THE CEREBELLUM

RECENT DEVELOPMENTS IN CEREBELLAR RESEARCH

ANNALS OF THE NEW YORK ACADEMY OF SCIENCES
Volume 978

THE CEREBELLUM

RECENT DEVELOPMENTS IN CEREBELLAR RESEARCH

Edited by Stephen M. Highstein and W. Thomas Thach

The New York Academy of Sciences
New York, New York
2002

Library of Congress Cataloging-in-Publication Data

The cerebellum : recent developments in cerebellar research / edited by
Stephen M. Highstein and W. Thomas Thach.
p.; cm. — (Annals of the New York Academy of Sciences; v. 978)
Includes bibliographical references and index.
ISBN 1-57331-436-6 (hardcover : alk. paper) — ISBN 1-57331-437-4 (paper: alk. paper)
1. Cerebellum--Congresses.
[DNLM: 1. Cerebellum--physiology--Congresses. WL 320 C4146 2002] I.
Highstein, Stephen M. II. Thach, W. Thomas. III. Series.]
Q11.N5 vol. 978
[QP379]
500 s—dc21
[612.8/ 2002014858

GYAT / PCP
Printed in the United States of America
ISBN 1-57331-436-6 (cloth)
ISBN 1-57331-437-4 (paper)
ISSN 0077-8923

ANNALS OF THE NEW YORK ACADEMY OF SCIENCES

Volume 978
December 2002

THE CEREBELLUM

RECENT DEVELOPMENTS IN CEREBELLAR RESEARCH

Editors
STEPHEN M. HIGHSTEIN AND W. THOMAS THACH

This volume is the result of a conference entitled **Recent Developments in Cerebellar Research**, sponsored by the McDonnell Center for Higher Brain Function of Washington University and held on November 15–18, 2001 in St. Louis, Missouri.

CONTENTS

Part III. Parallel Fiber Beams: What Do They Do?

Part IV. Climbing Fiber Inputs to the Cerebellum: What Is Signaled?

Part V. Cognitive Processing by the Cerebellum

Part VI. LTD, LTP, and Molecular Mechanisms of Learning and Memory

Part VII. Poster Papers

Financial assistance was received from:

- **McDONNELL CENTER FOR HIGHER BRAIN FUNCTION**
- **NATIONAL INSTITUTES OF HEALTH**
- **NATIONAL SCIENCE FOUNDATION**
- **WASHINGTON UNIVERSITY CONTINUING EDUCATION**
- **WASHINGTON UNIVERSITY DEPARTMENTS OF ANATOMY AND NEUROBIOLOGY, NEUROLOGY, OTOLARYNGOLOGY**

Introduction

STEPHEN M. HIGHSTEIN AND W. THOMAS THACH

Washington University School of Medicine, 4566 Scott Avenue, St. Louis, Missouri 63110, USA

Progress in cerebellar research and in the understanding of cerebellar function has proceeded at a rapid pace with the advent of new tools, techniques, and ideas. There is a wealth of new data on the organization of cerebellar microzones and on the function and operation of parallel-fiber beams. Optical techniques have been applied to the study of the propagation of parallel-fiber beams in several preparations. Multiple single-cell records have now been taken from the cerebellar cortex in awake, behaving mammals. These experimental results indicate that the cortex may be able to dynamically rearrange itself depending upon the behavioral context. Molecular techniques have been applied to the study of cerebellar cortical function. In a recent elegant study Golgi cells were ablated in an adult animal. Subsequently the resultant behavioral deficits were documented, and the time course of rccovery and its cellular and molecular mechanisms studied. Optical recordings from cortical slices in combination with patch recordings and pharmacology indicated a specific deficit, implying the normal function of Golgi cells. Further, physiologists have now learned how to record from cortical interneurons, for example, Golgi cells in awake, behaving animals, and new roles for these cells are emerging. A new cell type, the unipolar brush cell has been described, and its function is under study. Molecular mechanisms of cerebellar plasticity have emerged and transgenic animals and a host of neurobiological techniques have been employed in their study. An entire new generation of mathematical models of cerebellar function has emerged and has been influential in suggesting new physiological experiments. Studies of skeletal motor and oculomotor movements have recently been revealing of cerebellar function. These include studies of the influence of the cortex and nuclei on the initiation and termination of movement. There have also been numerous experiments on a role for the cerebellar cortex in motor learning.

Yet cerebellar research remains prominent for the many obvious and important *un*answered questions, and the continuing and vigorous controversies. What the essential mechanisms are underlying the cerebellum's role in coordinating movements are still debated. These include the relative roles of the timing of muscle contraction, the control of interaction torques, and the role of motor learning; whether the parallel fiber "beam" is a timer for motor sequences, a linker of synergic muscle ensembles, or a much weaker facilitator of the actions of distance Purkinje cells; whether the cerebellum is directly concerned with sensory perception and cognitive non-motor functions over and beyond motor control; whether the role of the inferior olive is to periodically "clock" elements of movement, to "teach" new combinations of elements, to register sensory events, motor events, or motor errors; and whether the

Ann. N.Y. Acad. Sci. 978: xi–xii (2002).

cerebellum contains explicit models of the motor system elements and behavior in order to implement predictive motor control.

The papers contained in this volume are the report of a forum on the cerebellum held in November 2001 at Washington University in St. Louis, Missouri. The proceedings of this forum are being published in the hope that they will be of interest not only to cerebellar neuroscientists but also to others in broad areas of the neurosciences including developmental biology, physiology, and pharmacology, space biology, and the medical fields of neurology, otolaryngology, and neurosurgery.

Disorders in Timing and Force of Finger Opening in Overarm Throws Made by Cerebellar Subjects

J. HORE,[a] D. TIMMANN,[b] AND S. WATTS[a]

[a]*Department of Physiology, University of Western Ontario, London, Ontario, Canada*

[b]*Department of Neurology, University of Essen, Essen, Germany*

ABSTRACT: Although there is agreement that an important sign of cerebellar dysfunction is disorder in timing of movement, it appears that authors who study different behaviors mean different things when they use the term "timing," and that the underlying mechanisms are likely to be different. For overarm throwing, skilled throwers can time ball release with a precision of less than 7 ms, whereas cerebellar subjects show a large variability of 50 ms or more in this timing. Furthermore, cerebellar patients show a larger variability in the amplitude of finger opening which could either reflect a disorder in force, or result indirectly from the increased variability in timing. To determine whether timing and force of finger opening were dependent variables, the time of ball release was plotted against the amplitude of finger opening. In control subjects these two parameters were related, with early (mistimed) throws having smaller finger amplitudes. However, in cerebellar subjects the increased variability in finger amplitude could not be accounted for by the increased variability in timing. Similarly, the increased timing windows could not be explained by disorder in force at the fingers. It is concluded that the abnormal finger opening that occurs in cerebellar patients when making overarm throws results from increased variability in both the timing and force of finger extension. Whether the increased variability in timing is a disorder in triggering finger opening at the right moment, or is due to a failure to combine finger opening appropriately with the hand trajectory, remains to be determined.

KEYWORDS: timing; overarm throwing; variability; finger force; finger amplitude; skill; cerebellar lesions; hand opening

CEREBELLAR DISORDERS OF TIMING MAY RESULT FROM BREAKDOWN IN DIFFERENT MECHANISMS

Although it is generally accepted that cerebellar lesions lead to disorders in timing, it is clear that this term means different things to authors who study different behavioral paradigms. For example, for some authors, the term disorders in timing

Address for correspondence: J. Hore, Department of Physiology, University of Western Ontario, London, Ontario N6A 5C1, Canada. Voice: 519-661-3817; fax 519-661-3827.
jon.hore@fmd.uwo.ca

Ann. N.Y. Acad. Sci. 978: 1–15 (2002). © 2002 New York Academy of Sciences.

means changes in the onset and offset of agonist and antagonist muscle contractions that produce a fast and accurate movement. This idea was clearly formulated 40 years ago by Braitenberg and colleagues when they proposed that a parallel fiber delay-line could be used to time bursts of agonist and antagonist muscle activity (see Refs. 1 and 2 for recent versions of this model). However, others have used disorders in timing to explain the degradation in performance that occurs in cerebellar subjects in a range of motor tasks such as rhythmic tapping[3] and sequencing of movements.[4] Similarly, lesions of the cerebellar cortex changed the timing of learned responses in the classically conditioned nictitating reflex and the vestibular ocular reflex.[5,6] And disorders in timing have been used as an explanation for the abnormalities cerebellar patients have in performing perceptual tasks such as judgment of the duration of brief tones,[7] determination of the velocity of moving objects,[8] and estimation of time intervals in a line bisection task.[9] It seems likely that the disorders in "timing" in these situations are due to a breakdown in different mechanisms.

PRECISION OF TIMING IN OVERARM THROWS MADE BY CONTROLS AND CEREBELLAR SUBJECTS

One movement where precise timing is required to achieve accuracy is opening the fingers to release the ball in overarm throwing. To investigate this timing we used five search coils taped to different finger and arm segments and sampled data at 1000 Hz. Subjects were asked to throw fast and accurately at a 6-cm^2 target 3 m in front of them. Using this technique, we showed that when the fingers open early on the flattened hand-path arc, the ball goes high, whereas when they open late, the ball goes low.[10] The precision of timing of ball release in skilled (accurate) throwers is impressive. If the moment of ball release from the fingertip is measured with respect to the moment in the forward throw when the hand is vertical in space, then recreational ball players have a timing variability of only a few milliseconds. This variability, which was determined in anywhere from 30 to 100 throws, was expressed statistically as a timing window for 95% of the throws by multiplying the SD of the mean time of ball release by 3.92.[11] The mean timing window for recreational ball players was 10 ms,[12] whereas for skilled throwers who had played competitive baseball or cricket it was 7 ms.[13] In both cases subjects threw from a seated position with their chest fixed by straps over the shoulder (so we could reconstruct hand translation). The timing window for our two most accurate throwers, who had been competitive baseball pitchers, was 2–3 ms when they threw fast and accurately from a standing position. Similarly, extrapolating our target size and distance to baseball reveals the remarkable fact that to hit the strike zone consistently in a series of throws requires a timing precision of 1–2 ms.

As expected, less accurate throwers had longer timing windows. Male recreational ball players had a mean timing window of 22 ms when throwing with their non-dominant arm from the seated position,[14] and a group of mainly female unskilled throwers had a mean timing window of 27 ms when throwing with their dominant arm.[13]

A striking finding in cerebellar subjects was they were much less accurate in both the high–low direction on the target[15] and the left–right direction.[16] Neither the instruction to throw at a slow, medium, or fast speed, or the actual speed that was

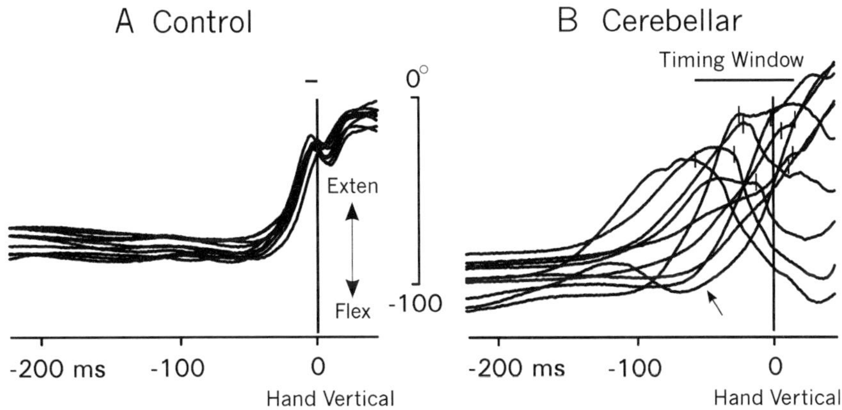

FIGURE 1. Timing of finger opening and ball release is more variable in a cerebellar subject. Angular position of the fingertip with respect to the hand for 10 throws aligned on the moment when the hand was vertical in space (0 time). (**A**) Matched control (Mi); (**B**) cerebellar subject (O1). *Short vertical lines* in (B) indicate moment of ball release. *Horizontal lines* give timing windows for ball release for all 10 throws. *Arrow* in (B) marks a throw in which the finger extended, flexed, then extended again.

achieved, affected accuracy. Although there was greater variability in the direction and height of the hand trajectory than for control throwers, the major cause of the inaccuracy was increased variability in the timing of ball release. This, in turn, was caused by increased variability in the timing of finger opening. FIGURE 1 shows superimposed traces of finger opening (extension) from 10 throws made by a matched control subject (FIG. 1A) and a cerebellar subject (FIG. 1B) aligned on the moment in the throw when the hand was vertical in space. The cerebellar subject (O1) was a 15-year-old male who had a surgical lesion in the region of the cerebellar nuclei. For each throw, the short vertical line gives the moment of ball release from the fingertip. The upper horizontal bars give the timing window for ball release for these throws. Compared to the control, the cerebellar subject shows a marked increase in the timing window for ball release. Across nine cerebellar subjects the mean timing window for ball release was 51 ms, whereas it was 14 ms for the matched controls.[13] Interestingly, some cerebellar subjects who showed large timing windows had relatively minor disorders in arm movements on clinical examination. For example, two subjects (O1 and Ls), who had timing windows of 98 and 97 ms, had clinical scores for upper arm ataxia of only 6/36 and 5/36, respectively.[13] Both subjects had surgical lesions in the region of the lateral cerebellar nuclei.

One problem with measuring timing of ball release in an overarm throw is that the release window depends on the variability of the reference point as well as the variability of ball release. As such, there is no one correct point that timing can be measured with respect to. We chose the moment in the throw when the hand was vertical in space for three reasons: because hand angle in space arises from the summation of rotations at shoulder, elbow, and wrist, and therefore finger opening must be timed with respect to the summation of these proximal joint rotations; because in accurate throwers, ball release occurs within a few milliseconds of this moment; and

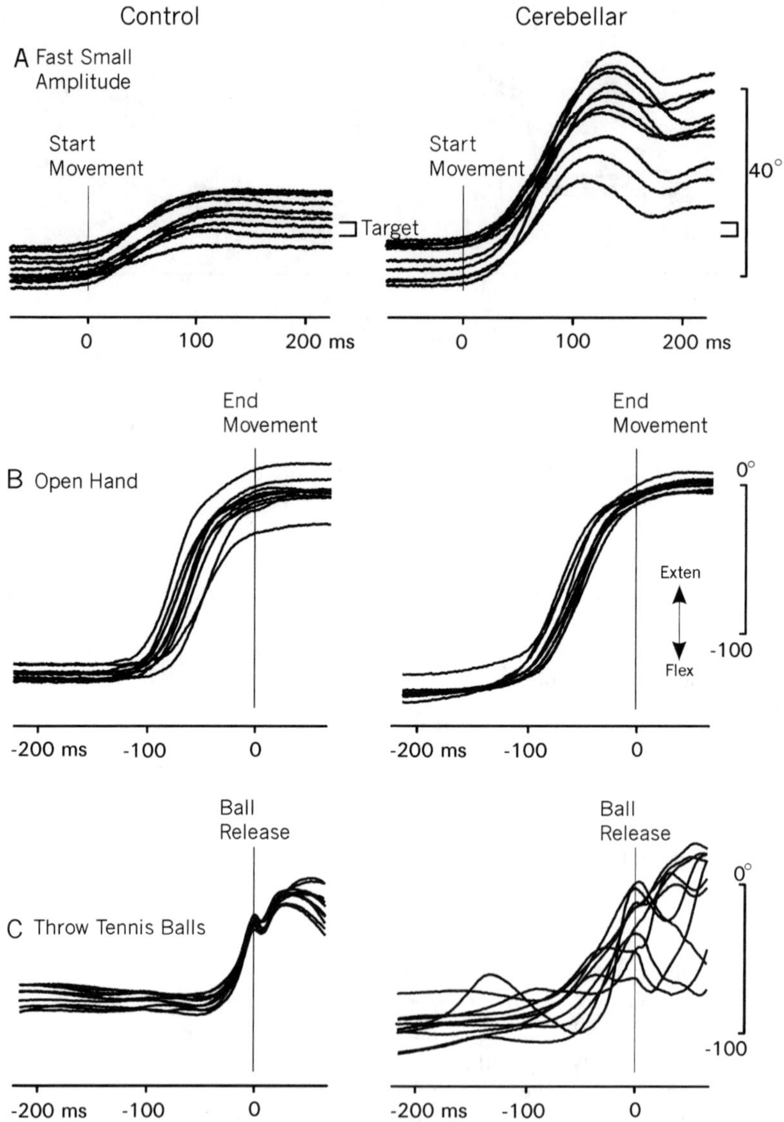

FIGURE 2. Cerebellar subject shows disorder in some hand-opening tasks but not others. Each set of records are 10 superimposed traces of angular position of fingertip with respect to hand aligned on a particular parameter (*vertical line*). Cerebellar subject shows (**A**) hypermetria in fast small-amplitude finger movement, (**B**) no disorder when opening finger to drop ball, and (**C**) increased variability in finger position while gripping and releasing ball in overarm throw. Note different scale in (A). Different throws than in FIGURE 1. Control Mi; cerebellar Ol.

because a relatively strong relation occurs between high–low ball accuracy on the target and the time of ball release measured this way. However, a problem with the hand vertical reference point is that its time will change if the direction of the hand path changes. For example, if the hand-path direction is more upward, the hand will take longer to get to vertical. To check the validity of timing windows, we measured them in nine cerebellar subjects and matched controls with respect to the times of three different parameters: (1) when the hand was vertical in space, (2) when elbow extension crossed a relatively high (robust) velocity threshold of 500°/s, and (3) when wrist flexion crossed a threshold of 300°/s. Mean values for the cerebellar patients were to hand vertical, 50 ms; to onset of elbow extension, 55 ms; and to onset of wrist flexion, 52 ms, and for the matched controls 13, 15, and 16 ms, respectively. These values show that timing windows for the three different parameters were within a few milliseconds of each other. Consequently, the use of hand vertical as a reference point for timing appears to be valid.

Two conclusions can be drawn from the preceding discussion. First, irrespective of the method of measurement, skilled throwers release the ball from throw to throw with a variability of only a few milliseconds. Second, compared to matched controls, cerebellar patients cannot throw accurately because they cannot precisely time release of the ball.

FINGER FORCE IN OVERARM THROWS MADE BY CONTROL AND CEREBELLAR SUBJECTS

As with timing, there is agreement that cerebellar lesions lead to disorders in control of force at the fingers, though the issue is also confused. One reason for the confusion is that there is a difference between acute and chronic patients; another is that there are differences between patients with lateral and medial cerebellar lesions. For the fingers, Holmes[17] described grasp strength in acute patients as being half that of the normal limb, whereas in chronic patients finger grip strength was within the normal range.[18,19] However, both acute and chronic patients with evidence of a lesion in the lateral cerebellum (i.e., infarction of superior cerebellar artery; surgical damage of lateral nuclei; arm ataxia) show decreases in the rate of change of force (or decreases in peak acceleration and peak agonist EMG activity).[17–22] This may contribute to the awkwardness in hand movements[23] and the impaired ability to pinch in a precision grip[24] observed in monkeys with dentate lesions.

We investigated the ability of cerebellar subjects with evidence of lateral cerebellar lesions to open their fingers in three situations that required different control of finger force.[13] In the first situation, they were asked to make fast and accurate finger extension movements of 10° moving all the fingers together. As shown in FIGURE 2A (right side), the cerebellar subject could not perform accurately in this task. Instead he showed the classic disorder of hypermetria. We have previously shown that hypermetria results from an inability to generate early antagonist muscle activity.[21] In the second situation, subjects were asked to hold the hand at about the position in an overarm throw where the ball was released and open the hand fully and drop the ball. FIGURE 2B shows that one representative cerebellar subject showed no obvious disorder in this task, presumably because he braked finger extension by viscoelastic forces. In contrast, in the third situation, in which the task was to open the fingers and release the ball in an overarm throw, the cerebellar subject showed major differences compared to the matched control, including more variable amplitudes of finger extension at the moment of ball release (FIG. 2C). In this figure traces are aligned on the moment of ball release. One likely possibility is that the increased variability in finger position throughout the throw resulted from a deficit in the more complex control of force that is required in an overarm throw.

We have studied the forces that occur on the fingers in an overarm throw in recreational ball players.[25,26] FIGURE 3 illustrates ball release from the fingers (FIG. 3A)

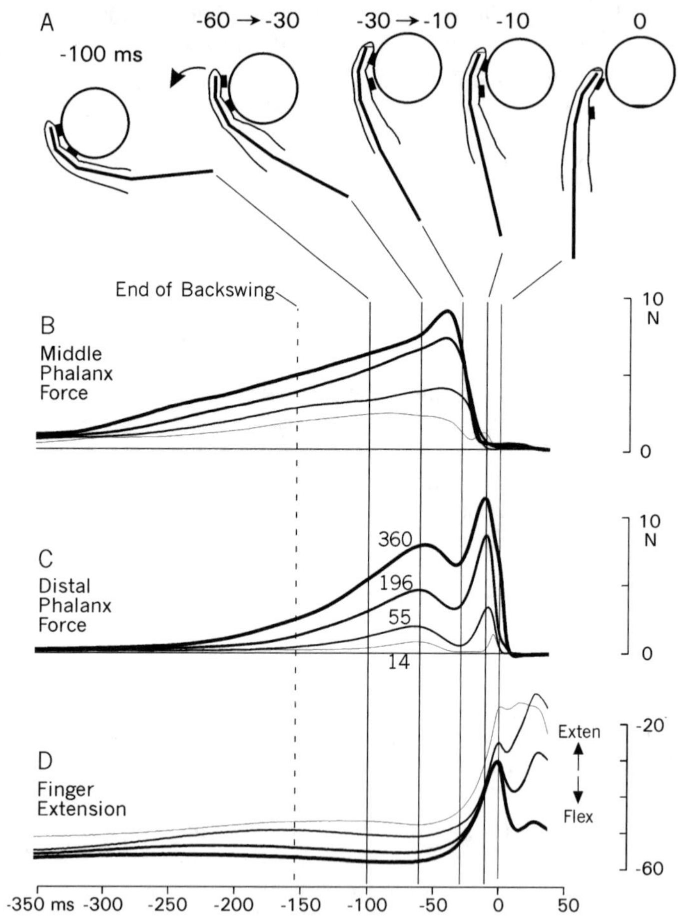

FIGURE 3. Ball release from the finger and the accompanying finger forces and finger positions in a skilled thrower. Times are with respect to the moment of ball release (time 0). *Black squares* indicate force transducers on distal and middle phalanges of middle finger. Each trace is the average of 10 throws made with each ball of a particular weight (14–360 g). Subject De.

and the accompanying finger forces (FIG. 3B, 3C) and finger positions (FIG. 3D) for throws made with balls of four different weights (14–360 g). Each trace represents the average from 10 throws. This subject threw with a slow backswing and had very small hand accelerations and decelerations during the backswing (not shown). Two important results can be seen: first, when allowance is made for their different grip positions during the throw, the amplitude of finger extension (FIG. 3D) was similar for all four balls in spite of their different weights, and second, at the end of the backswing (dashed vertical line), there was a force on the middle phalanx (FIG. 3B) that was proportional to ball weight, which could not be accounted for by passive back forces or gravity.[26] From this and other results in which back forces were measured directly, it was concluded that recreational ball players increased their active grip force throughout the backswing and forward throw in an anticipatory fashion to counteract the progressively increasing back force from the ball.

It follows that in cerebellar subjects, loss of precision in the ability either to predict back forces from the ball on the finger, or to produce the correct amount of grip force for the ball weight, could lead to the disorder in finger position that was observed throughout the throw (FIG. 2C). In other words, the disorder in finger position could have resulted from disorder in control of finger force.

DOES VARIABILITY IN FINGER FORCE CAUSE VARIABILITY IN FINGER TIMING?

Given that evidence has been found in cerebellar subjects that disorders occur in both timing and force of finger opening, the question arose whether in overarm throws these are independent disorders or whether one causes the other. For the case of a single joint movement, it is clear that a decrease in development of force would cause an increase in onset of timing measured to a set position or velocity threshold. But in overarm throws, cerebellar subjects showed increased variability in timing of finger opening rather than a delay. Could increased variability in force on the fingers have caused the increased variability in timing? One way this could occur would be if there was a failure in cerebellar subjects to produce a large enough grip force, such that the back force from the ball pushed the fingers open and released the ball at variable times. This is unlikely, because the evidence indicates that in control and cerebellar subjects finger opening is caused by a central command and not by back forces. Compelling evidence in control subjects for a central command comes from experiments in which recordings were made of finger position (with a search-coil) and finger force (with a force transducer), both taped to the distal phalanx. If finger opening was caused by the back force, it would be expected that onset of finger extension would be associated with an increase in force on the finger. In fact, records in all subjects showed that onset of finger extension (FIG. 3D) was associated with an initial decrease (and not an increase) in force on the distal phalanx (FIG. 3C). The explanation is that opening of the fingers is initiated by a central command that causes liftoff of the distal phalanx from the ball (Fig. 3A, –60 to –30) and a decrease in force on the distal phalanx. In agreement, a second piece of evidence comes from experiments in which back forces were changed by mechanically perturbing elbow extension. Irrespective of whether elbow extension was slowed, blocked, or replaced by elbow flexion, which in all cases markedly decreased back forces, finger opening

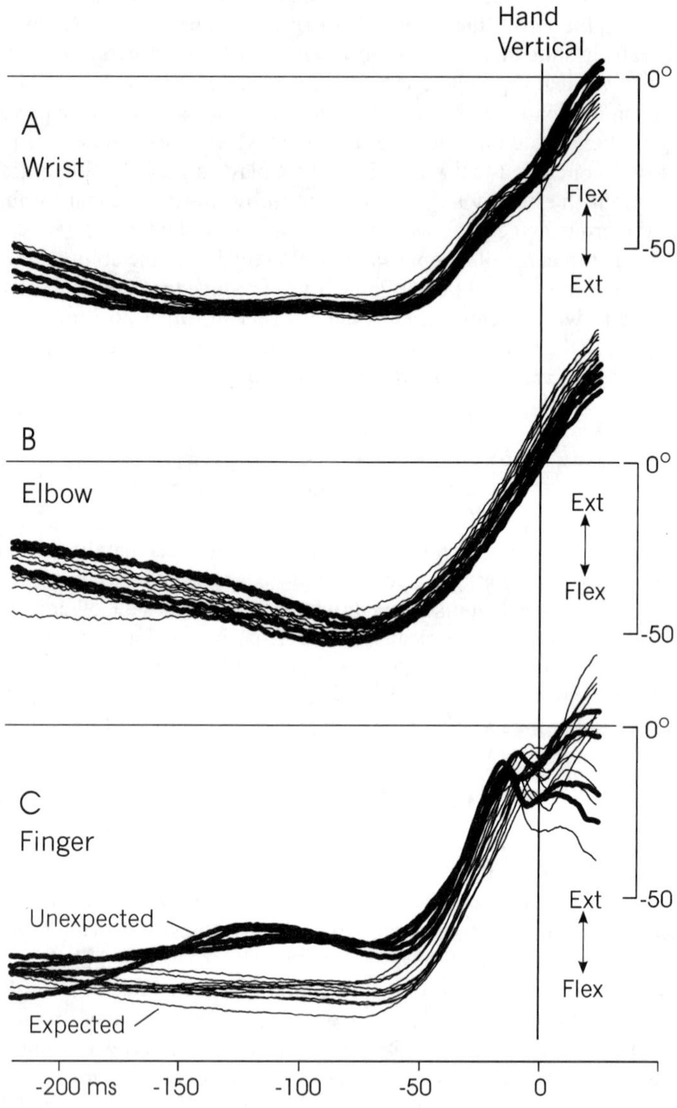

FIGURE 4. Effect of unexpected change in back force from the ball on the finger. *Thin lines* (expected): 12 throws when heavy ball was expected. *Thick lines* (unexpected): 4 throws when subject expected 55-g tennis ball and received the 220-g heavy tennis ball. (**A**) Angular rotation at wrist joint; (**B**) angular rotation at elbow joint; (**C**) angular rotation of distal phalanx of middle finger with respect to hand. Subject My.

always occurred and at the same latency as for normal throws.[27] This would not have occurred if back forces caused finger opening.

A third piece of evidence that back forces do not cause finger opening comes from the finding that when control subjects underestimated the back force by a factor of 4, the inappropriately large back force did not cause finger opening and ball release. This could not be tested in recreational ball players by changing ball weight in normal throws because they adjust finger amplitude on the very first throw with a ball of a new weight.[25] However, we overcame this problem by minimizing the time subjects had to handle the ball. This was achieved by having six experienced throwers hold their hand open with the arm stationary at the end of the backswing. The experimenter then placed a ball in the subject's hand, which the subject had to immediately grip and throw forward. To investigate the effect of an unexpectedly large back force, subjects were told they would throw tennis balls. After about 10 throws with tennis balls (55 g), unknown to the subjects, a switch was made and a tennis ball filled with concrete (220 g) was placed in their hand. All subjects reported that especially for the first unexpected heavy ball, they were completely unaware of the switch until well into the forward throw. At the end of the experiment subjects also made 10 throws with the heavy ball when it was expected. When the heavy tennis ball was unexpected, there was a small finger extension, but the finger did not open to release the ball. FIGURE 4 shows kinematic records from one representative subject of the last 200 ms of forward throws made when the heavy ball was expected (thin traces) and unexpected (thick traces). Records are aligned at the moment when the hand was vertical in space (0 ms) which occurred within 10 or so ms of the time of ball release from the fingertip. The duration of these throws, from initial ball contact with the hand to ball release from the fingertip, was between 450 and 750 ms. In these throws the initial forward acceleration of the upper arm caused the hand to first rotate backwards and downwards in space (not shown), which was associated with wrist extension (FIG. 4A) and elbow flexion (FIG. 4B). About 75 ms before the time of hand vertical, the hand began to rotate forward and up in space (not shown), which was associated with wrist flexion (FIG. 4A) and elbow extension (FIG. 4B). FIGURE 4C shows that starting about 200 ms before hand vertical in throws when the heavy ball was expected, there was a slight flexion at the finger for 150 ms (rotation of distal phalanx with respect to hand). However, when the heavy ball was unexpected, there was an extension of the finger of about 15°, consistent with it being pushed back by the unexpectedly large back force. This extension was followed first by finger flexion (which may be a reflex response) and then finger opening (extension), which released the ball. Similar results were found for five of the six subjects. In the sixth subject, changes in kinematics when the heavy ball was unexpected were more obvious at the wrist and elbow than the fingers. In summary, when subjects were tricked into underestimating the back forces from the ball, a small finger extension occurred as they gripped the ball during the forward throw, but it did not cause premature ball release.

In keeping with these findings there is evidence that finger opening in cerebellar patients is not produced by back forces. For back forces to push the fingers open, it would be necessary for cerebellar patients to have low finger grip forces. However, the data indicate that cerebellar patients can scale-up finger grip force to the weight of the ball,[13] but do so with increased variability compared to controls. Even if it was argued that cerebellar subjects have a highly variable generation of finger grip force,

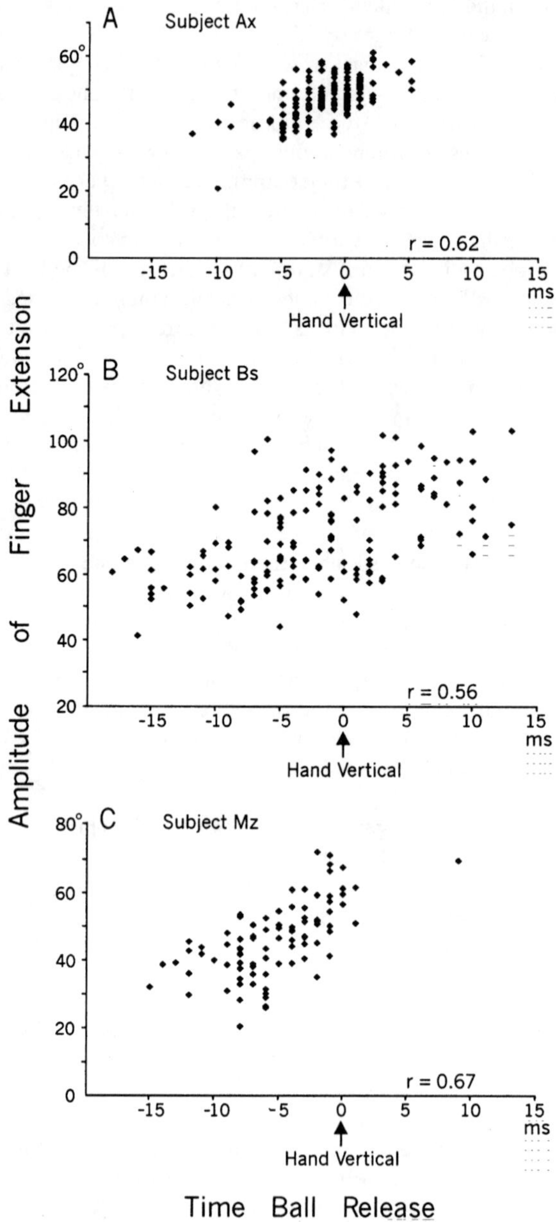

FIGURE 5. Relation between amplitude of finger extension and time of ball release. (**A**) One hundred fifty throws with right (dominant) arm in a recreational ball player (subject Ax). (**B**) One hundred fifty throws with left (nondominant) arm in a recreational ball player (subject Bs). (**C**) One hundred throws with right (dominant) arm in a female recreational ball player (subject Mz). Time of ball release was measured with respect to the moment in the throw when the hand was vertical in space. All throws made from a sitting position.

which results in the back force pushing the fingers open at various times in different throws, it would be expected that early finger opening (produced in throws that had extremely low finger grip force) would have a large initial velocity of finger opening because the back force was much greater than the grip force. Conversely, late finger opening (produced in throws that had relatively high finger grip forces) would have a small initial velocity of finger opening. The same argument would apply if variably excessive back forces added together with the central command for finger extension. However, as seen in FIG. 1B, and reported previously,[15] the records of finger opening do not show the predicted properties, that is, initial finger velocity (rate of change of initial finger extension) is variable throughout the throw.

Although we have not found evidence across cerebellar subjects for force affecting timing, the possibility still exists that this occurs, especially in cases of extreme disorder. For example, consider the throw (FIG. 1B) marked with an arrow. One explanation for the pattern of finger changes is that decreased grip force early in the throw allowed the back force to push the finger toward extension (cf. FIG. 4), that this was counteracted by a reflex that produced finger flexion, which in turn, delayed the onset of finger extension produced by the central command for finger opening. In this case, the timing of finger opening would be delayed by the abnormality in the control of force. Although this pattern occurred to various extents in up to 25% of throws in cerebellar subject Ol, it was not the cause of most of the timing variability, and it was not seen consistently in other cerebellar subjects. In summary, the evidence indicates that, considering all cerebellar subjects, the variable timing of ball release results, not from variable force, but from variable timing of central commands.

DOES VARIABILITY IN FINGER TIMING CAUSE VARIABILITY IN FINGER FORCE?

As shown in FIGURE 2C, a characteristic of throws made by cerebellar patients is that they show an abnormally large variability (1) in the amplitude of the change in finger position when gripping, (2) in the amplitude of finger opening, and (3) in the amplitude of the change in finger position 10 ms after ball release.[13] These changes were suggested to be due to a disorder in control of force. But given that change in back force from the ball on the fingers occurs throughout an overarm throw, and that in cerebellar subjects there is increased variability in the timing of ball release, the question arose whether these disorders in finger position were in fact due to disorders in force or whether they resulted from the increased variability in timing. The idea is first, that the back force, although not causing finger opening, may nevertheless contribute to the amplitude of finger extension, and second, that the earlier the ball is released, the smaller will be the back force (because hand acceleration is smaller) and the smaller will be the amplitude of finger extension. This was tested by examining the relation between timing of ball release and the amplitude of finger opening (which was taken as a measure of force on the finger). Two difficulties in investigating this relation in control subjects are that recreational ball players have narrow timing windows for ball release (about 10 ms) and that back forces do not vary greatly over a 10-ms period. Consequently, 50 throws may not be enough to observe the predicted relation, given that there are other factors that affect finger amplitude, for example, the inherent variability of central signals that control the amplitude of finger extension. We therefore analyzed data from a previous series of experiments in which six recreational ball players made 150 throws with a tennis ball, both with their dominant right arms and their nondominant left arms.[14] Time of ball release was measured with respect to the moment in the throw when the hand was vertical in space. Across subjects mean timing windows for ball release for 95% of the throws were 9 ms for the right arm and 22 ms for the left arm.

FIGURE 5 shows plots of the amplitude of finger extension against time of ball release. FIGURE 5A shows a subject throwing with his right (dominant) arm. Although ball release occurred over a relatively short time duration (the timing window was 11 ms), a statistically significant relation was found with early ball release associated with smaller finger amplitudes. For all six subjects this relation for the right arm was significant at the $p < .01$ level. Throws with the left arm showed greater variabil-

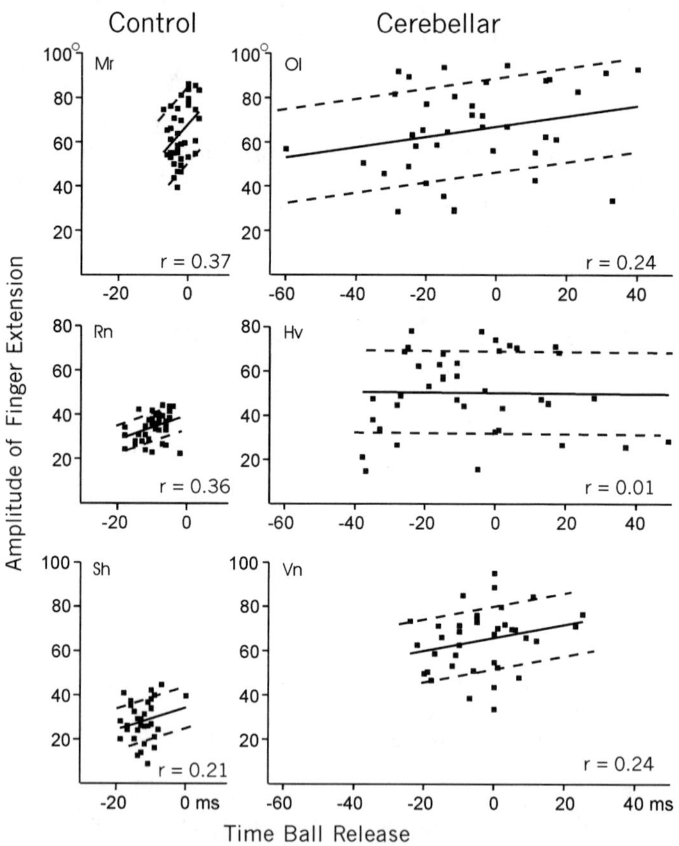

FIGURE 6. Increased variability in amplitude of finger extension is not caused by increased variability in timing. Time of ball release plotted with respect to the moment in throw when hand was vertical in space (time 0). *Continuous lines* are regression lines; *dashed lines* give standard error. The relation between finger amplitude and timing is not statistically significant in any of the three cerebellar subjects.

ity in timing of ball release and in amplitude of finger opening. FIGURE 5B shows for a second subject that in spite of this variability a similar relation occurred for throws with the left arm. This relation was significant for throws with the left arm in three of the six subjects at the $p < .01$ level, one at the $p < .05$ level, and it was not significant in two subjects. We also analyzed throws made with the dominant arm from four female recreational ball players, because they had relatively long timing windows (range 15–35 ms). Again in three cases a significant relation was found at the $p < .01$ level (see FIG. 5C for relation in subject Mz).

It follows from the preceding that in control subjects longer timing windows would be associated with greater variability in finger amplitude. Given that cerebellar subjects have longer timing windows, the possibility arose that this was the cause of their greater variability in finger amplitude. To investigate this we plotted finger amplitude against timing of ball release. FIGURE 6 shows these plots for three cerebellar subjects and their controls. The continuous line is the regression line; the dashed lines give the standard error (68% of the points lie between the dashed lines). Three things can be observed. First, two of the three control subjects showed the expected relation (as shown in FIG. 5) at the $p < .05$ level (r critical is .30). Second, none of the cerebellar subjects showed the relation at the $p < .05$ level. Third, cerebellar subjects showed larger standard errors, that is, they had greater variability between the two variables. Across the five cerebellar patients with the largest timing windows, the mean standard error was 17.0, whereas it was 9.6 for the matched controls. It can be concluded that in cerebellar subjects, the greater variability in finger amplitude is due to independent variability, that is, it cannot be attributed to greater variability in the timing of finger opening. Rather, the evidence indicates that it is due to greater variability in the control of finger force.

CEREBELLAR ROLE IN TIMING: TRIGGERING OR COMBINING

Although the evidence is convincing that cerebellar subjects cannot time finger opening precisely, it is not clear whether the fundamental disorder is a failure to trigger finger opening at the right moment in a throw, or is a failure to combine appropriately finger opening with some global kinematic parameter such as hand path in space. The latter possibility follows from the ideas of Thach *et al.*,[2] who proposed that the cerebellum normally acts to help combine the relative motions of different body parts into a single organized movement. In keeping with this, cerebellar patients had a breakdown in the combining of reach and grasp movements.[28] Two different schemes for finger opening based on these ideas of triggering or combining can be proposed for throws of different speeds. FIGURE 7A and 7B show a schematic angular hand path and finger extension for a fast throw. Note that the fast rate of change of angular hand position in the forward direction is associated with a fast rate of change of finger position during finger extension. Consider what the pattern might be for a slow throw (FIG. 7C). One interpretation of the triggering hypothesis is that finger extension with the same kinematics as for the fast throw will be triggered at a different time (FIG. 7D). An alternative possibility is that the slow hand path will be combined appropriately with a slow finger extension (FIG. 7E). These possibilities are currently being investigated.

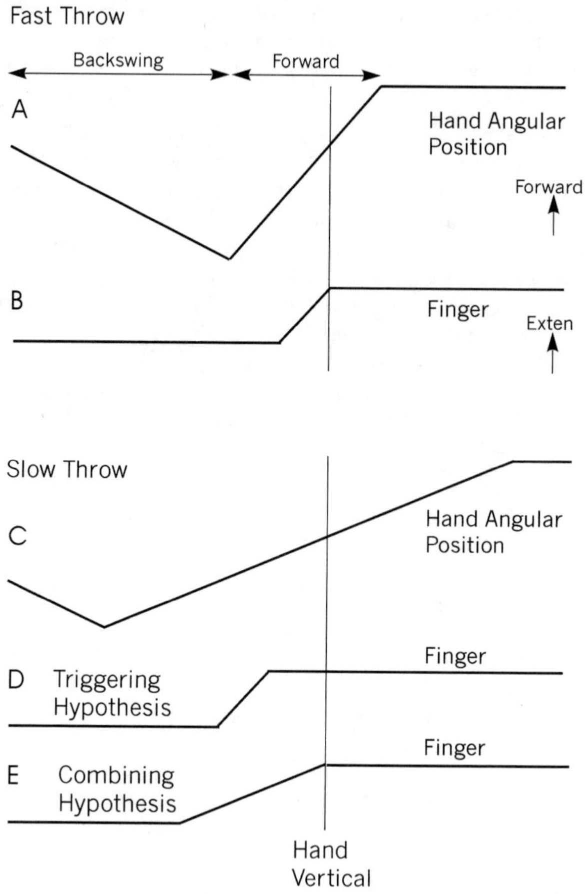

FIGURE 7. Schematic diagram of angular hand positions and finger extensions for fast and slow throws. The triggering hypothesis (D) and combining hypothesis (E) predict different finger kinematics for the slow throw.

REFERENCES

1. Braitenberg, V., D. Heck & F. Sultan. 1997. The detection and generation of sequences as a key to cerebellar function: experiments and theory. Behav. Brain Sci. **20:** 229–277.
2. Thach, W.T., H.P. Goodkin & J.G. Keating. 1992. The cerebellum and the adaptive coordination of movement. Annu. Rev. Neurosci. **15:** 403–442.
3. Ivry, R.B., S.W. Keele & H.C. Diener. 1988. Dissociation of the lateral and medial cerebellum in movement timing and movement execution. Exp. Brain Res. **73:** 167–180.
4. Inhoff, A.W., H.C. Diener, R.D. Rafal & R. Ivry. 1989. The role of cerebellar structures in the execution of serial movements. Brain **112:** 565–581.

5. RAYMOND, J.L., S.G. LISBERGER & M.D. MAUK. 1996. The cerebellum: a neuronal learning machine? Science **272:** 1126–1131.
6. PERRET, S.P. 1998. Temporal discrimination in the cerebellar cortex during conditioned eyelid responses. Exp. Brain Res. **121:** 115–124.
7. IVRY, R. 1997. Cerebellar timing systems. Int. Rev. Neurobiol. **41:** 555–573.
8. IVRY R.B. & H.C. DIENER. 1991. Impaired velocity perception in patients with lesions of the cerebellum. J. Cognit. Neurosci. **3:** 355–366.
9. NICHELLI, P., D. ALWAY & J. GRAFMAN. 1996. Perceptual timing in cerebellar degeneration. Neuropsychologia **34:** 863–871.
10. HORE, J., S. WATTS & D. TWEED. 1996. Errors in the control of joint rotations associated with inaccuracies in overarm throws. J. Neurophysiol. **75:** 1013–1025.
11. SOKAL, R.R. & F.J. ROHLF. 1981. Biometry. The Principles and Practice of Statistics in Biological Research. Freeman. San Francisco.
12. HORE, J., S. WATTS, J. MARTIN & B. MILLER. 1995. Timing of finger opening and ball release in fast and accurate overarm throws. Exp. Brain Res. **103:** 277–286.
13. TIMMAN, D., R. CITRON, S. WATTS & J. HORE. 2001. Increased variability in finger position occurs throughout overarm throws made by cerebellar and unskilled subjects. J. Neurophysiol. **86:** 2690–2702.
14. HORE, J., S. WATTS, D. TWEED & B. MILLER. 1996. Overarm throws with the nondominant arm: kinematics of accuracy. J. Neurophysiol. **76:** 3693–3704.
15. TIMMANN, D., S. WATTS & J. HORE. 1999. Failure of cerebellar patients to time finger opening precisely causes ball high-low inaccuracy in overarm throws. J. Neurophysiol. **82:** 103–114.
16. TIMMANN, D., S. WATTS & J. HORE. 2000. Causes of left-right ball inaccuracy in overarm throws made by cerebellar patients. Exp. Brain Res. **130:** 441–452.
17. HOLMES, G. 1939. The cerebellum of man. Brain **62:** 1–30.
18. MAI, N., P. BOLSINGER, M. AVARELLO, *et al.* 1988. Control of isometric finger force in patients with cerebellar disease. Brain **111:** 973–998.
19. MÜLLER, F. & J. DICHGANS. 1994. Dyscoordination of pinch and lift forces during grasp in patients with cerebellar lesions. Exp. Brain Res. **101:** 485–492.
20. HALLETT, M., A. BERARDELLI, J. MATHESON, *et al.* 1991. Physiological analysis of simple rapid movements in patients with cerebellar deficits. J. Neurol. Neurosurg. Psychiatry **53:** 124–133.
21. HORE, J., B. WILD & H.C. DIENER. 1991. Cerebellar dysmetria at the elbow, wrist and fingers. J. Neurophysiol. **65:** 563–571.
22. WILD, B., T. KLOCKGETHER & J. DICHGANS. 1996. Acceleration deficit in patients with cerebellar lesion. A study of kinematic and EMG-parameters in fast wrist movements. Brain Res. **713:** 186–191.
23. BOTTERELL, E.H. & J.F. FULTON. 1938. Functional localization in the cerebellum of primates III Lesions of the hemispheres (neocerebellum). J. Comp. Neurol. **69:** 63–87.
24. THACH, W.T., S.A. KANE, J.W. MINK & H.P. GOODKIN. 1992. Cerebellar output: multiple maps and motor modes in movement coordination. *In* The Cerebellum Revisited. R. Llinas and C. Sotelo, Eds. Springer-Verlag. New York.
25. HORE, J., S. WATTS & D. TWEED. 1999. Prediction and compensation by an internal model for back forces during finger opening in an overarm throw. J. Neurophysiol. **82:** 1187–1197.
26. HORE, J., S. WATTS, M. LESCHUK & A. MACDOUGALL. 2001. Control of finger grip forces in overarm throws made by skilled throwers. J. Neurophysiol. **86:** 2678–2689.
27. HORE, J., R. RITCHIE & S. WATTS. 1999. Finger opening in an overarm throw is not triggered by proprioceptive feedback from elbow extension or wrist flexion. Exp. Brain Res. **125:** 302–312.
28. ZACKOWSKI, K.M., W.T. THACH & A.J. BASTIAN. 2002. Cerebellar subjects show impaired coupling of reach and grasp movements. Exp. Brain Res. **146:** 511–522.

Cerebellar Limb Ataxia

Abnormal Control of Self-Generated and External Forces

AMY J. BASTIAN

Department of Neurology, Johns Hopkins Medical School, Baltimore, Maryland 21205, USA, and Kennedy Krieger Institute, Baltimore, Maryland 21205, USA

ABSTRACT: Our work has been focused on understanding the mechanism of movement abnormalities associated with cerebellar ataxia. The hypothesis tested is that the cerebellum acts to modulate muscle activity across multiple joints in anticipation of the mechanical interaction torques generated by one's own movement and by external forces. Individuals with cerebellar damage were studied in two sets of experiments. In the first experiment, we studied how cerebellar subjects' movement changed when interaction torques were present, and then reduced via mechanically constraining movement to a single joint. Consistent with the hypothesis, it was found that cerebellar endpoint errors were greatly improved when interaction torques were reduced. We also found that cerebellar deficits in the unconstrained condition were not explained by a general failure of torque timing or magnitude scaling. This supports the idea that the cerebellum plays a specific and important role in adjusting for the dynamics of one's own body movements. In the second experiment, we studied how well cerebellar subjects could adapt arm movements to external loads. Cerebellar subjects were tested as they adapted a catching movement to balls of different weight. It was found that they were slow or unable to adapt through practice and did not show evidence of storage of the adaptation. This suggests that the cerebellum is needed for rapid adaptations for loads in movement. Given these findings, we think that the cerebellum is important in anticipating and adjusting for the mechanical demands of movement though trial-and-error practice.

KEYWORDS: human; cerebellum; ataxia; interaction torque; dysmetria; adaptation; motor learning

INTRODUCTION

Controlling movement of the human body is a difficult task. From a mechanical perspective, the body can be thought of as a large set of linked limb segments, which have complicated dynamic interrelationships. For example, movement of any limb segment inevidibly causes forces at all limb segments that are linked to it. Despite

Address for correspondence: Amy J. Bastian, Kennedy Krieger Institute, 707 North Broadway, G05, Baltimore, MD 21205. Voice: 443-923-2718; fax: 443-923-2715.
bastian@kennedykrieger.org

Ann. N.Y. Acad. Sci. 978: 16–27 (2002). © 2002 New York Academy of Sciences.

this complexity, the nervous system is able to produce relatively invariant, rapid movements that are smooth and accurate. In addition, movement patterns can be rapidly changed to account for novel dynamic situations, such as interacting with objects in the environment.

Many different regions of the central nervous system are important for planning, executing, and adjusting movement. However, it is likely that distinct brain regions participate in different aspects of normal motor control. This is evidenced by the fact that damage of different motor regions results in specific types of movement deficits. For example, damage of the cerebellum causes a characteristic deficit in movement "coordination." When reaching, people with cerebellar damage often have deficits that include abnormally curved hand paths, abnormal coupling of elbow and shoulder movements, and over- and undershooting targets (dysmetria). These features, collectively known as ataxia, typically occur in the absence of any overt sensory deficits or weakness.

Our research has focused on understanding the mechanism of movement abnormalities associated with cerebellar ataxia. We have suggested that the cerebellum plays an important role in adjusting movements at a given limb segment for mechanical forces generated by movement of other limb segments. This is supported by work in humans[1–3] and in cats.[4] Specifically, it has been shown that cerebellar ataxia during multijointed reaching may be due to an inability to adjust for interaction torques. During a reach, elbow movement causes an interaction torque to occur at the shoulder, and shoulder movement causes an interaction torque to occur at the elbow. Cerebellar subjects have difficulty offsetting interaction torques, especially during fast, multijointed reaching movements.[1,3] This can result in an abnormal pattern of reaching, with the elbow and shoulder joints moving at inappropriate rates relative to one another and the fingertip overshooting the target.[1,2]

In addition to dealing with the dynamics of multijointed movements, the nervous system has to adjust for external loads and forces. We think that the cerebellum also plays an important role in anticipating and adjusting for changing external mechanical loads and forces on the body. These adjustments are likely made through practice, but would have to occur fairly rapidly in order to make movements with different loads efficient and precise. We have found that cerebellar damage reduces a person's ability to adjust for novel loads through trial-and-error practice.[5,6]

Therefore, our working hypothesis is that the cerebellum acts to modulate muscle activity across multiple joints in anticipation of the mechanical forces generated by one's own movement and by external forces. This aspect of control is critical for allowing individuals to produce movement of the appropriate trajectory and magnitude in the face of constantly varying circumstances. We have tested this hypothesis by studying individuals with cerebellar damage in two sets of experiments. In the first experiment, we studied how cerebellar subjects' movements change when interaction torques were present, and then reduced via mechanical fixation of a joint.[7] Consistent with our hypothesis, we found that cerebellar endpoint errors (dysmetria) were greatly improved when movement was mechanically constrained to a single joint. We also found that cerebellar deficits in the unconstrained condition were not explained by a general failure of torque timing or magnitude scaling. This supports the idea that the cerebellum plays a specific and important role in adjusting for the dynamics of one's own body movements. In the second experiment, we studied how well cerebellar subjects could adapt arm movements to external loads. We tested cerebellar subjects as they adapted a catching movement to balls of different weight. We found that they were slow or unable to adapt through practice and could not generate the appropriate anticipatory muscle activity for a given ball weight.[5,6] This suggests that the cerebellum is needed for rapid adaptations for loads in movement. All results have been previously published.[5,7]

EXPERIMENT 1: CEREBELLAR CONTROL OF INTERACTION TORQUES

Methods

In the first experiment, we tested (1) whether deficits in interaction torque compensation could be explained by torque scaling or timing deficits, and (2) whether reducing interaction torques improved movement performance of cerebellar subjects. Eight people with cerebellar cortical atrophy and eight matched controls participated in this study. The presence of cerebellar damage was assessed by neurological exam and MRI scan. All cerebellar subjects were clinically examined to ensure that the arm to be tested showed signs of ataxia only. Informed consent was acquired prior to testing all subjects.

All subjects were videotaped performing elbow flexion movements in the sagittal plane under shoulder-free and shoulder-fixed conditions (FIG. 1A). Arm kinematics

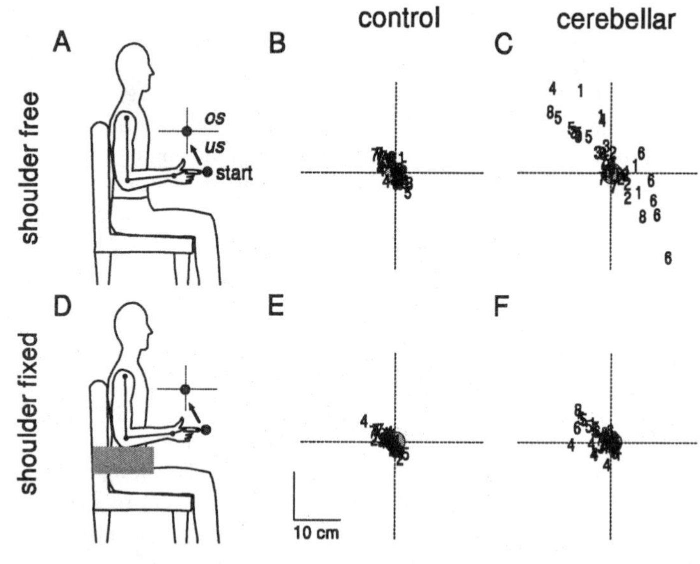

FIGURE 1. (**A**) Schematic of the shoulder-free condition. (**B**) Control subjects' endpoint errors in the shoulder-free condition. (**C**) Cerebellar subjects' endpoint errors in the shoulder-free condition. (**D**) Schematic of the shoulder-fixed condition. (**E**) Control subjects' endpoint errors in the shoulder-fixed condition. (**F**) Cerebellar subjects' endpoint errors in the shoulder-fixed condition. For all plots, each number indicates the errors from a different subject; 5 trials are plotted for each subject.

and muscle activity were recorded at 60 and 600 Hz, respectively. A target was located at a position that required ~35 degrees of elbow flexion and no shoulder movement. In the shoulder-free condition, seated subjects touched a target as fast as they could, while trying to stop on the target. The shoulder-fixed condition was identical to the shoulder-free condition, except that the elbow rested on a flat surface to mechanically constrain movement at the shoulder (FIG. 1D). For both conditions, three to five practice trials were given prior to 10 test trials.

Five test trials were analyzed for each condition from "start" of movement (time that wrist velocity exceeded 5% of its peak) to the "end" of the first phase of movement (time that wrist velocity dropped below 5% of its peak). Each trial performed by cerebellar subjects was classified into different endpoint error categories by assessing the magnitude and direction of endpoint error as: No Error (within 2 SD of control errors), an overshoot (OS = above target), or an undershoot (US = below target; FIG. 1A). Joint angles were also measured at the end of movement.

We then used videotaped kinematic data, estimates of the inertia of the limb segments,[8] and 3-segment inverse dynamics equations to calculate the time series of the (1) net torque, (2) gravitational torque, (3) interaction torque, and (4) muscle (residual) torque about both the elbow and shoulder joints.[1,7] We estimated the net torque for a given joint by taking the product of the moment of inertia of the involved segments (including the segment under consideration and all segments distal to it) and the angular acceleration around a given joint. The net torque would therefore be proportional to acceleration of a given joint. The gravitational torque is the torque produced by the force of gravity acting on the limb segments of the arm. The interaction torque is the "passive" mechanical torque generated when two or more linked segments move on one another. The muscle (residual) torque is produced by muscle and passive tissues; it is calculated by taking the difference between the net torque and the sum of the interaction and gravitational terms. All torque traces were normalized to each subject's body weight. Measures of interest for both shoulder and elbow joints were (1) the magnitude of muscle and interaction torque peaks, (2) the time of muscle and interaction torque peaks, and (3) the rate of muscle torque production.

EMG was rectified and low-pass filtered at 100 Hz. We then determined differences in the muscle activity levels between shoulder-free and shoulder-fixed conditions. We focused on the agonist muscles (biceps and brachioradialis) and the shoulder muscle needed to initially stabilize the shoulder joint (anterior deltoid). EMG signals for these muscles were integrated for the first 100 ms of movement. Percent change in integrated muscle activity from the shoulder-free to the shoulder-fixed conditions was then calculated.

All kinematic and kinetic measures were tested using ANOVA and Duncan's *post hoc* test. Student's t-tests were used to test whether the percent change in muscle activity was different between the control and cerebellar groups.

Results

Control subjects made small endpoint errors in both conditions; cerebellar subjects often produced large endpoint errors in the shoulder-free condition, but were able to perform at nearly control levels in the shoulder-fixed condition (FIG. 1). The majority of cerebellar errors in the shoulder-free condition were overshoots (40% OS errors, 17% US errors, and 43% classified as No Error). The main reason that

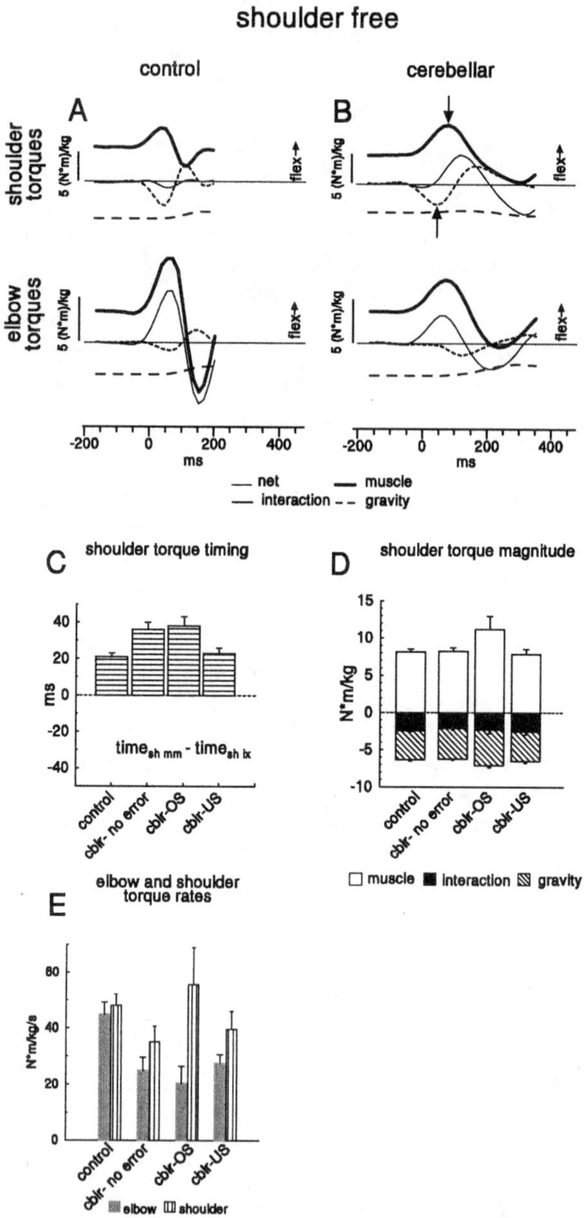

FIGURE 2. A single trial of normalized elbow and shoulder torque components from (**A**) a control and (**B**) a cerebellar subject, both moving in the shoulder-free condition. The cerebellar subject shown overshot the target in this trial. Data are from movements of similar peak velocity (control 161 cm/s, cerebellar 167 cm/s). Trials are aligned on the start of movement (time 0) and drawn to the end of movement, prior to corrections. *Arrows* indicate

cerebellar subjects produced OS errors in the shoulder-free condition was because they flexed the shoulder excessively. US errors were less common, and were due to too little elbow flexion. Cerebellar endpoint errors were dramatically reduced in the shoulder-fixed condition (15% OS errors, 7% US errors, and 78% classified as No Error).

The cerebellar group produced the most abnormal torque profiles in the shoulder-free condition. The major abnormality was at the shoulder joint when they made OS errors. FIGURE 2 shows single trials of torques (normalized to body weight) from a control subject and from a cerebellar subject that produced an OS error. Trials are matched for peak tangential wrist velocity. The control subject's shoulder-muscle torque is nearly equal and opposite the interaction torque in shape, and offset by an amount that counteracts the nearly constant gravitational torque (FIG. 2A). Because of this, the control subject produced a fairly flat net shoulder torque and did not move much at the shoulder. The cerebellar subject produced a peak shoulder-muscle torque that was too large and slightly too late to counter the interaction torque (FIG. 2B). Because of this early imbalance, cerebellar subjects produced a substantial net shoulder torque, and flexed excessively at the shoulder.

We then tested whether this deficit was due to a problem with timing, magnitude scaling, and/or rate of torque production. FIGURE 2C shows the difference in timing of the peak shoulder muscle and interaction torque (see arrows in FIG. 2B) for the controls and cerebellar group in the different error categories. The peak shoulder-muscle torque occurred later than the interaction torque in all groups, but most substantially for the cerebellar OS and cerebellar No Error groups. Thus, the abnormalities in timing did not appear to be specific to the cerebellar OS errors. Instead, we saw a specific abnormality in the magnitude of the shoulder torques. FIGURE 2D shows that the shoulder-muscle torque was larger in the OS category than in all other categories (all $p < .02$). Shoulder interaction and gravitational torques were not different between categories. When we analyzed the rate of flexor-muscle torque production at the elbow and shoulder, we found that the control group produced comparable torque rates, but the cerebellar groups produced higher rates of torque at the shoulder compared with the elbow. Statistical comparison of the difference between shoulder and elbow torque rates showed that only the cerebellar OS group was significantly different than controls. This was due to the fact that the cerebellar OS group tended to produce slightly higher shoulder torque rates and lower elbow torque rates compared with controls.

As stated earlier, cerebellar endpoint error was reduced to near control levels when they moved in the shoulder-fixed condition (78% of trials were within control variability). Cerebellar subjects also showed a striking difference in the magnitude of EMG activity in the anterior deltoid (AD) muscle. FIGURE 3 shows a single EMG trace from a control subject and a cerebellar subject moving in both conditions. The cerebellar subject shown is highly representative of the group, with prolonged AD activity (either tonic or multiple phasic bursts) in the shoulder-free condition, and

---

where timing and magnitude measurements were made. (**C**) Timing of the shoulder interaction torque peak relative to the shoulder-muscle torque peak. (**D**) Magnitude of shoulder torque components. (**E**) Elbow and shoulder-muscle torque rates (calculated from time 0 to the peak torque at each joint).

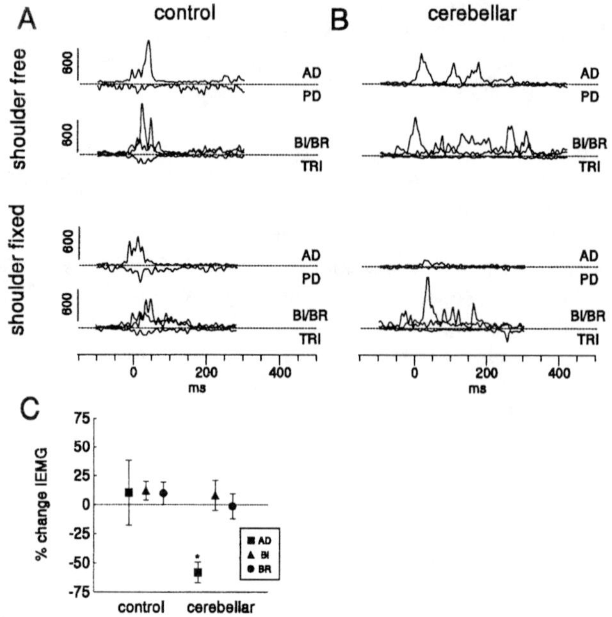

FIGURE 3. Single trial of EMG from (**A**) a control and (**B**) a cerebellar subject, moving under the shoulder-free condition (*top*) and shoulder-fixed condition (*bottom*). Trials are aligned on the start of movement (time 0) and drawn to the end of movement, prior to corrections. All trials are matched for peak tangential wrist velocity (control, shoulder free = 103 cm/s; control, shoulder fixed = 101; cm/s; cerebellar, shoulder free = 99 cm/s; cerebellar, shoulder free = 116 cm/s). (**C**) Percent change (mean ± SE) in integrated EMG from the shoulder-free to the shoulde-fixed conditions. Negative changes indicate decreased muscle activity in the shoulder-fixed condition.

nearly absent AD activity in the shoulder-fixed condition. Control subjects strongly activated the AD in both conditions, despite shoulder fixation in the second condition. FIGURE 3C shows the percent change in integrated EMG activity of the AD, biceps (BI), and brachioradialis (BR) from the shoulder-free to shoulder-fixed conditions. Control subjects increased the magnitude of activity in all three muscles. Cerebellar subjects increased EMG magnitude in the BI muscles, showed little change in the BR, and greatly decreased activity in the AD. AD activity was decreased by 58% in the cerebellar group and increased by 10% in the control group ($p < .05$).

EXPERIMENT 2: CEREBELLAR ADAPTATION TO NOVEL LOADS

Methods

In experiment 2, we determined whether cerebellar subjects show a deficit in adapting an arm movement (ball catching) to novel loads. To make a successful catch, an individual relies primarily on anticipatory muscle activity timed to occur

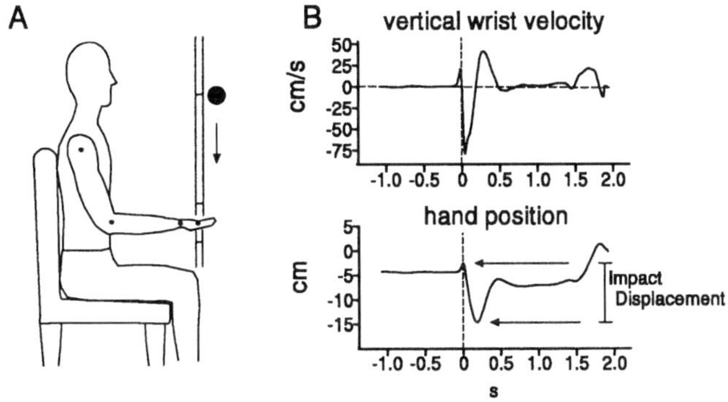

FIGURE 4. (**A**) Schematic of the catching task. (**B**) Typical vertical wrist velocity and position traces demonstrating impact selection and impact displacement calculation. The trial is aligned on impact, marked by the *dashed vertical line.*

before the moment of impact between the hand and the ball, and relies little on stretch reflex gain.[9] Nine subjects with cerebellar lesions and nine control subjects participated in this study. Cerebellar lesions were confirmed by imaging (CT or MRI). All cerebellar subjects underwent a neurological examination to determine the severity of the movement deficit and to check for signs of involvement of other motor structures. Informed consent was obtained from all subjects prior to testing.

All subjects caught two balls that were different weights, but identical in appearance. FIGURE 4A illustrates the experimental setup. A vertical pole was positioned next to the subject's outstretched hand, with markings indicating the top, bottom, and middle of a 10-cm "window." Subjects were instructed to catch the ball while attempting to maintain hand position within the window as they caught it. Each subject repeatedly caught one " light" ball and one "heavy" ball. Arm kinematics were recorded at 100 Hz and recorded trials were blocked as follows. The *baseline phase* consisted of 10–20 trials with one ball, the *adaptation phase* consisted of 25–35 trials with the other ball, and the *postadaptation phase* consisted of 15 additional trials with the first ball (to test an aftereffect). Subjects were not told when the balls would be switched, and thus had to adapt to the new ball weight. All subjects performed with light ball at baseline, the heavy ball during adaptation, and the light ball during postadaptation (light–heavy–light). Five of the nine subjects also performed with the heavy ball at baseline, the light ball during adaptation, and the heavy ball during postadaptation (heavy–light–heavy). After completing all of the catching trials, subjects were tested to determine if they could perceive the weight difference between the two balls.

In our analysis, we first determined impact (time of ball contact with the hand) using the vertical velocity of the wrist (FIG. 4B; velocity trace). For each trial, we then calculated impact displacement, which was defined as the vertical distance traveled by the wrist from impact to the first reversal in direction (FIG. 4B; position trace). Impact displacement was then plotted as a function of trial for each subject. Finally, we measured the rate of adaptation, or how quickly a subject's impact displacement values changed after repeated exposures to the new ball during the adaptation phase. The adaptation process for each individual was mathematically modeled using an exponential decay function.[10–12] The exponential decay time constant was taken as our measure of the rate of adaptation. This value represents the number of catches that it would take to obtain $(1 - e^{-1})$ or ~63.2% of the final adaptation.[10–12] Student's *t*-tests were used to assess differences between the control and cerebellar groups for adaptation coefficients.

Results

We found that cerebellar subjects adapted much more slowly than controls to the changed ball weight, regardless of whether it was heavy or light. FIGURE 5 shows examples of impact displacement plotted as a function of trial and the corresponding curve fits for both sets of adaptations (light–heavy–light and heavy–light–heavy). Typical control subjects are shown for both conditions (FIG. 5A and 5C). The control group adapted rapidly to the heavy ball (group mean 1.7 ± 0.2 trials) and light ball (1.6 ± 0.2 trials). Note that the controls shown also produced a large error in the op-

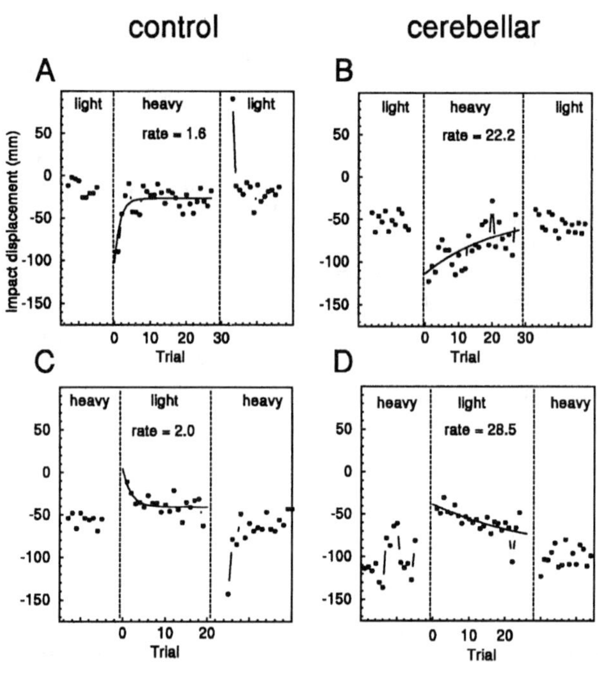

FIGURE 5. Impact displacement values plotted as a function of trial. Examples of (**A**) a control subject adapting to the heavy ball; (**B**) a cerebellar subject adapting to the heavy ball; (**C**) a control subject adapting to the light ball; (**D**) a cerebellar subject adapting to the light ball. Negative-impact displacements indicate displacement in the downward direction, while positive values indicate displacement upward. *Vertical dashed lines* separate baseline, adaptation, and postadaptation phases. The *small lines* connecting the impact displacement values are provided to show sequential order of the trials. The *large solid line* through the data points in the adaptation phase represents the exponential curve fit.

posite direction of the adaptation (negative aftereffect) on the first trial of the postadaptation phase, indicating that they had stored the adaptation. FIGURE 5B and 5D are typical cerebellar subjects who adapted much more slowly than controls to both the heavy and light ball. As a group, cerebellar subjects adapted more slowly to the heavy (30.9 ± 8.7) versus light balls (24.6 ± 8.1), though their adaptations to either were much slower that controls ($p < .001$). Cerebellar subjects also tended not to produce a robust negative aftereffect, as evidenced by the lack of a large error on the first trial of the postadaptation phase. Finally, neither group had any difficulty perceiving which ball was the heavier one at the end of the task.

DISCUSSION

Our results support the idea that the cerebellum acts to modulate muscle activity across multiple joints in anticipation of the mechanical forces generated by one's own movement and by forces in the environment. In our first experiment, we found that people with cerebellar damage had difficulty controlling shoulder movement when performing a fast elbow flexion. The cerebellar subjects could make nearly normal elbow movements when the shoulder was mechanically constrained. Reduction of the most prevalent error type (overshoot) was due to a decrease in unwanted shoulder flexion movements caused by overcompensation for interaction torques. We also found that, unlike controls, cerebellar subjects substantially reduced shoulder-muscle activity when moving in the shoulder-fixed condition. We speculate that the improvement in the cerebellar groups performance in the shoulder-fixed condition could be due to the absence of interaction torques, reduction of the number of muscles to be controlled, or a combination of these two factors.

In our second experiment, we found that cerebellar subjects adapted their catching much more slowly than controls. This was demonstrated by increased adaptation rates and by lack of negative aftereffects. We know from prior work that people must rely on anticipatory muscle activity timed to occur *before* the moment of impact between the hand and the ball to make this adaptation.[5,6,9] These data suggest the cerebellum is needed to acquire the appropriate anticipatory adjustments for environmental forces (loads) through trial-and-error practice.

Based on this and other studies, we suggest that the cerebellum is involved in scaling the relative activities of multiple muscles to account for various forces (including interaction torques and external forces) that oppose or assist the intended movement.[1,13,14] We think that the cerebellum may be constantly adapting motor output for this purpose, possibly through error feedback from previous movements. This contribution would be critical in ensuring that the limb segments are moved at the correct rate and extent relative to one another. This type of control would also have to be predictive in nature, given that many movements are too fast for on-line feedback processes to operate. We think that several lines of evidence are consistent with these ideas. First, cerebellar subjects have difficulty scaling the relative activities of agonist and antagonist muscles to move a single segment.[15–17] These single-joint deficits worsen when inertia is added to the moving segment.[18] Second, much more striking deficits are observed when multiple muscles are used to control a multijointed reaching movement. Multijointed movement deficits are more than one would expect given the single-joint deficits, implying that the single-joint deficits are not simply summing. We think that the additional mechanical complexity of interaction torques during multijointed movements may account for this difference. Finally, we have shown that people with cerebellar damage are slow or unable to adapt arm movements to different external forces through trial-and-error practice and do not show evidence of storing any improvement via a negative aftereffect. We suggest that the cerebellum plays an important role in combining the appropriate anticipatory activity across several muscles and modifying that muscle activity in response to changing demands through trial-and-error practice.

ACKNOWLEDGMENTS

This work was supported in part by NIH grants HD01199, HD07434, the Foundation for Physical Therapy, and the Neurology Section of the APTA. Thanks to Tom Thach, Tod Martin, Jeff Keating, Kathy Zackowski, and Catherine Lang for contributions to these experiments.

REFERENCES

1. Bastian, A.J., T.A. Martin, J.G. Keating & W.T. Thach. 1996. Cerebellar ataxia: abnormal control of interaction torques across multiple joints. J. Neurophysiol. **76:** 492–509.
2. Topka, H., J. Konczak & J. Dichgans. 1998. Coordination of multi-joint arm movements in cerebellar ataxia: analysis of hand and angular kinematics. Exp. Brain Res. **119**(4)**:** 483–492.
3. Topka, H. *et al.* 1998. Multi-joint arm movements in cerebellar ataxia: abnormal control of movement dynamics. Exp. Brain Res. **119**(4)**:** 493–503.
4. Cooper, S.E., J.H. Martin & C. Ghez. 2000. Effects of inactivation of the anterior interpositus nucleus on the kinematic and dynamic control of multijoint movement. J. Neurophysiol. **84**(4)**:** 1988–2000.
5. Lang, C.E. & A.J. Bastian. 1999. Cerebellar subjects show impaired adaptation of anticipatory EMG during catching. J. Neurophysiol. **82:** 2108–2119.
6. Lang, C.E. & A.J. Bastian. 2001. Additional somatosensory information does not improve cerebellar adaptation during catching. J. Clin. Neurophysiol. **112**(5)**:** 895–907.
7. Bastian, A.J., K.M. Zackowski & W.T. Thach. 2000. Cerebellar ataxia: torque deficiency or torque mismatch between joints? J. Neurophysiol. **83**(5)**:** 3019–3030.
8. Winter, D.A. 1990. Biomechanics and Motor Control of Human Movement, 2nd ed.: 11–102. Wiley-Interscience. New York.
9. Bennett, D.J., M. Gorassini & A. Prochazka. 1994. Catching a ball: contributions of intrinsic muscle stiffness, reflexes, and higher order responses. Can. J. Physiol. Pharmacol. **72:** 525–534.
10. Keating, J.G. 1994. The role of the cerebellar cortex in the performance and adaptation of a targeted ballistic movement of the wrist. Ph.D. Thesis, Washington Univ., St. Louis, MO.
11. Deuschl, G. *et al.* 1996. Adaptation motor learning of arm movements in patients with cerebellar disease. J. Neurol. Neurosurg. Psychiatry **60:** 515–519.
12. Martin, T.A. *et al.* 1996. Throwing while looking through prisms: I. Focal olivocerebellar lesions impair adaptation. Brain **119:** 1199–1211.
13. Schweighofer, N., M.A. Arbib & M. Kawato. 1998. Role of the cerebellum in reaching movements in humans. I. Distributed inverse dynamics control. Eur. J. Neurosci. **10**(1)**:** 86–94.
14. Schweighofer, N., J. Spoelstra, M.A. Arbib & M. Kawato. 1998. Role of the cerebellum in reaching movements in humans. II. A neural model of the intermediate cerebellum. Eur. J. Neurosci. **10**(1)**:** 95–105.

15. HALLETT, M., B.T. SHAHANI & R.R. YOUNG. 1975. EMG analysis of patients with cerebellar deficits. J. Neurol. Neurosurg. Psychiatry **38:** 1163–1169.
16. FLAMENT, D. & J. HORE. 1986. Movement and electromyographic disorders associated with cerebellar dysmetria. J. Neurophysiol. **55:** 1221–1233.
17. HORE, J., B. WILD & H.-C. DIENER. 1991. Cerebellar dysmetria at the elbow, wrist, and fingers. J. Neurophysiol. **65:** 563–571.
18. MANTO, M., E. GODAUX & J. JACQUY. 1994. Cerebellar hypermetria is larger when the inertial load is artificially increased. Ann. Neurol. **35:** 45–52.

The Nodulus and Uvula: Source of Cerebellar Control of Spatial Orientation of the Angular Vestibulo-Ocular Reflex

BERNARD COHEN,[a] PADMORE JOHN,[a] SERGEI B. YAKUSHIN,[a]
JEAN BUETTNER-ENNEVER,[b] AND THEODORE RAPHAN[c]

[a]*Department of Neurology, Mount Sinai School of Medicine, New York, New York, USA*

[b]*Department of Anatomy, University of Munich, Munich, Germany*

[c]*Department of Computer and Information Science, Brooklyn College of CUNY, Brooklyn, New York, USA*

ABSTRACT: The nodulus and rostral-ventral uvula of the vestibulo-cerebellum play a critical role in orienting eye velocity of the slow component of the angular vestibulo-ocular reflex (aVOR) to gravito-inertial acceleration (GIA). This is done by altering the time constants of "velocity storage" in the vestibular system and by generating "cross-coupled" eye velocities that shift the eye velocity vector from along the body yaw axis to the yaw axis in a spatial frame. In this report, we show that eye velocity generated through the aVOR by constant velocity centrifugation in the monkey orients to the GIA in space, regardless of the position of the head with respect to the axis of rotation. We also show that, after removal of the nodulus and rostral-ventral uvula, the spatial orientation of eye velocity to the GIA is lost and that eye velocity is then purely driven by the semicircular canals in a body frame of reference. These findings are further confirmation that these regions of the vestibulo-cerebellum control spatial orientation of the aVOR.

KEYWORDS: vestibulo-ocular reflex; velocity storage; otoliths; monkey; eye velocity; centrifugation

INTRODUCTION

The otolith organs make a major contribution to spatial orientation by detecting the direction of the head vertical with respect to a gravitational reference. From activity arising in these end-organs, the vestibular system generates appropriate commands to align the body, head, and eyes and their angular velocities to that

Address for correspondence: Bernard Cohen, M.D., Mount Sinai School of Medicine, Department of Neurology, Box 1135, 1 East 100th Street, New York, NY 10029. Voice: 212-241-1943; fax: 212-831-1610.
bernard.cohen@mssm.edu

Ann. N.Y. Acad. Sci. 978: 28–45 (2002). © 2002 New York Academy of Sciences.

reference. One aspect of this orientation is induced by deviations of the head with regard to gravity, which cause the vertical axes of the eyes to align with the direction of gravity or with gravito-inertial acceleration (GIA) in pitch, roll, and yaw through the linear vestibulo-ocular reflex (lVOR).[1,2] [Note that GIA is defined as the sum of all translational, centripetal, and gravitational linear accelerations acting on the head. When subjects are stationary, the upward acceleration of gravity is the GIA.]

There is another, perhaps more subtle vestibular mechanism that generates spatial orientation in the angular vestibulo-ocular reflex (aVOR) through a process termed velocity storage.[3–5] Velocity storage, which was originally modeled by an integrator with a time constant to explain responses in one dimension,[3] extends the low-frequency characteristics of the aVOR in three dimensions and encodes orientation information about the head in space.[4,5] The presence of a neural network that operates as an integrator with a time constant was inferred from observations that time constants of horizontal vestibular nystagmus during constant velocity rotation were considerably longer than the activity generated by the semicircular canals.[6,7] Average time constants of vestibular nystagmus in humans, for example, are approximately 15–20 s, whereas cupular activation that produced the nystagmus has a time constant of only about 4 s.[8] Activity stored by this network is utilized to cancel post-rotatory responses that occur at the end of rotations if the rotations last longer than 5–10 s.[3,9,10] Velocity storage maintains eye velocity at a constant level during optokinetic stimulation[6,11] and is responsible for optokinetic after-nystagmus (OKAN) that persists after optokinetic nystagmus (OKN) if subjects are in darkness.[3,6,11–14] Velocity storage also supports generation of compensatory eye velocities during otolith activation when rotating about axes tilted from the vertical (off-vertical axis rotation, OVAR)[15–18] and during circular locomotion.[19,20]

An analysis of the eye velocities in three dimensions produced by optokinetic and vestibular stimuli with the head tilted with regard to gravity has shown that an important aspect of velocity storage is that it is spatially oriented; it directs eye velocity toward the direction of gravity[4,5,21–23] or the GIA.[24,25] This spatial orientation is present for all modes of vestibular or optokinetic activation of velocity storage in all mammalian species that have been tested, including humans, monkeys, and rabbits (see Ref. 2 for review). Moreover, despite weak OKAN in humans,[26,27] the spatial orientation of OKN is robust.[28,29] For example, when subjects are upright so that their yaw or body axis is aligned with the acceleration of gravity, visual or vestibular stimuli around the yaw axis produce pure horizontal eye movements. If the subjects are tilted in pitch or roll, but the rotation of the visual surround is maintained around the yaw axis of the head, a cross-coupled vertical or roll eye velocity component appears that shifts the axis of rotation of the eyes. The consequence is that the eye velocity vector tends to align with the spatial vertical rather than with the yaw axis of the head and body.

Since velocity storage is common to optokinetic and vestibular responses,[3] eye velocity declines along an orientation vector during postrotatory nystagmus as well as during OKAN when the head is tilted with regard to gravity. For example, when monkeys are stopped in side-down positions after rotation about an axis tilted from gravity (OVAR), the post-OVAR eye velocity trajectory approaches an orientation vector similar to that established during OKAN at the same head orientation.[22] This shows that velocity storage is the primary mechanism for orienting eye velocity in the aVOR. This conclusion is supported by studies utilizing centrifugation,[30–32] as well as by the orientation of the eyes to gravity during caloric stimulation.[33] The spatial orientation of velocity storage was modeled by altering the yaw, pitch, and roll time constants and cross-coupling parameters of the velocity storage system matrix.[25,34] From the model, it would be predicted that the magnitude and the direction of the GIA encoded by otolith afferents determine the degree to which the time constants associated with the aVOR are modified by linear acceleration.

The orientation shifts associated with the spatial orientation of velocity storage during postrotatory nystagmus and in response to centrifugation with the head fixed in the upright appear to be mediated through the nodulus and uvula since the orientation of eye velocity to the GIA disappears when these regions are ablated.[31,34–37] The neural connections and organization of the nodulus and uvula are appropriate for control of spatial orientation. There is primary input from the labyrinths,[38,39] including the otoliths.[40] Secondary projections come from the vestibular nuclei[41–43] and tertiary pathways enter via climbing fibers from the inferior olive.[41,43,44] The cerebellar cortex across the vermis of the nodulus and uvula is organized into microzones that process information relative to movement in canal coordinates.[45] Nodulus Purkinje cells project to the non-VOR relay zones of medial vestibular nuclei[46] where neurons related to velocity storage are found.[47] The parametric modifications of the VOR by the GIA are likely mediated by GABAergic projections from Purkinje cells to the vestibular nuclei[48–50] since time constants of nystagmus can be reduced by administration of baclofen, a GABA-B receptor agonist,[51] or by electrical stimulation of the nodulus and uvula.[52] Thus, the nodulus and uvula have the appropriate connections and processing to sense the tilts of the GIA relative to the head vertical and to alter the reference frame of velocity storage.

From studies of the spatial orientation properties of velocity storage, we have concluded that information related to movement in a terrestrial environment, which activates the canals and otoliths, is processed in a head-based canal-coordinate frame that is rotated relative to the cardinal axes of the head.[2,53,54] Velocity storage, on the other hand, processes information in a spatially linked coordinate frame whose yaw axis (yaw eigenvector) is related to a combination of the head vertical and the net GIA.[2,4,5,33,55] In models of velocity storage, it was assumed that the pitch and roll axes of velocity always remain aligned with the head pitch and roll axes,[4,5] while the yaw axis is spatially oriented to the GIA. Together, these axes form a nonorthogonal coordinate frame. Experiments to date have been consistent with this model whether using OKN, rotational stimuli, postrotatory nystagmus, centrifugation that tilted the GIA relative to the rotation axis, or caloric stimulation.[4,5,22,28–30,32,33,56] However, in experiments using OVAR and centrifugation, the axis of rotation and the head yaw axis have always been aligned and it has not been possible to determine whether differences between the rotation axis and head yaw axis would alter the yaw eigenvector of velocity storage. Only with OKN was it possible to generate a stimulus axis that was different from the axes of the head and the GIA.[5,56] One purpose of this study was to determine whether the yaw axis eigenvector of velocity storage would be altered in an experimental condition in which static tilts of the head during centrifugation were used to provide disparities between the axis of rotation, the yaw axis of the head, and tilts of the GIA relative to the head. A second purpose was to determine whether control of the eye velocity vector and spatial orientation of the yaw eigenvector would be influenced by lesions of the nodulus and uvula as in previous studies.[34]

METHODS

Animal Preparation

Two cynomolgus monkeys were used in this study. The experiments conformed to the Principles of Laboratory Animal Care (1996) and were approved by the Institutional Animal Care and Use Committee. One (M9865) was a control animal and the other (M9358) was tested after nodulectomy. The animals were prepared under sterile conditions and general anesthesia. An acrylic ring was surgically implanted to stabilize the head so that the head could be held painlessly in the stereotaxic position during experiments,[57] and two search coils were attached to the left eye. A perilimbal coil measured the horizontal and vertical components of eye movements relative to the orbit[58,59] and a second coil, inserted under the superior oblique and superior rectus muscles, measured torsional movements relative to the orbit.[60]

Experimental Paradigm

Testing was conducted in darkness on a centrifuge composed of a beam mounted on a rotator (Neurokinetics, Pittsburgh, PA). Animals sat in a primate chair with their heads fixed in the center of a 53-cm^3 Plexiglas coil box that was attached to the primate chair (FIG. 1B). The primate chair and coil box were mounted rigidly with the animal centered 70 cm out on the arm of the centrifuge. The monkeys were oriented so that they faced the direction of motion with the left ear out. During each test, the head was either upright or tilted in roll ±15° or ±30° from the upright. In this paper, we utilized the upright position (0°) and head tilts 30° from the upright toward the axis of rotation[30] (FIG. 1B). The tilts of the head in roll were about an axis through the base of the skull, close to the labyrinths. Animals were tested with steps of angular velocity in darkness. Rotation began with an angular acceleration of $20°/s^2$ to a final velocity from 30°/s to 270°/s to produce 0.08 to 1.0 *g* of centripetal acceleration (A_c, FIG. 1C). This tilted the GIA 5° to 45° relative to the head. The trials ended when the per-rotatory nystagmus in darkness had declined to zero. The chair was then decelerated with the same stimulus profile and the room was illuminated for 5–10 min before the next rotation began. Putting the animals into light when they were stationary damped any remaining postrotatory nystagmus or secondary after-nystagmus. Levels of centripetal acceleration were varied in a pseudorandom sequence during testing. The duration of each trial generally lasted about 1 min in normal animals and 2–3 min in the nodulo-uvulectomy animal.

Coordinate Frame for the Head and Eyes

A head-fixed coordinate frame was defined according to a right-hand rule with the roll or naso-occipital (x) axis positive out of the nose, the interaural or pitch (y) axis positive out of the left ear, and the head-vertical or yaw (z) axis positive up (FIG. 1A). The coil box held two sets of 43-cm field coils that generated orthogonal oscillating magnetic fields at a frequency of 24 kHz. The axes of the field coils were aligned with the spatial y- and z-axes. When the head was either upright or tilted, the left eye, on which the coils were implanted, was at the center of the coil field. Since the field coils were fixed relative to space, eye movements induced by rotation were recorded in spatial coordinates, not in a head-centric frame of reference. Therefore,

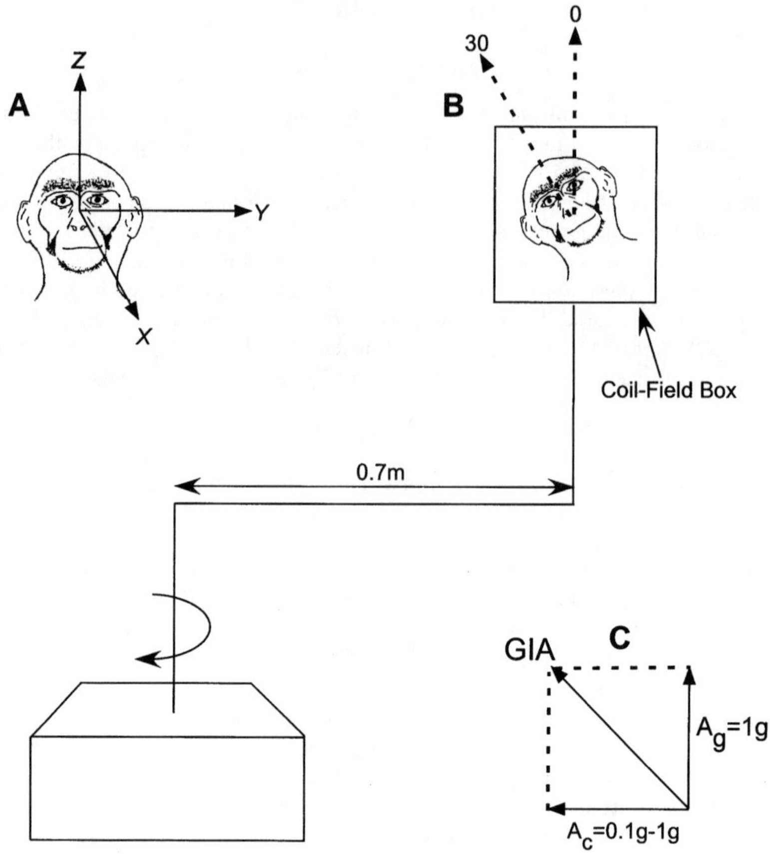

FIGURE 1. Experimental scheme. **(A)** Head coordinate frame. Positive directions of rotation around the *x*, *y*, and *z* axes, defined by the right-hand rule, are shown by the arrows. The x- or naso-occipital axis is positive out the nose, the y- or pitch axis is positive out the left ear, and the z- or yaw axis is positive out the top of the head. **(B)** During centrifugation, monkeys sat with their head in the middle of a coil-field box, 0.7 m from the axis of rotation. The animals were facing forward during rotation to the right with the left ear out. The head was either upright or tilted 30° from the upright, toward the axis of rotation and the tilt of the GIA. **(C)** Description of linear accelerations during centrifugation, as shown in part B. The upward acceleration of gravity (A_g) was 1 *g*, and the centripetal acceleration (A_c) ranged from 0.1 to 1 *g*. These accelerations combined with the acceleration of gravity to tilt the GIA vector in toward the axis of rotation from 5° to 45°. Tilt of the vector by 45°, as shown by the oblique upward arrow, which is 1.4 *g* in magnitude, represents a combination of the 1 *g* of centripetal and gravitational accelerations.

the horizontal and vertical gaze positions and velocities shown in FIGURES 3 and 5 (see below) represent eye rotations about spatial z- and y-axes, respectively. Eye velocities were calibrated by rotating the animals at 30°/s about the roll, pitch, and yaw axes of the head in light. It was assumed that horizontal and vertical gains were unity in this condition.[5,58,61] Since we will only report on horizontal and vertical movements, calibration of roll will not be considered further. In the figures, eye position changes and eye velocities are in spatial coordinates and are positive for left, down, and clockwise rotations of the eyes relative to the spatial vertical.

Data Collection and Processing

Data were recorded with amplifiers having a band-pass of dc to 40 Hz. The equipment was controlled and data were acquired with a computer. Voltages were digitized at 600 Hz/channel with 12-bit resolution and stored on disk. Eye position voltages were smoothed and digitally differentiated by finding the slope of the least-squares linear fit, corresponding to a filter with a 3-dB cutoff above 40 Hz, the cutoff frequency of the filters used for data acquisition. Saccades were detected and removed using maximum likelihood ratio criteria,[62] and the composite slow phase velocities were smoothed with a 2-Hz cutoff filter. To obtain eigenvectors of velocity storage, phase plots were constructed from the slow phase velocity of the per-rotatory nystagmus in the spatial y-z plane. This was done by combining the spatial yaw and pitch slow phase velocities, starting approximately 10 s after the angular acceleration had ended. By this time, the cupula would have returned to close to its original position and would no longer be activating central vestibular networks. Thus, the per-rotatory nystagmus would then be primarily driven by the activity that had been stored in the central velocity storage integrator,[3] and the eigenvectors of the induced nystagmus could be determined from the slope of the decline in eye velocity toward zero.[22,63] The phase plots were fitted with least-squares linear approximations to determine the average slope of the eigenvector in spatial coordinates. The average spatial slopes were then expressed as a function of the centripetal acceleration.

Cerebellar Ablations

After induction, the animal was intubated and surgery was performed under gas (isofluorane) anesthesia. The cerebellum was exposed, the secondary fissure was identified, and a dissection plane was carried into the vermis toward the floor of the fourth ventricle. The nodulus and uvula were removed to the lateral margins of the vermis by suction ablation (FIG. 2a–2c). Care was taken to remove the lateral rostral portions of the nodulus that lie close to the paramedian lobule (FIG. 2a and 2c). The cerebellar nuclei and surrounding white matter were not penetrated. The site of exposure was covered by artificial dura and the edges of the scalp approximated. Promethazine (12.5 mg) was used to reduce postoperative vomiting that frequently occurs after such ablations. The animal was allowed to recover for a month before testing was reinstituted.

Determination of Extent of Lesions

At the end of testing, the animal with the cerebellar lesion (M9358) was perfused through the heart under deep anesthesia with saline and a paraformaldehyde/formalin solution. The brain was removed, fixed, and processed for anatomical study. Upon histological examination, the cerebellar lesion was confined to the posterior vermis, which was removed up to the secondary fissure (FIG. 2b and 2c). Lobule VIIIB was partially damaged. As shown in the midsagittal and lateral sections in FIGURE 2a–2c,

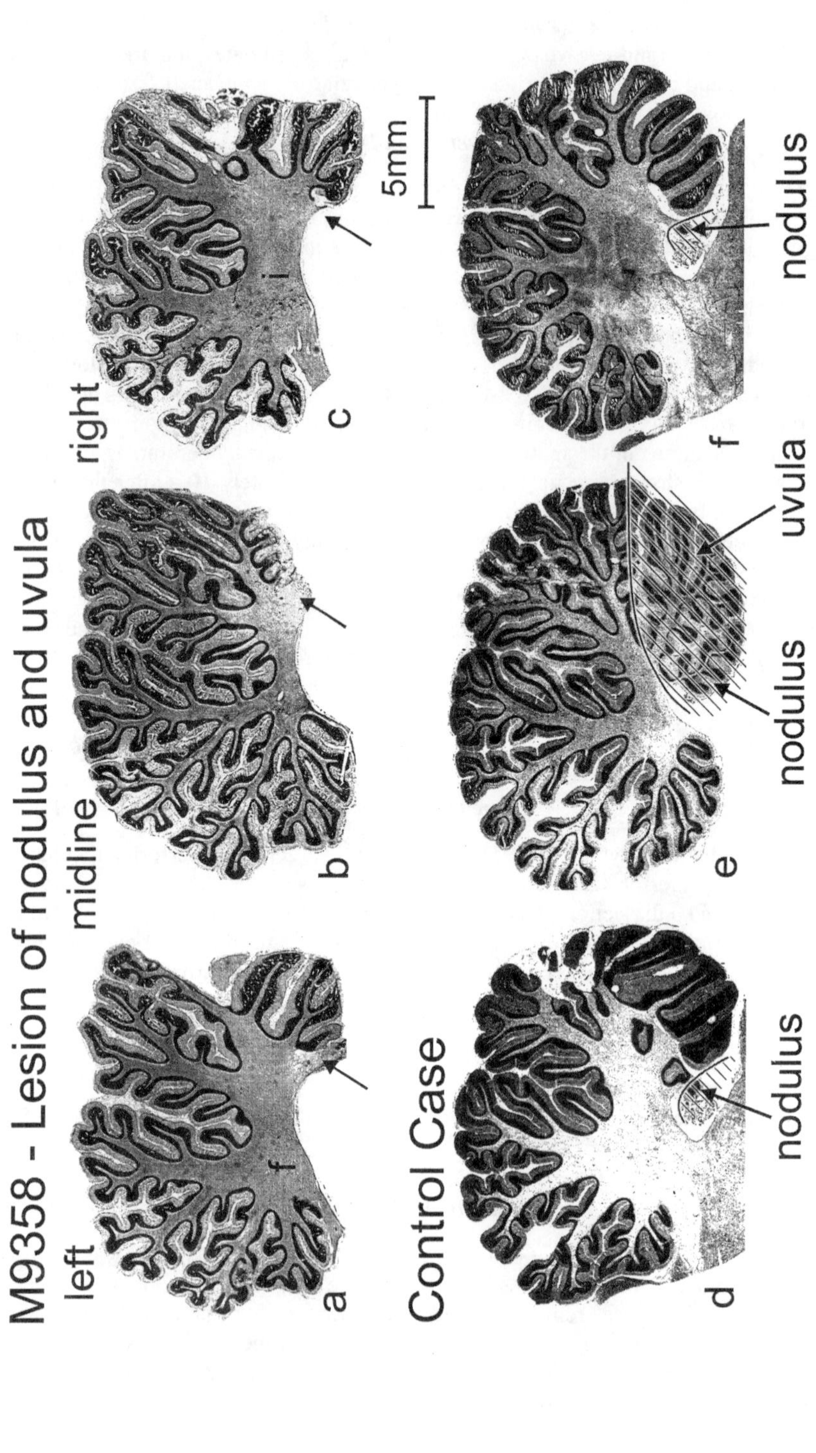

FIGURE 2. *See following page for legend.*

the nodulus and uvula were completely removed (compare with FIG. 2d–2f). The fastigial and interpositus nuclei and the surrounding white matter were intact, but there was slight damage to the stalk of the paraflocculus in the more rostral lateral portions on one side. The remainder of the cerebellum was intact.

RESULTS

Orientation of Eye Velocity to the GIA in Normal Animals

The spatial orientation of the slow phase velocity of per-rotatory nystagmus to the GIA during centrifugation with the head fixed in the upright position has been well demonstrated in previous publications.[24,34] Of interest here was whether the slow phase velocity of animals would show the same spatial orientation to the GIA with the head tilted 30° relative to the rotation axis. As previously reported in many studies, when animals are on-center and rotated about a vertical axis with the head upright, the nystagmus is predominantly spatially horizontal. Thus, for on-center rotation, the velocity vector of the nystagmus lay close to the spatial vertical, which in this instance was coincident with the dorsoventral (z) axis of the head. Small shifts in the velocity vector from the upright were produced by the presence of weak spontaneous upward slow phase velocity, which is frequently present in normal monkeys in darkness.

When the head was tilted, there was no shift in the spatial axis of eye rotation, that is, in the velocity vector of the nystagmus, during on-center rotation (not shown), and the velocity vector lay close to the spatial vertical, although the axis of eye rotation was now oblique with regard to the head. Findings were similar when the animal was moved off-axis and centrifuged at a low velocity that produced 0.08 *g* of centripetal acceleration, causing a 4.6° tilt of the GIA relative to the head (FIG. 3A). In this instance, the animal was rotated to its right (angular velocity along a −z direction), facing toward the direction of motion, with the left ear out and the head was tilted 30° toward the axis of rotation, as in FIGURE 1B. Consequently, the head was tilted 25.4° beyond the tilt of the GIA toward the axis of rotation. The induced peak spatial eye velocity was close to 30°/s. There was approximately 4°/s of vertical eye velocity in the 30° head-tilted position, and the eigenvector calculated from the tail of the response[22,33,34] was computed to be ≈13° (FIG. 3B). Orientation of the eye velocity vector toward the GIA plus spontaneous nystagmus could have contributed to the bias in the direction of the eigenvector beyond the GIA in this condition.

FIGURE 2. (a–c) Cresyl violet sections from the cerebellum of M9358 showing loss of the nodulus and uvula. **(d–f)** Corresponding sections from a normal cerebellum with regions of the nodulus and uvula in medial (e) and lateral (d, f) portions marked by the arrows. The structures that were removed were located in the hatched area below the solid curved lines. The upward arrow in panel a shows minor damage in the paramedian lobe, that in panel b shows associated damage of the rostral portion of lobule VIII, and that in panel c shows a small lesion in the white matter just next to the paramedian lobule on the opposite side. The nodulus and uvula were removed completely in both midline (b) and lateral sections. The calibration under panel c is 5 mm.

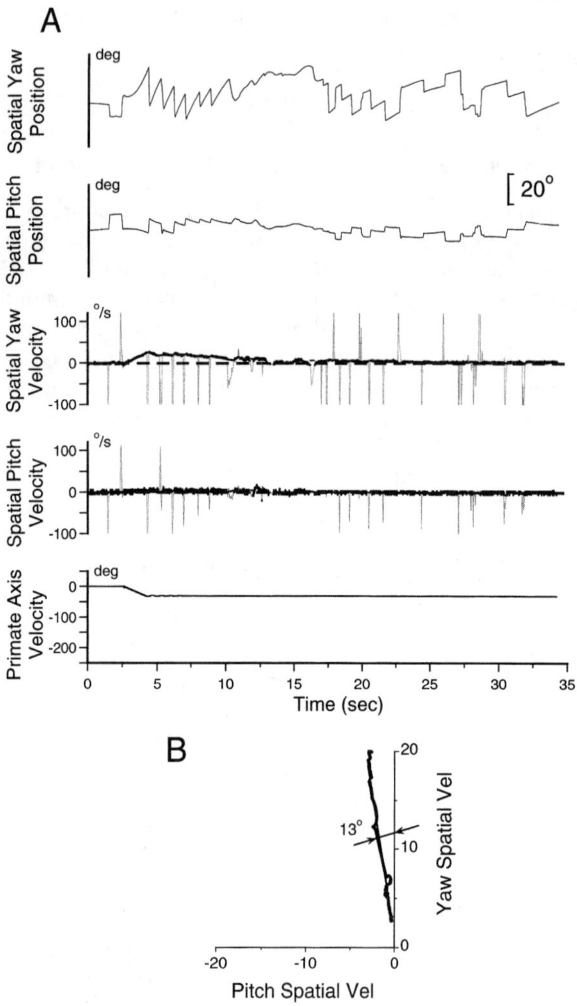

FIGURE 3. **(A)** Nystagmus induced by centrifugation at 30°/s, which produced 0.08 *g* of centripetal acceleration. The animal was positioned on the arm of the centrifuge facing the direction of motion with its head tilted to the right 30° toward the axis of rotation. Rotation was to the animal's right. Spatial horizontal and vertical eye positions (spatial yaw pos. and spatial pitch pos.) and velocities (spatial yaw vel. and spatial pitch vel.) are shown. The bottom trace shows stimulus velocity. Calibrations for position and velocity are shown on the ordinates and for time on the abscissa. **(B)** Eigenvector generated by combining yaw and pitch velocities. In this case, the velocities approached the origin along a slope of 13°, which was the eigenvector for this stimulus. Movements and velocities to the left and down are positive in this and succeeding figures.

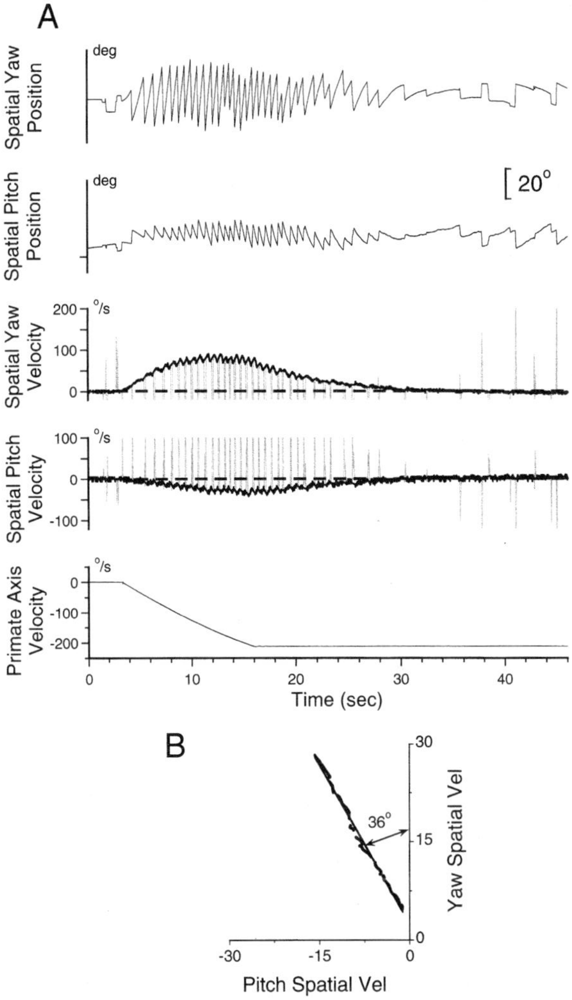

FIGURE 4. (A) Nystagmus induced by centrifugation at 214°/s, which produced 1 *g* of centripetal acceleration. Scheme as in FIGURE 3. The animal was positioned on the arm of the centrifuge facing the direction of motion with its head tilted to the right 30° toward the axis of rotation. Rotation was to the animal's right. The larger centripetal acceleration than in FIGURE 3 produced a relatively larger cross-coupled vertical component. **(B)** This shifted the eigenvector by 36° toward the GIA, which was tilted 45° relative to the axis of rotation.

When the animal was centrifuged at a higher velocity (218°/s), there was 1 *g* of centripetal acceleration and the GIA was tilted 45° toward the axis of rotation. The head was in the same position as shown in FIGURE 1B and, again, the animal was facing the direction of movement with the left ear out. Despite the fact that there was only a 15° difference between the yaw axis of the head and the tilted GIA, eye velocity developed a significant upward (−y) spatial pitch as well as a spatial yaw component (+z) (FIG. 4A). The increase in the upward vertical component rotated the eigenvector by 36° from the spatial vertical, 6° beyond the head yaw axis (FIG. 4B). If the nystagmus had been around the axis of rotation, the induced slow phase velocity would have been along the spatial vertical, oblique down and to the left for rotation to the animal's right. Instead it was almost head horizontal in this position. This could be explained by the fact that when the GIA is close to the head yaw axis, the time constant of the vertical response is short and the eigenvector is close to the head yaw axis. Tilts of the GIA relative to the head yaw axis will modulate the direction of the GIA, as demonstrated in FIGURE 5. As previously noted,[30,32,34] the time constant of the yaw component of nystagmus is shorter when the GIA is tilted relative to the head. Consistent with this, the time constant of the yaw component in FIGURE 3 was 7 s in response to a tilt of the GIA of 4.6°, while the time constant of the yaw component in FIGURE 4, where there was a 15° tilt of the GIA relative to the head vertical, was 4 s.

Tilts of the eye velocity vector in space, produced by varying amounts of centripetal accelerations from 0.08 to 1 *g*, are shown in FIGURE 5 (open circles).

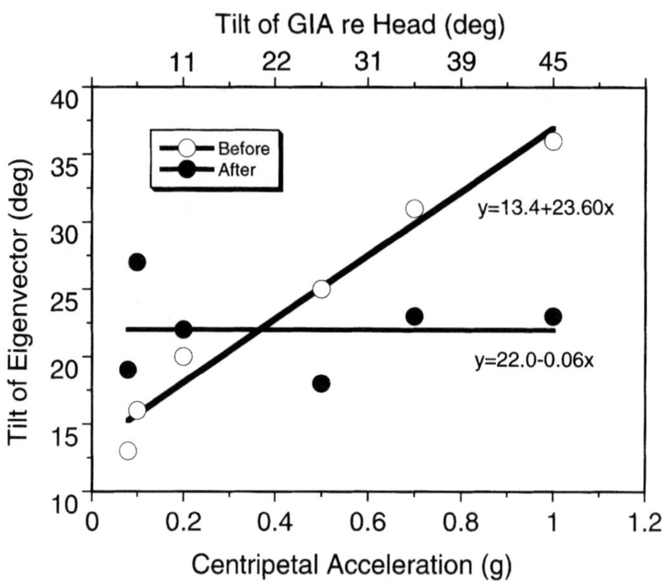

FIGURE 5. Relationship between centripetal linear accelerations from 0.08 to 1.0 *g* during rotation at different velocities that produced tilts in the GIA relative to the head up to 45° (abscissa). The angle of the eigenvector relative to the spatial vertical produced by each of these centrifugations is plotted on the ordinate. Normal monkey (open circles: include data from FIGS. 3 and 4): There was a linear increase in the tilt of the eigenvector for increases in the centripetal acceleration ($R = 0.986$). The sensitivity (slope) was 2.4° of tilt of the eigenvector for each increment of 0.1 *g* of centripetal acceleration. Nodulo-uvulectomy monkey (solid circles: include data from FIG. 6). The least-squares line of the data was flat, indicating there was no relation between the imposed centripetal acceleration and the tilt of the eigenvector.

There was a progressive increase in the tilt of the axis of eye velocity in space toward the GIA as the centripetal acceleration was increased. The increase in tilt angle as a function of centripetal linear acceleration was approximately linear with a sensitivity (slope) of 24° of tilt of the axis of eye velocity for 45° of tilt of the GIA relative to the acceleration of gravity and to the axis of rotation. The line crossed the ordinate at 13°, which was the background shift of the eigenvector produced by 4.6° of tilt of the GIA relative to the head in this animal (see FIG. 3B). Similar correlations between the spatial axis of eye rotation and the centripetal linear acceleration were found for head fixations in other tilt positions relative to the upright (not shown). These data indicate that the orientation of velocity storage to the GIA is maintained in three dimensions in the normal animal, even when the head is tilted in space.

Orientation of Eye Velocity after Removal of the Nodulus and Uvula

There was a striking difference in the eye velocities generated by centrifugation after the nodulus and uvula had been removed. As in previous studies,[36] the time constant of the nystagmus induced by rotation became longer. In the example shown in FIGURE 6A, the time constant of decline of the spatial horizontal component was 17 s, compared to a time constant of 4 s in the normal animal (FIG. 4) in response to the same stimulus. Consistent with the longer time constant, the animal tended to develop periodic alternating nystagmus after rotation or if left in darkness. In addition, the velocity storage component driving the induced nystagmus could not be discharged, that is, dumped, by exposure to a stationary visual stimulus as in normal animals.[36] Finally, there was striking direction-changing positional nystagmus in this animal, with the slow phase velocity being directed toward gravity for roll tilts as small as 10° from the upright.[65] This positional nystagmus was similar to that previously noted after muscimol injections into the nodulus[66] and will not be considered further here.

After nodulo-uvulectomy, the eye velocities that were induced by centrifugation were predominantly oriented to the axis of rotation rather than to the GIA. In response to the 1 *g* of centripetal acceleration, there was strong horizontal slow phase velocity (FIG. 6A, third trace), but much less vertical cross-coupled velocity (FIG. 6A, fourth trace). The eigenvector of the eye velocity in response to 1 *g* of centripetal acceleration was 23° (FIG. 6B) was close to the value obtained for centrifugation that produced a centripetal-acceleration of only 0.08 *g* (FIG. 5). The response to 1 *g* after nodulo-uvulectomy was also substantially different from the orientation of the eigenvector to the GIA in FIGURE 4B for a similar centripetal acceleration. Thus, the ability of the animal to orient its eigenvector to the GIA had been lost.

Data over a complete range of centripetal accelerations from 0.08 to 1 *g* were consistent with this finding; that is, there was no increase in the tilt of the axis of eye velocity despite increases in centripetal acceleration from 0.08 to 1 *g* (FIG. 5, solid circles, solid line). The same was true for rotation with the head tilted. Eye velocity still oriented to the axis of rotation as sensed by the semicircular canals, and there was little or no spatial orientation to tilts in the GIA. These findings demonstrate that the spatial orientation of velocity storage was lost after nodulo-uvulectomy and that, as during centrifugation when upright,[34] this loss was not altered by head tilt.

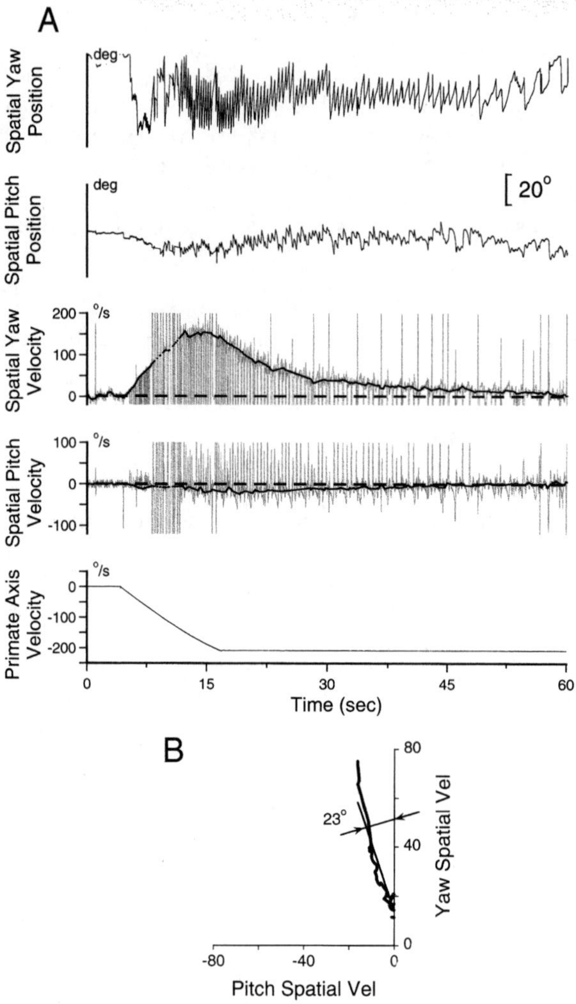

FIGURE 6. **(A)** Nystagmus induced by centrifugation at 218°/s, which produced 1 *g* of centripetal acceleration. Scheme and stimulus velocities as in FIGURE 4. The animal was positioned on the arm of the centrifuge facing the direction of motion with its head tilted to the right 30° toward the axis of rotation. Rotation was to the animal's right. Of note, the time constant of the yaw component was larger than in FIGURES 3 and 4, and there was relatively little cross-coupling to vertical. **(B)** Eigenvector generated by combining yaw and pitch velocities. In contrast to FIGURE 4 where 1 *g* of centripetal acceleration produced 36° of deviation of the eigenvector, eye velocities after nodulo-uvulectomy approached the origin along a slope of 23° from the vertical. As shown in FIGURE 5, this angle of deviation had no relationship to the magnitude of the centripetal acceleration.

DISCUSSION

The purpose of this study was to investigate whether the yaw eigenvector remains closely linked to the GIA when disparities are introduced between the yaw head axis and the axis of rotation. The data demonstrate the yaw eigenvector of velocity storage is spatially oriented and tends to align with the GIA in all conditions. Thus, tilts of the head that produce disparities between the rotation axis relative to the head yaw axis do not affect the direction of the yaw eigenvector. We also show that yaw orientation, a prominent feature of eye movements during head tilt and OVAR,[18] is lost following nodulo-uvulectomy. Consequently, following nodulus and uvular removal, rotations about an arbitrary head axis generate eye velocities whose trajectory in three dimensions follows the rotation axis and not the eigenvector. The findings support the proposition that the nodulus and uvula control the GIA-linked component of the yaw eigenvector and are consistent with the conclusion that the nodulus and uvula of the vestibulo-cerebellum form a center for controlling spatial orientation during low-frequency movements of the head.

The present results support the idea that spatial orientation of velocity storage orients the yaw axis eigenvector to the GIA, independent of the axis of rotation. When subjects are upright and rotated, the GIA is aligned with the rotation axis and head yaw. In this condition, all eye velocity trajectories are along the yaw axis. During centrifugation, the GIA tilts from being along the acceleration of gravity, and the eigenvector of velocity storage for upward cross-coupling tilts toward and is closely aligned with the GIA. Thus, a tilt of the GIA relative to the head will tilt the eigenvector by the amount of the tilt, demonstrating that not only does eye velocity orient to the tilted GIA when animals are upright, but that there is active orientation when the head is tilted and the axis of eye rotation is about other head axes. In every instance, the orientation to the GIA is maintained. Centripetal accelerations of 0.08 to 0.5 *g* are commonly encountered during turning.[67] In these experiments, the head was tilted 30° toward the axis of rotation from the upright. When the GIA was also tilted 30° toward the axis of rotation by imposing 0.5 *g* of centripetal acceleration, the GIA coincided precisely with the head yaw axis. Therefore, no tilt was sensed by the utricles, and the only stimulus to the otoliths was the increased *g* along the yaw axis, which would primarily affect the saccules. The semicircular canals signal in this case designated an axis of rotation about the spatial vertical. If the eyes had only responded to the semicircular canal stimulation, the axis of rotation would have been spatially vertical. Despite this, the eye velocity vector was tilted 26° from the spatial vertical toward the GIA (FIG. 5), close to the head horizontal and close to the yaw axis of the head. When the head yaw axis is aligned with the GIA, the time constant of the head pitch eye velocity is short and the yaw time constant is long. The implication of this is that changes in time constants of the pitch and yaw components in head coordinates were the critical elements in determining the orientation of the eigenvector in space. Consequently, the axis of eye rotation would be expected to align with the GIA under these circumstances. Such alignment could cause misalignment of vision with movement of the visual surround relative to the body. Presumably, suppressive mechanisms from the flocculus could discharge the stored activity and restore the axis of eye velocity to an angle closer to the axis of the rotational velocity.

Finally, the findings that the spatial orientation of velocity storage disappears after nodulo-uvulectomy provide strong support for the conclusion that the spatial orientation of velocity storage is dependent on the integrity of this region of the vestibulo-cerebellum. This is reasonable, given the direct input from the otolith organs, which is in marked contrast to the lack of such input to the flocculus, where the rapid components of eye velocity are controlled. Since eye velocity can no longer be oriented to the GIA after nodulo-uvulectomy, the input from the semicircular canals then becomes dominant. Although there was no effect of linear acceleration on the orientation of the eigenvector after operation, there were rather large deviations of the eye velocity vectors from the direction of the acceleration of gravity in every test condition (22°; FIG. 5). The reason for this is not clear, but it may be related to the fact that every test that was utilized in this study tilted the GIA relative to the head at least 5° or more. This level of static tilt has been shown to induce strong vertical and horizontal positional nystagmus after inactivation of the nodulus with muscimol[66] or after nodulo-uvulectomy.[65] These deviations could have been dependent on tilts of the head on the neck, which induced greater spontaneous vertical nystagmus. Since the time constant of vertical nystagmus is short after nodulectomy,[34,36] the vertical component is unlikely to have arisen from instability in the vertical mode of velocity storage. Such instabilities are present in the yaw component and lead to periodic alternating nystagmus.[34,36]

There are still many unanswered questions about orienting phenomena of the head and eyes. For example, orientation of the head to the GIA has been demonstrated in normal animals when subjected to centripetal acceleration during centrifugation[68] and in humans when turning corners,[69] indicating that it plays an important role in responding to tilts of the GIA while in motion. It is not known whether the head would continue to orient to the GIA after nodulo-uvulectomy, nor is it known whether body orientation to the GIA during turning would be affected. Furthermore, it is not known whether and how perception of the spatial vertical might be affected after such lesions since clean lesions of this portion of the nodulus and uvula rarely present themselves clinically. Regardless, the current data form a base for future study by confirming that the nodulus and uvula play a critical role in orienting the eyes to gravity and to the GIA.

ACKNOWLEDGMENTS

This work was supported by Grant Nos. EY11812, EY04148, DC03284, DC03787, and DC04996 and Core Center Grants DC05204 and EY01867. We thank Jun Maruta and Victor Rodriguez for assistance.

REFERENCES

1. COHEN, B., J. MARUTA & T. RAPHAN. 2001. Orientation of the eyes to gravito-inertial acceleration. Ann. N.Y. Acad. Sci. **942:** 241–258.
2. RAPHAN, T. & B. COHEN. 2002. The vestibulo-ocular reflex (VOR) in three dimensions. Exp. Brain Res. **145:** 1–27.
3. RAPHAN, T., V. MATSUO & B. COHEN. 1979. Velocity storage in the vestibulo-ocular reflex arc (VOR). Exp. Brain Res. **35:** 229–248.
4. RAPHAN, T. & D. STURM. 1991. Modelling the spatiotemporal organization of velocity storage in the vestibuloocular reflex by optokinetic studies. J. Neurophysiol. **66**(4): 1410–1420.

5. DAI, M., T. RAPHAN & B. COHEN. 1991. Spatial orientation of the vestibular system: dependence of optokinetic after nystagmus on gravity. J. Neurophysiol. **66**(4)**:** 1422–1438.
6. RAPHAN, T., V. MATSUO & B. COHEN. 1977. A velocity storage mechanism responsible for optokinetic nystagmus (OKN), optokinetic after-nystagmus (OKAN), and vestibular nystagmus. *In* In Control of Gaze by Brain Stem Neurons, pp. 37–47. Elsevier/North-Holland. Amsterdam/New York.
7. ROBINSON, D. 1977. Vestibular and optokinetic symbiosis: an example of explaining by modeling. *In* In Control of Gaze by Brain Stem Neurons, pp. 49–58. Elsevier/North-Holland. Amsterdam/New York.
8. DAI, M. *et al.* 1999. Model-based study of the human cupular time constant. J. Vest. Res. **9:** 293–301.
9. TER BRAAK, J.W.G. 1936. Untersuchungen ueber optokinetischen Nystagmus. Arch. Neerl. Physiol. **21:** 309–376.
10. MOWRER, O.H. 1937. The influence of vision during bodily rotation upon the duration of post-rotational vestibular nystagmus. Acta Otolaryngol. **25:** 351–364.
11. COHEN, B., V. MATSUO & T. RAPHAN. 1977. Quantitative analysis of the velocity characteristics of optokinetic nystagmus and optokinetic after-nystagmus. J. Physiol. (Lond.) **270:** 321–344.
12. WAESPE, W. & V. HENN. 1977. Neuronal activity in the vestibular nuclei of the alert monkey during vestibular and optokinetic stimulation. Exp. Brain Res. **27:** 523–538.
13. WAESPE, W. & V. HENN. 1977. Vestibular nuclei activity during optokinetic after-nystagmus (OKAN) in the alert monkey. Exp. Brain Res. **30:** 323–330.
14. WAESPE, W. & V. HENN. 1978. Conflicting visual vestibular stimulation and vestibular nucleus activity in alert monkeys. Exp. Brain Res. **33:** 203–211.
15. RAPHAN, T., B. COHEN & V. HENN. 1981. Effects of gravity on rotatory nystagmus in monkeys. Ann. N.Y. Acad. Sci. **374:** 44–55.
16. COHEN, B., J.I. SUZUKI & T. RAPHAN. 1983. Role of the otolith organs in generation of horizontal nystagmus: effects of selective labyrinthine lesions. Brain Res. **276:** 159–164.
17. MARUTA, J. *et al.* 2001. Orienting otolith-ocular reflexes in the rabbit during static and dynamic tilts and off-vertical axis rotation. Vision Res. **41:** 3255–3270.
18. KUSHIRO, K. *et al.* 2002. Compensatory and orienting eye movements induced by off-vertical axis rotation (OVAR) in monkeys. J. Neurophysiol. In press.
19. SOLOMON, D. & B. COHEN. 1992. Stabilization of gaze during circular locomotion in light: I. Compensatory head and eye nystagmus in the running monkey. J. Neurophysiol. **67**(5)**:** 1146–1157.
20. SOLOMON, D. & B. COHEN. 1992. Stabilization of gaze during circular locomotion in darkness: II. Contribution of velocity storage to compensatory eye and head nystagmus in the running monkey. J. Neurophysiol. **67**(5)**:** 1158–1170.
21. RAPHAN, T. & B. COHEN. 1986. Multidimensional organization of the vestibulo-ocular reflex (VOR). *In* Adaptive Processes in Visual and Oculomotor System, pp. 285–292. Pergamon. New York.
22. RAPHAN, T., M. DAI & B. COHEN. 1992. Spatial orientation of the vestibular system. Ann. N.Y. Acad. Sci. **656:** 140–157.
23. ANGELAKI, D.E. & B.J.M. HESS. 1994. Inertial representation of angular motion in the vestibular system of rhesus monkeys: I. Vestibuloocular reflex. J. Neurophysiol. **71:** 1222–1249.
24. WEARNE, S., T. RAPHAN & B. COHEN. 1997. Contribution of vestibular commissural pathways to velocity storage and spatial orientation of the angular vestibuloocular reflex. J. Neurophysiol. **78:** 1193–1197.
25. WEARNE, S. *et al.* 1997. Control of the three-dimensional dynamic characteristics of the angular vestibulo-ocular reflex by the nodulus and uvula. *In* Progress in Brain Research, pp. 321–334. Elsevier. Amsterdam/New York.
26. COHEN, B. *et al.* 1981. Velocity storage, nystagmus, and visual vestibular interactions in humans. Ann. N.Y. Acad. Sci. **374:** 421–433.
27. LAFORTUNE, S. *et al.* 1986. Human optokinetic afternystagmus: stimulus velocity dependence of two-component decay model and involvement of pursuit. Acta Otolaryngol. **101:** 183–192.

28. GIZZI, M. *et al.* 1994. Orientation of human optokinetic nystagmus to gravity: a model based approach. Exp. Brain Res. **99:** 347–360.
29. MOORE, S.T. *et al.* 2003. Ocular and perceptual responses to linear acceleration in microgravity: alterations in otolith function on the COSMOS and Neurolab flights. J. Vest. Res. In press.
30. RAPHAN, T., S. WEARNE & B. COHEN. 1996. Modeling the organization of the linear and angular vestibulo-ocular reflexes. Ann. N.Y. Acad. Sci. **781:** 348–363.
31. WEARNE, S., T. RAPHAN & B. COHEN. 1996. Nodulo-uvular control of central vestibular dynamics determines spatial orientation of the angular vestibulo-ocular reflex (aVOR). Ann. N.Y. Acad. Sci. **781:** 364–384.
32. WEARNE, S., T. RAPHAN & B. COHEN. 1999. Effects of tilt of the gravito-inertial acceleration vector on the angular vestibuloocular reflex during centrifugation. J. Neurophysiol. **81:** 2175–2190.
33. ARAI, Y. *et al.* 2002. Spatial orientation of caloric nystagmus in canal plugged monkeys. J. Neurophysiol. **88:** 914–928.
34. WEARNE, S., T. RAPHAN & B. COHEN. 1998. Control of spatial orientation of the angular vestibuloocular reflex by the nodulus and uvula. J. Neurophysiol. **79:** 2690–2715.
35. COHEN, H. *et al.* 1992. Habituation and adaptation of the vestibulo-ocular reflex: a model of differential control by the vestibulo-cerebellum. Exp. Brain Res. **90:** 526–538.
36. WAESPE, W., B. COHEN & T. RAPHAN. 1985. Dynamic modification of the vestibulo-ocular reflex by the nodulus and uvula. Science **228:** 199–201.
37. ANGELAKI, D.E. & B.J.M. HESS. 1995. Inertial representation of angular motion in the vestibular system of rhesus monkeys: II. Otolith-controlled transformation that depends on an intact cerebellar nodulus. J. Neurophysiol. **73:** 1729–1751.
38. KORTE, G. & E. MUGNAINI. 1979. The cerebellar projection of the vestibular nerve in the cat. J. Comp. Neurol. **184:** 265–278.
39. BARMACK, N.H. *et al.* 1993. Vestibular primary afferent projections to the cerebellum of the rabbit. J. Comp. Neurol. **327:** 521–534.
40. MAKLAD, A. & B. FRITZSCH. 2003. Developmental segregation of posterior crista and saccular fibers to the nodulus and uvula of the cerebellum in mice. Dev. Brain Res. In press.
41. BARMACK, N.H. & H. SHOJAKU. 1992. Representation of a postural coordinate system in the nodulus of the rabbit cerebellum by vestibular climbing fiber signals. *In* Vestibular and Brain Stem Control of Eye, Head, and Body Movement, pp. 331–338. Japan Sci. Soc. Press. Tokyo.
42. BARMACK, N.H. *et al.* 1992. Secondary vestibular cholinergic projection to the cerebellum of rabbit and rat as revealed by choline acetyltransferase immunohistochemistry, retrograde, and orthograde tracers. J. Comp. Neurol. **317:** 250–270.
43. SHOJAKU, H., N.H. BARMACK & K. MIZUKOSHI. 1991. Influence of vestibular and visual climbing fiber signals on Purkinje cell discharge in the cerebellar nodulus of the rabbit. Acta Otolaryngol. Suppl. **481:** 242–246.
44. TAN, H.S. & N.M. GERRITS. 1992. Laterality in the vestibulo-cerebellar mossy fiber projection to flocculus and caudal vermis in the rabbit: a retrograde fluorescent double-labeling study. Neuroscience **47:** 909–919.
45. WYLIE, D.R. *et al.* 1994. Projections of individual Purkinje cells of identified zones in the ventral nodulus to the vestibular and cerebellar nuclei in the rabbit. J. Comp. Neurol. **349:** 448–463.
46. SHOJAKU, H. *et al.* 1987. Topographical distribution of Purkinje cells in the uvula and the nodulus projecting to the vestibular nuclei in cats. Brain Res. **416**(1): 100–112.
47. REISINE, H. & T. RAPHAN. 1992. Neural basis for eye velocity generation in the vestibular nuclei during off-vertical axis rotation. Exp. Brain Res. **92:** 209–226.
48. HOLSTEIN, G.R., G.P. MARTINELLI & B. COHEN. 1999. Ultrastructure of GABA-immunoreactive vestibular commissural neurons related to velocity storage in the monkey. Neuroscience **93:** 171–181.
49. HOLSTEIN, G.R., G.P. MARTINELLI & B. COHEN. 1999. Ultrastructural features of non-commissural GABAergic neurons in the medial vestibular nucleus of the monkey. Neuroscience **93:** 183–193.
50. HOLSTEIN, G.R. *et al.* 1999. Ultrastructure of vestibular commissural neurons related to velocity storage in the monkey. Neuroscience **93:** 155–170.

51. COHEN, B., D. HELWIG & T. RAPHAN. 1987. Baclofen and velocity storage: a model of the effects of the drug on the vestibulo-ocular reflex in the rhesus monkey. J. Physiol. **393:** 703–725.
52. SOLOMON, D. & B. COHEN. 1994. Stimulation of the nodulus and uvula discharges velocity storage in the vestibulo-ocular reflex. Exp. Brain Res. **102:** 57–68.
53. YAKUSHIN, S.B. *et al.* 1995. Semicircular canal contribution to the three-dimensional vestibulo-ocular reflex: a model-based approach. J. Neurophysiol. **74:** 2722–2738.
54. YAKUSHIN, S.B. *et al.* 1998. Dynamics and kinematics of the angular vestibuloocular reflex in monkey: effects of canal plugging. J. Neurophysiol. **80:** 3077–3099.
55. RAPHAN, T. & B. COHEN. 1996. How does the vestibulo-ocular reflex work? *In* Disorders of the Vestibular System, pp. 20–47. Oxford University Press. London/New York.
56. DAI, M.J., T. RAPHAN & B. COHEN. 1992. Characterization of yaw to roll cross-coupling in the three-dimensional structure of the velocity storage integrator. Ann. N.Y. Acad. Sci. **656:** 829–831.
57. YAKUSHIN, S.B. *et al.* 2000. Functions of the nucleus of the optic tract (NOT): I. Adaptation of the gain of the horizontal vestibulo-ocular reflex. Exp. Brain Res. **131:** 416–432.
58. ROBINSON, D.A. 1963. A method of measuring eye movement using a scleral search coil in a magnetic field. IEEE Trans. Biomed. Eng. BME **10:** 137–145.
59. JUDGE, S.J., B.J. RICHMOND & F.C. CHU. 1980. Implantation of magnetic search coils for measurement of eye position: an improved method. Vision Res. **20:** 535–538.
60. DAI, M. *et al.* 1994. Effects of spaceflight on ocular counterrolling and spatial orientation of the vestibular system. Exp. Brain Res. **102:** 45–56.
61. CRAWFORD, J.D. & T. VILIS. 1991. Axes of eye rotation and Listing's law during rotations of the head. J. Neurophysiol. **65**(3)**:** 407–423.
62. SINGH, A. *et al.* 1981. Detection of saccades by a maximum likelihood ratio criterion. *In* Thirty-fourth ACEMB, Houston, p. 136.
63. RAPHAN, T. *et al.* 1992. Semicircular canal input and the spatial orientation of the vestibulo-ocular reflex. XVIIth Barany Society Meeting, Czechoslovakia, June 1–5, 1992.
64. YATES, B.J. *et al.* 2000. Responses of vestibular nucleus neurons to tilt following chronic bilateral removal of vestibular inputs. Exp. Brain Res. **130:** 151–158.
65. KUSHIRO, K. *et al.* 2001. Eye movements induced by low frequency linear acceleration in the monkey following nodulectomy and nodulus inactivation. Soc. Neurosci. Abstr. **31:** 403.10.
66. SHELIGA, B.M. *et al.* 1999. Control of spatial orientation of the angular vestibulo-ocular reflex by the nodulus and uvula of the vestibulocerebellum. Ann. N.Y. Acad. Sci. **871:** 94–122.
67. IMAI, T. *et al.* 2001. Interaction of the body, head, and eyes during walking and turning. Exp. Brain Res. **136:** 1–18.
68. JOHN, P. *et al.* 2001. Spatial orientation of eye velocity to gravito-inertial acceleration (GIA) in monkeys. Soc. Neurosci. Abstr. **31:** 403.9.
69. IMAI, T. *et al.* 1998. Stabilization of gaze when turning corners during overground walking. Soc. Neurosci. **24:** 162.5.

The Cerebellum as a Device for the Coordination of Movement

Summary of Presentations and of Discussion

ALBERT F. FUCHS, *Rapporteur*

This session featured contributions on both the somatomotor system and the oculomotor/vestibular system.

SUMMARY OF PRESENTATIONS

The papers relating to the somatosensory system, by Hore and by Bastian, were concerned with movements of the upper limb. Hore described the overarm throwing behavior of normal subjects and of patients with lesions of the dentate nucleus. The patients threw more slowly and had more variability in the timing of ball release and in the changes of finger position throughout the throw. However, the patients could, on average, scale finger force to the load just as normal subjects could. Although the variability in the timing of ball release for normal subjects ranged from 10 ms for recreational ball players to 27 ms for unskilled subjects, the variability for cerebellar patients was 50–100 ms. Hore concluded that during a throw the dentate nucleus is required for finger force and finger timing to be precise and consistent (low variability), but not for the scaling of force.

In her paper about upper limb movements of patients with less precise cerebellar lesions, Bastian seemed to reach different conclusions, at least with respect to interaction torques. During arm reaching, her patients with cerebellar disorder did not rotate the shoulder and elbow appropriately, causing curved reaching trajectories and dysmetria. With the shoulder fixed, this dysmetria was reduced. Bastian posited that patients cannot control interaction torques, which are torques produced across one joint in response to movement of another. Her view is that this is a deficit of torque scaling rather than timing because cerebellar patients that over- or undershoot can time their peak torques in nearly the same way that normal subjects can. In a second experiment, cerebellar patients adapted much more slowly (16–31 trials, time constant) than normal subjects (<2 trials, time constant) to catching a heavy falling ball after previously having learned to catch a light ball. If the switch in weights was made known, normal subjects improved their performance, whereas patients did not.

The two papers on the oculomotor system, by Cohen and by McCrea, were concerned with the role of the archicerebellum, but in two quite different vestibular situations. Cohen showed that when primates turn a corner, the body and eyes orient to the gravitoinertial acceleration vector; this orientation occurs with great reliability under a variety of tilts in the dark. This capability is lost in monkeys that have un-

Ann. N.Y. Acad. Sci. 978: 46–49 (2002). © 2002 New York Academy of Sciences.

dergone ablation of the nodulus and uvula, suggesting that these structures are important for this orienting response. After questioning by James Bower about why organisms should care about the vertical, Llinás opined that this orienting capability would allow us to align with a stationary world during all kinds of movement. Possibly to facilitate orientation, nodulus neurons that receive inputs from the otoliths also receive inputs from the neck.

To determine the role of the flocculus in modulation of the vestibulo-ocular reflex (VOR), McCrea recorded the activity of eye- and eye/head-velocity Purkinje cells (P-cells), as identified under passive head rotations, during a variety of eye/head interactions. Eye-velocity P-cells are not modulated during VOR suppression, whereas eye/head velocity P-cells are. Most P-cells did not respond when the body was rotated under a fixed head even though the trunk motion elicited the slow eye movements of the cervico-ocular reflex (COR); however, during COR suppression, firing rate again was modulated. This suggests to McCrea that eye-velocity P-cells might receive an as yet undisclosed input from neck proprioceptors. Floccular lesions impair a monkey's ability to suppress the VOR and cause a 50% reduction in the gain of smooth pursuit. However, such lesions have little effect on gaze shifts with the head unrestrained, a condition in which floccular P-cells show little modulation. Therefore, McCrea concludes that the flocculus only helps suppress the VOR and COR and is not crucial for gaze shifts.

SUMMARY OF DISCUSSION

In the spirit of the session title, The Cerebellum as a Device for the Coordination of Movement, the chair, Fuchs, wondered why the flocculus was not involved in the coordination of eye and head movements under conditions when both these movements can contribute to a gaze shift, whereas the first two speakers, Hore and Bastian, indicated that the cerebellum clearly is important in the coordination of certain somatic movements. McCrea suggested that he was recording the activity in a small "efferent zone" that is involved only with control of VOR gain; other zones could be involved with other components of a coordinated gaze shift. Presumably, only if *all* the efferent zones involved in the various components of a complex movement were compromised, as was the situation in Bastian's patients, would there be a deficit in coordinated movement. Bastian suggested that cerebellar areas other than the flocculus might be important in eye-head coordination by mentioning Thach's study in which cerebellar patients wearing prisms showed deficits in adapting eye, head, and trunk positions while throwing a ball. These patients had lesions in the posterior inferior cerebellar cortex.

Llinás questioned Bastian's conclusion that timing was not the principal deficit expressed by her cerebellar patients. He suggested that the issue of timing concerns not only when a movement or EMG begins and ends, but also the rate at which movement is generated. Bastian measured movement onsets and peak torques because measuring the timing of EMG bursts was more problematic. She agreed that consideration of the timing of other parts of the motor response might be revealing.

Ebner asked whether it was possible to extract a common message from the papers in this session as to the role of the cerebellum in the control of limb and eye/head movements. Hore was struck by the similarities in the deficits in saccades and limb movements produced by cerebellar damage. For both movements, there are deficits in movement accuracy and consistency or reliability which are thought to result from impairments of acceleration and braking. McCrea opined that, with regard to the oculomotor system, the cerebellum appears to be involved in some coordinate transformations and that investigators engaged in somatomotor studies might benefit from considering their complex reaching motions in the context of a coordinate transform. In his talk, McCrea suggested that smooth ocular tracking of images can be carried out in at least three different coordinate systems: spatial coordinates (VOR, pursuit of earth-referenced objects), head coordinates (VOR cancellation), and egocentric coordinates (hand-eye coordination). In his view, different classes of floccular P-cells may be related preferentially to each task. Cohen gave an example in the oculomotor system in which specific anatomical constraints require voluntary yaw head movements to rotate about an axis that is different from those of the semicircular canals, necessitating a coordinate transfer. It would have been useful to discuss details of the transformation(s) that were occurring in the flocculus/nodulus and to explore how consideration of coordinate transforms could provide specific insights in somatomotor experiments.

Stahl expressed surprise that the behavior of the flocculus with the head unrestrained seems to "break the general motif" of what the flocculus does with the head fixed. McCrea responded that floccular P-cells are relatively insensitive to head-free gaze shifts. However, rapid gaze shifts are accomplished by the saccadic system and the flocculus seems primarily involved with the slower smooth pursuit movements. On the other hand, neurons in both the oculomotor vemis and the caudal fastigial nucleus discharge vigorously during both head-fixed saccades (see the papers by Thier and Scudder in the following session) and head-unrestrained gaze shifts (see the poster by Brettler in this volume).

A common theme in the talks on the somatomotor system was that the cerebellum might play a role in anticipation (e.g., as in planning for interactive torques). Bastian indicated that cerebellar patients have anticipation deficits across several different types of movement. For example, they are unable to produce anticipatory movements to deal with motion of their own bodies, the presence of external objects, and standing. She proposed that the cerebellum is involved with predicting the muscle synergies that are necessary to make accomplished movements. Hore noted that in his studies with monkeys that were trained to overcome perturbations, cerebellar damage caused a delay in the antagonist activity that normally "brakes" a limb movement. He opined that the delay of antagonist activity represents a disorder of feedforward signals. In the absence of the cerebellum, the situation is dominated by feedback. Whereas pendular reflexes in cerebellar patients traditionally have been attributed to hypotonia, Hore would explain them by absence of a feedward signal controlling the antagonist.

Horak pointed out that muscle activity in the antagonist also is inappropriate in cerebellar patients required to adjust to predictable platform perturbations. When normal subjects are surprised with an unexpected perturbation, they too exhibit overactivity of the antagonist. Healthy subjects can adapt to the perturbations, whereas cerebellar patients have difficulty in adjusting. In particular, patients show considerable variability in their trial by trial adjustments, whereas healthy subjects do not. Considering all her data together, Horak concluded that in the postural system, a functioning cerebellum compensates by adjusting agonist gain and that the agonist hyperactivity in patients reflects compensation for poor antagonist control. Bower had an alternative explanation. For the first of many times in the meeting, he suggested that the real role of the cerebellum is to ensure that sensory systems work effectively. He observed that the organism with the largest cerebellum relative to its body size, the electric fish, sends most of its cerebellar output to a sensory, not a motor, structure. In his view, the major function of the cerebellum is not to control movement, but to make sure that the sensory signals upon which the organism depends are as good as they possibly can be. If the cerebellum makes vision better by reducing image-slip through regulation of VOR gain, Bower asked why it doesn't also have a role in improving the proprioceptive inputs that are operational during postural adjustments.

Lastly, Bell asked Hore how accurately movements could be timed and therefore, presumably, how accurately the cerebellum can accomplish that precision. Hore chose to study throwing because the end of the movement is ballistic in nature, that is, it is feedforward and not run by peripheral feedback. His fiduciary timing mark is the moment in a throw when the hand is vertical, although consideration of other kinematic points gives similar results. The average graduate student who is a good thrower has a variability in the timing of ball release of ~10 ms across 100 throws. The accuracy of the throw (above or below the target) is strongly correlated with this measure of timing. With regard to the absolute precision of this timing measure, the best throwers vary their timing by no more than 2 or 3 ms. Hore indicated that in order for a baseball pitcher to hit the strike zone consistently at 60 feet, he would need a timing precision of ~1 ms.

❧

In summary, this session was meant to explore whether the cerebellum is a device for the coordination of movement, with examples drawn from the somatomotor and oculomotor systems. The limb-movement studies used paradigms that involved the coordination of multiple joints, whereas the oculomotor studies used behaviors that required interactions between eye movements and movements of the trunk and head during locomotion and head-unrestrained gaze shifts. In throwing and reaching tasks, patients with rather large cerebellar lesions exhibited deficits in timing and in anticipation of the consequences of a movement, whereas monkeys with more circumspect lesions of the archicerebellum had deficits primarily in postural and oculomotor reflexes. Perhaps because of the diversity of experimental approaches in the two fields, the discussants indicated that it is currently impossible to discern a universal underlying principle for the cerebellar coordination of both multi-joint somatic movement and interactive eye, head, and trunk movements. As Cohen has suggested, perhaps we won't ever be able to attribute a single function to the cerebellum because it does *many* things. He observed that one would be equally frustrated when considering whether the cerebrum has a unitary role. This lack of closure was not unique to our session: later we learned that there is no consensus on even so fundamental a cerebellar issue as the role of climbing-fiber activity.

The Role of the Oculomotor Vermis in the Control of Saccadic Eye Movements

PETER THIER, PETER W. DICKE, ROMAN HAAS, CLAUS-DIETER THIELERT, AND NICOLAS CATZ

Department of Cognitive Neurology, University of Tübingen, Tübingen, Germany

ABSTRACT: The oculomotor vermis is a part of the posterior cerebellum, characterized by a low threshold (<10 μA) for evoked saccades. It comprises vermal lobuli VIc and VIIA. Many Purkinje cells in this area show eye position or saccade-related responses or combinations of the two and usually lack responses to the presentation of visual targets, guiding the oculomotor behavior. The saccade-related responses are directionally selective and show preferences for saccade amplitude or duration, which differ widely between cells. However, at the population level, these saccade-related Purkinje cells give a very precise account of the timing of the saccadic eye movement and, specifically, of the time it ends. This population signal might therefore contribute to determining the end of the saccadic eye movement. Furthermore, by changing the duration of the population response, the amplitude of the saccade could be changed. In other words, saccadic adaptation could be a consequence of changing a representation of time in the cerebellum.

KEYWORDS: oculomotor plant; Purkinje cell; cerebellum; adaptation; monkey

Several regions of the cerebellum house Purkinje cells that discharge in relation to saccadic eye movements. The region first identified and probably best understood is located in the posterior vermis, comprising vermal lobuli VI and VII, and occasionally also referred to as the oculomotor vermis or Noda's vermis, the latter name chosen to honor the late Hiroharu Noda, whose work has contributed considerably to our current view of this part of the cerebellum. A contribution of the posterior vermis and neighboring parts of the cerebellum to saccades was first suggested by experiments in which surgical lesions were carried out. For instance, Aschoff and Cohen[1] observed fewer saccades to the impaired hemifield after unilateral lesions of the cerebellar vermis, and Ritchie[2] described dysmetric saccades to visual targets after lesions of the posterior vermis. Dysmetric saccades have also been reported in the clinical literature as the consequences of cerebellar pathology involving the human vermis due to disease.[3,4] Ron and Robinson[5] showed that electric stimulation

Address for correspondence: P. Thier, Department of Cognitive Neurology, University of Tübingen, Hoppe Seyler-Str. 3, 72076 Tübingen, Germany. Voice: +49 7071 29 83057; fax: +49 7071 29 5326.

thier@uni-tuebingen.de

Ann. N.Y. Acad. Sci. 978: 50–62 (2002).

through electrodes placed in the posterior vermis and neighboring paravermis with currents of up to 1 mA was able to evoke saccadic eye movements. While this early work suggested a quite extended saccade representation in the posterior cerebellum, Noda and coworkers, by resorting to electric microstimulation, could show that the saccade representation was actually much smaller than hitherto assumed.[6–9] When stimulation currents were kept below 10 μA, saccades could only be evoked from lobuli VIc and VIIA of the posterior vermis, but not from the adjoining regions of the vermis and paravermis. Moreover, Noda and coworkers could show that the stimulation effects were a consequence of activating Purkinje cell axons and could rule out that the antidromic activation of vermal afferents, originating from the brainstem centers for saccades, contributed to the evoked saccades.[8] The oculomotor vermis projects to the saccade representation in the caudal fastigial nucleus, which in turn projects to the brainstem centers for saccades.[6,7] It is therefore very likely that the effects of activating Purkinje cells artificially by electric stimulation are mediated by this pathway. Electric microstimulation has clearly been very helpful in identifying the saccade-related parts of the posterior vermis and the pathway originating from there. On the other hand, its contribution to the characterization of the functional role of cerebellar cortex in the control of saccades has been limited. The early stimulation work described a dependency of the stimulation effect on the starting position of the eyes.[5] An analogous eye-position dependency has also been suggested by the lesion work, based on surgical ablations of the posterior cerebellar cortex[2] or inactivation of the deep cerebellar nuclei.[10] After these lesions, dysmetric saccades were observed with the amount of dysmetria being dependent on the starting position of the eyes. Electric stimulation and lesions might give rise to orbital position-dependent oculomotor responses because of a lack of the cerebellar functionality needed to compensate for the elastic forces that orbital tissue exerts on active eye movements (FIG. 1). When the eyes are held in a neutral, straight-ahead position, the sum of the elastic forces impinging on the eyes is zero. It increases linearly with increasing deviation from straight-ahead.[11,12] This force is supplemented by a second, velocity-dependent viscous force, mainly opposing high-velocity eye movements. Because of its centripetal orientation, the elastic force would assist centripetal eye movements and oppose centrifugal ones. In other words, in order to guarantee a saccade vector, whose size and direction are independent of the starting position of the eyes relative to the orbit, the neuronal signals controlling the eye muscles would have to take information on the starting position into account. If one assumes that the brainstem premotor circuits for saccades are ignorant of these elastic forces, there would be need for a corrective signal coming from somewhere else. The orbital position dependency of the effects of posterior vermal lesions and stimulation has suggested that this corrective circuitry might actually be offered by the oculomotor vermis. This idea is obviously a specific example of the concept suggesting that the function of the cerebellum is the generation (and adaptation) of internal models of the motor apparatus,[13] allowing the brain to translate representations of trajectories into real trajectories by compensating for the specific properties of the motor periphery.

If the idea that the oculomotor vermis sent a signal to the brainstem, taking care of the eye position-dependent elastic forces impinging on saccades, were correct, we might expect to observe a strong eye position dependency of saccade-related single-unit responses in the cerebellum. As exemplified in FIGURE 2A, saccade-related Purkinje cells in lobuli VIc and VIIA of the posterior vermis exhibit directionally se-

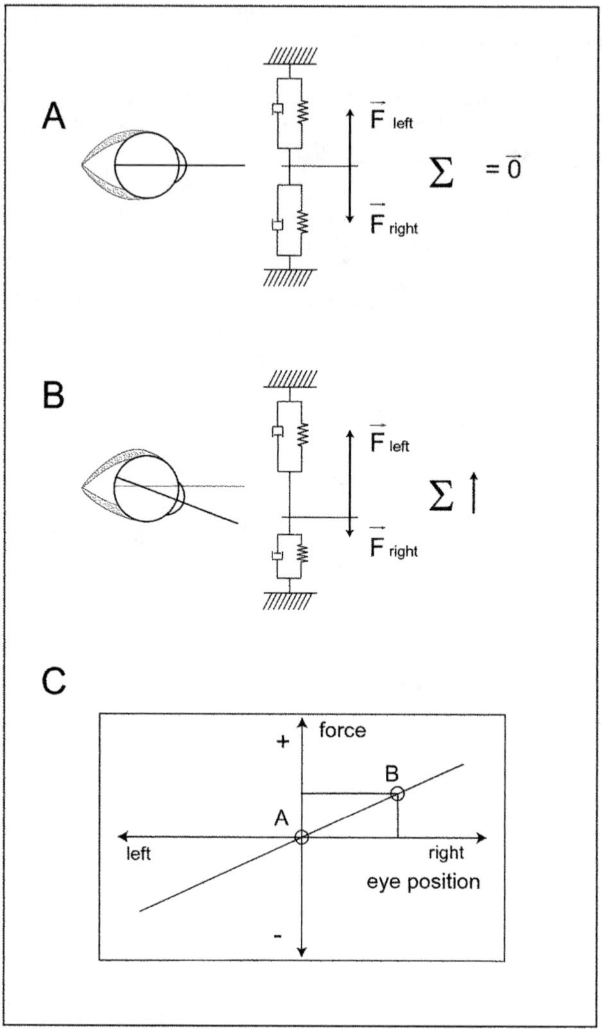

FIGURE 1. Active eye movements face elastic and viscous forces, based on the properties of orbital tissue and the eye muscle. The elastic force (symbolized by a spring) is position-dependent; the viscous force (symbolized by a piston) is velocity dependent. When the eyes are straight ahead (**A**), the elastic forces are balanced. However, if the eyes are moved into an eccentric position, the centripetally oriented forces are increased, while conversely, the centrifugally oriented forces are decreased, resulting in a net force back to straight-ahead (**B**). (**C**) Schematic relationship between net elastic forces and eye position. Only at the straight-ahead position do the elastic forces cancel out. Any deviation of the eyes away from straight-ahead evokes a net elastic force trying to move the eyes back. This force has to be compensated by muscle force in order to hold the eyes in the eccentric position.

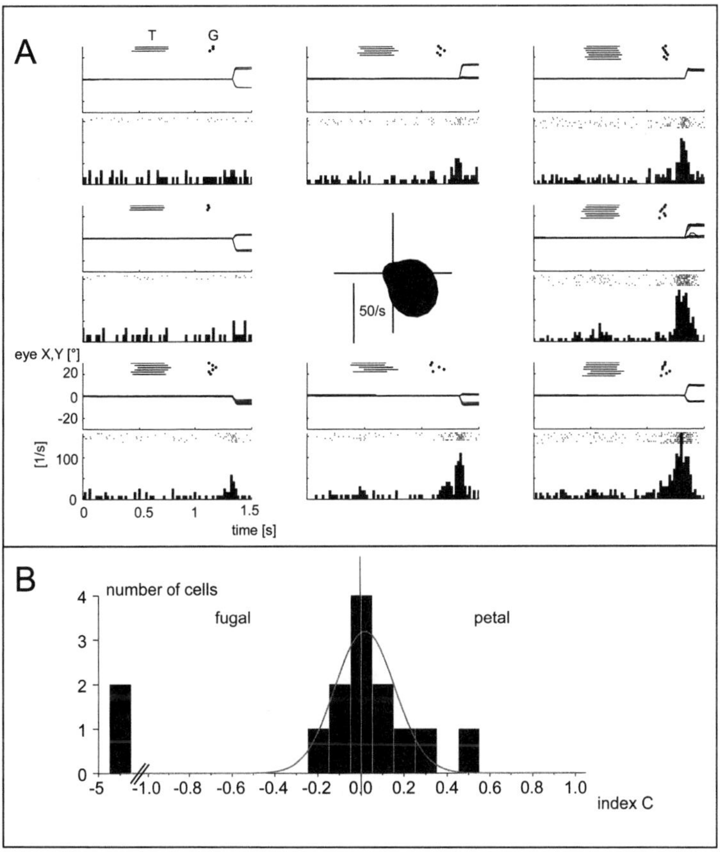

FIGURE 2. (**A**) Example of saccade-related Purkinje cell from the posterior vermis, preferring saccades to the lower right. The figure shows histograms (bin-width 5 ms) with accompanying raster plots of the discharge and records of the x and y components of eye position, characterizing responses to center-out "memory saccades" in 8 directions in the frontoparallel plane (peripheral target T on for 300 ms, duration of subsequent memory period 400 ms, amplitude 10°; G indicates the time the central fixation point was turned off, telling the monkey to make the saccade to the remembered peripheral location). In the center, a polar plot of the mean perisaccadic discharge rate as a function of direction is shown. (**B**) Distribution of an index $c = 1 - cf/cp$ capturing the amount of preference for centrifugal as opposed to centripetal saccades of posterior vermal Purkinje cells. cf is the mean perisaccadic activity for centrifugal saccades and cp the mean perisaccadic activity for centripetal saccades. Saccade-related responses were evoked by asking monkeys to make visually guided saccades of the same amplitude and direction in either a centrifugal or a centripetal manner. The distribution is centered on zero (sign-test, Dixon and Mood, $p > 0.05$).

lective bursts at the time the saccade is executed. In early experiments we compared the strength of these bursts for visually guided saccades of the same amplitude and direction, carried out in either a centrifugal or in a centripetal manner. FIGURE 2B plots the distribution of an index ***c***, comparing the responses to centrifugal and centripetal saccades. This index is defined as ***c*** = 1-cf/cp, with cf being the mean perisaccadic activity for centrifugal saccades and cp being the mean perisaccadic activity for centripetal saccades. ***c*** will be zero in the case of identical responses to centripetal and centrifugal saccades and deviate from zero in the one or the other direction in case of a preponderance of centripetal or centrifugal responses. The distribution of ***c*** shown in FIGURE 2B is centered on zero and does not significantly deviate from zero (sign-test, Dixon and Mood, $p > 0.05$). However, this finding is not necessarily at odds with the compensation hypothesis if one assumes that Purkinje cells with different types of eye position dependency may have contributed to the distribution, their different types of eye position dependencies canceling out at the population level. For instance one group of cells may discharge more strongly for centrifugal saccades because their response may be needed to increase the action of centrifugally pulling eye muscles. Others might conversely prefer centripetal saccades because they might dampen the signal for centripetal saccades. However, if this were actually the case, we would expect to find a clear, although possibly highly idiosyncratic orbital position dependency of individual saccade-related Purkinje cells in the oculomotor vermis. Our more recent work[14] shows that this is not the case. Of 59 cells, recorded from the oculomotor vermis of two monkeys, only six (= 10.2%) showed a significant influence of the starting position of the eyes, when fixed vector saccades in the preferred direction of a given cell were carried out from a set of nine different starting positions in the frontoparallel plane as explained in FIGURE 3C. Even in the few cells which showed a statistically significant dependency on orbital position, this effect was usually very weak. On the other hand, the overwhelming majority of cells showed identical saccade-related responses, independent of the starting position, as exemplified in FIGURE 3. These observations suggest that saccade-related Purkinje cells in the posterior vermis are not involved in compensating for the influence of the elastic forces on saccadic eye movement, at least not based on a rate code. This finding does not necessarily invalidate the compensation hypothesis in a more general form, since the posterior vermis contains a significant proportion of Purkinje cells, reminiscent of tonic or burst-tonic brainstem neurons, encoding eye position, occasionally interpreted as pursuit-related neurons because of their consistent activation during smooth-pursuit eye movements. As shown in FIGURE 4, these tonic and burst-tonic neurons are found intermingled with saccade-related neurons. An example of a tonic neuron is shown in FIGURE 5. This neuron discharged, whenever the eyes moved to the upper left, independent of the oculomotor strategy responsible for the position changes. Plot A was based on changes in eye position resulting from spontaneous, exploratory saccades, whereas plot B was derived from recordings of smooth-pursuit eye movements along a circular trajectory in the frontoparallel plane. In this case and many others, the eye-position sensitivities obtained were very similar. One speculative interpretation of the role of these position-dependent Purkinje cells could be that they take care of the passive elastic forces, impinging on the active eye movement. More specifically, they could be the basis of a differential signal, evening out the discrepancies between the actual brainstem signal driving the eyes and an ideal driving signal, discrepancies that probably keep changing continuously

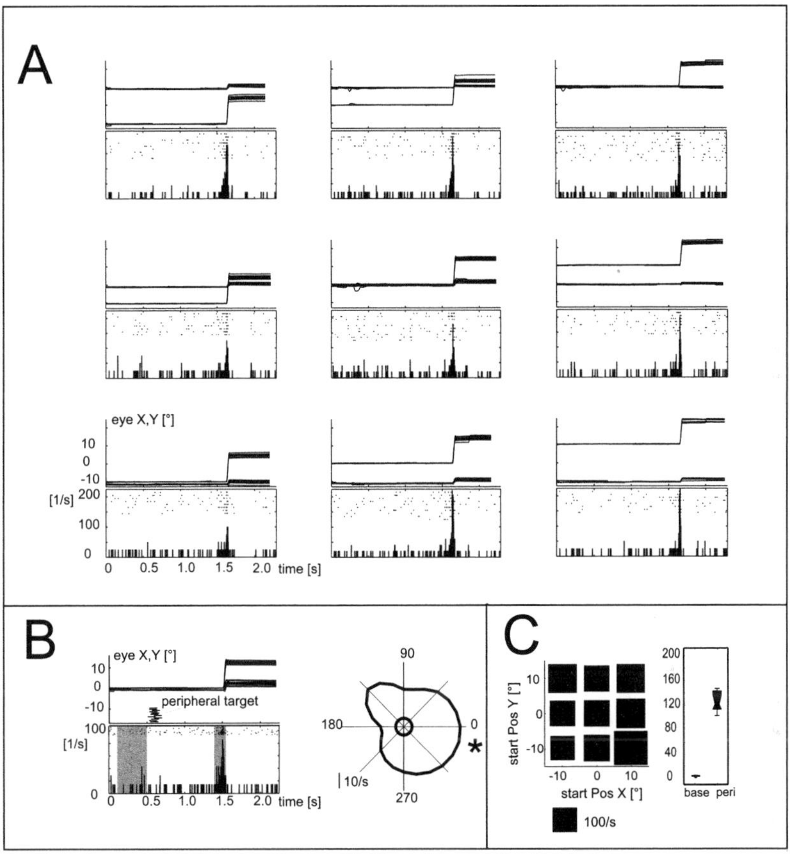

FIGURE 3. Lack of eye position dependency of the saccade-related responses of a posterior vermal Purkinje cell. (**A**) Responses for saccades to the right (visually guided saccades to the right, amplitude 15°), starting from nine different positions in the frontoparallel plane separated by 10° (same arrangement as in **C**). The cell discharged at the time the saccade was carried out with very little dependency on the starting position. (**C**) Plot of mean number of spikes in the burst (reflected by the size of the squares) as function of the starting position. A one-way analysis of variance with the factor starting position did not reveal any dependence of discharge rate on eye position. (**B**) Response of the same Purkinje cell in the memory saccade paradigm, in which the monkey made saccades to a position on the right which had been briefly cued a couple of 100 ms earlier (indicated by the short line segments "peripheral target"). Note the absence of a visual response and a lack of activity in the "memory" period between the offset of the cue and the offset of the central fixation point.

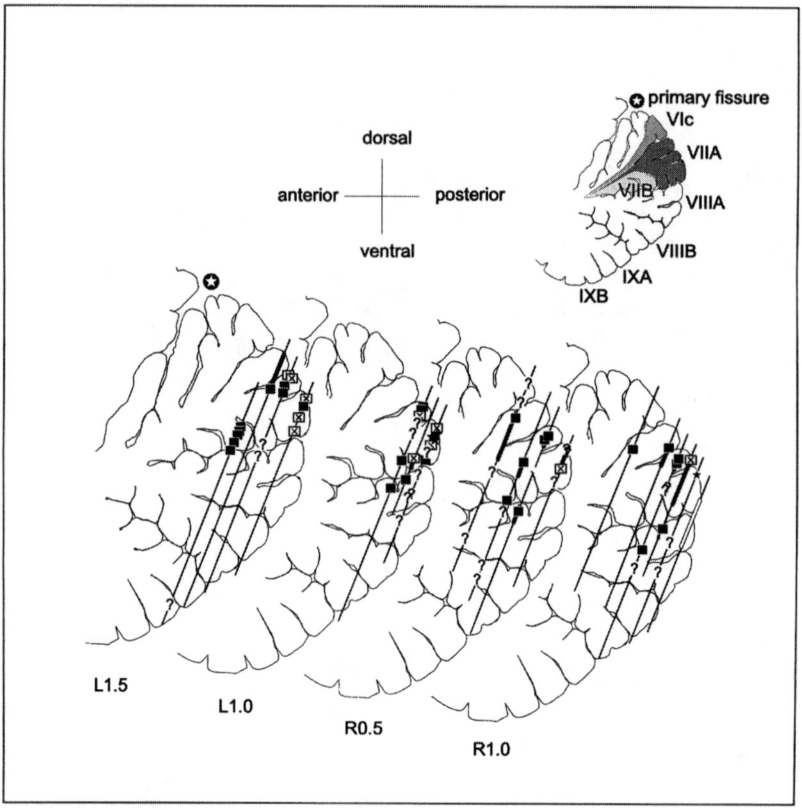

FIGURE 4. Parasagittal sections through the cerebellum of a rhesus monkey showing reconstructed electrode tracks through the posterior vermis. The various symbols represent specific functional types of Purkinje cells. *Black square*: "pursuit-related" neuron, showing sensitivity to eye positon and/or velocity during smooth pursuit, but usually also during other oculomotor behaviors; *asterisk*: saccade-related neuron, *square with x:* neuron showing mixed pursuit and saccade-related responses (= burst-tonic neuron); gray-shaded area: saccade-related background; *white area*: somatosensory background; ?: non-task-related neuron. The coordinates underneath the sections denote the position in millimeters relative to the midline (e.g., L1.5 = 1.5 mm to the left).

because of muscular fatigue and changes in the properties of orbital tissue. These eye position-dependent Purkinje cells represent one end of a spectrum of response preferences in the posterior vermis, whose other end is formed by the pure saccade-related Purkinje cells, with many Purkinje cells showing a mixture of eye position and saccade-related responses.

What is the role of the saccade-related Purkinje cells if not the compensation of the elastic forces? A possible answer is that these cells might be involved in compensating for the second passive force impinging on the eye movements, the velocity-dependent viscous force. Standard models of saccade generation assume

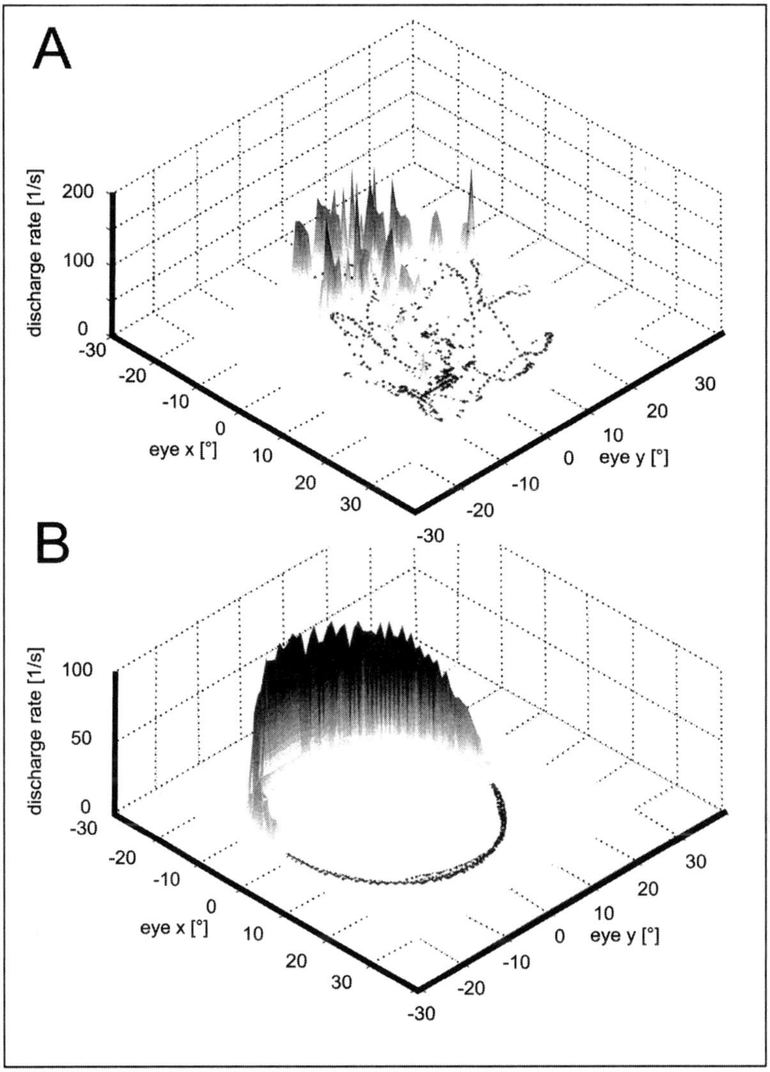

FIGURE 5. Example of burst-tonic Purkinje cell, whose responses to changes in static eye position and to smooth-pursuit eye movements are shown. Plot A was based on changes in static eye position, due to exploratory saccades. Estimates of mean eye position-related discharge were obtained for eye position bins of 1° × 1°. Discharge transients induced by the saccades were excluded by starting the analysis 155 ms after a saccade. Plot B was based on responses to circular smooth-pursuit in a counterclockwise manner in the frontoparallel plane (radius 20°; rate 0.2 Hz). Eye position bin width 1° × 1°. Note that this neuron discharged independent of the type of oculomotor behavior whenever the eyes moved to the left-upper-left.

that the velocity of the saccadic eye movement is determined by the size of an activity pulse generated in the brainstem premotor centers for saccades such as the paramedian pontine reticular formation.[15] The resulting saccade amplitude is the product of saccade velocity and duration, the latter determined by the duration of the pulse. Deviations of orbital viscosity from a standard would cause the normal relationship between pulse amplitude and eye velocity to change and thereby lead to dysmetric saccades. The saccades could be made normometric again either by changing the pulse amplitude or the duration. Our recent work on the posterior vermis[16] suggests that posterior vermal Purkinje cells might indeed provide a signal which could be used to adapt the pulse duration.

When tested in the memory-saccade paradigm, in which center-out saccades are made in darkness towards the remembered location of a cue, turned off a couple of 100 ms before the saccade is carried out, most saccade-related Purkinje cells exhibit pure, directionally selective saccade bursts (compare FIGURE 3B). They only rarely show visual responses or activity in the period of time when the monkey is waiting for the go-signal to start the saccade. Saccade duration and amplitude are closely linked. Saccade duration increases approximately linearly with amplitude for up to 40°, allowing one to vary saccade duration by simply asking monkeys to make saccades of different amplitudes. When saccades of different amplitudes are carried out in the preferred direction of a given cell, the amplitude dependency of the saccade-related bursts is highly idiosyncratic. Whereas some cells may show a monotonous increase in the number of spikes fired with increasing saccade amplitude, others show preferred amplitudes or no dependency on amplitude at all within a range of amplitudes up to 40°. In other words, one would most probably fail if one tried to determine the duration or amplitude of a saccade made by the monkey by monitoring the discharge pattern of individual cells. Unlike individual cells, though, larger groups of these saccade-related Purkinje cells provide a precise signature of saccade duration and amplitude. This is suggested by the conspicuous relationship between saccade duration and the duration of the population burst, the instantaneous discharge rate of a larger ($> n = 50$) group of saccade-related Purkinje cells, obtained by considering the timing of each spike fired by each cell in the sample. FIGURE 6A shows a plot of the resulting "population burst," based on 94 Purkinje cells from the posterior vermis as function of time around a saccadic eye movement. The population burst is plotted for three different saccade durations. It starts, independent of saccade duration, a couple of 10 ms before saccade onset and peaks exactly at the time the saccade starts, again independent of saccade duration. It is the decline of the population burst, which depends on saccade duration: it takes longer the longer the saccade lasts. This clear dependency of the time the population burst ends on the time the saccade ends is not restricted to the three saccade durations presented in FIG. 6A, but characterizes the full range of saccade durations tested (from less than 30 ms to almost 80 ms). This is shown in FIGURE 6B, which shows a pseudo 3-D plot of the population burst as function of saccade duration. In order to relate the time course of the population burst more precisely to the time course of the saccade, we measured the times of onset (a), peak (b), and offset (c) of the population burst relative to saccade onset of each saccade duration as well as the population burst duration (d), given by $c - a$. We determined population onset and offset times as the times when the population burst reached four times the baseline firing rate when building up and when declining. FIGURE 6C plots time t as function of a, b, c, and d, respec-

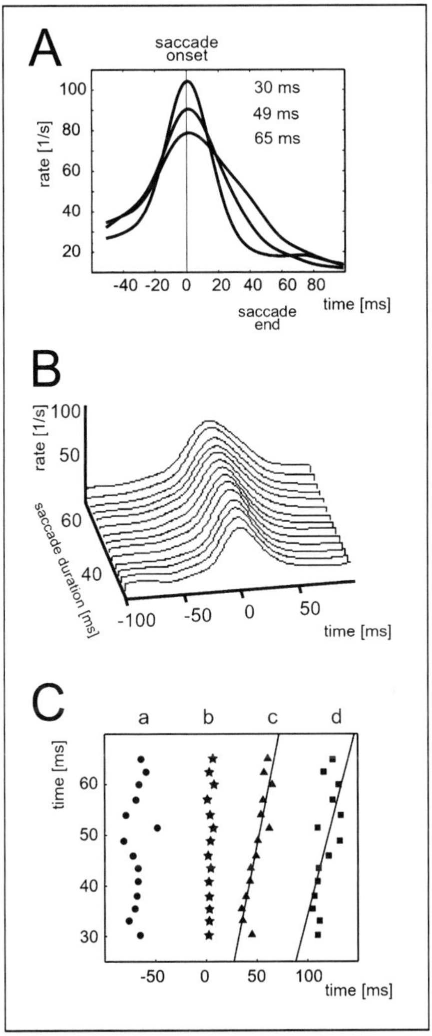

FIGURE 6. **(A)** Population burst profiles for 3 saccade durations of 30, 49, and 65 ms. **(B)** Dependence of population burst on saccade duration. The x-axis plots the time relative to saccade onset at 0 ms, the y-axis saccade duration, and the z-axis the mean instantaneous discharge rate of the population of 94 Purkinje cells. **(C)** Regression plots relating different parameters characterizing the population burst timing relative to the saccade. See the text for explanation. (Parts **A** and **C** taken from Thier *et al.*,[16] with permission.)

tively. In the case of *a, b* and *d, t* corresponds to saccade duration, in the case of *c* to the time of saccade termination. The plots are fitted by linear regressions. Both *c* and *d* increased linearly (c: $p < 0.00004$, d: $p < 0.01$) with the time of saccade termination and saccade duration, respectively, whereas neither *a* nor *b* depended significantly ($p > 0.05$) on saccade duration. The end of the population burst (*c*) as predicted by the regression corresponds very closely to the end of the saccades, whereas the population burst duration (*d*) underestimates saccade duration by 24%. This fact and the significantly higher coefficient of correlation for *c* as compared with *d* indicate that the population burst reflects the time of saccade termination. Individual cells fire their bursts at different times relative to the saccade, some reaching their maximum quite early, even before saccade onset, while others fire much later. None of the individual Purkinje cells reflect the termination of the saccade as accurately as the population response.[16]

Based on the number of Purkinje cells in rhesus monkeys[17] and the number of deep cerebellar nuclei neurons,[18] it can be estimated that on the order of 20 to 30 Purkinje cells converge on individual cells in the deep cerebellar nuclei. This means that a cell in the caudal part of the fastigial nucleus, the target of the posterior vermis, is probably influenced by a compound signal, not too different from the population burst as described before. In other words, the population burst is not a mathematical artifact, but most probably a direct functional consequence of the properties of the cerebello-nuclear projection. In view of the GABAergic nature of this projection, the population burst will deliver a strong hyperpolarizing signal to the recipient nuclear neuron, probably turning it off while the saccade is carried out. *In vitro* studies have shown that nuclear neurons fire strong rebound bursts upon cessation of hyperpolarization, as a consequence of hyperpolarization-activated mixed cation- and calcium-channels.[19,20] If such rebound bursts were also generated under *in vivo* conditions, we would expect to see saccade-related bursts close to the end of a saccade. Actually, many saccade-related nuclear neurons show such late saccade-related bursts,[21,22] a signal, which, if handed over to the brainstem machinery for saccades, might help to stop an ongoing saccade. Scudder[23] has recently shown that the timing of these late bursts can be changed by adapting saccade amplitude. For instance, if manipulations are carried out leading to longer-lasting, larger-amplitude saccades, these bursts occur later. If we assume that the timing of these later bursts fired by nuclear neurons is determined by the end of the vermal population burst, the conclusion is obviously that the manipulation leading to longer and larger saccades has increased the duration of the posterior vermal population burst. In other words, we propose that changes in the duration of the population burst might underlie saccadic learning. While we do not know whether the vermal population burst duration is indeed changed by saccadic adaptation, we do know that the posterior vermis is indispensable for saccadic adaptation. Lesioning vermal lobule VI and VII leads to an irreversible loss of short-term saccadic plasticity.[24] Short-term saccadic plasticity is the function that allows us to generate thousands of precise saccades despite the fact that the oculomotor periphery changes continously due to fatigue. We suggest that this is possible because of careful tuning of saccade duration, realized by optimizing a neuronal representation of saccade time at the level of the posterior vermis. We furthermore think that it may be worth considering that the principles of timing and of population coding which seem to characterize the role of saccade-related Purkinje cells in the posterior vermis may apply to other functional contributions of the cerebellum as well.

ACKNOWLEDGMENTS

This research was supported by grants from the Deutsche Forschungsgemeinschaft and by a grant from the Human Frontiers Science Programme (RGP 0023/2001-B).

REFERENCES

1. ASCHOFF, J.C. & B. COHEN. 1971. Changes in saccadic eye movements produced by cerebellar cortical lesions. Exp. Neurol. **32:** 123–133.
2. RITCHIE, L. 1976. Effects of cerebellar lesions on saccadic eye movements. J. Neurophysiol. **39:** 1246–1256.
3. BOTZEL, K., K. ROTTACH & U. BUTTNER. 1993. Normal and pathological saccadic dysmetria. Brain **116:** 337–353.
4. LEIGH, R.J. & D.S. ZEE. 1991. The Neurology of Eye Movements. F.A. Davis. Philadelphia, PA.
5. RON, S. & D.A. ROBINSON. 1973. Eye movement evoked by cerebellar stimulation in the alert monkey. J. Neurophysiol. **36:** 1004–1022.
6. NODA, H., S. SUGITA & Y. IKEDA. 1990. Afferent and efferent connections of the oculomotor region of the fastigial nucleus in the macaque monkey. J. Comp. Neurol. **302:** 330–348.
7. YAMADA, J. & H. NODA. 1987. Afferent and efferent connections of the oculomotor cerebellar vermis in the macaque monkey. J. Comp. Neurol. **265:** 224–241.
8. NODA, H. & T. FUJIKADO. 1987. Involvement of Purkinje cells in evoking saccadic eye movements by microstimulation of the posterior cerebellar vermis of monkeys. J. Neurophysiol. **57:** 1247–1261.
9. FUJIKADO, T. & H. NODA. 1987. Saccadic eye movements evoked by microstimulation of lobule VII of the cerebllar vermis of macaque monkeys. J. Physiol. **394:** 573–594.
10. VILIS, T., R. SNOW & J. HORE. 1983. Cerebellar saccadic dysmetria is not equal in the two eyes. Exp. Brain Res. **51:** 343–350.
11. ROBINSON, D.A. 1964. The mechanics of human saccadic eye movement. J. Physiol. **174:** 245–264.
12. ROBINSON, D.A. 1981. Control of eye movements. *In* Handbook of Physiology. Section I: The Nervous System. J.M. Brokkhart & V.B. Mountcastle, Eds.: 1275–1320. American Physiology Society. Bethesda, MD.
13. WOLPERT, D.M., R.C. MIALL & M. KAWATO. 1998. Internal models in the cerebellum. Trends Cognit. Sci. **2:** 338–347.
14. HAAS, R., P. THIER & P.W. DICKE. 1999. Saccade related responses of most posterior vermal purkinje cells do not depend on the starting position of the eyes. Soc. Neurosci. Abstr. **25:** 1652.
15. FUCHS, A.F., C.R.S. KANEKO & C.A. SCUDDER. 1985. Brainstem control of saccadic eye movements. Annu. Rev. Neurosci. **8:** 307–337.
16. THIER, P., P. DICKE, R. HAAS & S. BARASH. 2000. Encoding of movement time by populations of cerebellar Purkinje cells. Nature **405:** 72–76.
17. LANGE, W. 1975. Cell number and cell density in the cerebellar cortex of man and some other mammals. Cell Tiss. Res. **157:** 125–140.
18. GOULD, B.B. & P. RAKIC. 1981. The total number, time, or origin and kinetics of proliferation of neurons cmprising the deep cerebellar nuclei in the rhesus monkey. Exp. Brain Res. **44:** 195–206.
19. LLINÁS, R. & M. MÜHLETHALER. 1988. Electrophysiologie of guinea-pig cerebellar nuclear cells in the vitro brain stem-cerebellar preparation. J. Physiol. **404:** 241–258.
20. CZUBAYKO, U., F. SULTAN, P. THIER & C. SCHWARZ. 2001. Two types of neurons in the rat cerebellar nuclei as distinguished by membrane potentials and intracellular fillings. J. Neurophysiol. **85:** 2017–2029.
21. OHTSUKA, K. & H. NODA. 1990. Direction selective saccadic-burst neurons in the fastigial oculomotor region of the macaque. Exp. Brain Res. **81:** 659–662.

22. FUCHS, A.F., F.R. ROBINSON & A. STRAUBE. 1993. Role of the caudal fastigial nucleus in saccade generation I neuronal discharge patterns. J. Neurophysiol. **70:** 1723–1740.
23. SCUDDER, C.A. 1998. Discharge of fastigial nucleus neurons is altered during adaptive modification of saccade size. Soc. Neurosci. Abstr. **24:** 147.
24. BARASH, S., A. MELIKYAN, A. SIVAKOV, *et al.* 1999. Saccadic dysmetria and adaptation after lesions of the cerebellar cortex. J. Neurosci. **19:** 10931–10939.

Role of the Fastigial Nucleus in Controlling Horizontal Saccades during Adaptation

CHARLES A. SCUDDER

Departments of Otolaryngology and Neurobiology, University of Pittsburgh, Pittsburgh, Pennsylvania 15213, USA

ABSTRACT: Saccade size was adapted in rhesus monkeys using surgical weakening of the muscles of one eye combined with monocular viewing. Neurons in the caudal fastigial nucleus were recorded during the adaptation. Neuronal discharges changed in a way that could be interpreted as causing the changes in saccade size given our current knowledge of the projections of these neurons to the saccadic burst generator in the brain stem.

KEYWORDS: saccades; adaptation; fastigial nucleus; burst generator; extraocular muscle weakening; synaptic plasticity

INTRODUCTION

Considerable interest has been shown in the cerebellum, both as structure participating in the control of skilled movements and as a site of motor learning.[1–3] The present study uses saccadic eye movements as a model to investigate the parameters of cerebellar-neuron discharges that act to maintain movement accuracy during an adaptation paradigm. At the same time, the experiments help pinpoint the loci of the plastic changes that underlie adaptation, and address the nature of the signals controlled by the cerebellum to produce skilled movements

Saccades are both very brief (16–200 ms in a monkey) and very accurate,[4] but operate without concurrent visual feedback. For instance, moving the target during the saccade does not affect the amplitude of the current saccade.[5] Target location must be precisely assessed before movement onset, and the neural representation of that location must be translated into the appropriate amplitudes and durations of agonist and antagonist muscle forces.

Not surprisingly, this translation of target location into forces is not fixed, but is the product of continuous updating by the nervous system. A popular method for demonstrating this is to instantly displace the target of a saccade while the eye is moving.[6–9] At first, the saccades land where the target had been, but over time, change in amplitude or direction until they land near to the now-displaced target location. This is a genuine adaptation phenomenon, as shown by the fact that many human subjects are unaware of the intrasaccadic steps, and by the fact that

Address for correspondence: Charles A. Scudder, Ph.D., 5716 S.W. Hamilton Street, Portland, OR 97221. Voice: 503-297-3246.
scudder@pitt.edu

Ann. N.Y. Acad. Sci. 978: 63–78 (2002). © 2002 New York Academy of Sciences.

discontinuation of the intrasaccadic steps results in a gradual reversed adaptation instead of an immediate return to the pre-adaptation sensory-motor relationship.[6]

I use a different method for producing saccade adaptation in which the medial and lateral rectus extraocular muscles are surgically weakened in one eye. This procedure renders the horizontal eye movements in the "weakened" eye about half as large as those in the unoperated eye. The method has experimental advantages when working with monkey subjects that may not faithfully follow the target in the intrasaccadic-step task for the extended period needed to observe concomitant plastic changes occurring in the nervous system. It also has the advantage of producing large changes in the amplitude of saccades and presumably proportionately large plastic changes in the brain. Finally, it has the advantage of mimicking human pathology, as, in fact, it was modeled after human pathology. Abel *et al.*[10] and Kommerell *et al.*[11] showed that forcing human subjects to use a paretic eye, by patching the normal eye, produced gradual adaptive increases in the amplitude of saccades in the viewing paretic eye. Optican and Robinson[12] adapted this technique for use in monkeys, and Scudder *et al.*[5] showed that adaptive changes could occur rapidly if the number of target locations was restricted.

FIGURE 1 shows the locations where one might look for the plastic changes that mediate saccade adaptation. The command to make a saccade is issued by the supe-

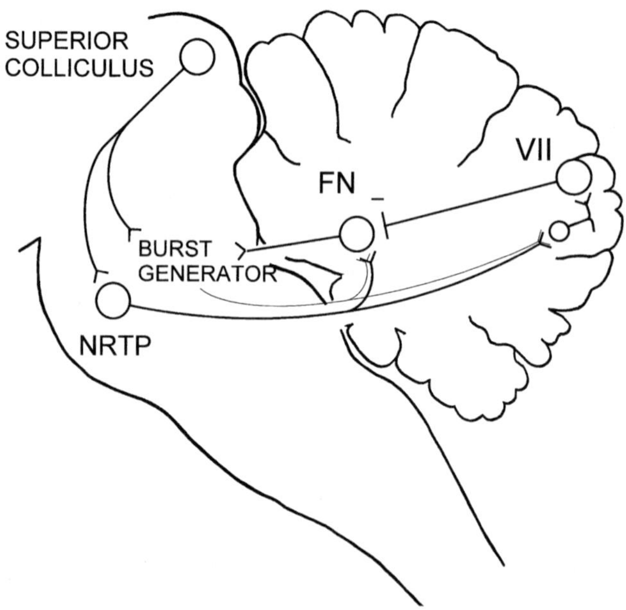

FIGURE 1. The major components of the saccadic system are superimposed on a parasagittal view of the monkey brain. The saccadic command from the superior colliculus traverses both a direct path to the saccadic burst generator in the medial pontine reticular formation, and an indirect pathway through the cerebellum. A weaker feedback path originates in the medial pontine reticular formation and terminates on fastigial nucleus (FN) and lobule VII neurons (*thin lines*). NRTP = nucleus reticularis tegmenti pontis.

rior colliculus and is conveyed directly to the "saccadic burst generator" located in the brainstem, which provides the immediate premotor signals to the motoneurons mediating horizontal saccades. The command is also conveyed to nucleus reticularis tegmenti pontis (NRTP), which gives rise to mossy fibers innervating vermal lobule VII and VIc and presumably the caudal fastigial nucleus via collaterals. After processing in the vermis, Purkinje cells send this signal back to the caudal fastigial nucleus, and finally, the fastigial nucleus projects to the same premotor neurons in the burst generator that receive direct and indirect input from the superior colliculus.[13] As a consequence, the cerebellum is well situated to both shape the command issued by the colliculus into a precise premotor command for the motoneurons, and to modify that command during motor learning. Consistent with this conclusion, neurons in the caudal fastigial nucleus discharge during saccades,[14–16] and microstimulation elicits saccades or modifies their size.[17–20]

The plastic changes probably do not occur in the superior colliculus.[21] This, together with the well-known evidence from *in vitro* experiments that the cerebellum contains plastic synapses at multiple levels[22] and evidence that adaptive changes don't occur without the midline cerebellum,[12,23–26] argues that the cerebellum probably contains the plastic synapses that mediate adaptation. As the output of the midline cerebellum, the fastigial nucleus offers the greatest opportunity to view the consequences of this plasticity. Therefore, the activity of fastigial-nucleus neurons was recorded as saccade size adaptively changed in an effort to reveal those plastic changes.

METHODS

Three rhesus monkeys were used in these experiments using procedures that conformed to University and NIH guidelines. Monkeys were surgically prepared for eye movement recording by implanting magnetic search coils in both eyes. Acrylic lugs were implanted for the painless immobilization of the head and to serve as attachments for a rigid patch that was used later to block vision in one eye. A recording chamber aimed at the fastigial nuclei was also implanted, and the medial and lateral recti of the right eye weakened. The tendons of these muscles were detached from the globe, the muscles cut, and the stumps were loosely sutured so that they reattached posteriorly on the globe with substantially less mechanical advantage. Further details of these and other methods are found in Scudder and McGee.[27]

For the experiments, monkeys were seated in a primate chair and placed in a dimly lit, sound-attenuating booth. To receive applesauce reward, monkeys were trained to fixate and track a target spot that was projected by a laser diode onto a white quarter cylinder situated in front of them. The target spot could be displaced using a mirror-galvanometer system under computer control. One experimental task was used to record the direction and size-dependencies of the fastigial neuron discharges. Monkeys initially fixated a central target that was extinguished as one of 16 possible peripheral targets was presented. Possible target loci were located along 8 equally-spaced directions on either an 8° or 15° circle. Usually, the central target was then re-presented to elicit a saccade in the opposite direction. A second, simpler, task was used during adaptation. There were only two potential targets separated by 20° horizontally, and the monkeys made saccades back and forth.

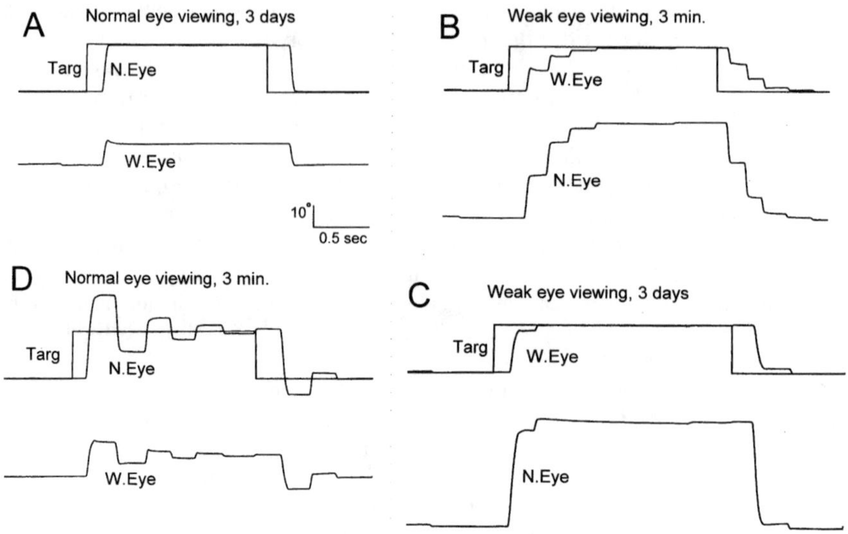

FIGURE 2. Long-term adaptation of saccade size produced by patching of either the normal or the weak eye. The four stages of adaptation (**A–D** in clockwise order) are shown for typical single trials. Horizontal position is illustrated for the target (Targ), normal eye (N.Eye), and weak eye (W.Eye). The vertical target position did not change. Initially, the patch covered the weak eye and the normal eye viewed the target (**A**). Adaptation was initiated by switching the patch to cover the normal eye so that saccades in the viewing weak eye were severely undershooting (**B**). After viewing with the weak eye for 3 days, saccade size increased in both eyes (**C**). Finally, switching the patch back to cover the weak eye resulted in severely hypermetric saccades in the viewing normal eye (**D**). Long-term adaptation to this situation produced decreases in saccade size and a return to normal-sized saccades in the normal eye (**A**).

The method and process of adaptation of saccade size is shown in FIGURE 2. Initially, monkeys wore a patch over the weakened eye and viewed the world monocularly using their normal, unoperated eye. FIGURE 2A shows that the monkeys made accurate saccades to a 20° horizontal target step, but the movement of the weakened eye was about half that needed to acquire the (non-visible) target. Immediately after switching the patch to the normal eye (FIG. 2B), the amplitudes of the initial saccade in each eye were about the same as in FIGURE 2A, but the small saccade in the viewing weakened eye fell well short of the target. The monkey had to make a series of hypometric saccades to get the weakened eye on target. At this point, the monkey either participated in an experiment (see below), or was returned to its cage with the rigid patch still covering the normal eye. After about three days, which I will call long-term, the monkey had significantly adapted (FIG. 2C). Saccade amplitudes in each eye had increased so that saccades in the viewing weakened eye were only slightly hypometric. Saccades in the patched normal eye had become about 80% larger than the target step. Switching the patch back to the weakened eye (FIG. 2D) again resulted in dysmetric saccades. This time, the viewing normal eye grossly overshot the target, and the monkey had to make a series of hypermetric saccades to acquire the target. Again, the monkey either participated in an experiment or was returned to its cage with the patch over the weakened eye. After about three days in his cage, adaptation had occurred so that saccades in the viewing normal eye had returned to orthometric (FIG. 2A).

Experiments began with the monkey adapted long-term to viewing with one eye. Metal microelectrodes (MPI) were driven through the caudal part of the fastigial nucleus in search of saccade-related neurons. When such a neuron was encountered and stable recordings obtained, saccades of different sizes and directions were elicited using the circular-target pattern, and the associated neuronal discharges recorded. When completed, the patch was switched to the opposite eye to induce short-term adaptation, the two-target task was started, and the eye movements and neuronal discharges were recorded again. Recordings were repeated every 15–20 minutes until isolation of the neuron was lost (20 to 180 minutes for the units included in this report). After the experiment, the patch was switched back to the original eye, and the monkey was returned to its cage. The next day, saccades had returned to approximately the same size as at the start of the previous day, so short-term adaptation was again possible. Typically, the cycle was repeated for a week, after which the patch was switched and the monkey adapted for three days. In so doing, neurons could be recorded during short-term adaptation in the opposite direction.

Neuronal discharges, target-movement, and eye-movement data were monitored on-line and saved to a VCR-based tape system (Vetter). Data were either directly digitized to a computer file, or were digitized off-line from the taped data. Digitized data were analyzed using an interactive program in which a user indicated relevant saccades and marked the start and end of the associated neuronal discharges. The computer automatically determined the start and end of the movement in both the horizontal and vertical eye-movement channels, fitting a second-order polynomial in order to interpolate the time when eye velocity departed from or returned to zero. The computer used the user's and the automatic marks to compute saccade metrics in both eyes and several discharge parameters. The count of number of spikes began at the user's mark, but was truncated at saccade end if the marked discharge ended later than the saccade. Peak discharge frequency was defined as the average frequency of the three consecutive spikes occurring before saccade end that occupied the shortest time interval. Discharge duration was computed as the time interval between the user's marks in order to obtain a measure independent of saccade duration.

In roughly a quarter of the neurons, discharges began and/or ended gradually, making it difficult to mark the beginning and/or end reliably. In these cases, plots of averaged instantaneous frequency were generated (like those in FIGS. 3 and 4), and the discharge onset and offset were determined graphically. Discharge onset was defined as the time at which averaged instantaneous frequency reached 33% of the difference between spontaneous and maximum rate, and offset was defined as the time when frequency fell to 40% of the same difference.

Saccade gain was computed as the amplitude of the horizontal component of the saccade in the viewing eye divided by the amplitude of the target step. Regression slopes obtained during adaptation (see RESULTS) were always referenced to saccade amplitude in the normal eye, regardless of which eye was viewing, so that slopes were computed on the same basis for both gain-increasing and gain-decreasing experiments.

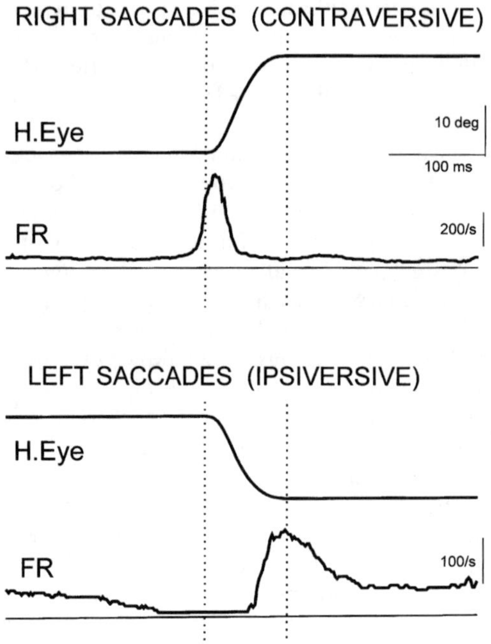

FIGURE 3. Averaged discharges of a typical fastigial nucleus neuron illustrated for rightward saccades (*top*) and leftward saccades (*bottom*). H.Eye = horizontal eye position; FR = firing rate. The neuron was recorded in the left nucleus, so the rightward movement (illustrated as upward) is contraversive, and the leftward movement is ipsiversive. For averaging, saccades were aligned on saccade onset (*dotted lines on left*). *Dotted lines on the right* mark the end of the longest-duration saccade.

RESULTS

Twenty-eight saccade-related neurons were recorded long enough that saccade gain changed by 5% or more. Their discharges exhibited characteristics similar to those reported by other authors.[14–16] FIGURE 3 illustrates data obtained from a neuron recorded in the left fastigial nucleus. Averaged horizontal eye position is shown in the top trace (H.Eye), and averaged instantaneous firing rate (FR) is shown below. For rightward saccades, which are contraversive for the left-sided recording (top panel), the neuron exhibited a burst of activity beginning before saccade onset and ending before saccade end (dotted lines). On average, the burst began 4.2 (±8.1 SD) ms before saccade onset, which is later than that reported by other authors. The difference may partly be accounted for by the use of a zero-velocity criterion for saccade onset, which was estimated to subtract ~2 ms from the latency when compared to a 20°/s criterion. In addition, there were a number of neurons, especially in one monkey, for which the burst followed saccade onset. In a third of the neurons, the burst ended before the saccade (as in FIG. 3). In another third, the burst usually ended

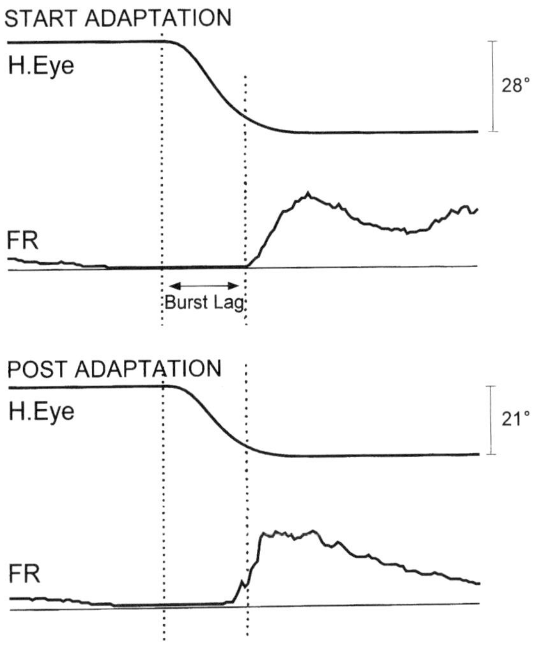

FIGURE 4. The change in the discharges of a fastigial nucleus neuron produced by adaptation of saccade size. H.Eye = horizontal eye position; FR = firing rate. Discharges are averaged across 25 ipsiversive (leftward) saccades aligned on saccade onset (*left dotted lines*). **Start Adaptation** data (*top panel*) were obtained with the monkey previously adapted to viewing with the weakened eye, but now viewing with the normal eye. Saccade size averaged 28°, and overshot the 20° target step (not illustrated). The onset of the neuronal discharge is marked with the *right dotted lines*. The very late resurgence in firing rate is associated with corrective saccades (not illustrated). **Post Adaptation** data (*bottom panel*) illustrate the smaller saccades (21° on average) produced by an hour of continued tracking with the normal eye. The neuronal discharge has advanced 12 ms relative to the control discharge, whose onset is marked by the *right dotted lines*, which is in the same location in both panels.

after the saccade ended, and in the last third, the burst and saccade ended roughly together.

For leftward, ipsiversive saccades (FIG. 3, lower panel), there was a suppression of the spontaneous firing rate followed by late burst. The delay of this burst, which I will call ipsiversive burst lag, averaged 23.9 (± 20.5 SD) ms after saccade onset, or about 47 ms before saccade end. Ohtsuka and Noda[15] maintained that there was a fixed 31 ms between burst onset and saccade end across saccades of different sizes, which led them to hypothesize that onset of the ipsiversive burst caused the termination of the saccade. Similar data obtained from the present sample before adaptation showed that this interval was not fixed, but for most neurons, there was still a statistically reliable relation between ipsiversive burst lag and saccade termination with a slope of 0.55 (± 0.27 SD, $N = 28$) ms-lag/ms-saccade on average.

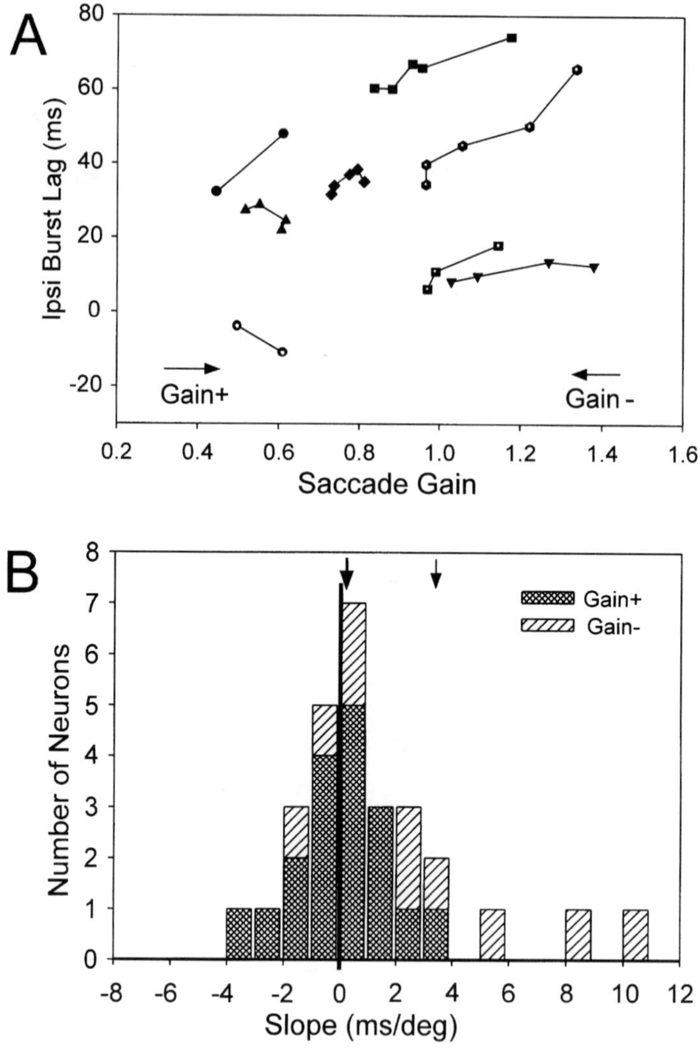

FIGURE 5. (**A**) Burst lag, the time from saccade onset to burst onset, associated with ipsiversive saccades is plotted as a function of saccade gain for eight representative neurons. Symbols plot average lag against average gain for 38–70 saccades collected at discrete intervals during adaptation. Traces to the left of gain = 0.9 plot data for experiments in which gain started low and increased as monkeys tracked the target with their weakened eye (Gain+ experiments). For traces to the right, gain started high and decreased as monkeys tracked the target with their normal eye (Gain– experiments). Most standard errors were smaller than the symbol size, which is ± 1.6 ms lag, or ± 0.014 gain units. (**B**) Histogram of the slope of the regression of ipsiversive burst lag against saccade size for 28 neurons. Slopes recorded during gain-increasing (Gain+) experiments clustered around zero, with a slight positive mean (*thick arrow*). Slopes recorded during gain-decreasing (Gain–) experiments were mostly positive with a significant positive mean (*thin arrow*).

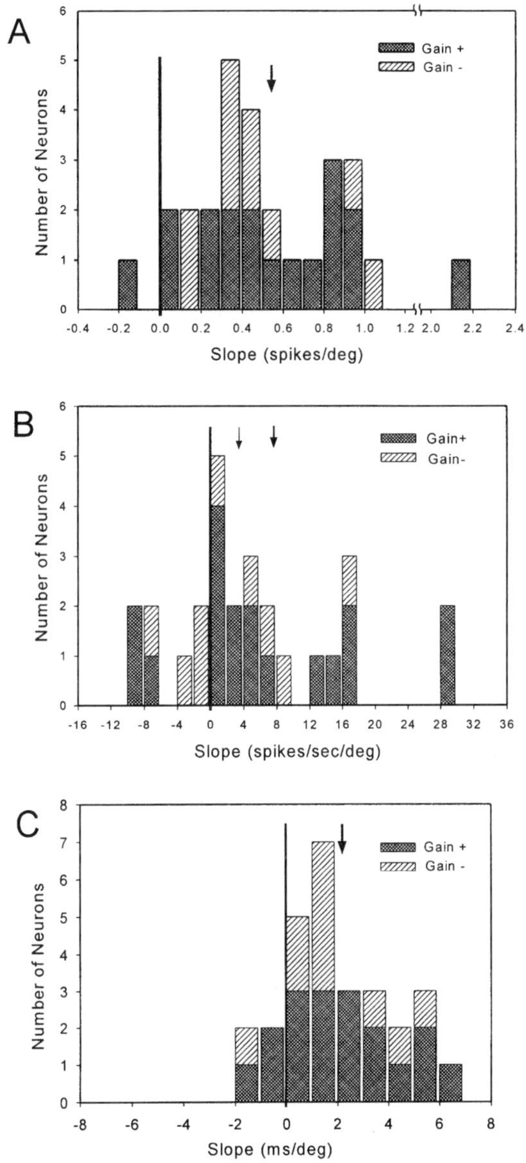

FIGURE 6. Histograms of the slope of the regression of number of spikes (**A**), peak frequency (**B**), and burst-end (**C**) against the size of contraversive saccades during adaptation. Mean values of slope for number of spikes and burst end are similar for Gain+ and Gain– experiments, so the average of the full sample is marked with a *single arrow*. Mean values of slopes for peak frequency are dissimilar, and so are marked separately for Gain– experiments (*thin arrow*) and Gain+ experiments (*thick arrow*).

Neural Correlates of Adaptation

A number of discharge parameters were observed to change as saccade size changed during adaptation. FIGURE 4 shows an example of a neuron in which ipsiversive burst lag decreased as saccade size decreased. The top traces show the averaged horizontal eye movement (H.Eye) and neuronal firing rate of a neuron in the right fastigial nucleus immediately after the patch was switched from the normal to the weak eye. The monkey generated 28° overshooting saccades to the 20° target step (as in FIG. 2D), which were associated with a late burst whose onset is indicated by the right dotted line. After an hour, saccade size had decreased to 21°, and ipsiversive burst lag had decreased by almost 12 ms. This rate of change (1.7 ms/deg) was close to average for the sample during gain-decreasing experiments.

FIGURE 5A plots ipsiversive burst lag against saccade gain for eight typical neurons. Symbols represent averaged data from 38–75 saccades obtained at different periods as adaptation progressed. Traces on the right illustrate burst lag as gain decreased (Gain–), as in FIGURE 4, while traces on the left illustrate burst lag as gain increased (Gain+). The important parameter is the slope of these traces, which we would expect to be positive for both gain-decreasing and gain-increasing experiments if the hypothesis is correct that the ipsiversive burst assists in the termination of the saccade. A histogram of these slopes obtained by linear regression is illustrated in FIGURE 5B for the sample of 28 neurons. The majority of neurons exhibited positive slopes, and on average the slope was positive and significantly different from zero (1.28 ± 3.0 ms/deg, $p < 0.01$, *t*-test). However, dividing the sample into those obtained during gain-decreasing and gain-increasing experiments revealed that the slopes were higher on average for neurons recorded during gain-decreasing experiments (3.22 ± 3.8 ms/deg, $N = 10$; thin arrow in 5B) than for those recorded during gain-increasing experiments (0.21 ± 1.7 ms/deg, $N = 18$; thick arrow). These two populations were significantly different ($p < 0.05$).

Seven neurons exhibited significant negative slopes, contrary to the hypothesis that the late ipsiversive burst terminates the saccade. However, four of these neurons had negative or nearly zero "lags" so that the ipsiversive burst actually preceded saccade onset (like the neuron in FIGURE 5A, lower left). These neurons were also unlike the majority of neurons in that they did not have a larger burst for contraversive saccades. Therefore, these neurons were never plausible candidates for the control of saccade termination, and could possibly serve some other role.

FIGURE 6 shows that there were also significant changes in the burst associated with contraversive saccades. Data were processed as above, and are summarized in a histogram of slopes as in FIGURE 5B. FIGURE 6A illustrates that for both gain-increasing and gain-decreasing experiments, number of spikes in the contraversive burst was positively correlated with saccade size during adaptation for all neurons but one. Accordingly, the mean slope was significantly different from zero (0.55 ± 0.44 spikes/deg; $p < 0.001$, $N = 28$) and was similar for both gain-increasing and gain-decreasing experiments (0.59 and 0.49, respectively). Because number of spikes is the product of burst frequency and duration, commensurate changes necessarily occurred in one or both of these parameters. FIGURE 6B illustrates that peak frequency of the burst was positively correlated with saccade size during adaptation for the majority of neurons. The mean value (6.09 ± 9.5 spikes/s/deg) was significantly above zero ($p < 0.002$), but differed for gain-increasing and gain-decreasing experiments (7.6 vs. 3.4, respectively). Finally, the time from saccade onset to burst-end was positively correlated with saccade size for most neurons (FIG. 6C), and the mean was significantly above zero (2.11 ± 2.17 ms/deg; $p < 0.001$, $N = 28$). Means for the two subpopulations were very similar (2.2 for gain+ and 2.0 for gain–). The positive number of spikes versus size correlations (FIG. 6A) also benefitted from a hidden contribution produced by the fact that the spike count was truncated at saccade end for some neurons, and that larger saccades tend to have longer durations.

Finally, there were two other changes of uncertain significance. The peak frequency of the burst during ipsiversive saccades increased slightly for both gain-increasing and gain-decreasing experiments, resulting in positive slopes for the former and negative slopes for the latter. Second, the lead of the burst during contraversive saccades was positively correlated with saccade size during adaptation (0.54 ± 1.06 ms/deg; $p < 0.05$, $N = 28$).

DISCUSSION

The results of this study show that several discharge parameters reliably change in fastigial nucleus neurons as adaptive mechanisms somewhere in the brain change saccade size. The most consistent change was in the number of spikes exhibited during contraversive saccades. This change was in turn produced by changes in peak frequency and burst duration. Recall that the measurement of burst duration was not truncated at saccade end, so the changes in this parameter surely reflect regulation by an independent adaptive mechanism. The increases in peak frequency during gain-increasing experiments were proportionally larger than the decreases in peak frequency during gain-decreasing experiments. A similar dichotomy was found for the changes in burst lag during ipsiversive saccades. Ipsiversive burst lag decreased during gain-decreasing experiments, but showed an insignificant increase during gain-increasing experiments. Overall, gain increases were relatively more dependent on increases in contraversive burst parameters and gain decreases were relatively more dependent on changes in ipsiversive burst lag.

There are at least two possible reasons why the neural correlates of gain+ and gain- adaptation might not be symmetric. One possibility is suggested by FIGURE 7, which shows the position and velocity profiles of saccades after long-term viewing with the weak eye. The figure shows the movements of the normal eye just after switching the patch back to the weak eye (as in FIG. 2D). Although control 30–35° saccades were not collected for comparison, these overshooting saccades appear to have an abnormal profile, with an excessively prolonged deceleration phase. The hypothesis is that these abnormally prolonged saccades were achieved by straining the limits of the adaptive mechanisms, which is supported by the finding that saccades adapted by viewing with the weakened eye never became orthometric.[5] The mechanism under the most strain, putatively that controlling saccade duration, might then be the first to revert when viewing with the normal eye allows recovery to occur. This hypothesis is consistent with the neurophysiology: gain decreases were largely accomplished by decreases in ipsiversive burst lag, which is thought to affect saccade duration.

More generally, adaptation may invoke a variety of plastic mechanisms which operate on different time scales. Indeed, the cerebellum alone is known to have several

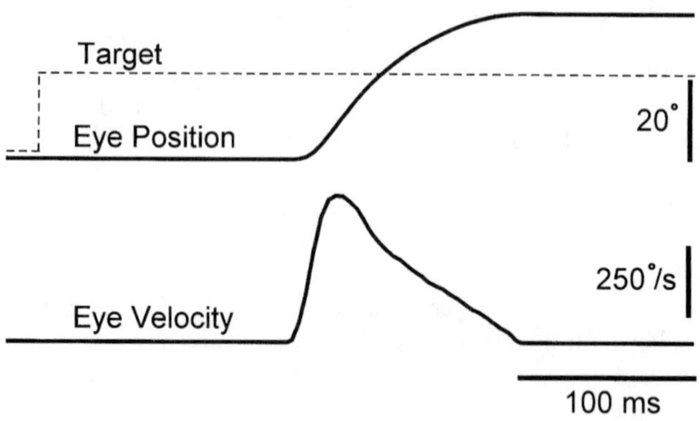

FIGURE 7. Eye position and eye velocity profiles of horizontal saccades in the normal eye after adaptation to viewing with the weak eye. The traces are the average of three saccades, each having the average velocity and duration for 55 other saccades recorded at this time period. Eye velocity has a highly asymmetric profile, with a very prolonged deceleration phase.

such mechanisms distributed along the pathway from the mossy fibers to the deep cerebellar nuclei.[22] In addition, long-term depression at a particular synapse is induced by a different pattern of stimuli than that which induces long-term potentiation, and each takes place at a different rate.[22,28,29] Therefore, it is quite plausible that short-term gain increases are mediated by different plastic mechanisms than are short-term gain decreases. These differences would be manifest as different parameters of fastigial nucleus discharges changing during gain-increasing and gain-decreasing experiments.

Locus of the Plastic Changes

The diagram of the saccadic system illustrated in FIGURE 1 suggests sites where plasticity could both occur and produce changes in the discharge of fastigial nucleus neurons. These sites include the superior colliculus, the brainstem burst generator, NRTP, and the cerebellum. Possible changes in the discharges of superior-colliculus neurons were sought by Frens and von Opstal[21] in an experiment using intrasaccadic steps to produce adaptation, and none were found. Plastic changes in the burst generator are possible, but it is unlikely that they could produce changes in the discharge of fastigial nucleus neurons having the magnitude reported here. This is because the projection to the cerebellum from the locus of the burst generator in the pontine reticular formation is regarded as minor in comparison to the heavy projection from NRTP.[30] Changes could also occur in NRTP, but because of the heavy projections from the fastigial nucleus to NRTP[31–33] they would probably also require the participation of the cerebellum. The cerebellum itself is the most promising site because of the demonstrated existence of plasticity *in vitro*[22] and because lesions of the cerebellum eliminate adaptation of saccade size.[12,23–26] In any case, whether the plastic changes occur in the cerebellum or in NRTP, the altered signals must pass through the fastigial nucleus to be effective in producing adaptation. Therefore, the signals

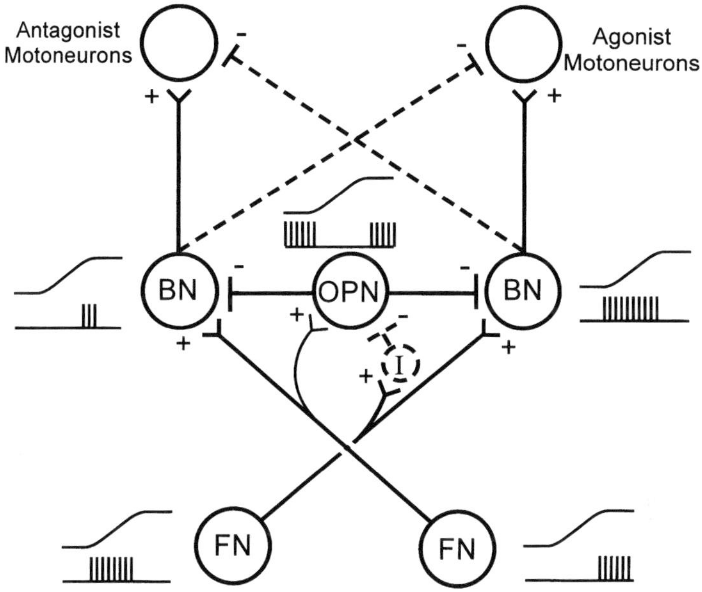

FIGURE 8. Schematized diagram of fastigial nucleus (FN) and burst generator connections. Excitatory connections are denoted by a +; inhibitory by a −. BN = burst neuron, representing both excitatory and inhibitory BNs. OPN = omnipause neuron; I = inhibitory interneuron. Schematized discharges are illustrated next to each neuron for a rightward saccade. BNs on the right side burst vigorously for the rightward saccade lightly before and during the movement. The burst excites agonist motoneurons on the right and inhibits antagonist motoneurons on the left.

that were recorded in the present study are probably those that caused the changes in saccade size. The discussion below will show that the changes in fastigial discharges reported above are mostly appropriate to produce saccade adaptation given the known projections of the fastigial nucleus to the saccadic burst generator.

Altered Fastigial Discharges Can Produce Saccade Adaptation

A more detailed diagram of the connections between the caudal fastigial nuclei and the saccadic burst generator is illustrated in FIGURE 8. Schematized discharge patterns associated with a rightward saccade are shown next to each neuron. Each fastigial nucleus (FN) projects contralaterally to both excitatory burst neurons and inhibitory burst neurons[13,31] (EBNs and IBNs, respectively, lumped together in the diagram as BNs). EBNs make excitatory connections with ipsilateral motoneurons (solid lines), and IBNs make inhibitory connections with contralateral motoneurons (dashed lines; evidence reviewed in Refs. 34 and 35). EBNs and IBNs each exhibit a vigorous burst before and during ipsiversive saccades, and exhibit a smaller later burst during contraversive saccades. The number of spikes in the EBN and IBN ipsiversive bursts are strongly correlated with saccade size, consistent with the tight control these neurons have on muscle force and the nature of orbital dynamics.[36, 37] Due to the above connections, the contraversive burst in fastigial nucleus neurons (left side FN) adds directly to the burst in the EBNs innervating agonist motoneurons (right side). Hence, the increases (decreases) in the number of spikes recorded during adaptation should translate directly into increases (decreases) in number of spikes in EBNs, and should produce increases (decreases) in saccade size. This predicted relation is, of course, what was observed.

Similarly, discharges in fastigial nucleus neurons during ipsiversive saccades (FN right side) affect the discharges of IBNs, and ultimately the forces produced by antagonist muscles. The earlier and stronger ipsiversive burst recorded in fastigial neurons should produce an earlier and stronger antagonist discharge that would act to decrease saccade size. Again, this relationship was observed in the present data. However, this mechanism was apparently not used to increase saccade size.

As shown in FIGURE 8, the fastigial nuclei also project to omnipause neurons (OPNs).[31] The OPNs discharge during fixation, and cease firing during saccades. By virtue of their direct inhibitory connections with EBNs and IBNs, OPN are thought to prevent firing in the burst neurons until the pause. Control of OPN firing can therefore have a powerful effect on EBN and IBN burst durations, and in turn, saccade size. Fastigial nucleus neurons are positioned to exert such control by means of a weak direct projection,[13] and an indirect inhibitory projection that has been hypothesized.[16] The inhibitory neuron (I) in the indirect pathway presumably receives other inputs so that it is only active during ipsiversive saccades (rightward on the right side). By way of this indirect inhibition, changes in the duration of fastigial nucleus discharges can cause commensurate changes in the pause of OPN firing. Thus an earlier onset or later end of the burst of fastigial neurons (left FN, FIG. 8) during contraversive saccades would be expected to increase OPN pause duration, EBN burst duration, and the size of saccades. In the present data, both fastigial burst lead and burst end changed in the expected manner as saccade size changed during both gain-increasing and gain-decreasing experiments.

Similarly, the late burst during ipsiversive saccades would be expected to affect OPN discharges via the direct excitatory connection from the opposite fastigial nucleus (right FN, FIG. 8). Specifically, the ipsiversive burst could promote the resumption of firing in the OPNs toward the end of the saccade. An earlier burst would therefore be expected to produce a shorter duration and smaller saccade, consistent with the present data. However, as above, the reciprocal effect does not appear to be used to produce increases in saccade size.

CONCLUSIONS

A motor-learning paradigm based on physiologically realistic muscle pathology was used to record both the adaptive changes in the size of saccadic eye movements and the associated changes in the discharge of neurons in the caudal fastigial nucleus. Evidence suggests that the observed changes in neuronal firing actually produced the changes in saccade size. First, the fastigial nucleus is in the output pathway of the cerebellum/NRTP, where the plastic synapses are probably located. Second, the projections of the fastigial nucleus to the neurons of the saccadic burst generator offer a number of mechanisms (not all of which were discussed) by which the fastigial nuclei could control saccade size. And finally, the observed changes in the fastigial nucleus neurons were appropriate to cause the observed changes in saccade size given the existence of these mechanisms. Although neither muscle forces nor motoneuron discharges were recorded, we can infer that the fastigial nucleus controlled the amplitudes and durations of the activity of the both the agonist and antagonist muscles.

ACKNOWLEDGMENTS

This work was supported by Grants R01-EY09210 and P30-EY08098 from the National Institutes of Health, and by the Eye and Ear Foundation of Pittsburgh.

REFERENCES

1. Hore, J., B. Wild & H.C. Diener. 1991. Cerebellar dysmetria at the elbow, wrist, and fingers. J. Neurophysiol. **65:** 563–571.
2. Ito, M. 2000. Mechanisms of motor learning in the cerebellum. Brain Res. **886:** 237–245.
3. Thach, W.T. 1998. A role for the cerebellum in learning movement coordination. Neurobiol.Learn.Memory **70:** 177–188.
4. Fuchs, A.F. 1967. Saccadic and smooth pursuit eye movements in the monkey. J. Physiol. (London) **191:** 609–631.
5. Scudder, C.A., E.Y. Batourina & G.S. Tunder. 1998. Comparison of two methods of producing adaptation of saccade size and implications for the site of plasticity. J. Neurophysiol. **79:** 704–715.
6. Deubel, H., W. Wolf & G. Hauske. 1986. Adaptive gain control of saccadic eye movements. Human Neurobiol. **5:** 245–253.
7. Deubel, H. 1987. Adaptivity of gain and direction in oblique saccades. *In* Eye Movements: From Physiology to Cognition. J.K. O'Regan & A. Levy-Schoen, Eds.: 181–190. Elsevier. Amsterdam.
8. Miller, J.M., T. Anstis & W.B. Templeton. 1981. Saccadic plasticity: parametric adaptive control by retinal feedback. J. Exp. Psychol. Hum. Percept Perform. **7:** 356–366.
9. Sraube, A., A.F. Fuchs, S. Usher & F.R. Robinson. 1997. Characteristics of saccadic gain adaptation in rhesus macaques. J. Neurophysiol. **77:** 874–895.
10. Abel, L.A., D. Schmidt, L.F. Dell'Osso & R.B. Daroff. 1978. Saccadic system plasticity in humans. Ann. Neurol. **4:** 313–318.
11. Kommerell, G., D. Olivier & H. Theopold. 1976. Adaptive programming of phasic and tonic components in saccadic eye movements: investigations in patients with abducens palsy. Invest. Ophthalmol. Vis. Sci. **15:** 657–660.
12. Optican, L.M. & D.A. Robinson. 1980. Cerebellar-dependent adaptive control of primate saccadic system. J. Neurophysiol. **44:** 1058–1076.
13. Scudder, C.A. & D.M. McGee. 2000. Connections of monkey saccade-related fastigial nucleus neurons revealed by anatomical and physiological methods. Soc. Neurosci.Abst. **26:** 971
14. Ohtsuka, K. & H. Noda. 1990. Direction-selective saccadic-burst neurons in the fastigial oculomotor region of the macaque. Exp. Brain Res. **81:** 659–662.
15. Ohtsuka, K. & H. Noda. 1991. Saccadic burst neurons in the oculomotor region of the fastigial nucleus of macaque monkeys. J. Neurophysiol. **65:** 1422–1434.
16. Fuchs, A.F., F.R. Robinson & A. Straube. 1993. Role of the caudal fastigial nucleus in saccade generation I. Neuronal discharge patterns. J. Neurophysiol. **70:** 1723–1740.
17. McElligott, J.G. & E.L. Keller. 1984. Cerebellar vermis involvement in monkey saccadic eye movements: microstimulation. Exp. Neurol. **86:** 543–558.

18. NODA, H. & T. FUJIKADO. 1987. Topography of the oculomotor area of the cerebellar vermis in macaques as determined by microstimulation. J. Neurophysiol. **58:** 359–378.
19. OHTSUKA, K. & H. NODA. 1991. The effect of microstimulation of the oculomotor vermis on discharges of fastigial neurons and visually directed saccades in macaques. Neurosci. Res. **10:** 290–295.
20. KELLER, E.L., D.P. SLAKEY & W.F. CRANDALL. 1983. Microstimulation of the primate cerebellar vermis during saccadic eye movements. Brain Res. **288:** 131–143.
21. FRENS, M.A. & A.J. VAN OPSTAL. 1997. Monkey superior colliculus activity during short-term saccadic adaptation. Brain Res. Bull. **43:** 473–483.
22. HANSEL, C., D.J. LINDEN & E. D'ANGELO. 2001. Beyond parallel fiber LTD: the diversity of synaptic and non-synaptic plasticity in the cerebellum. Nat. Neurosci. **4:** 467–475.
23. BARASH, S., A. MELIKYAN, A. SIVAKOV, *et al.* 1999. Saccadic dysmetria and adaptation after lesions of the cerebellar cortex. J. Neurosci. **19:** 10931–10939.
24. TAKAGI, M., D.S. ZEE & R.J. TAMARGO. 1998. Effects of lesions of the oculomotor vermis on eye movements in primate: saccades. J. Neurophysiol. **80:** 1911–1931.
25. GOLDBERG, M.E., S.Y. MUSIL, E.J. FITZGIBBON, *et al.* 1993. The role of the cerebellum in the control of saccadic eye movements. *In* The Role of the Basal Ganglia and Cerebellum in Voluntary Movements. N. Mano, I. Hamada, & M.R. DeLong, Eds.: 203–211. Elsevier. Amsterdam.
26. ROBINSON, F.R., A.F. FUCHS & C.T. NOTO. 2000. Role of the cerebellar caudal fastigial nucleus in saccade adaptation in monkey. Soc. Neurosci. Abs. **26:** 293
27. SCUDDER, C.A. & D.M. MCGEE. 2002. Adaptive modification of saccade size produces correlated changes in the discharges of fastigial nucleus neurons. J. Neurophysiol. Submitted for publication.
28. AIZENMAN, C.D., P.B. MANIS & D.J. LINDEN. 1998. Polarity of long-term synaptic gain change is related to postsynaptic spike firing at a cerebellar inhibitory synapse. Neuron 21:827–835.
29. ARMANO, S., P. ROSSI, V. TAGLIETTI & E. D'ANGELO. 2000. Long-term potentiation of intrinsic excitability at the Mossy fiber-granule cell synapse of rat cerebellum. J. Neurosci. **20:** 5208–5216.
30. THIELERT, C.-D. & P. THIER. 1993. Patterns of projections from the pontine nuclei and the nucleus reticularis tegmenti pontis to the posterior vermis in the rhesus monkey: a study using retrograde tracers. J. Comp. Neurol. **337:** 113–126.
31. NODA, H., S. SUGITA & Y. IKEDA. 1990. Afferent and efferent connections of the oculomotor region of the fastigial nucleus in the Macaque monkey. J. Comp. Neurol. **302:** 330–348.
32. BUISSERET-DELMAS, C., P. ANGAUT, C. COMPOINT, *et al.* 1998. Brainstem efferents from the interface between the nucleus medialis and the nucleus interpositus in the rat. J. Comp. Neurol. **402:** 264–275.
33. HOMMA, Y., S. NONAKA, K. MATSUYAMA & S. MORI. 1995. Fastigiofugal projection to the brainstem nuclei in the cat: an anterograde PHA-L tracing study. Neurosci. Res. **23:** 89–102.
34. FUCHS, A.F., C.R.S. KANEKO & C.A. SCUDDER. 1985. Brainstem control of saccadic eye movements. Annu. Rev. Neurosci. **8:** 307–337.
35. SCUDDER, C.A., C.R.S. KANEKO & A.F. FUCHS. 2002. The brainstem burst generator for saccadic eye movements: a modern synthesis. Exp. Brain Res. **142:** 439–462.
36. LUSCHEI, E.S. & A.F. FUCHS. 1972. Activity of brainstem neurons during eye movements of alert monkeys. J. Neurophysiol. **35:** 445–461.
37. ROBINSON, D.A. 1970. Oculomotor unit behavior in the monkey. J. Neurophysiol. **33:** 393–404.

Cerebellar Mechanisms in Eyeblink Conditioning

PHILIP J.E. ATTWELL, MAGNUS IVARSSON, LAURIE MILLAR, AND CHRISTOPHER H. YEO

Department of Anatomy and Developmental Biology, University College London, London WC1E 6BT, United Kingdom

ABSTRACT: A recent model of cerebellar learning in eyeblink conditioning predicts two sites of plasticity, the cerebellar cortex and cerebellar nuclei, which store information relating to timing and driving the movement, respectively. Consistent with this idea, lesions of the cortex or reversible "disconnections" of Purkinje cell output to the nuclei have been shown to disrupt response timing to produce short-latency conditioned eyeblinks. To better characterize potential cortical and nuclear plasticities, we analyzed the effects upon nictitating membrane (NM) and eyeblink conditioned responses (CRs) of different drugs administered to the cortex and to the nuclei. When either excitatory or inhibitory inputs to the cerebellar cortical lobule HVI were blocked by infusions of the AMPA receptor antagonist CNQX or the GABA-A receptor antagonists picrotoxin or SR95531, CRs were abolished. Similarly GABA-A receptor antagonists in the cerebellar nuclei abolished CRs. CR latencies were never shortened. However, blockade of AMPA/kainate receptor-mediated excitatory transmission to the nuclei had no effect upon CR frequencies or latencies. These results suggest that normal cortical and nuclear function is required for performance of NM and eyeblink CRs. We saw no evidence that CRs can be driven by AMPA/kainate receptor-mediated transmission from mossy fiber afferents to the cerebellar nuclei. So, although plasticity in the cerebellar nuclei is not ruled out, it is unlikely that a long-term change in AMPA receptor-mediated transmission from mossy fiber inputs to the nuclei is an essential mechanism in eyeblink conditioning. Our findings indicate that a fully functional olivo-cortico-nuclear loop is required to express all characteristics of associatively conditioned responses.

KEYWORDS: eyeblink; nictitating membrane; classical conditioning; cerebellar cortex; cerebellar nuclei; GABA

INTRODUCTION

The importance of the cerebellum for the acquisition and performance of eyeblink and nictitating membrane response (NMR) conditioning is now accepted, but we are still some way from understanding where and how these simple motor mem-

Address for correspondence: Dr. Christopher H. Yeo, Department of Anatomy and Developmental Biology, University College London, Gower Street, London WC1E 6BT, UK. Voice: +44 20 7679 7377; fax: +44 20 7679 3390.
c.yeo@ucl.ac.uk

Ann. N.Y. Acad. Sci. 978: 79–92 (2002). © 2002 New York Academy of Sciences.

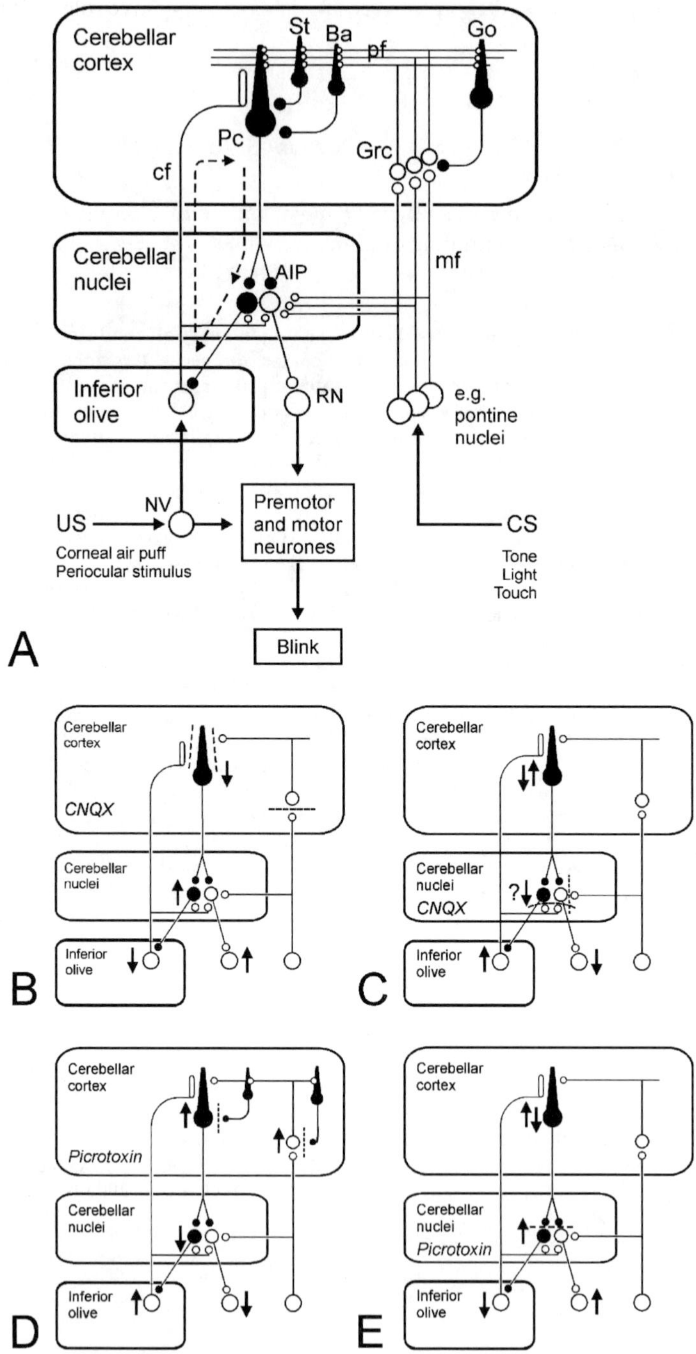

FIGURE 1. *See following page for legend.*

ories are formed. In particular, it is not clear what are the relative roles of the cerebellar cortex and cerebellar nuclei. Lesions of the cerebellar cortex, cerebellar nuclei, or inferior olive all prevent or impair expression of conditioned responses, and reversible inactivations of any of these structures prevent acquisition of conditioning (see Refs. 1–3 for reviews). Because of interconnections between olivary, cortical, and nuclear levels within the cerebellum, intervention at any level in this system has implications for information processing at the other levels. We have suggested that lesion and inactivation studies show that normal activity in the olivo-cortico-nuclear modules controlling eyeblink responses is necessary for acquisition and performance of conditioning, so plasticity essential for conditioning may reside at any, some, or all levels in this system[2,4,] (FIG. 1).

A mechanism for the storage of motor memories in the cerebellar cortex is strongly indicated in early theories[6,7] and in subsequent experimental studies of gain modification in the vestibulo-ocular reflex (VOR).[8,9] But electrophysiological analyses have also revealed learning-related changes at floccular cortical target neurons

---

FIGURE 1. Olivo-cortico-nuclear loops: inactivation effects. (**A**) A model of the cerebellum as a mediator of eyeblink conditioning. CS- and US-related information converges within the cerebellar cortex and within the cerebellar nuclei through mossy fiber and climbing fiber inputs, respectively (see Ref. 2). The olivo-cortico-nuclear loop (OCN) is indicated by *dashed arrows*. Conventions: Excitatory neurons and synapses are shown in white; inhibitory neurons and synapses in black. Abbreviations: AIP, anterior interpositus nucleus; Ba, basket cell; cf, climbing fiber; DAO, dorsal accessory olive; Go, Golgi cell; Grc, granule cell; HVI, cortical lobule HVI; NV, trigeminal nucleus; Pc, Purkinje cell; pf, parallel fibers; RN, red nucleus; St, stellate cell. (**B–E**) Simplified views of the circuitry shown in **A**, with cortical interneurons, multiple mossy fiber inputs, and some brainstem circuits omitted for clarity. Each panel shows how information transmission and excitabilities within the OCN may change after four different interventions. Excitability increases (↑) and decreases (↓) are indicated. (**B**) CNQX infusions in the cerebellar cortex block ionotropic, non-NMDA receptor-mediated transmission and block acquisition of NMR conditioning.[5] The main targets (shown as barred synapses) are parallel fiber inputs to Purkinje cells (and cortical interneurons: not shown), climbing fiber inputs to Purkinje cells, and mossy-fiber to granule cell synapses. The block of parallel-fiber synapses would reduce simple spike activity in Purkinje cells but may not abolish spontaneous activity. Cerebellar nuclear neurons would be partially disinhibited. (**C**) CNQX infusions in the cerebellar nuclei block any excitatory, AMPA/kainate-mediated input. Tonic excitatory inputs from mossy fibers and climbing fibers will be blocked (indicated by barred synapses). Dual-level learning models[3] place responsibility for driving the conditioned-blink response on the mossy fiber collateral, so CNQX infusions at this site should block this collateral input and abolish conditioned responses. (**D**) Blockade of GABA-A receptors in the cerebellar cortex (indicated by barred synapses) by local picrotoxin infusions will prevent feedforward inhibition of the Purkinje cell by the stellate and basket cells, as well as inhibitory feedback from Golgi cells to granule cells. The resultant increased excitability of both the granule cells and the Purkinje cells should lead to increased inhibitory output from Purkinje cells to the cerebellar nuclei. The decreased excitability of the nuclear cells would reduce their output to the olive and to their target output neurons. (**E**) Blockade of GABA-A receptors in the cerebellar nuclei (indicated by barred synapses) by local picrotoxin infusions disinhibits excitatory output neurons and the inhibitory nucleo-olivary projection. Increased inhibition in the inferior olive has effects similar to cutting or reversibly cooling climbing fiber inputs to produce a loss of complex spikes, but a significant increase in Purkinje cell simple spike frequencies. This blockade can also unmask short-latency, CS-driven eyelid responses.[15]

(FTNs) in the medial vestibular nucleus (MVN) during VOR modification,[10,11] leading to the suggestion that plasticity in the cerebellar cortex *and* at its efferent targets supports this learning. Perceived similarities in VOR modification and eyeblink conditioning have led to the proposal that dual-level learning processes in the cerebellar cortex and in the cerebellar nuclei mediate both forms of learning.[12]

The appeal of a common mechanism is clear, but it should be recognized that, at the behavioral level, there are important differences between these two different cerebellar-dependent forms of learning. The VOR needs speed and accuracy. Adaptive plasticity at FTNs in the medial vestibular nucleus ensures the fastest possible response in the vestibular reflex pathway, and cerebellar cortex can contribute further control at slightly longer latencies. Although the modification is velocity sensitive, the input cue is always vestibular. In contrast, conditioned eyeblink responses need not, and do not, occur at short latencies from conditional stimulus onset (CS). Through a learned association with the airpuff, or cutaneous electrical stimulus, that elicits the eyeblink reflex (the unconditional stimulus [US]), conditioned responses (CRs) can develop to CSs that include tactile, auditory, and visual events. Eyeblink conditioning cannot produce adaptive responses at CS–US intervals less than about 100 ms, but it can deal with a wide range of different cues. So, if the primary function of plasticity at cortical target neurons is to modify behaviour at the shortest possible latencies, then it may be less important in eyeblink conditioning than in modification of the VOR. However, if the primary purpose of plasticity at cerebellar nuclear or vestibular neurons is to allow adaptive *increases* in cerebellar output, as some models indicate, then it would be important in both of these forms of learning.

A Dual-Level Mechanism in Eyeblink Conditioning?

A formal model of dual-level cerebellar learning in eyeblink conditioning indicates that the cerebellar cortex is required for initial learning of conditioned eyeblink responses and then learning develops in the cerebellar nuclei.[13] The cerebellar cortex retains a memory related to the timing of the movement, whereas the cerebellar nuclei develop a memory sufficient to drive the movement from a CS instruction. There is indirect experimental evidence consistent with this idea. Lesions of anterior lobe cerebellar cortex output to the nuclei appear to disrupt the correct timing of conditioned eyeblinks by destroying the longer-latency, adaptively timed CRs and leaving small-amplitude, short-latency CS-driven responses.[14] Similar effects were seen after application of the GABA-A antagonist picrotoxin to the cerebellar nuclei, a procedure that has been likened to "disconnection" of the GABAergic Purkinje cell outputs from their nuclear targets.[15] It is suggested that these short-latency responses are driven by mossy fiber inputs to the cerebellar nuclei and that they are normally modulated by inhibitory drive from the cerebellar cortex.[16,17]

Experimental Support for a Dual-Level Learning Mechanism: Agreements and Disagreements

Experimental support for the dual-level model of eyeblink conditioning depends entirely upon the demonstration of short-latency responses after cortical lesions or cortical "disconnection." However, short-latency responses have not been seen in all lesion and inactivation studies. In our cortical lesion studies, we always saw complete losses of CRs or, where there were low levels of residual responses, they were erratically timed rather than of consistently short-latency.[18–20] Our lesions centred on lobule HVI, where eyeblink control microzones have been recorded in cats.[21,22] After cortical HVI inactivations with the non-NMDA ionotropic glutamate receptor antagonist 6-cyano-7-nitroquinoxaline-2,3-dione (CNQX), established nictitating membrane CRs gradually extended their latencies until the drug took full effect and CRs were abolished completely.[23] In related studies, HVI inactivation also blocked acquisition and consolidation of NMR conditioning.[5,24]

In contrast to our findings, those of Michael Mauk and colleagues demonstrated the short-latency response effect with *anterior lobe* lesions; and they found that HVI lesions were without effect. The reasons for the differences in the localization of effective lesion sites remains unclear. But, because of the major convergence of cortical inputs, it is unlikely that the "disconnection" experiments using GABA-A receptor blockers in the cerebellar nuclei would have discriminated between anterior lobe and HVI inputs, so issues related to the dual-level learning hypothesis can be resolved using this technique without further confounding the question of cortical localization.

In all of our previous analyses we have studied the NMR, but Mauk and colleagues have studied the external eyelid blink. These two response systems are mediated by different motoneuron pools in the accessory abducens and facial nucleus, respectively. There are likely to be differences in the tonic regulatory inflow to these pools, especially since the external eyelids are under voluntary control in most species. Short-latency responses might, therefore, be more easily evident in the eyeblink response system so it is important to test levels of expression of short-latency responses in both response systems.

CORTICAL AND NUCLEAR INACTIVATIONS DURING SIMULTANEOUS NMR AND EYEBLINK CONDITIONING: AN ANALYSIS OF CONDITIONED RESPONSE LATENCIES

We have now used simultaneous recording of NM and eyeblink conditioned responses during a range of cortical inactivations and functional "disconnections" in an attempt to reveal short-latency responses in either response system and to characterize possible cortical and nuclear plasticities more fully. Inactivation of cortico-nuclear inputs can be achieved by nuclear infusions of a GABA-A receptor antagonist such as picrotoxin, but there are problems associated with this drug.[15] Concentrations of picrotoxin above 0.5 mM induce depolarizing inactivation of the axonal spike-generating mechanisms[25] and this concentration has been exceeded in some recent studies.[16,26] Lower concentrations also have unwanted side effects since they will cause depolarization of axonal resting potential and decreased stimulation thresholds. To limit axon inactivation effects, we have used picrotoxin concentrations below the 0.5mM threshold, but in order to achieve optimal GABA-A receptor blockade we have also used an alternative, highly selective antagonist SR 95531 (Gabazine), which does not produce non-synaptic effects.

If functionally disconnecting the cortex from the nuclei is able to reveal short-latency responses, then they may also occur after reversible inactivation of the cerebellar cortex. We used cortical infusions of the AMPA/kainate receptor antagonist

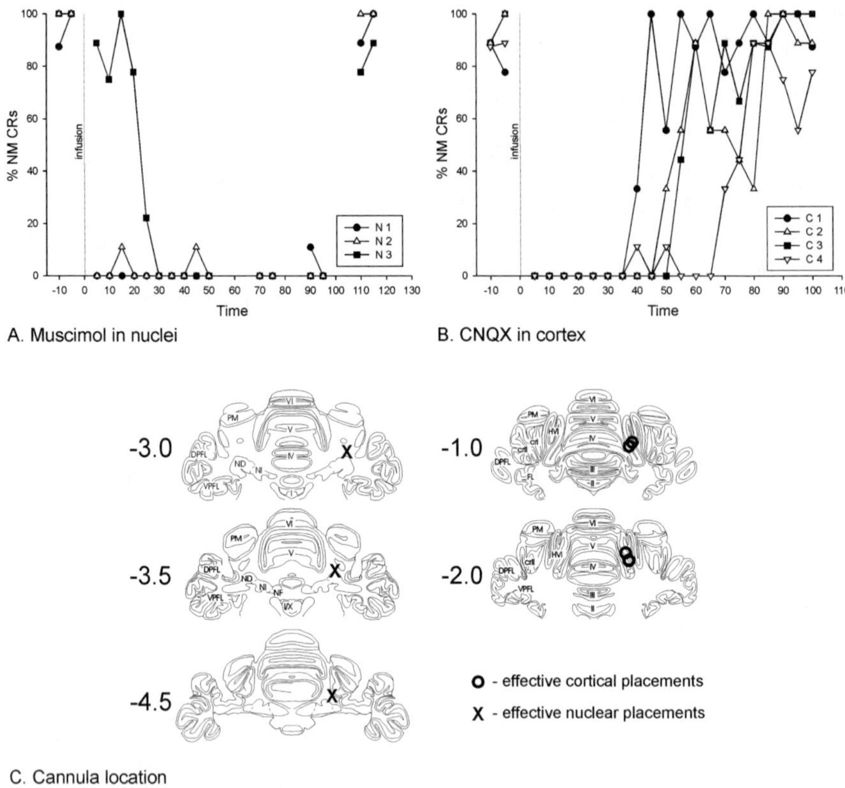

FIGURE 2. Optimal cannula locations and behavioral effects of drug infusions. The optimal locations for cortical and nuclear infusions were determined as previously described.[23] Initially, twenty training trials established a pre-infusion behavioral baseline. Subjects with cortical cannulae then received an infusion of CNQX (disodium salt; 3 mM, 2 µl infused over 2 min), and those with nuclear cannulae received muscimol (muscimol HBr; 1 mM, 2 µl). These initial infusions were made through injection cannulae (33G) projecting 1 mm beyond the end of the implanted guide cannulae. CR frequency (for definition see below) was then monitored during a 100-trial training session and, if CRs were not abolished, then the injection cannula was lowered by 0.5 mm and the infusion/testing procedure repeated. Further 0.5-mm increments, up to a maximum of 5 mm total, were tested. Subjects were excluded if CR performance was unaffected by the drug at all locations. The effectiveness of the drugs at the optimal cannulae locations are shown for muscimol in the nuclei **(A)** and CNQX in the cortex **(B)**. **CR definitions:** For the NM, a CR was defined as a response within the CS–US interval with amplitude greater than 0.5 mm and with an onset latency greater than 35 ms from CS onset.[19] An eyeblink EMG CR was defined as EMG activity during the CS–US interval (onset latency longer than 35 ms) at least 1.5 times the maximum amplitude of the background activity during the 500 ms preceding CS onset. CR frequency (%CRs) for both the NMR and eyeblink EMG were calculated for each block of 9 paired trials throughout the conditioning session. **(C)** Reconstructions of cannula tip positions from Nissl-stained, 50-µm serial sections.

CNQX to test again for this possibility. Furthermore, if short-latency responses were to be revealed by any of these procedures, then we could test whether they are driven by excitatory, collateral, AMPA receptor-mediated inputs to the cerebellar nuclei using infusions of CNQX within the nuclei.

Procedures

We tested NMR and eyeblink conditioning in 7 male Dutch belted rabbits. Before conditioning, each subject was implanted with a stainless steel guide cannula (26G) directed towards either the right cerebellar lobule HVI or cerebellar nuclei. All detailed procedures, including anaesthesia, surgery, and postoperative care, were as described in previous reports.[5] The cortical guide cannulae were implanted under visual guidance and the nuclear cannulae were positioned stereotaxically. Four stainless steel, Teflon-coated microwires were implanted into the orbicularis oculi muscle to record eyeblink EMG responses. The rabbits were trained using our standard techniques.[18] The CS was a 1-kHz sine wave tone of intensity of 85dbA. Background noise produced by ventilation fans was 57dbA. The US was periorbital electrical stimulation, the interstimulus interval (ISI) was 500 ms, and the intertrial interval was randomly selected between 25 and 35 s. NM position was monitored with an isotonic transducer and EMG activity in the orbicularis oculi muscle was recorded through the implanted microwires.

Subjects then received ten consecutive daily sessions of acquisition training. Each session consisted of 100 trials: In 90 trials the CS and US were paired, but on every tenth trial the CS was presented alone. Then, CNQX was infused in the cortically cannulated subjects and muscimol was infused in the nuclear-cannulated subjects and optimal sites were established by incrementally increasing the infusion cannula depth until a maximum behavioral impairment was seen (FIG. 2). Having established the optimal location for the infusions, subjects received daily sessions of training and various cortical or nuclear infusions were tested. Pre-infusion CRs were assessed over 20 trials and then 2μl infusions (all in 0.01M PBS, pH 7.4 vehicle over 2 min) of each drug were tested. The infusions were: CNQX (3 mM), picrotoxin (0.5 mM) or SR95531 (cortex, 0.1 mM; nuclei, 1mM). A training session commenced immediately after the injection, and the drug effects upon NM and eyeblink EMG CRs were monitored throughout the session to assess the time course for the onset and recovery of drug effects. Each subject was tested on each drug on different days and group mean data were obtained for each drug treatment. Where appropriate, changes in CR frequency were measured by comparing performance over 10 minutes (20 trials) before the infusion and over the time interval 20–30 minutes (20 trials) post infusion. Other comparisons, CR frequency, and latency measures were obtained as described in FIGURES 2 and 4.

GABAergic Blockade in Cerebellar Nuclei Abolishes CRs but CR Latencies Are Not Reduced: Blockade of AMPA/Kainate Receptor-Mediated Transmission in the Cerebellar Nuclei Has No Effect upon CR Frequencies or Latencies

Nuclear infusions of the GABA-A receptor antagonists picrotoxin and SR95531 both impaired NM and eyeblink EMG CR performance (FIG. 3A and B). When picrotoxin was infused, it completely abolished conditioned NM responses in one sub-

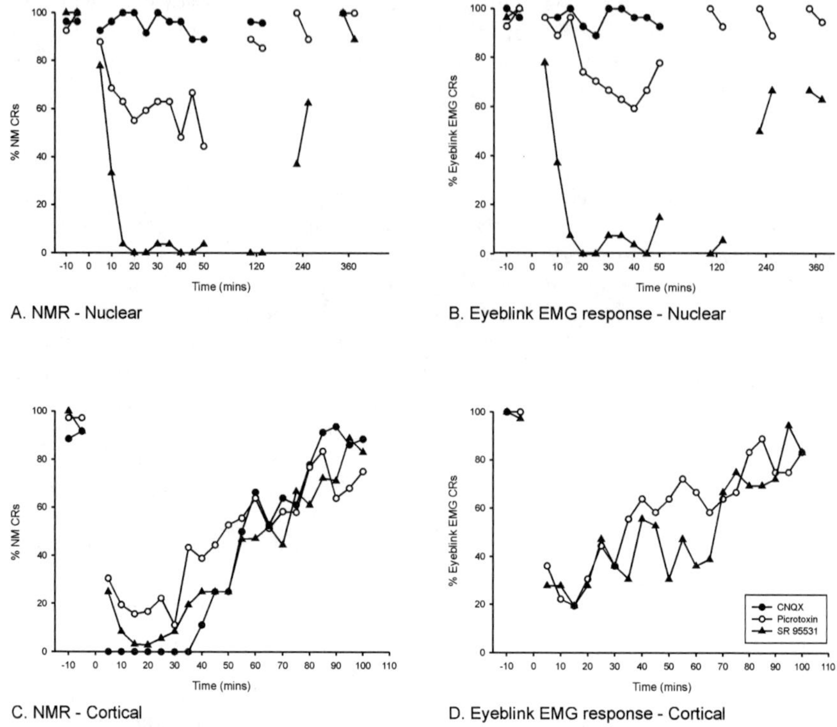

FIGURE 3. Effects of cortical and nuclear inactivations on CR frequency. Group mean frequency of conditioned responses before and after infusions of CNQX (3 mM, 2 μl; *solid circles*), picrotoxin (0.5 mM, 2 μl; *open circles*) and SR 95531 (nuclei: 1 mM, 2 μl; cortex: 0.1 mM, 1 μl; *solid triangles*). Data points are the mean percentage of CRs observed during each block of 9 paired trials before and after infusions at time 0 min (for definition of CRs see FIG. 2). (**A**) The effects of cerebellar nuclear drug infusions on NM CRs and (**B**) eyeblink EMG CRs. (**C**) The effect of cortical drug infusions on NM CRs and (**D**) eyeblink EMGs CRs.

ject, and partially blocked responses in the other two. SR 95531 fully abolished CRs in all three cases. However, blockade of the excitatory input to the nuclear cells with CNQX infusions produced no significant changes in CR frequency (Mann-Whitney U test: NMR, $p > 0.05$; eyeblink EMG, $p > 0.05$) (FIG. 3A and B), suggesting that these AMPA/kainate receptor-mediated transmission from these collaterals plays a minor role in the production of CRs.

Where drug infusions abolished CRs, CR latency measures could not be obtained. However, in all cases, there were time periods within the session where the drugs took effect or the effects wore off, so CR latency data were collected throughout the post-infusion period and compared with those from the pre-infusion period. Nuclear picrotoxin and nuclear SR95531 produced no significant changes in onset latencies and no significant changes in latency-to-peak for NM and eyeblink EMG CRs (FIG. 4).

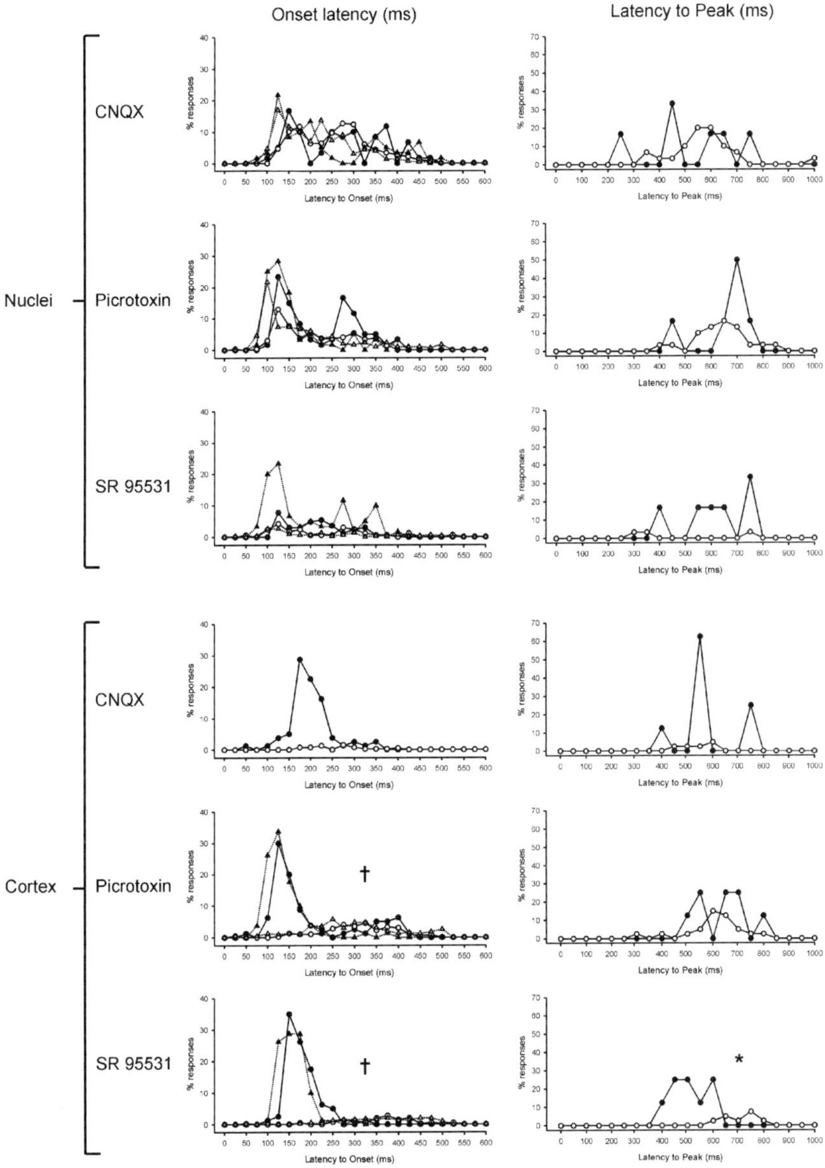

FIGURE 4. Drug effects upon the distribution of CR onset latencies and latencies-to-peak. CR latencies during administration of, and recovery from, infusions of CNQX (3 mM, 2 μl), picrotoxin (0.5 mM, 1 μl), and SR 95531 (nuclei: 1 mM, 2 μl; cortex: 0.1 mM, 1 μl). Frequency histograms (bin width 25 ms for onset latency, 50 ms for latency-to-peak) are represented in line format for clarity, but intermediate values between the bins are not implied. Points represent the mean percentage of trials that exhibit a response with a latency that falls within that particular bin. NM CRs before (*solid circles*), and after (*open circles*) drug infusions and eyeblink EMG CRs before (*solid triangles*) and after (*open triangles*) drug infu-

Nuclear CNQX, which did not affect CR frequency, produced no significant effects on onset latency in either the NM or eyeblink EMG CRs (NMR, 254.4 ± 13.3 ms compared to 242.6 ± 4.9 ms baseline; eyelid EMG responses, 198.0 ± 4.7 ms compared to 226.0 ± 15.5 ms baseline) and no significant changes in latency-to-peak of the NMR (505.6 ± 29.8 ms compared to 560.3 ± 42.74 ms) (FIG. 4).

GABAergic and Glutamatergic Blockade in the Cerebellar Cortex Significantly Impair CR Frequency, but Neither Releases Short-Latency Responses

When infused into the cerebellar cortex, CNQX completely and reversibly abolished fully established NM CRs (eyeblink EMG responses were not recorded during infusion of this drug) within a training session (FIG. 3C). Cortical picrotoxin abolished ($n = 3$) or impaired ($n = 1$) NM CRs (FIG. 3C) and eyeblink EMG CRs (FIG. 3D). SR 95531 had similar effects (FIG. 3C and D). There were no significant differences between the post-infusion CR frequencies for these three drugs (NMR: Friedman's statistic = 2.8 ($n = 3$); EMG: Mann Whitney U = 159.0), suggesting that GABAergic and glutamatergic transmission are of equal, and essential, importance for CR performance.

The cortical drug infusions mostly abolished CRs so effects on CR latency could only be studied as the drug took effect or wore off. Detailed analysis of these responses failed to reveal any shortening in the latency-to-onset or latency-to-peak of the NM or eyeblink EMG CRs with any of the drugs tested (FIG. 4). Only latency increases were ever seen. After picrotoxin infusions, onset latencies were increased significantly for both the NMR (298.3 ± 87.8 ms compared to 187.0 ± 98.9 ms baseline) and eyelid EMG responses (231.5 ± 98.5 ms compared to 122.9 ± 44.4 ms baseline). With SR 95531 infusions, the onset latency was increased in both NMR (374.2 ± 62.8 ms compared to 163.4 ± 38.0 ms baseline) and eyelid responses (231.5 ± 98.5 ms compared to 122.9 ± 44.4 ms baseline). After CNQX infusions NMR onset latencies showed a small, but insignificant, increase (233.1 ± 96.3 ms compared to 174.6 ± 42.8 ms baseline). The latency-to-peak of the NMR was also significantly increased following SR95531 infusions (694.3 ± 24.88 ms compared to 476.5 ± 694.3 ms baseline). There were no significant effects on latency-to-peak after either cortical picrotoxin or CNQX infusions.

---

sions are shown. * indicates a statistically significant difference ($p < 0.05$) between the mean NM CR latency before and after drug infusions; † indicates a statistically significant difference ($p < 0.05$) between the mean NM and eyeblink EMG CR latencies before and after drug infusions (t-test with Welch's correction for unequal variance where appropriate). **CR onset latency and latency to peak:** The onset latency of every NM CR was measured as the first displacement of 0.05 mm beyond baseline after CS onset. The onset latency of the eyelid EMG response was measured as the earliest time at which a spike reached an amplitude 1.5 times that of the maximum pre-CS baseline amplitude. These latencies were measured on all CS–US, paired and CS-alone unpaired trials. The latency-to-peak of the NM movement was also determined. Latency-to-peak of the eyeblink EMG CRs was not estimated.

CEREBELLAR CORTICAL AND NUCLEAR INACTIVATIONS: EVIDENCE FOR OR AGAINST A "DUAL-LEVEL" THEORY OF CEREBELLAR LEARNING?

We failed to find any evidence for short-latency, CS-driven responses in the NM and in the eyeblink EMG after any cerebellar cortical or cerebellar nuclear intervention with GABA or AMPA/kainate receptor antagonists. In all cases, wherever these drugs were effective, CR frequencies were reduced and latencies were extended and, finally, CRs were abolished. These findings are consistent with the suggestion that normal function is required both in the cerebellar cortex and in the cerebellar nuclei for performance of the conditioned NMR/eyeblink. Inactivation of either site is sufficient fully to abolish conditioned responses. But we found no evidence to support the idea that the CS can drive a CR through mossy fiber collaterals to the nuclei because disconnecting cortical outputs to the nuclei, by blocking GABAergic transmission at nuclear synapses with either picrotoxin or SR95531, failed to give rise to any short latency responses, but abolished CRs completely. Furthermore, blocking nuclear AMPA/kainate receptors did not produce any clear changes in the CRs. In no cortical or nuclear inactivation have we dissociated mechanisms for controlling associative and temporal aspects of the conditioned response. Our findings are consistent with our previous suggestion that a fully functional olivo-cortico-nuclear loop is required to express all characteristics of associatively conditioned responses.[5]

Now that we have tested eyeblink EMG CRs, our finding that GABA-A receptor blockers fail to release short-latency CS-driven responses is clearly at variance with findings by Mauk and colleagues.[15,16] However, it is consistent with a very recent report by Thompson and colleagues[27] in which nuclear picrotoxin also abolished NM CRs. However, in an extremely elegant experimental design, they showed that, by using successive nuclear infusions of muscimol (a competitive, GABA-A receptor agonist) and picrotoxin (a non-competitive GABA-A receptor antagonist), cortical input to the nuclei could be blocked by the antagonist while general excitability levels could be balanced to a relatively normal level by the agonist. Under these special conditions, short-latency responses could be detected. However, it should be noted that these responses were seen in subjects that gave quite high levels of short-latency responses in the no-drug condition. In our studies, the CR latencies were almost exclusively adaptively timed (long latency) in the no-drug condition.

The appearance of short-latency, CS-driven responses by manipulation of cerebellar nuclear excitability has been taken as evidence for "dual-level" plasticity. Specifically, it has been suggested that the cerebellar cortex may learn and control temporal aspects of the CR while cerebellar nuclear learning encodes a simpler CS–US association. These assumptions will need to be tested. Firstly, there is no evidence to suggest that these short-latency responses are produced from any part of the cerebellar circuitry. Since they are only seen when nuclear excitabilities are raised, they may be produced by extracerebellar mechanisms expressing CS-driven activation through disinhibited premotor and motor neurons. Secondly, there is no evidence to suggest that these short-latency, CS-driven responses are produced by CS–US association—the hallmark of a learned response. They have been seen at levels of around 15–20%[26] before conditioning begins and there have been no controls for non-associative effects, using non-paired presentations of the CS and US, during nuclear inactivation. Thirdly, our finding that blocking AMPA-kainate–mediated transmission input to the cerebellar nuclei has no significant impact on CR expression indicates that a CR control mechanism involving long-term changes in AMPA receptor-mediated transmission from mossy fiber collateral inputs is unlikely.

Recent evidence indicates that NMDA receptors can provide strong, dynamic inputs to cerebellar nuclear neurons[28] and blockade of cerebellar nuclear NMDA receptors with the selective antagonist D,L-2-amino-5-phosphonopentanoic acid (D,L-AP5) impaired the acquisition of eyeblink conditioning.[29] Intriguingly, nuclear infusions of the NMDA receptor antagonist did not significantly impair the amplitudes of established CRs, but they did delay CR execution because CR onset latencies and latencies-to peak were significantly extended.[29] These findings are consistent with the suggestion that the acquisition and production of accurate CRs is dependent upon NMDA receptor function in the cerebellar nuclei, but that the timing and activation of CRs are not clearly dissociable processes that can be ascribed to cortical and nuclear processing, respectively. Because performance of CRs was disturbed, D,L-AP5 infusions would have produced disturbances of excitability around the olivo-cortico-nuclear loop. The impairments in acquisition could, therefore, derive from changes in information processing at any level in the loop and do not confirm that there is essential plasticity in the cerebellar nuclei.

If, however, there is essential plasticity at the level of the cerebellar nuclei, then we do not know whether it is produced associatively. If the short-latency responses revealed by removal of cortical inhibition are produced non-associatively, but within the cerebellum, then they may be similar to the changes in intrinsic membrane properties seen in MVN neurons after labyrinthectomy[30] or similar to synaptically driven changes in cerebellar nuclear neuron intrinsic excitabilities seen *in vitro*.[31] If the mechanisms are extracerebellar, then a very wide range of possible mechanisms would need to be investigated.

In summary, does the absence of short-latency, CS-driven responses with nuclear GABA-A receptor manipulations rule out plasticity at the cerebellar nuclei in eyeblink/NMR conditioning? No. Such plasticity is not necessarily revealed by these manipulations. However, because nuclear infusions of CNQX have no effect on the production of established CRs, any essential plasticity at mossy fiber to cerebellar nuclear synapses cannot involve essential changes in AMPA receptor-mediated transmission. If, as has been proposed, there are critical changes at these synapses, they would have to involve a novel form of plasticity. Finally, would the presence of short-latency CS-driven responses have implicated plasticity at the cerebellar nuclei? No. The critical dependence of short-latency responses upon the cerebellar nuclei has yet to be established.

Although we now have evidence for memory consolidation processes mediated by the cerebellar cortex,[24] such evidence does not rule out memory storage in the cerebellar nuclei. The relative contributions of cerebellar cortex and nuclei to motor learning will continue to be questioned. With localized, reversible interventions as powerful tools for analysis, it seems that the question is nearing resolution.

REFERENCES

1. KIM, J.J. & R.F. THOMPSON. 1997. Cerebellar circuits and synaptic mechanisms involved in classical eyeblink conditioning. Trends Neurosci. **20:** 177–181.

2. YEO, C. & G. HESSLOW. 1998. Cerebellum and conditioned reflexes. Trends. Cogn. Sci. **2:** 322–330.
3. MEDINA, J.F., C.J. REPA, M.D. MAUK & J.E. LEDOUX. 2002. Parallels between cerebellum- and amygdala-dependent conditioning. Nat. Rev. Neurosci. **3:** 122–131.
4. YEO, C.H., D.H. LOBO & A. BAUM. 1997. Acquisition of a new-latency conditioned nictitating membrane response: major, but not complete, dependence on the ipsilateral cerebellum. Learn. Mem. **3:** 557–577.
5. ATTWELL, P.J., S. RAHMAN & C. H. YEO. 2001. Acquisition of eyeblink conditioning is critically dependent on normal function in cerebellar cortical lobule HVI. J. Neurosci. **21:** 5715–5722.
6. MARR, D. 1969. A theory of cerebellar cortex. J. Physiol (Lond) **202:** 437–470.
7. ALBUS, J. 1971. A theory of cerebellar function. Math.Biosci. **10:** 25–61.
8. ITO, M. 1972. Neural design of the cerebellar motor control system. Brain Res. **40:** 81–84.
9. ITO, M. 1982. Cerebellar control of the vestibulo-ocular reflex: around the flocculus hypothesis. Annu. Rev. Neurosci. **5:** 275–296.
10. MILES, F.A. & S.G. LISBERGER. 1981. Plasticity in the vestibulo-ocular reflex: a new hypothesis. Annu.Rev.Neurosci. **4:** 273–299.
11. RAYMOND, J.L. & S.G. LISBERGER. 1998. Neural learning rules for the vestibulo-ocular reflex. J. Neurosci. **18:** 9112–9129.
12. RAYMOND, J.L., S.G. LISBERGER & M.D. MAUK. 1996. The cerebellum: a neuronal learning machine? Science **272:** 1126–1131.
13. MEDINA, J.F. & M.D. MAUK. 2000. Computer simulation of cerebellar information processing. Nat. Neurosci. **3** (Suppl)**:** 1205–1211.
14. PERRETT, S.P., B.P. RUIZ & M.D. MAUK. 1993. Cerebellar cortex lesions disrupt learning-dependent timing of conditioned eyelid responses. J. Neurosci. **13:** 1708–1718.
15. GARCIA, K.S. & M.D. MAUK. 1998. Pharmacological analysis of cerebellar contributions to the timing and expression of conditioned eyelid responses. Neuropharmacology **37:** 471–480.
16. OHYAMA, T. & M. D. MAUK. 2001. Latent acquisition of timed responses in cerebellar cortex. J.Neurosci. **21:** 682–690.
17. GARCIA, K.S., P.M. STEELE & M.D. MAUK. 1999. Cerebellar cortex lesions prevent acquisition of conditioned eyelid responses. J. Neurosci. **19:** 10940–10947.
18. YEO, C.H. & M.J. HARDIMAN. 1992. Cerebellar cortex and eyeblink conditioning: a reexamination. Exp. Brain Res. **88:** 623–638.
19. HARDIMAN, M.J. & C.H. YEO. 1992. The effect of kainic acid lesions of the cerebellar cortex on the conditioned nictitating membrane response in the rabbit. Eur. J. Neurosci. **4:** 966–980.
20. GRUART, A. & C.H. YEO. 1995. Cerebellar cortex and eyeblink conditioning: bilateral regulation of conditioned responses. Exp. Brain Res. **104:** 431–448.
21. HESSLOW, G. 1994. Inhibition of classically conditioned eyeblink responses by stimulation of the cerebellar cortex in the decerebrate cat. J. Physiol. (Lond). **476:** 245–256.
22. HESSLOW, G. 1994. Correspondence between climbing fiber input and motor output in eyeblink-related areas in cat cerebellar cortex. J. Physiol. (Lond). **476:** 229–244.
23. ATTWELL, P.J., S. RAHMAN, M. IVARSSON & C.H. YEO. 1999. Cerebellar cortical AMPA-kainate receptor blockade prevents performance of classically conditioned nictitating membrane. J. Neurosci. **19:** RC45.
24. ATTWELL, P.J., S. COOKE & C.H. YEO. 2002. Cerebellar function in consolidation of a motor memory. Neuron **34:** 1011–1020.
25. FREEMAN, A.R. 1973. Electrophysiological analysis of the actions of strychnine, bicuculline and picrotoxin on the axonal membrane. J. Neurobiol. **4:** 567–582.
26. MEDINA, J.F., K.S. GARCIA & M.D. MAUK. 2001. A mechanism for savings in the cerebellum. J. Neurosci. **21:** 4081–4089.
27. BAO, S., L. CHEN, J.J. KIM & R.F. THOMPSON. 2002. Cerebellar cortical inhibition and classical eyeblink conditioning. Proc. Natl. Acad. Sci. USA **99:** 1592–1597.
28. ANCHISI, D., B. SCELFO & F. TEMPIA. 2001. Postsynaptic currents in deep cerebellar nuclei. J. Neurophysiol. **85:** 323–331.

29. CHEN, G. & J.E. STEINMETZ. 2000. Inra-cerebellar infusion of NMDA receptor antagonist AP5 disrupts classical eyeblink conditioning in rabbits. Brain Res. **887:** 144–156
30. CAMERON, S. A. & M. B. DUTIA. 1997. Cellular basis of vestibular compensation: changes in intrinsic excitability of MVN neurones. NeuroReport **8:** 2595–2599.
31. AIZENMAN, C. D. & D. J. LINDEN. 2000. Rapid, synaptically driven increases in the intrinsic excitability of cerebellar deep nuclear neurons. Nat. Neurosci. **3:** 109–111.

Role of the Nuclei in Eyeblink Conditioning

DAVID G. LAVOND

Department of Psychology and Neuroscience Program, University of Southern California, Los Angeles, California 90089-2520, USA

ABSTRACT: Evidence to date supports the strong conclusion that the cerebellum learns. Classical conditioning of the eyeblink response is critically dependent upon the cerebellum. The issue addressed here is whether cerebellar cortex or deep nuclei form the basic association. Learning occurs with large cerebellar cortical aspirations in rabbits and with a Purkinje-cell-deficient mutation in mice. The learned response is poorly timed, small in amplitude, and inconsistent in its occurrence. Learning nevertheless occurs. Lesions of the interpositus, on the other hand, prevent new learning and abolish previously learned conditioned responses. Small electrolytic lesions, kainic acid lesions, and temporary inactivation (cooling, muscimol, anisomycin) localize learning to the dorsolateral anterior interpositus nucleus. Learning-related unit activity—the signature of the engram—recorded throughout the brain depends on the interpositus. Electrical stimulation of interpositus afferents are needed for conditioning, and the conditioned interpositus has a lowered threshold. Finally, a recent anatomical study with electron microscopy shows synaptic changes in the excitatory inputs to the interpositus with conditioning. The interpositus is responsible for making the basic association between conditioned and unconditioned stimuli, which in turn allows ancillary learning to occur in cerebellar cortex, and possibly brainstem and forebrain.

KEYWORDS: cerebellum; classical conditioning; eyeblink; learning; deep nuclei; interpositus

INTRODUCTION

More than twenty years of research has shown that classical conditioning of the eyeblink response is critically dependent upon learning in the cerebellum.[1–8] At the beginning of the 1980s when this research started there was no preconception about the involvement of any particular neural structure, although it was known that animals could learn without cerebral cortex,[9,10] without hippocampus,[11,12] and indeed

Address for correspondence: David G. Lavond, Department of Psychology and Neuroscience Program, University of Southern California, Los Angeles, CA 90089-2520. Voice: 213-740-4041; fax: 213-740-5687.
lavond@neuro.usc.edu.

Ann. N.Y. Acad. Sci. 978: 93–105 (2002). © 2002 New York Academy of Sciences.

without any tissue above the midbrain.[13] Rather, the evidence of neural recordings of learning-related activity during conditioning suggested several possible locations for the association, including brainstem,[14] trigeminal,[15] and cerebellum.[1] Two decades of studies involving permanent or reversible lesions, recordings, and electrical stimulation, support the conclusion that the cerebellum is the site of this type of motor learning. At the same time, we ourselves have shown that the lateral cerebellum and interpositus nucleus of the cerebellum are not necessary for heart-rate classical conditioning[16] and are not needed for operant conditioning.[17] There has never been a claim that the cerebellum accounts for all types of learning. Others have shown that reference memory and semantic memory involve the hippocampus[18,19] and that amygdala is involved in emotional learning.[20,21] The evidence for involvement of the cerebellum in eyeblink classical conditioning is extensive.

The issue here is not whether the cerebellum is a learning machine, but whether the evidence supports the conclusion that eyeblink learning occurs in the cerebellar cortex or in the deep cerebellar nuclei, specifically in the interpositus nucleus, or both. In the accompanying article, Chris Yeo presents his evidence for the idea that cerebellar cortex is the site of learning. Since at least 1985,[22] we have suggested the alternative idea that learning in the interpositus nucleus can account for all the observed results of experiments on conditioning using lesions, recordings, stimulation, and pharmacology. In this article, we present evidence that the cortex learns something during classical conditioning, but does not learn the fundamental association, that the cortex is involved but is not essential, and that the interpositus nucleus is the likely site of learning and memory for the classically conditioned eyeblink response.

POINTS OF AGREEMENT

Before we begin with the issue of cortex versus nuclei, it is worthwhile to explicitly identify those points in which we think that Yeo and ourselves are in agreement. First, we probably both agree that cerebellum is critically important for fine motor coordination. One of the misconceptions, without any supporting evidence of which we are aware, is that we deny this role of the cerebellum. This attribution is completely false. The notion of cerebellar plasticity being related to learning does not exclude its role in fine motor coordination any more than the reverse. Second, there is clear evidence that cerebellar cortex is important for the proper timing and amplitude of the classically conditioned learned motor response. Subjects without cerebellar cortex are poor performers: they take longer to learn, their learned responses are smaller in amplitude and have longer (HVI lesion) or shorter (anterior lobe lesion) onset latencies, their peaks are not well timed to coincide with the unconditioned stimulus, and they are often inconsistent (conditioned responses are present on some trials and not on others). Third, cerebellar cortex learns during classical conditioning. What exactly it learns is the question. The debate is whether the fundamental association occurs in the cerebellar cortex and expresses itself through the interpositus, or whether plasticity in cerebellar cortex is merely supportive of learning the association that occurs in the interpositus. Fourth, we both agree that the interpositus is the important output pathway for the learned response, for lesions completely and selectively block the learned behavior. The difficulty is to separate out the performance from the learning effects.

POINTS OF CONTENTION

The point of departure is the question of what exactly is learned by interpositus and cerebellar cortex. We argue for the following conditional relationship: *If* the interpositus has learned the association between the conditioned and unconditioned stimuli, *then* cerebellar cortex can learn to perform the response better. The most frequent alternative solution is to argue that the cerebellar cortex itself forms the association, and the interpositus nucleus is merely an output relay or is sensitized to enhance the behavior. There are other possibilities, including that learning occurs in both places (cortex and nucleus) independently (and usually but not necessarily assumed to be equally potent), or that learning occurs in both places (cortex and nucleus) in a dependent manner such that both loci must be functioning. The evidence below supports neither of the latter two suggestions.

A second potential point of departure is whether learning occurs in one location (e.g., cerebral or cerebellar cortex) and somehow transfers to the deep nuclei after it is well learned. For example, learning how to ride a bicycle may initially be cognitive (cortical) learning, which then settles down to lower nervous regions when it becomes an unconscious habit. In this form of classical conditioning there is no evidence that learning and remembering occur in different places, nor, if it did, is it clear what mechanism could account for such a process. Rather, the evidence is that learning occurs and remains in the interpositus nucleus. Thus, in the example, learning to ride a bicycle always occurs at the lower nervous system, and initially the cortex provides strategies that make interpositus learning more efficient. No postulate of different types of learning or magical transfer mechanisms are needed.

INITIAL EVIDENCE FOR INTERPOSITUS

The first evidence for the involvement of the cerebellum in classical conditioning involved large cerebellar lesions that included both cortex and deep nuclei.[1] Clark and colleagues showed that large electrolytic lesions of the deep nuclei were sufficient to completely, selectively, and permanently abolish the learned behavioral response.[23] Interestingly, they also showed that neural unit activity recorded in the hippocampus was also abolished with the lesion, an observation that we will return to later. Because electrolytic lesions destroy fibers of passage and are still relatively large, we showed that kainic acid lesions as small as one cubic millimeter in the dorsolateral interpositus were sufficient to abolish conditioned responding.[22] Interestingly, in a study of the effects of bicuculine injections, Mamounas and colleagues found that mere insertion of the injection cannula could sometimes cause enough damage to the interpositus to completely abolish conditioning.[24] These studies not only argue for a well-localized effect, but caution against drawing wrong conclusions from large cerebellar cortical aspirations that could damage or disrupt the normal functioning of the interpositus nucleus.

Evidence Against Cerebellar Cortex as the Locus for Learning

Previous observations of McCormick and others have shown that small cerebellar cortical lesions that spared the deep nuclei did not abolish conditioning. Subsequently it was shown that large cerebellar cortical lesions did not abolish a previously learned conditioned response [25] and did not prevent initial learning.[26] These lesions included *at a minimum* the HVI area identified by Yeo and colleagues as being critical,[2,3] plus large areas of ansiform and paramedian lobes. In some animals the lesion extended into HV and other anterior cortical regions, into the vermis, and with nearly bilateral lesions in a few animals. We got so frustrated at the discrepancy between our results and Yeo's that we even tried duplicating their training parameters and strain of rabbit.[27] But the general observations held that very large lesions did not prevent or abolish the conditioned response, that they caused severe motor deficits that the animals would recover from, and that the performance of the conditioned response was delayed and of poor quality, but that learning did occur. The effects on learning included small amplitudes, poor timing (usually longer latencies with posterior cortical lesions, and sometimes shorter latencies with anterior cortical lesions), inconsistent responding, and taking much longer to reach criterion. On average, animals take seven times longer to learn than to relearn after cerebellar cortical lesions. Logan in her dissertation confirmed and extended these findings, showing in addition that discrimination training did not depend upon cerebellar cortex.[28] Cerebellar cortical lesions are accompanied by partial but not complete degeneration of the inferior olive, as one would expect.

Because the rate of relearning the eyeblink response after cortical lesion is similar to the rate of learning by a normal group of animals, it is sometimes assumed that relearning is actually new learning, that is, that cerebellar cortex did the original learning and after its removal that another part of the brain must now be learning. However, this is not the proper comparison. The correct comparison of relearning is to how long it takes a naive animal without cerebellar cortical to acquire (FIG. 1). The answer is about seven times as long as normal. Thus, in fact, as we have previously pointed out, subjects in the relearning study actually show considerable retention. Examination of the data in the relearning experiment also show individual animals with good retention when compared to their original learning rate. Similar observations have been made for the effects of cortical lesions on brightness discriminations in rats in explaining the classic observations of Lashley.[29,30]

More recently, Yeo and colleagues reported that bilateral lesions of HVI were necessary to show cortical dependence.[31] However, as we just noted, we have previously reported in the retention and acquisition experiments several instances of nearly bilateral lesions with learning—as long as the interpositus is not damaged.[25,26] Nevertheless, recently we again created bilateral cortical lesions, but subsequently used serial lesions of cerebellar cortex to minimize trauma and enhance recovery in removing cerebellar cortex bilaterally, and replicated our finding that it took longer to learn but that the animals did learn.[32]

The overriding lesson to be learned is that one should immediately suspect that interpositus damage or disruption can account for the loss of learned responses whenever a cortical lesion is made. Remember this observation: Very large lesions of cerebellar cortex do not prevent learning or retention, whereas very small lesions of the interpositus are sufficient to cause complete loss of learned responses.

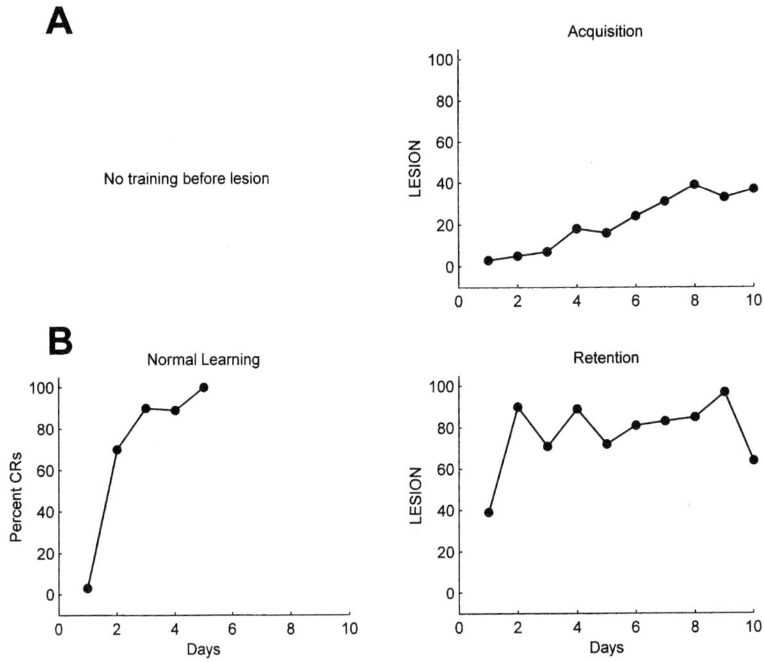

FIGURE 1. Comparison of the rates of learning and relearning with cerebellar cortical lesions measured as percent conditioned responses per day of training. (**A**) Learning for a group of animals with a cerebellar cortical lesion is prolonged, being about seven times the normal learning rate (the data are taken from Ref. 25, Figure 1). The learned responses are also of poor quality (poorly timed and small amplitudes) and not consistently present. (**B**) Relearning after an intervening cerebellar cortical lesion for one animal (the data are taken from Ref. 26, Figure 1C) shows retention compared to initial training, and also shows considerable retention when compared to naive learning with a cerebellar cortical lesion, as seen in (A). Along with the transfer of training experiment,[48] this figure illustrates behavioral evidence that cerebellar cortical lesions do not destroy a previously existing memory.

PROBLEMS IN INTERPRETATION OF CEREBELLAR CORTICAL LESIONS

There are a couple of problems with the interpretations of the effects of cerebellar cortical lesions on classical conditioning as we have presented them. Foremost among these problems is that, in our experience, it is nearly impossible to completely remove cerebellar cortex by aspiration (which we have never done) without damaging the interpositus nucleus (which we have done), either directly by removing the interpositus or indirectly by the trauma associated with pressing or pulling of the aspirator on the medulla of the cerebellum. Since Mamounas showed that very little damage caused by the mere presence of a cannula is sufficient to cause complete abolition, it behooves one to suspect first that a small amount of interpositus damage accounts for the loss rather than the larger cortical damage. On the other hand, in cases where there is little suspicion of deep nuclear damage, and hence conditioned responses are present, there is the strong possibility—*no, the certainty*—that there is cortical sparing that can be invoked as the explanation for learning. Oddly, the explanation of sparing is sometimes invoked even when the sparing is in the "wrong" place, that is, even with complete HVI removal and when there is certain to be some cortical sparing elsewhere. This condition has led to suspicions of the essential involvement of anterior cerebellar cortex or contralateral HVI. The fact that we have previously provided cases of large lesions that include these and nearly the entire cerebellar cortex leads us to doubt that. Despite these observations, cortical sparing remains an argument.

A Better Cerebellar Cortical Lesion

Probably the best evidence that cerebellar cortex is not essential for the basic association between the conditioned and unconditioned stimuli is the mutant mouse experiment by Thompson and his colleagues.[33] The pcd mutant mouse loses all its Purkinje cells by about day 20 without developing abnormal cortical connections. Sparing is not a problem here. The results of training are in complete agreement with the aspiration experiments: It takes about seven times as long to learn as normals, and the learning is not good in terms of latency and amplitude or as consistent. To be fair, as with all mutant mouse experiments of this type, however, it cannot be ruled out that some compensatory mechanism is engaged in recovery. Nevertheless, learning without Purkinje cells is clearly established. Lesion of the interpositus, however, wiped out conditioned responding in the pcd mutant mouse.[34]

Evidence for the Interpositus as the Locus for Learning

The evidence for the central role of the interpositus in learning the classically conditioned eyelid response includes the following. First, as given earlier, very small lesions of the interpositus completely and permanently abolish already learned responses or prevent learning of a new response, whereas very large lesions of cerebellar cortex disrupt but do not prevent learning and remembrance. Second, temporary, reversible lesions of the interpositus prevent any learning in any other part of the nervous system. Third, temporary inactivation abolishes all learning-related activity (the "signature" of the engram) everywhere in the nervous system that has been observed so far. Fourth, the lack of "transfer" to the contralateral cerebellum when training shows that there is no surviving memory after interpositus lesion. Fifth, stimulation of cerebellar afferents localize learning to the cerebellum, and decreases in threshold in the interpositus occur with conditioning. Sixth, synaptic anatomical changes are associated with conditioning-induced changes in excitatory synapses in the interpositus. We have already discussed the first point, the lesion results. The following succinctly examines the other points of evidence.

Reversible Lesions of the Interpositus Prevent Learning Anywhere Else

The second piece of evidence for the central role of the interpositus in classical conditioning comes from studies of the effects of reversible lesions on acquisition. We first showed that cooling the interpositus to suppress synaptic transmission during acquisition training prevented any evidence of later retention,[35] whereas there was excellent retention following similar suppression of the red nucleus.[36] The importance of the interpositus during acquisition was confirmed using the GABA agonist muscimol[37] and the protein synthesis inhibitor anisomycin.[38] Together these findings with reversible lesions argue strongly that the essential plasticity for classical eyelid conditioning occurs in the interpositus. Had the treatments merely interrupted learning that occurred prior to the interpositus, for example, in cerebellar cortex, then removing the inactivation would have shown immediate retention, as in the case of red nucleus inactivation.

Reversible Lesions of the Interpositus Abolish the Signature Everywhere Else

The third piece of evidence for the central role of the interpositus in classical conditioning comes from physiological studies of learning-related unit activity. Thompson and colleagues had previously shown that hippocampal activity predicts the presence and form of the learned behavioral response.[39] This neural activity, which we call the "signature" of the engram, can be recorded throughout much of the brain.[40–42] However, removal of cerebral cortex, hippocampus, and indeed all tissue above the midbrain does not prevent learning. As we previously noted, Clark and colleagues reported that the hippocampal signature is abolished with interpositus lesions.[23] In extending these findings, we have found that cooling of the interpositus removes this signature everywhere we have recorded (pons,[43] trigeminal,[44] and also in reticular formation, reticular tegmental nucleus, HVI, anterior cerebellar cortex, and facial nucleus[45]). In agreement, Bao and colleagues concluded more recently that lateral pons receives the engram from the interpositus, rather than the other way around, after stimulating the lateral reticular nucleus as a conditioned stimulus.[46] Thus, it seems probable that the interpositus is the source of this signature throughout the brain. The abolition of the signature in cerebellar cortex is a key observation for the argument here. Significantly, sensory-evoked activity in cerebellar cortex remains with interpositus inactivation. Thus, rather than originating in cerebellar cortex, the signature originates in the interpositus and is projected to the cortex. One of the major omissions in the thinking about models of the cerebellar circuit has been the projection from interpositus back to cerebellar cortex.

No Transfer of Training

The fourth piece of evidence for the central role of the interpositus in classical conditioning comes from the early observation that training on the unlesioned side is very rapid after contralateral interpositus lesion.[1,45] After normal training, animals given unilateral cerebellar lesion show no learned responses when retrained on that side. Subsequent training on the contralateral side, however, produces rapid learning. One way to interpret this is to suggest that the memory survived the lesions. However, we showed that if the retraining step is omitted after the lesion, that subsequent contralateral training proceeds as with naive animals.[48] Clearly, the *attempt* to retrain on the lesioned side is the key factor. By attempting to train, even though no learned responses appear on that side, new learning is induced on the contralateral side. This result shows quite convincingly that no memory survives the lesion.

Stimulation Lowers the Threshold of the Interpositus

The fifth piece of evidence for the central role of the interpositus in classical conditioning comes from a series of stimulation studies conducted by Thompson and colleagues. Briefly, these experiments show that direct stimulation of cerebellar afferents can be substituted for the conditioned stimulus (lateral pons and middle cerebellar peduncle stimulation),[49–51] or unconditioned stimulus (inferior olive stimulation),[52] or both,[53] that direct stimulation of the interpositus can substitute for the conditioned stimulus,[54] but not the unconditioned stimulus,[55] that the threshold of the pontine-to-interpositus connection is lowered with pontine stimulation used as the conditioned stimulus,[56] and that learning induced by lateral reticular stimulation occurs in the cerebellum.[57] Together, these studies show that electrical stimulation of interpositus afferents is necessary for conditioning, and that as a result of this conditioning the interpositus has a lowered threshold. Since an early observation by McCormick that stimulation of the interpositus elicited eyeblinks in untrained animals, it has been recognized that this form of learning is a modification of a preexisting, hardwired neural circuit.

Training Induces Anatomical Changes in Synapses in the Interpositus

The sixth piece of evidence for the central role of the interpositus in classical conditioning comes from the recent electron microscopic examination by Freeman and his colleagues.[57] Rabbits given paired but not control conditioning trials (pseudoconditioning and also home-cage conditioning) show increased excitatory synapses in the interpositus, whereas the inhibitory projections from the cerebellar cortex remain unchanged. Changes in the excitatory synapses are exactly what one would expect for learning in the interpositus. The inhibitory synapses were unchanged, ruling out a significant role for cerebellar cortex. For the excitatory synapses, further research will be needed to see if the mossy fiber projections to the interpositus are selectively affected over the climbing fiber projections, as one would expect for modification in the conditioned stimulus pathway.

DISCUSSION

Since the original electrolytic experiment of Clark and colleagues[23] and our own study using kainic acid lesions,[22] we have thought, as a working hypothesis, that the interpositus could be responsible for learning the basic association in classical conditioning and that other regions, like the hippocampus, could act to modify the basic pattern of learning. The involvement of the hippocampus in trace conditioning as opposed to delay conditioning[58] serves as a model for that idea. As a consequence, we have been skeptical of and tested the notion that cerebellar cortex is the locus of learning for this type of motor memory. It has not helped that most if not all of the computer models that are based on eyeblink conditioning experiments have simulated cerebellar cortex alone. In the few models where the interpositus is included, its role has been minor, usually as a mere relay. That has not prevented these models from obtaining computer-generated results that agree with behavioral observations. In general, the results of computer modeling deserve a great deal of skepticism.

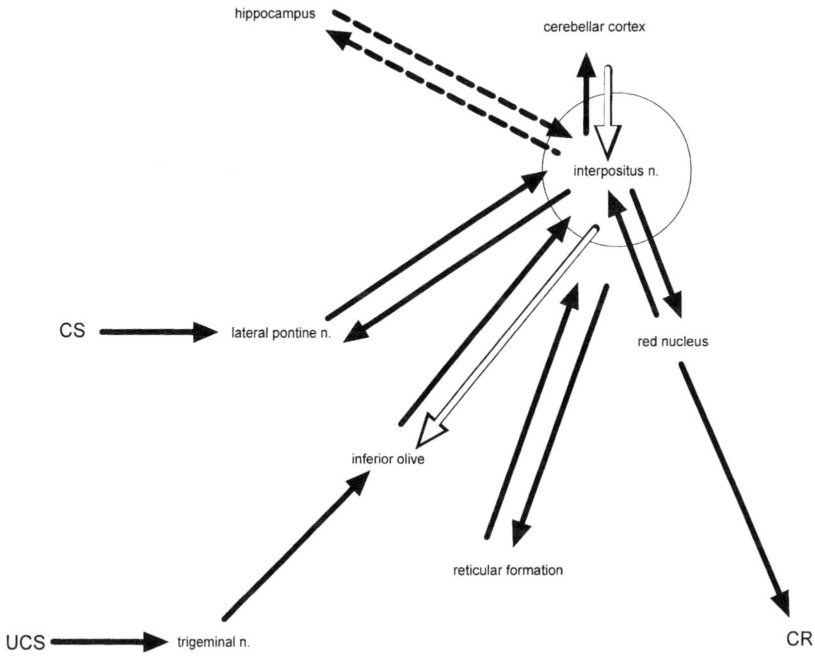

FIGURE 2. Diagram to emphasize the central nature of the interpositus in learning the association between the conditioned and unconditioned stimuli for eyeblink conditioning. Once the association has been formed in the interpositus, feedback to other neural structures allows ancillary learning. For example, feedback to the lateral pons allows learning about the stimulus qualities of the tone-conditioned stimulus. As another example, feedback to the cerebellar cortex allows learning about the timing and amplitude of the learned response.

A good model of the circuitry for classical conditioning of the eyelid response should include those features of connectivity that we have previously summarized (Ref. 6, Figure 1). But more, an essential feature that we wish to emphasize here is the central position of the interpositus (FIG. 2): *If* the interpositus has learned, *then* we have suggested that other areas—cerebellar cortex, reticular formation, lateral pons, trigeminal, red nucleus, hippocampus—can learn some different aspect of the conditioning, learn to enhance the timing and amplitude of the learned response, and learn to selectively attend to the conditioning stimuli, as we have suggested previously.[5]

The recent anatomical study by Freeman and colleagues[57] exemplifies the kind of study of the interpositus that is needed. The anatomy of the interpositus is not complex in comparison to cerebellar cortex, and the interpositus has probably been overlooked because of that simplicity. But "simplicity" is not synonymous with "incapable." One should bear in mind that the mammalian spinal cord is capable of conditioning.[59–61] There is no reason to believe that classical conditioning of the eyelid response is any more complicated than the spinal cord so that it would require distribution of the same memory between cortex and interpositus. Further, if structure is associated with function, then one should expect that *different* memories are subsumed in the cortex and interpositus, rather than redundancy of the same memory. More detailed investigation of the anatomical, physiological, and pharmacological changes within the interpositus during conditioning, during the recall of memories, and during extinction, should be pursued, as well as analogous studies of the contributions of cerebellar cortex and other structures that receive the signature of the engram and contribute to the whole behavioral experience.

SUMMARY

The cerebellum is critically involved in learning the classically conditioned eyeblink response. Cerebellar cortex learns to effect good timing and amplitude of the learned response. The interpositus nucleus learns the basic association between the conditioned and unconditioned stimuli, and coordinates different aspects of learning in cerebellar cortex, brainstem, and hippocampus. In interpretations of the effects of cerebellar lesions, it bears remembering the observations that very large lesions of cerebellar cortex do not abolish or prevent learning, whereas very, very small lesions of the interpositus can permanently and selectively block conditioned responses. Models of classically conditioned motor responses need to reflect these differences.

ACKNOWLEDGMENT

This work was supported by the Office of Naval Research under Contract N00014-00-1-0923. I thank Richard F. Thompson for helpful comments.

REFERENCES

1. McCormick, D.A., D.G. Lavond, G.A. Clark, *et al.* 1981. The engram found? Role of the cerebellum in classical conditioning of nictitating membrane and eyelid responses. Bull. Psychon. Soc. **18:** 103–105.
2. Yeo, C.H., M.J. Hardiman & M. Glickstein. 1984. Discrete lesions of the cerebellar cortex abolish the classically conditioned nictitating membrane response of the rabbit. Behav. Brain Res. **13:** 261–266.
3. Yeo, C.H., M.J. Hardiman & M. Glickstein. 1985. Classical conditioning of the nictitating membrane response of the rabbit. II. Lesions of the cerebellar cortex. Exp. Brain Res. **60:** 99–113.
4. Lavond, D.G. 2000. Reflections on R.F. Thompson's discovery of the engram. Festschr. Symp.: The Engram Found: A Celebration of the Scientific Contributions of Richard F. Thompson (Organized by T.W. Berger, M. Baudry, and L.W. Swanson), Univ. of Southern California, Los Angeles, CA.
5. Lavond, D.G. & S.A. Kanzawa. 2001. Inside the black box. *In* Model Systems and the Neurobiology of Associative Learning: A Festschrift in Honor of Richard F. Thompson. J.E. Steinmetz, M. Gluck, and P. Solomon, Eds.: 245–269. Erlbaum. Manwah, NJ.
6. Lavond, D.G. & S.A. Kanzawa. 2001. Reflections on R.F. Thompson's discovery of the engram. Engram **1:** 6–14.
7. Lavond, D.G. & M.C. Cartford. 2000. Eye blink conditioning circuitry: tracing, lesion, and reversible lesion experiments. *In* Eyeblink Classical Conditioning: Animals. D.S. Woodruff-Pak and J.E. Steinmetz, Eds.: 51–80. Kluwer. Boston.

8. THOMPSON, R.F., S. BAO, L. CHEN, *et al.* 1997. Associative learning. Int. Rev. Neurobiol. **41:** 151–189.
9. OAKLEY, D.A. & I.S. RUSSELL. 1972. Neocortical lesions and classical conditioning. Physiol. Behav. **8:** 915–926.
10. OAKLEY, D.A. & I.S. RUSSELL. 1977. Subcortical storage of Pavlovian conditioning in the rabbit. Physiol. Behav. **18:** 931–937.
11. KIM, J.J., R.E. CLARK & R.F. THOMPSON. 1995. Hippocampectomy impairs the memory of recently, but not remotely, acquired trace eyeblink conditioned responses. Behav. Neurosci. **109:** 195–203.
12. SCHMALTZ, L.W. & J. THEIOS. 1972. Acquisition and extinction of a classically conditioned response in hippocampectomized rabbits (Oryctolagus cuniculus). J. Comp. Physiol. Psychol. **79:** 328–333.
13. ENSER, D. 1976. A study of classical nictitating membrane conditioning in neodecorticate, hemidecorticate and thalamic rabbits. Ph.D. Thesis, Univ. of Iowa, Iowa City.
14. LAVOND, D.G., D.A. MCCORMICK, G.A. CLARK, *et al.* 1981. Effects of ipsilateral rostral pontine reticular lesions on retention of classically conditioned nictitating membrane and eyelid responses. Physiol. Psychol. **9:** 335–339.
15. DESMOND, J.E. & J.W. MOORE. 1983. A supratrigeminal region implicated in the classically conditioned nictitating membrane response. Brain Res. Bull. **10:** 765–773.
16. LAVOND, D.G., J.S. LINCOLN, D.A. MCCORMICK & R.F. THOMPSON. 1984. Effect of bilateral lesions of the dentate and interpositus cerebellar nuclei on conditioning of heart-rate and nictitating membrane/eyelid responses in the rabbit. Brain Res. **305:** 323–330.
17. HOLT, L., M.D. MAUK & R.F. THOMPSON. 1984. Effects of lateral cerebellar lesions on classical and operant conditioning. Unpublished manuscript.
18. OLTON, D.S., J.T. BECKER & G.E. HANDELMANN. 1979. Hippocampus, space and memory. Behav. Brain Sci. **2:** 313–365.
19. SQUIRE, L.R. 1982. The neurophysiology of human memory. Annu. Rev. Neurosci. **5:** 241–273.
20. DAVIS, M., D.S. GENDELMAN, M.D. TISCHLER & P.M. GENDELMAN. 1982. A primary acoustic startle circuit: lesion and stimulation studies. J. Neurosci. **2:** 791–805.
21. KIM, J.J., R.A. RISON & M.S. FANSELOW. 1993. Effects of amygdala, hippocampus, and periaqueductal gray lesions on short- and long-term contextual fear. Behav. Neurosci. **107:** 1093–1098.
22. LAVOND, D.G., T.L. HEMBREE & R.F. THOMPSON. 1985. Effect of kainic acid lesions of the cerebellar interpositus nucleus on eyelid conditioning in the rabbit. Brain Res. **326:** 179–182.
23. CLARK, G.A., D.A. MCCORMICK, D.G. LAVOND & R.F. THOMPSON. 1984. Effects of lesions of cerebellar nuclei on conditioned behavioral and hippocampal neuronal responses. Brain Res. **291:** 125–136.
24. MAMOUNAS, L.A., R.F. THOMPSON & J. MADDEN. 1987. Cerebellar GABAergic processes: evidence for critical involvement in a form of simple associative learning in the rabbit. Proc. Natl. Acad. Sci. USA **84:** 2101–2105.
25. LAVOND, D.G., J.E. STEINMETZ, M.H. YOKAITIS & R.F. THOMPSON. 1987. Reacquisition of classical conditioning after removal of cerebellar cortex. Exp. Brain Res. **67:** 569–593.
26. LAVOND, D.G. & J.E. STEINMETZ. 1989. Acquisition of classical conditioning without cerebellar cortex. Behav. Brain Res. **33:** 113–164.
27. CLARK, R.E. & D.G. LAVOND. 1994. Reacquisition of eyeblink classical conditioning following large cerebellar cortical lesions in Dutch belted rabbits. Behav. Brain Res. **61:** 101–106.
28. LOGAN, C.G. 1991. Cerebellar cortical involvement in excitatory and inhibitory classical conditioning. Ph.D. Thesis, Stanford Univ., Palo Alto, CA.
29. MEYER, D.R. & P.M. MEYER. 1977. Dynamics and bases of recoveries of functions after injuries to the cerebral cortex. Physiol. Psychol. **5:** 133–165.
30. MEYER, D.R. & P.M. MEYER. 1984. Bases of recoveries from perinatal injuries to the cerebral cortex. *In* Behavioral Biology of Early Brain Damage. S. Finger and C.R. Almli, Eds.: 211–227. Academic Press. New York.

31. GRUART, A. & C.H. YEO. 1995. Cerebellar cortex and eyeblink conditioning: bilateral regulation of conditioned responses. Exp. Brain Res. **104:** 431–448.
32. MOJTAHEDIAN, S., S. KANZAWA, D. LAVOND & R. THOMPSON. 2001. Effects of bilateral lesions of the cerebellar cortex on eyeblink conditioning. Soc. Neurosci. Abstr. **27:** 201.
33. CHEN, L., S. BAO, J.M. LOCKARD, *et al.* 1996. Impaired classical eyeblink conditioning in cerebellar-lesioned & Purkinje cell degeneration (pcd) mutant mice. J. Neurosci. **16:** 2829–2838.
34. CHEN, L., S. BAO & R.F. THOMPSON. 1999. Bilateral lesions of the interpositus nucleus completely prevent eyeblink conditioning in Purkinje cell-degeneration mutant mice. Behav. Neurosci. **113:** 204–210.
35. CLARK, R.E., A.A. ZHANG & D.G. LAVOND. 1992. Reversible lesions of the cerebellar interpositus nucleus during acquisition and retention of a classically conditioned behavior. Behav. Neurosci. **106:** 879–888.
36. CLARK, R.E. & D.G. LAVOND. 1993. Reversible lesions of the red nucleus during acquisition and retention of a classically conditioned behavior in rabbits. Behav. Neurosci. **107:** 264–270.
37. KRUPA, D.J., J.K. THOMPSON & R.F. THOMPSON. 1993. Localization of a memory trace in the mammalian brain. Science **260:** 989–991.
38. BRACHA, V., M.L. WEBSTER, M.K. STACHOWIAK & J.R. BLOEDEL. 1995. Injections of anisomycin into the interposed nuclear region affect consolidation of the conditioned eyeblink response. Soc. Neurosci. Abstr. **21:** 1222.
39. BERGER, T.W., B. ALGER & R.F. THOMPSON. 1976. Neuronal substrate of classical conditioning in the hippocampus. Science **192:** 483–485.
40. MCCORMICK, D.A., D.G. LAVOND, N.H. DONEGAN & R.F. THOMPSON. 1982. Neuronal responses of the rabbit brainstem and cerebellum during performance of the classically conditioned nictitating membrane/eyelid response. Brain Res. **271:** 73–88.
41. MCCORMICK, D.A., D.G. LAVOND & R.F. THOMPSON. 1983. Neuronal responses of the rabbit brainstem during performance of the classically conditioned nictitating membrane (NM)/eyelid response. Brain Res. **271:** 73–88.
42. MCCORMICK, D.A. & R.F. THOMPSON. 1984. Neuronal responses of the rabbit cerebellum during acquisition and performance of a classically conditioned nictitating membrane/eyelid response. J. Neurosci. **4:** 2811–2822.
43. CLARK, R.E., E.B. GOHL & D.G. LAVOND. 1997. The learning related activity which develops in the pontine nuclei during classical eyeblink conditioning is dependent upon the interpositus nucleus. Learn. Mem. **3:** 532–544.
44. CLARK, R.E. & D.G. LAVOND. 1996. Neural unit activity in the trigeminal complex with interpositus or red nucleus inactivation during classical eyeblink conditioning. Behav. Neurosci. **110:** 1–9.
45. CLARK, R.E., A.A. ZHANG & D.G. LAVOND. 1997. The importance of cerebellar cortex and facial nucleus in acquisition and retention of eyeblink/NM conditioning: evidence for critical unilateral regulation of the conditioned response. Neurobiol. Learn. Mem. **67:** 96–111.
46. BAO, S., L. CHEN & R.F. THOMPSON. 2000. Learning- and cerebellum-dependent neuronal activity in the lateral pontine nucleus. Behav. Neurosci. **114:** 254–261.
47. MCCORMICK, D.A., G.A. CLARK, D.G. LAVOND AND R.F. THOMPSON. 1982. Initial localization of the memory trace for a basic form of learning. Proc. Natl. Acad. Sci. USA **79:** 2731–2742.
48. LAVOND, D.G., S.A. KANZAWA, D. IVKOVICH & R.E. CLARK. 1994. Transfer of learning but not memory after unilateral cerebellar lesion in rabbits. Behav. Neurosci. **108:** 284–293.
49. STEINMETZ, J.E., D.G. LAVOND & R.F. THOMPSON. 1985. Classical conditioning of the rabbit eyelid response with mossy fiber stimulation as the conditioned stimulus. Bull. Psychon. Soc. **23:** 245–248.
50. STEINMETZ, J.E., D.J. ROSEN, P.F. CHAPMAN, *et al.* 1986. Classical conditioning of the rabbit eyelid response with a mossy fiber stimulation CS. I. Pontine nuclei and middle cerebellar peduncle stimulation. Behav. Neurosci. **100:** 871–880.

51. STEINMETZ, J.E., D.J. ROSEN, D.S. WOODRUFF-PAK, *et al.* 1986. Rapid transfer of training occurs when direct mossy fiber stimulation is used as a conditioned stimulus for classical conditioning. Neurosci. Res. **3:** 606–616.
52. MAUK, M.D., J.E. STEINMETZ & R.F. THOMPSON. 1986. Classical conditioning using stimulation of the inferior olive as the unconditioned stimulus. Proc. Natl. Acad. Sci. USA **83:** 5349–5353.
53. STEINMETZ, J.E., D.G. LAVOND & R.F. THOMPSON. 1989. Classical conditioning in rabbits using pontine nucleus stimulation as a conditioned stimulus and inferior olive stimulation as an unconditioned stimulus. Synapse **3:** 225–233.
54. POULOS, A.M. & R.F. THOMPSON. 2001. Extracellular stimulation of the interpositus nucleus as a conditioned stimulus during interstimulus interval shifts in delay eyeblink conditioning. Soc. Neurosci. Abstr. **27:** 1689.
55. CHAPMAN, P.F., J.E. STEINMETZ & R.F. THOMPSON. 1988. Classical conditioning does not occur when direct stimulation of the red nucleus or cerebellar nuclei is the unconditioned stimulus. Brain Res. **442:** 97–104.
56. TRACY, J., J.K. THOMPSON, D.J. KRUPA & R.F. THOMPSON. 1998. Evidence of plasticity in the ponto-cerebellar CS pathway during classical conditioning of the eyeblink response in the rabbit. Behav. Neurosci. **112:** 267–285.
57. KLEIM, J.A., J.H. FREEMAN, R. BRUNEAU, *et al.* 2002. Synapse formation is associated with memory storage in the cerebellum. Proc. Natl. Acad. Sci. USA **99:** 13228–13231.
58. SOLOMON, P.R., E.F. VANDER SCHAFF, R.F. THOMPSON & D.G. WEISZ. 1986. Hippocampus and trace conditioning of the rabbit's classically conditioned nictitating membrane response. Behav. Neurosci. **100:** 729–744.
59. PATTERSON, M.M. 1975. Effects of forward and backward classical conditioning procedures on a spinal cat hindlimb flexor nerve response. Physiol. Psychol. **3:** 86–91.
60. PATTERSON, M.M., C.F. CEGAVSKE & R.F. THOMPSON. 1973. Spinal conditioning. Bull. Psychon. Soc. **1:** 139–141.
61. PATTERSON, M.M., J.E. STEINMETZ, A.L. BEGGS & A.G. ROMANO. 1982. Associative processes in spinal reflexes. *In* Conditioning: Representation of Involved Neural Functions. C.D. Woody, Ed.: 637–650. Plenum. New York.

An Experimental Study of Posterior Interpositus Involvement in the Genesis and Control of Conditioned Eyelid Responses

LYDIA JIMÉNEZ-DÍAZ,[a] AGNÈS GRUART,[a] F. JAVIER MIÑANO,[b] AND JOSÉ M. DELGADO-GARCÍA[a]

[a]*División de Neurociencias, Laboratorio Andaluz de Biología, Universidad Pablo de Olavide, Sevilla-41013, Spain*

[b]*Unidad de Investigación, Laboratorio Experimental de Farmacología, Hospital Universitario de Valme, Universidad de Sevilla, Sevilla-41014, Spain*

ABSTRACT: The participation of the rostral–dorsal part of the posterior interpositus nucleus in the genesis of classically-conditioned eyelid responses in alert cats was studied. Cats were prepared for the recording of left eyelid movements with the search-coil-in-a-magnetic-field technique and of the electromyographic activity of the orbicularis oculi muscle. A stimulating electrode was implanted in the contralateral red nucleus and a guide cannula in the ipsilateral interpositus. The cannula allowed unitary recordings, electrical microstimulation, drug injection, and push-pull perfusion of the selected interpositus site. Perfusate was used for amino acid analysis with an HPLC-EC method. Animals were conditioned with a delay (CS: tone, US: air puff) paradigm. Antidromically identified interpositus neurons fired during eyelid conditioned responses (CRs) with a variable delay to CR start. Some neurons (type A) increased and other (type B) decreased their firing during CR performance. Electrical microstimulation of posterior interpositus was able to increase CR amplitude when applied during CS presentation. Muscimol decreased CR amplitude in well-trained animals. The analysis of push-pull perfusate showed a significant increase in glutamate, glycine, and taurine, but not GABA levels, during conditioning in comparison with habituation and extinction periods. Results indicate an involvement of the rostral-dorsal part of the posterior interpositus nucleus in the genesis of eyelid conditioned responses, playing an enhancing or controlling role in CR performance, rather than being its initiator.

KEYWORDS: cats; cerebellum; classical conditioning; electrical stimulation; GABA; glutamate; glycine; interpositus nucleus; muscimol; neuronal activity; nictitating membrane/eyelid responses; push-pull perfusion; taurine

Address for correspondence: Prof. José M. Delgado-García, División de Neurociencias, Laboratorio Andaluz de Biología, Universidad Pablo de Olavide, Ctra. de Utrera, Km. 1, Sevilla-41013, Spain. Voice: + 34-954-349374; fax: + 34-954-349375.
jmdelgar@dex.upo.es

Ann. N.Y. Acad. Sci. 978: 106–118 (2002). © 2002 New York Academy of Sciences.

INTRODUCTION

The classical conditioning of nictitating membrane/eyelid responses has been used for the past five decades as a useful tool for the study of neuronal processes underlying motor learning in mammals.[1,2] It is now well established that eyelid conditioned responses (CRs) are different in latency, profile, peak velocity, and other kinetic properties from reflexively evoked blinks, a fact indicating a different neural origin and/or peculiar generation mechanism.[3–8] As an indicative example of these kinetic differences, reflex blink responses in cats following trigeminal stimulation are characterized by a fast (up to 2000 deg/s) upper lid downward displacement of a fixed rise time (about 20 ms), while CRs are ramplike responses, with peak velocities never surpassing 300 deg/s.[5]

Both the neural sites where this particular type of motor learning could take place and the subcellular and molecular events involved in it are currently a matter of continuous debate.[8–12] One of the favorite suggested sites is a region not well located in the interpositus nucleus.[12–17] Perhaps the problem of the precise location of the interpositus site related to blink conditioning is that most of the early proposals have been supported mainly by lesion and pharmacological studies.[12–14,16,18] In a recent contribution to this dilemma, specific portions of the lateral margin of both anterior and posterior interpositus neurons following attenuated virus application to the upper eyelid in rats have been labeled.[19] Indeed, neuronal recording in alert-behaving cats during classical conditioning of eyelid responses has confirmed the involvement of a specific portion of the rostral–dorsal zone of the posterior interpositus nucleus in the performance of eyelid CRs.[15]

Our aim in this study was to present a useful experimental approach to the study of the specific contribution of selected posterior interpositus sites in the genesis and control of conditioned eyelid responses. This experimental approach allows the simultaneous study of the electrical activity of identified interpositus units,[15] the effects of local elctrical microstimulation,[15] of drug microinjection,[13–15] and the push-pull perfusion of the selected site in a search for neurotransmitters and related substances involved in the acquisition of conditional eyelid responses.[20–22] Obviously, this experimental model can be successfully applied to other sites (e.g., cerebellar cortex, hippocampus, red nucleus) also related to this sort of motor learning.

EXPERIMENTAL PROCEDURES

Subjects

Experiments were carried out on eight adult female cats weighing 2–2.5 kg. Animals were obtained from an authorized supplier (Iffa-Credo). All of the experimental procedures were performed following guidelines of the European Union (86/609/EU) and Spanish regulations (BOE/67:8509/1988) for the use of laboratory animals in chronic experiments.

Surgical Procedures

Under deep anesthesia (35–40 mg/kg, i.p. of sodium pentobarbital and 0.5 mg/kg i.m. of atropine sulfate) and sterile conditions, animals were implanted with a search coil in the center and lower end of the upper eyelid. The coil was made from Teflon-coated multistranded stainless-steel wire with an external diameter of 50 μm, of five turns, with a diameter of 3 mm. Animals were also implanted with two hook electrodes, bared 1 mm at their tips, in the orbicularis oculi muscle. Although not illustrated here, EMG records were mostly used to confirm latency data obtained from eyelid coil records.[5] A stimulating bipolar (200 μm, enamel-coated silver wire) electrode was implanted in the magnocellular division of the contralateral red nucleus for antidromic activation of recorded interpositus neurons. A recording window (5 × 5 mm) was opened in the occipital bone to allow access to the deep cerebellar nuclei. The dura mater was removed and an acrylic chamber was constructed around the opening. The cerebellar surface was protected with a piece of silicone sheet and sterile gauze, and capped with a hermetic plastic cover. A head-holding system, perpendicular to the horizontal stereotaxic plane, was attached to the skull. Wires were soldered to a socket fixed to the holding system. A detailed description of this chronic preparation has been presented elsewhere.[5,15,23]

Guide Cannula Implantation

Recording sessions started 2 weeks after surgery and lasted < 3 h for a maximum of 20 days. During some preliminary sessions, deep cerebellar nuclei were systematically explored with glass microelectrodes (2M NaCl, 3–6 MΩ of resistance) in order to find the posterior interpositus area related to eyelid responses.[15] Once this area was found, a guide cannula consisting of a 21-gauge stainless-steel needle was implanted up to 1 mm over the desired site. The cannula was maintained sterile and protected by a removable 25-gauge stainless-steel rod during nonrecording periods. Following cannula implantation, the cerebellar surface was protected with a sterile silicone sheet, and the chamber sealed with acrylic cement (see FIG. 1).

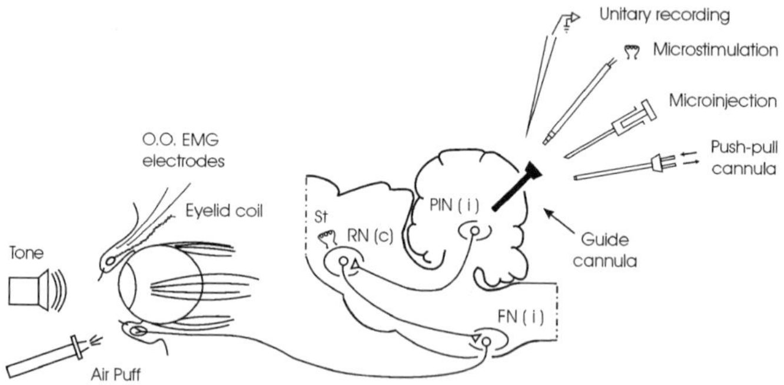

FIGURE 1. Experimental design. Alert-behaving cats were presented with tones and air puffs to evoke reflex and conditioned eyelid responses. The upper eyelid position was recorded with the search coil in a magnetic-field technique. The electromyographic activity of the orbicularis oculi muscle (O.O. EMG) was also recorded. A 21-gauge guide cannula was implanted 1 mm over the selected recording site in the rostral–dorsal part of the posterior interpositus using a transcerebellar approach. Recording or stimulating electrodes, an injection needle, or a push-pull cannula could be lowered through the guide into the nucleus, depending on the ongoing experiment. FN, facial nucleus; PIN, posterior interpositus nucleus; RN, red nucleus; i, c, ipsi- and contralateral to the left (recorded) eyelid.

Classical Conditioning

Conditioning sessions started one day after the guide cannula was implanted. Conditioning consisted of two habituation sessions, eight conditioning sessions, and six extinction sessions. A delay (tone–air-puff) conditioning paradigm was used. The CS consisted of a 370-ms, 600-Hz, 90-dB tone. The US started 270 ms after CS onset and consisted of a 100-ms, 3-kg/cm^2 air puff directed to the left cornea. The US coterminated with the CS. As illustrated in FIGURE 2, a conditioning session consisted of 12 blocks separated by a variable (5 ± 1-min) interval. These intervals were used to change the experimental design for unitary recording, electrical microstimulation, drug microinjection, or push-pull perfusion (see FIG. 1) when needed. Each block consisted of 10 trials separated by 30 ± 10-s intervals, at random. The CS was present alone in the first trial of each block. A complete conditioning session lasted for about 2 h. The CS was presented alone during habituation and extinction sessions for the same number of blocks per session and trials per block, and with a similar random interblock and intertrial distribution.[5]

Unitary Recordings, Electrical Stimulation, Drug Injection and Push-Pull Perfusion of Selected Posterior Interpositus Sites

Unitary recording during conditioning was carried out with tungsten electrodes of 0.01-in. external diameter, and 5-MΩ resistance (AM Systems, Inc.). The tungsten electrode was introduced with the help of a Canberra micromanipulator (Narishige, Japan) through the guide cannula. The recording electrode never projected >2 mm from the cannula tip. Neurons were identified by their antidromic activation from the contralateral red nucleus. Electrical stimuli in the red nucleus consisted of single- or double- (1- to 2-ms interval) cathodal, 50-μs^2 pulses with current intensities <0.2 mA. Neuronal activity was filtered in a bandwidth of 10 Hz to 10 kHz.[15,23]

For microstimulation of the selected posterior interpositus site, we used trains (370 ms) of cathodal square pulses (50 μs) at 20 Hz, with a current intensity 50 μA. This selective pattern of stimulation was carried out following the indications of a previous study.[23] Microstimulation was carried out with the help of home-made, bipolar, concentric electrodes, with an external diameter of 25-gauge. The stimulating electrode was allowed to project 1.5 mm beyond the guide cannula tip.

Injections of muscimol, an agonist of the $GABA_A$ receptor,[16] were carried out with the help of a Hamilton syringe (1-μL) connected by calibrated plastic tubing to a 25-gauge stainless- steel needle inserted inside the guide tube. Injections were carried out at a rate of 0.25 μL/min. A total of 1.25 μg/kg of muscimol (Sigma) dissolved in 1 μL of artificial cerebrospinal fluid (pH = 7.4) was injected 20 min before selected conditioned sessions.

Push-Pull Perfusion

A concentrically configured push-pull cannula system described previously[21,22] was used to perfuse the posterior interpositus nucleus. The push-pull cannula consisted of a 33-gauge stainless-steel inner (push) cannula inserted into a 25-gauge outer (pull) cannula. The separation of the tips of the push and the pull cannulae was 0.3 mm. The push-pull cannula system protruded from the guide cannula by 1 mm. An artificial cerebrospinal fluid (ref. no. 59-7316 from Harvard Apparatus, Inc.,

Massachusettes) was perfused through the push-pull cannula in pulses of 5 min at a constant flow rate of 20 μL min, using a peristaltic pump (Isco Wiz, Nebraska), during the conditioning session, and samples corresponding to the 12 conditioning blocks of each session were collected.[20] The system was allowed to equilibrate, and fractions of 5 min were collected in tubes containing 10 μL of perchloric acid 0.1 M and injected into an HPLC for subsequent electrochemical detection.

Determination of Amino Acids

Each sample of perfusate was analyzed by using a sensitive and selective isocratic method with precolumn derivation of amino acids with *o*-phthaldehyde (OPA)/2-mercaptoethanol, prior to detection by a Coulochem Electrochemical Detector modified from the early description of Donzanti and Yamamoto.[24] The perfusate was analyzed for its content in alanine, gamma-amino butyric acid (GABA), glycine, glutamate, glutamine, serine, taurine, and threonine. Aliquots (20 μL) of each sample were injected directly into the HPLC system (System Gold; Beckman Instruments, Fullerton, California), consisting of a programmable solvent pump (model 126 Beckman) with a pulse-dampener, an automatic refrigerated microsample injector (model CMA 200, Carnegie Medicine), and an analytical 3-μm C18 HR-80 reverse-phase analytical column (8 cm × 4.6 mm i.d.; ESA Inc., Bedford, Massachusettes). The ESA Coulochem II electrochemical detection system consisted of a model 5020 guard cell (detector setting, +650 mV) and a 5010 high-sensitivity dual analytical cell (cell 1, –400 mV; cell 2, +600 mV). The detection limit of the system was 200 fM.

The mobile phase for the separation of amino acids consisted of 0.1 M $Na_2 HPO_4$ in 25% ethanol adjusted to pH 6.75 with phosphoric acid (85%, HPLC grade). The flow rate was set to 1.2 mL/min, and background pressure was approximately 2500 psi.

Chemicals and Reagents

Amino acids and 2-mercapto-ethanol were purchased from Sigma (St. Louis, Missouri). All other reagents were the highest analytical-grade available (Fisher Scientific, Pennsylvania). OPA was purchased from Eastman-Kodak (Rochester, New York).

Histology

At the end of the experiments, animals were deeply re-anesthetized (sodium pentobarbital, 50 mg/kg, i.p.). Electrolytic marks were placed at the recording sites with the help of a tungsten electrode and at stimulating sites with the implanted electrode (1 mA for 10 s). Animals were perfused transcardially with saline and phosphate-buffered formalin. Serial sections (50-μm) of the brain stem and cerebellum were mounted on glass slides and stained with Cresyl violet for confirmation of stimulating and recording sites.

Data Collection and Analysis

Eyelid position, EMG and neuronal activity, and 1-V rectangular pulses corresponding to CS and US presentations were stored digitally on a videotape recording

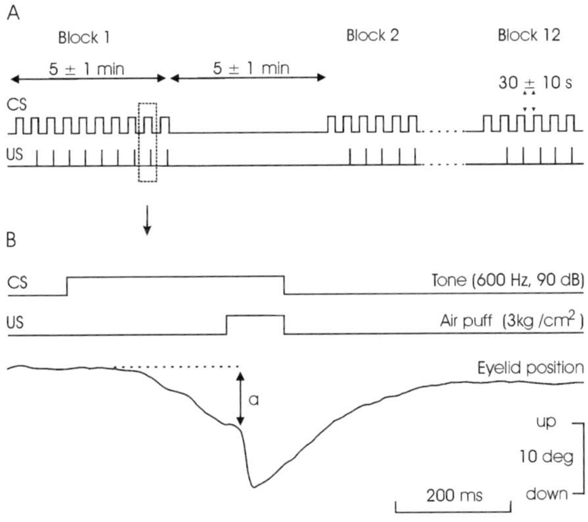

FIGURE 2. Conditioning paradigm. A whole experiment consisted of two habituation, eight conditioning, and six extinction sessions. (**A**) Each conditioning session (one per day) consisted of 12 blocks of 10 trials each. The conditioned stimulus (CS) was presented alone in the first trial of each block. Mean duration for blocks, interblock intervals, and intertrial intervals is indicated. (**B**) The CS consisted of a 370-ms, 600-Hz, 90-dB tone. The unconditioned stimulus consisted of a 100-ms, 3-kg/cm^2 air puff presented to the left cornea. A typical eyelid conditioned response (CR), recorded during the fifth conditioning session from cat #3, is illustrated. Calibrations for the eyelid response are indicated in (B). a, CR peak amplitude.

system at a sampling frequency of 11 kHz. Data were transferred off-line through an analog/digital converter (CED 1401 Plus) to a computer for later analysis. Action potentials were fed into a window discriminator, and the corresponding Schmidt trigger pulses stored on the same computer. Commercial computer programs (Spike 2 and SIGAVG from CED) were modified to represent instantaneous firing rate (in spikes/s) and eyelid position (see FIGS. 2 and 3). Statistical differences of mean values were determined with the help of Student's *t*-test. The program System Gold 2.01 (Beckman) was used to calculate amino acid levels in each sample. Neurochemical data were analyzed using analysis of variance with time as repeated measure coupled with Student's *t*-test when appropriate. The criterion for statistical significance was $P < .05$.

POSTERIOR INTERPOSITUS NEURONAL ACTIVITY DURING CONDITIONING EYELID RESPONSES

It was shown in an earlier study that neurons located in specific areas of the fastigial, interpositus, and dentate nuclei fire in association with reflexively evoked eyelid responses.[23] Those neurons were classified as type A and B. Although functionally different, type A and B neurons are usually recorded intermingled in tight patches of deep cerebellar nuclei. Type A neurons have a mean firing rate of

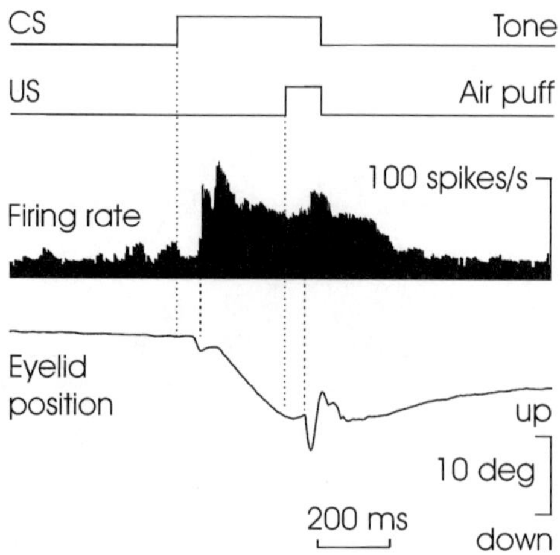

FIGURE 3. Activity of a representative posterior interpositus neuron, classified as type A, and recorded during the fifth conditioning session of cat #2. From *top to bottom* are illustrated the delay conditioning paradigm (CS, tone; US, air puff), the neuron firing rate, and eyelid position. The recorded neuron was identified by its antidromic activation from the contralateral red nucleus. The illustrated firing rate corresponds to the average of neuronal activity across a recording session (n = 120 trials). The illustrated eyelid position was averaged from the same recording session. Calibrations as indicated.

10–45 spikes/s during alertness, and increase their firing during reflex blinks, regardless of the sensory modality of the stimulus evoking the blink (visual, acoustic, trigeminal). The activation of type A neurons always lags the beginning of the blink by a few ms, but the latency of the neural response is usually time-locked to the beginning of the reflex blink.[23] In contrast, type B neurons exhibit a high mean firing rate (20–80 spikes/s) and decrease their firing during the downward, closing phase of reflex blinks. This partial, or total, inactivation of their spontaneous firing also starts slightly after the beginning of the blink.

We decided to carry out an in-depth study of the rostral–dorsal part of the posterior interpositus, because this specific zone appeared intensely labeled following attenuated rabies virus injection in the upper eyelid of rats.[19] In fact, both type A and B neurons can be easily recorded in this area, firing in relation to both reflex and conditioned eyelid responses.[15] In addition, neurons related to face and/or eye have been recorded in the rostral posterior interpositus in monkeys.[25,26] FIGURE 3 illustrates an example of type A neuron recorded during the fifth conditioning session of cat #2. The neuron started to fire at 23 ms after CR onset and maintained a higher firing rate than control average across CR performance. Accordingly, type A neurons located in the posterior interpositus exhibit discharge properties for CRs similar to those presented during reflex blinks.[15,23] In the same way, type B neurons recorded here inhibited or even stopped their firing during CR performance (not illustrated). Type B neurons started their inhibitory response >20 ms following the beginning of the CR.

Recently, it has been proposed that type A and B neurons can play an important role in the push-pull system represented by eyelid kinematics.[27] Thus, type A neurons could activate orbicularis oculi motoneuron firing across the red nucleus relay, while type B cells could, directly or indirectly, disfacilitate levator palpebrae motoneuron firing to allow lid closing during the CR. In this regard, it has been shown that some posterior interpositus neurons project to premotor nuclei of the extraocular motor system,[28] and that type B neurons can be antidromically activated from the perioculomotor areas.[23]

EFFECTS OF THE ELECTRICAL MICROSTIMULATION OF THE RECORDING SITES ON EYELID CONDITIONED RESPONSES

The finding of both type A and type B neurons firing in relation to CRs in the rostral–posterior interpositus prompted us to check the effects of electrical microstimulation, muscimol injection, and push-pull perfusion of the same recording area.

A systematic study of the effects of electrical microstimulation of the rostral–dorsal part of the posterior interpositus nucleus was carried out at the end of the sixth–eighth conditioning sessions in cats #3 and #4. Electrical stimulation of the posterior interpositus evoked a small, ramplike eyelid closing, similar in profile to a true CR, but with a wavier appearance (see Ref. 15, figure 7). As shown in FIGURE 4, the peak amplitude of the eyelid response evoked by microstimulation of the posterior interpositus was 50% smaller ($p < .001$) than the CR evoked by the presentation of CS alone. Nevertheless, the joint presentation of the CS and of the electrical stimulus evoked significantly larger eyelid responses (about 120%, $p < .001$) than those resulting from the simple addition of CS and electrical microstimulation isolated effects (see FIG. 4). It should be noted that, as already reported,[15] the pairing of the electrical train of stimuli and CS presentation modified CR amplitude, but not its latency (81.2 ± 9 ms for CS alone presentation and 79.7 ± 11 ms for paired CS and electrical microstimulation; $n = 30$) or duration (217 ± 17 ms for CS alone presentation and 209 ± 21 ms for paired CS and electrical microstimulation; $n = 30$). Accordingly, the extra activation of posterior interpositus in the presence of an ade-

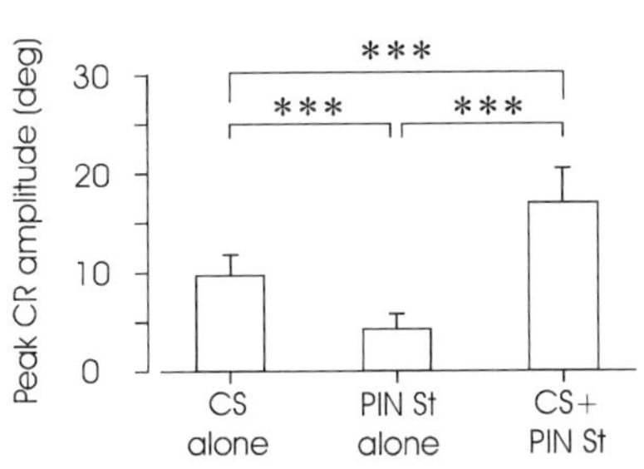

FIGURE 4. Effects of the electrical microstimulation of the posterior interpositus nucleus on eyelid conditioned responses. These data were collected at the end of the sixth to eighth conditioning session in cat #4. Peak eyelid amplitudes (mean ± S.E.; $n = 30$, in degrees) evoked by CS alone, interpositus stimulation (PIN St) alone, and the combined presentation of both stimuli (CS + PIN St) are illustrated. Note that the combined presentation of both stimuli produced a larger response than the sum of the effect of each stimulus presented alone. The electrical stimulus consisted of 50-μs cathodal square pulses of 200 μA, presented at 20 Hz for 370 ms. *** Indicates that $P < .001$.

quate CS is able to increase CR amplitude, but not to modify its latency, suggesting that the motor command generating the CR is originated elsewhere.[27,29]

EFFECTS OF MUSCIMOL MICROINJECTION IN THE RECORDING SITES

The injection of muscimol (1.25 μg/kg dissolved in 1 μL of artificial cerebrospinal fluid) in the rostral–dorsal part of the posterior interpositus nucleus during any session across conditioning decreased the amplitude of the CR by 20–30%. The effects of muscimol were systematically studied during its application in the sixth conditioning session in cats #5 and #6 (see FIG. 5). Peak amplitude of CRs during the conditioning session when muscimol was injected was 6 ± 0.9 deg, a significantly ($p < .001$) lower value than those obtained during the preceding (fifth session: 10.5 ± 1.7 deg) and the following (seventh session: 15 ± 2.4 deg) conditioning sessions. As already indicated for electrical microstimulation of the posterior interpositus, the injection of muscimol in the same zones did not modify CR latencies (fifth conditioning session: 78 ± 12 ms; sixth conditioning session: 74 ± 14 ms; and seventh conditioning session: 70 ± 8 ms).

It has been reported that chemical inactivation of the anterior interpositus nucleus with muscimol in rabbits prevents the acquisition of nictitating membrane CRs.[10,12] In the present experiments, as well as in a previous report,[15] both in cats, the CRs were already acquired when muscimol was applied, but it has been reported that muscimol injection in the posterior interpositus does not modify the learning curve.[15] In this latter case, a significant decrease–but not a complete abolition–of the CR was observed. In fact, a different susceptibility of rabbits[14,16] and cats[13,15] has been reported. In any case, the results obtained with muscimol infusion in the pos-

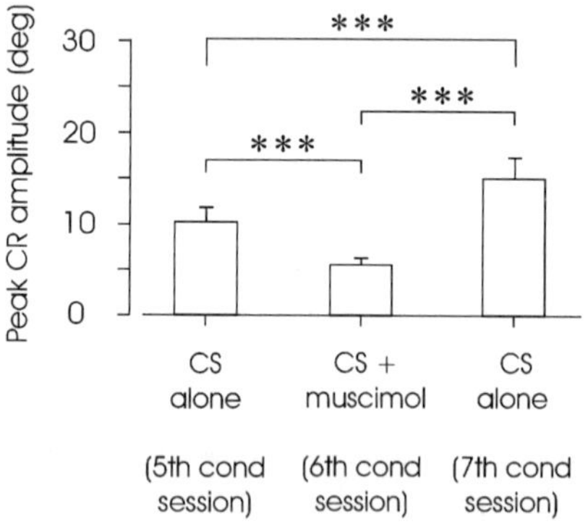

FIGURE 5. Effects of muscimol microinjection in the posterior interpositus nucleus on eyelid conditioned responses. Muscimol was injected during the sixth conditioned session in cats #5 and #6, as a single application (1.25 μg/kg), 20 min prior to the beginning of the conditioning session. Peak eyelid amplitudes (mean ± S.E.; $n = 240$, in degrees) evoked by CS presentation without (fifth and seventh conditioning sessions) and with muscimol injection into the interpositus nucleus (sixth conditioning session) are illustrated. *** Indicates that $P < .001$.

terior interpositus indicate that this structure is certainly involved in CR performance. One still-open issue is whether muscimol injections in the posterior interpositus are also able to modify the amplitude of eyelid reflex responses[27,30] or their effects are restricted to CRs.

PUSH-PULL PERFUSION STUDY OF EXTRACELLULAR AMINO ACID LEVELS IN THE RECORDING SITES

The push-pull perfusion system was successfully checked in different animals, but a systematic, complete study was carried out in cats #7 and #8. Five-minute samples, corresponding to conditioning blocks were collected during habituation, conditioning, and extinction sessions. Each sample contained a total of 100 μL of perfusate. Samples were analyzed with the HPLC-EC method[24] for the eight amino acids indicated above, in the Experimental Procedures section. The area of the peak corresponding to each amino acid in the chromatogram indicated its concentration (in pg/μL) in the sample. FIGURE 6 shows the mean values obtained for glutamate, glycine, taurine, and GABA for the 12 blocks/session from the two animals. Data shown correspond to the second habituation session, sixth conditioning session, and fourth extinction session. According to these results, a significant increase in the release of glutamate, glycine, and taurine was observed during conditioning when compared with concentration levels of the same amino acids during habituation and extinction sessions. Concentration values corresponding to alanine, serine, and threonine were not modified by conditioning (not illustrated). Surprisingly, values col-

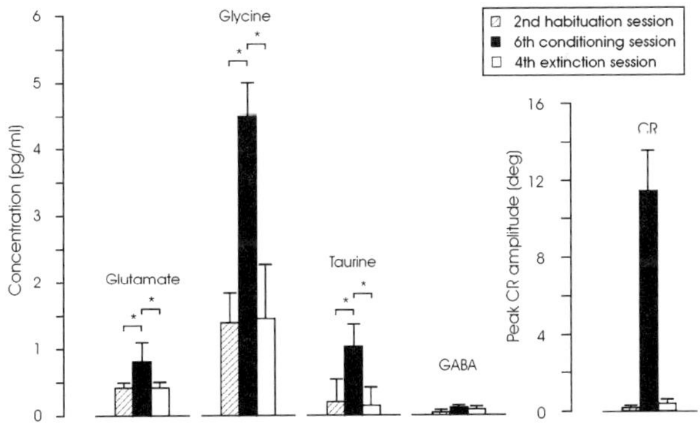

FIGURE 6. Extracellular levels of glutamate, glycine, taurine, and GABA in the rostral–dorsal part of the posterior interpositus nucleus across conditioning. Illustrated data were collected from cats #7 and #8. Five-minute samples containing 100 μL of perfusate were collected during each conditioning block (see FIG. 2) and averaged ($n < 12$ blocks/conditioning session/cat) during the second habituation session, the sixth conditioning session, and the fourth extinction session. For comparative purposes, peak amplitudes of eyelid responses for the same sessions in the same animals are also illustrated. Mean values ± SE are indicated. Note a significant ($P < .05$ at least) increase in collected samples for glutamate, glycine, and taurine. Surprisingly, values for GABA release were not significantly modified across conditioning.

lected for GABA did not significantly change across conditioning, although a slight increase for values obtained during the sixth conditioning session was observed when compared with habituation values (see FIG. 6). This latter result was not related to the HPLC-EC method, because a significant peak for GABA was detected using control samples. Thus, taking into account the rapid turnover of GABA, a feasible explanation is that this amino acid was rapidly degraded and/or retaken up following its release.[31] Further experiments seem necessary to clarify the specific role of GABA in the genesis and control of conditioned eyelid responses in the posterior interpositus nucleus.

The significant increase in the release of glutamate, glycine, and taurine at the rostral–dorsal part of posterior interpositus nucleus reinforces their involvement in the classical conditioning of eyelid responses, as previously documented for these three aminoacids.[32–36] Glutamate seems to act on NMDA receptors at the interpositus nucleus in rabbits.[33] The putative mechanism of glycine in deep cerebellar nuclei could be associated to a cooperative action on the NMDA channel or, as recently proposed, by acting on presynaptically located Cl^- channels, enhancing glutamate release.[33] Nevertheless, an inhibitory role for glycine, paralleling that of GABA neurotransmitter, cannot be ruled out.[37] Finally, taurine has been reported to produce a dose-dependent depression in the firing rate of neurons of deep cerebellar nuclei, that can be blocked by bicuculline and picrotoxin, but not by strychnine. These results suggest that taurine could have a GABA-like effect on these nuclei.[36] Obviously, these preliminary data introduce a larger order of complexity regarding the molecular and subcellular events taking place in cerebellar nuclei during classical conditioning of eyelid responses, and open new research strategies on the role of specific amino acid neurotransmitters in the genesis and modulation of these eyelid conditioned responses.

CONCLUSIONS AND FUTURE DEVELOPMENTS

Present results confirm a direct involvement of neurons located in a specific zone of the rostral–dorsal division of the posterior interpositus nucleus in relation with the acquisition of eyelid conditioned responses. According to data presented here for cats, and to earlier reports collected from the same species,[15,27] it cannot be conclusively proposed that the interpositus nucleus is the site where eyelid conditioned responses are generated. Available data, collected from recording, microstimulation and drug-injection studies, rather support a role of cerebellar nuclei as an enhancer of learned motor commands originated in a different place, such as the motor cortex.[29] The technical developments proposed here could represent a first attempt to use a multidisciplinary approach for the study of selected neuronal sites during motor learning in alert-behaving animals.

REFERENCES

1. GORMEZANO, I., E.J. KENOE & B.S. MARSHALL. 1983. Twenty years of classical conditioning research with rabbits. Prog. Psychobiol. Physiol. **21:** 1068–1070.
2. THOMPSON, R.F. & D.J. KRUPA. 1994. Organization of memory traces in the mammalian brain. Annu. Rev. Neurosci. **17:** 519–549.

3. DOMINGO, J.A., A. GRUART & J.M. DELGADO-GARCÍA. 1997. Quantal organization of reflex and conditioned eyelid responses. J. Neurophysiol. **78:** 2518–2530.
4. EVINGER, C. 1995. A brain stem reflex in the blink of an eye. News Physiol. Sci. **10:** 147–153.
5. GRUART A., P. BLÁZQUEZ & J.M. DELGADO-GARCÍA. 1995. Kinematics of spontaneous, reflex, and conditioned eyelid movements in the alert cat. J. Neurophysiol. **74:** 226–248.
6. TRIGO, J.A., A. GRUART & J.M. DELGADO-GARCÍA. 1999. Discharge profiles of abducens, accessory abducens, and orbicularis oculi motoneurons during reflex and conditioned eyeblinks in alert cats. J. Neurophysiol. **81:** 1666–1684.
7. WELSH, J.P. 1992. Changes in the motor pattern of learned and unlearned responses following cerebellar lesions: a kinetic analysis of the nictitating membrane reflex. Neuroscience **47:** 1–19.
8. WOODY, C.D. 1986. Understanding the cellular basis of memory and learning. Annu. Rev. Physiol. **37:** 433–493.
9. BLISS, T.V.P. & G.L.A. COLLINGRIDGE. 1993. A synaptic model of memory: long-term potentiation in the hippocampus. Nature **361:** 31–39.
10. KIM J.J. & R.F. THOMPSON. 1997. Cerebellar circuits and synaptic mechanisms involved in classical eyeblink conditioning. Trends Neurosci. **20:** 1177-1181.
11. MALENKA, R.C. 1995. LTP and LTD: dynamic and interactive processes of synaptic plasticity. Neuroscientist **1:** 35–42.
12. MAUK, M.D. 1997. Roles of cerebellar cortex and nuclei in motor learning: contradictions or clues? Neuron **18:** 343–346.
13. BRACHA, V. *et al.* 1999. Inactivation of interposed nuclei in the cat: classically conditioned withdrawal reflexes, voluntary limb movements and the action primitive hypothesis. Exp. Brain Res. **126:** 77–92.
14. GARCIA, K.S. & M.D. MAUK. 1998. Pharmacological analysis of cerebellar contributions to the timing and expression of conditioned eyelid responses. Neuropharmacology **37:** 471–480.
15. GRUART, A. *et al.* 2000. Cerebellar posterior interpositus nucleus as an enhancer of classically conditioned eyelid responses in alert cats. J. Neurophysiol. **84:** 2680–2690.
16. KRUPA, D.J., J.K. THOMPSON & R.F. THOMPSON. 1993. Localization of a memory trace in the mammalian brain. Science **260:** 989–991.
17. SCHREURS, B.G. *et al.* 1998. Intracellular correlates of acquisition and long-term memory of classical conditioning in Purkinje cell dendrites in slices of rabbit cerebellar lobule HVI. J. Neurosci. **18:** 5498–5507.
18. YEO, C.H., M.J. HARDIMAN & M. GLICKSTEIN. 1985. Classical conditioning of the nictitating membrane response of the rabbits. I. Lesions of the cerebellar nuclei. Exp. Brain Res. **600:** 87–98.
19. UGOLINI, G., S. MORCUENDE & J.M. DELGADO-GARCÍA. 1999. Brainstem and cerebellar centers mediating neural control of orbicularis oculi motoneurons: retrograde transneuronal tracing with rabies virus. Soc. Neurosci. Abstr. **25:** 1403.
20. JIMÉNEZ-DÍAZ, L. *et al.* 2001. A push-pull study of extracellular amino acid levels in the posterior interpositus nucleus during classical conditioning of eyelid responses in cats. Soc. Neurosci. Abstr. **27:** 295.7.
21. MYERS, R.D., J.M. PEINADO & F.J. MIÑANO. 1988. Monoamine transmitter activity in lateral hypothalamus during its perfusion with insulin or 2-DG in sated and fasted rats. Physiol Behav. **44:** 633–643.
22. MYERS, R.D., A. ADELL & M.F. LANKFORD. 1998. Simultaneous comparison of cerebral dialysis and push-pull perfusion in the brain of rats: a critical review. Neurosci. Biobehav. Rev. **22:** 371–387.
23. GRUART, A. & J.M. DELGADO-GARCÍA. 1994. Discharge of identified deep cerebellar nuclei neurons related to eye blinks in the alert cat. Neuroscience **61:** 655–681.
24. DONZANTI, B.A. & B.K. YAMAMOTO. 1988. An improved and rapid HPLC-EC method for the isocratic separation of amino acid neurotransmitters from brain tissue and microdialysis perfusates. Life Sci. **43:** 913–922.
25. MASON, C.R. *et al.* 1998. Organization of reaching and grasping movements in the primate cerebellar nuclei is revealed by focal muscimol inactivations. J. Neurophysiol. **79:** 537–554.

26. ROBINSON, F.R. 2000. Role of cerebellar posterior interpositus nucleus in saccades. I. Effects of temporary lesions. J. Neurophysiol. **84:** 1289–1302.
27. DELGADO-GARCÍA, J.M. & A. GRUART. 2002. The role of interpositus nucleus in eyelid conditioned responses. Cerebellum **2:** In press.
28. GONZALO-RUIZ, S. & G.R. LEICHNETZ. 1990. Connections of the caudal cerebellar interpositus complex in a New World monkey (*Cebus apella*). Brain Res. Bull. **25:** 919–927.
29. AOU, S., C.D. WOODY & D. BIRT. 1992. Changes in the activity of units of the cat motor cortex with rapid conditioning and extinction of a compound eye blink movement. J. Neurosci. **12:** 549–559.
30. WIKGREN, J. & T. KORHONEN. 2001. Interpositus nucleus inactivation reduces unconditioned response amplitude after paired but not explicitly unpaired treatment in rabbit eyeblink conditioning. Neurosci. Lett. **308:** 181–184.
31. FREEMAN, M.E., J.D. LANE & J.E. SMITH. 1983. Turnover rates of amino acid neurotransmitters in regions of rat cerebellum. J. Neurochem. **40:** 1441–1447.
32. GARDETTE, R., M. KRUPA & F. CREPEL. 1987. Differential effects of serotonin on the spontaneous discharge and on the excitatory amino acid-induced responses of deep cerebellar nuclei neurons in rat cerebellar slices. Neuroscience **23:** 491–500.
33. CHEN, G. & J.E. STEINMETZ. 2000. Intra-cerebellar infusion of NMDA receptor antagonist AP5 disrupts classical eyeblink conditioning in rabbits. Brain Res. **887:** 144–156.
34. SATO, K. *et al.* 1991. Localization of glycine receptor alpha 1 subunit m RNA-containing neurons in the rat brain: an analysis using in situ hybridization histochemistry. Neuroscience **43:** 381–395.
35. VITTEN, H. & J.S. ISAACSON. 2001. Synaptic transmission: exciting times for presynaptic receptors. Curr. Biol. **11:** R695–R697.
36. BILLARD, J.M. 1990. Taurine in deep cerebellar nuclei of the rat. In vivo comparison to GABA inhibitory effect. Brain Res. **514:** 155–158.
37. CHEN, S. & D.E. HILLMAN. 1993. Colocalization of neurotransmitters in the deep cerebellar nuclei. J. Neurocytol. **22:** 81–91.

Relative Roles of the Cerebellar Cortex and Nuclei

Summary of Presentations and of Discussion

RICHARD THOMPSON AND JENNIFER RAYMOND, *Rapporteurs*

The discussion was introduced by the moderator, J. DelGado Garcia, who posed several questions. The first concerned timing: are there oscillations in the olive and cerebellum with different frequencies, and how do they interact to yield precise timing of behavioral responses? The second question concerned the proposed diagrams of the cerebellum with regard to learning, as in eyeblink conditioning. What happens when two different CS stimuli are used, particularly of the same modality? can the proposed diagram of the system account for the result? The third question concerned latencies. In recordings, interpositus unit responses occur with both the unconditioned stimulus/reflex and the condition stimulus/responses, that is, they fire in relation to both types of stimuli. Can this be accounted for in the current models?

SUMMARY OF DISCUSSION

Each of the discussants raised several related issues for discussion. Thompson argued that the "basic" memory trace for eyeblink conditioning is stored in the interpositus nucleus, but that the evidence also strongly supports the view that a memory trace is formed in the cerebellar cortex, concerned with timing of the conditioned response, modulation of amplitude, and so on. Evidence from his laboratory, Lavond's laboratory, and Yeo's laboratory strongly support these views. In particular, reversible inactivation localized to the anterior interpositus nucleus completely prevents learning and subsequent learning shows no savings, whereas inactivation of the output from the cerebellum, the superior cerebellar peduncle, during training does not prevent learning at all.

Thompson showed new data presented at the fall Society for Neuroscience meeting by John Freeman and associates using eyeblink conditioning in the freely moving rat. The trained animals showed rapid learning and the unpaired control showed little or no sensitization. An EM analysis showed a marked and highly significant increase in the number of excitatory synapses in the interpositus nucleus (presumably mossy fiber synapses) and no increase in inhibitory synapses (presumably from Purkinje neurons) in the trained animals and no changes in synapse numbers in the unpaired control animals. This constitutes very strong morphological evidence for memory storage in the interpositus nucleus.

Work from the laboratories of Thompson, Yeo, Mauk, Steinmetz, and others using eyeblink conditioning supports the hypothesis that a memory trace(s) is(are)

Ann. N.Y. Acad. Sci. 978: 119–121 (2002).

formed in cerebellar cortex, modulating timing and amplitude of the CR. Thompson's work with mutant and KO mice shows a consistent positive correlation between the occurrence and degree of cerebellar cortical LTD and degree of eyeblink conditioning, arguing that LTD is a mechanism of memory trace formation in the cerebellar cortex, consistent with and strongly supportive of the elegant studies of Ito and associates on LTD as a mechanism of memory formation and storage in adaptation of the VOR and other forms of cerebellar memory.

Several participants echoed Thompson's view that there is more than one site of plasticity associated with cerebellum-dependent learning, and that these sites may mediate distinct aspects of learning. In the past, there was considerable debate about whether learning is mediated by the cerebellar cortex or nucleus. This either/or debate has yielded to a new discussion: Exactly how does each of these structures contribute to learning? Timing appears to be one organizing principle. In addition, Yeo discussed the evidence that the cerebellar cortex and nuclei have distinct roles in memory consolidation, extinction, and savings, citing work from his own lab and that of Mauk.

Raymond raised the question of whether we should expect the cerebellar cortex and nuclei to be performing similar signal-processing functions for all tasks that depend on the cerebellum. The anatomy and physiology of the cerebellum are very regular across its extent, and so many motor and perhaps cognitive tasks rely on very similar neural circuits through the cerebellum. This has led to the suggestion that the cerebellum is performing a similar computation for a wide range of tasks. Nevertheless, there are several differences between regions of the cerebellum, for example, in the expression of molecular markers like zebrin II. These regional variations may simply provide fine "tuning" of the cerebellar circuit for particular tasks, but it is also possible that the seemingly small regional variations can endow the cerebellar circuit architecture with very different signal-processing properties. Thier expressed confidence that the computational principle is the same throughout the cerebellum.

Much of the general discussion with the audience focused on an issue raised by Llinás, namely, the extent to which the evidence to date has demonstrated that memory traces, or at least neuronal/synaptic plasticity is established in the cerebellum as a result of learning. There appears to be a concensus of all the speakers, Thier, Scudder, Yeo, Lavond, and Delagado-García that this is the case, at least for the establishment of neuronal/synaptic plasticity. Llinás expressed some skepticism. He was asked what he felt would be conclusive evidence for memory storage in the cerebellum. Llinás did not answer the question directly, but said he would recognize compelling evidence when he saw it. Several other participants offered criteria for establishing the cerebellum as a site of memory storage, and cited experimental studies of eyeblink conditioning that fulfill most of the criteria. Controls for surgical and genetic lesion studies were suggested: it should be shown that there is no effect of the lesion on sensory processing or motor performance that prevents the animal from performing the learning task, and it should be shown that the lesion has not disrupted the transmission of neural signals that are necessary for plasticity at a downstream site. One participant expressed the opinion that some of the strongest evidence for a site of memory in the cerebellum is the work of Thompson and colleagues showing that the CS and US can be replaced with direct activation of the mossy fiber and climbing fiber pathways. Another participant thought that the most compelling evidence would be a direct measurement of changes in particular synapses after learning.

The discussion in this session had considerable participation by the audience and a number of issues were raised but not necessarily settled. The tenor of this session and the discussion was spirited but generally quite positive.

The Olivocerebellar System as a Generator of Temporal Patterns

Y. YAROM AND D. COHEN

Department of Neurobiology, Life Science Institute and the Center for Computational Neuroscience, Jerusalem , Israel 91904 and
Department of Neurobiology, Duke University Medical Center, Durham, North Carolina 27710, USA

Abstract: The large number of diverse functions attributed to the cerebellum appears to be inconsistent with its simple, homogeneous and evolutionary preserved structure. A homogeneous structure that participates in a variety of functions implies that a common denominator underlies all of them. Since the concept of precise timing can be recognized in almost all cerebellar functions, it is likely, therefore, that the basic cerebellar circuit is capable of generating temporal patterns. Of the different mechanisms that can generate temporal patterns, two are suggested by the functional anatomy of the cerbellum: transmission lines or oscillators. Our recent experimental observations indicate that the olivary oscillatory property is more likely to serve this function. We propose that interactions between the cerebellum and the inferior olive endow the system with the ability to generate complex temporal patterns. These temporal patterns can be used for fine adjustment of motor output, sensory expectation, or shifting attentions.

Keywords: inferior olive; parallel fibers; oscillations; imaging; electrical coupling; timing

INTRODUCTION

Deciphering the cerebellar mystery has been a major focus of neurosciences endeavors for over a century. Recent studies demonstrate that in addition to the traditional role of the cerebellum in motor control, it is also involved in sensory acquisition and in cognitive function. The wide breadth of cerebellar function is somewhat surprising in view of the simplicity and homogeneity of the cerebellar cortex, which is composed of few cell types that are connected in an orderly fashion. This rather simple structure has been preserved through evolution, although the cerebellum itself has expanded and acquired new functions. Since the anatomical and physiological properties of a circuit define its computational capabilities, it is reasonable to assume that the basic cortical circuit performs the same computation in

Address for correspondence: Y. Yarom, Department of Neurobiology, Life Science Institute, Hebrew University, Jerusalem, Israel 91904. Voice: 972-2-6585172; fax: 972-2-6586296.
Yarom@vms.huji.ac.il

Ann. N.Y. Acad. Sci. 978: 122–134 (2002).

the various cerebellar areas as well as across different species. One should, therefore, search for a theme, a common denominator to all cerebellar functions.

The concept of precise timing, or the prediction of time-related events, can be easily recognized in almost all the different functions attributed to the cerebellum. Indeed, it has been previously suggested that timing is the common denominator of cerebellar functions. Numerous cerebellar models deal with different aspects of time. Some focus on the parallel-fiber system, which conveys information to the Purkinje cells along its path at a certain conduction velocity. Others focus on cerebellar involvement in learning, where precise timing of the two input systems—the parallel fibers and climbing fibers—is essential. Our recent observations on the functional organization of the cerebellar cortex and of the inferior olivary nucleus lead us to suggest that it is the interactions between these structures that endow the system with the ability to coordinate a variety of context-dependent inputs by generating complex temporal patterns. These temporal patterns, in turn, can be used in different functions, such as fine adjustment of motor output, sensory expectation, or shifting attentions.

TIMING: THE COMMON DENOMINATOR OF CEREBELLAR FUNCTION

Neurological symptoms of cerebellar patients typically include lack of coordination, difficulties in maintaining balance, postural abnormalities, and intention tremor. Because these symptoms often occur in the absence of substantial sensory impairment, scientists generally view the cerebellum as a motor structure. All these motor deficits could be manifestations of a damaged timing mechanism. For example, in order to properly coordinate multijoint movements, all participating muscles must be activated sequentially and at precise times. Similarly, lack of balance and appearance of tremor may reflect maladjustment in the timing of muscle activation.

This view is supported by numerous studies on the relationship between cerebellar activity and motor performance.[1–8] Often, the dependence of performance on timing is explicit.[9,10] Unfortunately, relatively few studies have directly addressed the issue of timing. One of these is an investigation of ball throwing,[11] in which cerebellar patients fail to hit a target because they failed to precisely coordinate the time of finger opening. Ivry and colleagues used a finger-tapping task to show another timing-related cerebellar malfunction.[12] In this study, cerebellar patients performed poorly when they had to estimate precise time intervals without external cues. Several such studies led Ivry to formulate a "Cerebellar Timing System" hypothesis, which proposes that the cerebellum generates "a near-infinite set of hourglass or interval-type timers."

This temporal theme can also explain the participation of the cerebellum in nonmotor functions.[12] Recently, it was demonstrated that the human cerebellum predicts the occurrence of self-administered sensory stimuli.[13] This implies that when the subject moves, the cerebellum predicts the precise time of consequent sensory input. The authors of this study reviewed their finding in the concept of "efference copy," whereby the cerebellum receives both a copy of the motor command and sensory information on body position. It has been suggested that the cerebellum uses efference copy to improve motor performance.[14,15] However, a study by Tesche and Karhu[16] directly shows that efference copy is not required for sensory prediction, since the cerebellum can predict rhythmic sensory input that is unrelated to motor performance. The cerebellum is thus capable of producing temporal patterns that correspond to forthcoming input and that are independent of motor behavior.

An ability to predict sensory inputs that are unrelated to motor command may imply that the cerebellum provides the sensory cortex with information about the importance of an incoming signal. Such an alerting rule of the cerebellum is in agreement with the suggestion that the cerebellum is involved in attention.[17] Courchesne[17] claimed that just as the cerebellum plays a control role in coordination of movements, it also serves as a controller of moment-to-moment shifts in mental focus of attention. This would explain the variety of cognitive dysfunctions that have been attributed to the cerebellum, such as dyslexia[18] and autism.[19]

We conclude that the only way to bind these diverse cerebellar functions with a single computational process is to assume that the cerebellum provides the temporal information that enables precise performance of various brain functions. Accordingly the role of the cerebellum is to translate this temporal information into firing patterns. The duration of each firing sequence, as well as its pattern, is uniquely designed for a particular task. For example, when a goal-directed movement is to be executed, the cerebellum generates a pattern that encodes the precise temporal sequence of muscle activation. The cerebellar activity modulates the command descending from the motor cortex, leading to the execution of a smooth movement. Since these patterns must be constantly adjusted for changes in the external environment or internal conditions, an almost infinite number of patterns are required. Because the number of available neurons is finite, the mechanism that underlies the pattern generation must be at once precise and modifiable.

MECHANISMS OF TIMING

Of the potential neuronal generators that might underlie the production of temporal patterns, two are suggested by the functional anatomy of the cerebellum and the inferior olive. The generator may use transmission lines that convey signals at a constant velocity by measuring the appearance of signals at different locations along the line. In this case, the conduction velocity and the spatial resolution dictate the temporal sensitivity of the system. Alternatively, the generator may use an oscillator that functions as a clock, with temporal sensitivity being determined by the time period of the oscillator.

THE PARALLEL FIBER SYSTEM: HYPOTHETICALLY AN IDEAL TRANSMISSION LINE

Morphologically, the most prominent organizational feature of the cerebellar cortex is the arrangement of the longitudinally arrayed parallel fibers with respect to the perpendicularly arrayed Purkinje cells. This organization resembles the structure of a delay line that is capable of generating accurate time intervals.[20] Assuming that the conduction velocity of the parallel fibers is 20 cm/s and that two adjacent Purkinje cells located along that path are, on average, 50 microns apart, the minimal time interval that the system can generate is a quarter of a millisecond. However, evidence suggests that this highly time-sensitive system does not utilize its transmission line properties to produce accurate time intervals. Were parallel fibers to provide a significant input to Purkinje cells, we would expect all Purkinje cells along a beam of parallel fibers to respond to focal activation of mossy fibers; such a response should generate a bandlike organization of the receptive fields. Experimental observations, however, reveal a patchlike organization of mossy fibers receptive fields.[21]

In order to explain this discrepancy, we studied the functional properties of the parallel-fiber system, using a high-resolution optical imaging system that enabled simultaneous monitoring of neuronal activity in 128 (or 464) locations covering an area of 0.5 to 4 mm^2 (for details see Ref. 22). We studied the spread of electrical activity in the cerebellar cortex following direct activation of the parallel fibers (stim. 1 in FIG. 1) and compared it to the response elicited by mossy-fiber stimulation. The latter was achieved by inserting a stimulating electrode into the cerebellar white matter (stim. 2 in FIG. 1). As shown in FIGURE 1, direct stimulation of parallel fibers elicited a beam of activity that propagated in the mediolateral direction at a velocity of 20 cm/s. A completely different response was observed when a stimulus was applied to the white matter. In this case, the response was a patch of synchronized activity that appeared at the center of the field.

In previous studies, we demonstrated that these signals reflect only postsynaptic activity and that spikes generated in parallel fibers are not detected.[22] Therefore, the absence of propagating activity in the response to mossy-fiber stimulation does not necessarily mean that the parallel fibers are themselves inactive. It is possible that the parallel fibers are activated, but that their activation is insufficient to pass the threshold required to activate Purkinje cells along their path. To ascertain that the lack of propagation is not a result of conduction failure in the parallel fibers, we examined the interaction between the responses to parallel- and mossy-fiber stimulations. The intensity of the parallel-fiber stimulation was adjusted to barely evoke a response along the beam (FIG. 2A). Mossy-fiber stimulation evoked a patch of activity that partially overlapped the band of activity evoked by the parallel-fiber stimulation (FIG. 2B). Synchronous activation of both stimuli generated a supralinear response in an area outside the patch and along the parallel fibers (FIG. 2C), indicating that the two stimuli together evoked a detectable postsynaptic response that cannot be detected following either stimulus alone. We conclude that the parallel-fiber–Purkinje-cell synapses are functional during the natural mode of activation, but that their efficiency is low. Therefore, the parallel-fiber system does not serve as an accurate time measuring device, and rather than considering it as a transmission line, its function should be reevaluated in the context of weak interactions between adjacent cortical areas.

THE OLIVOCEREBELLAR SYSTEM: A PRECISE, ACCURATE, MODIFIABLE TIMER

Although there is general agreement as to the importance of the powerful input to the cerebellar cortex provided by the climbing fibers, there are fierce debates regarding its role in cerebellar function. One focus of debate is the question of rhythmicity of the climbing-fiber input. Complex spike activity, as recorded continuously from a single Purkinje cell *in vivo*, does not appear to be rhythmic.[23] However, such activity recorded simultaneously from a population of Purkinje cells exhibits a clear rhythm of 5–10 Hz.[24] The rhythmic nature of climbing-fiber input is also supported by the finding that harmaline induces rhythmic activity in the inferior olive, which in turn, produces a 10-Hz muscle tremor.[25,26] Further support for the rhythmic nature of the climbing-fiber activity comes from *in vitro* studies, where sub- and suprathreshold rhythmic activity has been documented.[27–30] Subthreshold rhythmic activity is expressed as sinusoidal oscillations of membrane potential (FIG. 3A).

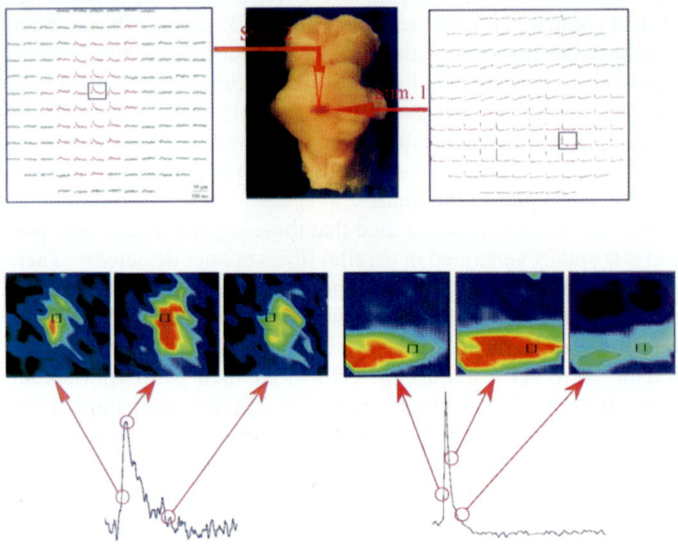

FIGURE 1. Beams and patches: Two modes of activity in the cerebellar cortex. Voltage-sensitive dye was applied to image the activity in the cerebellar cortex from an isolated guinea pig cerebellum (*top middle*). The dye was injected into one of the cerebellar folia in the vermis (*red spot*) and two stimulating electrodes (*red arrows*) were used to evoke either the parallel-fiber input (*right*) or the mossy-fiber input (*left*). The latter was achieved by inserting the stimulating electrode into the cerebellar white matter underneath the recorded area. A 40× objective allowed monitoring of changes in fluorescence from an area of 0.25 mm^2 when an array of 128 photodiodes was used (each diode covers an area of 50-by-50 microns). The results are presented either as a change in fluorescence as a function of time at each location (**top left** and **right panels**) or as color-coded distribution of fluorescence at three different times (**middle panels**). **Lower panels**: the times of each of the color pictures are marked on the traces recorded at the designated location (*black square*). The response to surface stimulation (*top right*) is characterized by a beam of activity propagating along the parallel-fiber axis. The locations of highest depolarization were marked by red. Note that as expected from the anatomical organization of the cerebellar cortex, the excitatory beam was accompanied by lateral inhibition. A patch of synchronized excitatory activity characterizes the response to white matter stimulation (*top left*). The colored presentations show that the surface-evoked activity propagated along the beam of parallel fibers, whereas the white-matter-evoked activity occurred synchronously.

These oscillations occur throughout the nucleus, as demonstrated by simultaneous intracellular recordings from two olivary neurons (FIG. 3B) or when a larger population is monitored using optical imaging (FIG. 3C).

The oscillations are unstable both in time and in space. In the time domain, they are composed of epochs of sustained oscillations lasting about 20 s and interrupted by quiescent periods of various durations (FIGS. 3 and 4). The phase of the oscillations in two sequential oscillating periods was unrelated. In addition, the frequency changed between 4 and 8 Hz during the periods of sustained oscillations (FIG. 4B). Blocking the synaptic inputs reversibly blocked the shift in frequency, whereas the characteristic intermittent periods of oscillations were not affected.[31] In space, the ensemble activity alternated between standing waves (FIG. 3C) and propagating waves.[31] In the first mode, neurons oscillated in-phase (FIG. 3B), while a phase difference characterized the second mode. Indeed, a phase difference of up to 50 ms (depending on the distance) was occasionally observed (FIG. 4A).

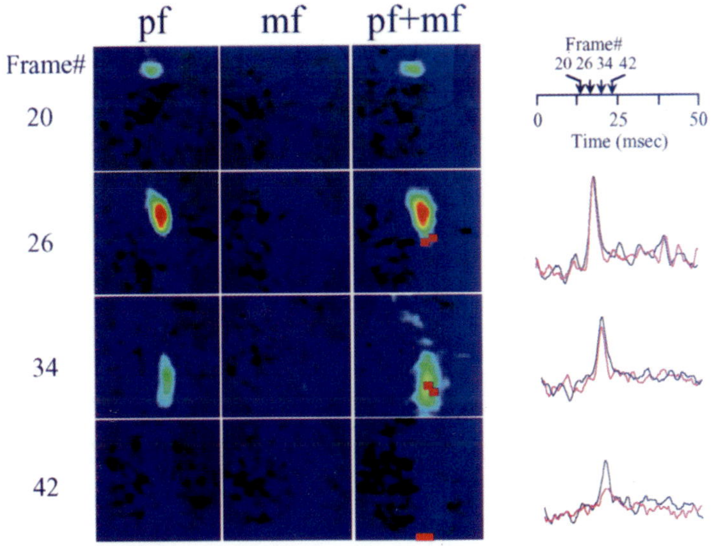

FIGURE 2. Nonlinear summation of the responses evoked by parallel-fiber (pf) and mossy-fiber (mf) stimulation suggests that pf's were activated by white-matter stimulation. The responses were measured with a 464-photodiode array, digitized at 2.5 kHz, using ×20 objective. The color-coded presentations show the responses to pf and mf stimulations and the response to both stimuli (pf + mf), in which the mf stimulation preceded that of the pf's by 4 ms. The number of the frame is shown to the left of each row, and their relative position along the time axis is shown in the *upper right schematic*. Note that when both stimuli were delivered, the beam, which propagated from top to bottom, is longer than that evoked by pf's alone. The *right column* shows the responses at the designated location (*red squares*). Each record is an average of the responses measured at the two locations. *Black traces* are the actual recordings, and *red traces* are the arithmetic sum of the pf and the mf responses. Note that at the center of the field (*second row*) the response to both stimuli is similar to the arithmetic summation of the two individual responses. At the distal end of the beam (*bottom of the field*) a prominent supralinear summation is evident.

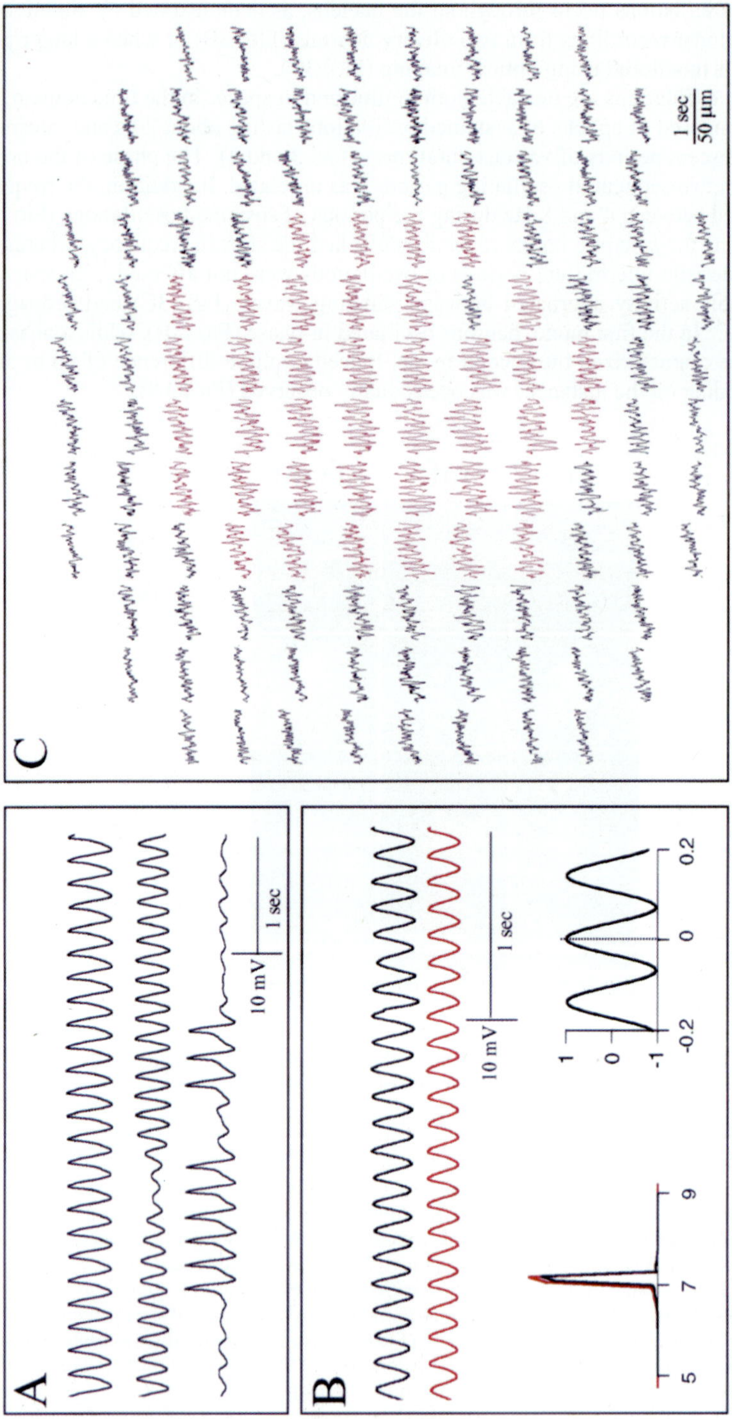

FIGURE 3. *See following page for legend.*

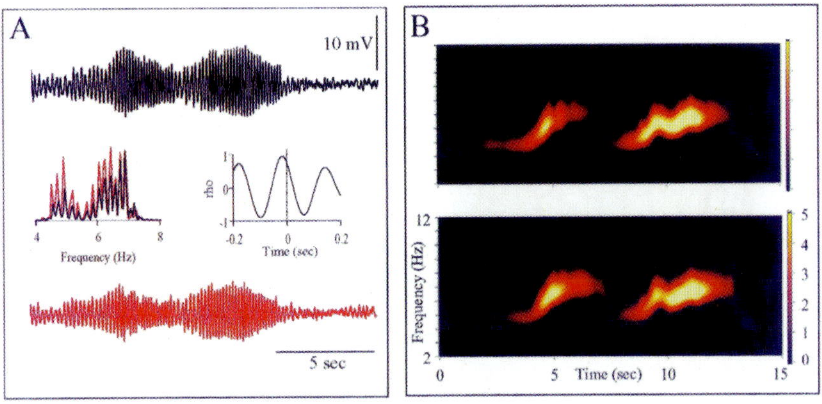

FIGURE 4. The frequency of the subthreshold oscillations undergoes spontaneous changes. (**A**) A prolonged epoch of subthreshold oscillations recorded simultaneously from two olivary neurons (*top black* and *bottom red traces*). Power spectra (*middle left panel*) and cross-correlation (*middle right panel*) analyses show that both neurons oscillated at a similar frequency and with a 20-ms phase shift. (**B**) Time-frequency analysis of the oscillatory activity reveals identical and synchronous changes in frequency. The frequency (*y*-axis) is presented as a function of time (*x*-axis), and its amplitude is color-coded (see scale to the right). The signal was convoluted with a Gabor wavelet for the time-frequency analysis (see Ref. 48).

In view of these observations, and the assumption that the subthreshold oscillations underlie the rhythmic activity of the climbing fibers, it is not surprising that it is difficult to disclose an underlying rhythm when recording from a single Purkinje cell. The frequency of the subthreshold oscillations ranges from 5 Hz to 10 Hz, whereas the average firing rate of a climbing fiber is close to 1 Hz, suggesting that individual olivary neurons produce action potential only once every fifth to tenth cycle of the oscillation (see FIG. 5). In order to uncover rhythmic behavior when the firing rate is so low, autocorrelation analysis of a prolonged recording is required. However, during such a recording, several oscillating epochs may occur, each starting at a different phase. Moreover, in each of them, the frequency may shift. Therefore, it would be nearly impossible to detect the oscillations by analyzing a single

---

FIGURE 3. Subthreshold oscillations of olivary neurons. Intracellular recordings (**A, B**) and optical imaging of voltage-sensitive dye (**C**) were performed from olivary neurons in rat brain-stem slices. (A) Three examples of spontaneous subthreshold oscillations, each recorded from a different neuron. Several types of oscillations were observed, ranging from continuous stable oscillations (*top*) to short epochs of oscillatory activity (*bottom*). (B) Simultaneous recordings from two olivary neurons (*red* and *black traces*) reveal identical frequency (*lower left, power spectra*) and in-phase oscillatory activity (*lower right, cross-correlation*). (C) Optical imaging of oscillatory activity in the inferior olive nucleus. The slice was incubated in a voltage-sensitive dye (RH-414) for 15 minutes prior to recording and imaged with a 128-photodiode array using a ×40 objective. The results are presented as a change in absolute fluorescence as a function of time at the relative locations. The *traces labeled in red* were those that showed significant oscillations. Note that the activity is confined to an area in the middle of the imaged field.

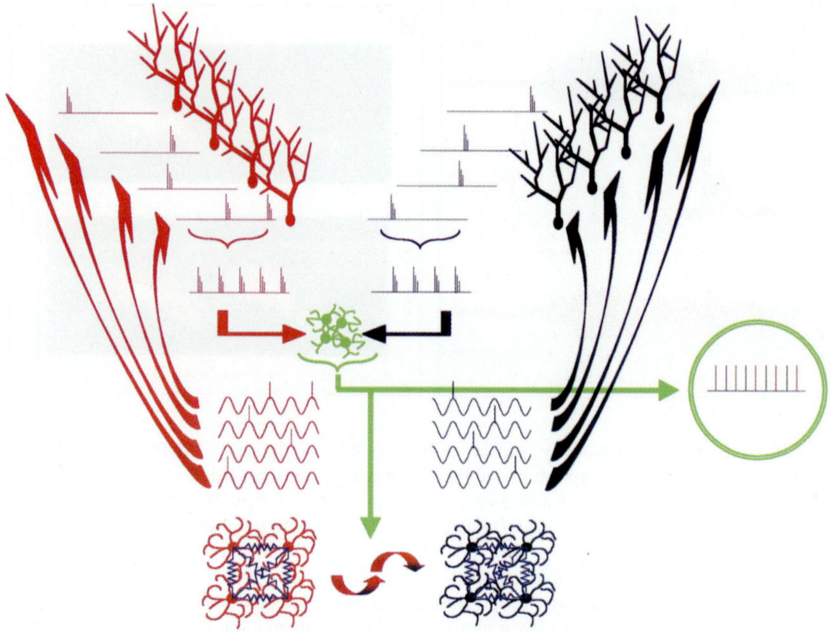

FIGURE 5. Schematic presentation of the olivocerebellar system as a generator of temporal patterns. Two groups of electrically coupled olivary neurons are shown at the bottom. Each group generates synchronous subthreshold oscillations that occasionally elicit action potentials. Since these oscillations propagate from the red neurons to the black, a phase shift exists between the two groups. In this example, a 180-degree phase shift is presented. Each neuron in the group projects to a single Purkinje cell (*top*), occasionally triggering a complex spike. Since the subthreshold oscillations in each olivary group is synchronized, the summation of the corresponding group of Purkinje cells shows rhythmic complex spike activity. Both groups of Purkinje cells (*red* and *black*) project to the deep cerebellar nuclei (*green*). The 180-degree phase difference between these two groups will double the output of the deep cerebellar nuclei (*green circle*).

cell's activity. On the other hand, simultaneously monitoring the activity from a population of Purkinje cells should reveal the underlying rhythmicity (see FIG. 5), and the larger the population, the higher the likelihood that it will be detected. Indeed, simultaneous recording of climbing-fiber activity from a large number of Purkinje cells in either an anesthetized or awake rat showed a strong tendency for rhythmic behavior.[32,33]

PROPOSED MECHANISM OF TIMING

We have briefly described the results from two sets of experiments. The first showed that input arriving via the mossy fibers to the cerebellar cortex does not substantially activate the parallel fibers. This observation led us to conclude that the parallel fibers do not behave like delay lines, and to rule out a potential timing mechanism based on their activity. The second showed that the inferior olive is capable of generating subthreshold oscillations whose properties can explain some of the controversial observations regarding the rhythmic nature of climbing-fiber activity. We conclude that climbing-fiber activity is indeed rhythmic, and that its oscillatory nature can only be detected when a whole population of Purkinje cells is considered. These synchronized subthreshold oscillations and the resultant complex spike rhythmicity in an ensemble of Purkinje cells, can potentially serve as a timing mechanism that generates precise temporal patterns.

Is there a morphological substrate for a functional organization of a population of Purkinje cells that receive input from functionally related olivary neurons? The well-known sagittal bands of the cerebellar cortex seem to be the perfect basic time unit. The functional sagittal organization of the olivary projection to the cerebellar cortex is well established, and the conservation of this organization in the projection to the deep cerebellar nuclei (DCN) has been documented.[34,35] The idea that a sagittal band of Purkinje cells and their corresponding olivary input can serve as a basic clock unite was previously suggested by Llinas and his collogues, who reported that the organization of climbing-fiber activity on the cerebellar cortex is "discrete and occurs as independent bandlike coherent states."[36] The isochronicity of this band is ensured by synchrony at the level of the inferior olive, and by climbing-fiber morphology, which compensates for the distance by altering the conduction velocity.[37] Such isochronicity is essential if the precision of the basic clock is to depend on a population of neurons.

What is the accuracy of the olivocerebellar clock? The simplest assumption would be that accuracy depends on the period of the oscillation. The 5–10-Hz oscillations recorded in the inferior olive impose an accuracy of 100–200 ms. This is far too low to account for the millisecond-range precision attributed to cerebellar function. However, as shown earlier, when the oscillations in the olive propagate in space, phase lags occur between the subthreshold oscillations of neighboring cells. These phase lags, which determine the temporal resolution of the clock, could be as low as 10 ms.[31] FIGURE 5 demonstrates how a phase difference of 180 degrees between two groups of oscillating neurons doubles the rate of the cerebellar output. In this figure, the subthreshold oscillations propagated from the red group of olivary cells to the black group. The oscillations in the black group lagged the red group by 180 degrees. The sum of complex spike activity in the corresponding black Purkinje cells reveals the basic frequency of the subthreshold oscillation. If the two groups of Purkinje cells converge at the same location in the DCN, or at any remote structure down the stream of information flow, the overall frequency will be doubled. Thus, the downstream convergence of Purkinje cell axons extends the accuracy of the clock beyond the limits imposed by the period of the oscillation.

To be effective, a clock should have a resetting mechanism, it should be recruited as needed by the system, and it should begin to tick at the appropriate point in time. In neuronal terms, the onset, occurrence, and termination of the oscillations should be under neuronal control. Following experimental[30,38,39] and theoretical studies[40,41] on the origin of the subthreshold oscillations, we formulated a "network oscillations" hypothesis. According to this hypothesis, each neuron by itself cannot oscillate; it is only when a network of electrically coupled neurons is formed that the membrane potential of each component will begin to oscillate. This hypothesis implies that interneuronal coupling controls the onset and termination of the oscilla tions, and that if the cerebellum can control the coupling between olivary neurons, it can theoretically serve as a resetting mechanism.

The inferior olive nucleus is densely innervated by inhibitory synapses originating from the DCN.[42] Furthermore, the inhibitory DCN cells that project to the inferior olive, and the excitatory cells that project to the thalamus and premotor nuclei, receive input from the same Purkinje cells.[43] Hence, the same cortical output modulates the olivary nucleus and is transmitted to other brain structures. The inhibitory inputs to the olive are strategically positioned next to the gap junctions. It has been proposed that such an arrangement can serve as an uncoupling mechanism,[44,45] by redirecting the current to leak outside instead of passing from one cell to the other. We have shown that olivary neurons are coated with GABAergic receptors of various properties.[46] We also demonstrated that GABA application can profoundly affect the subthreshold oscillations and occasionally switch an oscillating epoch to a quiescent period.[47] Inhibitory input from the DCN can thus control the onset and termination of the olivary clock, while at the same time controlling the size of the network and the spatial propagation of the oscillations.

We hypothesize that the role of the olivocerebellar system is to generate a large repertoire of temporal patterns. Modifiable interactions between the olivary nucleus, the cerebellar cortex, and deep nuclei generate these temporal patterns. The olivary subthreshold rhythmic activity contains all the components necessary to act as precise, yet modifiable, oscillators, and the cerebellar cortex, via the DCNs, translates this oscillatory activity into firing patterns. There are several fundamental questions that we have not addressed in this brief manuscript. For example, how is rhythmic activity translated into a sequential firing pattern? What determines that amplitude of the response? How are these patterns learned, stored, and recruited by a specific input? These questions should be addressed by studies specifically motivated by this hypothesis. Despite these open questions, one can envision a scenario in which the cerebellum, when in need of a specific temporal pattern, assembles a group of olivary neurons by pausing the activity in the DCN, and thereby removing the inhibition from appropriate olivary gap-junctions. The size and location within the olive at which the inhibition is removed determines the size of the functional ensemble, as well as its spatial organization. Once the inhibition is removed, the ensemble starts oscillating, thus generating a temporal pattern that is uniquely suited to the task at hand.

REFERENCES

1. JOHNSON, M.T., C.R. MASON & T.J. EBNER. 2001. Central processes for the multiparametric control of arm movements in primates. Curr. Opin. Neurobiol. **11**(6)**:** 684–688.
2. ITO, M. 1984. The Cerebellum and Neurnal Control. Raven Press. New York.
3. LLINAS, R.R. & J.I. SIMPSON. 1981. Cerebellar control of movement. *In* Handbook of Behavioral Neurobiology. A.L. Towe & E.S. Luschei, Eds.: 231–302. Plenum. New York
4. BLOEDEL, J.R. & V. BRACHA. 1995. On the cerebellum, cutaneomuscular reflexes, movement control and the elusive engrams of memory. Behav. Brain Res. **68**(1)**:** 1–44.
5. MIALL, R.C. & G.Z. RECKESS. 2002. The cerebellum and the timing of coordinated eye and hand tracking. Brain Cogn. **48**(1)**:** 212–226.

6. SCHWEIGHOFER, N. *et al.* 1998. Role of the cerebellum in reaching movements in humans. II. A neural model of the intermediate cerebellum. Eur. J. Neurosci. **10**(1): 95–105.
7. THACH, W.T., H.P. GOODKIN & J.G. KEATING. 1992. The cerebellum and the adaptive coordination of movement. Annu. Rev. Neurosci. **15:** 403–442.
8. RAMNANI, N. *et al.* 2001. The cerebellum and parietal cortex play a specific role in coordination: a PET study. Neuroimage **14**(4): 899–911.
9. DIENER, H.C. *et al.* 1993. Cerebellar dysfunction of movement and perception. Can. J. Neurol. Sci. **20**(Suppl 3): S62–S69.
10. WELSH, J.P. *et al.* 1995. Dynamic organization of motor control within the olivocerebellar system. Nature. **374**(6521): 453–457.
11. TIMMANN, D., S. WATTS & J. HORE. 1999. Failure of cerebellar patients to time finger opening precisely causes ball high-low inaccuracy in overarm throws. J. Neurophysiol. **82**(1): 103–114.
12. IVRY, R. 1997. 1997. Cerebellar timing systems. Int. Rev. Neurobiol. **41:** 555–573.
13. BLAKEMORE, S.J., D.M. WOLPERT & C.D. FRITH. 1998. Central cancellation of self-produced tickle sensation. Nat. Neurosci. **1**(7): 635–640.
14. WITNEY, A.G., S.J. GOODBODY & D.M. WOLPERT. 1999. Predictive motor learning of temporal delays. J Neurophysiol. **82**(5): 2039–2048.
15. KAWATO, M. 1999. Internal models for motor control and trajectory planning. Curr. Opin. Neurobiol. **9**(6): 718–727.
16. TESCHE, C.D. & J.J. KARHU. 2000. Anticipatory cerebellar responses during somatosensory omission in man. Hum. Brain Mapp. **9**(3): 119–142.
17. COURCHESNE, E., *et al.* 1994. Impairment in shifting attention in autistic and cerebellar patients. Behav. Neurosci. **108**(5): 848–865.
18. NICOLSON, R.I., A.J. FAWCETT & P. DEAN. 2001. Developmental dyslexia: the cerebellar deficit hypothesis. Trends Neurosci. **24**(9): 508–511.
19. AKSHOOMOFF, N.A., E. COURCHESNE & J. TOWNSEND. 1997. Attention coordination and anticipatory control. Int. Rev. Neurobiol. **41:** 575–598.
20. BRAITENBERG, V. 1967. Is the cerebellar cortex a biological clock in the millisecond range? Prog. Brain Res. **25:** 334–346.
21. BOWER, J.M. & D.C. WOOLSTON. 1983. Congruence of spatial organization of tactile projections to granule cell and Purkinje cell layers of cerebellar hemispheres of the albino rat: vertical organization of cerebellar cortex. J. Neurophysiol. **49**(3): 745–766.
22. COHEN, D. & Y. YAROM. 1999. Optical measurements of synchronized activity in isolated mammalian cerebellum. Neuroscience **94**(3): 859–866.
23. KEATING, J.G. & W.T. THACH. 1995. Nonclock behavior of inferior olive neurons: interspike interval of Purkinje cell complex spike discharge in the awake behaving monkey is random. J. Neurophysiol. **73**(4): 1329–1340.
24. LLINAS, R. & K. SASAKI. 1989. The functional organization of the olivo-cerebellar system as examined by multiple Purkinje cell recordings. Eur. J. Neurosci. **1:** 587–602.
25. LLINAS, R. & R. VOLKIND. 1973. The olivo-cerebellar system: functional properties as revealed by harmaline-induced tremor. Exp. Brain Res. **18:** 69–87.
26. WELSH, J.P. 1998. Systemic harmaline blocks associative and motor learning by the actions of the inferior olive. Eur. J. Neurosci. **10**(11): 3307–3320.
27. LLINAS, R. & Y. YAROM. 1986. Oscillatory properties of guinea-pig inferior olivary neurones and their pharmacological modulation: an in vitro study. J. Physiol. **376:** 163–182.
28. LLINAS, R. & Y. YAROM. 1981. Properties and distribution of ionic conductances generating electroresponsiveness of mammalian inferior olivary neurones in vitro. J. Physiol. **315:** 569–584.
29. LLINAS, R. & Y. YAROM. 1981. Electrophysiology of mammalian inferior olivary neurones in vitro. Different types of voltage-dependent ionic conductances. J. Physiol. **315:** 549–567.
30. LAMPL, I. & Y. YAROM. 1997. Subthreshold oscillations and resonant behavior: two manifestations of the same mechanism. Neuroscience **78**(2): 325–341.
31. DEVOR, A. & Y. YAROM. 2002. Generation and propagation of subthreshold waves in a network of inferior olivary neurons. J Neurophysiol. In press.

32. LANG, E.J. *et al.* 1999. Patterns of spontaneous Purkinje cell complex spike activity in the awake rat. J. Neurosci. **19**(7)**:** 2728–2739.
33. SUGIHARA, I., E.J. LANG & R. LLINAS. 1995. Serotonin modulation of inferior olivary oscillations and synchronicity: a multiple-electrode study in the rat cerebellum. Eur. J. Neurosci. **7**(4)**:** 521–534.
34. VOOGD, J. & M. GLICKSTEIN. 1998. The anatomy of the cerebellum. Trends Neurosci. **21**(9)**:** 370–375.
35. VOOGD, J. & T.J. RUIGROK. 1997. Transverse and longitudinal patterns in the mammalian cerebellum. Prog. Brain Res. **114:** 21–37.
36. FUKUDA, M., T. YAMAMOTO & R. LLINAS. 2001. The isochronic band hypothesis and climbing fibre regulation of motricity: an experimental study. Eur. J. Neurosci. **13**(2)**:** 315–326.
37. SUGIHARA, I., E.J. LANG & R. LLINAS. 1993. Uniform olivocerebellar conduction time underlies Purkinje cell complex spike synchronicity in the rat cerebellum. J. Physiol. **470:** 243–271.
38. YAROM, Y. 1991. Rhythmogenesis in a hybrid system—Interconnecting an olivary neuron to an analog network of coupled oscillators. Neuroscience **44**(2)**:** 263–275.
39. MANOR, Y. *et al.* 2000. To beat or not to beat: a decision taken at the network level. J. Physiol. Paris **94**(5–6)**:** 375–390.
40. MANOR, Y. *et al.* 1997. Low-amplitude oscillations in the inferior olive: a model based on electrical coupling of neurons with heterogeneous channel densities. J. Neurophysiol. **77**(5)**:** 2736–2752.
41. LOEWENSTEIN, Y., Y. YAROM & H. SOMPOLINSKY. 2001. The generation of oscillations in networks of electrically coupled cells. Proc. Natl. Acad. Sci. USA **98**(14)**:** 8095–8100.
42. DE ZEEUW, C.I. *et al.* 1989. Ultrastructural study of the GABAergic, cerebellar, and mesodiencephalic innervation of the cat medial accessory olive: anterograde tracing combined with immunocytochemistry. J. Comp. Neurol. **284**(1)**:** 12–35.
43. TEUNE, T.M. *et al.* 1998. Single Purkinje cell can innervate multiple classes of projection neurons in the cerebellar nuclei of the rat: a light microscopic and ultrastructural triple-tracer study in the rat. J. Comp. Neurol. **392**(2)**:** 164–178.
44. SPIRA, M.E. & M.V. BENNETT. 1972. Synaptic control of electrotonic coupling between neurons. Brain Res. **37:** 294–300.
45. LLINAS, R. 1974. 18th Bowditch Lecture: motor aspects of cerebellar control. Physiologist **17:** 19–46.
46. DEVOR, A., J.M. FRITSCHY & Y. YAROM. 2001. Spatial distribution and subunit composition of GABA(A) receptors in the inferior olivary nucleus. J. Neurophysiol. **85**(4)**:** 1686–1696.
47. DEVOR, A. & Y. YAROM. 2000. GABAergic modulation of olivary oscillations. Prog. Brain Res. **124:** 213–220.
48. PALVA, J.M. *et al.* 2000. Fast network oscillations in the newborn rat hippocampus in vitro. J. Neurosci. **20**(3)**:** 1170–1178.

The Organization of Cerebellar Cortical Circuitry Revisited

Implications for Function

JAMES M. BOWER

The Research Imaging Center at The University of Texas Health Science Center–San Antonio, and the Cajal Neuroscience Research Center at the University of Texas–San Antonio, Texas 78284-6240, USA

ABSTRACT: For more than 35 years there has been experimental evidence that parallel fiber activity does not generate the beams of activated Purkinje cells hypothesized on the basis of cortical anatomy and assumed by most theories of cerebellar cortical function. This paper first reviews the evidence for and against the parallel fiber beam hypothesis, and then discusses the findings of our recent experimental and model-based investigations intended to better understand parallel fiber effects on Purkinje cells. A principal conclusion of these studies is that the excitatory effects of parallel fibers on Purkinje cell dendrites are modulating and must be considered in the context of a balancing inhibitory influence provided by molecular layer interneurons to these same dendrites. It is proposed that this association of excitation and inhibition can account for the lack of beam-like effects on Purkinje cells. The paper concludes by considering the consequences of this new interpretation of cerebellar cortical circuitry for current theories of cerebellar function.

KEYWORDS: cerebellum; Purkinje cell; parallel fibers; granule cells; cerebellar circuitry; modeling; theory

INTRODUCTION

Starting with the work of Ramon y Cajal at the turn of the last century[1] the nearly crystalline structure of the cerebellar cortex has allowed investigators to develop a highly detailed description of its neurons and network architecture.[2–9] This knowledge, in turn, has meant that functional speculations concerning the cerebellum have been more closely linked to real anatomy and physiology than is often the case for brain theories.[0–17] Because of this close linkage, these speculations have had an unusually direct influence on physiological, anatomical, and behavioral investigations.[18–21]

While it is laudable that cerebellar theories and cortical structure have so often been closely linked, it is the thesis of this article that a fundamental assumption con-

Address for correspondence: James M. Bower, The Research Imaging Center, The University of Texas Health Science Center–San Antonio, 7703 Floyd Curl Drive, San Antonio, TX 78284-6240. Voice: 210-587-8080; fax: 210-587-8152.
bower@uthscsa.edu

Ann. N.Y. Acad. Sci. 978: 135–155 (2002). © 2002 New York Academy of Sciences.

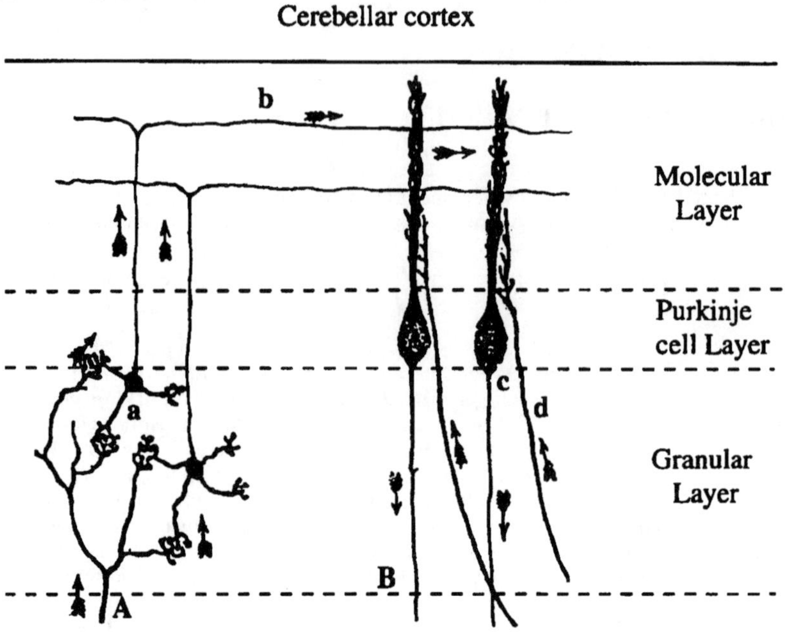

FIGURE 1. Connections of the cerebellar cortex as drawn by Ramon y Cajal[1] and modified slightly by Ito.[21] Following the arrows on the left side of the diagram, mossy fiber inputs (**A**) make excitatory connections with granule cells (**a**), whose axons ascend through the molecular layer, bifurcate, and then all course as parallel fibers (**b**) through the isoplanar dendritic tree of the Purkinje cell (**c**). The Purkinje cell axon (**B**) then provides the sole output of the cerebellar cortex. Also shown in this diagram is the other major afferent input to the cerebellar cortex, the climbing fiber projection (**d**) arising in the inferior olive. (From Ito.[21] Reproduced by permission.)

cerning the physiological organization of cerebellar circuitry is, in fact, incorrect. Specifically, at the base of many contemporary ideas about cerebellar function is a classic description and understanding of the influence of the excitatory granule cells which provide many thousands of synapses on the Purkinje cells which, in turn, provide the sole output of cerebellar cortex (FIG. 1). This classical view assumes that the primary excitatory drive on Purkinje cells comes from synapses associated with the parallel fiber segment of the granule cell axon and that this strong positive excitatory dendritic input directly results in the generation of somatic action potentials. As indicated by the arrows in FIGURE 1, this relationship was implied by the earliest descriptions of cerebellar cortical circuitry.[1] In this and all subsequent diagrams of cerebellar cortical architecture, granule cell axons arising from the cell body (a) are shown to bifurcate in the molecular layer and course as parallel fibers (b) contacting distant Purkinje cells (c).

While the assumption that parallel fiber input is directly associated with Purkinje cell output is in keeping with classical descriptions of synaptic integration in dendrites,[22] it is not supported by numerous studies which have consistently failed to demonstrate any direct excitatory parallel fiber effects on Purkinje cells.[23–27] Instead, it appears as though the largest granule cell drive on Purkinje cells comes from synapses associated with the granule cell axon as it ascends radially through the molecular layer.[25,28–30] In the classically drawn diagrams of cerebellar circuitry (FIG. 1) this relationship is obscured by the fact that the ascending segment of the granule cell is placed nowhere near the overlying Purkinje cells. Perhaps as a consequence, the influence of synapses associated with the ascending segment of the granule cell axon has been little considered when speculating about cerebellar cortical function. The strong influence of the ascending segment synapses, when coupled with the apparently nonclassical influences of parallel fiber synapses on Purkinje cells, requires a major revision of our traditional view of cerebellar cortical processing.

THE TRADITIONAL VIEW

The highly regular geometrical relationship between the parallel fiber segment of the granule cell axon and the isoplanar dendrites of the overlying Purkinje cells have intrigued experimentalists and theorists for more than 100 years (FIG. 1). We now know that each individual parallel fiber makes a single or at most two synapses on perhaps half the Purkinje cells it encounters along its several millimeter path.[31] However, because there are so many granule cells, each Purkinje cell is estimated to receive a total of between 150,000 and 175,000 granule cell synapses,[31] most from the parallel fibers.[32] This excitatory synaptic convergence is the largest known in the nervous system. The importance of this geometrical relationship between parallel fibers and Purkinje cells is reinforced by the fact that a similar geometry is found in all of the many and diverse animals that have a cerebellum[33] as well as in all regions of a single individual's cerebellum.[3]

Given the size and unusual geometry of the parallel fiber to Purkinje cell projection, it is not surprising that its significance has been the focus of a great deal of functional speculation.[29] In general, these ideas can be grouped into either one of two categories or some combination of both. The first interprets the fact that single parallel fibers sequentially contact Purkinje cells over several millimeters to suggest that cerebellar circuitry performs a timing function.[19,34–40] While traditionally proposed in the context of motor control,[19,34,36,37,39,41] a presumed timing function has been applied more recently to sensory and even cognitive tasks.[42]

The second category of parallel fiber-related hypothesis focuses more on the number of parallel fibers making contact with each Purkinje cell. Typically, these ideas have emphasized some synaptic learning process, which selects a subset of the total number of parallel fibers to be active.[14,43] Although usually not stated explicitly, the assumption in many of these models is that much less than 150,000 synapses is necessary to drive Purkinje cell output, and therefore the full complement of parallel fiber synapses represents a potential field of connections from which the appropriate subset is selected. In the most famous and influential examples of this interpretation, it was proposed independently by Marr[11] and Albus[12] that parallel fiber synapses are either enhanced[11] or suppressed[12] in response to coincident activity in the other afferent pathway to Purkinje cells, climbing fiber projections from the inferior olive (d in FIGURE 1). As with pure timing theories, these ideas have traditionally been applied to motor-related learning,[18,44] but more recently have also been applied to cerebellar contributions to sensory/cognitive tasks.[45] Several ideas about cerebellar function combine both timing and learning functions.[17,46]

Whichever feature of parallel fiber projections onto the Purkinje cells is emphasized, all of these ideas assume that parallel fiber synapses provide the primary granule cell–related excitatory drive on Purkinje cells.[19,34–40] Most also assume that Purkinje cell dendrites function like classical integrate-and-fire devices, summing synaptic inputs until some threshold for activation of the Purkinje cell soma is surpassed.[17,47] As described in the next section, quite a bit of experimental evidence exists to suggest that this is not the case.

TO BEAM OR NOT TO BEAM, THAT IS THE QUESTION

The idea that parallel fibers provide a strong excitatory drive on successively activated Purkinje cells would seem to follow naturally from our 100-year-old understanding of cerebellar circuitry (FIG. 1) as well as firmly established principles of dendritic integration.[22] In fact, this idea, which is generally referred to as "the beam hypothesis,"[10] was proposed on almost purely anatomical grounds well before there was any direct supporting physiological evidence."[34] However, subsequent physiological studies of the parallel fiber system by Sir John Eccles and his colleagues seemed to fully support the beam hypothesis.[48] Typical of the work of the Eccles laboratory,[10] great care was taken in these now classical experiments to experimentally isolate parallel fiber influences from other aspects of the circuitry. To do so, large surface stimulating electrodes (the so-called "LOC" stimulus) were used to directly activate parallel fibers while field potential recordings were obtained along the resulting "beam" running down their length. This direct electrical stimulation technique, still in remarkably wide use today,[49,50] results in evoked responses with a temporal delay with distance from the site of stimulation that can be directly related to the conduction velocity of the parallel fibers.[48] These experiments, when combined with the existing understanding of cerebellar cortical architecture, resulted in the beam hypothesis shown in its original formulation in FIGURE 2.

So, what is the problem? Put as simply as possible, experimental investigations in which the parallel fiber system is activated other than by direct application of electrical current have consistently failed to reveal strong excitatory beam-like parallel fiber effects on Purkinje cells. The first study of this type was actually reported more than 33 years ago, in 1969, by Bell and Grimm.[23] In that experiment, pairs of Purkinje cells presumably sharing parallel fiber inputs were recorded simultaneously and examined for correlated discharge. Were shared parallel fibers driving the activity of these Purkinje cells, these authors should have found a temporal correlation in firing patterns with delays corresponding to the distance between the recorded Purkinje cells. While these authors did find correlations between some pairs of Purkinje cells, correlated cells were separated by less than 100 μm, and the correlation delay was 1 millisecond or less. As the authors themselves point out, this result implies that these cells were receiving some source of simultaneous inputs rather than delayed inputs from parallel fibers.[23] Given that parallel fibers in the cats used in these experiments extend for upwards of 4 millimeters,[31] parallel fiber inputs should have resulted in correlations between cells separated by much larger distances. Clearly concerned by the failure to obtain this result, the authors explained their results by proposing that

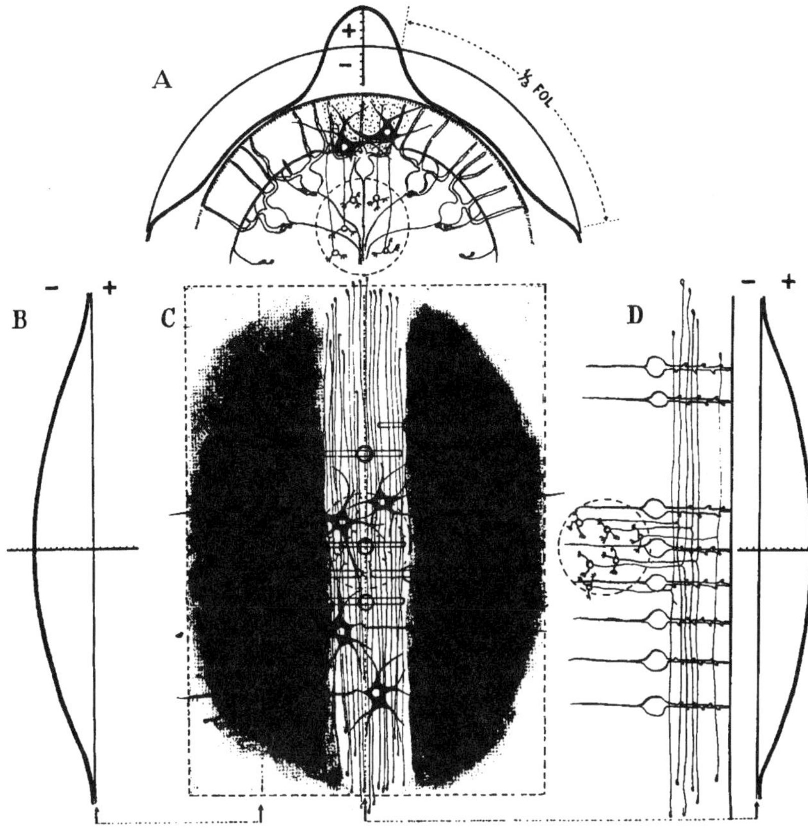

FIGURE 2. Diagrammatic representation of the beam hypothesis, as originally presented by Eccles *et al.*[10] The central diagram (**C**), shows the expected pattern of Purkinje cell excitation (light central area) and inhibition (surrounding dark region) following activation of the parallel fibers. (From Eccles *et al.*[10] Reproduced by permission.)

more distant Purkinje cells had a much lower probability of sharing parallel fibers than did nearby cells. The bottom line, however, is that these experiments failed to reveal any beam-like behavior of Purkinje cells driven by parallel fibers.

This first indication that all was not right with the beam hypothesis was followed by experiments published by the Eccles laboratory itself in 1972.[24] In these experiments, the locations of Purkinje cells responding to tactile activation of the cat forepaw were mapped and found to occur in spatial "clumps" or "patches" (FIG. 3). Again, the authors clearly expected to find activated Purkinje cell beams, and when they did not, they proposed that the stimuli they had presented were either too weak to activate the beams, or that the clusters of activated Purkinje cells were produced by converging subthreshold beams.[24] The idea that Purkinje cell activation depends on complex patterns of temporal convergence of parallel fiber beam-like activity has more recently formed the basis for both theoretical[40] and *in vitro* physiological[51]

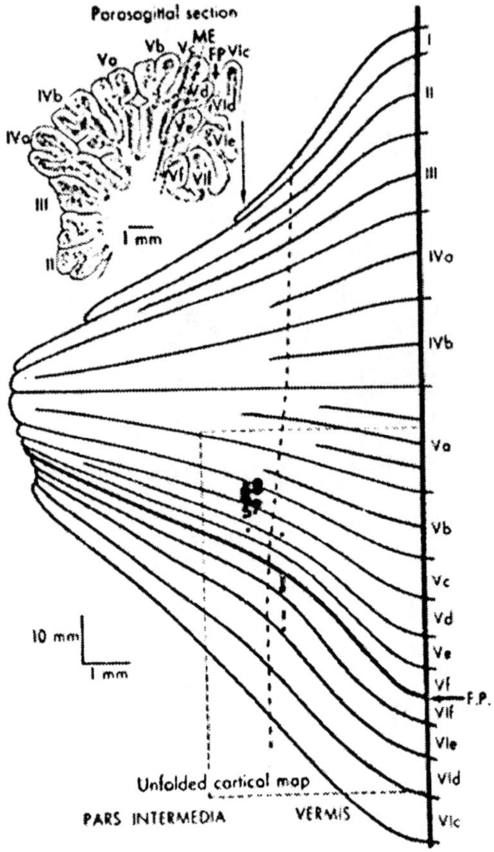

FIGURE 3. Results of cerebellar mapping studies by Eccles *et al.*[24] indicating the location of Purkinje cells responding to touch of the cat's forepaw. The region of activated Purkinje cells is much less than would be expected from the parallel fibers which course several millimeters medial and lateral (approximately left and right in this diagram) along each folium. (From Eccles *et al.*[24] Reproduced by permission.)

studies in the laboratory of Valentino Braitenberg. The fact remains, however, that Eccles and his colleagues themselves showed that natural activation of peripheral inputs did not produce beams of activated Purkinje cells.

Unfortunately, at the time of the Eccles experiments, little was known about the fine spatial pattern of afferent mossy fiber projections to the cerebellum, so it was not possible to test the idea that overlapping subthreshold parallel fiber beams produced the clusters of Purkinje cell activity. Those afferent input maps that existed relied on low spatial-resolution surface field potential recordings primarily obtained using the electrical stimulation of peripheral nerves.[52,53] It was not until the late

1970s that Welker and colleagues, using fine-grained micromapping techniques and tactile peripheral stimulation, demonstrated that there was also a "patchy" pattern of tactile afferent projections to the granule cell layer.[54–56] When these patchy patterns of granule cell layer activation were compared to tactilely evoked responses in the overlying Purkinje cells,[25] the patchy pattern of activated Purkinje cells reported by Eccles *et al.*[24] was found to directly reflect the patchy organization of afferent activation of the granule cell layer.

THE RADIAL ORGANIZATION OF CEREBELLAR CORTICAL CIRCUITRY

As just described, by the early 1980s several different studies suggested that parallel fibers might not have a classical excitatory influence on cerebellar Purkinje cells.[23–25] Instead, it appeared as though parallel fiber beams were at best subthreshold in their effects on Purkinje cells. Furthermore, our own studies directly comparing activation of the granule cell layer with overlying Purkinje cells suggested that the granule cell layer immediately underlying each Purkinje cell had the most substantial influence on the cell's spiking output.[25] This datum, however, raised two important questions: first, why don't parallel fiber inputs generate beam-like Purkinje cell output?; and second, why does the granule cell layer underlying each Purkinje cell seem to have so much influence?

Shortly after our original description of the vertical (radial) organization of cerebellar cortex,[57] Llinas[58] proposed an explanation for both the lack of obvious effects of the parallel fibers, and the apparently special influence of the granule cells underlying each Purkinje cell. Referring to the previous anatomical observation by Mugnaini[59] that granule cell axons make synapses as they ascended through the molecular layer prior to bifurcating into parallel fibers, Llinas suggested that the radial organization of cerebellar cortex could be the influence of these synapses (FIG. 4). Ascending segment granule cell axons were ideally oriented to make multiple synaptic contacts on the overlying Purkinje cell, and this input was likely to be temporally synchronous after activation of the granule cell layer. In contrast, each parallel fiber segment made one or at most two contacts on a Purkinje cell,[60] and variations in the conduction velocities of parallel fibers[61] could be expected to result in a growing input asynchrony as they coursed away from an activated region of the granule cell layer. Thus it was proposed that the number of granule cell synaptic inputs, and their degree of synchrony, accounted for the radial, and not beam-like activation of Purkinje cells.[25,58]

CONTRARY EVIDENCE

In the years following the proposal that the ascending segment of the granule cell axon might provide a more potent input to Purkinje cells than the parallel fibers, several concerns have been raised by both cerebellar anatomists and physiologists. The first involved the number of synapses expected to be associated with the ascending segment.[60] While quantifying the parallel fiber system of the rat, Napper and Harvey[60] used indirect means to estimate that ascending segment synapses were

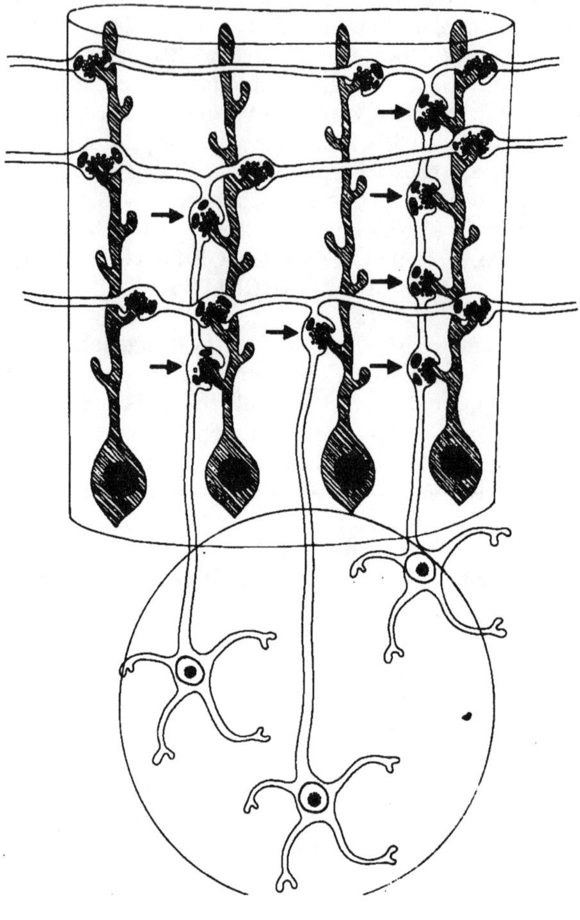

FIGURE 4. First proposal by Llinás that the radial organization of the cerebellar cortex was due to synapses associated with the ascending segment of the granule cell axon. (From Llinás.[58] Reproduced by permission.)

likely to represent at most 3% of the total granule cell synaptic contacts on any one Purkinje cell. While these authors did subsequently note that any change in their assumptions would require a modification of this estimate,[31] this estimate stood until we ourselves undertook a direct serial electron micrographic study of the ascending segment synapses.[32] Using two different approaches, we were able to calculate that ascending segment synapses were likely to constitute up to 20% of the total granule cell synaptic input to Purkinje cells. Further, these studies demonstrated that ascending segment synapses were both structurally distinct from parallel fibers and contacted a different region of the Purkinje cell dendrite. These anatomical results therefore suggested that there were more than enough ascending granule cell synapses to produce strong excitatory effects on Purkinje cells and that the synapses associated with these two populations of synapses might have distinctly different physiological effects or functions.

While there would seem to be an anatomical basis for a strong effect of ascending axon synapses on Purkinje cells, in 1994 Vranesic *et al.*[62] published physiological data suggesting that these synapses might not have a strong effect on Purkinje cells. Using an *in vitro* brain slice preparation and measuring neuronal activity using voltage-sensitive dyes, these authors claimed to find no evidence for an antidromic activation of Purkinje cells via the ascending granule cell axon following a LOC-type parallel fiber electrical stimulation. This experiment was performed in part in response to our previous caution that a LOC-type stimulus could be expected to activate not only parallel fibers orthodromically, but also ascending segment axons antidromically.[25] These authors concluded with support for strong parallel fiber effects on Purkinje cells. However, a closer look at the data, and their interpretation leads to several concerns. First, the signal obtained using voltage-sensitive dies almost certainly measures dendritic responses and not the somatic output of the Purkinje cell on which the results of Bower and Woolston[25] are based. Second, the *in vitro* slicing procedures used in these experiments interfere with inhibitory mechanisms in the cerebellum, which we have reason to believe play an important role in regulating parallel fiber effects (see below). Finally, their analysis failed to adequately consider the contribution of parallel fibers to the postsynaptic signals recorded. In our own *in vitro* study published in the same year,[28] we used the more direct technique of intracellular electrical recording to demonstrate that substantial granule cell–induced excitatory Purkinje cell responses remained after parallel fiber inputs had been removed.

While most cerebellar physiologists seemed willing to accept some influence of ascending axon segments on Purkinje cells,[40] considerable resistance has remained to the idea that parallel fibers do not provide strong excitatory effects on Purkinje cells. This may not be surprising, given that most hypotheses regarding cerebellar function assume strong parallel fiber effects. However, this general resistance was bolstered by the publication of a paper in 1992 by Garwicz and Andersson[62] that purported to present direct evidence contradicting the results of Bower and Woolston[25] as well as those of Eccles *et al.*[24] While this paper is often sited as evidence that parallel fibers are strongly excitatory,[21,62,63] the methods used by these authors actually precluded a test of either of the previous results.

First, Garwicz and Andersson designed their experiments around cortical field potentials rather than recordings made directly from single Purkinje cells. By their own acknowledgement, those few single neuron recordings made in the course of their study were not obtained systematically. Unfortunately, like voltage-sensitive dyes,[62] field potential recordings reflect synaptic effects in dendrites averaged over many neurons, rather than the spiking behavior of individual cells.[64] Thus, the finding of beam-like field potential propagation by Garwicz and Andersson only indicates that parallel fiber inputs influence Purkinje cell dendrites. This has never been in doubt. Instead, the single-unit recording experiments of Bower and Woolston,[25] Eccles *et al.*,[24] as well as Bell and Grimm[23] all suggest that this dendritic influence fails to directly produce spiking in the Purkinje cell soma. Oddly enough, in the discussion of their paper, Garwicz and Andersson actually warn against single-unit recording, noting that field potentials are less variable in nature.[62] Again, however, field potentials are transparent to the spiking activity of individual neurons, and it is the spiking output of the Purkinje cell that is of interest here (and of interest to subsequent neurons).

A second concern with respect to the results of Garwicz and Andersson is that they were obtained using electrical stimulation of peripheral nerves,[62] rather than natural stimulation of peripheral receptors, as was the case for Bower and Woolston[25] and Eccles *et al.*[24] While the electrical stimulation of peripheral nerves is again a procedure that has persisted since the 1940s,[52,53] Kolb *et al.*[26] have shown that this type of electrical stimulation produces different patterns of cerebellar activity than those induced with more natural stimuli. Further, while also using field potential recordings, the study by Kolb *et al.* sampled cerebellar responses at higher spatial and temporal resolution and more completely analyzed the depth.[62] Kolb *et al.* concluded that there was "no strong evidence" for beam-like propagation of Purkinje cell activity.[26] These authors also controlled for arterial blood pressure and continually measured the viability of parallel fiber conduction, both concerns raised by Garwicz and Andersson[62] with respect to the original reports by Bower and Woolston[25] and Eccles *et al.*[24]

Despite the evidence from multiple laboratories over more than 33 years that parallel fibers don't do what they are supposed to do, a very recent study undertaken in the cat by Ekerot and Jorntell[63] again purports to support strong beam-like parallel fiber effects on Purkinje cells. While these authors did use single-unit Purkinje cell recordings and natural peripheral stimulation, their conclusion, unfortunately is not based on a direct comparison of granule cell layer and Purkinje cell responses. Instead the receptive field data on which they draw this conclusion were obtained from different experiments using different animals (see their Figure 3, for example[63]). Further, these authors interpreted their results as supporting the existence of strong parallel fibers inputs to Purkinje cells despite the fact that they also noted the Purkinje cells' receptive fields were smaller than would be expected from beam-like activation, and instead tended to resemble mossy fiber receptive fields. In this case, the small size of the Purkinje cell receptive fields was taken as evidence for prior parallel fiber synaptic modification of the type hypothesized by Albus,[65] as the authors noted a similarity between the climbing fiber and simple spike receptive fields of individual Purkinje cells.[63] We have also recently shown a correspondence between climbing fiber and simple spike receptive fields in comparable cerebellar regions of the rat.[66] However as already discussed, we have also shown using direct comparisons of granule cell and Purkinje cell responses that the simple spike receptive fields of Purkinje cells are similar to the receptive fields of the underlying granule cell layer.[25] Thus, the receptive fields of Purkinje cells are due to a direct granule cell layer projection via ascending axon synapses, and do not require any form of climbing fiber-directed learning.

Finally, there is also an important methodological difficulty with the report of Ekerot and Jorntell.[63] In their studies, Purkinje cell receptive fields are determined by listening to spiking activity on an auditory monitor and/or observing spike sequences on an oscilloscope while manually palpating the skin of the animal. Given the ongoing background spontaneous activity of Purkinje cells, we have found that the careful determination of receptive fields is entirely dependent on the use of modern spike-triggered averaging techniques and the use of repetitive mechanically controlled tactile stimuli.[25,66] Without such techniques, in our experience, it is very difficult to make the fine distinctions necessary to compare receptive field properties.

In concluding this review of evidence against a radial organization of cerebellar cortex, it is worth mentioning that several other reports since 1983 have supported the original ressults of Bower and Woolston[25] and Eccles *et al.*[24] We have already mentioned the careful electrophysiological recording study of Kolb *et al.*[26] However, support for a strong influence of underlying granule cells on Purkinje cells was also indirectly provided by the results of Sharp *et al.*,[67] who mapped the activity of c-fos gene expression in the cerebellum and demonstrated a radial alignment of Fos-positive granule and Purkinje cells. More recently, Cohen and Yarom[27] have shown, using voltage-sensitive dye recordings in an isolated cerebellar/midbrain preparation, that molecular-layer activation is radially restricted to a region directly overlying activated regions of the granule cell layer.

NO BEAMS, WHAT THEN?

It is our view that the clear evidence from multiple laboratories collected over almost 35 years that parallel fibers do not produce Purkinje cell beams requires a reexamination and reinterpretation of how the cerebellum works. It is also clear, however, that the complexity of what was once considered a relatively straightforward cerebellar circuitry requires the use of more sophisticated tools to uncover function.

The approach we have taken to unravel these structure/function relationships involves an interplay between what we have referred to as structurally realistic models[68] and experimental investigations. Structurally realistic models are defined as models that are fundamentally based on what is known of the anatomical and physiological structure of the biological object of interest, and that explicitly do not assume a particular function in their construction. A primary advantage of the use of models of this type is that new functional ideas often emerge from the exploration of the model, while the intrinsic realism of the model allows these ideas to be directly related back to experimental investigations.[68] Exactly this process has resulted in the slow emergence of a new view of cerebellar cortical circuitry in our laboratory.[29,30] While space does not allow a detailed review of this previous work, the final sections of this paper will outline the main features of this new view and discuss the resulting reinterpretation of traditional ideas about the function of this circuitry. Readers interested in additional technical details are referred to the referenced papers.

The foundation and base for our modeling work has been the development and exploration of a structurally realistic model of the cerebellar Purkinje cell.[69–73] This model, which is based on the actual 3-D anatomy of reconstructed Purkinje cells,[74] was first tuned to replicate previously described responses of Purkinje cells to current and voltage clamp recordings *in vivo* and *in vitro*.[69,70] Again, the reader is referred to the original papers for details of the model's structure, and of these initial experiments to test and tune model performance. Probably, however, the most important understanding to emerge from this work is that synaptic integration in this neuron bears little resemblance to the classical summation of synaptic currents leading to a supra-threshold somatic action potential assumed to take place in most neurons. Instead, the dynamics of the Purkinje cell model are dominated by a complex interaction of synaptic conductances, large intrinsic voltage-dependent conductances found in the dendrite and soma of this cell,[75–78] and the cell's morphology.

The knowledge that Purkinje cell dynamics is complex and related to voltage-dependent ionic currents was, of course, well established experimentally prior to our model's construction.[75–87] What the model allowed us to explore more easily than could be done with pure physiological techniques was how these electrical components might contribute to Purkinje cell output.[88,89] This exploration produced several new ideas concerning the control of Purkinje cell dynamics. First, we found that the generation of *in vivo*–type activity patterns in the model required the presence of continuous low levels of background synaptic input. In fact, we could toggle the behavior of the model back and forth between *in vivo* and *in vitro* types of activity by simply including or removing background synaptic inputs.[88] Perhaps more importantly, realistic Purkinje cell behavior required not only background excitatory "parallel fiber" inputs, but also background inhibitory synaptic inputs of the sort contributed by molecular layer interneurons.[88,89] While these types of inhibitory interneurons have been described from the earliest analysis of cerebellar cortical circuitry,[1,90] they have been much less considered in ideas of cerebellar function.[30] On the basis of our models, we now believe that the influence of parallel fibers on Purkinje cells is intimately tied to the molecular-layer inhibition also converging on Purkinje cell dendrites (see below).

The second set of relationships the model has allowed us to explore are those between the large voltage-dependent calcium, sodium and potassium conductances found in the Purkinje cell's dendrite and soma.[88,89] Surprisingly, we found in the model that the interplay of somatic and dendritic currents results in a net current flow from the soma to the dendrites[89] rather than the other way around as is usually assumed in neurons. On average, the modeled Purkinje cell dendrite was actually a current sink not a source, and most often served as a break on intrinsic somatic firing, rather than a drive on the soma. Again, the Purkinje cell appeared to be anything but a "normal" neuron. The consequences of this unusual reversal of somatic and dendritic roles were most interesting when examining the synaptic effects of background parallel fiber and molecular-layer inputs.[89] While the large voltage-dependent conductances dominate the control of somatic spiking, the state of these conductances were found to be regulated in subtle ways by the background synaptic inputs. Specifically, parallel fiber excitation when combined with molecular-layer inhibition produced what is, in essence, a local membrane voltage clamp controlling the activation state of the large conductances.[89] Working in a push–pull relationship, the balance between parallel fiber excitation and molecular-layer inhibition substantially changes the activation state of intrinsic Ca and K channels, which in turn influences the current flow between dendrite and soma and the spiking output of the Purkinje cell. These rather remarkable results were subsequently supported by experimental studies.[91] Because this synaptic influence is mediated through the intrinsic voltage-dependent conductances, it is also highly indirect,[89] consistent with physiological results already discussed that have failed to show a direct influence of spontaneous parallel fiber activity on Purkinje cell spiking.[23]

The modeling results described so far only consider the state of the model when generating the type of spontaneous firing characteristic of Purkinje cells *in vivo*.[92] However, we have recently begun using the model to try to better understand the response of Purkinje cells to activation of the granule cell layer.[25] Specifically, we first used the model to test the hypothesis that the radial organization of cerebellar cortex was a result of the synchronous activation of ascending granule cell synapses.[25,58] To our surprise, the model tuned previously only on current and voltage clamp data[70] already contained a mechanism that specifically amplified synchronous synaptic inputs.[93] This mechanism was dependent on the same intrinsic Ca and K conductances responsible for regulating the ongoing spontaneous activity of the cell soma. In this case, however, synchronous activation of excitatory granule cell inputs produced a Ca-dependent subspiking threshold propagation of membrane depolarization which, on reaching the soma, could directly influence the generation of action potentials. With multiple simulated stimuli, the short latency responses of modeled and real peri-stimulus time histograms were remarkably similar.

The discovery in the model of a Ca-dependent mechanism that amplified synchronous synaptic input appeared to strongly support the proposal by Llinas[94] that the special influence of the granule cells underlying a particular Purkinje cell was a consequence of the synchronized synaptic inputs for ascending segment axons, although, of course, the model accomplished this feat through a novel mechanism. Accordingly, we then set about testing the other component of the original hypothesis, that the lack of Purkinje cell beams was due to the progressive desynchronization of the parallel fiber volley. This test, however, first required the construction of a network model of cerebellar cortex to generate realistic patterns of parallel fiber input.[96] Here again the model produced surprising results suggesting that desynchronization was, in fact, not enough to suppress beam-like Purkinje cell activity. Instead, even with the maximum likely conduction delays, parallel fiber inputs could also evoke the Ca-dependent amplification mechanism with a subsequent direct influence on somatic spiking. Thus, some mechanism other than or in addition to desynchronization appeared to be responsible for the lack of Purkinje cell beams *in vivo*.

Remarkably, once again, the proper behavior of the parallel fibers required the introduction of inhibition from molecular-layer interneurons. Specifically, when added to the model, the inhibition from molecular-layer inteneurons counterbalanced the effects of parallel fiber activation of the dendrite suppressing the beam.[95] At first this behavior might seem counterintuitive because inhibitory neurons are themselves activated by parallel fibers, and therefore the arrival of inhibition on the dendrite would be expected to be temporally delayed. However, the temporal dispersion of parallel fiber activity, originally evoked by Llinas to account by itself for the lack of parallel fiber beams,[58] instead assures that the one synapse delay in inhibition is inconsequential, allowing inhibition to keep pace with arriving parallel fiber excitation. While the feed-forward nature of this pathway has been noted for many years,[96] it has to our knowledge never been proposed as a means to control or interact with parallel fiber excitation. Instead, molecular-layer inhibition has been viewed more as an independent influence on Purkinje cells.[10] The reader should also note that, in the model, the large nearly synchronous activation of the Purkinje cell dendrite by the ascending segment synapses happens too fast to be counterbalanced by inhibition. Thus, ascending segment synapses are free to invoke the dendritic amplification mechanism and directly influence somatic spiking. Finally, ironically, evidence that inhibition may play such a role is already available in the studies reporting strong parallel effects in *in vitro* brain slice preparations.[51,61,97] The slicing procedures used to process the tissue directly interfere with inhibition, and, in addition, inhibition is often pharmacologically blocked in these experiments.

IMPLICATIONS FOR FUNCTION

As summarized in FIGURE 5, the view of cerebellar cortical circuitry that is emerging from these studies is very different from that assumed by most contemporary models of cerebellar function. Whereas the traditional view of granule cell ef-

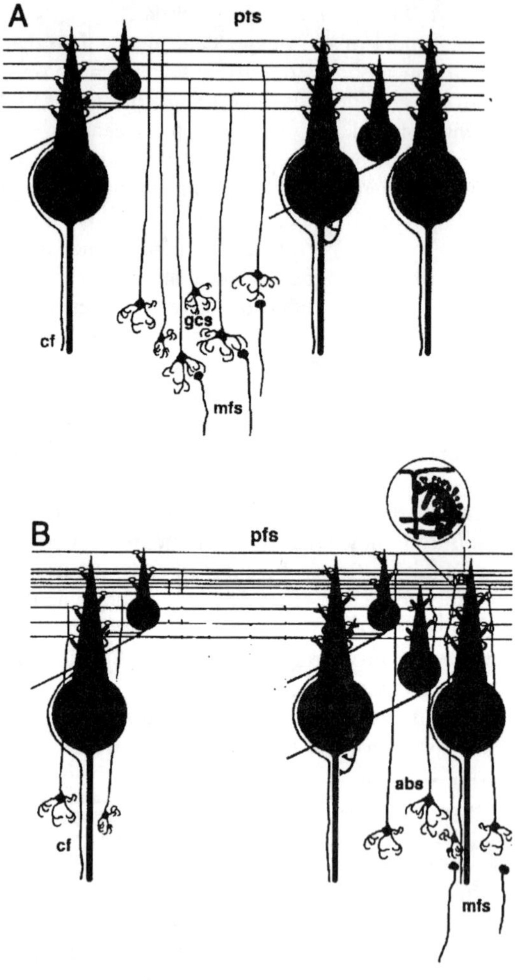

FIGURE 5. The traditional view of cerebellar cortical circuitry (**A**) is contrasted with the ideas emerging from our model-based studies (**B**). The *inset* in B indicates the fact that ascending segment and parallel fiber synapses terminate on different regions of the Purkinje cell dendrite (From Bower.[30] Reproduced by permission.)

fects in cerebellar cortex primarily emphasized the parallel fibers (FIG. 5A), the new view both distinguishes between the ascending and the parallel fiber segment of the granule cell axon, and also provides an important role for molecular-layer inhibition (FIG. 5B). Each of these changes has important functional implications.

The proposal that parallel fiber excitation operates in the context of molecular-layer inhibition substantially changes the way one thinks about cerebellar cortical circuitry. For example, balancing parallel fiber excitation with inhibition solves a conundrum that has bothered cerebellar theorists and physiologists for many years, namely, how can a Purkinje cell cope with an excitatory input with a potential dynamic range of a few to 150,000 excitatory synaptic inputs? If, as our model suggests, cerebellar cortical circuitry assures a balance between parallel fiber excitation and molecular-layer inhibition, then any increase in parallel fiber activity would be naturally counterbalanced. In this way it is not necessary to postulate mechanisms for reducing the total number of effective parallel fiber inputs,[11,12,98] and Purkinje cell behavior at any instant can be influenced by any number of parallel fiber inputs.

The model has also opened up new ways of thinking about the interaction between the different synaptic influences on cerebellar Purkinje cells. When the radial organization of cerebellar circuitry was first described, we specifically proposed that the parallel fiber synapses might be modulatory in nature.[25] However, we had no firm ideas about what type of modulation was involved. The model, however, suggests how ascending and parallel fiber inputs may interact. Specifically, the same Ca channels under local voltage clamp regulation of parallel fibers/inhibition are also responsible for the subthreshold amplification mechanism that relays the synchronous activation of the ascending segment synapses to the soma. Because parallel fiber synapses, in addition, are situated between the distal ascending segment inputs and the more proximal parts of the Purkinje cell dendrite and its soma,[32] there is a clear opportunity for the parallel fiber/inhibitory system to affect the response of the cell to ascending segment inputs. We have recently reported modeling results that demonstrate just such a modulation.[72]

While this paper has not really considered the climbing fiber afferent system, our hypothesis concerning the interaction between ascending and parallel fiber/inhibitory inputs also suggests a quite different role for input from the inferior olive. Modeling results suggest that the long time constants of the intrinsic Ca and K conductances influenced by the parallel fibers result in a Purkinje cell dendrite whose current state is influenced by 10s of msecs of previous parallel fiber activity. The "memory" for this influence is found in the dynamics of the intrinsic calcium-related conductances. The massive increases in free calcium associated with climbing fiber activation will almost certainly reset these dendritic dynamics, wiping out the influence of previous parallel fiber/molecular-layer interneuron patterns. While a resetting function for the climbing fiber has been suggested previously,[99,100] the current modeling results provide a specific mechanism.

CONSEQUENCES FOR EXISTING IDEAS OF CEREBELLAR FUNCTION

If our emerging view of the physiological relationships between granule cell input and Purkinje cell output are correct, it will be necessary to reconsider both existing models, as well as the interpretation of several important strains of cerebellar experimentation. Perhaps one of the most important involves the much studied phenomena of cerebellar long-term depression (LTD).[21] Inspired by the Marr/Albus hypothesis of parallel fiber synaptic learning,[11,12] physiologists have worked for years to demonstrate that parallel fiber synaptic efficacy can be changed under the right physiological conditions.[21] The resulting discovery of LTD has been taken as explicit evidence for cerebellar learning theories.[21] However, as has been pointed out by De Schutter,[101] the apparent need for a balance between parallel fiber excitation and molecular-layer inhibition raises the possibility that LTD is not related to classical ideas of learning at all, but instead provides a mechanism to slowly adjust the overall strength of parallel fiber synaptic inputs in order to maintain this balance. Having such a mechanism dependent on intracellular calcium, as is the case for LTD,[102] would tie its induction directly to the interaction between synaptic and intrinsic dendritic conductances that the model suggests are critically important to function. In fact, while most advocates of LTD's involvement in a Marr/Albus type mechanism remain focused on induction by climbing fiber activity,[17,103,104] it has been shown that granule cell synaptic input alone is capable of increasing intracellular free calcium[105] and at least one form of LTD. The fact that LTD takes many minutes to hours to induce,[106] is most easily induced when inhibition is blocked,[19] and can spread between many nearby parallel fiber synapses[107] all sound more like a long-term "metabolic type" regulation in response to overactive parallel fibers than a synapse-specific learning mechanism. In addition it has been shown that intracellular calcium also potentiates inhibition by molecular-layer interneurons.[108] In other words, Purkinje cells appear to contain Ca-related mechanisms both for decreasing the strength of parallel fiber synapses and increasing the potency of inhibitory inputs. This is exactly the type of regulation one would expect were Purkinje cell behavior dependent on an overall balance between parallel fiber excitation and molecular-layer inhibition. If this view is correct, it means that the many investigators, driven to study LTD by an interest in establishing a Marr/Albus type cerebellar learning mechanism,[109–116] may have actually, in effect, appropriated a fundamentally different mechanism for this purpose.

Finally, we believe this emerging view of cerebellar circuitry pushes strongly in the direction of thinking of this structure as a real-time processing device, whose output is governed strictly by the pattern of inputs it receives. Over the last several years, driven in part by these physiological and modeling results, we have proposed that cerebellar cortex implements an algorithm that evaluates the quality of sensory data being obtained for use by the rest of the nervous system.[29,30] Such an evaluation is likely to intrinsically require that the sensory information being received from one set of peripheral receptors be considered in the context of data simultaneously being received from other sensory receptors involved in the same sensory, motor, or cognitive task. It is our view that this context is provided by the parallel fiber/molecular layer interneuron system, which then modulates Purkinje cell responses produced by ascending segment synapses activated by afferent information arising from specific sensory receptors. Many further modeling and experimental studies will be required, however, to determine the exact nature of the algorithm.

ACKNOWLEDGMENT

This research was funded by NSF Grant RO1 NS37109.

REFERENCES

1. CAJAL, S. 1911. Histologie du System Nerveux de l'Homme et des Vertebres. Maloine. Paris.
2. BLOEDEL, J.R. & J. COURVILLE. 1981. Cerebellar Afferent Systems *In* Handbook of Physiology: The Nervous System, vol II. V.B. Brooks, Ed.: 735–829. American Physiological Society. Bethesda, MD.
3. PALAY, S.L. & P. CHAN. 1974. Cerebellar Cortex: Cytology and Organization. Springer. Berlin.
4. PALKOVITS, M., P. MAGYAR & J. SZENTAGOTHAI. 1971. Quantitative histological analysis of the cerebellar cortex in the cat. I. Number and arrangement in space of the Purkinje cells. Brain Res. **32:** 1–13.
5. PALKOVITS, M., P. MAGYAR & J. SZENTAGOTHAI. 1971. Quantitative histological analysis of the cerebellar cortex in the cat. II. Cell numbers and densities in the granular layer. Brain Res. **32:** 15–30.
6. PALKOVITS, M., P. MAGYAR & J. SZENTAGOTHAI. 1971. Quantitative histological analysis of the cerebellar cortex in the cat. III. Structural organization of the molecular layer. Brain Res. **34:** 1–18.
7. PALKOVITS, M., P. MAGYAR & J. SZENTAGOTHAI. 1972. Quantitative histological analysis of the cerebellar cortex in the cat. IV. Mossy fiber-Purkinje cell numerical transfer. Brain Res. **45:** 15–29.
8. SCHILD, R.F. 1980. Length of the parallel fibers in rat cerebellar cortex. J. Physiol. **303:** 25P.
9. MUGNAINI, E. 1972. The histology and cytology of the cerebellar cortex. *In* The Comparative Anatomy and Histology of the Cerebellum: The Human Cerebellum, Cerebellar Connections, and Cerebellar Cortex. O.L.A.J. Jansen, Ed.: 201–251. The University of Minnesota Press. Minneapolis, MN.
10. ECCLES, J.C., M. ITO & J. SZENTAGOTHAI. 1967. The Cerebellum as a Neuronal Machine. Springer. Berlin.
11. MARR, D. 1969. A theory of cerebellar cortex. J. Physiol. (London) **202:** 437–471.
12. ALBUS, J.S. 1971. A theory of cerebellar function. Math. Biosci. **10:** 25–61.
13. FUJITA, M. 1982. Adaptive filter model of the cerebellum. Biol. Cybern. **45:** 195–206.
14. ITO, M. 1984. The Cerebellum and Neural Control. Raven Press. New York.
15. KAWATO, M. & H. GOMI. 1992. A computational model of 4 regions of the cerebellum based on feedback-error learning: biological cybernetics. **68(2):** 95–103.
16. KILLEEN, P. & J. FETTERMAN. 1993. The behavioral theory of timing-transition analyses: a journal of the experimental analysis of behavior. **59(2):** 411–422.
17. MAUK, M.D. & N.H. DONEGAN. 1997. A model of pavlovian eyelid conditioning based on the synaptic organization of the cerebellum. Learn. Mem. **3:** 130–158.
18. THOMPSON, R.F. 1988. The neural basis of basic associative learning of discrete behavioral responses. Trends Neurosci. **11:** 152–155.
19. KARACHOT, L., R. KADO & M. ITO. 1994. Stimulus parameters for induction of long-term depression in in vitro rat Purkinje-cells. Neurosci. Res. **21(2):** 161—168.
20. DESCHUTTER, E. & R. MAEX. 1996. Cerebellum: cortical processing and theory. Curr. Opin. Neurobiol. **6:** 759–764.
21. ITO, M. 2001. Cerebellar long-term depression: characterization, signal transduction, and functional roles. Physiol. Rev. **81:** 1143–1195.
22. RALL, W. 1964. Theoretical significance of dendritic trees for neuronal input-output relations. *In* Neuronal Theory and Modeling. R.F. Reiss, Ed.: 73–97. Stanford University Press. Stanford, CA.
23. BELL, C. & R. GRIMM. 1969. Discharge properties of Purkinje cells recorded on single and double microelectrodes. J. Neurophysiol. **32:** 1044–1055.
24. ECCLES, J.C. *et al.* 1972. Investigation on integration of mossy fiber inputs to Purkyne cells in the anterior lobe. Exp. Brain Res. **13:** 54–77.
25. BOWER, J.M. & D.C. WOOLSTON. 1983. Congruence of spatial organization of tactile projections to granule cell and Purkinje cell layers of cerebellar hemispheres of the albino rat: vertical organization of cerebellar cortex. J. Neurophysiol. **49:** 745–766.
26. KOLB, F.P. *et al.* 1997. Spatial distribution of field potential profiles in the cat cerebellar cortex evoked by peripheral and central inputs. Neuroscience **81:** 1155–1181

27. COHEN, D. & Y. YAROM. 1998. Patches of synchronized activity in the cerebellar cortex evoked by mossy-fiber stimulation: questioning the role of parallel fibers. Proc. Natl. Acad. Sci. USA **95:** 15032–15036.
28. JAEGER, D. & J.M. BOWER. 1994. Prolonged responses in rat cerebellar Purkinje cells following activation of the granule cell layer: an intracellular in vitro and in vivo investigation. Exp. Brain Res., **100:** 200–214.
29. BOWER, J.M. 1997. Is the cerebellum sensory for motor's sake, or motor for sensory's sake? The view from the whiskers of a rat. Progr. Brain Res. **114:** 483–516.
30. BOWER, J.M. 1997. The cerebellum and the control of sensory data aquisition. *In* International Review of Neurobiology, vol 41: The Cerebellum and Cognition. J. Schmahmann, Ed.: 489–513. Academic Press. San Diego, CA.
31. HARVEY, R.J. & R.M.A. NAPPER. 1991. Quantitative studies of the mammalian cerebellum. Progr. Neurobiol. **36:** 437–463.
32. GUNDAPPA-SULUR, G., E. DESCHUTTER & J.M. BOWER. 1999. The ascending branch of the granule cell axon: a significant component in cerebellar cortical circuitry. J. Comp. Neurosci. **406:** 580–596.
33. LLINÁS, R. 1969. Neurobiology of Cerebellar Evolution and Development. American Medical Association/Education Research Foundation. Chicago, IL.
34. BRAITENBERG, V. & R.P. ATWOOD. 1958. Morphological observations on the cerebellar cortex. J. Comp. Neurol. **109:** 1–33.
35. FREEMAN, J.A. & C.N. NICHOLSON. 1970. Space-time transformaiton in the frog cerebellum through an intrinsic tapped delay-line. Nature (Lond.) **226:** 640–642.
36. ECCLES, J.C. 1973. The cerebellum as a computer: patterns in space and time. J. Physiol. (Lond.) **229:** 1–32.
37. PELLIONISZ, A. & R. LLINÁS. 1980. Tensorial approach to the geometry of brain function: cerebellar coordination via a metric tensor. Neuroscience **5:** 1125–1138.
38. LLINÁS, R. 1984. Functional significance of the basic cerebellar circuit in motor coordination. *In* Cerebellar Functions. J.R. Bloedel, Ed.: 171–185. Springer Verlag. New York.
39. GABBIANI, F., J. MIDTGAARD & T. KNÎPFEL. 1994. Synaptic integration in a model of cerebellar granule cells. J. Neurophysiol. **72:** 999–1009.
40. BRAITENBERG, V., D. HECK & F. SULTAN. 1997. The detection and generation of sequences as a key to cerebellar function: experiments and theory. Behav. Brain Sci. **20.**
41. KEELE, S. & R. IVRY. Does the cerebellum provide a common computation for diverse tasks: a timing hypothesis. Ann. N.Y. Acad. Sci. 1990. **608:** 179–211.
42. IVRY, R.B. *et al.* 2002. The cerebellum and event timing. This volume.
43. BRINDLEY, G.S. 1964. The use made by the cerebellum of the information it receives from the sense organs. IBRO Bull. **3:** 80.
44. THACH, W.T. 1998. A role for the cerebellum in learning movement coordination. Neurobiol. Learn. Mem. **70:** 177–188.
45. THACH, W.T. 1998. What is the role of the cerebellum in motor learning and cognition? Trends Cogn. Sci. **2:** 331–337.
46. PAULIN, M.G. 1997. The role of the cerebellum in neural representation of moving systems. *In* The Cerebellum and Cognition. J. Schmahmann, Ed.: 515–533. Academic Press. New York.
47. HOUK, J.C. 1988. Schema for motor control utilizing a network model of the cerebellum. *In* Neural Information Processing Systems. D.Z. Anderson, Ed.: 367–376. American Institute of Physics. New York.
48. ECCLES, J.C., R. LLINÁS, & K. SASAKI. 1966. Parallel fibre stimulation and the responses induced thereby in the Purkinje cells of the cerebellum. Exp. Brain. Res. **1:** 17–29.
49. HANSON, C.L., G. CHEN & T.J. EBNER. 2000. Role of climbing fibers in determining the spatial patterns of activation in the cerebellar cortex to peripheral stimulation: an optical imaging study. Neuroscience **96:** 317–331.
50. CHEN, G., C.L. HANSON & T.J. EBNER. 1998. Optical responses evoked by cerebellar surface stimulation *in vivo* using neutral red. Neuroscience **84:** 645–668.
51. HECK, D. 1999. Sequential stimulation of rat and guinea pig cerebellar granular cells in vitro leads to increasing population activity in parallel fibers. Neurosci. Lett. **263:** 137–140.

52. SNIDER, R.S. & A. STOWELL. 1942. Evidence of a representation of tactile sensibility in the cerebellum of the cat. Fed. Proc. Fed. Am. Soc. Exp. Biol. **1:** 82–83.
53. SNIDER, R.S. & A. STOWELL. 1944. Receiving areas of the tactile, auditory, and visual systems in the cerebellum. J. Neurophysiol. **7:** 331–357.
54. WELKER, W. 1987. Spatial organization of somatosensory projections to granule cell cerebellar cortex: functional and connectional implications of fractured somatotopy (summary of Wisconsin studies). *In* New Concepts in Cerebellar Neurobiology. J.S. King, Ed.: 239–280. A.R. Liss. New York.
55. SHAMBES, G.M., J.M. GIBSON & W. WELKER. 1978. Fractured somatotopy in granule cell tactile areas of rat cerebellar hemispheres revealed by micromapping. Brain Behav. Evol. **15:** 94–140.
56. SHAMBES, G.M., D.H. BEERMANN & W. WELKER. 1978. Multiple tactile areas in cerebellar cortex: another patchy cutaneous projection to granule cell columns in rats. Brain Res. **157:** 123–128.
57. BOWER, J., D. WOOLSTON & J. GIBSON. 1980. Congruence of spatial patterns of receptive field projections to Purkinje cell and granule cell layers in the cerebellar hemispheres of the rat. Soc. Neurosci. Abstr. **6:** 511.
58. LLINÁS, R. 1982. General discussion: radial connectivity in the cerebellar cortex: a novel view regarding the functional organization of the molecular layer. *In* The Cerebellum: New Vistas (Exp. Brain Res. Suppl. Vol. 6). S.L. Palay & V. Chan-Palay, Eds.: 189-194. Springer Verlag. New York.
59. MUGNAINI, E. 1972. The histology and cytology of the cerebellar cortex. *In* The Comparitive Anatomy and Histology of the Cerebellum: The Human Cerebellum, Cerebellar Connections, and Cerebellar Cortex. O. Larsell & J. Jansen, Eds.: 201-262. University of Minnesota Press. Minneapolis, MN.
60. NAPPER, R.M.A. & R.J. HARVEY. 1988. Number of parallel fiber synapses on an individual Purkinje cell in the cerebellum of the rat. J. Comp. Neurol. **274:** 168–177.
61. VRANESIC, I. *et al.* 1994. Signal transmission in the parallel fiber Purkinje-cell system visualized by high-resolution imaging. Proc. Natl. Acad. Sci. USA **91(26):** 13014–13017.
62. GARWICZ, M. & G. ANDERSSON. 1992. Spread of synaptic activity along parallel fibres in cat cerebellar anterior lobe. Exp. Brain Res. **88:** 615–622.
63. EKEROT, C.F. & H. JORNTELL. 2001. Parallel fibre receptive fields of Purkinje cells and interneurons are climbing fibre-specific. Eur. J. Neurosci. **13:** 1303–1310.
64. KETCHUM, K.L. & L.B. HABERLY. 1993. Membrane currents evoked by afferent fiber stimulation in rat piriform cortex. I. Current source-density analysis. J. Neurophysiol. **69:** 248-260.
65. ALBUS, J.S. 1971. A theory of cerebellar function. Math. Biosci. **10:** 25–61.
66. BROWN, I.E. & J.M. BOWER. 2000. Congruence of mossy fiber and climbing fiber tactile projections in the lateral hemispheres of the rat cerebellum. J. Comp. Neurol. **429:** 59–70.
67. SHARP, F.R. *et al.* 1989. C-fos expression and [^{14}C]2-deoxyglucose uptake in the caudal cerebellum of the rat during motor/sensory cortex stimulation. J. Comp. Neurol. **284:** 621–636.
68. BOWER, J.M. 1990. Reverse engineering the nervous system: an anatomical, physiological, and computer based approach. *In* An Introduction to Neural and Electronic Networks, S. Zornetzer, J. Davis & C. Lau, Eds.: 3–24. Academic Press. New York.
69. DE SCHUTTER, E. & J.M. BOWER. 1994. An active membrane model of the cerebellar Purkinje cell II. Simulation of synaptic responses. J. Neurophysiol. **71:** 401–419.
70. DE SCHUTTER, E. & J.M. BOWER. 1994. An active membrane model of the cerebellar Purkinje cell I. Simulation of current clamp in slice. J.Neurophysiol. **71:** 375–400.
71. SANTAMARIA, F. & J.M. BOWER. 2002. Why don't parallel fibers activate beams of Purkinje cells? A combined computational and experimental investigation. J. Neurosci. Manuscript under review.
72. SANTAMARIA, F. *et al.* 2002. Modulatory effects of parallel fiber and molecular layer interneuron synaptic activity on Purkinje cell responses to ascending segment input: a modeling study. J. Comp. Neurosci. In press.
73. JAEGER, D., E. DE SCHUTTER & J.M. BOWER. 1997. The role of synaptic and voltage-gated currents in the control of Purkinje cell spiking: a modeling study. J. Neurosci. **17:** 91–106.

74. RAPP, M., I. SEGEV & Y. YAROM. 1994. Physiology, morphology and detailed passive models of guinea-pig cerebellar Purkinje cells. J. Physiol. (Lond.) **474:** 101–118.
75. LLINÁS, R. & M. SUGIMORI. 1980. Electrophysiological properties of in vitro Purkinje cell somata in mammalian cerebellar slices. J. Physiol. (Lond.) **305:** 171–195.
76. LLINÁS, R. & M. SUGIMORI. 1992. The electrophysiology of the cerebellar Purkinje cell revisited. *In* The Cerebellum Revisited. R. Llinas & C. Sotelo, Eds.: 167–181. Springer Verlag. New York.
77. SUGIMORI, M., A.R. KAY & R. LLINÁS. 1994. The persistent Na+ current in cerebellar Purkinje cells has a single channel conductance distinct from the inactivating current. Soc. Neurosci. Abstr. **20** [part 1]: 63.
78. SUGIMORI, M. & R. LLINÁS. 1992. Dual patch-clamping of mammalian Purkinje cells in cerebellar slices. Soc. Neurosci. Abstr. **18** [part 2]: 1358.
79. HOUNSGAARD, J. & C. YAMAMOTO. 1979. Dendritic spikes in Purkinje cells of the guinea pig cerebellum studied in vitro. Exp. Brain Res. **37:** 387–398.
80. LLINAS, R. & M. SUGIMORI. 1980. Electrophysiological properties of in vitro Purkinje cell dendrites in mammalian cerebellar slices. J. Physiol. (Lond.) **305:** 197–213.
81. EKEROT, C., F. & O. OSCARSSON. 1981. Prolonged depolarization elicited in Purkinje cell dendrites by climbing fibre impulses in the cat. J. Physiol. (Lond.) **318:** 207–221.
82. CRÇPEL, F., J. DUPONT & R. GARDETTE. 1983. Voltage clamp analysis of the effect of excitatory amino acids and derivatives on Purkinje cell dendrites in rat cerebellar slices maintained in vitro. Brain Res. **279:** 311–315.
83. CAMPBELL, N.C. & G. HESSLOW. 1984. Plateau potentials evoked by climbing-fibre stimulation are restricted to the Purkinje cell dendrites of the cat. Neurosci. Lett. **45:** 187–192.
84. HIRANO, T. & H. OHMORI. 1986. Voltage-gated and synaptic currents in rat Purkinje cells in dissociated cell cultures. Proc. Natl. Acad. Sci. USA **83:** 1945–1949.
85. ROSS, W.N. & R. WERMAN. 1987. Mapping calcium transients in the dendrites of Purkinje cells from the guinea-pig cerebellum in vitro. J. Physiol. **389:** 319–336.
86. CONNOR, J.A. & H.Y. TSENG. 1988. Measurement of intracellular Ca2+ in cerebellar Purkinje neurons in culture: resting distribution and response to glutamate. Brain Res. Bull. **21:** 353–361.
87. TANK, D.W. *et al.* 1988. Spatially resolved calcium dynamics of mammalian Purkinje cells in cerebellar slice. Science **242**: 773–777.
88. DESCHUTTER, E. & J. BOWER. 1994. An active membrane model of the cerebellar purkinje-cell 2 simulation of synaptic responses. J. Neurophysiol. **71**(1): 401–419.
89. JAEGER, D., E. DESCHUTTER & J.M. BOWER. 1996. The role of synaptic and voltage-gated currents in the control of Purkinje cell spiking: a modeling study. J. Neurosci. In press.
90. SULTAN, F. & J.M. BOWER. 1998. Quantitative Golgi study of the rat cerebellar molecular layer interneurons using principal component analysis. J. Comp. Neurol. **393:** 353–373.
91. JAEGER, D. & J.M. BOWER. 1999. Synaptic control of spiking in cerebellar Purkinje cells: dynamic current clamp based on model conductances. J. Neuroscience. **19:** 6090–6101.
92. WATERHOUSE, B.D. & J.G. MCELLIGOTT. 1980. Simple spike activity of Purkinje cells in the posterior vermis of awake cats during spontaneous saccadic eye movements. Brain Res. Bull. **5:** 159–168.
93. DESCHUTTER, E. & J.M. BOWER. 1994. Simulated responses of cerebellar Purkinje cells are independent of the dendritic location of granule cell synaptic inputs. Proc. Natl. Acad. Sci. USA **91:** 4736–4740.
94. LLINÁS, R. 1982. General discussion: radial connectivity in the cerebellar cortex: a novel view regarding the functional organization of the molecular layer. *In* The Cerebellum: New Vistas (Exp. Brain Res. Suppl. Vol. 6), S.L. Palay & V. Chan-Palay, Eds.: 189–194. Springer Verlag. New York.
95. SANTAMARIA, F. & J.M. BOWER. 1997. A realistic cerebellar network: The temporal and spatial, direct and indirect influences of granule cell axons on Purkinje cells. Soc. Neurosci. Abst. **23:** 2364.

96. ECCLES, J.C., R. LLINÁS & K. SASAKI. 1966. The inhibitory interneurons within the cerebellar cortex. Exp. Brain. Res. **1:** 1–16.
97. BARBOUR, B. 1993. Synaptic currents evoked in Purkinje cells by stimulating individual granule cells. Neuron **11:** 759–769.
98. ITO, M. 1996. Cerebellar long-term depression. Trends Neurosci. **19**(1):11–12.
99. GIBSON, A.R. & R.S. GELLMAN. 1987. Functional implications of inferior olivary response properties. *In* Cerebellum and Neuronal Plasticity. M. Glickstein, C. Yeo & J. Stein, Eds.: 119-140. Plenum Press. New York.
100. BLOEDEL, J.R. & W.J. ROBERTS. 1971. Action of climbing fibers in cerebellar cortex of the cat. J. Neurophysiol. **34:** 17–31.
101. DESCHUTTER, E. 1995. Cerebellar long-term depression might normalize excitation of Purkinje-cells: a hypothesis. Trends Neurosci. **18**(7): 291119-140.295.
102. KONNERTH, A., J. DREESSEN & G.J. AUGUSTINE. 1992. Brief dendritic calcium signals initiate long-lasting synaptic depression in cerebellar Purkinje cells. Proc. Natl. Acad. Sci. USA. 1992. **89:** 7051–7055.
103. CHAPEAUBLONDEAU, F. AND G. CHAUVET. 1991. A neural network model of the cerebellar cortex performing dynamic associations. Biological cybernetics. **65**(4): 267–279.
104. SCHREURS, B. & D. ALKON. 1993. Rabbit cerebellar slice analysis of long-term depression and its role in classical-conditioning. Brain Res. **631**(2): 235–240.
105. HARTELL, N. 1996. Strong activation of parallel fibers produces localized calcium transients and a form of ltd that spreads to distant synapses. Neuron **16**(3): 601–610.
106. KARACHOT, L. *et al.* 2001. Induction of long term depression in cerebellar Purkinje cells requires a quickly turned over protein. J. Neurophysiol. **86:** 280–289.
107. REYNOLDS, T. & N.A. HARTELL. 2000. An evaluation of the synapse specificity of long term depression in rat cerebellar slices. J. Physiol. (Lond.) **527:** 563-577.
108. KANO, M. 1994. Calcium-induced long-lasting potentiation of gabaergic currents in cerebellar purkinje-cells. Jpn. J. Physiol. **44:** S131–S136.
109. ITO, M. Long-term depression. Annu. Rev. Neurosci. 1989. **12**: 85-102.
110. IBA, A. *et al.* 1994. Deficient cerebellar long-term depression and impaired motor learning in mglur1 mutant mice. Cell **79**(2): 377–388.
111. EKEROT, C. & M. KANO. 1985. Long-term depression of parallel fibre synapses following stimulation of climbing fibres. Brain Res. **342:** 357–360.
112. KANO, M. & M. KATO. 1998. Mode of induction of long-term depression at parallel fibre-Purkinje cell synapses in rabbit cerebellar cortex. Neurosci. Res. **5**: 544—556.
113. LINDEN, D. 1994. Long-term synaptic depression in the mammalian brain. Neuron **12**(3): 457–472.
114. SAKURAI, M. 1989. Depression and potentiation of parallel fiber-Purkinje cell transmission in in vitro cerebellar slices. Exp. Brain Res. **17:** 220–230.
115. SCHREURS, B.G. & D.L. ALKON. 1993. Rabbit cerebellar slice analysis of long-term depression and its role in classical conditioning. Brain Res. **631:** 235–240.
116. SHIBUKI, K., *et al.* 1996. Deficient cerebellar long-term depression, impaired eyeblink conditioning, and normal motor coordination in gfap mutant mice. Neuron. **16**(3): 587–599.

Dynamic Correlation of Neuronal Activity in Rat Cerebellar Cortex Modulated by Behavior

DETLEF HECK,[a] FELIX KÜMMELL,[a] WILLIAM T. THACH,[b] AND AD AERTSEN[a]

[a]*University of Freiburg, Institute of Biology III, Neurobiology & Biophysics, 79104 Freiburg, Germany*

[b]*Washington University School of Medicine, Department of Anatomy and Neurobiology, St. Louis, Missouri, USA*

ABSTRACT: Sprague-Dawley rats (2–4 months old) were trained to perform a reaching–grasping task while their head was fixated and multielectrode recordings were performed in the ipsilateral cerebellar hemisphere. Multiunit (MU) activity was recorded with 2–3 electrodes in the Purkinje cell layer at various depths (1.5–4.5 mm) from the pial surface while the animal either performed the reaching–grasping task or was at rest. Recording sites were visually aligned along the transverse or the sagittal axis of Crus IIa. Excess correlations of MU spike activity were calculated at 10-ms time resolution using the joint peristimulus time histogram (JPSTH), to reveal the dynamics of cross-correlations corrected for nonstationarities in spike rates. Peak correlation amplitudes and areas were calculated separately for a time period preceding the movement and for a succeeding period during which the movement occurred. Correlations were compared across different paradigms: transversal versus sagittal alignment of recordings sites and behavior versus rest. No significant differences were found between transversally and sagittally aligned recording sites. Significant differences in peak correlation amplitude and/or peak area, however, were found both between premovement and movement time, as well as between both these time periods and periods while the animal was at rest.

KEYWORDS: cerebellum; correlation dynamics; timing; parallel fibers; multielectrode recording; awake rat; reaching-grasping

INTRODUCTION

The cerebellar cortex is wired in such a way that all intrinsic excitatory fibers, that is, the parallel fibers, run in the transversal direction, whereas the inhibitory axons extend in the sagittal plane.[1] The functional significance of this arrangement, especially the effect of parallel fibers on the spike firing of Purkinje cells, is still under debate.[2–7] Cells aligned along the axis of parallel fibers receive largely overlapping

Address for correspondence: Dr. Detlef Heck, University of Freiburg, Institute of Biology III, Neurobiology & Biophysics, Schaenzlestr. 1, 79104 Freiburg, Germany. Phone: +49 761 203 2862; fax: +49 761 203 2860.

heck@biologie.uni-freiburg.de

Ann. N.Y. Acad. Sci. 978: 156–163 (2002).

excitatory input,[8,9] which, if sufficiently strong, could drive the cells to correlated spike firing. We investigated the temporal correlation of neuronal spike activity in the Purkinje cell layer of awake-behaving rats using multielectrode recordings and dynamic correlation analysis.[10] Multiunit (MU) spike activity was recorded simultaneously from 2–3 sites that were either aligned with the transversal axis of Crus IIa or perpendicular to that axis. The temporal correlation of spike activity was calculated at 10-ms resolution for time periods preceding the movement and during the movement, and for a control period during which the animal was at rest. Multiunit spike activities at different sites exhibited temporally loose correlations, with peak widths of cross-correlograms in the order of 100 ms. These correlations were strongly modulated during behavior, but did not differ significantly between transversally and sagittally aligned recording sites. The independence of these loose temporal correlations off the highly asymmetric cerebellar cortical anatomy suggests that they may be caused by external sources providing common input to the observed neurons.

METHODS

Surgery and Behavior

Sprague-Dawley rats (2–4 months old) were trained to reach for a small food pellet.[11] Rats were kept in a training cage with a grid wall. Food pellets were offered outside the cage in front of the grid wall. Rats learned to reach for the food pellets in one to three training sessions, each session lasting 4 h. After successful training, screws for head fixation and a recording chamber were mounted on the skull using acrylic cement. On day 3 after surgery, the animals were set in a head-fixating device, and trained to reach for and grasp food pellets during head fixation. The pellets were manually set down on a capacitance switch mounted in front of the rat (about 1 cm below the mouth). The switch was set off by the rat's paw touching the food pellet, and the status of the capacitance switch (open or closed) was recorded together with the spike data and later used for event-triggered spike-train analysis (see below). The behavioral performance was taped on video for off-line classification of reaching–grasping movements. A light-emitting diode, visible in the video recordings, was connected to the capacitance switch, and lit up when the paw touched the pellet. Only successful trials, that is, grasping of the pellet in one reach, were included in the event-triggered spik-train analysis. Because of the manual delivery of food pellets, the rats were able to predict pellet presentation by the experimenter's movement, and typically reached for the pellet as soon as it was set down on the switch.

Electrophysiology

Multiunit neuronal activity in cerebellar cortex was recorded with 2–3 electrodes simultaneously using a multielectrode recording system (System Eckhorn, Uwe Thomas Recording, Germany). Electrodes, aligned in a row and spaced at 305 μm, were lowered into the ipsilateral hemisphere 1.5 to 4.5 mm below the pial surface (Crus 1), aiming at the paramedian lobe. Penetration depth was controlled and registered for each electrode individually with a resolution of 1 μm. Recordings from the Purkinje cell layer were accepted when complex spikes could be clearly identified.[12]

All electrodes were placed in the same lobule as confirmed by penetration depth, and the number of times the Purkinje cell layer was penetrated. The assembly of electrodes was inserted either in a transversal (n = 14) or a sagittal (n = 9) orientation. Orientation of the electrodes was adjusted visually in relation to the longitudinal axis of Crus IIa. Deviations from alignment with the parallel fibers or the perpendicular axis, however, may have occurred in cases when the electrodes were advanced more posteriorly. Multiunit activity was recorded while the animal performed the reaching–grasping task or while at rest, that is, with no food pellet presented. About 10 min of resting data were recorded at least 5 min after completion of a series of reaches. Animals were awake and alert during the resting period, but no food pellets were presented. Data obtained from four different experimental paradigms: recordings along or perpendicular to the transverse axis, while the animal was at rest or behaving, were analyzed separately. Spike data traces were bandpass filtered (0.1 Hz–5 kHz) and digitized at 20 kHz (CED 1401, Cambridge Electronic Design, UK).

Data Analysis

To calculate excess correlation in data with nonstationary rates, as typically observed in behaving animals, we used the joint-peristimulus time histogram (JPSTH)[10] (FIG. 1). With this technique, correlation values are normalized for rate modulations within trial and rate variability across trails.[10,13,14] To perform event-related correlation analysis on data from resting periods without behavioral events, artificial event triggers were generated at pseudorandom intervals (minimum interval 4.2 s; maximum interval 5 s). JPSTH analysis of the behavioral data revealed on average a high correlation 0.5 s and earlier before the touching of the food pellet, and a clearly reduced correlation around the time of the touch. We therefore calculated the average correlations separately for the time periods −1.5 to −0.5 s before and ± 0.5 s around the touch event. Since the delay between movement onset and touching of the food pellet was between 100 and 300 ms, we considered the two periods representing time before and during the reaching–grasping movement, respectively.

Average cross-correlograms were smoothed with a Gaussian (σ = 20 ms). After smoothing the peak value, the deviation of peak time from zero time lag and the area under the correlation peak were measured within the boundaries defined by the average width at half maximum.

RESULTS

Neuronal activity modulated during the reaching–grasping task was recorded at 30 different sites at various depths in three animals. Behavior-related modulation of MU firing rate was most pronounced between 3.5- and 4.5-mm penetration depth, that is, presumably in the paramedian lobe. The most common observation was an increase in firing rate around the time of the grasping movement (FIG. 1). Rate increases typically lasted several hundred milliseconds, and occurred between about 1 s before and 1 s after the paw touched the food pellet. Time of occurrence of the rate change relative to the touch varied with recording sites.

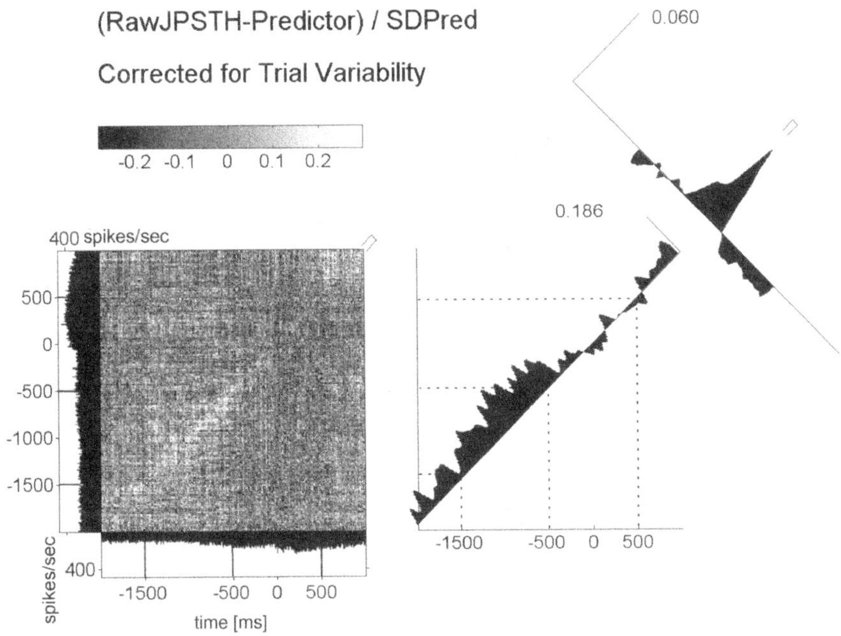

FIGURE 1. Joint-peri-stimulus-time histogram (JPSTH) reveals the dynamics of excess spike correlation after correcting for rate modulations within trial and rate variability across trials. The abscissa and ordinate each show a standard peri-stimulus-time histogram (PSTHs) of multiunit spike recordings aligned on the behavioral trigger (time zero = touching of the food pellet). Each element in the JPSTH matrix gives in gray scale the correlation value for a given time delay between the two spike trains. The elements aligned along the main diagonal (*bottom left to upper right*) hold the values for correlations around zero time delay, that is, synchronous firing. The main diagonal is shown again in histogram format to the right of the matrix. As clearly visible in the matrix as well as in the histogram of the main diagonal, excess correlation of neuronal activity was higher during the period preceding the touch than during the period around the touch. Correlation analysis was therefore performed separately on two time intervals (−1.5 to −0.5 and −0.5 to 0.5 s), as indicated by the *lines* in the two PSTHs and the *dashed lines* in the graph showing the smoothed diagonal. The *correlation histogram* on the upper right is the result of averaging over all elements of the matrix in a direction parallel to the main diagonal. It represents the average excess correlation across the entire 3 s trial duration. Its peak amplitude and the area under the peak (calculated within the limits of the full width at half maximum) were measured separately for the premovement and the movement period, and compared across experimental paradigms. Binwidth in the PSTH and JPSTH matrix is 10 ms. Binwidth in the average cross-correlogram is 5 ms. Correlation histograms were smoothed with a Gaussian (sigma = 20 ms).

Correlation Analysis

Excess correlation of multiunit activity recorded on two electrodes separated by 305 or 610 mm was calculated using the JPSTH with a binwidth of 10 ms (FIG. 1). Peak area and maximum of the smoothed average cross-correlograms were analyzed and compared across paradigms and time periods relative to the touching of the food pellet (FIG. 2). Neuronal activity was clearly correlated during both the behavioral

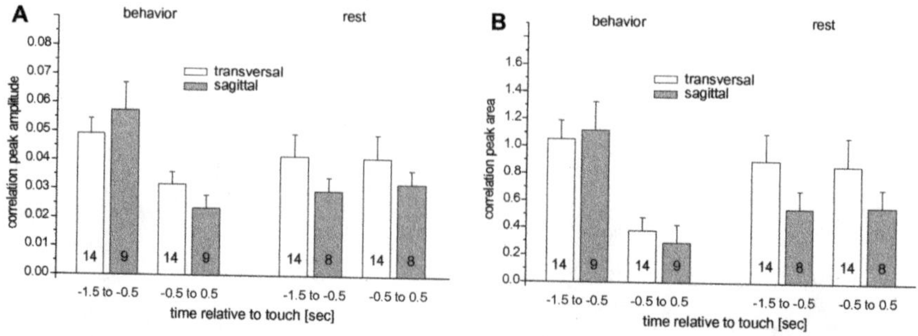

FIGURE 2. Correlation of neuronal activity recorded from sites aligned with the parallel fibers (transversal) or perpendicular to that axis (sagittal). Data were analyzed separately for the period preceding the reaching for the food pellet (−1.5 to −0.5 s) and the time around the touch (−0.5 to 0.5 s). Event-triggered analysis of data recorded during rest was based on pseudorandom event triggers. Average excess correlations for the two time periods were calculated using the JPSTH (cf. FIG. 1). After smoothing the correlation histogram, the (**A**) peak amplitude and (**B**) the area under the peak were measured. No significant differences (Wilcoxon) in peak amplitude or peak area could be found comparing transversally with sagittally aligned recording sites. Numbers of measurements are shown in each *bar*. *Error bars* show S.E.M.

and the resting periods. The temporal precision of the correlation, however, was rather low, with an average width of the correlation peak at half maximum of 204 ± 105 ms.

Neuronal correlation was strongly modulated during behavior. The delay between movement onset and the touching of the food pellet was about 100 to 300 ms. Visual inspection of the JPSTH results revealed a markedly stronger correlation of MU spike activity during the period −1.5 to −0.5 s before the rat touched the food pellet (premovement) than during the period −0.5 to 0.5 s around the touch (during the reaching–grasping movement) (FIG. 1). Therefore, all correlation analyses were performed separately on data obtained during the premovement and the movement intervals.

We measured the correlation peak amplitude (FIG. 2A) and the correlation peak area (FIG. 2B) for sagittally and transversally aligned recording sites and compared the values across three different sets of data: (1) data recorded before the reaching movement, (2) during the reaching–grasping movement, and (3) data recorded while the animal was at rest, but awake. No significant differences between transversal or sagittal alignment of recording sites were found (FIG. 2A and 2B). We also tested for a dependence of the correlation on distance between recording sites. Neither the peak amplitude nor the peak area were significantly different when comparing recording sites separated by 305 or 610 μm (data not shown).

Highly significant differences, however, in both correlation peak area and correlation peak amplitude were found when comparing the periods before and during the reach (FIG. 3A and 3B). Both correlation measures were significantly smaller ($p < 0.01$) during the reaching–grasping movement than during the time immediately preceding the movement.

A comparison of behavioral and control (rest) data revealed that neuronal activity recorded during the resting period was on average less correlated than during the pe-

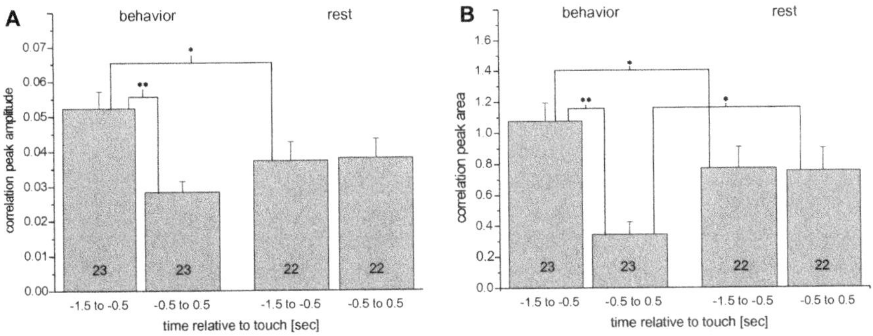

FIGURE 3. Comparison of the (**A**) excess correlation peak amplitude and (**B**) correlation peak area of neuronal activity measured during the time periods preceding and during the reaching–grasping movement, with the correlation during rest. In the behavioral paradigm, excess correlation peak amplitude as well as peak area were significantly higher during the period preceding the movement (−1.5 to −0.5 s) than during movement (−0.5 to 0.5 s). Peak correlation preceding the touch was significantly higher than the peak correlation during a period of identical duration at rest. Compared to the control situation at rest, correlation peak area was significantly higher during the period preceding the touch and significantly lower during the period around the touch ($**p < 0.01$; $*p < 0.05$). Numbers of measurements are shown in each *bar*. *Error bars* show S.E.M.

riod preceding the reaching–grasping movement in the behavioral paradigm. Both the mean correlation peak area and correlation peak amplitude were significantly higher in the period preceding the reach than during rest ($p < 0.05$; FIG. 3A and 3B). On the other hand, the correlation during the reaching movement was significantly lower than during rest. This difference, however, was significant only for the correlation peak area ($p < 0.05$; FIG. 3B), but not significant when comparing the correlation peak amplitude (FIG. 3A).

DISCUSSION

We have shown here that the correlation of neuronal activity in the Purkinje cell layer of the cerebellum undergoes dynamic changes during a reaching–grasping task. Comparing the excess spike correlation during behavior with data obtained during rest, we found a higher than resting correlation during the time immediately preceding the movement and a lower than resting correlation during the actual movement.

In our experiments we used electrodes spaced 305 μm apart to record simultaneously from 2–3 sites along the transverse axis, that is, in line with the parallel fibers or along the sagittal axis, that is, aligned with the inhibitory axons. Typically, neurons in the Purkinje cell layer mostly increased their firing rate in relation to the reaching–grasping movement. Other than expected from a "beam"-like organization,[15] where lateral inhibition would occur between neighboring areas in the sagittal direction, there was no obvious difference in rate change patterns between transversal and sagittal recordings.[11]

In the present study we have shown that the temporal correlation of neuronal activity at a 10-ms time resolution during rest or behavior was not different when comparing recording sites aligned along the transversal or along the sagittal axis. The precision, however, with which we could align the electrodes with the anatomical axes of the cerebellar cortex was limited, especially in deep recordings. Vos and colleagues[16] have reported differences in the correlation of Golgi cell activity (expressed in *z*-scores[17]) when comparing recordings along the transversal and the sagittal axis. They recorded from anaesthetized animals and verified the alignment of their recording sites anatomically. Using the same *z*-score analysis as Vos *et al.* on our data, however, we also found significantly higher peak correlations along the transversal compared to the sagittal axis, both during behavior and during the resting period. This difference was no longer significant, however, when we calculated excess correlation using the joint-PSTH correcting for rate nonstationarities in the data (both rate modulations within trial and rate variability across trials). It is thus possible that nonstationarities of rates are responsible for the differences between the transversal and sagittal recordings reported by Vos *et al.*

The excess correlation of multiunit activity was strongly modulated during the reaching–grasping movement. The movement, that is, lifting the paw off the ground, starts between 100 and 300 ms before the touch occurs. The correlation of neuronal activity was significantly higher immediately before than during the reaching–grasping movement. The correlation during the premovement period was also significantly higher than during the control or resting period, indicating that the time shortly before the movement was not equivalent to a resting state. The increased correlation measured during the premovement time may reflect a state of anticipation, similar to what has been observed by Riehle *et al.*[18] in the primary motor cortex of behaving monkeys. We supplied the food pellets manually. Therefore the experimenters' movements were most likely an early preparatory signal (see methods). During the actual reaching–grasping movement, excess correlation was significantly reduced, both compared to the correlation measured in the premovement period and also compared to the resting state.

SUMMARY

Neurons in the Purkinje cell layer of awake rats showed a behavior-related modulation of spike correlation at a 10-ms time resolution. The dynamics of this correlation were independent of the alignment of the recording sites in the transverse or sagittal direction, and were also independent of the distance between recording sites. The rather loose temporal correlations observed in this study could be brought about either by intrinsic cerebellar cortical connections, equally affecting sagittal and transversal recording sites, or by external sources. However, given the highly asymmetric wiring of the cerebellar cortex, the first possibility appears unlikely. We therefore suggest that the observed correlations and their dynamics are due to activity in the cerebellar afferents, that is, the mossy-fiber system, possibly affecting loose spike correlation through the rising parts of granule cell axons.[2,3] We have, however, also observed temporal correlations in cerebellar activity occurring at submillisecond precision.[19] The dependence of these highly precise correlations on recording-site alignment and their dynamics during behavior are to be investigated.

ACKNOWLEDGMENTS

This research was supported by NS RO1-1777 and by the Deutsche Forschungsgemeinschaft (DFG). We thank Dr. R. Clark for considerable help with animal training and video analysis of behavior, and Drs. Fahad Sultan and Jeff Keating for helpful comments on earlier versions of the manuscript.

REFERENCES

1. PALAY, S.L. & V. CHAN-PALAY. 1974. Cerebellar Cortex. Cytology and Organization. Springer-Verlag. Berlin.
2. LLINÁS, R. 1982. Radial connectivity in the cerebellar cortex: a novel view regarding the functional organization of the molecular layer. *In* The Cerebellum, New Vistas. S.L. Palay and V.Chan-Palay, Eds.:189–192. Springer-Verlag. New York.
3. BOWER, J.M. & D.C. WOOLSTON. 1983. Congruence of spatial organization of tactile projections to granule cell and Purkinje cell layers of cerebellar hemispheres of the albino rat: Vertical organization of cerebellar cortex. J. Neurophysiol. **49:** 745–766.
4. GARWICZ, M. & G. ANDERSSON. 1992. Spread of synaptic activity along parallel fibres in cat cerebellar anterior lobe Exp. Brain Res. **88:** 615–622.
5. HECK, D. 1995. Sequential input to guinea pig cerebellar cortex *in vitro* strongly affects Purkinje cells via parallel fibers Naturwissenschaften **82:** 201–203.
6. BRAITENBERG, V., D. HECK & F. SULTAN. 1997. The detection and generation of sequences as a key to cerebellar function. Experiments and theory. Behav. Brain Sci. **20:** 229–245.
7. HECK, D. & F. SULTAN. 2002. Cerebellar structure and fuction: making sense of parallel fibers Hum. Mov. Sci. In press.
8. HARVEY, R.J. & R.M.A. NAPPER. 1991. Quantitative studies on the mammalian cerebellum. Prog. Neurobiol. **36:** 437–463.
9. NAPPER, R.M.A. & R.J. HARVEY. 1988. Number of parallel fiber synapses on an individual Purkinje cell in the cerebellum of the rat. J. Comp. Neurol. **274:** 168–177.
10. AERTSEN, A.M.H.J., G.L. GERSTEIN, M.D. HABIB & G. PALM. 1989. Dynamics of neuronal firing correlation: modulation of "effective connectivity." J. Neurophysiol. **61:** 900–917.
11. HECK, D., R. CLARK & W.T. THACH. 1999. Multi-electrode recording in rat cerebellar cortex during a reaching-grasping task [Abstr.]. Soc. Neurosci. Abstr. **25:** 149.5.
12. THACH, W.T. 1968. Discharge of cerebellar Purkinje and nuclear neurons during rapidly alternating arm movements in the monkey J. Neurophysiol. **31:** 785–797.
13. VAADIA, E., I. HAALMAN, M. ABELES, *et al.* 1995. Dynamics of neuronal interactions in monkey cortex in relation to behavioural events Nature **373:** 515–518.
14. BRODY, C.D. 1999. Disambiguating Different Covariation Types. Neural Comput. **11:** 1527–1535.
15. ECCLES, J.C., M. ITO & J. SZENTÁGOTHAI. 1967. The Cerebellum as a Neuronal Machine. Springer-Verlag Berlin.
16. VOS, B., R. MAEX, A. VOLNY-LURAGHI & E. DE SCHUTTER. 1999. Parallel fibers synchronize spontaneous activity in cerebellar golgi cells. J. Neurosci. **19:** RC6.
17. MELSSEN, W.J. & W.J.M. EPPING. 1987. Detection and estimation of neural connectivity based on crosscorrelation analysis. Biol. Cybern. **57:** 403-–414.
18. RIEHLE, A., S. GRÜN, M. DIESMANN & A. AERTSEN. 1997. Spike synchronization and rate modulation differentially involved in motor cortical function. Science **278:** 1950–1953.
19. BELL, C.C. & T. KAWASAKI. 1972. Relations among climbing fiber responses of nearby Purkinje cells J. Neurophysiol. **35:** 155–169.

Temporal Organization of Activity in the Cerebellar Cortex: A Manifesto for Synchrony

PHILIPPE ISOPE, STÉPHANE DIEUDONNÉ, AND BORIS BARBOUR

Laboratoire de Neurobiologie (CNRS UMR 8544), Ecole Normale Supérieure, 75230 Paris Cedex 05, France

ABSTRACT: The issues of temporal coding and the temporal organization of activity have aroused a great deal of interest in sensory systems, cortex, thalamus, and hippocampus. Strangely, despite the important timing roles attributed to the cerebellum, little consideration has been given to the organization of activity within the cerebellar circuitry. In fact, there is evidence of a remarkable temporal patterning of activity in even the earliest cerebellar recordings. The evidence for the existence of high-frequency oscillations in the cerebellar cortex is reviewed and possible mechanisms are discussed; one involves the synchrony of parallel fiber inputs to Purkinje cells. It is shown how synchronous and oscillatory activity can enable extremely precise timing and also how they can maximize the information storage capacity of the cerebellar cortex.

KEYWORDS: cerebellum; cerebellar cortex; Purkinje cells; information storage; high-frequency oscillations; parallel fibers; climbing fibers

EVIDENCE FOR SYNCHRONY

In one of the first ever electrophysiological recordings of the cerebellum, Adrian[1] placed a silver-ball electrode upon the cerebellar cortical surface and recorded the signals reproduced in FIGURE 1. Well-organized oscillations of potential were observed, with characteristic frequencies between 150 and 250 Hz (in the figure the frequency is about 180 Hz). Particularly telling is the comparison with potentials recorded from the cerebral cortex, for which the now well-known oscillations of around "40 Hz" were observed (60 Hz in the figure). Given the type of recording, it is certain that the oscillatory field potentials reflect synchronous activity in a population of neurons. In light of the knowledge we have accumulated about the cerebellar cortex in the intervening 68 years, we shall try to explain the possible origin and function of these high-frequency signals.

A few studies reproduced the findings of Adrian,[2–5] but overall it seems that the phenomenon was forgotten during the war and in the postwar excitement of single-

Address for correspondence: Boris Barbour, Laboratoire de Neurobiologie (CNRS UMR 8544), Ecole Normale Supérieure, 46 rue d'Ulm, 75230 Paris Cedex 05, France. Voice: +33 1 44 32 37 36; fax: +33 1 44 32 38 87.

barbour@ens.fr

Ann. N.Y. Acad. Sci. 978: 164–174 (2002). © 2002 New York Academy of Sciences.

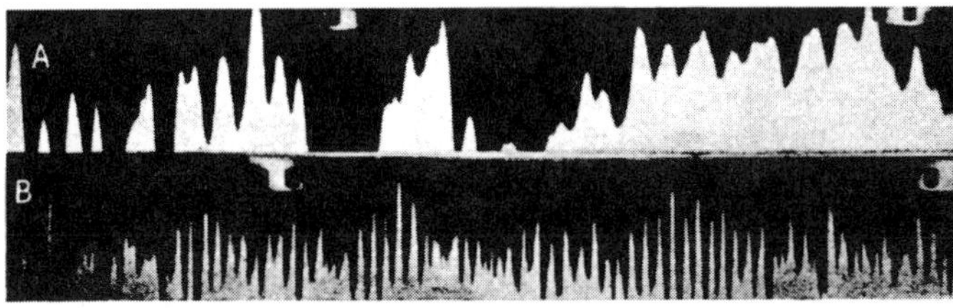

FIGURE 1. The first evidence for high-frequency field potential oscillations in the cerebullar cortex. Cat under chloroform and ether. (**A**) Spontaneous potential waves from cerebral cortex (parietal), about 60/s. (**B**) From cerebellar cortex (vermis), about 180s. White marks at top of records give ¼-s interval. [Reproduced from Adrian (1935) with the kind permission of The Physiological Society.]

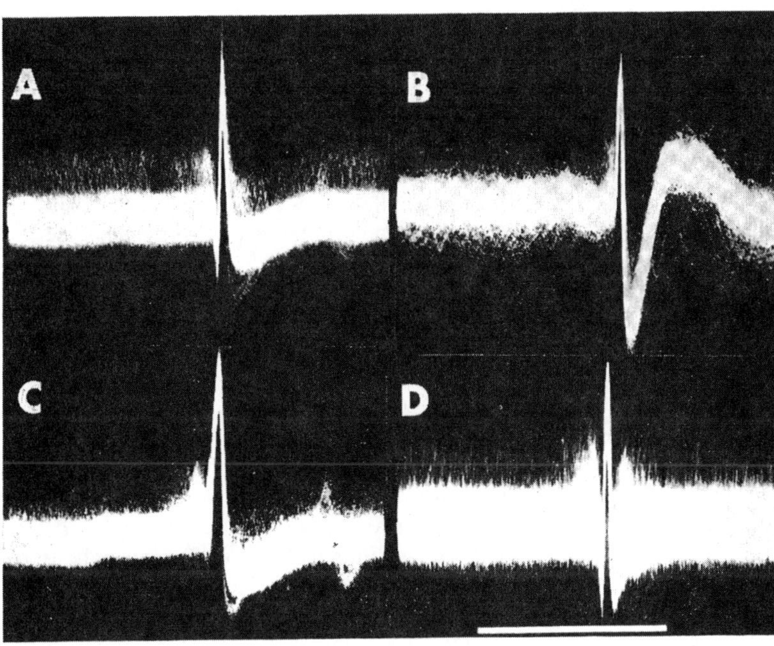

FIGURE 2. Evidence for synchrony and oscillations in multiunit recordings. Relationship between small and large amplitude spikes recorded on the same electrode. Each picture is made by superimposing sweeps triggered 10 ms before the large spike. Calibration mark is 10 ms. [Reproduced from Bell and Grimm (1969) with the kind permission of the *Journal of Neurophysiology*.]

unit and intracellular recordings. The next important findings relating to these oscillations were studies of correlated activity between Purkinje cells.[6,7] FIGURE 2 shows some specimen data from one of these articles.[6] In each recording, action potentials from a pair of cells were recorded on a single electrode. The two cells can be distinguished on the basis of action-potential amplitude. The superimposed traces were triggered on the larger spike. In each case, a clear correlation exists between the two spikes, so there is synchronous activity in the cerebellar cortex. Its precision is remarkable: the correlations are easily exact to within a millisecond. In panels A and B, all action potentials were identified as arising from Purkinje cells, on the basis of complex spikes, so the activity of Purkinje cells, at least, can be precisely correlated. In panels C and D, only the larger spike in panel C was identified as a Purkinje cell, so it is possible that other cells also participate in the synchronous activity. A striking observation can be made in panel C: precisely timed occurrences of the smaller spike appear about 6 ms after the larger spike (and about 7 ms after the position of the synchronous smaller spikes). There can be little doubt that this well-defined interval is related to the high-frequency population oscillations recorded by Adrian.

What circuit elements could underlie such oscillatory activity? Purkinje cells obviously contribute. Because Adrian was recording at the cerebellar cortical surface and it is generally accepted that field potentials tend to reflect synaptic activity rather than action potentials, one's attention is immediately drawn to the parallel fiber system. Parallel fiber–Purkinje cell synapses are present at very high densities, even at the surface of the cerebellar cortex. Parallel fibers also contact molecular layer interneurons (stellate cell and basket cells) and Golgi cell dendrites. Stellate and basket cells make inhibitory connections upon Purkinje cells, themselves, and Golgi cells. Purkinje cells can also make intracortical inhibitory connections. If we ignore the various cell types for a moment, we can see that in the molecular layer we still have a very simple circuit. Excitatory parallel fibers excite all cells, which are all inhibitory. The essential question is what type of interplay exists between the excitatory parallel fibers and the inhibitory systems.

It should be pointed out that in trying to understand the high-frequency oscillations, we can safely ignore the climbing fiber system, because we know that climbing fibers are only active at very low frequencies.

In the hippocampus, brief "ripples," field-potential oscillations of similar high frequency (200 Hz), have been described.[8–11] In that circuit the principal element is thought to be an electrically coupled network of interneurons that inhibit pyramidal calls at the somatic level.[12–14] During periods of intense excitatory input, the interneurons become active (synchronously) at high frequencies and essentially "chop up" action potential emission via pyramidal cell somata. Recently, it has been suggested that electrical coupling between pyramidal cell axons may also play a role.[15]

To what extent can this proposed mechanism of a synchronously active interneuron network be transplanted to the cerebellar cortex? Basket cells provide somatic and even axonal inhibition of Purkinje cells. It has also long been known that they are coupled by gap junctions in the "pinceau." Although it has proved difficult to demonstrate physiologically any electrical coupling of molecular layer interneurons in the rat, convincing evidence has been provided for the guinea pig.[16,17] The cerebellum (of the guinea pig at least) therefore possesses an analogous interneuron system, driven in a feed-forward manner by the parallel fibers. This network is possibly sufficient to explain the correlations observed between Purkinje cells (and other cells). However, because the synchronizing event in this mechanism is the onset of inhibition, the most difficult events to synchronize are the action potentials, which occur some time later, following recovery from inhibition. It is unclear whether this indirect method of synchronization could really generate the precise correlations observed.

THE HYPOTHESIS OF PARALLEL FIBRE SYNCHRONY

In light of this uncertainty, we propose an additional mechanism: *synchronisation of the parallel fiber inputs themselves*. Although this is somewhat speculative, some evidence for it and a plausible generating mechanism exist. The best evidence for synchronized parallel fiber inputs is provided by the paper of Bell and Grimm[6] and is reproduced in FIGURE 3. Two Purkinje cells were recorded with two electrodes separated by a short distance in the parallel fiber direction; traces were triggered on the spikes of one of the Purkinje cells (lower traces). The second Purkinje cell (upper traces) clearly fires synchronous spikes, but these have a bimodal timing distribution. A natural explanation (suggested by the authors) is that the excitatory input can be propagated in either direction along the parallel fibers, altering which of the Purkinje cells is excited first. This would explain the precise timing, the bimodal distribution, and accounts reasonably for the timing difference (cf. the Discussion sec-

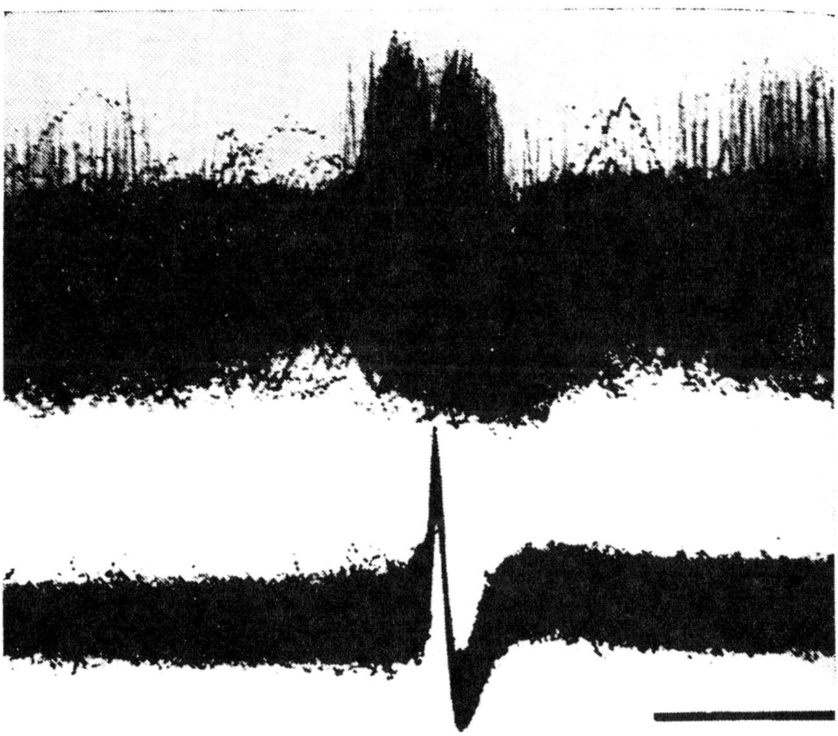

FIGURE 3. Invertible correlations between Purkinje cells are consistent with phasic drive from parallel fibers. Bimodal character of interaction between two Purkinje cells. *Superimposed traces* in which the sweeps were locked to the occurrence of a spike in the lower channel. Calibration mark is 4 ms. [Reproduced from Bell and Grimm (1969) with the kind permission of the *Journal of Neurophysiology*.]

tion of the original paper). However, if we accept this explanation, we must also accept that rather precisely timed events are communicated by the parallel fibers, rather than some diffuse excitation that would have neither a clear onset nor an obvious directionality.

Additional support for the synchronization of parallel fiber inputs comes from a plausible mechanism for its generation. Golgi cells inhibit many if not all of the granule cells in a region of cerebellar cortex. Thus, in addition to regulating the average levels of granule cell activity, as proposed by Marr[18] and Albus,[19] Golgi cells will tend to interrupt periodically granule cell transmission from the region they innervate. This provides an additional method for synchronizing parallel fiber activity. This possibility has been recognized.[20,21] However, both groups seemed to consider a frequency range of around 20 Hz, rather than the frequencies an order of magnitude higher that we seek to explain here. One of the reasons for this discrepancy is that the average frequency of Golgi cell activity *in vivo* is around 20 Hz.[22] However, the De Schutter group has provided evidence that Golgi cells *in vivo* can be active at much higher frequencies, at least for short bursts.[23] These bursts of high-frequency activity (comparable to those under discussion here) were elicited in response to whisker stimulation.

If parallel fibers are active synchronously, why aren't large compound EPSPs observed in intracellular recordings? There are two answers to this question. The first is that it may be necessary to use particular stimulation protocols (cf. the Golgi cell recordings) and possibly anesthetic conditions. The second is that maybe they have been observed. Or rather, depolarizing events occurring at high frequencies in Purkinje cells have been reported. Unfortunately, neither of the two reports is at all conclusive. In some of the first intracellular recordings of Purkinje cells, Granit and Phillips[24] described the occurrence of prepotentials (FIG. 4A) that clearly have too high a frequency to be due to the climbing fiber. Unfortunately, however, Granit and Phillips used KCl electrodes, so it is possible that the intracellular recording was taken after a sufficiently long time for IPSPs to have become inverted. Nevertheless, they also recorded extracellular "giant spikes" (presumably when their microelectrode was between the glial cell sheath and the Purkinje cell membrane). It is stated that these extracellular action potentials were preceded by "prepotentials," which were sometimes subthreshold. These were probably not complex spikes, which were first identified in another part of this work, and the prepotentials may therefore represent large EPSPs recorded in the confined extracellular space. Eccles and coworkers[25,26] also describe such intracellular "potentials;" they are clearly visible in the recording shown in FIGURE 4B. However, the explanation given by Eccles for these potentials is that they "are generated by a rhythmic firing locus on some injured area of the cell, for example a small dendrite damaged by the micro electrode."[26] A clear recording of EPSPs consistent with synchronous parallel fiber activity would constitute an important element of proof of our hypothesis.

A frequent finding is that Purkinje cells *in vivo* show peaky autocorrelograms, indicating a tendency to rhythmic activity. This was reported by Bell and Grimm,[6] and has been observed more recently.[27,28] The relation between this mode of activity and any population synchrony/oscillation in the cerebellar cortex has not been elucidated, although it is obviously suggestive. In general, however, the reported frequencies for the rhythmic activity are somewhat lower than for the population oscillation. Moreover, Purkinje cells *in vitro* appear to be spontaneously active in the absence of

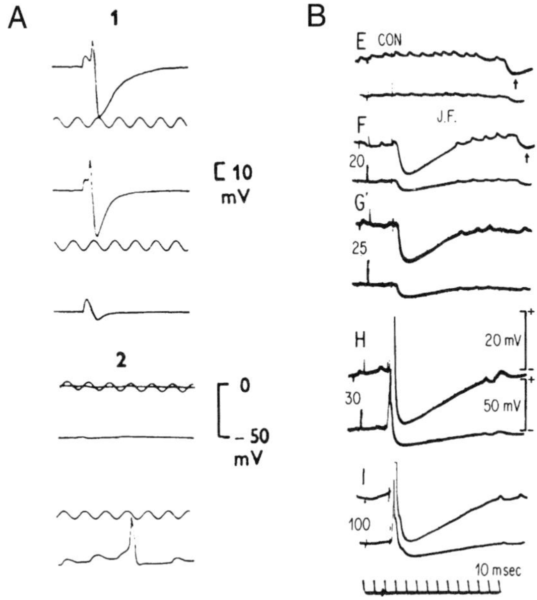

FIGURE 4. Possible intracellular correlates of synchronized parallel fiber inputs recorded in Purkinje cells. (**A**) (1) Giant spike with prepotential, and (*lowest sweep*) prepotential alone, initially positive-going. a.c. amplification. (2) Penetration of the same cell with brief survival. d.c. record. Spontaneous positive-going prepotentials, one of which generates an impulse. (**B**) Inhibitory postsynaptic potentials evoked by LOC and J.F. stimulations. Intracellular recording with a 10 MΩ potassium citrate microelectrode at 800 μ depth and with –50-mV resting potential. All intracellular records are shown at the same high and low amplification. E is a control record of base line. High frequency potential oscillations are observed throughout the panel, but particularly in E. (Reproduced from Granit and Phillips (1956) and Eccles *et al.* (1967) with the kind permission of The Physiological Society and Springer-Verlag.)

all synaptic input.[29] It is thus possible that when deprived of (suprathreshold) synaptic input, Purkinje cells enter a "holding" mode of spontaneous activity, during which the activity of the population of Purkinje cells is presumably not tightly correlated. Such a mechanism would appear quite reasonable, if only to prevent complete disinhibition of neurons in the deep nuclei.

Before discussing the possible benefits to cerebellar function of synchronizing parallel fiber inputs, it is worth making a final observation related to oscillatory activity in the cerebellum. The synapse between parallel fibers and Purkinje cells is thought to be a major site for storing information acquired during motor learning. Storage is thought to involve synaptic plasticity, with most study so far being directed toward long-term depression (LTD).[30] The parallel fiber–Purkinje cell synapse has a number of peculiarities. Of relevance here are the metabotropic receptors (mGluRs) found just outside the synaptic cleft on the postysnaptic (Purkinje cell) membrane[31] and the presynaptic NMDA receptors expressed in a similar location on the parallel fiber varicosity.[32] Activation of both receptor types has been shown to be necessary for the induction of LTD *in vitro*.[33,34]

Most intriguingly, activation of both receptor types requires repetitive parallel fiber activity.[34,35] For the metabotropic receptors, this has simply been attributed to their perisynaptic location: several action potentials would be required to build up enough glutamate to activate them. For the presynaptic NMDA receptors, an additional mechanism comes into play. In order to open, they must bind glutamate, which can occur rapidly, but also change conformation, which only happens slowly (a few milliseconds).[36] In addition, the receptor must be exposed to a depolarized membrane potential in order to relieve a voltage-dependent channel block by external magnesium ions.[37] Because of the presynaptic location of NMDA receptors, the only sufficient depolarization available is an action potential. However, by the time an action potential has released glutamate that has bound to the NMDA receptor and the slow opening reaction of the NMDA receptor has occurred, the parallel fiber will have completely repolarized. A second action potential is therefore necessary to relieve the voltage-dependent block by magnesium ions. To sum up, current studies of plasticity *in vitro* indicate a necessity for at least two action potentials in the same parallel fiber. This suggests that the kind of oscillatory activity just discussed might selectively enable synaptic plasticity at the parallel fiber–Purkinje cell synapse.

POTENTIAL BENEFITS OF SYNCHRONOUS ACTIVITY

In order to illustrate some of the potential benefits to cerebellar operation of synchronous activity, we have constructed a very simple model that contains 200 parallel fibers, 50 interneurons, and 1 Purkinje cell. The parallel fibers all contact the Purkinje cell and contact the interneurons in a probabilistic manner. Sufficient excitation of an interneuron causes it to induce an inhibitory conductance a short while later in the Purkinje cell. Further details are given in the legend to FIGURE 5. We shall compare the behavior of the Purkinje cell as the synchronicity of the parallel fiber input is varied. The times of each action potential (one per parallel fiber) were drawn randomly from a Gaussian probability density function. When the standard deviation of the timing distribution was 0.25 ms, a very reproducible EPSP-IPSP sequence was obtained (FIG. 5A), similar to those recorded by Eccles and coworkers following direct stimulation of parallel fibers. With larger standard deviations (S.D. = 1 ms), the EPSP amplitude became smaller and more variable (FIG. 5B). When the parallel fiber action potentials were spread out even more in time, the IPSP was also affected (not shown).

The variation of mean EPSP amplitude with timing standard deviation is shown in FIG. 5C. As expected, the quick succession of EPSP and IPSP imposes strict coincidence detection upon the Purkinje cell. Truly synchronous inputs are much more effective than even mildly desynchronized inputs. This is very interesting from the point of view of the limits of timing precision in the cerebellum. Such behavior (which could plausibly occur in real tissue) would allow a single cell to distinguish between very similar parallel fiber inputs with slightly different timing properties. Even a slight desynchronization of one or two milliseconds could easily make the difference between subthreshold and suprathreshold. In learning a precise behavior, a Purkinje cell therefore has every interest in exploiting synchronous inputs: they enable precise temporal definition of the learnt behavior controlled by the Purkinje cell; they occupy the least time, allowing more calculations per unit time; they are

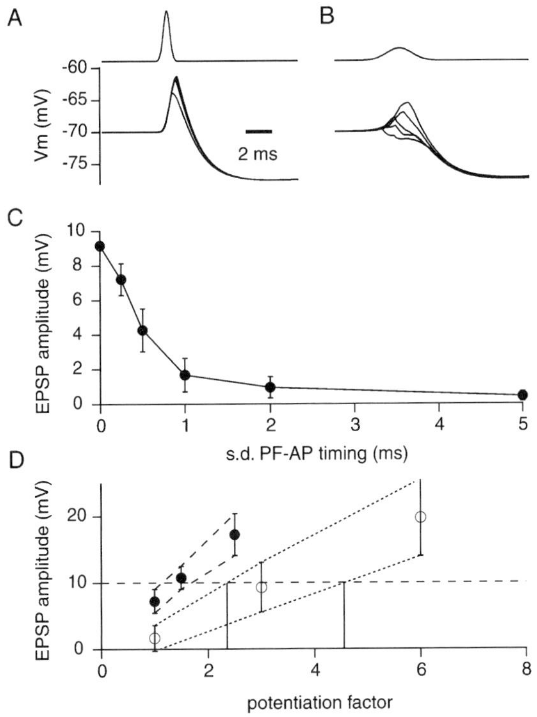

FIGURE 5. A simple model of monosynaptic and di-synaptic (via interneurons) inputs to a Purkinje cell. (**A**), (**B**) Two hundred parallel fibers, each firing a single action potential, induce a biphasic reponse in the Purkinje cell: a monosynaptic EPSP truncated by a di-synaptic IPSP. The degree of synchrony among parallel fibers affects the EPSP amplitude and reliability (parallel fiber action potential timing-distribution is diagrammed at the top of each panel). Highly synchronous inputs (A: S.D. = 0.25 ms) are more effective than more spread-out inputs (B: S.D. = 1 ms). (**C**) Full relationship for several other parameter values. (**D**) Average EPSP amplitudes ±2 S.D. for timing-S.D. = 0.25 ms (*filled circles*) and timing-S.D. = 1 ms (*open circles*) resulting when the standard excitatory conductance of the parallel fiber–Purkinje cell synapses was multiplied by the given potentiation factor. The *dashed horizontal line* represents threshold. In order to give a reliable response, the mean EPSP for an input must be at least 2 S.D. from the threshold. For the less synchronous input (*open circles*), the potentiation factors giving the largest reliable subthreshold response (95% correct) and the smallest reliable suprathreshold response are indicated by *vertical lines*. A potentiation of 92% would be required to make the transition. A smaller value (47% potentiation) applies for the more synchronous input. Methods: 200 parallel fibers all contacted a single Purkinje cell. Each parallel fiber had a 1/20 chance of contacting each of 50 interneurons. All interneurons inhibit the Purkinje cell. The Purkinje cell was modeled as a 1500 pF capacitor resting at –70 mV, while interneurons were 10 pF capacitors resting at –65 mV. Excitatory conductances were exponential with a peak of 2.85 nS (reversal potential at zero) and a time constant of 0.5 ms. The peak inhibitory conductance (per interneuron) was 20 nS (reversal potential at –80 mV). This is unrealistically large, unless interneurons do synchronize their firing via gap junctions (which was not implemented here). A more complex model of a Purkinje cell would implement somatic/axonal inhibition, which might allow lower conductances to be effective. The values used here were adjusted so as to reproduce approximately the EPSP–IPSP sequences reported by Eccles and coworkers following

the most economical in terms of numbers of fibers necessary to evoke a response. In a comprehensive study of di-synaptic inputs in the hippocampus, Pouille and Scanziani[38] concluded that they enforce precise timing of the output. The proposed benefit here is subtly different and nearer the classic notion of coincidence detection: we suggest that they subserve the selection of synchronous inputs.

We wish to draw attention to another less obvious potential benefit of synchronous inputs. Its usefulness depends upon Purkinje cell (and interneuron) behavior being determined by a few parallel fibers. The suggestion that at any given time most parallel fibers are inactive was made both by Marr[18] and Albus.[19] Indirect support has come from work showing that only small numbers of parallel fiber inputs are necessary to excite Purkinje cells and interneurons, at least in young animals.[39] At this conference, we presented our work (Isope and Barbour, unpublished), in which we determined that as few as 150 average parallel fiber inputs will suffice to excite a Purkinje cell in the adult rat. This represents about 0.1% of the total number synapsing on a Purkinje cell. If the assumption is made that a number of parallel fibers of this order is active at any given time, it turns out that the responses in the Purkinje cell will be subject to stochastic noise and that this effect worsens as inputs become desynchronized.

This has consequences for the learning of stereotyped behaviors by the Purkinje cell. It is probable that the role of the Purkinje cell is to learn to associate a particular input with a specific output. If an input must excite the Purkinje cell, the synapses involved must be potentiated to ensure this. Conversely, if an input should not lead to the emission of an action potential, it will be necessary to depress the synapses involved, presumably using LTD. If the responses of the Purkinje cell are to be reliable, the mean potentiated response must always be suprathreshold, while the depressed response must always be subthreshold. It is here that the dependence of variability upon synchronization could be important. The degree of potentiation/depression necessary to make a highly variable response reliable in terms of action potential threshold is greater than for an invariant response (FIG. 5 D). This in turn has repercussions for the overall information storage capacity of the Purkinje cell. Problems arise when synapses are shared between inputs that must give opposite responses. For noisy responses, the learning of one input requires strong depression, while learning the other calls for powerful potentiation. There comes a point of saturation where the conflicting demands made of synapses cannot be met. If responses have low noise, a greater number can be learnt before problems of overlap and saturation arise. Since EPSPs induced by synchronized inputs appear to be less variable than those elicited by desynchronized inputs, it follows that their use would maximize Purkinje cell storage capacity.

("LOC") stimulation of a parallel fiber beam. An interneuron was deemed to fire an action potential when its potential reached –60 mV, inducing its inhibitory conductance in the Purkinje cell with a delay of 0.7 ms. The decay of the inhibitory conductance was set to a slow value (10-ms time constant) in order to allow concentration upon the leading phase of the synaptic response in the Purkinje cell. The modeling was fairly insensitive to the value of resting membrane resistance. The timing of parallel fiber activations was determined by drawing from a Gaussian random distribution with a predefined standard deviation. Knowing the conductance time course for the cells, it was simple to calculate their membrane potential using trapeziodal integration of the differential equation.

Some of the earliest cerebellar recordings provided strong evidence for a high degree of temporal organization of activity in the cerebellar cortex. More recent work suggests that neighboring Purkinje (and other) cells tend to be active synchronously, as part of a high-frequency oscillation in the molecular layer. Feed-forward inhibition, electrical coupling of interneurons, and synchronization of granule cells by Golgi cells are among the possible mechanisms. The cerebellar circuitry is well-adapted to exploit synchronous inputs in precise timing tasks. The synchronization of parallel fiber inputs can also augment the information storage capacity of Purkinje cells.

REFERENCES

1. ADRIAN, E.D. 1935. Discharge frequencies in the cerebral and cerebellar cortex. J. Physiol. (London) **83:** 33P–33P.
2. DOW, R.S. 1938. The electrical activity of the cerebellum and its functional significance. J. Physiol. (London) **94:** 67–86.
3. BROOKHART, J.M., G. MORUZZI & R.S. SNIDER. 1958. Origin of cerebellar waves. J. Neurophysiol. **14:** 181–190.
4. CREPAX, P. & F. INFANTELLINA. 1955. L'attività elettrica del limbo isolato di corteccia cerebellare di gatto. Boll. Soc. Ital. Biol. Sper. **31:** 1229–1231.
5. MORRUZI, G. 1957. Esperimenti e considerazioni sull'elettronarcosi cerebellare. Arch. Sci. Biol. (Bologna) **41:** 91–104.
6. BELL, C.C. & R.J. GRIMM. 1969. Discharge properties of Purkinje cells recorded on single and double microelectrodes. J. Neurophysiol. **32:** 1044–1055.
7. EBNER, T.J. & J.R. BLOEDEL. 1981. Correlation between activity of Purkinje cells and its modification by natural peripheral stimuli. J. Neurophysiol. **45:** 948–961.
8. O'KEEFE, J. 1976. Place units in the hippocampus of freely moving rats. Exp. Neurol. **51:** 78–109.
9. YLINEN, A. *et al.* 1995. Sharp wave-associated high-frequency oscillation (200 Hz) in the intact hippocampus: network and intracellular mechanisms. J. Neurosci. **15:** 30–46.
10. CHROBAK, J.J. & G. BUZSAKI. 1996. High-frequency oscillations in the output networks of the hippocampal-entorhinal axis of the freely behaving rat. J. Neurosci. **16:** 3056–3066.
11. CSICSVARI, J. *et al.* 1999. Fast network oscillations in the hippocampal CA1 region of the behaving rat. J. Neurosci. **19:** RC20.
12. GIBSON, J.R., M. BEIERLEIN & B.W. CONNORS. 1999. Two networks of electrically coupled inhibitory neurons in neocortex. Nature **402:** 75–79.
13. BEIERLEIN, M., J.R. GIBSON & B.W. CONNORS. 2000. A network of electrically coupled interneurons drives synchronized inhibition in neocortex. Nat. Neurosci. **3:** 904–910.
14. GALARRETA, M. & S. HESTRIN. 1999. A network of fast-spiking cells in the neocortex connected by electrical synapses. Nature **402:** 72–75.
15. SCHMITZ, D. *et al.* 2001. Axo-axonal coupling. A novel mechanism for ultrafast neuronal communication. Neuron **31:** 831–840.
16. MANN-METZER, P. & Y. YAROM. 1999. Electrotonic coupling interacts with intrinsic properties to generate synchronized activity in cerebellar networks of inhibitory interneurons. J. Neurosci. **19:** 3298–3306.
17. MANN-METZER, P. & Y. YAROM. 2000. Electrotonic coupling synchronizes interneuron activity in the cerebellar cortex. Prog. Brain Res. **124:** 115–122.
18. MARR, D. 1969. A theory of cerebellar cortex. J. Physiol. **202:** 437–470.
19. ALBUS, J.S. 1971. A theory of cerebellar function. Math. Biosci. **10:** 26–51.
20. DIEUDONNÉ, S. 1998. Submillisecond kinetics and low efficacy of parallel fibre-Golgi cell synaptic currents in the rat cerebellum. J. Physiol. (London) **510:** 845–866.
21. MAEX, R. & E. DE SCHUTTER. 1998. Synchronization of Golgi and granule cell firing in a detailed network model of the cerebellar granule cell layer. J. Neurophysiol. **80:** 2521–2537.

22. EDGELY, S.A. & M. LIDIERTH. 1987. The discharges of cerebellar Golgi cells during locomotion in the cat. J. Physiol. (London) **392:** 315–332.
23. VOS, B.P., A. VOLNY-LURAGHI & E. DE SCHUTTER. 1999. Cerebellar Golgi cells in the rat: receptive fields and timing of responses to facial stimulation. Eur. J. Neurosci. **11:** 2621–2634.
24. GRANIT, R. & C.G. PHILLIPS. 1956. Excitatory and inhibitory processes acting upon individual Purkinje cells of the cerebellum in cats. J. Physiol. (London) **133:** 520–547.
25. ECCLES, J.C., R. LLINAS & K. SASAKI. 1966. Intracellularly recorded responses of the cerebellar Purkinje cells. Exp. Brain Res. **1:** 161–183.
26. ECCLES, J.C., M. ITO & J. SZENTAGOTHAI. 1967. The Cerebellum as a Neuronal Machine. Springer-Verlag. Berlin, Heidelberg, New York.
27. GOOSSENS, J. *et al.* 2001. Expression of protein kinase C inhibitor blocks cerebellar long-term depression without affecting Purkinje cell excitability in alert mice. J. Neurosci. **21:** 5813–5823.
28. DE ZEEUW, C.I. *et al.* 1997. Association between dendritic lamellar bodies and complex spike synchrony in the olivocerebellar system. J. Neurophysiol. **77:** 1747–1758.
29. HAUSSER, M. & B.A. CLARK. 1997. Tonic synaptic inhibition modulates neuronal output pattern and spatiotemporal synaptic integration. Neuron **19:** 665–678.
30. ITO, M. 1989. Long-term depression. Annu. Rev. Neurosci. **12:** 85–102.
31. MATEOS, J.M. *et al.* 2000. Immunolocalization of the mGluR1b splice variant of the metabotropic glutamate receptor 1 at the parallel fibre-Purkinje cell synapse in the rat cerebellar cortex. J. Neurochem. **74:** 1301–1309.
32. CASADO, M., S. DIEUDONNE & P. ASCHER. 2000. Presynaptic N-methyl-D-aspartate receptors at the parallel fiber- Purkinje cell synapse. Proc. Natl. Acad. Sci. USA **97:** 11593–11597.
33. ICHISE, T. *et al.* 2000. mGluR1 in cerebellar Purkinje cells is essential for long-term depression, synapse elimination and motor coordination. Science **288:** 1832–1835.
34. CASADO, M., P. ISOPE & P. ASCHER. 2002. Involvement of presynaptic N-methyl-D-aspartate receptors in cerebellar long-term depression. Neuron **33:** 123–130.
35. TAKECHI, H., J. EILERS & A. KONNERTH. 1998. A new class of synaptic response involving calcium release in dendritic spines. Nature **396:** 757–760.
36. Lester, R.A. *et al.* 1990. Channel kinetics determine the time course of NMDA receptor-mediated synaptic currents. Nature **346:** 565–567.
37. NOWAK, L. *et al.* 1984. Magnesium gates glutamate-activated channels in mouse central neurones. Nature **307:** 462–5.
38. POUILLE, F. & M. SCANZIANI. 2001. Enforcement of temporal fidelity in pyramidal cells by somatic feed-forward inhibition. Science **293:** 1159–1163.
39. BARBOUR, B. 1993. Synaptic currents evoked in Purkinje cells by stimulating individual granule cells. Neuron **11:** 759–769.

In Defense of the Cerebellum

VALENTINO BRAITENBERG

Max-Planck-Institute for Biological Cybernetics, D-72076 Tuebingen, Germany

ABSTRACT: The very special intrinsic connectivity of the cerebellar cortex plays but a minor role in present-day theories of cerebellar function, and it is hardly used as a source of inspiration for experiments. It is argued here that a direct translation of structure into physiological relations inescapably leads to some propositions about cerebellar function that could be tested experimentally.

KEYWORDS: parallel fibers; spatiotemporal patterns; vestibulo-ocular integration

INTRODUCTION

In no other part of the nervous system is the internal connectivity as thoroughly known as in the cerebellar cortex. Moreover, although other cortices are not known in comparable detail, one can confidently assert that the pattern in the cerebellum is unique. The anatomical knowledge, mainly due to Ramon y Cajal,[1] was complemented by an equally exhaustive definition of the basic physiological properties of the various types of cerebellar neurons thanks to the work of Eccles and his associates.[2]

This would seem to make it possible to go directly from the elementary mesh of the cerebellar network to a definition of its global operation, and hence to an explanation of the "functions" of the normal cerebellum and of the "symptoms" of its derailment as they appear to the clinical neurologist. Nobody has succeeded in building this bridge, in spite of some proposals that were seductive in their generality, but too general to serve as an explanation of the uniqueness of the cerebellum.

The stagnation of our theorizing is not caused by lack of experimental findings, which have been forthcoming at an impressive rate in recent years. Rather, it seems that most of the experiments were not so much aimed at an elucidation of the special kind of computation typical for the cerebellum, as at questions that apply to the nervous system everywhere, such as membrane physiology and plasticity on the one hand, and the mapping of input and output connections on the other hand.

In this situation it seems legitimate to make a fresh start by reproposing the level of analysis where the cerebellum is most characteristically itself, the level intermediate between cytology and fiber-bundle tracing—that of the geometry of the intracortical fiber felt.

Address for correspondence: Valentino Braitenberg, Max-Planck-Institute for Biological Cybernetics, Spemannstr. 38, D-72076 Tuebingen, Germany.
valentino.braitenberg@tuebingen.mpg.de

Ann. N.Y. Acad. Sci. 978: 175–183 (2002). © 2002 New York Academy of Sciences.

SOME NUMBERS

The remarkable divergence and convergence of signals inherent in the numerical relations between the various types of neurons, in the ramifications of their dendrites and axons, and in their synaptic connections, have sometimes been invoked as one of the basic features of cerebellar anatomy. However, this cannot be the whole, or even the main, story. Comparison with the cerebral cortex is illuminating.

The number of output fibers originating in the Purkinje cells of the (human) cerebellar cortex is of the order 10^7.[3,4] The same order of magnitude can be assumed for the input fibers (climbing and mossy). Input is relayed to output through 10^{11} interneurons, the vast majority of them granular cells.[3,4]

In the telencephalic cortex the sensory input fibers are a few times 10^6 (of which the visual system has the lion's share). Again, the motor output (the pyramidal tract) is of the same order. All the other (not corticospinal) pyramidal cells, of the order of 10^{10}, are interneurons, with a small (15%) admixture of inhibitory neurons.

Thus, in both cortices the divergence between input and interneurons, as well as the convergence from interneurons to output, is by a factor of 10^4.

The number of synapses in the cerebellar cortex, a few times 10^2 on the axons of 10^{11} granular cells, or a few times 10^5 on the dendrites of 10^7 Purkinje,[5] plus a similar number on the (more numerous) basket, stellate, Golgi cells, turns out to be of the order of 10^{13}.

In the cerebral cortex, the number of synapses is usually assumed to be of the order of 10^{14}. There are 10^4 synapses on each of the 10^{10} pyramidal cells, and fewer on the smaller number of stellate cells.[6] Given the rough rule of a constant density of synapses everywhere in the gray substance, the greater number of synapses in the cerebral cortex, which is about four times as thick as the cerebellar cortex and double in area, is not surprising.

The proportion of synapses residing on spines—85% in the cerebellum[5] and 75% in the cerebral cortex[6]—may be taken as a measure of plasticity, if we interpret spines as embodying the learning element, as some do.

The real surprise is provided by the percentage of inhibitory neurons among all the neurons of the two cortices. Although of the five neuronal types in the cerebellum, four are inhibitory and only one (the granular cell) is excitatory, because of the great numerical preponderance of granular cells, the percentage of inhibitory neurons (Purkinje, plus basket, stellate, and Golgi cells) turns out to be only about 1%. In the telencephalic cortex the estimate for the number of inhibitory (nonpyramidal) neurons is around 15%.[6]

On the whole, these numbers (except perhaps the last pair) are not sufficiently different in the two cortices to serve as an explanation of the striking differences in their geometry.

SOME STRUCTURAL TRAITS

(1) The histology of the cerebellar cortex is uniform throughout. The only possible local exception—quantitative, not qualitative—may be in the tiny flocculus, where deviations from the general scheme have been described both in the density of granular cells and in the length of parallel fibers.[7]

(2) In any small fraction of the cerebellar cortex two directions of the cortical plane are defined by totally different fiber populations oriented at right angles to each other. The only excitatory connections, the parallel fibers, run in the transversal (x) direction, while the axons running in the longitudinal (y) direction are all inhibitory (x, y are intrinsic coordinates of the cerebellar cortex, not everywhere coincident with the macroscopic directions right–left and anteroposterior).

(3) There are no cortico–cortical connections in the white substance underlying the cerebellar cortex, and no intrinsic fibers in the cerebellar–cortical gray substance longer than a few millimeters in both the x and y directions.

(4) All intracortical connections are symmetrically arranged: for every fiber running in the $+x$-direction, there is one in the $-x$-direction. The same goes for $+y$ and $-y$.

(5) The two systems of input fibers distribute their contacts in radically different ways. Climbing fibers target individual neurons; mossy fibers reach extended, overlapping regions, each fiber selecting its sites of termination (the so-called rosettes) within the region of its distribution, in an apparently haphazard way.

PHYSIOLOGICAL DEDUCTIONS

It would be doing the physiologists an injustice if we did not take their findings seriously. Axons conduct at certain speeds, synapses excite or inhibit, dendrites somehow add up their inputs and ultimately cause all or none spikes to be generated in the axon of their own neuron. If we accept all of this, we get an inescapable translation of the neuroanatomy into statements about function. In the case of the cerebellar cortex, this reads as follows.

(1) The basic transformation of input into output signals is the same in the different contexts (motor, vestibular, etc.) in which the cerebellum operates in its various specialized regions.

(2) Activation of neuronal elements relayed through mossy fibers–granular cells spreads in the $+x$ and $-x$ directions along parallel fibers. What spreads in the $+y$ and $-y$ directions is inhibition generated by inhibitory interneurons, one synaptic step farther away from the input. (Spread of inhibition in the y-direction is secondary to spread of excitation along x.)

(3) All interactions are local. The functional principle of the cerebellar cortex must be embodied in a mesh no larger than a few millimeters in any direction.

(4) Interactions in opposite directions along both x and y have the same effect.

(5) Climbing-fiber input reaches the output directly in a point-to-point fashion, bypassing the internal cortical machinery. On the other hand, each mossy fiber excites an extended, discontinuous pattern of separate points in the input to the network.

AN APOLOGY OF PARALLEL FIBERS

End of the Beam, Endless Grooves

Without any doubt, if we take neuroanatomy at face value, shifting signals to the right and left in the parallel-fiber system must be at the root of cerebellar information handling. However, the concept of a "beam" of parallel fibers exciting Purkinje (and other) cells along its way fell into disgrace, perhaps rightly so, if for no other reason but that the term suggests something strong and inflexible. The length of the beam is limited to the length of parallel fibers, which is short compared to the width of the cortical regions in which the relevant interactions take place. Even more importantly, the bulk of parallel fibers reaching an individual Purkinje cell is composed of fibers taking their origin from different places along the "beam" up to a distance corresponding to the length of the parallel-fiber branches. Therefore, activation of granular cells at a restricted site can only produce excitation in a small percentage of all the fibers, perhaps not enough to meet the threshold of Purkinje cells. On the other hand, synchronous activation of a more extended set of granuular cells might also not be effective, for it will be dispersed in time (by up to about 10 ms) due to varying delays in parallel fibers of different length.

The solution to this riddle of an apparent ineffectiveness of the parallel-fiber–Purkinje-cell junction[8,9] comes from the idea of spatiotemporal summation in the parallel-fiber system. Given no more than the anatomy and the elementary tenets of neurophysiology, it is straightforward to define the dynamics of stimuli that will necessarily produce a high density of active parallel fibers in the input to a Purkinje cell. Imagine an active mossy fiber that sets off a few spikes in a group of granular cells a certain distance away from a Purkinje cell. The spikes will travel toward the cell (and away from it, but this is irrelevant) at the characteristic conduction velocity of parallel fibers. Imagine another mossy fiber a little closer to that Purkinje cell activating some more granular cells a little later, at the right time so that their spikes will be synchronized with those of the first group in their parallel travel along the parallel fiber, and so on. Clearly, successive excitation delivered to a transversal row of granular cells, spaced and timed in the right way so as to match conduction velocity in parallel fibers, will be the optimal stimulus.

This mode of operation,[3,10] which makes of the cerebellar cortex a *movement* (*or sequence*) *detector,* has been shown by Heck[11,12] to work in surviving slices of the rat cerebellum. It would be surprising if it did not work in the intact brain, and with natural stimulation, but the experimental verification is still outstanding.

COMMENTS

(a) In the idealized situation of a patch of excitation moving at the right speed across the cerebellum (say, in the "lobulus simplex," where the folia span the whole width of the cerebellum), we expect the density of activity in the parallel fibers, and presumably in their postsynaptic partners, to increase linearly up to a distance corresponding to the length of one parallel-fiber branch. Beyond that, saturation is reached and, if the input keeps "moving" at the right speed, the density remains the same to the end of the movement, perhaps all the way to the margin of the cerebellar

cortex. Thus the effect of "moving" input is not limited to the length of parallel fibers, corresponding to delays of about 10 ms, but to that of the folia, several centimeters in larger cerebella corresponding to times of 100 or more milliseconds. The density at saturation is probably higher for *patterns* of excitation moving in a coherent fashion than for a single moving patch.

(b) To avoid confusion with the outdated concept of a "beam of parallel fibers," we prefer to talk about a *groove* in which spatiotemporal summation takes place along the x-direction of the cerebellar cortex. The term groove seems appropriate not only because, unlike the "beam," it does not suggest a limited length, but also, because of the layout of inhibitory neurons in the y-direction, that is, across the direction of parallel fibers, which can be expected to form walls of inhibition flanking the groove of traveling excitation and keeping the (postsynaptic) excitation in the groove.

(c) The density of active parallel fibers in a groove, once it has reached saturation level, is the same for input "moving" in opposite directions. This is not so for the initial stretch where the build up of activity to saturation occurs. Here obviously "movements" of input excitation in opposite directions produces skewed distributions of activity in the parallel-fiber system that are mirror images of each other.

(d) The interpretation of the cerebellar cortex as a detector of spatiotemporal ("moving") pattern implies sharp tuning for a certain velocity—that of conduction in parallel fibers. This, as far as we know, is the same in all parts of the cerebellar cortex, about 0.5 m/s. If real movement is represented there in various sensory or motor contexts (sequences of muscle activations, retinal slip, moving somatosensory stimuli, etc.), the very different velocities of these movements must be scaled to the intrinsic velocity of the cerebellum by an appropriate magnification factor in the projections.

(e) Afferent excitation "moving" in the input to the cerebellum does not necessarily reflect real movement in some outside space. It may well be that sequences of disparate events (signals from the cortex, proprioceptive feedback, sensory input, motor control signals, etc.) are represented in the mossy-fiber system in such a way that each sequence presented in the right order and with the right timing produces orderly "moving" input in a transversally oriented row of granular cells.

DIFFICULTIES AND OPEN QUESTIONS

If patterned input moving in the direction of the folia is the key to cerebellar function, the question is, in what context is such movement detection relevant? And, are we sure it occurs at all?

The easy answer invokes one of the senses in which outside space is projected in an orderly manner onto a sensory surface in such a way as to transform continuous movement outside in continuous movement of nervous activity across some part of the brain. What comes to mind is the sense of vision. Here indeed movement of the panorama (e.g., due to a shift of gaze) is well known to be one of the variables that enters a computation in which the other important partner is the vestibular system, and for which the cerebellum is largely held responsible. The trouble is that the detection of visual movement (in the cerebral cortex or in the mesencephalon) seems to precede the arrival of the signals at the cerebellum (at least in the flocculus[13]), so that the role of the cerebellar cortex as a movement detector would seem to be superfluous. Besides, in order for the cerebellum to detect movement in the visual field by means of its inherent movement-detecting mechanism, the visual input would have to be displayed somewhere in a topographically correct fashion. Such exact retinotopy to my knowledge has not yet been ascertained anywhere in the cerebellar cortex.

There is a conceptual difficulty in the otherwise very seductive idea of the cerebellum as the site where inputs from different modalities, but referring to the same events, are compared—for example, rotation of the body perceived through the labyrinth and through vision. The dimensionality of the vestibular and of the visual signals is very different. From our knowledge of the periphery of the vestibular sense it would seem that the vestibular channel conveys a total of 10 quantities, referring to three rotations and two linear accelerations measured by the labyrinth on each side (why there are 15,000 fibers in the vestibular nerve is another question). In contrast, the visual channel has an enormous number of dimensions, of the order of millions in the periphery. Where is the dimensionality of the visual information reduced to the few dimensions that can be matched to the vestibular information? Only in the vestibular nuclei, where the two are known to converge? Or in the pretectum? Or in the cerebellar cortex, where the relevant information on movement could be extracted from the visual signal?

If one had invented the cerebellum and tried to put it to good use, one might like to feed the two modalities to be matched to the two sets of input fibers, mossy and climbing. The detection of movement in the visual field could be the business of the mossy-fiber–parallel-fiber channel, and the vestibular signals could come in through the climbing fibers, the two of them reaching their compromise on the Purkinje cells. Unfortunately, the opposite seems to be true, at least in the celebrated flocculus, where the vestibular input is said to be mossy[14] and the visual input climbing.[15]

However, the flocculus is not all there is to the vestibulocerebellum. Actually, a place to look for a detector of a shifting visual panorama would be in a region close to the midline, where the two visual half-fields could be expected to be represented without loss of continuity in the mossy-fiber–parallel-fiber system, which is equally continuous across the midline. Nodulus, uvula, and posterior vermis would be good candidates also for another reason. There the folia, or if you wish, the grooves in the parallel-fiber system, are limited in length to a few millimeters, since there is no anatomical continuity between the medial backward extension of the cerebellar cortex (the posterior vermis, uvula, etc.) and its two lateral backward extensions (the hemispheres, paraflocculus, etc.[3,16,17] The shortness of the grooves, as we have seen [comments (*a*) and (*c*) above], makes movement detection directional, as it should be if the shifting panorama is to be compared to the clearly directional vestibular signal. Over a stretch of a few millimeters "moving" input would sweep up, as it were, spikes on one side or the other of a limited corridor, depending on the direction. To make use of this asymmetry, the output signals of the two halves of the nodulus (posterior vermis or uvula) to the right and left of the midline would have to be conveyed separately to a pair of cerebellar nuclei, which then represent differentially movement in the two opposite directions (with the low dimensionality matching that of the vestibular signals). This much, at least, is supported by anatomy.

But there is another difficulty of a quantitative nature. If the grooves in the medial–posterior cerebellum are not much longer than the parallel fibers, the events they monitor cannot last much longer than 10 ms. Translated into visually perceived movement due to a shift of gaze of about 200°/s, this means that visual movement is detected in a narrow window of about 2°. This makes sense in some contexts of visual–vestibular–optomotor interactions, but not in others.

Two more troubling questions. The macroscopical layout of the (flattened) cerebellar cortex[3,16–18] is very similar in many mammalian species: an anterior lobe, continuous between right and left and generally tapered towards it front end, and three separate backward extensions, the middle one narrow and of fairly uniform width (the posterior vermis), the two lateral ones wider but tapered posteriorly (paraflocculus and flocculus). In larger mammals the anteroposterior extension of the whole cortex is *at least ten times its width*, a fact that clearly demonstrates that it is not the surface area that counts in the evolution of the cerebellum, but something related to the x and y dimensions separately. The strange shape of the cerebellar cortical sheet must be isomorphic to something, but what it is, we don't know. It certainly is not the distribution of masses in the body, nor is it the vector space of movements or any such intuitive notion.

Finally, our idea of the cerebellum or, rather, of the mossy-fibers–parallel-fibers–Purkinje-cells junction as a movement (or sequence) detector does not contribute to our understanding of the role of climbing fibers. Whether they are involved in learning, or whether they have an ongoing role in movement control as a direct route overriding the cerebellar cortical computation in situations of emergency, is an open but legitimate question. One thing is certain: projection of input spaces onto the cerebellum through one or the other channel must have radically different meanings.

CONCLUSION

The inescapable translation of essential features of cerebellar anatomy into statements about function has not yet resulted in a generally accepted view of the basic role of the cerebellum. It may, however, eventually lead to a coherent interpretation of a vast body of experimental facts. For this to happen, fact-finding would have to be increasingly directed toward proof or disproof of clearly stated hypotheses, not about ionic channels and plasticity, but about what makes the cerebellum so different from other parts of the brain. Some of the ideas in this paper may serve this purpose.

If I may express three wishes to the experimental community (to which, alas, I do not belong any more), here they are:

(1) The mapping of external spaces onto the surface of the cerebellum (sensory, proprioceptive, motor homunculus, continuous, or fractured) should concentrate on the mossy-fiber input, with an eye on the rows of input points aligned in the direction of the parallel fibers. It should be studied at a resolution much finer than in many previous studies.

(2) The results of all mapping studies should be projected onto the unfolded surface of the cerebellum in its intrinsic x,y-coordinate system. On such maps we could hope to understand the meaning of the geometry and reciprocal arrangement of the various representations. It is not enough to squash the cerebellum somewhat and pull its lobes apart, as in some illustrations in the literature.

(3) The logic, if there is any, of the apparently haphazard distribution of input points ("rosettes") served by a single mossy fiber should by investigated and its possible origin in an early ontogenetic learning process should be considered.

Finally, to the theoreticians who are willing to apply their mental and possibly mathematical powers to a general theory of the cerebellum, I offer some food for thought.

What sequences of events (internal or external), lasting a few tenths of a second, require precisely timed sequences of motor responses, and different sequences of responses when they are presented in reverse? This question presents itself when we consider the responses of selected Purkinje cells along a groove as timed by "moving" excitation in the mossy-fiber–parallel-fiber system. The selection may be due to synapses of different strength by which they are coupled to the parallel fibers. It is very possible that a particular Purkinje cell receives strong synapses mainly from the left branches of parallel fibers, and another one from right branches, making the selection of active Purkinje cells different for "movement" in opposite directions. HINT: Could this have something to do with the sequence of extensions and flexions in the chain of articulations between the heel and the head (or between the shoulder and the hand, etc.) that corrects for perturbations applied to one end or the other end of the chain?

In terms of mechanics, what is the meaning of the "smoothing of movements," which to some neurologists is the main function of the cerebellum? HINT: It may have something to do with counteracting the inertial (centrifugal, Coriolis, etc.) parassitary forces that necessarily accompany movements in any multijoint system.

What kind of manifold is mapped on the surface of the cerebellar cortex with its two orthogonal coordinates, x and y? In particular: What is the meaning of the anteroposterior dimension? What is represented in the very different laterolateral dimension?

At present we have only fragments of truth about the cerebellum. But before we try to put them together, we should have some idea about what the complete truth might look like.

ACKNOWLEDGMENT

The competent advise of Dr. Fahad Sultan is very gratefully acknowledged.

REFERENCES

1. CAJAL, R. Y. 1911. Histologie du Système Nerveux de l'homme et des Vertébrés. Malone. Paris.
2. ECCLES, J.C., M. ITO & J. SZENTÁGOTHAI. 1967. The Cerebellum as a Neuronal Machine. Springer-Verlag. New York.
3. BRAITENBERG, V. & R.P. ATWOOD. 1958. Morphological observations on the cerebellar cortex. J. Comp. Neurol. **109:** 1–34.
4. ANDERSEN, B.B., L. KORBO & B. PAKKENBERG. 1992. A quantitative study of the human cerebellum with unbiased stereological techniques. J. Comp. Neurol. **326:** 549–560.

5. Harvey, R.J. & R.M.A. Napper. 1991. Quantitative studies on the mammalian cerebellum. Prog. Neurobiol. **36:** 437–463.
6. Braitenberg, V. & A. Schüz. 1998. Cortex: Statistics and Geometry of Neuronal Connectivity. Springer-Verlag. Heidelberg.
7. Llinas, R. 1982. Radial connectivity in the cerebellar cortex: a novel view regarding the functional organization of the molecular layer. *In* The Cerebellum, New Vistas. S.L. Palay and V. Chan-Palay, Eds. Springer-Verlag. New York.
8. Bower, J.M. & D.C. Woolston. 1983. Congruence of spatial organization of tactile projections to granule cell and Purkinje cell layers of cerebellar hemispheres in the albino rat: vertical organization of cerebellar cortex. J. Neurophysiol. **49:** 745–766.
9. Braitenberg, V. 1983. The cerebellum revisited. J. Theor. Neurobiol. **2:** 237–241.
10. Heck, D. 1993. Rat cerebellar cortex in vitro responds specifically to moving stimuli. Neurosci. Lett. **157:** 95–98.
11. Braitenberg, V., D. Heck & F. Sultan. 1997. The detection and generation of sequences as a key to cerebellar function: experiments and theory. Behav. Brain Sci. **20:** 229–277.
12. Noda, H. 1986. Mossy fibres sending retinal-slip, eye and head velocity signals to the flocculus of the monkey. J. Physiol. **379:** 39–60.
13. Ito, M., M. Sakurai & P. Tongroach. 1982. Climbing-fibre induced depression of both mossy-fibre responsiveness and glutamate sensitivity of cerebellar Purkinje cells. J. Physiol. **324:** 113–134.
14. Maekawa, K. & J.I. Simpson. 1973. Climbing fiber responses evoked in vestibulocerebellum of rabbit from visual pathway. J. Neurophysiol. **36:** 649–666.
15. Fahle, M. & V. Braitenberg. 1985. Some quantitative aspects of cerebellar anatomy as a guide to speculation on cerebellar functions. *In* Cerebellar Functions, J.R. Bloedel, J. Dichgans, and W. Precht, Eds. Springer-Verlag. Berlin.
16. Sultan, F. & V. Braitenberg. 1993. Shapes and sizes of different mammalian cerebella. A study in quantitative comparative neuroanatomy. J. Hirnforschung **34:** 79–92.
17. Van Essen, D.C. 2002. Surface-based atlases of cerebellar cortex in human, macaque, and mouse. This volume.

Parallel-Fiber Beams: What Do They Do?

Summary of Presentations and of Discussion

JAMES M. BOWER, *Rapporteur*

The primary issue considered in the formal presentations made prior to this discussion session was the nature of the excitatory influence of the parallel fibers on Purkinje cells. In each presentation, evidence was provided that this influence was more complex than previously assumed by the highly influential "beam hypothesis" formulated more than 50 years ago by Eccles, Ito, and Szentagothai (*The Cerebellum as a Neuronal Machine*, Springer-Verlag, Berlin, 1967).

The overall tone of the discussion section was set well by the moderator, Steve Highstein, who asked the panel, "If the parallel fibers don't produce Purkinje-cell beams, by what mechanism is their influence limited?" Much of the subsequent discussion revolved around this question.

SUMMARY OF DISCUSSION

The panel discussion began by considering the results presented to the meeting by Barbour. Bower started that discussion by wondering how it was that the amplitude of granule-cell-induced responses in Purkinje cells were the same whether the Purkinje cell was close to the activated region of the granule cell layer, or at a distance. He noted that ascending granule-cell axon segments almost certainly provide multiple synaptic inputs to overlying Purkinje cells, so it was not clear why multiple synaptic inputs would produce the same amplitude event as that produced by the one parallel-fiber synapse, presumably contacting a distant Purkinje cell. Braitenburg provided one possible explanation, noting that the experimental procedure used by Barbour left only the superficial-most granule cell layer intact, and that maybe these granule cells branched at the lowest depths of the molecular layer, and therefore only provided one or a few synaptic contacts on overlying Purkinje cells. In response, it was noted by Barbour that experiments labeling individual granule cells in his laboratory and others, found no strict relationship between the position of a granule cell in the granule-cell layer, and the depth of its parallel-fiber bifurcation in the molecular layer.

The discussion then turned to the possibility that the voltage changes measured by Barbour and colleagues were a secondary consequence of the activation of calcium channels as suggested by Bower's Purkinje-cell models, and might not therefore directly reflect the amplitudes of synaptic currents. Barbour stated he thought this

Ann. N.Y. Acad. Sci. 978: 184–187 (2002). © 2002 New York Academy of Sciences.

explanation was unlikely because the potentials they recorded were small. Bower then suggested that this would probably take a model to resolve, as the currents controlling Purkinje-cell membrane potentials are complex.

The first comment from the audience was made by Bell, who asked whether the larger likelihood of getting a Purkinje-cell response from stimulating a granule cell below it might be related to the larger number of synapses it could be expected to make on each Purkinje cell as compared to the parallel fibers. He went on to point out that when the granule-cell layer as a whole was considered, rather than just the 35 microns remaining intact in Barbour's experiments, the influence of the parallel fibers could still be substantial.

Bower then raised concerns about the fact that much of the data presented by Barbour was obtained when inhibition had been pharmacologically blocked, and the modeling and the physiological results Bower had presented suggested that the parallel-fiber system might only function normally in the presence of inhibition. He went on to point out that experiments performed by Barbour in which inhibition was not blocked resulted in large inhibitory effects on Purkinje cells.

With respect to the question of inhibition, DeZeeuw asked Bower from the audience whether his EM studies of the granule-cell axon synapses shed any light on whether the parallel fibers might have a relative preference for contacting molecular-layer inhibitory neurons rather than Purkinje-cell dendrites. Bower responded by saying that his laboratory had not yet looked at this important question experimentally, but he would not be surprised if a considerable number of the parallel-fiber contacts were made on inhibitory interneurons. Braitenburg then pointed out that, while no one had done this experiment, the fact that there are about 50 times more inhibitory neurons in the molecular layer than Purkinje cells would lead one to suspect that there might be a parallel-fiber preference for molecular-layer interneurons, even though Purkinje cells have 10 times more synapses overall.

Concurring with both DeZeeuw and Braitenburg, Bower then speculated that the lack of parallel-fiber inputs found in many of Barbour's paired granule-cell/Purkinje-cell recordings, which Barbour had suggested were synapses made silent by learning, might actually reflect a lack of synaptic contact to begin with. Barbour responded that after extensive light and EM analysis, Napper and Harvey had concluded that 85% of parallel-fiber varicosities contacted Purkinje-cell spines. Bower pointed out that Napper and Harvey had also calculated from the same data that less than 5% of granule-cell inputs to Purkinje cells should come from the ascending segments, and this number was off by a factor of 5.

From the audience it was noted that the data presented by Barbour reflected a fairly small sample size and that, perhaps, a larger sample size would reveal granule cells with more influence on overlying Purkinje cells. Barbour suggested that this was not impossible, but that another important point involved the spatial selectivity of the mossy-fiber system, which he suggested might be somewhat diffuse. Braitenburg stated subsequently that his analysis of the anatomy suggests that there is not much spatial smearing of mossy-fiber inputs.

Garwicz then raised the possibility from the audience that the lack of Purkinje-cell beams might be a result of experimental interference with the viability of parallel fibers at their bifurcation, as suggested by experiments on the cat he and his colleagues had conducted in 1992. In response, Bower pointed out that it was critical in experiments of this type to directly measure the spatial distribution of granule-cell-layer activity, and the field-potential techniques used in these previous experiments may not have been adequate for that purpose. In addition, Bower pointed out that the data he had presented earlier in the meeting clearly showed that one could switch between no beams and beams simply by blocking inhibition and that the parallel fibers were therefore working well throughout the experiment. Bower's offering that perhaps the anterior lobe of the cat, where Garwicz had performed his experiments, might be more sensitive to experimental conditions than the lateral hemispheres of the rat was challenged by Yarom, who was quick to point out that the two areas were morphologically uniform.

Stepping back from the details for a moment, Yarom then asked the question whether at least it was possible to leave this meeting with an agreement about whether there was a Purkinje-cell beam or not, noting that this was not as complicated as whether there was learning or not. After chuckles from the audience, Barbour commented that, in fact, resolving the influence of parallel fibers on Purkinje cells was probably going to require that we use stimuli that had meaning to the animal, and thus one could not so easily avoid the larger question of cerebellar function.

Following up on this point, Mauk then asked a question from the audience that was related to Bower's claim in his earlier talk that the lack of Purkinje-cell beams posed problems for most if not all cerebellar learning theories. Specifically, Mauk made the point that learning theories, including Marr's original theory, did not, in fact, expect to find long beams of activated Purkinje cells, but instead proposed that Purkinje-cell responses to parallel-fiber inputs were a result of competition between parallel-fiber excitation and molecular-layer inhibition, a mechanism he felt was supported by Bower's data. Bower responded that the issue with respect to learning theories was not the existence of a beam of Purkinje-cell activation, but was instead the assumption made by these theories that Purkinje-cell activity was driven by parallel fibers. Bower asserted that the balance of the evidence presented in the previous session had suggested that it was the granule cells underneath Purkinje cells, and not those contacting Purkinje cells via parallel fibers, that provided the driving input to these cells.

Braitenburg then asked for clarification on the difference between synapses "driving" Purkinje cell output and "modulating" that output. Bower then attempted to clarify the distinction by stating that driving means that synaptic input results directly in time-locked Purkinje-cell output, while modulation means that parallel fibers do not directly affect somatic output, but instead influence the output of other synapses, in this case, those of the ascending segment of the granule-cell axon.

A question was then raised by a member of the audience as to whether Yarom could take advantage of the spatial organization of vestibular inputs to the cerebellar nodulus to counteract the suggestion that Purkinje-cell beams do not exist. Yarom first stated that in his *in vitro* isolated cerebellar preparation, the vestibular system was no longer intact, so these experiments could not be performed. However, he went on to clarify his position with respect to the Purkinje-cell beams by stating that he was in agreement with Bower that there were no such beams, but that he differed with Bower on the mechanistic reason they do not exist, with Bower emphasizing the role of inhibition, and his group leaning more toward relative differences between the number of synapses activated on each Purkinje cell by parallel fibers as compared to the synapses associated with the ascending granule cell axon segments.

Yarom then again said that he hoped that in this meeting we could at least agree that mossy-fiber input does not produce Purkinje-cell beams regardless of the mechanism. He pointed out that the data presented in this meeting, as well as most of the data in the literature, all point to the conclusion that there is no beam, so, we should decide on that and continue with a discussion of the mechanisms responsible.

❧

The final statement made by Valentino Braitenburg seemed particularly appropriate:

> I think the organizers invited us because they hoped to have a good aggressive show. (Audience laughter.) Jim (Bower) you started your talk by saying the beam is dead, I have always liked the idea of a beam going across the cerebellum. The other thing is, is the cerebellum a machine that is prewired, or is it one that is set up by learning? We must admit we have all been beaten, we don't have to beat each other, because we have all been beaten, our pet ideas have been taken apart by people who have made beautiful experiments, and we learned something in the process. So, I think we should go back to the essentials, to build up some model of the cerebellum, because what we don't know is why it is made the way it is. Why, even a child looking at the cerebellum can tell it is different from the cerebral cortex, macroscopically. But we don't know why. I am fanatically convinced, however, that the cerebellum's structure can be explained and must be explained...perhaps by thinking some more, and going back to basics, we can arrive at a beautiful synthesis.

Functional Significance of Climbing-Fiber Synchrony

A Population Coding and Behavioral Analysis

JOHN P. WELSH

Neurological Sciences Institute, Oregon Health and Science University, Beaverton, Oregon 97006, USA

ABSTRACT: Population coding and behavioral approaches were taken toward analyzing the functional significance of the climbing-fiber system. Analyses of neuronal interaction using the joint peristimulus time histogram showed that given a low rate of firing, the climbing-fiber system organizes itself to fire synchronously during movement—a feature that bears little or no relationship to the modulation in firing rate during movement or whether olivary neurons respond to a sensory stimulus. Moreover, the climbing-fiber system avoids synchrony during a passive sensory response but actively makes a transition into synchrony as a movement is initiated. Thus, from a functional viewpoint, the active feature of the climbing-fiber system to organize into synchronously firing cell ensembles is uniquely motor. Analyses of behaving rats without an inferior olive revealed that the climbing-fiber system optimizes the timing of skilled movement by reducing reaction time and the interval between repetitive movements by 100 milliseconds. Finally, using classical delay eyeblink conditioning, it was found that the inferior olive is essential for learning about rapid sequences of events but not the same event sequence when given more slowly. It was concluded that the climbing-fiber system exerts its function through synchrony, which provides a 100 ms advantage in movement speed and the ability to learn about events that are rapidly presented in time.

KEYWORDS: climbing-fiber system; cerebellum; delay eyeblink conditioning; mossy-fiber system

INTRODUCTION

It is not uncommon for a paper on the cerebellum to begin by stating that despite knowing much about its morphology, connectivity, and synaptology, there is no clear consensus about how the cerebellum operates or about its function. Of course, that statement is continuously reinforced by the sharp contrast between the unique spatial constancy of the cerebellar microcircuitry and the relative temporal inconstancy of hypotheses about how the cerebellum actually operates. The inability to resolve

Address for correspondence: John P. Welsh, Neurological Sciences Institute, Oregon Health and Science University, 505 NW 185th Avenue, Beaverton, OR 97006. Voice: 503-418-2645; fax: 503-418-2501.
welshj@ohsu.edu

Ann. N.Y. Acad. Sci. 978: 188–204 (2002). © 2002 New York Academy of Sciences.

cerebellar function from the spectacular array of available facts has led to the implicit view that the operating principles of the cerebellum are far more complicated than previously thought, and perhaps even more complicated than for other parts of the brain. This view assumes that the facts in hand are sufficient and all that is needed is a better operating framework into which those facts can be placed. However, over the 30 years since the cerebellar microcircuit was elucidated, other areas of the brain have become similarly well described, but their operating principles and functions also have remained elusive. Thus, the facts so far obtained with the methodologies employed may not be sufficient to understand function, and new approaches toward investigating the cerebellum are required. One such approach, which is now providing new insight into the function of numerous brain areas, is that of measuring the activity of neuronal ensembles in an attempt to understand how function is coded at the population level.[1] Thus, the purpose of this chapter is to highlight recent advances in understanding population coding within the climbing-fiber system and its possible functional meaning for movement control and learning as revealed in behaving animals.

It is important to define what it meant by population coding at the outset. At its most superficial level, population coding refers to the information that is conveyed in the relative amounts of activity across multiple neurons or neural systems. This simplest level of analysis can be elucidated with techniques such as functional metabolic imaging, electro- and magnetoencephalography, and inferred with sequential single electrode recordings. However, a deeper type of population coding refers to the information conveyed in the specific temporal patterns of neurons' action potentials with respect to one another. Approaches toward this level of analysis are electrophysiological recording of many single neurons using multiple microelectrodes or high-resolution voltage imaging. It is important to emphasize that the latter approaches allow an analysis of population codes beyond the simple summing of simultaneously recorded activities, but allow the interactions between neurons to be related to a particular behavior. For the field of cerebellar electrophysiology, like most fields, the "one neuron–one experiment" approach has dominated and has related the spikes of individual cerebellar neurons to specific movements or movement parameters. On the other hand, a population coding approach seeks to elucidate the relationship between two or more neurons' interaction with a particular movement parameter. As we shall see below, the single-neuron approach has led to a myriad of conflicting views on climbing-fiber function, whereas the population coding approach is beginning to reveal new insights.

BACKGROUND: A DIVERSITY OF DATA AND VIEWS OF THE CLIMBING-FIBER SYSTEM

The ability to relate the activity of the climbing-fiber system to motor control has been hampered by the low (1–2 Hz) average firing frequency of single olivary neurons, which is incongruent with their role in the control of fast movement. Indeed, it has been difficult to understand how a skilled and fast movement like playing rapid arpeggios on the piano can be controlled by neurons that fire, on average, only once or twice a second. Clearly, even large groups of such slowly firing olivary neurons make virtually no numeric contribution to the aggregate spike activity of the cerebellar, striatal, and cortical systems that simultaneously influence movement. This quandary has led to a view that the inferior olive may have no role in the real time control of movement.[2]

On the other hand, electrophysiological experiments have found robust responses of olivary neurons to peripheral sensory events such as somatic and visual stimuli. The robustness is defined by the ratio of the magnitude of the evoked response to the uniquely low baseline rate, which yields captivating stimulus-locked peaks of activity that can appear much like a primary sensory neuron but at longer latency.[3,4] Such sensory robustness has led to the characterization of the climbing-fiber system as a unique type of accessory sensory system,[5] has brought forth views of climbing fibers as sensory comparators,[6] and more recently, as providing a central guide for optimizing the acquisition of sensory input.[7] A corollary of the view of olivary neurons as "sensory neurons" is that their sensory fidelity may trigger synaptic modification of Purkinje cells underlying motor and even nonmotor associative learning.[8]

Coming full circle, recording experiments using behaving monkeys have found significant modulations of climbing-fiber activity that are specifically related to the generation of movements.[9–11] Although such modulations are often weak in comparison to neurons of the motor cortex or striatum, repeatability and statistical significance have helped to reinforce the view that the inferior olive does play some important, but undefined role, in movement generation.

In summary, the current state of climbing-fiber research using the single neuron recording approach has led to the view that olivary neurons are either "sensory" or "motor," a state of affairs that is particularly unsatisfying. It seems even less satisfying to adopt the compromise position—that olivary neurons are "sensory-motor"—because, although undoubtedly true, changing the label does not generate any greater insight. A critical issue is how olivary neurons are to be defined functionally. Should their function be defined on the basis of their input response or on the basis of the consequence of their output? Indeed, the impasse lies in the fact that the available data have not allowed us to understand the latter, while the former may not be very discriminating, because even motoneurons can show tight stimulus-locked responses to peripheral events, and we do not think of them as sensory. Recent electrophysiology data obtained using population coding techniques offer the prospect of breaking this impasse.

BREAKING THE IMPASSE: POPULATION CODING BY CLIMBING-FIBER SYNCHRONY

FIGURES 1 and 2 show complex spike data obtained in a skilled movement paradigm that was developed for cerebellar multineuron electrophysiology. Here, the complex spikes of 32 Purkinje cells were recorded from crus IIa with as many microelectrodes while a rat received a tone that signaled it to protrude its tongue out of the mouth. The details of the training paradigm[12] and electrophysiology[13] have been published. FIGURE 1A shows raster plots of the complex spikes of 13 representative Purkinje cells during two 500 ms epochs of movement. In the two epochs, the rat performed 7 tongue protrusions, and each of the 13 Purkinje cells fired at least 1 complex spike. The initial qualitative impression of the rasters reinforces a most common finding in climbing-fiber electrophysiology: each Purkinje cell fires com-

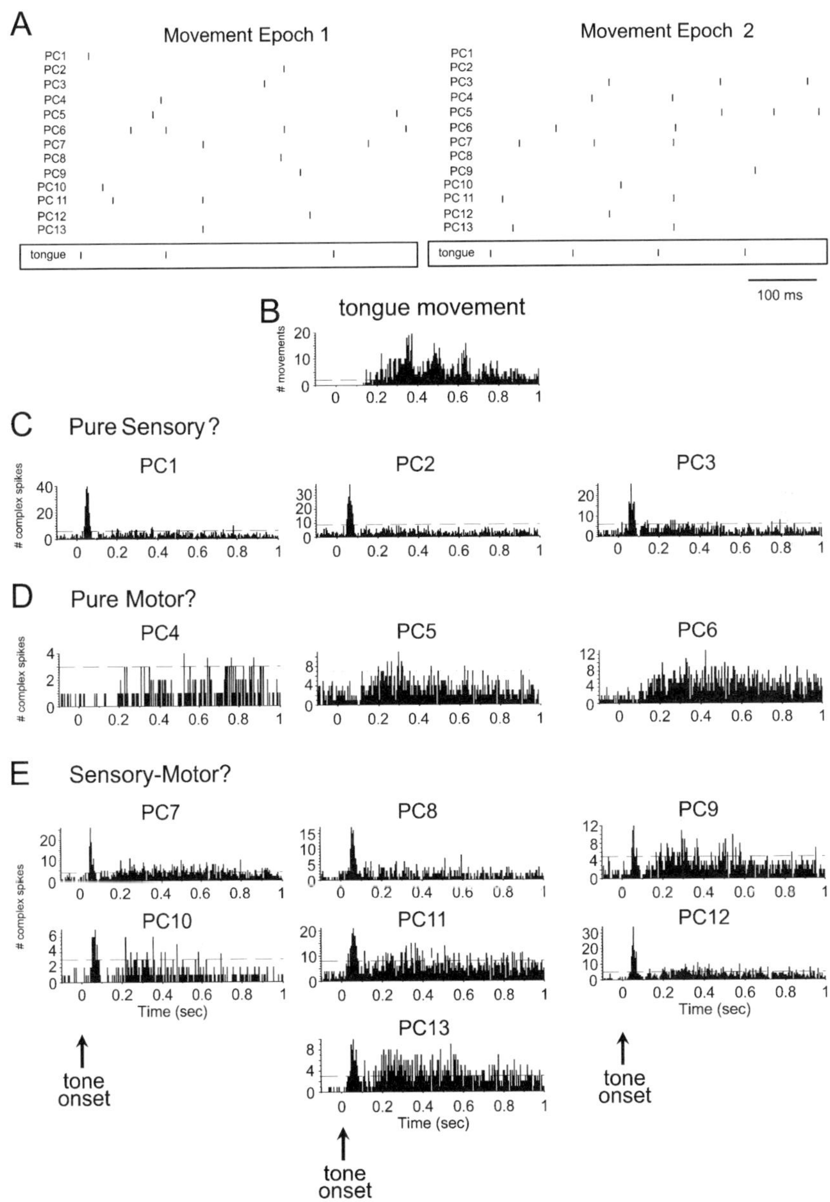

FIGURE 1. Single neuron analysis of complex spikes. (**A**) Complex spike rasters of 13 Purkinje cells recorded simultaneously during conditioned tongue protrusion. **(B)** Peristimulus time histogram of tongue protrusion. In all plots, time 0 represents the onset of a 750-ms tone that signaled the rat to protrude its tongue. **(C–E)** Peristimulus time histograms of the complex spikes of the 13 Purkinje cells shown in **A**. The neurons were grouped on the basis of their response to the tone alone (**C**), movement alone (**D**), or both (**E**). Dotted lines represent 99% confidence intervals, and bin widths equal 5 ms.

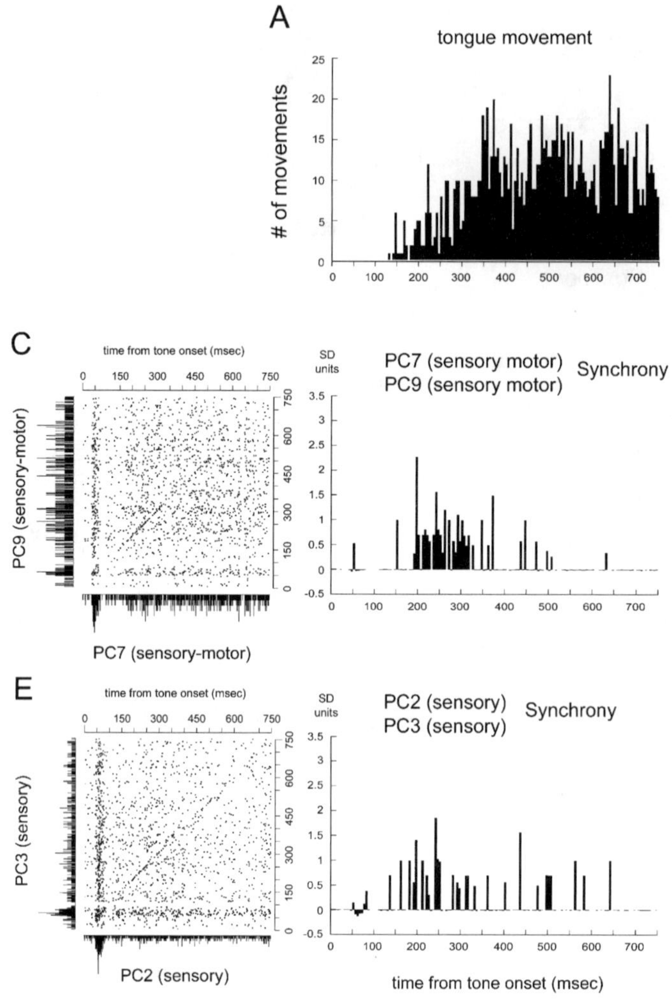

FIGURE 2. Interaction analysis of the data from FIG. 1. In all histograms, time 0 represents onset of the tone that triggered a tongue protrusion. **(A)** Perievent histogram of tongue movement. **(B–F)** JPSTH analyses of Purkinje cell pairs. JPSTHs are presented as raw occurrences in 1-ms bins for various combinations of cell types defined by the single neuron analysis. Note the prominent line of points along the diagonal of each JPSTH (except **F**), indicating the occurrence of complex spike synchrony. To the right of each JPSTH is the normalized poststimulus time synchrony histogram presented in standard deviation units and 5-ms bins.

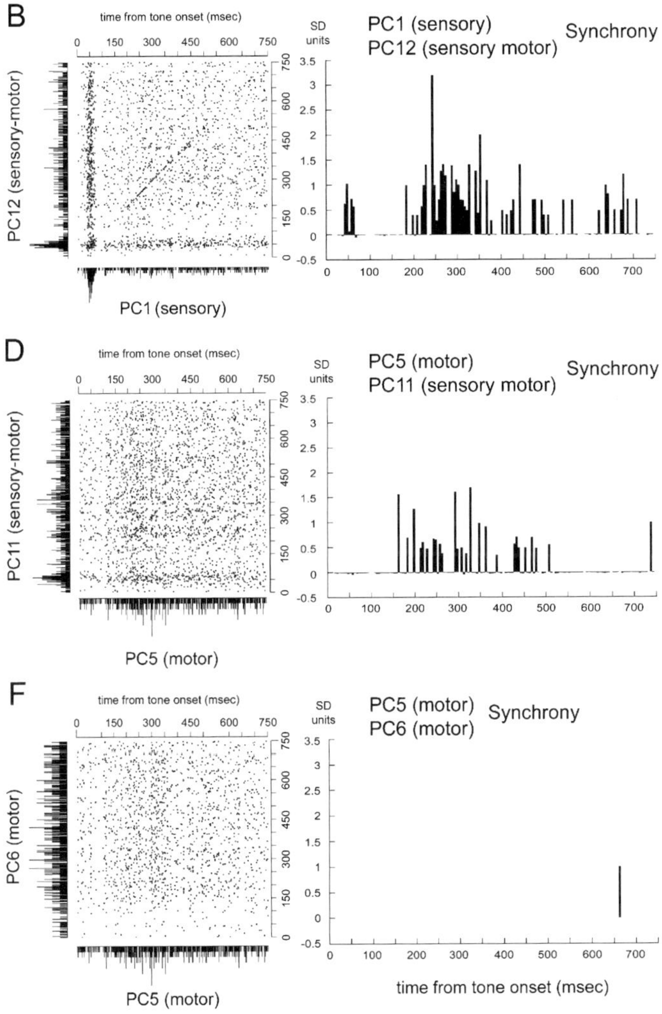

FIGURE 2. *Continued.*

plex spikes infrequently and with very weak, if any, relationship to the ongoing movement. Moreover, any relation that one might observe on a given instance of movement appears not to occur subsequently and is not easily identifiable in previous instances of the movement. Thus, the raw data give the impression of an almost random relationship between a Purkinje cell's complex spikes and movement.

However, statistical treatment of the data using standard perievent histograms reveals that complex spikes do not occur randomly but bear significant and robust relationships to specific aspects of the paradigm. FIGURES 1B–E present the movement and complex spike data as perievent histograms in relation to the onset of the tone-conditioned stimulus. As seen in FIGURE 1B, the tongue movement had an earliest latency of approximately 200 ms after tone onset, and movements occurred from 0.2 to 1 s after onset of the tone. The histograms clearly separated the 13 Purkinje cells into three groups. The first group (FIG. 1C) showed a brief and robust complex spike response to the tone but only baseline activity during the movement. One could classify such Purkinje cells as showing a pure sensory response to the tone. The second group (FIG. 1D) did not show the sensory response to the tone, but showed a significant modulation during the movement. A sensible classification of such Purkinje cells is as motor. The third group (FIG. 1E) combined the sensory and motor features just described and could be classified as sensory-motor. Such classification of the Purkinje cells into sensory, motor, and sensory-motor on the basis of the histograms is reasonable and replicates findings of previous investigators in a variety of paradigms. Yet, with such a classification in hand, one can refer back to the rasters to try to find differences in the moment-to-moment complex spiking of the three groups of Purkinje cells. Indeed, no clear difference can be seen among the sensory, motor, and sensory-motor types of Purkinje cells on specific instances of the movement. Thus, it is difficult to know how (indeed, whether) any of the Purkinje cells' complex spiking shown in the histograms relates to instances of the behavior.

Because the 13 Purkinje cells were recorded simultaneously, it can be determined whether particular temporal interactions among the neurons' complex spike activity relate to the performance of the movement. The joint peristimulus time histogram (JPSTH) was employed, a method that elucidates dynamic interactions between two neurons' spike activities relative to a trigger event.[14,15] The JPSTH is a two-dimensional matrix formed from two perievent histograms that are laid out on the ordinate and abscissa. In the construction of a JPSTH, the data are "played into" the matrix so that all logical ANDs of spike occurrences at various delays are plotted in a temporal plane. Features of the temporal interaction among the two sets of spikes produce regions of increased point density within the matrix. There are three patterns that are readily revealed by the JPSTH. First, a broad stripe of point density parallel to the axes indicates a response to the stimulus. Second, stripes of point density that run parallel to the lower-left to upper-right diagonal indicate a consistent temporal offset between the two neurons' spikes. Third, using a 1 ms bin width, a very thin stripe of high point density precisely on diagonal reveals spike synchrony. Fluctuations in the density of points along the diagonal indicate the averaged, stimulus-locked modulation of synchrony. In order to correct for "accidental" occurrences of coincidence due to increases in firing rate during the trial, the matrix is normalized by subtracting a product matrix divided by the number of trials and is divided again by the product of the standard deviations of the two perievent histograms. This converts the matrix into correlation values whose significance is expressed in standard deviation units. Slices of data extracted from the diagonal of the corrected JPSTH matrix indicate the occurrence of synchrony statistically greater than would be expected from that produced by independent comodulation of firing rate during the trial.

JPSTH analyses were performed on the complex spike data of FIG. 1. In the JPSTH plots of these data, the origin is the time of tone onset, and the spike times are plotted in 1 ms bins. For ease of presentation, the JPSTHs are shown as raw spike occurrences, while a histogram of data along the main diagonal is presented to the right of each matrix in standard deviation units after correction for changes in firing rate that occur due to the trial. Because the bin width of the JPSTH matrix is 1 ms, extraction of the main diagonal generates a poststimulus time synchrony histogram. It is important to reemphasize that the analysis reveals the occurrence of complex spike synchrony that is statistically greater than that expected simply by the increased firing rate, per se.

JPSTH analyses were performed on various combinations of Purkinje cells characterized as sensory, motor, and sensory-motor on the basis of the poststimulus time analysis shown in FIGURE 1. FIGURE 2B shows the coincident complex spike firing of a pure sensory Purkinje cell (PC1) with a sensory-motor Purkinje cell (PC12). The striking feature of this JPSTH was the prominent diagonal line through poststimulus time, indicating moments of synchrony during the trial. The poststimulus time synchrony histogram revealed that significant synchrony occurred throughout the trial and was greatest between 175–500 milliseconds. A second moment of synchrony occurred at 650 ms into the trial. Thus, the two neurons fired synchronous complex spikes conditionally, at discrete moments of the trial. A similar result was obtained from the JPSTHs of two sensory-motor Purkinje cells (FIG. 2C) and between a motor and sensory-motor Purkinje cell (FIG. 2D). However, the synchrony between the two pairs was confined to 150–500 ms of the trial. What is remarkable about the poststimulus time synchrony histograms is that the statistical significance of the synchrony was greatest during the movement, when overall rates of complex spiking were lowest, and insignificant during the initial sensory response to the tone when corrected for the change in firing rate.

Even more surprising relationships were seen in the JPSTHs between two sensory and two motor Purkinje cells. In the JPSTH of two sensory Purkinje cells (FIG. 1E), robust complex spike synchrony occurred not during the response to the tone, but during the movement. Here, again, a prominent diagonal line formed through the JPSTH matrix, revealing that the complex spikes among two sensory neurons occurred synchronously during the movement, long after their sensory response was over. This remarkable feature contrasted with the firing rate during the movement, which was not different than baseline. Lastly, the JPSTH of two motor Purkinje cells revealed no synchrony at any time during the movement, even though the single-neuron analysis indicated a significant motor-related modulation.

The significance of these findings is very clear: given a low rate of firing, the climbing-fiber system organizes itself to fire synchronously during movement—a feature that bears little or no relationship to single olivary neurons' firing rate or whether or not the neuron responds to a stimulus. So-called sensory Purkinje cells, defined on the basis of their robust and time-locked complex spike response to a peripheral stimulus, are, in fact, motor when one considers their tendency to fire synchronously specifically during movement. Because the corrected JPSTH analysis reveals the occurrence of complex spike synchrony that is statistically greater than that expected simply by the increased firing rate, it reveals an active process that reorganizes the temporal interactions among olivary neurons in the movement latency period. Although many Purkinje cells show a short-latency response to a tone before movement onset, synchrony during the sensory response is not greater than that produced by the increased firing rate. The population coding analysis reveals a remarkable principle: the climbing-fiber system avoids synchrony during its passive sensory response but actively organizes into synchronous ensembles during movement. Thus, from a functional viewpoint, the active features of the climbing-fiber system are uniquely motor.

POPULATION CODING OF COMPLEX-SIMPLE SPIKE INTERACTIONS: AMPLIFYING CLIMBING-FIBER FUNCTION

A major question is whether the climbing- and mossy-fiber systems act independently or in a concerted way to control movement. Numerous studies have found that a complex spike on a given Purkinje cell modulates the subsequent rate of simple spiking for a brief period—the well-described complex spike–simple spike interaction. Using single microelectrode recordings, some investigators found an inhibitory effect of a complex spike on simple spike rate,[16,17] while others found an enhancing effect.[18,19] Such interactions have led to the hypothesis that the sole function of the climbing-fiber system is to modulate the rate of simple spiking. A circumstantial argument against this hypothesis can be made on numerical grounds: given the infrequent occurrence of complex spikes (1–2 Hz), the high frequency of simple spikes (40–300 Hz), and the brief (50–100 ms) modulation of the former on the latter, such modulation would have a negligible impact on the aggregate mossy-fiber system throughput and would, thus, seem to be a rather weak way for such a synaptically powerful afferent to exert function. Indeed, simple spike modulation may be epiphenomenal—an unavoidable consequence of the two systems sharing a final path through which they exert function. Yet, population coding data support a third and perhaps more intriguing possibility: the climbing-fiber system uses synchrony to modulate simple spike rate in order to potentiate its own independent action.

We developed a paradigm for examining interneuronal interaction within the cerebellar cortex using microelectrode recordings of Purkinje cell complex and simple spikes[20] with a new technology that allowed the independent manipulation of microelectrodes in a high density array.[21] A major finding was that mossy-fiber system throughput, as recorded in the simple spikes of multiple simultaneously recorded Purkinje cells, was robustly modulated by synchrony within the climbing-fiber system. This was observed as the well-described transient pause in simple spiking after a complex spike in the same Purkinje cell.[22] However, using a multielectrode approach, we found that the modulation of simple spiking did not require a given Purkinje cell to fire a complex spike, but was on those occasions highly correlated to the synchronous complex spikes of other Purkinje cells.

FIGURE 3 shows the fundamental finding. Stimulation of the motor cortex evoked simple and complex spikes in the contralateral crus IIa cerebellar cortex, as recorded with an array of 16 microelectrodes (FIG. 3A). FIGURE 3B shows an autocorrelation analysis of the simple spikes of 2 simultaneously recorded Purkinje cells (of 10 re-

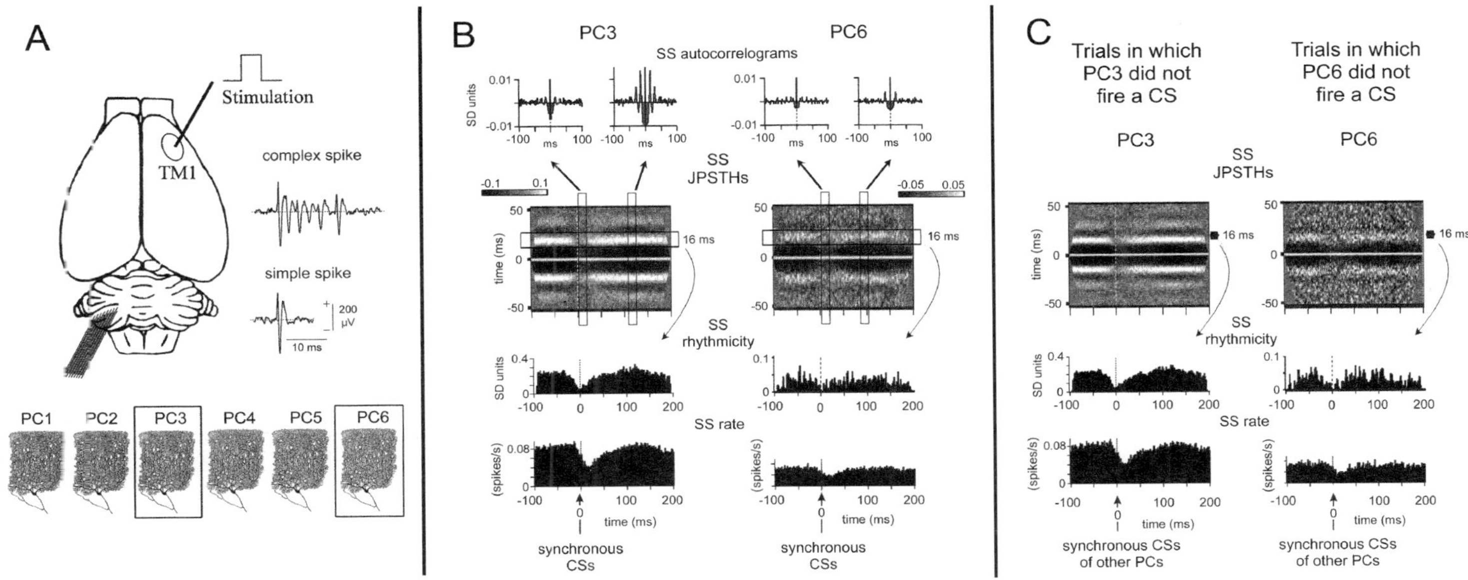

FIGURE 3. Population coding of complex spike-simple spike interactions. (**A**) Stimulation of the right tongue motor cortex (TM1) evokes complex spikes (CS) and simple spikes (SS) in the left crus IIa. (**B**) Climbing-fiber synchrony determines SS rhythm dynamics. The SSs of 2 Purkinje cells were recorded simultaneously with the CSs of other Purkinje cells. Times of CS synchrony serve as trigger events for an SS peristimulus time autohistogram. (**C**). Same analysis as in **B**, but excluding trials in which the target Purkinje cells fired a CS. Note that the SS rate and rhythm modulation is just as profound. (From Schwarz & Welsh.[20] Reproduced by permission.)

corded at the same time). Time 0 in the bottom three plots is the time in which the complex spikes of Purkinje cells in the array occurred synchronously after the motor cortex was stimulated. The trials selected for analysis in FIGURE 3B included trials in which the complex spikes of the target Purkinje cells (PC3 and PC6) contributed to the population response, although they were not limited to such trials. The occurrence of complex spikes in the population produced a prominent inhibition of simple spike rate in the two target Purkinje cells, which lasted about 50 ms (FIG. 3B, bottom row). The times of complex spikes were used to trigger normalized auto-JPSTHs of the Purkinje cells' simple spikes, which are shown rotated by 45 degrees (FIG. 2B, second row). The patterns in the auto-JPSTHs indicated that the rhythmicity in the two Purkinje cells' simple spikes was robustly modulated by the occurrence of complex spike synchrony. This is clearly seen in the normalized autocorrelograms of simple spiking calculated at specific times after complex spikes and in the modulation of the perievent time correlation histogram shown under the two auto-JPSTHs. The prominent peak at 16 ms in the autocorrelograms indicates rhythmic simple spiking of 60–70 Hz. However, the strength of the simple spike rhythmicity changed dynamically, such that it was transiently reduced in the 50 ms after the complex spikes and transiently enhanced 50 ms later.

In FIGURE 3C, the same analysis was performed but excluding trials in which the two target Purkinje cells fired a complex spike; time 0 in these plots represents a moment of complex spike synchrony that did not involve the two target cells. On these trials, the modulation of simple spike rate and rhythmicity by the climbing fibers was just as robust as on trials when the target Purkinje cells participated in the population response. Because each Purkinje cell receives only one climbing fiber, the data demonstrated that the modulation of simple spike rate and rhythmicity by the climbing-fiber system is determined by network interactions within the cerebellar cortex that are driven by complex spike synchrony. Such a network interaction would necessarily involve the cerebellar cortical interneurons.[20]

What is the possible functional significance of the network interaction between the climbing- and mossy-fiber systems? Dr. Cornelius Schwarz and I have proposed that the modulation of the mossy-fiber system facilitates the ability of climbing-fiber synchrony to influence movement.[20] As shown earlier,[23] synchronous activation of the inferior olive produces a sequence of excitation-inhibition-excitation in deep nuclear neurons, respectively produced by excitatory climbing-fiber collaterals, synchronously firing inhibitory terminals, and rebound excitation. The duration of the hyperpolarization of deep nuclear neurons induced by synchronous olivary firing has a minimum of about 60 ms—a duration that is nearly identical to the duration of the simple spike rate decrease immediately following a moment of complex spike synchrony. This implies that the duration of the pause in simple spiking firing induced by complex spike synchrony is tuned to the membrane properties of the deep nuclear neurons and reduces synaptic inhibition for the time required for the emergence of rebound excitation. The subsequent facilitation of simple spikes would restore deep nuclear inhibition and cancel rebound excitation, in order to limit the rapid spike bursts.

The results from this population coding approach suggest a view in which climbing fibers act as an independent, real-time control system that periodically has a strong entraining effect on the mossy-fiber system by virtue of spike synchrony. This view supports the hypothesis that climbing fibers exert important effects on movement in their own right[24,25] but adds the feature that each episodic bout of synchrony in the inferior olive is self-facilatory, in that it briefly coopts the mossy-fiber system in order to amplify its own effect on deep nuclear neurons. Such a takeover of mossy-fiber system function may represent a unique example of exaptation in the nervous system.[26] Given that complex spikes occur synchronously during specific epochs of movement (FIG. 2), the data indicate that the two independent systems of the cerebellum become transiently bound into a single system at critical moments during movement.

NECESSITY OF CLIMBING FIBERS FOR FAST MOVEMENT AND LEARNING ABOUT RAPID EVENT SEQUENCES

Three behavioral paradigms were used to determine the contribution of the climbing-fiber system to the generation of movement and associative learning. The first paradigm was similar to the one described above, in which restrained rats were trained to protrude their tongue out of their mouth. The conditioning procedure was a hybrid classical-operant paradigm in which rats first associated a 1.5-s, 90-dB, 2-kHz tone conditioned stimulus (CS) with intraoral water as an unconditioned stimulus (US). The drop of water elicited a reflex response of rhythmic licking and swallowing, which was then conditioned to the tone by repeated presentations of the two stimuli at a 1.5-s CS-US interval. After acquisition of a conditioned licking response, its occurrence was reinforced on operant trials in which the rat was given the water upon making the response. The conditioned response was then shaped to reach a target 6 mm away from the mouth by making delivery of the water contingent upon gradually longer tongue protrusions. The second paradigm trained rats to make repeated extension-flexion movements of the arm about the elbow joint. Restrained rats were trained under free operant conditions to press a lever under a fixed ratio 10 schedule (FR10), which delivered a water drop after every 10 lever presses. The paradigm was very robust and supported thousands of lever presses in a single 90-min session. The third paradigm was classical eyeblink conditioning. The eyeblink of restrained rats was measured electromyographically using a bipolar recording electrode in the superior orbicularis oculi muscle. The eyeblink was conditioned to an auditory white noise stimulus that was followed by a 100-ms, 3-mA, 60-Hz shock to the lateral margin of the eye. Two CS-US intervals were employed in separate rats: 350 and 700 ms in classical delay procedures, and 110 trials per day were given at an average 30-s intertrial interval.

The effect of removing the inferior olive on the learning and motor performance supported by these paradigms was assessed. The inferior olive was removed by systemic administration of 3-acetylpyridine, harmaline, and niacinamide.[27] This procedure kills 96 ± 0.1% of olivary neurons,[12,28] sparing only a small population in the medial aspect of the caudal medial accessory olive and beta nucleus. FIGURE 4B shows the complete loss of neurons at a midlevel of the olive as revealed by immunostaining neuronal-specific enolase. As measured by the immunostaining and a Nissl stain, damage to other parts of the brain was undetectable. Recovery for at least three weeks after the lesion was allowed before testing.

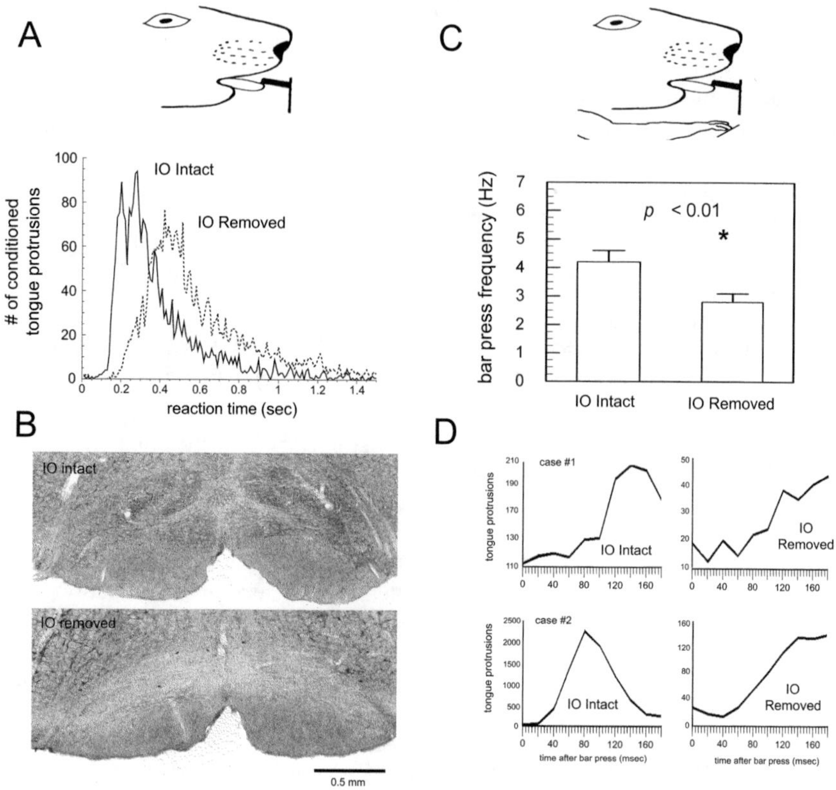

FIGURE 4. Inferior olive is required for rapid movement. (**A**) Onsets of conditioned tongue protrusions are delayed by 100 ms in the absence of the inferior olive (n=5 normal, 6 lesioned rats). (**B**) Immunostaining for neuron-specific enolase verifies that 3-acetylpyridine kills inferior olivary neurons. (**C**) Frequency of lever presses is reduced after removing the inferior olive (n=5 rats tested before and after the lesion). (**D**) The latency of tongue movements after lever presses is prolonged by removing the inferior olive.

Operant Paradigms

Rats were able to acquire conditioned tongue protrusion responses in the absence of the inferior olive and showed acquisition curves that were not different from normal.[12] Nevertheless, olivary-lesioned rats showed a significantly prolonged onset latency of the conditioned tongue protrusion (FIG. 4A). Onset latencies of normal rats had an earliest value of approximately 100 ms and a mode of 275 milliseconds. The earliest onset latency of olivary-lesioned rats was 200 ms, and the mode was 425 milliseconds. The difference in the onset latency distributions was significant ($p < 0.01$). Thus, rats without climbing fibers were 100–150 ms slower in initiating the conditioned response.

In the lever-pressing paradigm, normal rats showed two interesting behaviors. First, lever pressing was highly rhythmic and exhibited a steady frequency of 4.2 ±

0.3 Hz. Second, the rats frequently exhibited tongue protrusions as they were pressing the lever to meet the response contingency for the water. The tongue protrusions became entrained to the rhythm of the lever pressing and reached maximal extension at a stereotypic latency after the press. This behavior was acquired with experience in the paradigm and represented the acquisition of a learned motor synergy. Removing the inferior olive had two profound effects. First, the mean frequency of lever pressing was reduced to 2.8 ± 0.3 Hz ($p < 0.01$, FIG. 4C), representing an increase of 119 ms in the latency to execute a repeated movement of the arm. Second, the modal latency between lever press and tongue extension was increased by 40–105 ms (FIG. 4D), representing a significant temporal prolongation of the motor synergy.

Classical Eyeblink Conditioning

FIGURE 5 shows an experiment in which four rats were given eyeblink conditioning (FIG. 5A). Two rats were conditioned using a 350 ms CS-US interval, while two were conditioned under a 700 ms CS-US interval. In each pair, one had its inferior olive removed five weeks prior to the conditioning. FIGURE 5B shows typical orbicularis oculi records of conditioned eyeblink responses (CRs) at the two intervals. The CRs of normal rats acquired at the 700 ms CS-US interval had longer onset latencies than those conditioned under the shorter interval, a feature of eyeblink conditioning.[29] Removing the inferior olive prevented the acquisition of CRs at the 350-ms interval (filled squares, FIG. 5C), replicating previous reports.[30,31] However,

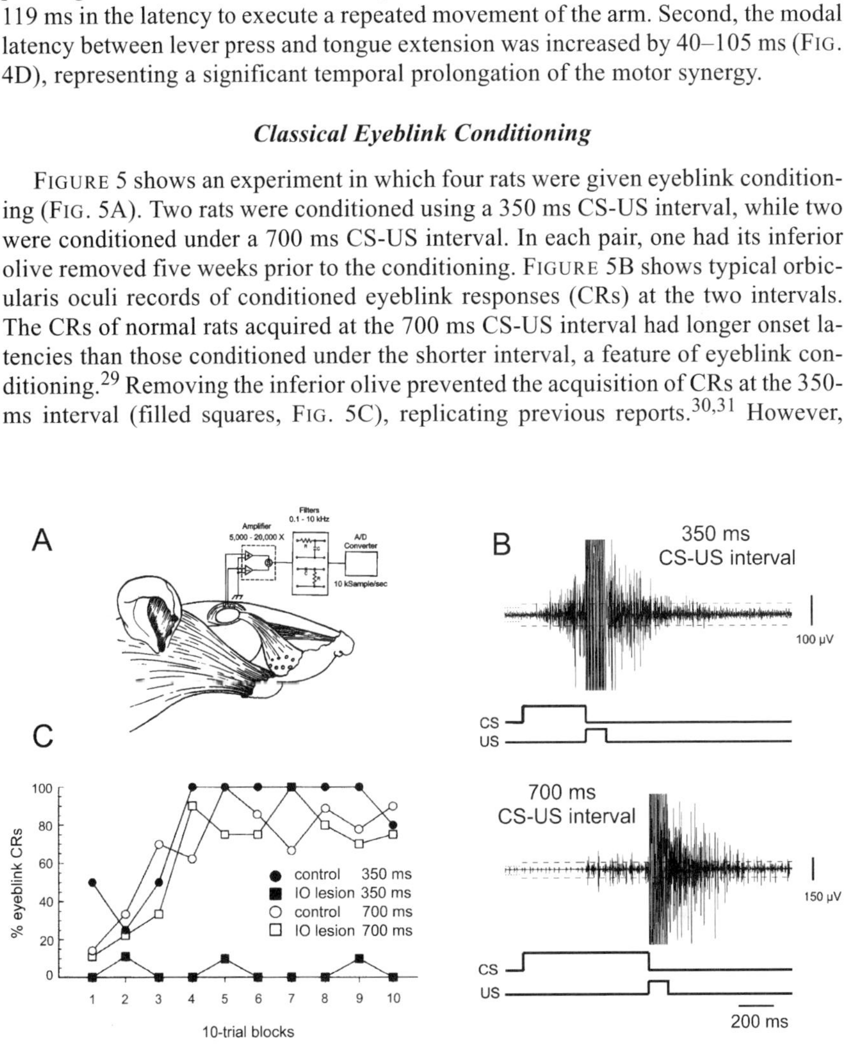

FIGURE 5. Eyeblink conditioning after removing the inferior olive. (**A**) Setup for measuring orbicularis oculi EMG. (**B**) Examples of orbicularis oculi responses conditioned under a 350-ms and 700-ms CS-US interval. (**C**) Inferior olive lesion prevents acquisition of conditioned eyeblink responses under a 350-ms CS-US interval (*filled squares*) but not under a 700-ms CS-US interval (*open squares*). Ahn and Welsh, unpublished.

there was no difference in the CR acquisition between the olivary lesioned and normal rats conditioned at the 700 ms CS-US interval (compare open circles and open squares in FIG. 5C). Quantitative analysis revealed that the lesioned rats had 96–97% of their olivary neurons removed. The experiment demonstrated that the inferior olive is required for eyeblink conditioning only at a short CS-US interval and not at a longer CS-US interval.

SUMMARY AND CONCLUSIONS

Four points emerge from the population coding and behavioral analysis of the climbing-fiber system. First, the dynamic interaction among olivary neurons that allows them to trigger complex spikes synchronously in ensembles of Purkinje cells is a motor property of this system. Purkinje cells that would be defined as sensory on the basis of a single neuron analysis show strong modulation in complex spike synchrony during movement that is unrelated to changes in firing rate or the sensory response. In fact, the climbing-fiber system actively avoids assembling into synchronously firing ensembles during its response to a sensory event. Second, the climbing-fiber system controls the temporal dynamics of simple spike firing in populations of Purkinje cells by using synchrony to modulate mossy-fiber system throughput. The simple spike modulation has a time constant that suggests that it augments climbing fiber–related effects on deep nuclear neurons in order to maximize the occurrence of rebound excitation produced by synchronously firing Purkinje cells. Third, the climbing-fiber system is essential for rapid movement. Three separate analyses of skilled movement indicate a 100-ms slowing of movement onset and speed in the absence of climbing fibers. Fourth, the climbing-fiber system is essential for learning about short-time intervals as measured by classical eyeblink conditioning. This finding raises the intriguing possibility that the speed-engendering properties of the climbing-fiber system that permit rapid motor performance also contribute to the ability to associate events that are closely spaced in time. Because synchrony is such a prominent characteristic of the climbing-fiber system, it will be important to understand the contribution of electrotonic coupling for both motor control and for the learning about rapid event sequences.

ACKNOWLEDGMENTS

I thank Drs. E. Ahn, E. Lang, C. Schwarz, I. Sugihara, and Mr. R. Marquez for invaluable assistance with the collection of various parts of these data. This research was supported by a grant from the United States National Institute for Neurological Disorders and Stroke (NS31224-11).

REFERENCES

1. NICOLELIS, M.A.L. 2001. Prog. Brain Res. Adv. Neural Popul. Coding. Vol. 130.
2. THACH, W.T. 1968. Discharge of Purkinje and cerebellar nuclear neurons during rapidly alternating arm movement in the monkey. J. Neurophysiol. **31:** 785–797.

3. GELLMAN, R., J.C. HOUK & A.R. GIBSON. 1983. Somatosensory properties of the inferior olive of the cat. J. Comp. Neurol. **215:** 228–243.
4. BLOEDEL, J.R. & T.J. EBNER. 1984. Rhythmic discharge of climbing fibre afferents in response to natural peripheral stimuli in the cat. J. Physiol. **352:** 129–146.
5. SIMPSON, J.I. 1984. The accessory optic system. Annu. Rev. Neurosci. **7:** 13–41.
6. OSCARSSON, O. 1980. Functional organization of olivary projection to the cerebellar anterior lobe. *In* The Inferior Olivary Nucleus: Anatomy and Physiology. J. Courville, C. de Montigney & Y. Lamarre, Eds. Raven Press. New York.
7. BOWER, J.M. 1997. Is the cerebellum sensory for motor's sake, or motor for sensory's sake: The view from the whiskers of a rat? *In* Progress in Brain Research. The Cerebellum: From Structure to Control. C.I. DeZeeuw, P. Strata & J. Voogd, Eds.: **114:** 463–496. Elsevier. Amsterdam.
8. ITO, M. 1972. Neural design of the cerebellar motor control system. Brain Res. **40:** 81–84.
9. MANO, N.I. 1974. Simple and complex spike activities of the cerebellar Purkinje cell in relation to selective alternate movement in intact monkey. Brain Res. **70:** 381–393.
10. FU, Q.G., C.R. MASON, D. FLAMENT, *et al.* 1997. Movement kinematics encoded in complex spike discharge of primate cerebellar Purkinje cells. Neuroreport **8:** 523–529.
11. KITAZAWA, S., T. KIMURA & P.B. YIN. 1998. Cerebellar complex spikes encode both destinations and errors in arm movements. Nature **392:** 494–497.
12. WELSH, J.P. 1998. Systemic harmaline blocks associative and motor learning by the actions of the inferior olive. Eur. J. Neurosci. **10:** 3307–3320.
13. WELSH, J.P., E.J. LANG, I. SUGIHARA & R. LLINÁS. 1995. Dynamic organization of motor control within the olivocerebellar system. Nature **374:** 453–457.
14. AERTSEN, A.M.H.J., G.L. GERSTEIN, M. HABIB & G. PALM. 1989. Dynamics of neuronal firing correlation: modulation of "effective connectivity." J. Neurophysiol. **61:** 900–917.
15. GERSTEIN, G.L. 1999. Correlation-based analysis methods for neural ensemble data. *In* Methods for Neural Ensemble Recording. M.A.L. Nicolelis, Ed.: 157–177. CRC Press. Boca Raton.
16. BELL, C.C. & R.J. GRIMM. 1969. Discharge properties of Purkinje cells recorded on single and double microelectrodes. J. Neurophysiol. **32:** 1044–1055.
17. BLOEDEL, J.R. & W.J. ROBERTS. 1971. Action of climbing fibers in cerebellar cortex of the cat. J. Neurophysiol. **34:** 17–31.
18. EBNER, T.J. & J.R. BLOEDEL. 1981. Role of climbing fiber afferent input in determining responsiveness of Purkinje cells to mossy fiber inputs. J. Neurophysiol. **45:** 962–971.
19. MCDEVITT, C.J., T.J. EBNER & J.R. BLOEDEL. 1982. The changes in Purkinje cell simple spike activity following spontaneous climbing fiber inputs. Brain Res. **237:** 484–491.
20. SCHWARZ, C. & J.P. WELSH. 2001. Dynamic modulation of mossy fiber system throughput by inferior olive synchrony: A multielectrode study of cerebellar cortex activated by motor cortex. J. Neurophysiol. **86:** 2489–2504.
21. WELSH, J.P. & C. SCHWARZ. 1999. Multielectrode recording from the cerebellum. *In* Methods for Neural Ensemble Recording. M.A.L. Nicolelis, Ed.: 79–100. CRC Press. Boca Raton.
22. SIMPSON, J.I., D.R. WYLIE & C.I. DEZEEUW. 1996. On climbing fiber signals and their consequence(s). Behav. Brain Sci. **19:** 384–398.
23. LLINÁS, R. & M. MÜHLETHALER. 1988. Electrophysiology of guinea-pig cerebellar nuclear cells in the in vitro brain stem-cerebellar preparation. J. Physiol. (Lond.) **404:** 241–258.
24. LLINÁS, R. 1970. Neuronal operations in cerebellar transactions. *In* The Neurosciences: Second Study Program. F.O. Schmitt, Ed.: 409–426. Rockefeller University Press. New York.
25. WELSH, J.P. & R. LLINÁS. 1997. Some organizing principles for the control of movement based on olivocerebellar physiology. *In* Progress in Brain Research. The Cerebellum: From Structure to Control. C.I. DeZeeuw, P. Strata & J. Voogd, Eds.: Vol. **114:** 449–461. Elsevier. Amsterdam.

26. GOULD, S.J. & E.S. VBRA. 1982. Exaptation—a missing term in the science of form. Paleobiology **8:** 4–15.
27. LLINÁS, R., K. WALTON, D.E. HILLMAN & C. SOTELO. 1975. Inferior olive: Its role in motor learning. Science **190:** 1230–1232.
28. WELSH, J.P., B. CHANG, M.E. MENAKER & S.A. AICHER. 1998. Removal of the inferior olive abolishes myoclonic seizures associated with a loss of olivary serotonin. Neuroscience **82:** 879–897.
29. SMITH, M.C., S.R. COLEMAN & I. GORMEZANO. 1969. Classical conditioning of the rabbit's nictitating membrane response at backward, simultaneous, and forward CS-US intervals. J. Comp. Physiol. Psychol. **69:** 226–231.
30. MINTZ, M., D.G. LAVOND, A.A. ZHANG, *et al.* 1994. Unilateral inferior olive NMDA lesion leads to unilateral deficit in acquisition and retention of eyelid classical conditioning. Behav. Neural Biol. **61:** 218–224.
31. WELSH, J.P. & J.A. HARVEY. 1998. Acute inactivation of the inferior olive blocks associative learning. Eur. J. Neurosci. **10:** 3321–3332.

What Do Complex Spikes Signal about Limb Movements?

TIMOTHY J. EBNER, MICHAEL T. V. JOHNSON, ALEXANDER ROITMAN, AND QINGGONG FU

Department of Neuroscience, University of Minnesota, Minneapolis, Minnesota 55455, USA

ABSTRACT: Deciphering the information or signals carried by the complex spike discharge of Purkinje cells has proven to be problematic, primarily because of low frequency discharge and lack of adequate analytical techniques. This problem is particularly acute for studies of limb movements. To this end the relationship of cerebellar Purkinje cell complex spike discharge to direction and speed were studied in a manual-tracking task. Two monkeys were trained to pursue track targets moving in one of eight directions and at one of four speeds. An analysis based on Poisson regression modeling fitted the complex spike counts during single movement trials to target direction and/or speed. Using single trial data, the Poisson modeling demonstrated that the complex spike discharge for a majority of the Purkinje cells was significantly fit to tracking direction and speed. A second analysis based on the directional distribution of position and speed errors and a Poisson regression model of complex spike discharge to tracking position and speed errors found little relationship to movement error. Comparison of the preferred direction of the complex spike discharge with that of the simple spike activity revealed a reciprocal relationship for many cells. Thus, the complex spike discharge signals both tracking direction and speed but not movement errors. Furthermore, treating complex spike counts as a Poisson process provides a powerful tool for analyzing these events in single trials, without the need for extensive averaging.

KEYWORDS: climbing fiber; complex spike; Purkinje cells; direction; speed; pursuit tracking; limb movements

INTRODUCTION

The question of what motor or sensory signals are carried by the complex spike discharge of Purkinje cells remains unresolved. Produced by the powerful excitatory action of the climbing-fiber afferents, the low frequency discharge of complex spikes has made quantitative analyses difficult and may limit the signal content. It has been argued that the complex spike discharge is inherently random,[1] and it is the downstream intracellular signaling effects that are of primary concern.[2] Others have em-

Address for correspondence: Timothy J. Ebner, M.D., Ph.D., Department of Neuroscience, University of Minnesota, Lions Research Building, 2001 Sixth St. S.E., Room 421, Minneapolis, MN 55455. Voice: 612-626-9200; fax: 612-626-9201.
ebner001@tc.umn.edu

Ann. N.Y. Acad. Sci. 978: 205–218 (2002).

phasized the effects of complex spike discharge on the Purkinje cell's simple spike discharge, either in the short term[3] or long term.[2,4] Still others have stressed the inherent rhythmicity and coupling of inferior olivary neurons, the source of the climbing-fiber afferents, and that the information in the complex spike discharge is encoded in the population dynamics.[5,6] Lastly, and the subject of this study, others postulate that complex spikes carry specific movement-related signals.[7–11]

The relationship and contribution of complex spike activity to limb movement control is not as well understood as for the oculomotor system.[12] Again, a range of perspectives exists. Work in the cat suggests that inferior olive neurons or Purkinje cell complex firing does not encode movement parameters but rather somatosensory information.[13–16] In fact, the inferior olive can be gated off during volitional limb movements in the cat.[13,14,16] In contrast, complex spike discharge modulates in relation to specific parameters of limb movement in the monkey, as has been found for oculomotor behaviors.[7,8,10] Occurring primarily around movement onset[9,17–19] complex spike discharge is correlated with the direction and amplitude of reaching movements[20] and covaries with the speed of wrist movements.[18] Several studies have emphasized a role for complex spike discharge in signaling movement error.[11,19,21]

Unresolved is how to analyze complex spike discharge. One approach is simply to average a large number of trials,[10,11,21] using statistical or linear regression techniques to examine for correlations with task parameters. Another approach is to record from a population of Purkinje cells, extracting the signal from the ensemble.[6,22,23] However, analyses based on the analysis of single trial data generally have not been used, since complex spikes are discrete events that occur relatively rarely (~1 Hz). This suggests that methods based on "count" data rather than continuous data may be more useful. In this report, we introduce the use of Poisson regression modeling to the analysis of complex spike discharge in single trials. During a manual visuomotor tracking task, direction and speed modulate complex spike discharge, supporting the postulated role of complex spike discharge in signaling specific movement parameters.

MATERIALS AND METHODS

Tracking Tasks and Recording Procedures

Only a brief recounting of the methods is included, as the behavioral paradigm and recording procedures were described fully in a previous publication.[24] Two female rhesus monkeys (*Macaca mulatta*, 4–6 kg) were trained to use a two-joint manipulandum to make visually guided arm tracking movements over a horizontal workspace. In the *bell task*, the animal tracked targets that moved with bell-shaped speed profiles. The animal first superimposed the cursor (0.5 cm diameter) over a centrally positioned start target (1.2 cm diameter) for a randomized period of 0.75–1.5 s ("hold 1"). The target then began to move in one of eight directions (0°–315° in 45° intervals) at one of four bell-shaped speed profiles (peak speeds 2, 3, 4, and 5 cm/s and track durations 5.8, 4.5, 3.5, and 2.5 s). The animal was required to track the moving target by maintaining the cursor within the confines of the target for a distance of 5 centimeters. A "hold 2" period followed the track period. In the *constant speed task*, the animal also positioned the cursor in the central start box to start a trial. Following the randomized hold period (1–2 s), the "cue period" began with a square target box (1.2 cm diameter) appearing at one of the eight target positions (0°–315° in 45° increments), each 5 cm distant from the start box. The target moved at one of four constant speeds along a straight line to intersect the start box. During the cue period, the animal was required to maintain the cursor in the start box while watching the moving target. When the target reached the central start box, the start box was extinguished and the target continued to move along the same direction and speed for another 5 centimeters. During this "track period" the animal was required to keep the cursor within the moving target box. Four speeds of 2, 3, 4, and 5 cm/s were used, resulting in cue and track periods of 2.5, 1.7, 1.25, and 1.0 s in duration, respectively. The trial sequence was repeated over the eight directions and four target speeds in a randomized block design. For both tasks each direction-speed combination was presented 5–10 times.

Details of the chronic Purkinje cell recordings, including surgery, electrophysiological methods, and data acquisition, including eye movements and electromyographic activity (EMG) are included in an earlier report.[24] Most recordings were either anterior to the primary fissure in lobule V or posterior in lobule VI, extending mediolaterally from the intermediate region well into the hemisphere (HV and HVI).

Data Analyses

Complex spike data were analyzed using single-trial data using Poisson regression models.[25,26] The Poisson model, also called a log-linear model, was necessary for single-trial analysis of the complex spike discharge because the statistical errors in low-frequency data may not be independently and identically distributed, a necessary assumption for linear regression.

The first analysis evaluated the relationships of complex spike discharge rate to movement direction by fitting the mean complex spike discharge rate (μ/τ) at each target movement direction (θ) to a cosine tuning model for each target speed:[27,28]

$$\ln(\mu/\tau) = \alpha + \left(\sum_{j=1}^{5-10} \beta_j \text{trial}\right) + \xi \sin\theta + \eta \cos\theta + \varepsilon \quad (1)$$

This is expressed alternatively as:

$$\ln\mu - \ln\tau = \alpha + \left(\sum_{j=1}^{5-10} \beta_j \text{trial}\right) + \gamma[\cos\theta - \theta_{pd}] + \varepsilon \quad (2)$$

The logarithm of the trial duration (τ), called the *offset* term, was included in the model to correct for the unequal task durations. The trial (j) term was included as a categorical variable to account for trial-to-trial trends; however, tracking direction and speed were presented pseudorandomly. The peak of the cosine function corresponds to the direction in which cell discharge rate was highest and is referred to as the preferred direction (PD; θ_{pd} in Eq. 2) of the cell.[27] The regression coefficients α and γ (Eq. 2) represent the mean discharge rate over all directions and the depth of discharge modulation, respectively. The statistical error is represented by ε. An index of the depth of cell discharge modulation relative to the baseline firing was given as $I_{dir} = \gamma/\alpha$.[27] A Chi-Square statistic is used to test whether the fit was significant ($p<0.5$).

The mean complex spike discharge rate (μ/τ) for each peak (bell task) or constant speed (constant speed task) was also analyzed over each of the eight directions using a Poisson regression model with speed (s) as the predictor of interest:

$$\ln(\mu/\tau) = \alpha + \left(\sum_{j=1}^{5-10} \beta_j \text{trial}\right) + \delta s + \varepsilon \quad \textbf{(3)}$$

The value of δ represents the change in mean discharge rate with unit increases in speed along a given direction, in units of spikes/s per cm/s.

Another set of analyses determined the position and speed errors for the track period and correlated the complex spike count discharge with those two error signals. The position error, E_p, was the absolute difference between the animal's actual position and the required tracking position at each time bin, averaged throughout the entire track period. The speed error, E_s, was the difference between the animal's actual speed and the required tracking speed at each time bin, averaged throughout the track period. For each error signal, a strong directional dependency was observed, and, therefore, the preferred direction (PD) was determined using a standard cosine tuning model. In addition, the complex spike discharge rate (μ/τ) during the track period was analyzed using a Poisson regression model with E_p and E_s as the predictors:

$$\ln(\mu/\tau) = \alpha + \left(\sum_{j=1}^{5-10} \beta_j \text{trial}\right) + \xi_p E_p + \xi_s E_s + \varepsilon \quad \textbf{(4)}$$

RESULTS

Cerebellar Purkinje cell discharge was recorded from 154 cells in two monkeys during the performance of one of the two tracking tasks. Of these 154 cells, the complex spike discharge of 37 cells in the bell task and of 48 cells in the constant speed task was adequately isolated and discriminated throughout the entire set of trials and selected for further analyses.

The first analysis evaluated whether the complex spike discharge could be modeled as a Poisson process, using the counts per trial as the random variable. As shown in FIGURE 1 for the complex spike discharge of two Purkinje cells, the Poisson distribution provides an excellent approximation of the sampled probability density function. To estimate the Poisson probability density distribution (i.e., fitted distribution), the mean complex spike count per trial was used. The actual distributions were found to fit a Poisson distribution (Chi-square goodness of fit test) for these two cells and for the vast majority of cells (71%). Therefore, the complex spike counts can be modeled using a Poisson process.

The histograms and contour plots of a Purkinje cell for which the complex spike discharge was tuned to movement direction and modulated by movement speed during the track period of the constant speed task is shown in FIGURE 2. There was little

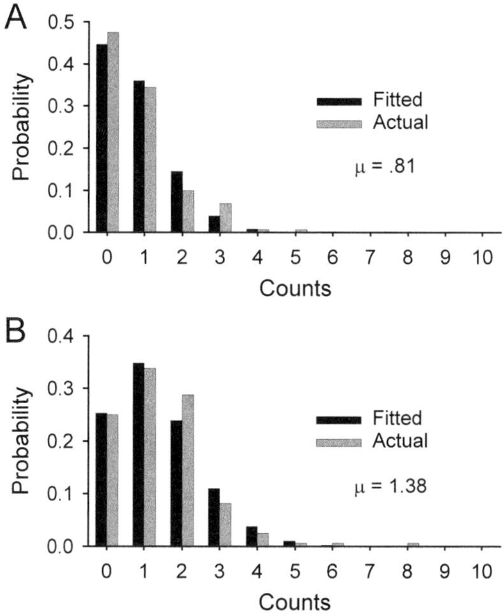

FIGURE 1. Two examples of the sample and estimated probability distributions of complex spike counts for two Purkinje cells. **A** is from the track period in the bell task, and **B** is from the track period in the constant speed task. Actual data is *open bars*, and the fitted distribution is *filled bars*. The mean μ of the Poisson distribution, was estimated from the sample average counts per trial and is given in each plot.

observable modulation in the complex spike discharge during the hold and cue periods. During the track period, however, complex spike discharge was greatest for 225°–270° tracking and increased with tracking speed. The complex spike discharge was significantly tuned to direction for movements at speeds of 5 cm/s (PD of 244° and I_{dir} of 0.88). The Poisson regression model also revealed significant speed modulation for tracking along 225° (δ = 0.30) and 270° (δ = 0.31).

Direction and Speed-Related Complex Spike Discharge Modulation

The spatial distributions of the complex spike PDs for the two tracking tasks are shown in FIGURE 3. The complex spike discharge of 21 of the 37 cells (57%) recorded during the bell task was significantly modulated with tracking direction, and the average I_{dir} was 1.27. The distribution of the PDs was not uniform (Watson U^2 test, $p<0.05$). In the cue period of the constant speed task (FIG. 3B), the complex spike discharge of 23/48 cells (48%) was tuned to movement direction (mean $I_{dir} = 1.38$). The distribution of PDs was not uniformly distributed ($p<0.05$). The complex spike discharge of 35/48 cells (75%) was tuned to movement direction during the track period of the constant speed task (FIG. 3C). Again, the PDs were not uniformly distributed ($p<0.05$), and the mean I_{dir} was 1.41. The large percentage of Purkinje cells with cue- and/or track-related modulation and similar depth of modulation suggests

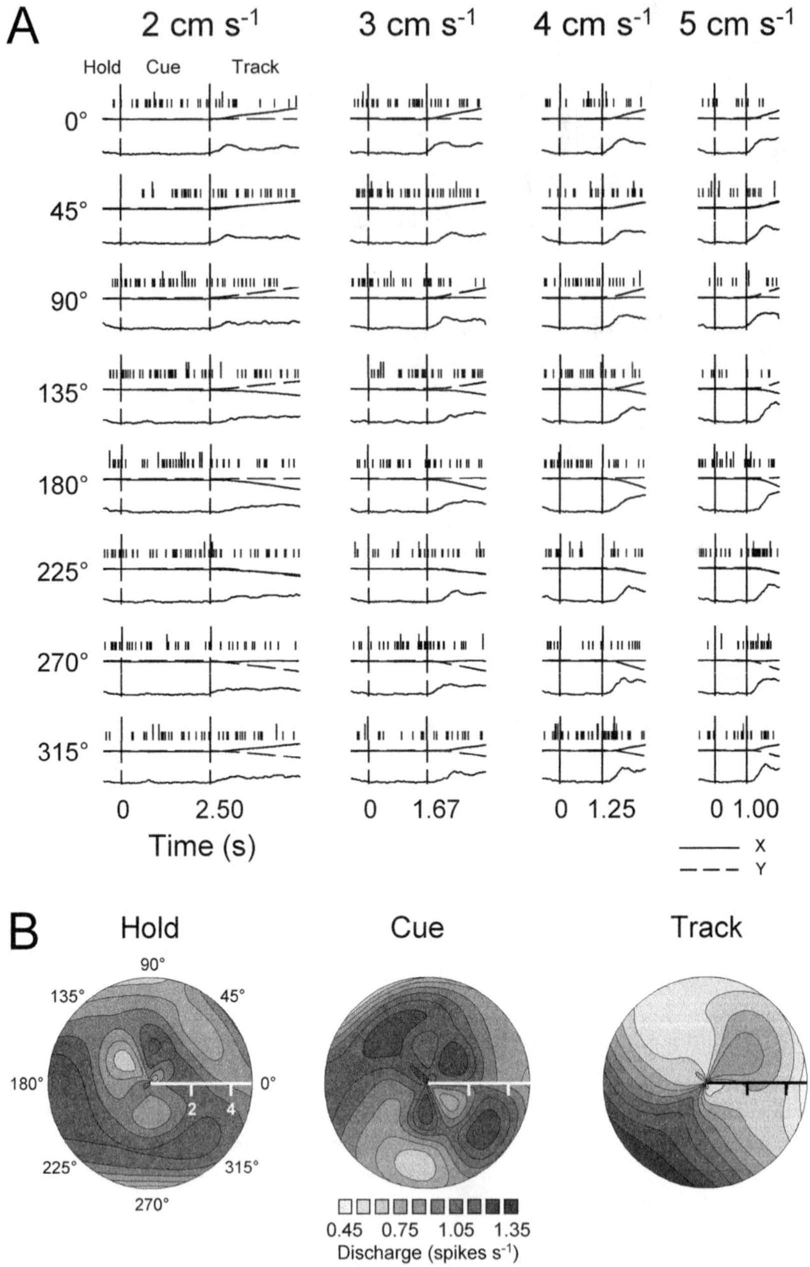

FIGURE 2. Complex spike discharge of a Purkinje cell tuned to direction and modulated by speed during the track period of the constant speed task. **(A)** Complex spike discharge is shown for all eight directions (0–315°) and all four speeds (2–5 cm/s) in spike

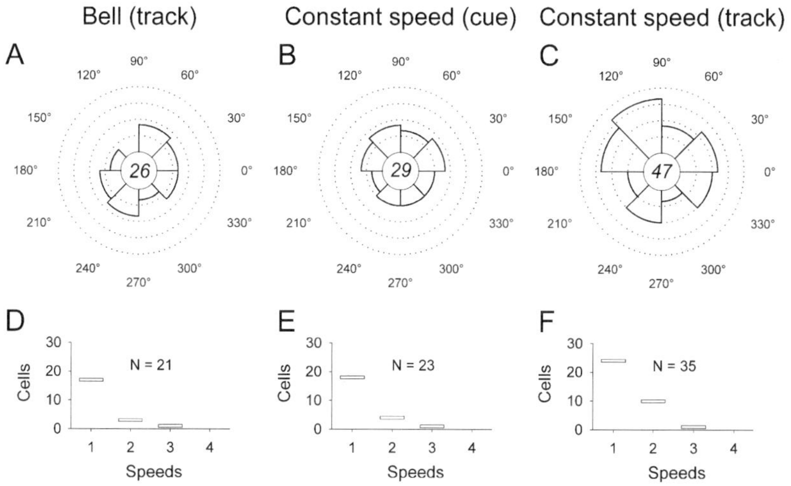

FIGURE 3. Summary of the directional complex spike discharge. (**A**–**C**) Rose plot of the distribution of PDs for all Purkinje cells with significant directional modulation irrespective of movement speed for the bell (**A**) and constant speed (**B** and **C**) tasks. Numbers of statistically significant PDs are given at the center of each plot. Length of pedals projecting outward from the center is the number with each dotted line equal to 3. (**D**–**F**) Histograms of the total numbers of cells (N) with significant directional discharge at each movement speed.

that visual and/or eye movements were equally effective as the actual arm movements in modulating the complex spike discharge.

The complex spike discharge was tuned to movement direction at usually only one of the four speeds, a common feature in either tracking or task period task (FIGS. 3D–F). In the bell task (FIG. 3D), 17 of the 33 cells were tuned to direction at just one speed. Similarly, for the constant speed task, during the cue period, 18 of the 23 cells were tuned to direction at only one speed and for the track period, 24 of 35 cells. This specificity of the directional tuning to just one speed mirrors what was previously described for the simple spike discharge.[24]

The complex spike discharge was also significantly modulated by tracking speed (Eq. 3). Twenty-one of the 37 cells (70%) recorded during the bell task were significantly modulated by movement speed along one or more directions (FIGS. 4A and D). Speed modulation occurred over all directions of the workspace with mainly positive speed slopes (mean absolute $\delta = 0.51$). Most cells had significant relation-

histogram form (20-ms time bins, average of 10 movement trials). (**B**) The three contour plots indicate the averaged firing as a function of all 32 direction-speed combinations. Direction is mapped on the circumference, and speed is mapped on the radius. From *left to right*, the polar plots refer to the hold, cue, and track periods of the trial sequence. The *legend* below the plots indicates the color mapped density of firing in complex spikes/second.

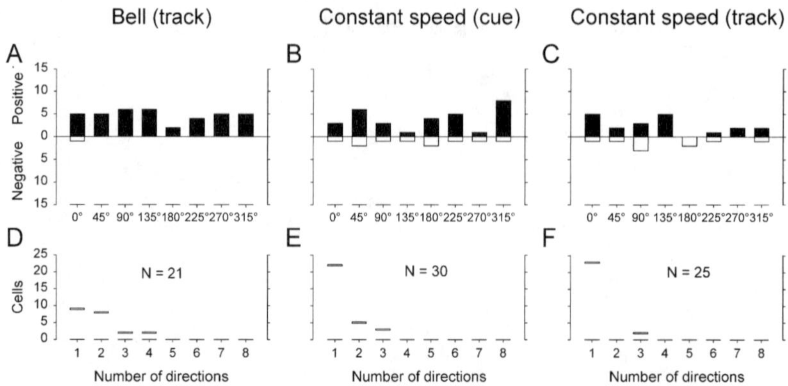

FIGURE 4. Distribution of directions along which speed-related modulation occurred. (**A–C**) Distribution of positive (*upward filled bars*) and negative (*downward open bars*) speed slopes for cells with significant speed-related modulation in the track period of the bell task (**A**), and in the cue (**B**) and track (**C**) periods of the constant speed task at each direction. The number of histogram bars exceeds the number of Purkinje cells with significant relationships to movement speed (indicated in each histogram in **D–F**), since some cells were modulated with speed along one or more directions. (**D–F**) Histograms of the total numbers of cells (N) with significant speed-related discharge at one or more directions for the track period of the bell task (**D**), and the cue (**E**) and track (**F**) periods of the constant speed task.

ships to speed along only one or two directions (FIG. 4D). In the cue period of the constant speed task, the complex spike discharge of 30/48 cells (62.5%) was significantly modulated with movement speed (mean absolute $\delta = 0.55$) and were distributed over all directions (FIGS. 4B and E). Similar findings were made for the 25 cells (45.6%) whose complex spike discharge was modulated by speed in the track period (mean absolute $\delta = 0.50$). As in the bell task, most cells had speed-related modulation for movements along only one direction (FIGS. 4E and 4F), and the slopes were mostly positive. Therefore, the finding that the complex spike discharge of most cells was directionally tuned usually at only one speed (FIG. 4) has a complement in the speed results. Also, the complex spike speed modulation in the cue period was similar to the modulation during the track periods.

Relationship of Complex Spikes to Tracking Errors

Both the PDs of the position and speed errors were dependent on tracking direction (FIGS. 5A and 5B), a result not unexpected when one considers the anisotropic properties of the arm and its interaction with the manipulandum.[29] The distributions of the PDs of the position and speed errors were highly nonuniform (Watson U^2, $p<0.5$), being more prevalent in the contralateral half of the workspace (FIG. 5). When compared to the distribution of the complex spike PDs during tracking (FIGS. 3A and C), this suggests that complex spikes were not correlated with these kinematic errors.

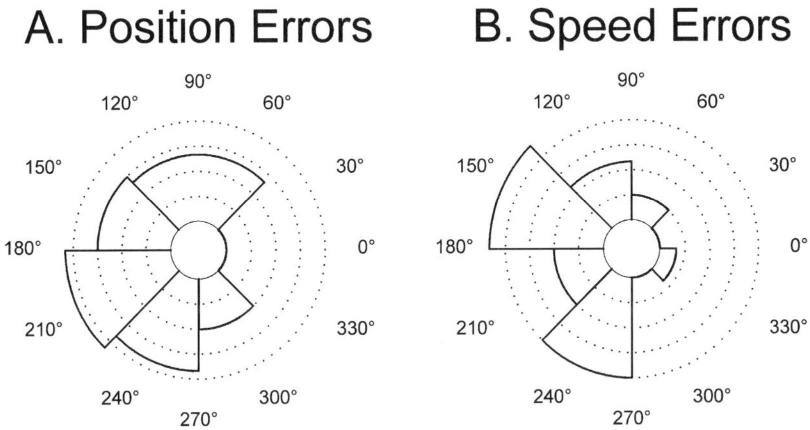

FIGURE 5. (**A**) Distribution of the PDs for tracking direction error. (**B**) Distribution of the PDs for tracking speed error. These distributions should be compared with the distributions of the PDs for the tracking periods of the bell task (FIG. 3A) or constant speed task (FIG. 3C).

A second explicit test of this question was undertaken, using the Poisson regression model to fit the complex spike counts to positional and/or directional error (Eq. 4). The complex spike discharge of only five cells had a significant correlation with position error, and only five cells had a significant correlation with direction error. Representing only 6% of the population, this occurrence rate is not unexpected by chance. Therefore, the complex spikes are not related to these two estimates of the tracking error.

Relationship between Simple and Complex Spikes in Individual Purkinje Cells

A reciprocal relationship between the complex spike and simple spike discharge has been found during visuo-vestibular behaviors.[7,8,10] A reciprocal relationship was also observed during manual tracking in this study. As shown in FIGURE 6 for two Purkinje cells, the simple spike and complex spike discharge are nearly 180° out-of-phase. This was observed in the bell task and constant speed task, and in the cue and track periods. Of the 14 Purkinje cells with significant directional tuning in both the simple spike and complex spike discharge, 71% ($n = 9$) showed this inverse firing relationship. However, for several Purkinje cells the simple spike and complex spike discharge were in-phase. Therefore, while a reciprocal relationship was common, it was not invariant.

DISCUSSION

A major analytical problem in relating climbing-fiber activity to higher-order movement parameters stems from the low firing rate (~1 Hz) of complex spikes. Several studies[10,18–20] have attempted to circumvent this problem by averaging or

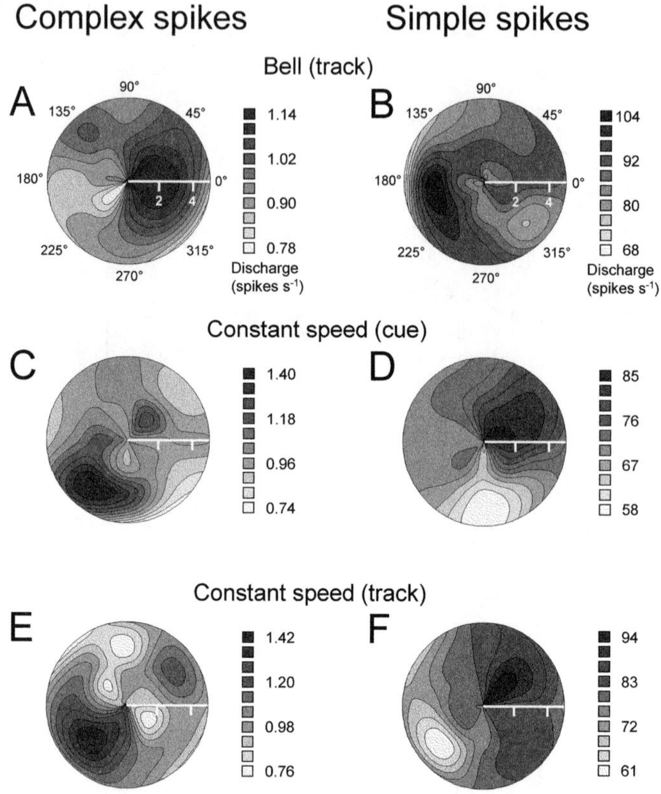

FIGURE 6. Contour plots of the complex spike and simple spike discharge profiles for two Purkinje cells in the bell tracking task (**A** and **B**) and the constant speed task (**C**–**D** cue period and **E**–**F** track period). Conventions as in FIG. 1.

analyzing complex spike activity over a large number of trials, and thereby treated complex spike discharge as though it were a continuous variable. A more sensitive approach is to use methods designed to relate quantal events or counts to continuous parameters. An information-theoretic approach has been used to relate climbing-fiber discharge to amplitude and error on a single-trial basis.[10,11] Kobayashi *et al.* introduced the use of logistic regression in the analysis of climbing-fiber activity, but relied extensively on averaged data in the description of their findings. In ordinary least-squares regression, as was used in the study of simple spike activity,[24,30] statistical errors were assumed to be independently and identically distributed, an assumption that cannot be made when the response variable (i.e., complex spike discharge) is quantal. Logistic regression is designed for use with a binary response variable (taking values of 0 or 1). Similarly, Poisson regression is used when the response takes on a limited number of interval counts (0, 1, 2, 3, ...) similar to the discharge properties of complex spikes. Examination of the number of complex spikes reveals a remarkable similarity to a Poisson process. Therefore, the Poisson regression approach seems well suited for the analysis of the relationship between complex spike discharge rate and task parameters on a single trial basis. The high percentage of cells with a significant correlation between the complex spike discharge and tracking direction and speed confirms the utility of this approach.

Correlation of Complex Spike Discharge to Direction and Speed

The complex spike discharge, like that of the simple spike discharge,[24] was modulated in a majority of cells by direction and speed. This result complements that of a previous study of complex spike activity during whole-arm reaching in primates, which also described directionally tuned and amplitude-modulated discharge.[20] Directionally tuned complex spike discharge was observed most commonly at one movement speed, and speed-related discharge modulation occurred most often along one movement direction (FIGS. 3 and 4). This is similar to that found for the relationship of simple spike discharge to these parameters,[24] and suggests that complex spikes may also be tuned to a preferred velocity. Overall, the findings show that complex spike discharge is modulated by specific task parameters and can be interpreted to encode these parameters.

Evolution of Complex Spike Discharge from Visual Instruction to Tracking

The constant speed task incorporated a visual instruction (cue) and a movement (track) period. Thus, comparisons concerning the likelihood that a complex spike occurred during the visual as compared to the actual movement phase of the task could be made. In the cue period, complex spike discharge was related to direction or speed in approximately 50% of the Purkinje cells—only slightly less than the track period. The mean I_{dir} and δ for the cue period were similar to that found for the track period. Therefore, direction and speed-related complex spike modulation was common when the animal was only viewing the moving target.

Possible sources of climbing-fiber activity in the cue period are eye movements, which were uncontrolled in this task, and visual stimuli. Climbing-fiber activity is evoked by the movement of the visual field[7] and during smooth pursuit eye movements.[8,10] The latter is the most likely source of the cue period modulation. Oculomotor behavior was similar over the trial sequence from cue to track periods in the face of changes in the direction and speed correlations.[24] While it is possible that eye movements and visual processing account for the kinematic correlations found through the entire trial sequence, the authors are not aware of any description of climbing-fiber modulation due to eye movements or visual stimuli for the areas of cerebellar cortex recorded in this study. A study explicitly examining Purkinje cell responses to these two variables is needed.

Comparisons of Complex and Simple Spike Discharge in Individual Cells

A reciprocal relationship for directional tuning between these two classes of action potentials has been found for floccular Purkinje cells for the movement of the visual field[7] and for smooth pursuit eye movements.[8,10] The preferred direction of the complex spike discharge was out of phase from the preferred direction of the simple spike discharge for the majority of cells (FIG. 6). A similar reciprocal relationship in the directional tuning occurs in reaching movements.[20,30]

The preferred directions of floccular Purkinje cells are discretely distributed and aligned along the response axes of the semicircular canals or the rotational axes of the extraocular muscles.[7,8] Modest deviations from reciprocal simple spike and complex spike preferred directions are attributed to admixtures of retinal slip and efference copy inputs to the Purkinje cell.[7] This rather fixed arrangement suggests that simple and complex spike discharge is organized in the same reference frame and a common reference frame should simplify the needed sensorimotor transformations. However, the reciprocal relationship found for limb movements in this and a previous report[20] is not as rigid as that found in the floccular Purkinje cells. Recent reports suggest that in-phase simple spike and complex spike modulations occur during head and eye movements.[31] Nor does a clear set of coordinate axes emerge. The preferred directions of the simple spike activity do show a distinct preference for the anterior-posterior movements.[24,30,32] Although the distribution of preferred directions for the complex spike discharge was not uniform, there is not a dramatic directional preference. This may reflect that multiple afferent systems (proprioceptive, exteroceptive, and visual) and effector systems (joints and muscles) are involved. The reduced stereotypy in the relationship between simple and complex spike directionality may reflect the complexities of the sensorimotor transformations needed to control a multiple degree-of-freedom arm.

The final question is whether the directionality in complex spike discharge determines or contributes to the directionality in the simple spike discharge. Possibly the less frequent complex spikes shape the directional tuning of the simple spikes. Both short-term mechanisms[3,33] and/or a long-term mechanisms[2,10,34,35] are possible. Climbing-fiber input might spatially focus modulation of the Purkinje cells by modifying the responsiveness to mossy-fiber input. This may select or weight the recruitment of specific directionally responsive cells to match the direction of a volitional movement. Variations in the ongoing kinematic performance (i.e., direction, velocity, and acceleration) would be signaled by the more frequent simple spike discharge.

ACKNOWLEDGMENTS

We wish to thank M. McPhee for assistance with graphics and histology and B. Swanson and L. King for help with manuscript preparation. We also thank J. Coltz for assistance with the early phases of this study. E. Ebner assisted in programming. This work was supported by National Institutes of Health Grant 5R01-NS-18338, and National Science Foundation Grant IBN-9873478.

REFERENCES

1. KEATING, J.G. & W.T. THACH. 1995. Nonclock behavior of inferior olive neurons: interspike interval of Purkinje cell complex spike discharge in the awake behaving monkey is random. J. Neurophysiol. **73:** 1329–1340.
2. ITO, M. 2001. Cerebellar long-term depression: characterization, signal transduction, and functional roles. Physiol. Rev. **81:** 1143–1195.

3. Ebner, T.J. & J.R. Bloedel. 1987. Climbing fiber afferent system: intrinsic properties and role in cerebellar information processing. *In* New Concepts in Cerebellar Neurobiology. J.S. King, Ed.: 371–386. Alan R. Liss, Inc. New York.
4. Ito, M., M. Sakurai & P. Tongroach. 1982. Climbing fibre induced depression of both mossy fibre responsiveness and glutamate sensitivity of cerebellar Purkinje cells. J. Physiol. **324:** 113–134.
5. Sasaki, K., J.M. Bower & R. Llinás. 1989. Multiple Purkinje cell recording in rodent cerebellar cortex. Eur. J. Neurosci. **1:** 572–586.
6. Welsh, J.P., E.J. Lang, I. Sugihara, *et al.* 1995. Dynamic organization of motor control within the olivocerebellar system. Nature **374:** 453–457.
7. Graf, W., J.I. Simpson & C.S. Leonard. 1988. Spatial organization of visual messages of the rabbit's cerebellar flocculus. II. Complex and simple spike responses of Purkinje cells. J. Neurophysiol. **60:** 2091–2121.
8. Stone, L.S. & S.G. Lisberger. 1990. Visual responses of Purkinje cells in the cerebellar flocculus during smooth-pursuit eye movements in monkeys. I. Simple spikes. J. Neurophysiol. **63:** 1241–1261.
9. Ojakangas, C.L. & T.J. Ebner. 1992. Purkinje cell complex and simple spike changes during a voluntary arm movement learning task in the monkey. J. Neurophysiol. **68:** 2222–2236.
10. Kobayashi, Y., K. Kawano, A. Takemura, *et al.* 1998. Temporal firing patterns of Purkinje cells in the cerebellar ventral paraflocculus during ocular following responses in monkeys II. Complex spikes. J. Neurophysiol. **80:** 832–848.
11. Kitazawa, S., T. Kimura & P.B. Yin. 1998. Cerebellar complex spikes encode both destinations and errors in arm movements. Nature **392:** 494–497.
12. West, G.A., D.R. Haynor, R. Goodkin, *et al.* 1994. Magnetic resonance imaging signal changes in denervated muscles after peripheral nerve injury. Neurosurgery **35:** 1077–1086.
13. Apps, R. 2000. Gating of climbing fibre input to cerebellar cortical zones. Prog. Brain Res. **124:** 201–211.
14. Gellman, R., A.R. Gibson & J.C. Houk. 1985. Inferior olivary neurons in the awake cat: detection of contact and passive body displacement. J. Neurophysiol. **54:** 40–60.
15. Rushmer, D.S., W.J. Roberts & G.K. Angter. 1976. Climbing fiber responses of cerebellar Purkinje cells to passive movement of the cat forepaw. Brain Res. **106:** 1–20.
16. Horn, K.M., P.L. Van Kan & A.R. Gibson. 1996. Reduction of rostral dorsal accessory olive responses during reaching. J. Neurophysiol. **76:** 4140–4151.
17. Bauswein, E., F.P. Kolb, B. Leimbeck, *et al.* 1983. Simple and complex spike activity of cerebellar Purkinje cells during active and passive movements in the awake monkey. J. Physiol. **339:** 379–394.
18. Mano, N.-I., I. Kanazawa & K.-I. Yamamoto. 1986. Complex-spike activity of cerebellar Purkinje cells related to wrist tracking movement in the monkey. J. Neurophysiol. **56:** 137–158.
19. Ojakangas, C.L. & T.J. Ebner. 1994. Purkinje cell complex spike activity during voluntary motor learning: relationship to kinematics. J. Neurophysiol. **72:** 2617–2630.
20. Fu, Q.-G., C.R. Mason, D. Flament, *et al.* 1997. Movement kinematics encoded in complex spike discharge of primate cerebellar Purkinje cells. NeuroReport **8:** 523–529.
21. Wang, J.-J., J.H. Kim & T.J. Ebner. 1987. Climbing fiber afferent modulation during a visually guided, multi-joint arm movement in the monkey. Brain Res. **410:** 323–329.
22. Llinás, R. & K. Sasaki. 1989. The functional organization of the olivo-cerebellar system as examined by multiple Purkinje cell recordings. Eur. J. Neurosci. **1:** 587–602.
23. Lang, E.J., I. Sugihara & R. Llinás. 1996. GABAergic modulation of complex spike activity by the cerebellar nucleoolivary pathway in rat. J. Neurophysiol. **76:** 255–275.
24. Coltz, J.D., M.T.V. Johnson & T. Ebner. 1999. Cerebellar Purkinje cell simple spike discharge encodes movement velocity in primates during visuomotor arm tracking. J. Neurosci. **19:** 1782–1803.
25. Cameron, A.C. 1998. Regression Analysis of Count Data. Cambridge. New York.
26. Agresti, A. 1996. An Introduction to Categorical Data Analysis. Wiley. New York.
27. Georgopoulos, A.P., J.F. Kalaska, R. Caminiti, *et al.* 1982. On the relations between the direction of two-dimensional arm movements and cell discharge in primate motor cortex. J. Neurosci. **2:** 1527–1537.

28. FU, Q.-G., J.I. SUAREZ & T.J. EBNER. 1993. Neuronal specification of direction and distance during reaching movements in the superior precentral premotor area and primary motor cortex of monkeys. J. Neurophysiol. **70:** 2097–2116.
29. SAINBURG, R.L., M.F. GHILARDI, H. POIZNER, *et al.* 1995. Control of limb dynamics in normal subjects and patients without proprioception. J. Neurophysiol. **73:** 820–835.
30. FU, Q.-G., D. FLAMENT, J.D. COLTZ, *et al.* 1997. Relationship of cerebellar Purkinje cell simple spike discharge to movement kinematics in the monkey. J. Neurophysiol. **78:** 478–491.
31. BELTON, T., M. SUH & J.I. SIMPSON. 2001. The non-visual complex spike signal in the flocculus responds to challenges to the vestibulo-ocular reflex gain. Soc. Neurosci. Abstr. **27:** 70.9.
32. FORTIER, P.A., J.F. KALASKA & A.M. SMITH. 1989. Cerebellar neuronal activity related to whole-arm reaching movements in the monkey. J. Neurophysiol. **62:** 198–210.
33. EBNER, T.J. & J.R. BLOEDEL. 1981. Role of climbing fiber afferent input in determining responsiveness of Purkinje cells to mossy fiber inputs. J. Neurophysiol. **45:** 962–971.
34. ITO, M. 1989. Long-term depression. Annu. Rev. Neurosci. **12:** 85–102.
35. GILBERT, P.F.C. & W.T. THACH. 1977. Purkinje cell activity during motor learning. Brain Res. **128:** 309–328.

Inhibitory Control of Olivary Discharge

ALAN R. GIBSON, KRIS M. HORN, AND MILTON PONG

Barrow Neurological Institute, St. Joseph's Hospital, Phoenix, Arizona 85013, USA

ABSTRACT: The inferior olivary nucleus is the sole source of an entire afferent system to the cerebellum, the climbing-fiber system. Inferior olivary neurons are very sensitive to the appropriate sensory stimuli, such as light contact to the paw. Yet, when animals move about, olivary cells show little change in discharge rate. Apparently some mechanism prevents the cells from discharging to stimuli generated by the animal's own movement. The inferior olive receives a massive inhibitory input from small cells in the cerebellar and vestibular nuclei. This article reviews the results from several experiments that suggest that the inferior olive is specifically targeted by inhibitory inputs that prevent responses to stimuli resulting from self-produced movement. Oscarsson proposed that the inferior olive provides the cerebellum with information about errors of motor performance and about spinal reflexes. We argue that it is unlikely that the inferior olive provides information about movement errors, although the olive may signal the occurrence of sensory events that are likely to elicit reflex movements. Another popular theory of climbing-fiber action argues that the climbing fibers play a role in altering the strength of the parallel fiber–Purkinje cell synapse. The cerebellum is important for the formation of classically conditioned responses, and input generated by the unconditioned stimulus does provide effective stimulation of olivary neurons. Although the olive does not generate the unconditioned response, it may provide the cerebellum with information necessary for the formation of conditioned responses.

KEYWORDS: inferior olive; inhibition; cerebellum; red nucleus; cuneate nucleus; interpositus

INTRODUCTION

Cells in the inferior olive (IO) provide climbing-fiber (CF) input to the cerebellum.[6] The physiological and anatomical properties of IO neurons are unusual, and understanding the conditions that elicit olivary discharge should help in unraveling both olivary and cerebellar function. Although olivary neurons discharge at low rates, their discharge generates a complex spike in the Purkinje cell, which has a strong effect on the cell. The complex spike is characterized by a powerful depolarization of the target Purkinje cell and is followed by a prolonged period of hyperpolarization.[7,8] Low rates of CF discharge can completely preclude simple spike discharge of the Purkinje cell.[4,9] Most cells in the IO are responsive to some form of somatosensory stimuli, and, even in anesthetized animals, many cells are highly sen-

Address for correspondence: Alan R. Gibson, Barrow Neurological Institute, St. Joseph's Hospital, Phoenix, AZ 85013. Voice: 602-406-3732; fax: 602-406-4172.
agibson@chw.edu

Ann. N.Y. Acad. Sci. 978: 219–231 (2002). © 2002 New York Academy of Sciences.

sitive to stimuli such as light touches to the hair and taps to the skin.[10–13] High sensitivity is not surprising, since neurons in the spinal cord as well as in the trigeminal, gracile, and cuneate nuclei terminate on olivary neurons.[14,15] These input pathways involve only a small number of synapses from peripheral receptors to the IO, so they are not strongly depressed by anesthesia.

In the awake animal, olivary responses can be elicited by the same stimuli that are effective in the anesthetized animal.[16] However, if the animal is actively moving, olivary cells show no pronounced increase in discharge rate, even when the movement results in contact to parts of the body appropriate for activating the neurons.[16–19] This lack of responding suggests that active movement suppresses olivary discharge, either by inhibiting olivary neurons or neurons in its afferent pathways (or both). We will review several of our findings that indicate that inhibition of responses during movement is specific to the olive. Since IO neurons cannot modify cerebellar activity when their responses are suppressed, modulation of olivary sensitivity places constraints on the contribution that IO discharge can make to cerebellar function.

SOURCES OF OLIVARY INHIBITION

Anatomic data support the possibility that active movement inhibits the IO. The inhibitory neurotransmitter gamma-aminobutyric acid (GABA) is marked by the presence of its synthetic enzyme, glutamic acid decarboxylase (GAD). The IO in a variety of species, including humans, contains a large number of boutons that immunostain for GAD.[20,21] A large source of this inhibitory input arises from small fusiform cells in the cerebellar nuclei,[22,23] which are known to be GABAergic.[22–26] Cells in the cerebellar nuclei that project to nonolivary targets, such as the thalamus and red nucleus, are excitatory and have large increases in discharge during movement.[27] Since a single Purkinje cell can innervate both olivary and nonolivary projecting cells,[28] it is likely that increased nuclear activity during movement is associated with increased inhibition in the IO.

In the anesthetized cat, increasing cerebellar nuclear activity produces a profound depression of olivary activity. FIGURE 1 illustrates evoked responses in the rostral dorsal accessory division of the olive (rDAO) to electrical stimulation. The receptive field was large, and cutaneous shock anywhere in the receptive field elicited a robust response (FIG. 1B). We increased activity in the cerebellar interpositus nucleus, which projects to rDAO, by injecting bicuculline, an antagonist to the inhibitory neurotransmitter GABA (collaborative research with Dr. Tom Ruigrok, Erasmus University). The bicuculline injection produced high-bursting activity in interpositus neurons (FIG. 1A), which was associated with a total suppression of rDAO responses to cutaneous shock (FIG. 1B). As the effect of the bicuculline wore off, rDAO neurons again became responsive to electrical stimulation.

Although cerebellar nuclei are probably the largest source of inhibitory input to the IO, other brain stem regions, such as ventral areas of the cuneate nucleus, contain GABAergic neurons that project to the IO.[20] These cells may allow activity of descending motor pathways, such as the rubrospinal tract (RST) and corticospinal tract, to inhibit IO neurons. In the anesthetized preparation, trains of stimulus pulses to the red nucleus (RN) or motor cortex produce a prolonged period of suppression of olivary responses to peripheral stimulation.[29,30] At least for the RN, the inhibitory effect does not

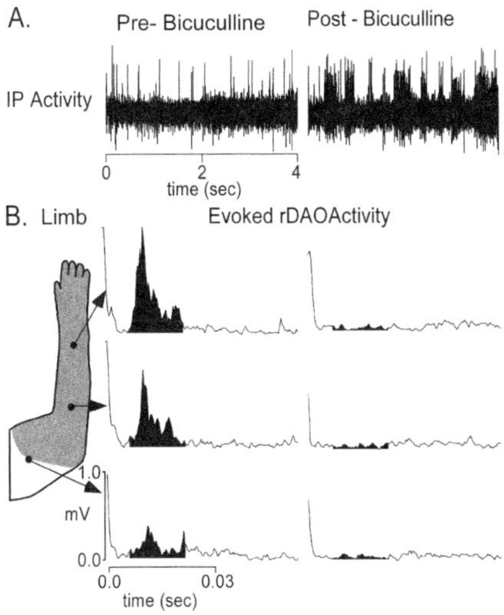

FIGURE 1. Increased cerebellar activity eliminates cutaneous sensitivity in the inferior olive. (**A**) Multiunit activity in the cerebellar interpositus nucleus before (*left*) and after (*right*) injection of GABA antagonist bicuculline into the cerebellar interpositus nucleus (1 μL at 0.1% in saline). Bursting activity increases after injection. (**B**) Evoked potentials measured in rDAO before (*left*) and after (*right*) injection of GABA in interpositus. Cutaneous electric shocks were applied to three different points across the receptive field as indicated on the diagram to the left. Response traces start at the onset of the shock. Potentials are reduced after increasing activity in the cerebellum.

depend on the cerebellum, since inhibition following stimulation occurs even after the cerebellum has been ablated.[29] However, this effect might be mediated by inhibition of afferent pathways to the IO, since other reports have indicated that stimulation of RN (or motor cortex) can inhibit sensory responses in the cuneate nucleus.[31,32] The RST and corticospinal tract have extensive terminations in regions of the cuneate and gracile nuclei that project to the IO.[33]

The hind limb region of the rDAO receives monosynaptic input from spinal afferents that travel in the ventral funiculus (VF) of the spinal cord.[34] Therefore, it is possible to test if inhibition produced through RN or RST stimulation is specific to the IO by stimulating the VF rather than the skin.[35] If RN-mediated inhibition occurs at an extraolivary site, then IO activation via stimulation of the VF should not be suppressed by RN stimulation. FIGURE 2 illustrates evoked responses in rDAO to VF stimulation with and without conditioning stimulation to the RN. The rDAO response to VF stimulation is greatly attenuated following RN conditioning stimulation, thereby indicating that inhibition is occurring within the IO.

However, RN-mediated inhibition might not be specific to the IO since afferent pathways to other structures may also be inhibited. We examined this possibility by recording evoked responses in ascending sensory pathways. Shocking the hind limb produces an evoked potential from the fibers in the VF and the dorsal column (DC) at mid-thoracic levels of the cord. FIGURE 3 illustrates the evoked response in the VF and DC with and without conditioning stimulation to the RN. As previously illustrated, the

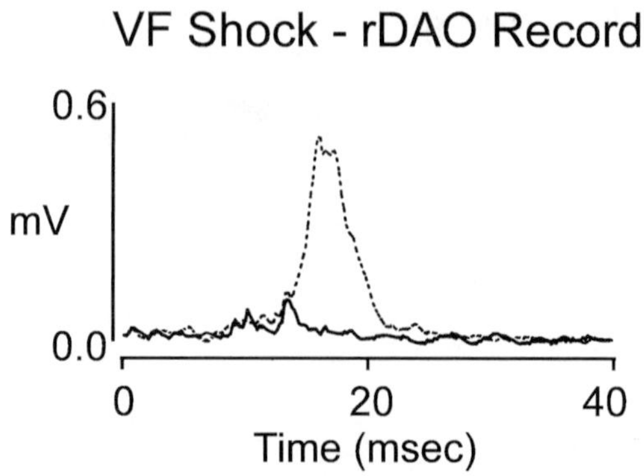

FIGURE 2. RN stimulation reduces IO sensitivity by inhibition within the olive. Stimulation of the ventral funiculus (VF) excites IO neurons via monosynaptic input. Two averages (n=20) of rDAO potentials recorded after shock (40 μA, 0.5-ms pulse) to ventral funiculus at time zero, with (*solid*) and without (*dashed*) RN conditioning stimulation (60 μA, 0.1-ms pulse, 400 Hz, 100-ms train). Adapted from Horn *et al.*[35]

evoked rDAO response is greatly reduced following RN stimulation, but neither the VF nor DC response is attenuated. Additionally, sensory evoked potentials recorded from the thalamus are unaffected by RN conditioning (FIG. 4). Therefore, RN stimulation produces inhibition specific to the olivocerebellar pathway. At least some of the inhibition occurs within the IO, although it is possible that olivary-projecting cells in various sensory nuclei are also inhibited.

OLIVARY INHIBITION DURING BEHAVIOR

Although experiments with anesthetized preparations indicate that activity in descending motor pathways can inhibit the IO, the interplay of excitatory and inhibitory influences in the behaving animal may not result in periods of IO response inhibition. To test olivary sensitivity during behavior, we trained cats to reach out, grasp, and retrieve a lever (FIG. 5) and then recorded IO activity during the task.[36] Many of our experiments focused on the activity of the rDAO division of the IO, since light cutaneous contact, which is easy to generate as a stimulus, strongly activates rDAO neurons. Additionally, the well-defined and highly organized cutaneous receptive fields of rDAO neurons[11] allow the localization of cells responding to stimulation of specific parts of the body, such as the distal forelimb.

During the undisturbed course of the reaching task, rDAO neurons discharge at a low rate in an apparently random fashion. FIGURE 6A illustrates recordings from a rDAO neuron with a receptive field on the toes during several reach-to-grasp trials. The rDAO neuron discharges throughout the trial with no obvious association to any particular behavioral event. FIGURE 6B illustrates discharges from the same cell on trials where a force perturbation has been introduced to the feeder handle during retrieval (dashed line). The perturbation is strong enough to momentarily reverse the motion of the limb. Again, there is no rDAO discharge related to the perturbation. The same cell,

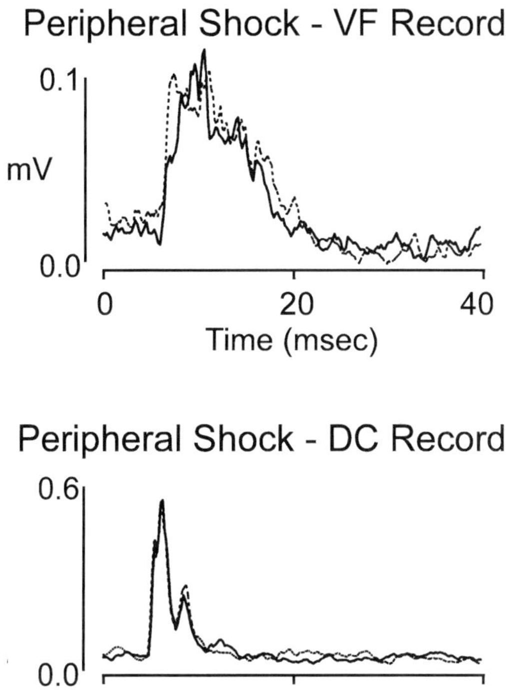

FIGURE 3. RN stimulation does not inhibit spinal sensory pathways. Averaged responses (n=20) to peripheral shock on hind limb at time zero, with (*solid*) and without (*dashed*) RN stimulation. Recordings show evoked responses from the VF (*top*) and dorsal columns (*bottom*). RN stimulation had no effect on the potentials at either location. Adapted from Horn *et al.*[35]

however, responds to light taps on the platform during stance periods between reaching trials (FIG. 7).

The rDAO cell illustrated in the previous FIGURES is typical of all rDAO cells. Sensitivity is high during stance but rapidly decreases at the beginning of the reach. Another way to stimulate rDAO cells is to use subcutaneous shock through implanted electrodes. rDAO neurons can be activated with single low current stimulus pulses, and the cats show no aversive response to the stimulation and continue to reach without interruption. The advantage of the electrical stimulation over mechanical stimulation is that the electrical stimuli can be presented at precise times throughout the task. The added precision allows measuring the amplitude of the evoked rDAO response during different periods of the reaching task. FIGURE 8A plots the amplitude of the rDAO evoked response as a function of time, with time 0 representing the beginning of reach. The rDAO response amplitude is high and constant throughout the stance period but rapidly decreases at the onset of reach. The response amplitude remains low until the cat returns to a standing position. The suppression of rDAO responses is not solely dependent upon movement, as different limb postures are associated with different amplitudes of response (hold versus stance). Similar findings

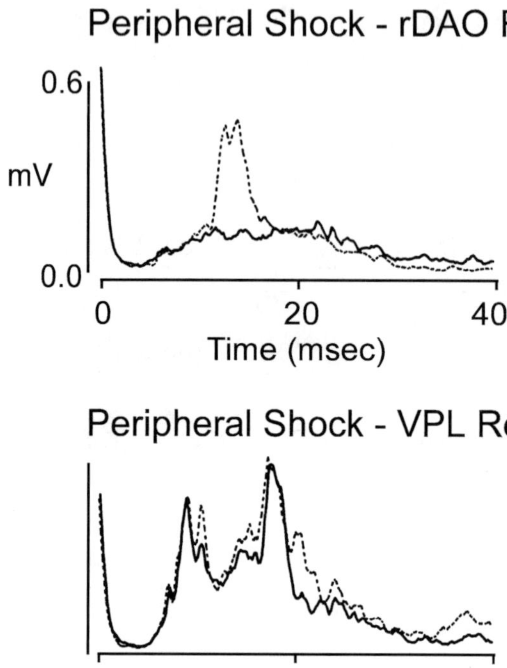

FIGURE 4. RN stimulation inhibits olivary but not thalamic sensory responses. Averaged responses ($n=20$) to peripheral shock on hind limb at time zero, with (*solid*) and without (*dashed*) RN stimulation. Recordings show responses in rDAO (*top*) and in ventral posterior lateral nucleus (VPL) of the thalamus (*bottom*). RN stimulation inhibits the evoked potentials in rDAO but not VPL. Adapted from Horn *et al.*[35]

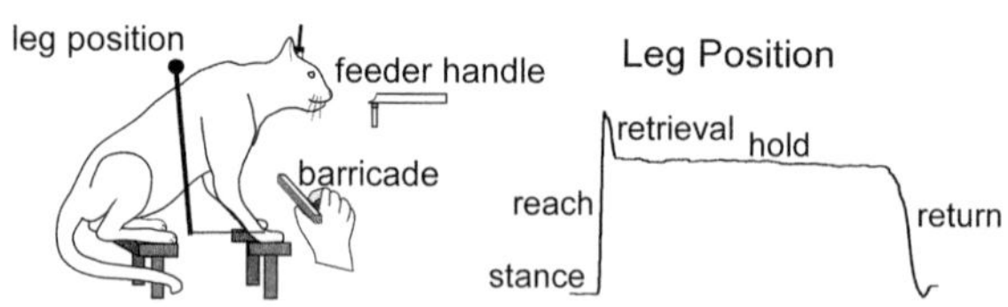

FIGURE 5. Reach-to-grasp task provides a behavior during which olivary activity can be monitored in the awake animal. Diagram on *left* shows the general setup, including a barricade to perturb the animal's movements. Trace on *right* is a measure of leg position from a potentiometer attached to a lever arm that is attached to the cat's wrist by a string. The cat initially has its front paws resting on the pedestals. After a tone sounds, the cat reaches out to the feeder handle, grasps the handle, and then pulls back toward its mouth. As the animal holds the handle, it licks off a food reward. The cat then releases the handle and returns the forelimb to the pedestal.

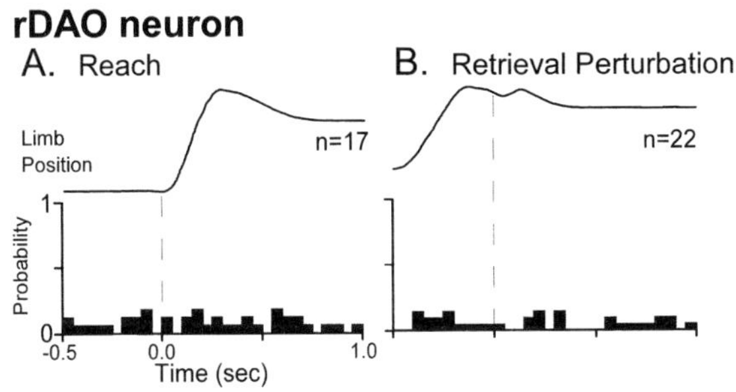

FIGURE 6. rDAO neurons do not respond during the reach-to-grasp task. Average limb position traces (*top*) and probability of spike firing (50-ms bins) recorded from an rDAO cell with a receptive field on the toes. (**A**) Data aligned to the time of forelimb lift from the platform (*vertical dashed line*). The rDAO cell showed no increase in firing during any phase of the task. (**B**) Data aligned on the occurrence of a perturbation that pulls the handle away from the animal during retrieval. The rDAO cell failed to respond to the perturbation. Adapted from Horn *et al.*[36]

of reductions in rDAO sensitivity during reaching have been reported by Apps *et al.*[37] using measures of evoked responses from the C1 and C3 cerebellar cortical zones, which receive input from rDAO.[38]

Several papers have reported reductions in sensory responses during movement,[39,40] and the reduction in the rDAO response may not differ from that of other sensory systems. To test this, we (in collaboration with Dr. John Stein of Oxford University) performed an experiment similar to the preceding rDAO experiment by measuring responses in the main cuneate nucleus. The main cuneate nucleus receives cutaneous afferents from the forelimb and supplies input to the rDAO.[15,41] During recording, it was immediately apparent that the cuneate neurons differed from those of rDAO. Cells in the cuneate had large increases in discharge during the task, and the increases corresponded to specific events, such as lifting the foot or grasping the feeder handle. FIGURE 9A illustrates discharge of a cuneate cell with a receptive field on the paw. The cell discharged at a high rate when the cat lifted his foot at the beginning of the reach. FIGURE 9B illustrates discharge from the same cuneate cell with the discharge records aligned to a torque pulse applied to the feeder handle during retrieval. In contrast to the rDAO cell shown in FIGURE 6B, there was a strong response coinciding with the onset of the perturbation.

The discharge associated with movement suggested that cuneate neurons maintain high sensitivity during the reaching task. To quantify the sensitivity at different times during the task, we recorded the average evoked potential from the main cuneate to a cutaneous electrical shock, as we did from rDAO (FIG. 7). FIGURE 8B illustrates the amplitude of the evoked response from a recording site within the main cuneate. When the stimulation was applied during stance, the cuneate showed a high-amplitude evoked response, just as the rDAO did. However, during the reaching task, the evoked response from the cuneate showed no reduction in amplitude. Clearly, response modulation is much stronger in rDAO than in the main cuneate nucleus. In fact, the IO modulation is so strong that rDAO neurons are essentially gated on or off by the behavioral state.

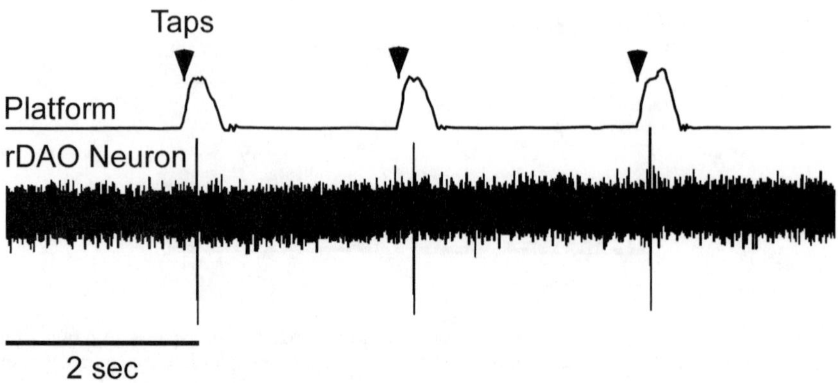

FIGURE 7. Rostral DAO cells respond to stimulation during stance. Same cell as in FIG. 6. *Top trace* shows pressure record of the foot pedestal. The *arrows* indicate the onset of taps applied by the experimenter to the cat's toes. The *bottom trace* shows the activity of the rDAO cell. Each tap elicits one spike. Adapted from Horn *et al.*[36]

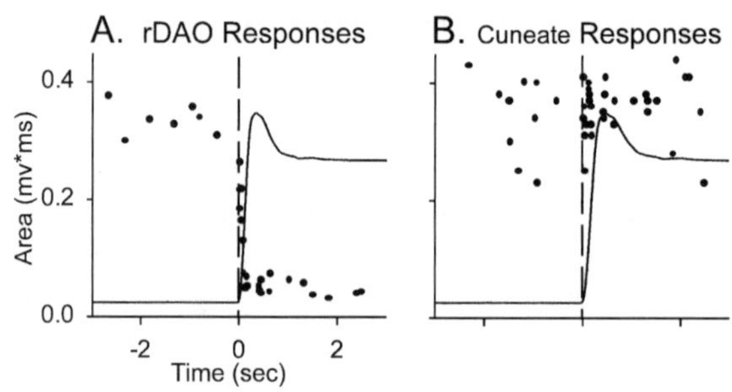

FIGURE 8. Rostral DAO and cuneate cells respond differently during movement. Graphs plot amplitude of evoked responses in rDAO and cuneate to subcutaneous shock at times before and after the onset of movement (*dashed line*). Each point is the integration of the rectified response measured across an 8-ms window centered on the peak of the response. *Black trace* shows corresponding forelimb movement. Evoked responses in rDAO are strongly reduced shortly after onset of movement, but movement has little effect on cuneate responses. Adapted from Horn *et al.*[36]

EXPECTED CONTACT

Oscarsson[1] hypothesized that the IO acts as a comparator to signal perturbations of movement due to spinal reflexes or unexpected changes in conditions, such as load or resistance. To do this, the IO compares inputs based on motor commands to feedback from receptors and generates an error signal when these inputs do not match. This "error detection" model has become the most commonly held theory of olivary function (i.e., see Ref. 4), and several experiments have produced data that

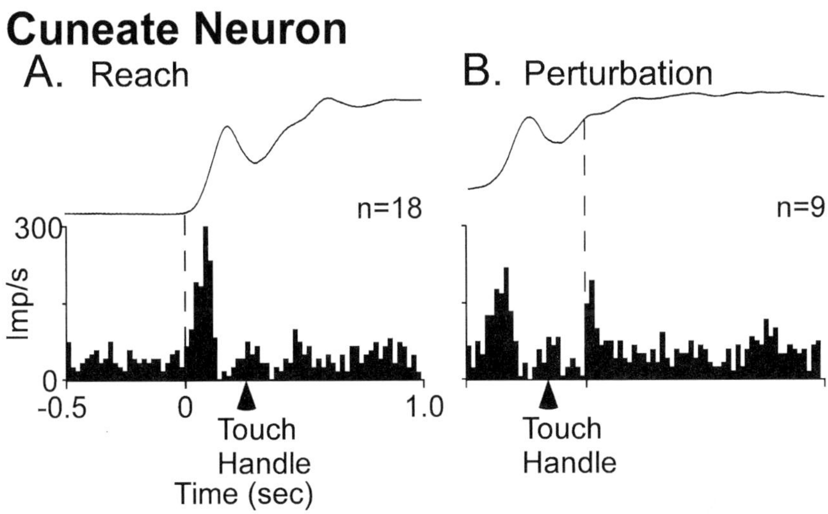

FIGURE 9. Cuneate cells respond during movement. Averaged forelimb traces (showing measured elbow angle) and averaged discharge rate for a cuneate cell. The cuneate neuron had a receptive field on the toes. In contrast to rDAO cells, the cuneate cell responds to both the task (**A**, aligned on foot lift, *dashed line*) and to perturbations during the task (**B**, aligned on perturbation, *dashed line*).

appear to be consistent with the model. For example, if a rung is unexpectedly allowed to drop while a cat is walking on a ladder, Purkinje cells often (but not always) respond with a complex spike.[42] Similarly, complex spikes are generated when the load on a manipulandum is changed, and these responses wane as the animal gains experience (learns) with the new load.[43] Presumably, the motor program is modified, and the IO ceases to discharge when the motor command produces the expected movement.

In our reaching paradigm, the cat has a great deal of experience with reaching out and grasping the lever, so he is "expecting" contact with the handle at the end of the reach. Therefore, IO neurons do not respond to stimulation produced by contact with the handle. How would the IO respond if a cat reached for a handle that wasn't present? The act of reaching suggests that the cat is "expecting" contact with the handle, so the IO should fire at the time when the cat would normally touch the handle. Furthermore, the forelimb would continue to extend when the handle wasn't contacted, so actual limb position would fail to match the planned movement and neurons in the IO should be activated.

To test this, we recorded IO neurons while a well-trained cat reached out and grasped the handle. We then moved the handle to a new location and, for several trials, the cat reached to the accustomed handle position and made several attempts to grasp the missing handle. Regardless of the division of the IO, no cells responded at the time of "expected" contact with the handle.[44] However, some of the cells did respond when the cat bumped into various pieces of apparatus during the grasping attempts.

Apparently, IO cells require peripheral stimulation for activation, since self-generated limb movements in the absence of contact do not generate IO discharge, even when the movement results in an "error." With repeated attempts, the cats do learn to reach to the new location in order to retrieve the handle. At least this form of learning can occur without an error signal from the IO. The absence of discharge to the missed grasp is not particularly surprising, since no contact to the limb occurs during these attempts. The primary information about forelimb movement during the missed reach probably arises from Ia muscle afferents, which provide a massive mossy-fiber input to cerebellar cortex via the external cuneate nucleus.[45–47] In contrast, the IO receives no input from the external cuneate.[48]

Response properties of IO neurons are diametrically opposed to the properties of a signal that one might expect for a movement error signal: (1) IO neurons do not discharge to self-produced movement, even when the movement results in an error. (2) IO discharge requires an appropriate stimulus, and many IO cells are most sensitive during quiet stance. (3) A single olivary neuron can respond to a variety of inputs such as light touch, group II afferents, and c-fibers,[49] but not to Ia muscle afferents, which provide parametric information about movement. (4) Olivary pathways are slowly conducting in comparison with other afferent pathways. Although these properties suggest that the IO does not provide an error signal for correcting movement, they may be consistent with Oscarsson's[1] proposal that the IO reports the altering of descending commands by reflex activity. Many IO response properties are consistent with stimuli that might elicit reflex action. For example, touch to the paw as the cat places his foot back on the pedestal appears to elicit a reflex to avoid the obstruction, and cells in some divisions of the IO fire to these touches. Similarly, during stance, light cutaneous stimuli are apt to elicit changes in limb position, and noxious stimuli elicit withdrawal of the limb. However, reflexes are also triggered by afferents, such as Ia muscle afferents and visual stimuli that are poorly represented in the IO. Why would only some stimuli that elicit reflexes be effective for stimulating olivary cells?

The dendritic tree of the Purkinje cell is electrically isolated from the soma,[50] and it is likely that the climbing fiber climbs in order to influence membrane properties throughout the Purkinje cell. One proposed function of the climbing fiber is to alter the strength of synapses on the Purkinje cell, thereby providing a mechanism for plasticity.[2–4] A great deal of research supports a role for the cerebellum in motor learning, but most models of cerebellar plasticity view the climbing-fiber input as a "teaching" signal to adjust movement.[4] This model is similar to Oscarsson's original hypothesis[1]: the IO provides error feedback by comparing actual to commanded movement. To do this, climbing-fiber activity must be related to movement, which does not seem to be the case.

However, olivary input to the cerebellum is not necessarily "teaching" about movement performance. The olivary information may be relevant for movement selection rather than correction. The cerebellum has been shown to be important in the formation of stimulus–response associations in the classical conditioning paradigm.[5,51,52] In this learning model, climbing-fiber discharge signals the cerebellum about the occurrence of the unconditioned stimulus (US). Several studies support the view that climbing fibers carry information that is necessary for the formation and retention of a conditioned response (CR),[53–56] which is consistent with a US function.

The typical experimental US, such as an air puff to the eye or an electrical shock to a limb, provides a potent stimulus for olivary activation. The US provides no information about movement and does not depend upon movement. Olivary discharge reflects the somatic consequences of the US, which are likely to include stimulation involving multiple modalities, such as touch, pressure, and pain. The learned association is between the conditioned stimulus (CS), an arbitrary stimulus such as a tone or light, and the US. This association allows the animal to predict the likely somatic consequences of an initially neutral stimulus (which is signaled by the rapidly conducting mossy-fiber pathways to the cerebellum). After conditioning, the CS will elicit a variety of additional behavioral responses from the dog. The CS may provide information useful for organizing, or selecting, motor responses rather than adjusting mechanics of movement.

In summary, most olivary cells with receptive fields on the limb are inhibited during movement but are sensitive to a variety of stimuli during rest. The properties are not suitable for providing information about self-produced movement or movement error. Olivary response properties are consistent with providing information about the occurrence of an unconditioned stimulus: one possible function may be to form associations that allow the appropriate selection of responses to normally arbitrary environmental cues.

REFERENCES

1. Oscarsson, O. 1980. Functional organization of olivary projection to the cerebellar anterior lobe. *In* The Inferior Olivary Nucleus: Anatomy and Physiology. J. Courville, C. De Montigny & Y. Lamarre, Eds.: 279–289. Raven Press. New York.
2. Marr, D. 1969. A theory of cerebellar cortex. J. Physiol. **202:** 437–470.
3. Albus, J.S. 1971. A theory of cerebellar function. Math. Biosci. **10:** 25–61.
4. Ito, M., M. Sakurai & P. Tongroach. 1982. Climbing fibre induced depression of both mossy fibre responsiveness and glutamate sensitivity of cerebellar Purkinje cells. J. Physiol. **324:** 113–134.
5. McCormick, D.A. *et al.* 1982. Initial localization of the memory trace for a basic form of learning. Proc. Natl. Acad. Sci. USA **79:** 2731–2735.
6. Courville, J. & F. Faraco-Cantin. 1978. On the origin of the climbing fibers of the cerebellum. An experimental study in the cat with an autoradiographic tracing method. Neuroscience **3:** 797–809.
7. Eccles, J.C., R. Llinas & K. Sasaki. 1966. The excitatory synaptic action of climbing fibres on the Purkinje cells of the cerebellum. J. Physiol. **182:** 268–296.
8. Martinez, F. E., W.E. Crill & T.T. Kennedy. 1971. Electrogenesis of cerebellar Purkinje cell responses in cats. J. Neurophysiol. **34:** 348–356.
9. Rawson, J.A. & K. Tilokskulchai. 1982. Climbing fibre modification of cerebellar Purkinje cell responses to parallel fibre inputs. Brain Res. **237:** 492–497.
10. Eccles, J.C. *et al.* 1972. Cutaneous mechanoreceptors influencing impulse discharges in cerebellar cortex. III. In Purkinje cells by climbing fiber input. Exp. Brain Res. **15:** 484–497.
11. Gellman, R., J.C. Houk & A.R. Gibson. 1983. Somatosensory properties of the inferior olive of the cat. J. Comp. Neurol. **215:** 228–43.
12. Rushmer, D.S., W.J. Roberts & G.K. Augter. 1976. Climbing fiber responses of cerebellar Purkinje cells to passive movement of the cat forepaw. Brain Res. **106:** 1–20.
13. Bauswein, E. *et al.* 1983. Simple and complex spike activity of cerebellar Purkinje cells during active and passive movements in the awake monkey. J. Physiol. **339:** 379–394.
14. Berkley, K.J. & P.J. Hand. 1978. Projections to the inferior olive of the cat. II. Comparisons of input from the gracile, cuneate, and the spinal trigeminal nuclei. J. Comp. Neurol. **180:** 253–64.
15. Boesten, A.J. & J. Voogd. 1975. Projections of the dorsal column nuclei and the spinal cord on the inferior olive in the cat. J. Comp. Neurol. **161:** 215–237.

16. GELLMAN, R., A.R. GIBSON & J.C. HOUK. 1985. Inferior olivary neurons in the awake cat: detection of contact and passive body displacement. J. Neurophysiol. **54:** 40–60.
17. DUGAS, C. & A.M. SMITH. 1992. Responses of cerebellar Purkinje cells to slip of a hand-held object. J. Neurophysiol. **67:** 483–95.
18. HARVEY, R.J., R. PORTER & J.A. RAWSON. 1977. The natural discharges of Purkinje cells in paravermal regions of lobules V and VI of the monkey's cerebellum. J. Physiol. **271:** 515–536.
19. ARMSTRONG, D.M., S.A. EDGLEY & M. LIDIERTH. 1988. Complex spikes in Purkinje cells of the paravermal part of the anterior lobe of the cat cerebellum during locomotion. J. Physiol. **400:** 405–414.
20. NELSON, B.J. & E. MUGNAINI. 1988. The rat inferior olive as seen with immunostaining for glutamate decarboxylase. Anat. Embryol. **179:** 109–127.
21. NELSON, B.J. *et al.* 1989. Comparative study of glutamate decarboxylase immunoreactive boutons in the mammalian inferior olive. J. Comp. Neurol. **286:** 514–539.
22. MCCREA, R.A., G.A. BISHOP & S.T. KITAI. 1978. Morphological and electrophysiological characteristics of projection neurons in the nucleus interpositus of the cat cerebellum. J. Comp. Neurol. **181:** 397–419.
23. LEGENDRE, A. & J. COURVILLE. 1987. Origin and trajectory of the cerebello-olivary projection: an experimental study with radioactive and fluorescent tracers in the cat. Neuroscience **21:** 877–891.
24. ANGAUT, P. & C. SOTELO. 1987. The dentato-olivary projection in the rat as a presumptive GABAergic link in the olivo-cerebello-olivary loop. An ultrastructural study. Neurosci. Lett. **83:** 227–231.
25. ANGAUT, P. & C. SOTELO. 1989. Synaptology of the cerebello-olivary pathway. Double labelling with anterograde axonal tracing and GABA immunocytochemistry in the rat. Brain Res. **479:** 361–365.
26. DE ZEEUW, C.I. *et al.* 1988. A new combination of WGA-HRP anterograde tracing and GABA immunocytochemistry applied to afferents of the cat inferior olive at the ultrastructural level. Brain Res. **447:** 369–375.
27. THACH, W.T. 1978. Correlation of neural discharge with pattern and force of muscular activity, joint position, and direction of intended next movement in motor cortex and cerebellum. J. Neurophysiol. **41:** 654–676.
28. TEUNE, T.M. *et al.* 1998. Single Purkinje cell can innervate multiple classes of projection neurons in the cerebellar nuclei of the rat: a light microscopic and ultrastructural triple-tracer study in the rat. J. Comp. Neurol. **392:** 164–178.
29. Weiss, C., J.C. Houk & A.R. Gibson. 1990. Inhibition of sensory responses of cat inferior olive neurons produced by stimulation of red nucleus. J. Neurophysiol. **64:** 1170–1185.
30. LEICHT, R., M.J. ROWE & R.F. SCHMIDT. 1973. Cortical and peripheral modification of cerebellar climbing fibre activity arising from cutaneous mechanoreceptors. J. Physiol. **228:** 619–635.
31. GRAY, B.G. & J.O. DOSTROVSKY. 1983. Modulation of the sensory responses of cat trigeminal and cuneate neurons by electrical stimulation of the red nucleus. Soc. Neurosci. Abstr. **9:** 247.
32. TOWE, A.L. & S.J. JABBUR. 1961. Cortical inhibition of neurons in dorsal column nuclei of cat. J. Neurophysiol. **24:** 488–498.
33. MCCURDY, M.L., A.R. GIBSON & J.C. HOUK. 1992. Spatial overlap of rubrospinal and corticospinal terminals with input to the inferior olive. Neuroimage **1:** 23–41.
34. OSCARSSON, O. & B. SJOLUND. 1977. The ventral spino-olivocerebellar system in the cat. III. Functional characteristics of the five paths. Exp. Brain Res. **28:** 505–520.
35. HORN, K.M., T.M. HAMM & A.R. GIBSON. 1998. Red nucleus stimulation inhibits within the inferior olive. J. Neurophysiol. **80:** 3127–3136.
36. HORN, K.M., P.L. VAN KAN & A.R. GIBSON. 1996. Reduction of rostral dorsal accessory olive responses during reaching. J. Neurophysiol. **76:** 4140–4151.
37. APPS, R., M.J. ATKINS & M. GARWICZ. 1997. Gating of cutaneous input to cerebellar climbing fibres during a reaching task in the cat. J. Physiol. **502:** 203–214.
38. GROENEWEGEN, H.J., J. VOOGD & S.L. FREEDMAN. 1979. The parasagittal zonation within the olivocerebellar projection. II. Climbing fiber distribution in the intermediate and hemispheric parts of cat cerebellum. J. Comp. Neurol. **183:** 551–601.

39. JIANG, W., C.E. CHAPMAN & Y. LAMARRE. 1991. Modulation of the cutaneous responsiveness of neurones in the primary somatosensory cortex during conditioned arm movements in the monkey. Exp. Brain Res. **84:** 342–354.
40. GHEZ, C. & M. PISA. 1972. Inhibition of afferent transmission in cuneate nucleus during voluntary movement in the cat. Brain Res. **40:** 145–155.
41. GROENEWEGEN, H.J. & J. VOOGD. 1975. Projections from the gracile and internal cuneate nuclei to the inferior olive in the cat as demonstrated by autoradiography. Acta Morphol. Neerl. Scand. **13:** 232–234.
42. ANDERSSON, G. & D.M. ARMSTRONG. 1987. Complex spikes in Purkinje cells in the lateral vermis (b zone) of the cat cerebellum during locomotion. J. Physiol. **385:** 107–134.
43. GILBERT, P.F. & W.T. THACH. 1977. Purkinje cell activity during motor learning. Brain Res. **128:** 309–328.
44. HORN, K.M. *et al.* 2001. Discharge of inferior olive neurons during reaching errors. Soc. Neurosci. Abstr. 577.4.
45. JASMIN, L. & J. COURVILLE. 1987. Distribution of external cuneate nucleus afferents to the cerebellum: II. Topographical distribution and zonal pattern—an experimental study with radioactive tracers in the cat. J. Comp. Neurol. **261:** 497–514.
46. JASMIN, L. & J. COURVILLE. 1987. Distribution of external cuneate nucleus afferents to the cerebellum: I. Notes on the projections from the main cuneate and other adjacent nuclei. An experimental study with radioactive tracers in the cat. J. Comp. Neurol. **261:** 481–496.
47. MACKIE, P.D., J.W. MORLEY & M.J. ROWE. 1999. Signalling of static and dynamic features of muscle spindle input by external cuneate neurones in the cat. J. Physiol. **519** (Pt **2**)**:** 559–569.
48. MCCURDY, M.L., J.C. HOUK & A.R. GIBSON. 1998. Organization of ascending pathways to the forelimb area of the dorsal accessory olive in the cat. J. Comp. Neurol. **392:** 115–133.
49. EKEROT, C.F. *et al.* 1987. Climbing fibres projecting to cat cerebellar anterior lobe activated by cutaneous A and C fibres. J. Physiol. **386:** 529–538.
50. HAUSSER, M., G. MAJOR & G.J. STUART. 2001. Differential shunting of EPSPs by action potentials. Science **291:** 138–141.
51. ATTWELL, P. J., S. RAHMAN & C. H. YEO. 2001. Acquisition of eyeblink conditioning is critically dependent on normal function in cerebellar cortical lobule HVI. J. Neurosci. **21:** 5715–5722.
52. YEO, C.H., M.J. HARDIMAN & M. GLICKSTEIN. 1984. Discrete lesions of the cerebellar cortex abolish the classically conditioned nictitating membrane response of the rabbit. Behav. Brain Res. **13:** 261–266.
53. MAUK, M.D., J.E. STEINMETZ & R.F. THOMPSON. 1986. Classical conditioning using stimulation of the inferior olive as the unconditioned stimulus. Proc. Natl. Acad. Sci. USA **83:** 5349–5353.
54. MCCORMICK, D.A., J.E. STEINMETZ & R.F. THOMPSON. 1985. Lesions of the inferior olivary complex cause extinction of the classically conditioned eyeblink response. Brain Res. **359:** 120–130.
55. MEDINA, J.F., W.L. NORES & M.D. MAUK. 2002. Inhibition of climbing fibres is a signal for the extinction of conditioned eyelid responses. Nature. **416:** 330–333.
56. YEO, C.H., M.J. HARDIMAN & M. GLICKSTEIN. 1986. Classical conditioning of the nictitating membrane response of the rabbit. IV. Lesions of the inferior olive. Exp. Brain Res. **63:** 81–92.

Complex Spike Activity in the Flocculus Signals More than the Eye Can See

J. I. SIMPSON, T. BELTON, M. SUH, AND B. WINKELMAN

Department of Physiology and Neuroscience, New York University School of Medicine, New York, New York 10016, USA

ABSTRACT: Modulation of the complex spike activity of Purkinje cells in the cerebellar flocculus can convey not only visual signals but also nonvisual signals. The nonvisual complex spike modulation, which is readily observed with vestibular stimulation of the awake rabbit in darkness, is approximately in-phase with the concomitant simple spike modulation. This nonreciprocal relationship contrasts to the reciprocal relationship found when the rabbit is afforded vision.

KEYWORDS: cerebellum; complex spike; climbing fiber; Purkinje cell; inferior olive; optokinetic; vestibulo-ocular reflex; rabbit; eye movement

INTRODUCTION

It is well known that the climbing-fiber input to the cerebellar flocculus carries signals of retinal image motion.[1–4] In contrast, the fact that the activity of these same climbing fibers can also be modulated by vestibular stimulation presented in darkness has received only scant attention.[2,5,6] To address this imbalance and also the question of what is signaled by the vestibularly induced modulation of the climbing fiber–driven complex spike (CS) activity of floccular Purkinje cells, we began by characterizing the nonvisual CS modulation using more robust vestibular stimuli than previously used. In this note, we report some of the salient features of the nonvisual CS signal and compare them with those of the visual CS signal.

METHODS

We recorded from the flocculus of awake, Dutch-belted rabbits restrained on a turntable. Our focus was those Purkinje cells whose CS activity was best-modulated optokinetically when a full-field textured pattern rotated about the vertical axis. This type of Purkinje cell we call a vertical axis Purkinje cell.[3] Using a range of stimulus frequencies (0.05–2.0 Hz) and velocities (3–67°/s), we investigated the modulation of the CS activity when the rabbit was sinusoidally oscillated in the dark and in the

Address for correspondence: Prof. J.I. Simpson, Department of Physiology and Neuroscience, NYU School of Medicine, 550 First Ave., New York, NY 10016. Voice: 212-263-5428; fax: 212-689-9060.

john.simpson@med.nyu.edu.

Ann. N.Y. Acad. Sci. 978: 232–236 (2002). © 2002 New York Academy of Sciences.

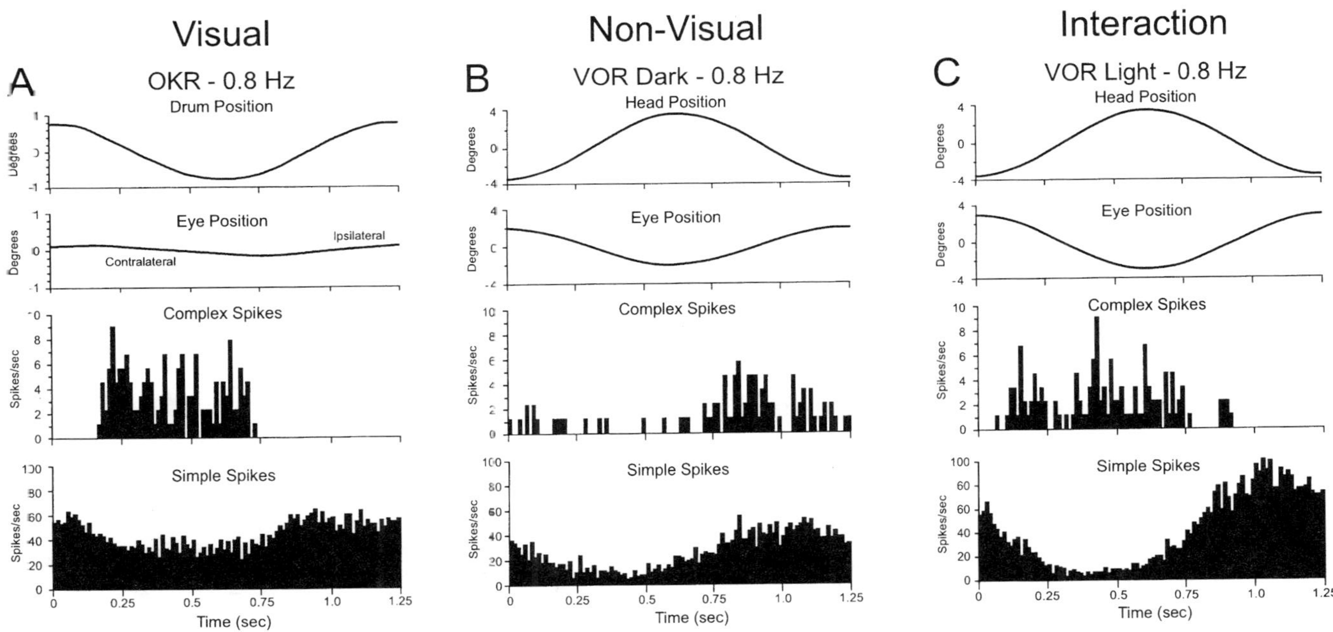

FIGURE 1. Illustration of the visually and vestibularly induced modulations of CS activity in the flocculus of the awake rabbit. During optokinetic stimulation at 0.8 Hz (**A**), the CS modulation was out-of-phase with the SS modulation, while during vestibular stimulation at the same frequency in darkness (**B**) it was in-phase with the SS modulation. In both conditions the SS activity increased with ipsilateral eye movement. When the image velocity during vestibular stimulation in the light (2.5°/s for **C**) was comparable to that during OKR (2.8°/s for **A**), the CS modulation was out-of-phase with the SS modulation, but the pattern of the CS modulation was altered compared to that during OKR, particularly with regard to its transient characteristics. The vestibular stimulus in the dark (**B**) was the same as that in the light (**C**).

light about the vertical axis. Movement of the ipsilateral eye was measured using the search coil technique[7] so as to determine the vestibulo-ocular response in darkness (VOR Dark) and in light (VOR Light). Since the CS activity recorded in darkness was found to increase approximately in-phase with contralateral head velocity, modulation depth was defined as the difference in the number of CSs occurring in each of the two halves of the stimulus cycle centered on peak head velocity divided by the total number of CSs in the whole stimulus cycle. We considered CS activity to be modulated when the ratio of the two half-cycle firing rates was significantly different from 1.0 (chi-square test, $p < 0.05$). To assess the interaction between the visually and vestibularly induced CS modulations, we also used an optokinetic drum to produce eye movements (optokinetic response, OKR) with retinal image velocities comparable to those accompanying the eye movements produced by vestibular stimulation in the light.

RESULTS

The modulation of the CS activity of vertical axis Purkinje cells in the rabbit flocculus often reflected not only retinal image motion, but also a nonvisual signal, as exemplified in FIGURE 1. While vestibular stimulation in the dark at low frequencies (such as 0.1 Hz) was not effective in evoking modulation of CS activity, vestibular stimulation in the dark at higher frequencies (0.4–1.0 Hz) was often effective. The nonvisual CS modulation was approximately in-phase with the concomitant simple spike (SS) modulation, and the CS activity increased when the head rotated contralaterally. This nonreciprocal relationship between the CS and SS modulation stands in marked contrast to the reciprocal relationship found when an image motion signal was present (FIG. 1). Of 141 vertical axis Purkinje cells recorded in two rabbits oscillated in the dark, 76% (107) had significant CS modulation. For the significantly modulated Purkinje cells (n=71) recorded in the rabbit presented with the greater range of vestibular stimuli (0.05–2.0 Hz and velocity amplitudes of 3–67°/s), the CS modulation depth in darkness peaked at 0.4–0.8 Hz, with stimulus amplitudes of 12.5–38°/second. That is, with frequency held constant, the relationship between the nonvisual CS modulation and head velocity was nonmonotonic, as exemplified in FIGURE 2.

DISCUSSION

The climbing fibers of the rabbit cerebellar flocculus that carry visual signals can also carry nonvisual signals, which are readily observed during vestibular stimulation in darkness. The nonmonotonic relationship between the nonvisual CS modulation and head velocity is reminiscent of the nonmonotonic relationship between the visual CS modulation and image velocity,[1,8] but other comparisons reveal differences. First, the visual and nonvisual CS modulations are, respectively, out-of-phase and in-phase with the concomitant modulations of the SS activity, which in both cases increases with ipsilateral eye movement. Second, the visual CS modulation is readily associated with a performance error,[4,9,10] but what is the error to be associated with the nonvisual CS modulation that occurs in darkness?

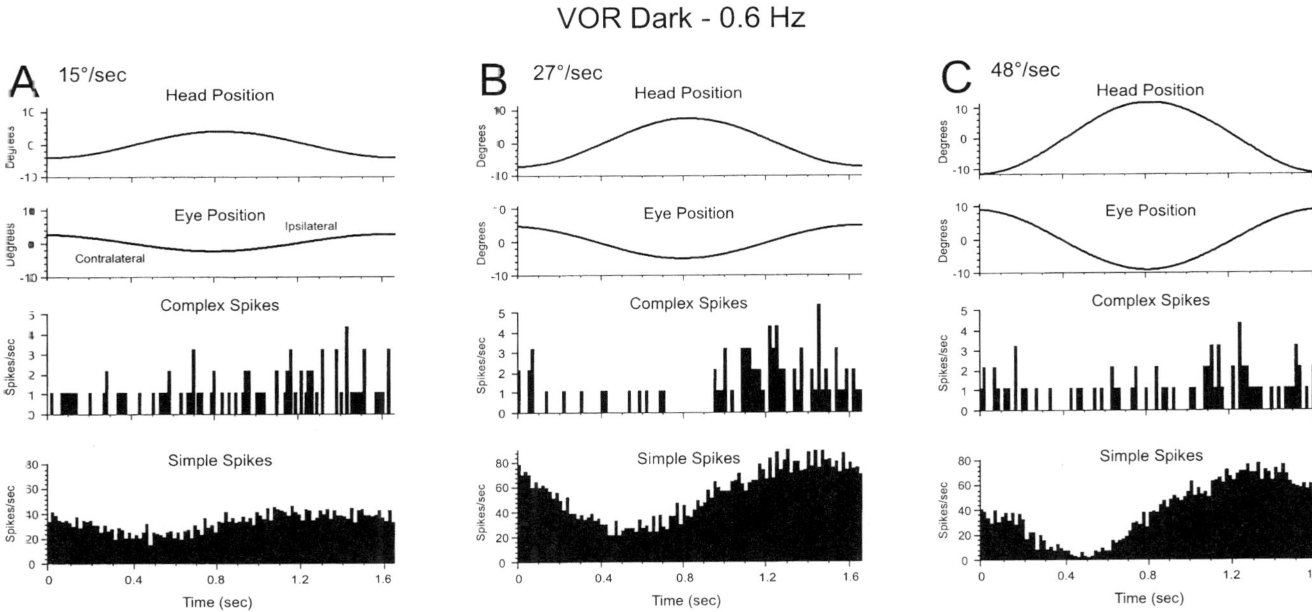

FIGURE 2. Illustration of the relationship between the amplitude of vestibular stimulation in the dark and the depth of the nonvisual CS modulation. At the stimulation frequency of 0.6 Hz, the CS modulation depth showed a nonmonotonic relationship with the amplitude of the sinusoidal vestibular stimulus. That is, the CS modulation was greater when the peak stimulus velocity was 27°/s **(B)** than when it was 15°/s **(A)** or 48°/s **(C)**. Note that at each amplitude the SS modulation was in-phase with the nonvisual CS modulation, but that, in contrast to the CS modulation, the amplitude of the SS modulation increased monotonically with the stimulus amplitude (and with the amplitude of the eye movement).

The projection from the nucleus prepositus hypoglossi to the caudal dorsal cap of the inferior olive[11–13] may be the source of the nonvisual CS modulation, in which case it might be derived from the vestibular and/or eye movement signals present in that nucleus.[14–16] However, a straightforward relationship between the amplitude of the head and/or eye movements and the nonvisual CS modulation was not apparent. What is apparent is that the "visual" climbing fibers can convey signals to the flocculus about more than the eye can see.

ACKNOWLEDGMENTS

This work was supported by NS-13742 and DC-03284.

REFERENCES

1. SIMPSON, J.I. & K.E. ALLEY. 1974. Visual climbing fiber input to rabbit vestibulo-cerebellum: A source of direction-specific information. Brain Res. **82:** 302–308.
2. GHELARDUCCI, B., M. ITO & N. YAGI. 1975. Impulse discharges from flocculus Purkinje cells of alert rabbits during visual stimulation combined with horizontal head rotation. Brain Res. **87:** 66–72.
3. GRAF, W., J.I. SIMPSON & C.S. LEONARD. 1988. Spatial organization of visual messages of the rabbit's cerebellar flocculus. II. Complex and simple spike responses of Purkinje cells. J. Neurophysiol. **60:** 2091–2121.
4. SIMPSON, J.I., D.R. WYLIE & C.I. DE ZEEUW. 1996. On climbing fiber signals and their consequence(s). Behav. Brain Sci. **19:** 380–394.
5. LEONARD, C.S. & J.I. SIMPSON. 1986. Simple spike modulation of floccular Purkinje cells during the reversible blockade of their climbing fiber afferents. *In* Adaptive Processes in Visual and Oculomotor System. E.L. Keller & D.S. Zee, Eds.: 429–434. Pergamon. Oxford, UK.
6. DE ZEEUW, C.I. *et al.* 1995. Phase relations of Purkinje cells in the rabbit flocculus during compensatory eye movements. J. Neurophysiol. **74:** 2051–2064.
7. ROBINSON, D.A. 1963. A method of measuring eye movement using a scleral search coil in a magnetic field. IEEE. Trans. Biomed. Electron. **10:** 137–145.
8. BARMACK, N.H. & D.T. HESS. 1980. Multiple-unit activity evoked in the dorsal cap of inferior olive in the rabbit. J. Neurophysiol. **43:** 151–163.
9. ITO, M. 1982. Cerebellar control of the vestibulo-ocular reflex—around the flocculus hypothesis. Annu. Rev. Neurosci. **301:** 275–296.
10. FRENS, M.A., A.L. MATHOERA & J. VAN DER STEEN. 2001. Floccular complex spike response to transparent retinal slip. Neuron **30:** 795–801.
11. MCCREA, R.A. & R. BAKER. 1985. Anatomical connections of the nucleus prepositus of the cat. J. Comp. Neurol. **237:** 377–407.
12. DE ZEEUW, C.I., P. WENTZEL & E. MUGNAINI. 1993. Fine structure of the dorsal cap of the inferior olive and its GABAergic and non-GABAergic input from the nucleus prepositus hypoglossi in rat and rabbit. J. Comp. Neurol. **327:** 63–82.
13. BARMACK, N.H., M. FAGERSON & P. ERRICO. 1993. Cholinergic projection to the dorsal cap of the inferior olive of the rat, rabbit, and monkey. J. Comp. Neurol. **328:** 263–281.
14. LOPEZ-BARNEO, J. *et al.* 1982. Neuronal activity in prepositus nucleus correlated with eye movement in the alert cat. J. Neurophysiol. **47:** 329–352.
15. MCCREA, R.A. 1988. The nucleus prepositus. *In* Reviews of Oculomotor Research. Neuroanatomy of the Oculomotor System. D.A. Robinson & H. Collewijn, Eds. 203–225. Elsevier. Amsterdam.
16. MCFARLAND, J.L. & A.F. FUCHS. 1992. Discharge patterns in nucleus prepositus hypoglossi and adjacent medial vestibular nucleus during horizontal eye movement in behaving macaques. J. Neurophysiol. **68:** 319–332.

Vestibularly Evoked Climbing-Fiber Responses Modulate Simple Spikes in Rabbit Cerebellar Purkinje Neurons

NEAL H. BARMACK AND VADIM YAKHNITSA

Neurological Sciences Institute, Oregon Health and Sciences University, Beaverton, Oregon 97006, USA

ABSTRACT: The nodulus receives a primary vestibular afferent input from the ipsilateral labyrinth and a vestibularly related climbing-fiber input originating from the contralateral labyrinth. Previously we demonstrated that increased discharge of vestibularly evoked climbing-fiber responses (CFRs) in nodular Purkinje cells was correlated with decreased discharge of simple spikes (SSs). This left unresolved the question of whether vestibularly evoked antiphasic behavior of CFRs and SSs reflects a common neural mechanism or the activation of two separate parallel pathways. We answered this question using natural vestibular stimulation to modulate the discharge of uvula-nodular Purkinje cells recorded extracellularly in unilaterally labyrinthectomized, chloralose urethane–anesthetized rabbits. In such animals, vestibular primary afferents projecting to the uvula-nodulus as mossy fibers remained intact on the side contralateral to the unilateral labyrinthectomy. The discharge of CFRs recorded in ipsilateral nodular Purkinje cells was increased by ipsilateral roll-tilt while the discharge of SSs was increased by contralateral roll-tilt. These polarities were reversed for Purkinje cells recorded in the contralateral uvula-nodulus. The polarity of SS discharge recorded from Purkinje cells on both sides of the nodulus was opposite to that of the vestibular primary mossy-fiber afferents. SSs continued to respond to contralateral roll-tilt even when the primary vestibular afferent mossy-fiber pathway was destroyed by the unilateral labyrinthectomy. Although the discharge of SSs recorded in the contralateral uvula-nodulus was increased by contralateral roll-tilt, this modulation was reduced relative to that observed in Purkinje cells recorded in the ipsilateral uvula-nodulus. We conclude that vestibularly evoked CFRs caused the modulation of SS discharge.

KEYWORDS: vestibular primary afferents; inferior olive; climbing-fiber response; mossy fiber; simple spike; parasolitary nucleus; nodulus; cerebellum; Golgi cell; parallel fiber; Purkinje cell

Address for correspondence: Neal H. Barmack, Neurological Sciences Institute, 505 NW 185th Avenue, Oregon Health and Sciences University, Beaverton, Oregon 97006. Voice: 503-418-2560; fax: 503-418-2501.

barmackn@ohsu.edu

Ann. N.Y. Acad. Sci. 978: 237–254 (2002).

INTRODUCTION

The discharge of cerebellar Purkinje cells is characterized by two distinctly different action potentials, termed "simple spikes" and "climbing-fiber responses." Simple spike (SSs) have waveforms of short duration (0.75–1.25 ms) and spontaneous discharge rates of 20–60 impulses per second. Climbing-fiber responses (CFRs) have broad (2–5 ms), multipeaked action potentials and discharge rates of 1–5 impulses per second.[1] CFRs are evoked in Purkinje cells by the multisynaptic contacts of a single climbing fiber as it wraps around the primary dendrite.[1,2] SSs are evoked by the excitatory drive of approximately 150,000 parallel fibers, the bifurcated axons of granule cells, that course through the Purkinje cell dendritic tree. Parallel fiber–evoked EPSCs are small (2–60 pA). At least 50 granule cells must discharge simultaneously to evoke a single SS.[3]

Modulation of granule cell activity is often ascribed to the activity of the mossy-fiber afferents that terminate on granule cell dendrites in the well-characterized glomerular synapse (FIG. 1B). However, each glomerulus also includes an inhibitory axon terminal from a Golgi inhibitory interneuron.[4] Granule cells also receive synaptic inputs from unipolar brush neurons (UBCs), found in highest density in the uvula-nodulus.[5] Like granule cells, UBCs receive excitatory inputs from vestibular primary afferents.[6]

SSs may reflect a combined excitatory drive from both mossy fibers and UBCs relayed through granule cells. It is equally possible that SSs reflect Golgi cell inhibition of granule cells at the glomerular synapse.[7,8] Since Golgi cells comprise the only cerebellar neuron, other than the Purkinje cell, on which climbing fibers synapse,[7,9–11] climbing fiber–evoked Golgi cell activity could play a role equal to that of mossy-fiber afferents or UBC afferents in modulating granule cell discharge.

In order to understand how cerebellar circuitry modulates CFRs and SSs, the signaling capacity of mossy fibers and climbing fibers must be tested separately. The vestibular circuitry projecting to the uvula-nodulus offers this possibility. More than 70% of primary vestibular afferents project to folia 9d and 10 in the ipsilateral uvula-nodulus as mossy fibers[12–16] (FIG. 1A). Vestibularly modulated climbing fibers originate from two subnuclei of the inferior olive (β-nucleus and dorsomedial cell column [dmcc]) and cross the midline to synapse on Purkinje neurons in the contralateral uvula and nodulus[17–21](FIG. 1C). The β-nucleus and dmcc receive secondary vestibular projections from the ipsilateral parasolitary nucleus (Psol), a small GABAergic nucleus that receives a primary vestibular afferent projection from the ipsilateral anterior and posterior semicircular canals as well as the ipsilateral utricular otolith.[22,23]

Both climbing-fiber and mossy-fiber systems are engaged by natural vestibular stimulation. Roll-tilt evokes *increased* discharge from ipsilateral primary vestibular afferents projecting as mossy fibers to the ipsilateral uvula-nodulus. Ipsilateral roll-tilt also evokes *increased* activity of neurons in the contralateral β-nucleus and dmcc, caused by release from GABAergic inhibition from the contralateral Psol. These disinhibited olivary neurons project to the ipsilateral uvula-nodulus.[24,25]

The vestibulo-cerebellar circuitry implies that modulated SSs originate from the ipsilateral labyrinth and modulated CFRs originate from the contralateral labyrinth. This offers the possibility of separating mossy-fiber and climbing-fiber signals before they enter the uvula-nodulus so that they can be tested independently. This sep-

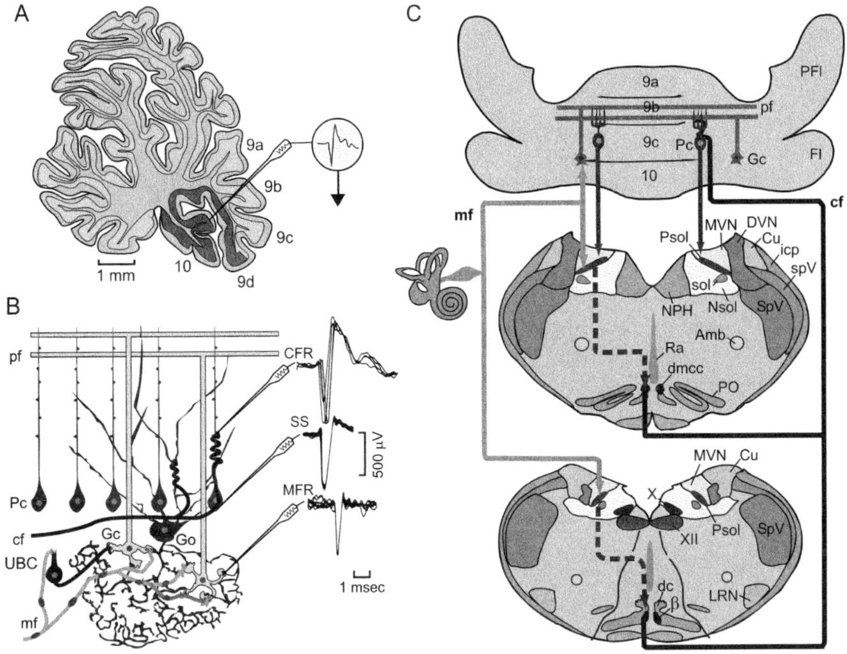

FIGURE 1. Cellular and system circuitry relevant to vestibular signals recorded in uvula-nodulus. (**A**) Sagittal view of rabbit cerebellum shows the angle at which microelectrode penetrates the different folia of the uvula-nodulus. The region of the uvula-nodulus receiving primary vestibular afferent mossy-fiber terminals is indicated by darker shading. (**B**) Microcircuitry of mossy-fiber and climbing-fiber projections to cerebellum. Climbing fiber (cf) makes synaptic contact with Purkinje cell (Pc) and Golgi cell (Go). Mossy-fiber (mf) contacts granule cell dendrites (Gc) in glomeruli at which Golgi cell axons also terminate. Three different extracellularly recorded waveforms are illustrated: (1) climbing-fiber response (CFR) is a multipeaked action potential recorded from the superficial aspect of the Purkinje cell layer, (2) simple spikes (SS) is recorded from the base of the Purkinje cell layer, and (3) mossy-fiber response (MFR) is short duration action potential recorded in the granule cell layer. (**C**) Vestibular mossy-fiber and climbing-fiber projections to uvula-nodulus. Vestibular primary afferents send collateral projections to the ipsilateral uvula-nodulus (folia 9d and 10), where they terminate upon granule cells as mossy fibers and to the ipsilateral vestibular complex where they synapse upon several vestibular nuclei, including the parasolitary nucleus (Psol). The Psol sends a descending GABAergic projection to the ipsilateral inferior olive (β-nucleus and dmcc). The axons of neurons in the β-nucleus and dmcc cross the midline and synapse as climbing fibers on Purkinje cells in the contralateral uvula-nodulus. Anatomical abbreviations: Amb, nucleus ambiguous; β, β-nucleus; cf, climbing fiber; Cu, cuneate nucleus; dc, dorsal cap of Kooy; dmcc, dorsomedial cell column; DVN, descending vestibular nucleus; Fl, flocculus; Gc, granule cell; Go, Golgi cell; icp, inferior cerebellar peduncle; LRN, lateral reticular nucleus; mf, mossy fiber; MVN, medial vestibular nucleus; NPH, nucleus prepositus hypoglossi; Nsol, nucleus solitarius; Pc, Purkinje neuron; pf, parallel fiber; PFl, paraflocculus; PO, principal olive; Psol, parasolitary nucleus; sol, tractus solitarius; Ra, nucleus of the raphe; SpV, spinal trigeminal nucleus; spV, spinal trigeminal tract; UBC, unipolar brush cell; X, dorsal motor nucleus of the vagus, XII, hypoglossal nucleus.

aration can be achieved surgically by making a unilateral labyrinthectomy. After a unilateral labyrinthectomy, the vestibular primary afferent mossy-fiber signal to the ipsilateral uvula-nodulus would be absent. However, the vestibularly modulated climbing-fiber signal to the ipsilateral uvula-nodulus would remain intact since it originates from the contralateral inferior olive. Conversely, the primary vestibular afferent mossy-fiber pathway to the contralateral uvula-nodulus would remain intact, while the vestibularly modulated climbing-fiber cross-projection from the ipsilateral inferior olive would be impaired.

We have used the different origins of vestibular climbing and mossy fibers to investigate whether intact mossy-fiber and climbing-fiber systems are necessary and/or sufficient for modulation of SSs and CFRs. Specifically we measured vestibularly evoked SSs and CFRs in the uvula-nodulus of rabbits immediately after one labyrinth was surgically destroyed. Using natural vestibular stimulation, we recorded from ipsilateral Purkinje cells deprived of primary vestibular mossy-fiber input and in contralateral Purkinje cells with reduced climbing-fiber signals.

METHODS

Anesthesia and Surgery

Fifteen pigmented rabbits (weight 0.8–1.7 kg) were anesthetized intravenously with α-chloralose (50 mg/kg) and urethane (500 mg/kg) or intramuscularly with ketamine hydrochloride (50 mg/kg), xylazine (6 mg/kg), and acepromazine maleate (1.2 mg/kg). Rectal temperature was monitored and maintained at 37° C. The adequacy of anesthesia was evaluated using the corneal reflex as an indicator.

In a preparatory operation, a dental acrylic plug was formed to the calvarium of each rabbit. This plug held two inverted stainless steel screws (8-32) to the dorsal surface of the calvarium. Five smaller stainless steel screws (2-56) were screwed into the calvarium and helped to anchor the acrylic plug. The larger inverted screws mated with a metal rod that maintained the head rigidly in the center of a three-axis vestibular rate table.

Unilateral Labyrinthectomy

The left labyrinth was destroyed surgically in eight rabbits. A retroauricular 5 mm incision was made to gain access to the middle ear through the tympanic membrane. The stapes was disinserted from the oval window. A fine dental drill was used to open the bony cochlea-labyrinth. The pigmented membranous labyrinth was visualized and removed with a fine forceps. To make certain that the entire vestibular apparatus was damaged, the dental drill was used to excavate the remaining bony labyrinth. Approximately 60 minutes following the completion of the operation, electrical recording was begun from Purkinje cells in the uvula-nodulus.

Microelectrode Recording

The uvula-nodulus was approached by reflecting the muscles overlying the cisterna magna and enlarging the dorsal aspect of the foramen magnum. The outer layers of the dura mater were stripped away exposing folia 9b and 9c through the remaining layer of dura mater. A hydraulic microdrive, attached to the head restraint bar, advanced tungsten microelectrodes through folia 9b and 9c towards folia 9d and 10 (FIG. 1A). The microelectrodes had an extended taper so that the diameter of the final 4 mm (the length necessary for a complete penetration of the uvula-nodulus) was less than 50 µm. The tip impedance was approximately 4 MΩ.

Action potentials were discriminated with a window-slope discriminator-Schmitt trigger and analyzed using Spike 2 software (Cambridge Electronic Design). Evoked single unit activity was displayed on-line as peristimulus histograms, constructed using different numbers of stimulus cycles at different stimulus frequencies. Each stimulus cycle was divided into 180 bins. Interspike intervals for spike occurrences were stored in these bins. The reciprocal of these interspike intervals, spike frequency, was averaged for the number of spike occurrences in each bin.

Histological Verification of Recording Sites

The location of each recorded neuron was marked electrolytically (–8 µA, 30 s). At the conclusion of each experiment, the rabbit was deeply anesthetized and perfused transcardially with 0.9% saline, followed by 10% paraformaldehyde. The brain was removed and cryoprotected with 10%, 20%, and 30% sucrose in 0.1 M PBS, pH 7.2. The posterior cerebellum was blocked sagittally, mounted onto cork with OCT compound, and frozen in isopentane cooled with dry ice. Sagittal frozen sections (35 µm) were mounted serially.

Vestibular Stimulation

The rate table was sinusoidally oscillated about its vertical axis (yaw), about its longitudinal axis (roll), or about its interaural axis (pitch) (±10°, 0.005–0.800 Hz). During vestibular stimulation, the vision of the rabbit was occluded by hemispherical ping pong balls.

Static roll-tilt stimulation tested the otolithic origin of vestibularly evoked discharges. The rabbit was tilted 5–10° about some combination of the roll-pitch axes. After an adaptation period of 20–30 s, the average discharge frequency was measured for the next 20–30 seconds. The rabbit was then tilted in the opposite direction. A difference in mean discharge frequency of more than 40%, evoked for tilts in the two opposite directions, was taken to indicate a climbing-fiber sensitivity to linear acceleration.

Vestibular **Null Plane** *Measurement*

The peripheral origins of vestibularly modulated signals were characterized using a "null technique." While the rabbit was roll-tilted sinusoidally, the angle of the rabbit's head was varied continuously from roll-to-pitch, until a minimal modulation of neuronal activity was attained. On either side of this null plane, the phase of modulated activity shifted by 180° with respect to the sinusoidal forcing function. The optimal plane, the plane of maximal neuronal modulation, was assumed to be orthogonal to the null plane.[26]

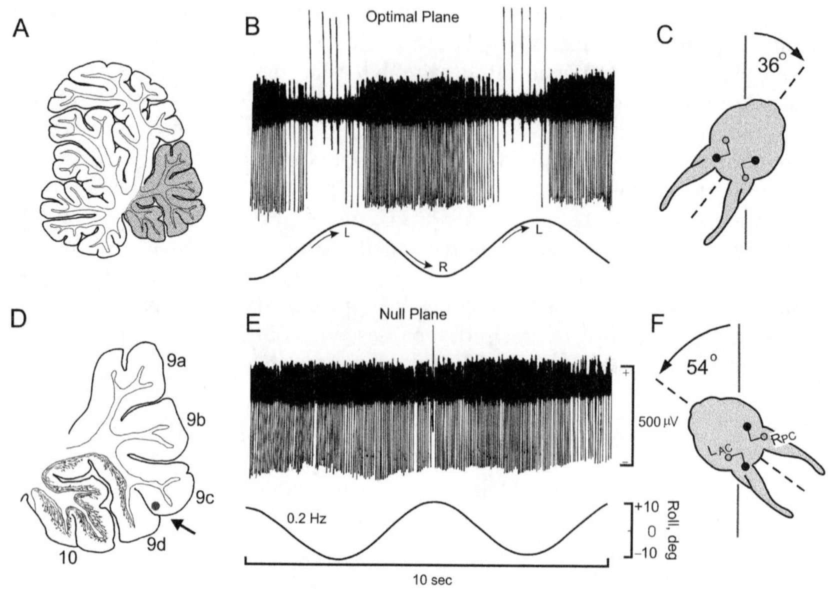

FIGURE 2. Climbing-fiber responses (CFRs) and simple spikes (SSs) recorded from a Purkinje cell during vestibular stimulation in the optimal and null planes. Sinusoidal vestibular stimulation, bounded by the roll and pitch axes, was used to evoke CFRs and SSs from a Purkinje cell localized to the junction of folia 9c and 9d. **(A)** Illustration of the uvula-nodulus from which microelectrode recordings were obtained (*shaded region*). This *shaded region* is illustrated at larger magnification in **(D)**. The distribution of primary vestibular afferents to folia 9d and 10 is illustrated. The location of the Purkinje cell from which the recordings illustrated in **(B)** and **(E)** were obtained is indicated by the arrow. **(B)** Stimulation in the *optimal plane* evoked positive-going CFRs and negative-going SSs. The sinusoidal trace indicates head position as the rabbit was rolled about an axis that was 36° clockwise with respect to the vertical axis, illustrated in **C**. In this Purkinje cell, the *optimal plane* corresponded approximately to the planes of the left posterior and right anterior semicircular canals, L_{PC}-R_{AC}. **(D)** This neuron was found 220 μm to the left of the midline. **(E)** Neither the CFRs nor the SSs were modulated when the head of the rabbit was oriented in the *null plane.* **(F)** The null plane was oriented 54° counterclockwise with respect to the vertical axis.

RESULTS

Antiphasic Behavior of CFRs and SSs in Intact Anesthetized Rabbits

When a rabbit rolled about the longitudinal axis, CFRs were evoked in Purkinje cells in the folia 9c, 9d, and 10. Increased CFR discharge was evoked when the rabbit was rolled onto the side ipsilateral to the recorded Purkinje neuron (FIG. 2B). In more than 150 Purkinje cells recorded from the uvula-nodulus, the CFRs of 80% were modulated by both static and dynamic roll-tilt. In these Purkinje cells, ipsilateral roll evoked an increase in CFRs and a decrease in SSs. Conversely, contralateral roll evoked a decrease in CFRs and an increase in SSs. We isolated ten cells in the granule cell layer that were identified as mossy-fiber terminals on the basis of the brevity of their action potential and by the fact that the electrode could be moved as much as 50 μm while still holding the unit. The discharge of mossy fibers increased during ipsilateral and decreased during contralateral roll, slightly leading the discharge of CFRs.

By positioning the head about the vertical axis it was possible to determine the head position at which the vestibular roll stimulus evoked no change in the discharge of either CFRs or SSs. This head position corresponded to the "null plane" (FIG. 2E and F). The optimal roll plane was orthogonal to the "null plane" (FIG. 2B and C). Most of the CFRs recorded from Purkinje cells in folia 9c, 9d, and 10 had optimal planes that were aligned with either the ipsilateral posterior or anterior semicircular canals. Purkinje cells having CFRs with optimal planes aligned with the ipsilateral posterior canal were located in a medial zone extending from the midline to about 1.2 mm lateral to the midline. Purkinje cells having CFRs aligned with the ipsilateral anterior semicircular canal were located in a more lateral zone starting at about 1 mm and extending to 2.2 mm lateral of the midline.

In 29 Purkinje cells located in folia 9c, 9d, and 10, in which both CFRs and SSs could be discriminated clearly, we determined the average modulation for vestibular roll stimulation in both the optimal and null planes (FIG. 3). In the optimal plane, there was a mean modulation of 3 impulses/cycle for CFRs and 14.5 impulses/cycle for SSs. This modulation was reduced to near zero in the null plane.

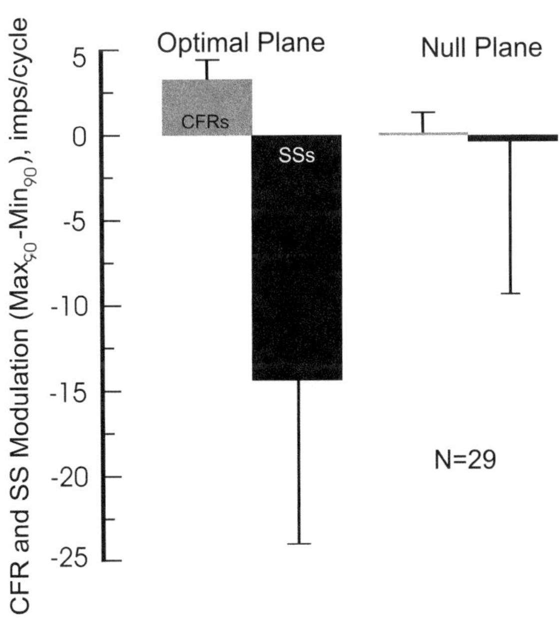

FIGURE 3. Modulation of CFRs and SSs during vestibular stimulation in *optimal* and *null planes.* CFRs and SSs were recorded in 29 uvula-nodular Purkinje cells during roll-pitch stimulation in the *optimal* and *null planes*. Vestibularly evoked increase in frequency of CFRs was related to the reduction in frequency of SSs in the *optimal plane*. During vestibular stimulation in the CFR null plane there was no modulation of SSs.

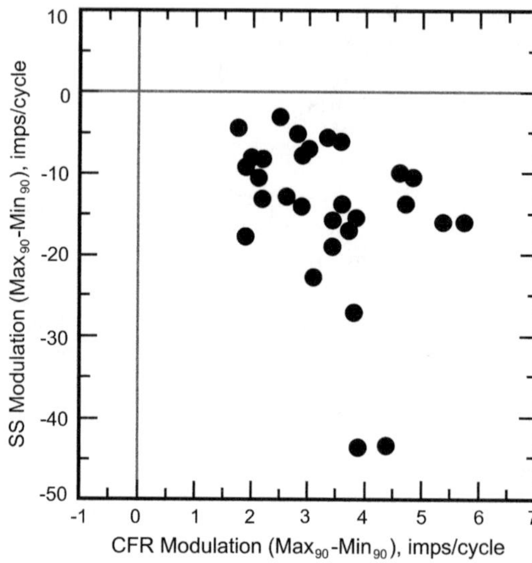

FIGURE 4. Peak modulation of CFRs and SSs during vestibular stimulation in *optimal plane.* CFRs and SSs were simultaneously recorded in 29 uvula-nodular Purkinje cells during vestibular roll stimulation in the *optimal plane*. The peak SS modulation is plotted as a function of peak CFR modulation for each Purkinje cell.

For these 29 Purkinje cells, the amplitude of modulation of the CFRs was correlated with the amplitude of modulation of SSs (FIG. 4). These data demonstrate that SSs have the same null plane as CFRs. The correlation of CFR modulation with SS modulation could indicate that both were determined by two or more parallel central circuits receiving identical vestibular information from the peripheral labyrinth. Alternatively, these data are consistent with the idea that CFRs induced the modulation of SSs.

After establishing the optimal plane for each recorded Purkinje cell, we determined the phase of vestibularly evoked peak discharge with respect to the sinusoidal forcing function. The mean phase of the peak discharge of CFRs led ipsilateral-down head position by 10.4°. The mean phase of peak discharge of SSs led ipsilateral-down head position by 182.4°. These data indicate that SSs were antiphasic with respect to CFRs. The mean phase of mossy-fiber responses led ipsilateral head position by 38.8°, a slightly greater phase lead than that of CFRs (FIG. 5).

Antiphasic Behavior of CFRs and SSs in Ipsilateral Uvula-Nodulus of Anesthetized Rabbits following Unilateral Labyrinthectomy

According to known central vestibular projections, a unilateral labyrinthectomy should deprive the ipsilateral uvula-nodulus of vestibular primary afferent mossy fibers, while at the same time reducing vestibularly modulated climbing-fiber signals to the contralateral uvula-nodulus (FIG. 1). If vestibular modulation of SSs in Purkinje cells were caused by modulated vestibular primary afferent mossy-fiber signals, then the left unilateral labyrinthectomy should cause a loss of SS modulation

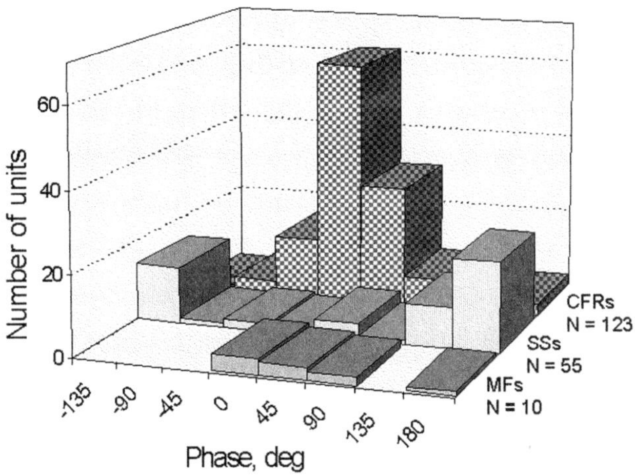

FIGURE 5. Phase of vestibularly evoked CFRs, SSs, and mossy fibers during sinusoidal roll in the *optimal plane*. The phase of vestibularly evoked peak CFRs, SSs, and mossy fibers were measured relative to head position during stimulation in the *optimal plane* at 0.1–0.2 Hz. The peak discharges of CFRs and mossy fibers (MFs) had phase leads of 10.4 and 38.8° with respect to head position during ipsilateral tilt. The peak discharge of SSs had a lag of 177.6° with respect to head position during ipsilateral tilt. A phase lead is indicated by positive values, and a phase lag is indicated by negative values.

in ipsilateral Purkinje cells while retaining modulation of CFRs that originate from the contralateral (intact) inferior olive. Both vestibularly modulated SSs and CFRs were present in the ipsilateral uvula-nodulus following a left unilateral labyrinthectomy (FIG. 6). Ipsilateral roll evoked increased discharges of CFRs and decreased discharges of SSs over a range of testing frequencies (FIG. 6D, E, and F). These vestibularly evoked responses were similar in all respects to those previously observed in intact rabbits.

Antiphasic Behavior of CFRs and SSs in Contralateral Uvula-Nodulus of Anesthetized Rabbits following Unilateral Labyrinthectomy

If the left labyrinth were the only source of vestibular primary afferents to the left parasolitary nucleus and if the left parasolitary nucleus were the only source of vestibularly modulated signals to the ipsilateral inferior olive, then a left labyrinthectomy would prevent vestibularly modulated climbing-fiber signals from reaching the contralateral uvula-nodulus. This is clearly not the case. Following a left labyrinthectomy, we recorded from 16 Purkinje cells in the contralateral uvula-nodulus that evinced vestibularly modulated CFRs. Increases in CFRs and decreases in SSs were evoked when the rabbit was rolled onto its right side (FIG. 7).

Following the left labyrinthectomy the discharge of SSs in both the left and right uvula-nodulus remained approximately antiphasic with respect to vestibularly evoked CFRs. CFRs evoked from Purkinje cells in the ipsilateral uvula-nodulus lead

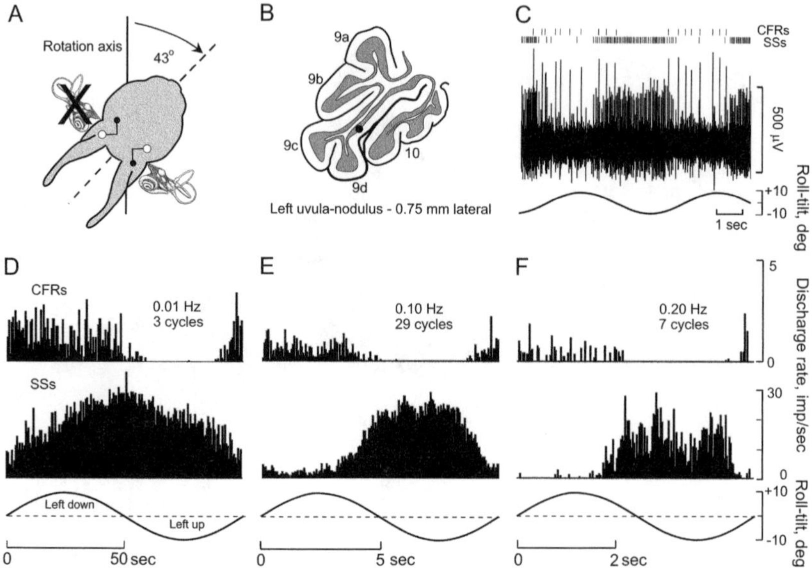

FIGURE 6. Vestibular modulation of CFRs and SSs in Purkinje cells ipsilateral to unilateral labyrinthectomy. Following a left labyrinthectomy, vestibularly evoked activity was recorded from a Purkinje cell in the left uvula. (**A**) The *optimal plane* for modulating CFRs was 43° clockwise with respect to the vertical axis. Rotation in this plane optimally stimulated the right anterior semicircular canal. (**B**) The Purkinje cell was located in folia 9d, 0.75 mm to the left of the midline. (**C**) CFRs and SSs were modulated antiphasically. CFRs increased and SSs decreased when the rabbit was rotated onto its left side. The *top two traces* indicate discriminated occurrences of CFRs and SSs. The *bottom trace* shows roll-tilt position. An upward deflection of this trace corresponds to left side down. (**D, E,** and **F**) Peristimulus histograms of CFRs and SSs were constructed for stimulus frequencies of 0.01, 0.10, and 0.20 Hz.

ipsilateral head position by 13.8°. CFRs evoked from Purkinje cells in the contralateral uvula-nodulus lead contralateral head position by 6.7°. SSs evoked from Purkinje cells in the ipsilateral uvula-nodulus lead ipsilateral head position by 189.3° and SSs recorded in the contralateral uvula-nodulus lead contralateral head position by 203.9° (FIG. 8A and B).

Although the antiphasic behavior of CFRs and SSs recorded from Purkinje cells in both the left and right uvula-nodulus remained intact, there was a difference in the extent of CFR-induced modulation on the two sides. Separate plots were made of the relation between CFR and SS modulation for Purkinje cells recorded in the left and right uvula-nodulus (FIG. 9). The slope of this function for Purkinje cells ipsilateral to the unilateral labyrinthectomy was not statistically different than the slope of the same function in Purkinje cells recorded in rabbits with intact labyrinths ($p = .87$, analysis of covariance) (FIG. 4 and FIG. 9A). However, the slope of the CFR-induced SS function was reduced significantly in Purkinje cells recorded from the right uvula-nodulus ($p < .05$, analysis of covariance) (FIG. 9B).

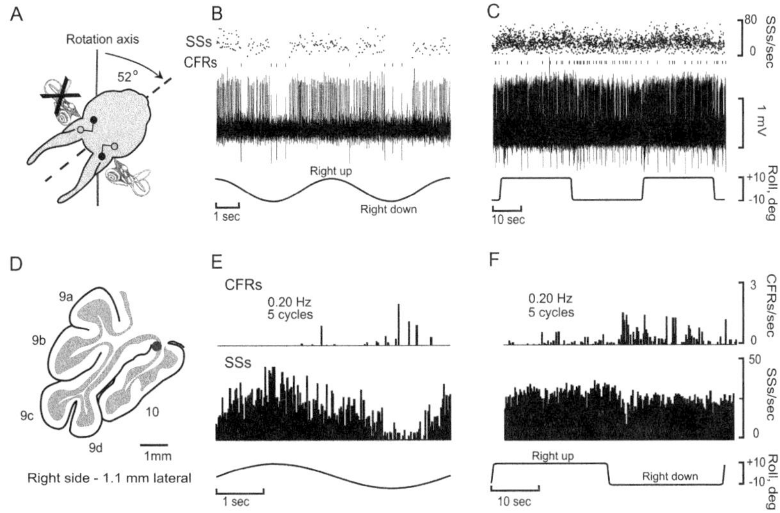

FIGURE 7. Vestibular modulation of CFRs and SSs in Purkinje cell contralateral to unilateral labyrinthectomy. Following a left labyrinthectomy, vestibularly evoked activity was recorded from a Purkinje cell in the right nodulus. (**A**) The *optimal plane* for modulating CFRs was 52° clockwise with respect to the vertical axis. Rotation in this plane optimally stimulated the right anterior semicircular canal. (**B**) CFRs and SSs were modulated antiphasically. CFRs increased and SSs decreased when the rabbit was rotated onto its right side. The *top two traces* indicate discriminated occurrences of CFRs and SSs. The *bottom trace* shows roll-tilt position. An upward deflection of this trace corresponds to right side up. (**C**) The antiphasic behavior of CFRs and SSs was maintained during static tilt. (**D**) The Purkinje cell was located in folia 10, 1.10 mm to the right of the midline. (**E** and **F**) Peristimulus histograms during sinusoidal roll (**E**) and static tilt (**F**) illustrate the weak antiphasic behavior of CFRs and SSs with greater clarity.

DISCUSSION

Vestibularly Evoked CFRs and SSs in Rabbits with Intact and Unilaterally Damaged Labyrinths

In normal rabbits, CFRs are evoked in the uvula-nodulus by roll onto the ipsilateral side. SSs are evoked by roll onto the contralateral side. If the SSs were evoked by the action of vestibular primary afferent mossy fibers relayed through granule cells, then one would expect that SSs would have the same polarity as do vestibularly evoked mossy-fiber signals. However, they have the opposite polarity. These findings could have several interpretations: SSs are caused by the action of mossy fibers, but the polarity of the signal is inverted at either the vestibular primary afferent mossy-fiber synapse on the granule cell or at the parallel fiber synapse on the Purkinje cell dendrite. However, glutamate appears to be the primary neurotransmitter released by both vestibular primary afferents[27,28] and cerebellar granule cells.[29–32] It is unlikely that the necessary sign inversion would be produced at the mossy fiber–

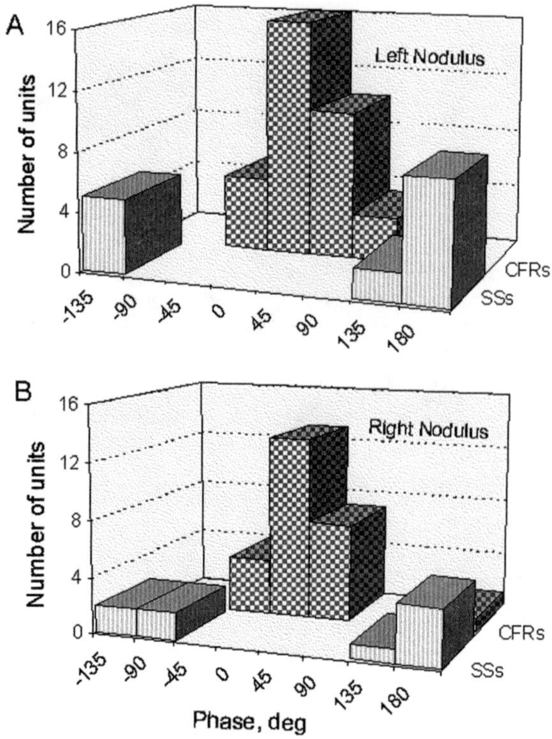

FIGURE 8. Phase of vestibularly evoked CFRs and SSs in uvula-nodulus Purkinje cells during sinusoidal roll in *optimal plane* following unilateral labyrinthectomy. The phase of peak vestibularly evoked CFRs and SSs were measured relative to head position during stimulation in the *optimal plane* at 0.1–0.2 Hz. The relative phases for CFRs and SSs were similar for Purkinje cells recorded ipsilaterally, (**A**), or contralaterally, (**B**), to the unilateral labyrinthectomy. A phase lead is indicated by positive values, and a phase lag is indicated by negative values.

granule cell synapse or parallel fiber–Purkinje cell synapse. The sign inversion could be caused by climbing fiber–induced excitation of Golgi cells and subsequent Golgi cell–induced inhibitory modulation of granule cells.

A unilateral labyrinthectomy prevents modulation of ipsilateral vestibular primary afferent mossy fibers. Since the vestibular primary afferent mossy-fiber projection to the uvula-nodulus is exclusively ipsilateral,[16] a unilateral labyrinthectomy prevents vestibular primary afferent mossy fibers from contributing to the vestibularly evoked modulation of SSs.

Although the vestibular modulation of SSs is not dependent on vestibular primary afferent mossy fibers, it is possible that these afferents could influence granule cell discharge in more subtle ways. Vestibular primary afferents could control the general excitability of granule cells and thereby influence the spontaneous rate and dynamic range of SS modulation evoked by CFRs.

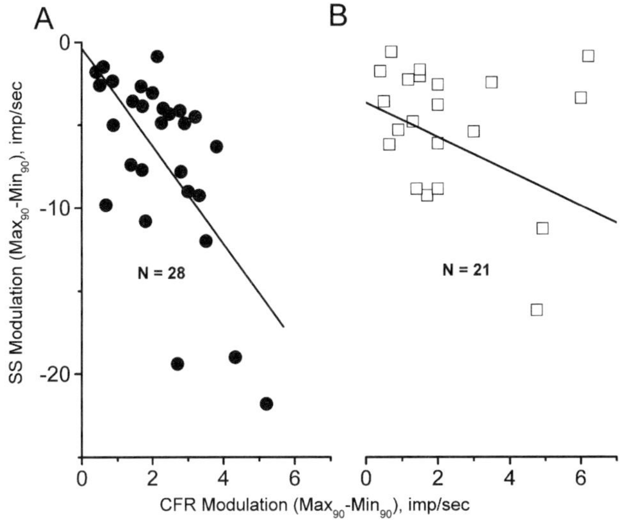

FIGURE 9. Left labyrinthectomy causes reduced modulation of SSs in right uvula-nodulus during vestibular stimulation in the *optimal plane.* (**A**) The peak modulation of SSs as a function of vestibularly evoked CFRs recorded from Purkinje cells in the left uvula-nodulus was unaffected by a left labyrinthectomy. (**B**) The peak modulation of SSs as a function of vestibularly modulated CFRs recorded from Purkinje cells in the right uvula-nodulus was reduced. The slopes of the fitted linear regression lines were statistically different (analysis of covariance, $p < .05$).

Possible Role of Secondary Vestibular Afferent Mossy Fibers in Regulation of Purkinje Cell SSs

While we *can* rule out the possibility that vestibular primary afferent mossy fibers contribute significantly to vestibularly evoked antiphasic behavior of CFRs and SSs, we *cannot* exclude the possibility that vestibular secondary afferents could, in part, account for the modulation of SSs. Unlike vestibular primary afferent projections, vestibular secondary afferent projections to folia 9d and 10 are bilateral.[33–39] A unilateral labyrinthectomy acutely reduces the spontaneous and vestibularly evoked activity in both the ipsilateral and contralateral vestibular complex. This activity remains depressed in the ipsilateral vestibular complex for 10 hours. Resting and stimulus-evoked activity of neurons in the contralateral nucleus returns to within 60% of normal within four hours.[40–43] Since our recordings extended to 8 hours postlabyrinthectomy, a unilateral labyrinthectomy would not necessarily have eliminated inputs to the uvula-nodulus from the contralateral vestibular complex. However, this would have required a fortuitous type I projection from the contralateral vestibular complex to the cerebellum.

On the basis of anatomical projections, it is unlikely that modulated primary or secondary vestibular afferents contributed to SS modulation. Vestibularly evoked antiphasic behavior of CFRs and SSs was obtained in all folia of the uvula-nodulus.

However, only folia 9c and 9d receive projections from primary and secondary vestibular afferents.[16,26,44]

Revision of Wiring Diagram for Vestibular Activation of Climbing Fibers

While a unilateral labyrinthectomy deprives the ipsilateral uvula-nodulus of its vestibular primary afferent input, it does not eliminate a descending vestibular projection to the ipsilateral inferior olive. We were able to record well-modulated CFRs in the contralateral uvula-nodulus following a unilateral labyrinthectomy. This suggests that the parasolitary nucleus is not the only source of vestibular information to the β-nucleus and dmcc.

One possible source of descending vestibular information is the Y-group. This nucleus is divided into dorsal and ventral segments. Each of these segments may have different projections and different functions. Cells in the dorsal Y-group receive primary and secondary vestibular afferents[15,45] and are retrogradely labeled by tracer injections into the contralateral inferior olive.[46,47] The Y-group also projects to the ipsilateral flocculus and nodulus as well as the oculomotor complex.[45] Cells in the dorsal Y-group are labeled immunohistochemically by an antibody to aspartate, rather than γ-amminobutyric acid.[48] This suggests that vestibularly modulated cells projecting to the inferior olive from the contralateral dorsal Y-group would have the same functional effect on olivary neurons as would GABAergic cells projecting to the inferior olive from the ipsilateral parasolitary nucleus.[22,23]

The existence of a vestibular pathway to the inferior olive, other than that from Psol, accounts for the observation that vestibularly evoked CFRs are recorded in the contralateral uvula-nodulus following a unilateral labyrinthectomy. While a unilateral labyrinthectomy clearly reduces the efficacy of climbing-fiber signals to these contralateral Purkinje cells (FIG. 9), a more direct demonstration of the effects of blocking vestibular climbing fibers, perhaps at the level of the inferior olive, would provide further insight into the role of CFRs in modulating SSs.

Induction of SS Modulation by Climbing-Fiber Activation of Golgi Neurons?

Most theories concerning climbing-fiber functions ascribe to them a more indirect role in regulating Purkinje cell activity. The relatively low and seemingly unpredictable discharges of CFRs have been rationalized theoretically by the attribution of rare functions to them. CFRs have been characterized as "phasic motor generators,"[49] "threshold devices,"[50] "event detectors,"[51] "error detectors,"[52] "frequency synchronizers,"[53] and "gain changers."[54] These theories assume incorrectly that the low discharge frequencies of CFRs are incapable of encoding sensory stimulation and contributing to the parametric control of movement. On the contrary, optokinetic and vestibular stimulation continuously modulate the discharge of climbing fibers.[26,55–57] In the uvula-nodulus, climbing fibers not only evoke low frequency CFRs in Purkinje cells, but they also regulate the higher frequency discharge of SSs. Consequently, it is unnecessary to resort to theories of climbing-fiber function that have no prospect of being rejected by experiment. Our results assign importance to the role of the cerebellar Golgi neuron as a cell that could allow climbing-fiber inputs to modulate SSs. This idea could be tested in transgenic conditional "knockouts" in which the Golgi cells of a fully developed cerebellum are rendered dysfunctional.[58]

Functions of CFRs

Climbing fibers could have both short-term and long-term functions in regulating the discharge of Purkinje cells. The short-term function could provide a spatial filter in which the climbing fiber, acting on Golgi cells, determines the discharge of SSs. In this role the topography of the climbing fiber rather than the topography of mossy fibers would determine the polarity and primary spatial characteristics of SSs.

"Long term depression" (LTD) provides empirical support for how CFRs can reduce the frequency of SSs.[59–62] The conjunctive coupling of parallel fiber activation with CFRs reduces the efficacy of parallel fiber synapses on Purkinje cell dendrites.[63] In LTD, the climbing fiber acts as a temporal filter in which simultaneously occurring parallel fiber inputs are selectively weakened, allowing a single climbing fiber to shape the synaptic influence of parallel fibers originating from spatially remote granule cells. Since parallel fibers attain lengths of up to 7 mm,[64,65] virtually every granule cell in a beam extending across the entire width of the nodulus potentially could synaptically contact every Purkinje cell. By depolarizing the entire Purkinje cell dendritic tree, a single CFR could potentially influence the synaptic efficacy of more than 150,000 parallel fibers.[4] Both Golgi-mediated granule cell inhibition and conjunctive LTD contribute to the reciprocal modulation of SSs by CFRs.

ACKNOWLEDGMENTS

This research was supported by a grant from the National Eye Institute (EY04778) to NHB. We thank Ms. Mary Westcott for her expert histological and technical assistance.

REFERENCES

1. Granit, R. & C.G. Phillips. 1956. Excitatory and inhibitory processes acting upon individual Purkinje cells of the cerebellum in cats. J. Physiol. (Lond.) **133:** 520–547.
2. Eccles, J.C., R. Llinás & K. Sasaki. 1966. The excitatory synaptic action of climbing fibers on the Purkinje cells of the cerebellum. J. Physiol. (Lond.) **182:** 268–296.
3. Barbour, B. 1993. Synaptic currents evoked in Purkinje cells by stimulating individual granule cells. Neuron **11:** 759–769.
4. Fox, C.A., D.E. Hillman, K.A. Siegesmund & C.R. Dutta. 1967. The primate cerebellar cortex: a Golgi and electron microscopic study. *In* Progress in Brain Research, Vol.25: The Cerebellum. C.A. Fox & R.S. Snider, Eds.: 174–225. Elsevier. New York.
5. Diño, M.R., R.J. Schuerger, Y.B. Liu, *et al.* 2000. Unipolar brush cell: A potential feedforward excitatory interneuron of the cerebellum. Neuroscience **98:** 625–636.
6. Diño, M.R., A.A. Perachio & E. Mugnaini. 2001. Cerebellar unipolar brush cells are targets of primary vestibular afferents: an experimental study in the gerbil. Exp. Brain Res. **140:** 162–170.
7. Hámori, J. & J. Szentágothai. 1966. Participation of Golgi neuron processes in the cerebellar glomeruli: an electron microscope study. Exp. Brain Res. **2:** 35–48.
8. Eccles, J.C., R. Llinás & K. Sasaki. 1966. The mossy fibre-granule cell relay of the cerebellum and its inhibitory control by Golgi cells. Exp. Brain Res. **1:** 82–101.
9. Desclin, J.C. 1976. Early terminal degeneration of cerebellar climbing fibers after destruction of the inferior olive in the rat. Synaptic relationships in the molecular layer. Anat. Embryol. (Berl.) **149:** 87–112.
10. Sugihara, I., H. Wu & Y. Shinoda. 1999. Morphology of single olivocerebellar axons labeled with biotinylated dextran amine in the rat. J. Comp. Neurol. **414:** 131–148.

11. Hámori, J. & J. Szentágothai. 1980. Lack of evidence of synaptic contacts by climbing fibre collaterals to basket and stellate cells in developing rat cerebellar cortex. Brain Res. **186:** 454–457.
12. Carpenter, M.B., B.M. Stein & P. Peter. 1972. Primary vestibulocerebellar fibers in the monkey: distribution of fibers arising from distinctive cell groups of the vestibular ganglia. Am. J. Anat. **135:** 221–250.
13. Korte, G.E. 1979. The brainstem projection of the vestibular nerve in the cat. J. Comp. Neurol. **184:** 279–292.
14. Carleton, S.C. & M.B. Carpenter. 1984. Distribution of primary vestibular fibers in the brainstem and cerebellum of the monkey. Brain Res. **294:** 281–298.
15. Kevetter, G.A. & A. Perachio. 1986. Distribution of vestibular afferents that innervate the sacculus and posterior canal in the gerbil. J. Comp. Neurol. **254:** 410–424.
16. Barmack, N.H., R.W. Baughman, P. Errico & H. Shojaku. 1993. Vestibular primary afferent projection to the cerebellum of the rabbit. J. Comp. Neurol. **327:** 521–534.
17. Barmack, N.H. 1996. GABAergic pathways convey vestibular information to the beta nucleus and dorsomedial cell column of the inferior olive. Ann. N.Y. Acad. Sci. **781:** 541–552.
18. Barmack, N.H., M. Fagerson, B.J. Fredette, E. Mugnaini & H. Shojaku. 1993. Activity of neurons in the beta nucleus of the inferior olive of the rabbit evoked by natural vestibular stimulation. Exp. Brain Res. **94:** 203–215.
19. Barmack, N.H. 1999. The central vestibular system. *In* Neuroscience Secrets. M.T.T. Wong-Riley, Ed.: 115–140. Hanley and Belfus. Philadelphia.
20. Kaufman, G.D. & A.A. Perachio. 1994. Translabyrinth electrical stimulation for the induction of immediate-early genes in the gerbil brainstem. Brain Res. **646:** 345–350.
21. Kaufman, G.D., M.J. Mustari, R.R. Miselis & A.A. Perachio. 1996. Transneuronal pathways to the vestibulocerebellum. J. Comp. Neurol. **370:** 501–523.
22. Barmack, N.H., B.J. Fredette & E. Mugnaini. 1998. Parasolitary nucleus: A source of GABAergic vestibular information to the inferior olive of rat and rabbit. J. Comp. Neurol. **392:** 352–372.
23. Barmack, N.H. & V. Yakhnitsa. 2000. Vestibular signals in the parasolitary nucleus. J. Neurophysiol. **83:** 3559–3569.
24. Barmack, N.H. & H. Shojaku. 1992. Vestibularly induced slow oscillations in climbing fiber responses of Purkinje cells in the cerebellar nodulus of the rabbit. Neuroscience **50:** 1–5.
25. Fushiki, H. & N.H. Barmack. 1996. Simple spike activity modulated by vestibularly-evoked climbing fiber responses in the rabbit cerebellar nodulus [abstract]. Soc. Neurosci. Abst. **22:** 1091.
26. Fushiki, H. & N.H. Barmack. 1997. Topography and reciprocal activity of cerebellar Purkinje cells in the uvula-nodulus modulated by vestibular stimulation J. Neurophysiol. **78:** 3083–3094.
27. Raymond, J., A. Nieoullon, D. Dememes & A. Sans. 1984. Evidence for glutamate as a neurotransmitter in the cat vestibular nerve: radioautographic and biochemical studies. Exp. Brain Res. **56:** 523–531.
28. Yamanaka, T., M. Sasa & T. Matsunaga. 1997. Glutamate as a primary afferent neurotransmitter in the medial vestibular nucleus as detected by in vivo microdialysis. Brain Res. **762:** 243–246.
29. Sandoval, M.E. & C.W. Cotman. 1978. Evaluation of glutamate as a neurotransmitter of cerebellar parallel fibers. Neuroscience **3:** 199–206.
30. Somogyi, P., K. Halasy, J. Somogyi, *et al.* 1986. Quantification of immunogold labelling reveals enrichment of glutamate in mossy and parallel fibre terminals in cat cerebellum. Neuroscience **19:** 1045–1050.
31. Kingsbury, A., V. Gallo & R. Balazs. 1988. Stimulus-coupled release of amino acids from cerebellar granule cells in culture. Brain Res. **448:** 46–52.
32. Ottersen, O.P., N. Zhang & F. Walberg. 1992. Metabolic compartmentation of glutamate and glutamine: morphological evidence obtained by quantitative immunocytochemistry in rat cerebellum. Neuroscience **46:** 519–534.
33. Brodal, A. & B. Hoivik. 1964. Site and mode of termination of primary vestibulocerebellar fibres in the cat. Arch. Ital. Biol. **102:** 1–21.

34. EPEMA, A.H., J.M. GULDEMOND & J. VOOGD. 1985. Reciprocal connections between the caudal vermis and the vestibular nuclei in the rabbit. Neurosci. Lett. **57:** 273–278.
35. MAGRAS, I.N. & J. VOOGD. 1985. Distribution of the secondary vestibular fibers in the cerebellar cortex. Acta Anat. **123:** 51–57.
36. SATO, Y., K.-I. KANDA, K. IKARASHI & T. KAWASAKI. 1989. Differential mossy fiber projections to the dorsal and ventral uvula in the cat. J. Comp. Neurol. **279:** 149–164.
37. THUNNISSEN, I.E., A.H. EPEMA & N.M. GERRITS. 1989. Secondary vestibulocerebellar mossy fiber projection to the caudal vermis in the rabbit. J. Comp. Neurol. **290:** 262–277.
38. EPEMA, A.H., N.M. GERRITS & J. VOOGD. 1990. Secondary vestibulocerebellar projections to the flocculus and uvulo-nodular lobule of the rabbit: A study using HRP and double fluorescent tracer techniques. Exp. Brain Res. **80:** 72–82.
39. BARMACK, N.H., R.W. BAUGHMAN, F.P. ECKENSTEIN & H. SHOJAKU. 1992. Secondary vestibular cholinergic projection to the cerebellum of rabbit and rat as revealed by choline acetyltransferase immunohistochemistry, retrograde and orthograde tracers. J. Comp. Neurol. **317:** 250–270.
40. PRECHT, W., H. SHIMAZU & C.H. MARKHAM. 1966. A mechanism of central compensation of vestibular function following hemilabyrinthectomy. J. Neurophysiol. **29:** 996–1010.
41. SMITH, P.F. & I.S. CURTHOYS. 1988. Neuronal activity in the ipsilateral medial vestibular nucleus of the guinea pig following unilateral labyrinthectomy Brain Res. **444:** 308–319.
42. RIS, L., B. CAPRON, C. DE WAELE, P.-P. VIDAL & E. GODAUX. 1997. Dissociations between behavioural recovery and restoration of vestibular activity in the unilabyrinthectomized guinea-pig. J. Physiol **500:** 509–522.
43. RIS, L. & E. GODAUX. 1998. Neuronal activity in the vestibular nuclei after contralateral or bilateral labyrinthectomy in the alert guinea pig. J. Neurophysiol. **80:** 2352–2367.
44. KORTE, G. & E. MUGNAINI. 1979. The cerebellar projection of the vestibular nerve in the cat J. Comp. Neurol. **184:** 265–278.
45. BLAZQUEZ, P., A. PARTSALIS, N. M. GERRITS & S. M. HIGHSTEIN. 2000. Input of anterior and posterior semicircular canal interneurons encoding head-velocity to the dorsal Y group of the vestibular nuclei. J. Neurophysiol. **83:** 2891–2904.
46. WENTZEL, P. R., D. R. WYLIE, T. J. RUIGROK & C. I. DE ZEEUW. 1995. Olivary projecting neurons in the nucleus prepositus hypoglossi, group y and ventral dentate nucleus do not project to the oculomotor complex in the rabbit and the rat. Neurosci. Lett. **190:** 45–48.
47. DE ZEEUW, C.I., N.M. GERRITS, J. VOOGD, *et al.* 1994. The rostral dorsal cap and ventrolateral outgrowth of the rabbit inferior olive receive a GABAergic input from dorsal group Y and the ventral dentate nucleus. J. Comp. Neurol. **341:** 420–432.
48. KUMOI, K., N. SAITO & C. TANAKA. 1987. Immunohistochemical localization of γ-aminobutyric acid- and aspartate-containing neurons in the guinea pig vestibular nuclei. Brain Res. **416:** 22–23.
49. BERTHOZ, A. & R. LLINÁS. 1974. Afferent neck projection to the cat cerebellar cortex. Exp. Brain Res. **20:** 385–401.
50. MURPHY, J.T., W.A. MACKAY & F. JOHNSON. 1973. Differences between cerebellar mossy and climbing fibre responses to natural stimulation of forelimb muscle proprioceptors. Brain Res. **55:** 263–289.
51. RUSHMER, D.S., W.J. ROBERTS & G.K. AUGTER. 1976. Climbing fiber responses of cerebellar Purkinje cells to passive movement of the cat forepaw. Brain Res. **106:** 1–20.
52. ANDERSSON, G. & D.M. ARMSTRONG. 1987. Complex spikes in Purkinje cells in the lateral vermis of the cat cerebellum during locomotion. J. Physiol. (Lond.) **385:** 107–134.
53. LLINÁS, R.R. & Y. YAROM. 1986. Oscillatory properties of guinea-pig inferior olivary neurones *in vitro.* J. Physiol. (Lond.) **376:** 163–182.
54. EBNER, T.J. & J.R. BLOEDEL. 1981. Role of climbing fiber afferent input in determining responsiveness of Purkinje cells to mossy fiber inputs. J. Neurophysiol. **45:** 962–971.
55. SIMPSON, J.I. & K.E. ALLEY. 1974. Visual climbing fiber input to rabbit vestibulocerebellum: A source of direction–specific information Brain Res. **82:** 302–308.
56. SIMPSON, J.L., W. GRAF & C. LEONARD. 1981. The coordinate system of visual climbing fibers to the flocculus. *In* Developments in Neuroscience, Vol 12: Progress in Oculomotor Research. Fuchs, A.F. & W. Becker, Eds.: 475–484. Elsevier. North Holland.

57. BARMACK, N.H. & D.T. HESS. 1980. Multiple-unit activity evoked in dorsal cap of inferior olive of the rabbit by visual stimulation J. Neurophysiol. **43:** 151–164.
58. WATANABE, D., H. INOKAWA, K. HASHIMOTO, *et al.* 1998. Ablation of cerebellar Golgi cells disrupts synaptic integration involving GABA inhibition and NMDA receptor activation in motor coordination. Cell **95:** 17–27.
59. ITO, M., M. SAKURAI & P. TONGROACH. 1982. Climbing fiber induced depression on both mossy fibre responsiveness and glutamate sensitivity of cerebellar Purkinje cells. J. Physiol. (Lond.) **324:** 113–134.
60. ITO, M. & L. KARACHOT. 1989. Long-term desensitization of quisqualate-specific glutamate receptors in Purkinje cells investigated with wedge recording from rat cerebellar slices. Neurosci. Res. **7:** 168–171.
61. SAKURAI, M. 1987. Synaptic modification of parallel fibre-Purkinje cell transmission in *in vitro* guinea-pig cerebellar slices. J. Physiol. (Lond.) **394:** 463–480.
62. CRÉPEL, F. & D. JAILLARD. 1991. Pairing of pre- and postsynaptic activities in cerebellar Purkinje cells induces long-term changes in synaptic efficacy *in vitro.* J. Physiol. (Lond.) **432:** 123–141.
63. NARASIMHAN, K. & D.J. LINDEN. 1996. Defining a minimal computational unit for cerebellar long-term depression. Neuron **17:** 333–341.
64. BRAND, S., A.L. DAHL & E. MUGNAINI. 1976. The length of parallel fibers in the cat cerebellar cortex. An experimental light and electron microscopic study. Exp. Brain Res. **26:** 39–58.
65. MUGNAINI, E. 1983. The length of cerebellar parallel fibers in chicken and Rhesus monkey. J. Comp. Neurol. **220:** 7–15.

Climbing Fiber Inputs to the Cerebellum: What Is Signaled?

Summary of Discussion

J. HOUK, *Rapporteur*

One problem in understanding cerebellar function has been how to conceptualize the role of the climbing fiber (cf) as distinct from that of the mossy fiber (mf). The climbing fiber originates exclusively from the inferior olive nucleus, projects with little divergence to excite powerfully a few Purkinje, Golgi and deep nuclear cells, but fires at relatively low rates of 1–2/seconds. The mossy fiber originates from many different sources, branches to excite many granule cell, which in turn branch to excite weakly many Purkinje (basket, stellate, and Golgi II) cells; each of these cells receive many parallel fibers, excited by each one parallel fiber (pf) weakly, but are influenced by the aggregate to fire at relatively high frequencies of up to several hundred spikes/second.

Llinás posed the question of whether one should think of the two input systems as being parallel and independent, each serving a uniquely different purpose, or instead working essentially together in combination. Several supported the view that the two systems are at least to some extent independent. Welch emphasized the power of the climbing fiber synapse, and the tendency, seen in his and others' work, to fire synchronously across the olive and thus across much of cerebellar cortex, and in many instances periodically. This could be expected to have a powerful driving effect on the excitatory output of the cerebellum, such as to cause and pace movements. This view had been suggested from the harmaline studies of Lamarre and Llinás and colleagues in which the drug produced a whole-body tremor in the range of 10 Hz, a periodic firing of IO neurons also in the range of 10 Hz, and the reduction of the tremor by ablation of the inferior olive. Thach pointed out that electric stimulation of the inferior olive by a number of workers including Gibson, Rawson, Ito, Barmack, and Mauk had not produced the sudden movements that might have been predicted by such a hypothesis. Barmack said that instead he had observed "slow" eye movments, which could have been due to a conditioning effect on cerebellar output as well as a straight-through drive. Barmack further observed that averaging eye movement on the occurrence of ION/PcCS discharge revealed very slight movements of the eye in a direction such as to increase retinal slip, with a latency of 18 ms and a magnitude of 10s of minutes of arc. Ebner asked whether others had observed the periodicity of olive cell discharge that had been seen in the early harmaline (in cat) studies and the more recent studies in rodents of Lang and Welch and

Commentary transcribed and edited from tape by Tom Thach.

Ann. N.Y. Acad. Sci. 978: 255–257 (2002). © 2002 New York Academy of Sciences.

colleagues, and whether the periodicity, when observed, was somehow due to the fact that the movements themselves (licking) were periodic. Simpson said that in his single-electrode studies in rabbit one could occasionally see periodic discharge in ION's and Pc complex spikes, but that it "came and went" over time. In single-unit recording in macaques, Ebner remarked he had seen little, and Gibson and Thach no detectable periodicity of Pc CS discharge (see abstract of Hakimian *et al.*). Gibson remarked that injections into the rostral inferior olive of macaques of lidocaine tended to slow upper limb movements and injections of bicuculline to speed the movements, but that the movement pattern was not obviously disrupted, as would have occurred with injections into the cerebellum proper, thus further questioning a major driving effect of olive discharge upon movement. Welch replied that in studies where the rats made non-periodic limb movements (lever press), the inferior olive–Pc complex spike discharge was much less periodic. Nevertheless, there was still a tendency to periodicity, as well as synchronicity across olive and Purkinje cells. Furthermore, with strong sensory or electrical stimulation, there was a tendency to a few beats of damped oscillatory discharge. He stressed that rhythmicity is a population event, and can only be fully appreciated in multiple (30–40) electrode recordings. Others felt that multiple cell recordings might indeed increase the sensitivity of detection of a tendency toward a periodic signal, but that the necessity of such a resort testified to the weakness of the signal. Bell raised the question as to whether the ION-cf discharge would be appropriate to serve as a "clock" or timing function of any sort, given that the interspike interval is of the order of 100–130 ms, yet the timing of coordinated movements is the range of tens of milliseconds.

Others took the tack that the function of the climbing fiber is more likely to be understood as cooperative *with* the mf in some complementary function. Bell raised the question posed previously by Harmon and Kado as to whether the climbing fiber might serve to "erase" the electrotonic memory of the Pc and thus "reset" it to some initial condition. Gibson cited work of Garwisz and Eckerot commenting that such a function might serve to bias the operation of the system: if triggered by a painful stimulus, the climbing fiber might serve to alter a locomotor command appropriate to the location and nature of the painful stimulus.

Bower commented that the strength of the climbing fiber terminal synapses, yet the low frequency of firing, would in itself suggest some sort of biasing function, as in a "refresh" operation, rather than an independent drive path through the nuclei. Yet Llinás pointed out that the cf-Pc-induced IPSP in the deep cerebellar nuclear cell is much more powerful than that caused by mf-Pc IPSP, which could be consistent with a drive-through function of disinhibition of downstream cerebellar targets.

Houk began the discussion period with his own summary of the evidence that the climbing fiber likely serves to condition the Purkinje cell pf synapse through LTD, and thus to play a pivotal role in the cerebellar contribution to motor learning (the topic of another session). Others argued as to whether the case had unequivocally been made that the ION-cf IS indeed involved in motor learning.

Llinás said that the more we know about the details of the operation, the more mysterious the essential operation becomes. Prompted by comment of Baker about the utility of comparative studies, Llinás pointed out that the cerebellar circuitry, with very little variation, is remarkably preserved in its basic plan across species. That, as he and many have concluded, bespeaks some uniformity and uniqueness of function. Yet, each of the experts emphasizes a particular feature—the adequate sensory stimulus, the synaptic effect, the behavioral effect of discharge, and so on. Like the six blind men of Hindustan, each attempting to comprehend through palpation the essence of the elephant, the experts at this meeting did not come up with one hypothesis.

> "And so these men of Indostan
> Disputed loud and long,
> Each in his own opinion
> Exceeding stiff and strong.
> Though each was partly right,
> All were in the wrong."
>
> –Godfrey Saxe

Nevertheless, there was consensus that in such a situation, it is good for the blind men to get together in congenial surroundings and talk!

On the Amazing Olivocerebellar System

R. LLINÁS, E. LEZNIK, AND V.I. MAKARENKO

Department of Physiology and Neuroscience, New York University School of Medicine, New York, New York 10016, USA

ABSTRACT: Over the last four decades elegant sets of a single-cell studies, originating from various research groups, have contributed significantly to our understanding of olivary and cerebellar physiology. Nevertheless questions relating to the dynamic properties of olivocerebellar network, as a system, remain unsolved. We may be reaching the limits of what can be learned using the single-cell recordings. Further research on this subject may require study of the spatiotemporal activity profiles of ensemble neuronal activity. This paper summarizes results obtained using voltage-sensitive dye imaging in inferior olive slices, and the use of mathematical modeling to address such activity profiles.

KEYWORDS: olivocerebellar system; inferior olive; voltage sensitive dyes, optical imaging; mathematical modeling

INTRODUCTION

The inferior olives, the origin of the olivocerebellar system, are bilateral nuclei located in the ventral aspect of the brain stem. The afferent type, known as climbing fibers,[1] forms one of the two major afferent pathways to the cerebellar cortex and represents the exclusive means of projection from the inferior olives (IO) to the contralateral cerebellar cortex.[2] Climbing fibers directly innervate all Purkinje cells (PC) in the cerebellar cortex generating the largest synaptic junction in the vertebrate central nervous system. Indeed, activation of a climbing fiber is followed by an all-or-nothing burst of spikes,[3] later known as a complex spike,[4] in all of its postsynaptic PCs. The total number of IO neurons is approximately one order of magnitude smaller than that of the PCs, indicating that on the average an IO neuron generates 10 climbing fibers.[5] In addition to contacting PCs, climbing fibers also produce collateral branches that terminate on all cerebellar nuclei.[6]

Inferior olive neurons fire spontaneously at 1–10 Hz and can exhibit rhythmic oscillatory activity near 10 Hz.[7,8] Likewise, the resulting complex spikes in the PCs display a 10-Hz rhythmicity. Furthermore, simultaneous electrode recordings from multiple PCs have shown that complex spikes occur synchronously in a group of PCs.[10–12] Such synchronous activation of multiple PCs by the IO nucleus suggests that clusters of IO neurons have the ability to fire rhythmically and in unison. The

Address for correspondence: Dr. Rodolfo Llinás, Department of Physiology and Neuroscience, New York University School of Medicine, 550 First Ave., New York, NY 10016.
llinar01@endeavor.med.nyu.edu

Ann. N.Y. Acad. Sci. 978: 258–272 (2002). © 2002 New York Academy of Sciences.

fact that IO neurons are electrically coupled to each other and tend to fire in groups[13,14] led to the proposal that climbing fibers may perform a timing function in motor coordination.[15] According to that hypothesis, IO activity functions as a motor timing signal by generating synchronous and rhythmic activation of cerebellar nuclear and descending and ascending connectivity to the spinal cord and thalamus. The timing of these cerebellar nuclei is being further controlled by the synchronized inhibitory barrage originated by climbing fiber activation of PC complex spikes. Evidence to support the motor timing proposal includes the following: synchronous firing of a population of PCs in anesthetized and awake animals,[10,16,17] recordings of rhythmic inhibitory potential in the deep cerebellar nuclei,[18] and temporal correlation between firing of the olivocerebellar system and execution of movements.[19]

The ability of the olivocerebellar system to generate synchronous rhythmic activity has been attributed to the intrinsic oscillatory properties of the IO neurons[8,20–25] and their electrotonic coupling.[8,13,14,26,27] In particular, several types of voltage-dependent calcium and potassium conductance, in addition to those involved in action potential generation, enable IO cells to oscillate and fire rhythmically at 1–10 Hz. These conductances include a high-threshold Ca^{2+} conductance, a somatic low-threshold Ca^{2+} conductance, a Ca^{2+}-activated K^+ conductance, and a hyperpolarization-activated cationic conductance.[8,20,21,23]

Concerning electrical coupling, gap junctions, as in other CNS structures,[28] constitute the main communication pathway between the IO neurons.[14,29] Such electrotonic coupling is proposed to play a crucial role in synchronizing IO oscillations and in generating groups of concurrently oscillating neurons[21] and is assumed to be controlled by return glomerular inhibition.[15] Inferior olive afferents were, in fact, found to modulate the efficiency of electrotonic coupling via inhibition at the IO glomeruli. The pathway turned out to be supported by a set of cerebellar nuclear GABAergic neurons[30] that, in addition, represent almost 50% of the total neuronal population in such nuclei, given some measure of the importance of this feedback inhibitory pathway. Indeed, it was experimentally determined that such inhibitory input can control the degree and distribution of synchronous oscillatory activity in the IO nucleus[31] and cerebellum.[11,32,33] Dynamically contained groups of IO neurons oscillating in-phase can in turn synchronously activate a population of PCs and thereby control patterns of synchronous activity in the cerebellum during motor coordination.[19]

Although synchronized IO oscillations are a neuronal ensemble event, they have mainly been studied on a single-cell level, and no information on their spatial profiles has been available. Thus, we attempted to address this issue by utilizing a voltage-sensitive dye optical imaging technique.[31] This technique is currently a methodology of choice in studying the geometrical distribution of activity in a large neuronal ensemble.[34] We have shown that ensemble oscillations in the IO emanate from clusters of synchronized activity, where each cluster is a localized functional event composed of hundreds of cells. Given the distribution of complex spike activity in the cerebellum cortex, we proposed that these clusters are very likely responsible for the synchronized activation of the Purkinje cells observed in previous *in vivo* multielectrode experiments.[31] Furthermore, when our experimental results were compared with those obtained by computational modeling of IO neuronal ensembles endowed with oscillatory electrical properties and electrotonic coupling,[35,36] we showed that neuronal oscillatory clustering is a direct consequence of the combined electrotonic/intrinsic properties of coupled IO neurons.

OPTICAL VOLTAGE-SENSITIVE DYE IMAGING RESULTS

Voltage-sensitive dye imaging of oscillatory activity was successfully implemented in rodent IO slices.[31] Thus, after IO electrical stimuli, which served to reset the phase of subthreshold oscillation and to entrain a large proportion of neurons to in-phase oscillations,[21] imaging of the spatiotemporal profiles of ensemble IO oscillations was unambiguously observed. Indeed, synchronization of oscillatory activity over the IO network increased the amplitude of the optical signal to a level that could easily be detected with our imaging setup. Such an oscillatory reset was also observed with intracellular recordings from IO neurons (FIG. 1A and 1B).

In agreement with previous intracellular results,[21] extracellular stimulation at the dorsal border of the IO nucleus generated a full action potential followed by membrane hyperpolarization in nearby neurons (FIG. 1A, boxed area). Our results also demonstrated that if the cell was oscillating at the time of the stimulus, its oscillations were momentarily stopped but then resumed with a different phase approximately 750 ms after the stimulation. Moreover, it was also determined that this extracellular stimulation reset the phase, but not the amplitude or frequency, of the subthreshold oscillation (FIG. 1B). The reset could repeatedly be observed in all cells recorded, and the resulting oscillation phase shift was remarkably constant (FIG. 1B, right panel).

Description of the Optical Signals

Our imaging responses consisted of an initial change in background fluorescence induced by direct electrical stimulation followed by several oscillatory waves that spread throughout the IO nucleus (FIG. 1C and 1D). These ensemble IO oscillations were calcium derived and were also observed in concurrently obtained intracellular recordings from IO neurons.[31]

The optically recorded IO oscillations consisted of clusters of coherent neuronal activity in which each cluster was a localized functional entity (FIG. 1D, second and third image panels). These oscillatory groupings of cells arose from several independent regions of the slice, but they had the same oscillation frequency and were phase coherent. The phase coherence was most clearly observed after the phase reset triggered by the electrical stimuli. The optically recorded oscillations were highly repeatable in time and space; the average of three oscillatory cycles had the same spatiotemporal profile as that of the individual oscillations (compare second and third image panels in FIG. 1D).

Within the temporal resolution used in these experiments (4.8 ms between frames), the optical oscillations started synchronously everywhere in the slice (see FIG. 1C [im] pixel 1, 2, 3, and 4). The frequency of optically recorded clusters varied from 1.6 to 6.5 Hz, which was in range for intracellularly recorded subthreshold oscillations.[21, 22, 37]

The optically recorded oscillatory clusters had a dynamic spatial organization. They changed their size depending on the oscillation phase, so that they embraced the largest area during the upward phase of the oscillations (FIG. 1D, second image panel, third image). Hence, each cluster consisted of a core region and the adjoining area. The core region demonstrated a close to constant size, but the extent of the adjoining area was found to be phase dependent. We calculated the core area and maximum area (i.e., core region plus adjoining area at its uppermost extent) for several representative clusters in each experiment. The mean core area and mean maximum area of a cluster occupied an area of 10^{-2}.[31]

Because most of the IO neurons are coupled via dendrodendritic gap junctions, which are not located within the same plane,[38,39] we assumed that the optically recorded IO clusters were three-dimensional structures. We measured the area of each cluster and then estimated its volume by multiplying the area of the cluster by the thickness of the live tissue in the slice. (This could be done because fluorescence can be detected up to 700 μm in depth,[40] and our slices were only 300 μm thick.) Next, we approximated the number of cells in each cluster by multiplying the density of IO cells in the rat (80,000 cells/mm^3)[41] by the volume of each cluster. Based on the foregoing assumptions, we found that both the core area and the maximum area of a cluster consisted of hundreds of cells.

Our optical data therefore indicate that at the network level, the IO nucleus is organized in functionally coupled activity clusters. Each cluster is comprised of several hundred cells, which may act in unison to simultaneously activate groups of cerebellar Purkinje cells, as was previously observed with multiple electrode recordings.[10,12,16,42]

Factors Affecting Cluster Dimensions

We considered the possibility that the level of functional coupling among the IO neurons defined dimensions of the optically recorded fluorescent clusters.[43] If this were the case, drugs that modulate electrical coupling among olivary neurons, such as picrotoxin,[11] would change cluster dimensions. Indeed, application of picrotoxin increased the size of the oscillatory clusters by partly merging them into larger areas of coherent activity and by partly activating additional neuronal regions (FIG. 2A and 2B). These results also demonstrate that clustering was not an artifact resulting from sparsity of neuronal survival, as neurons previously occupying quiescent sites became active after drug administration.

At the single cell level, the addition of picrotoxin decreased the frequency and amplitude of spontaneous subthreshold oscillations without affecting the membrane potential of the cell (FIG. 2C and 2D), which was consistent with an increased cluster size. That is, expansion of cluster size could have increased the electrical load of the network, resulting in a reduction of amplitude and frequency of oscillations. Modulation of the effective electrotonic coupling changed cluster dimensions; therefore, modulation of afferent GABAergic inputs on the IO glomeruli[29, 38] should produce the observed results.

In conclusion, dimensions of clusters are probably determined by the IO electrical coupling coefficient and therefore by the magnitude and distribution of the return inhibition from the cerebellar nuclear feedback, as was demonstrated in previous *in vivo* experiments[11,44] and supported with mathematical modeling.[31, 36]

Modeling Inferior Olive Clusters

As just indicated, the IO nucleus appears to play a significant role in the timing of organized movement. The anatomical and physiological properties of these neurons are sufficiently well established to warrant the development of quantitative computer modeling of both single cell and network properties.

Current knowledge indicates that the subthreshold membrane potential oscillations of IO neurons have an important functional role and complex dynamics. The latter implies that in addition to individual IO neuron dynamics, the electrotonic communication matrix woven by the dendrodendritic gap junction connectivity is crucial in the organization of such a timing signal. This ensemble activity is organized dynamically using neuronal synchronization as a way to rapidly engage or disengage a given neuron within a synchronized group. This arrangement leads to generation of spatiotemporal patterns of activity into dynamic clusters. These clusters depend on the modulation generated by the connectivity of the system with external stimuli and with other levels within the CNS. Indeed, such clustering results from interaction among IO neurons and external stimuli. Nonlinear mathematical analysis of experimental data from single neurons (FIG. 3) and from coupled pairs of IO neurons indicates that despite being almost perfectly periodic, IO oscillations have a weakly chaotic nature.[35] This property markedly increases the sensitivity of the dynamics independent of input origin. Surprisingly, in this neuronal system the chaotic degree of the dynamics does not increase (i.e., the Lyapunov spectrum remains practically the same) regardless of the number of elements in the system. Indeed, the synchronization itself was found to belong to a very specific type, known as chaotic phase synchronization, and observed in a notoriously robust dynamic system.[35]

If the IO nucleus is to play a role in movement coordination, it must be able to control functional parameters in time, locality, and number of degrees of freedom. We have proposed[35] that to meet these requirements the IO uses its intrinsic oscillatory properties for temporal organization, that is, its ability to organize, support, and

---

FIGURE 1. (**A**) Intracellular recording of spontaneous oscillations at 2 Hz interrupted by an extracellular stimulus. After extracellular stimulation (marked with an *arrowhead*), oscillations disappeared for 750 ms (*boxed area*) and then resumed. The membrane potential was at –60 mV. (**B**) Left, intracellular recordings of spontaneous (*dashed black line*) and stimulus-evoked (*solid black line*) oscillations from the same cell are superimposed. Their corresponding power spectra are shown below. Note that extracellular stimulation only modified the phase of spontaneous oscillations without affecting its amplitude and frequency. Right, six individual intracellular traces of stimulus-evoked oscillations from the same cell are superimposed. Each trace is shown in a different color. Their corresponding power spectra are displayed below. In every recording, the frequency of stimulation-evoked oscillation was the same (2.0 Hz). Note that in each trace the stimulation-induced shift in the cell's oscillatory rhythm is remarkably similar. Oscillations are clearly seen after the stimulus-induced reset, but they can barely be detected before stimulation. Calibration bar: 1 mV; A, 1 second; (B) and (C), 500 ms. (**C**) A complete time-course of optical responses for four pixels. Locations of the pixels are marked with circles of different color in (D). The optical response consisted of an initial response to a train of extracellular stimuli (two stimuli at 10 Hz) followed by three oscillatory cycles. The frequency of the optically recorded oscillations was 4 Hz. (**D**) Several multiframe displays for each of the regions marked in (C). Only the area of the inferior olive (IO) is shown. Position of the stimulating electrode is indicated. The color bar gives the color scale, with dark purple corresponding to lowest fluorescence and red corresponding to the highest fluorescence from 0 to 2.7×10^{-3} $\Delta F/F$. (1) Response of the IO nucleus to the first stimulus. (2) Location and spatial spread of one cycle of the ensemble oscillations. The oscillations emanate from several clusters of closely spaced cells throughout the IO (see the change in amplitude with time). These clusters are highly repeatable from cycle to cycle over an oscillation sequence. (3) An average of three oscillatory cycles further underlines the stability of fluorescent clusters. (Figure adapted from Leznik *et al.*[31])

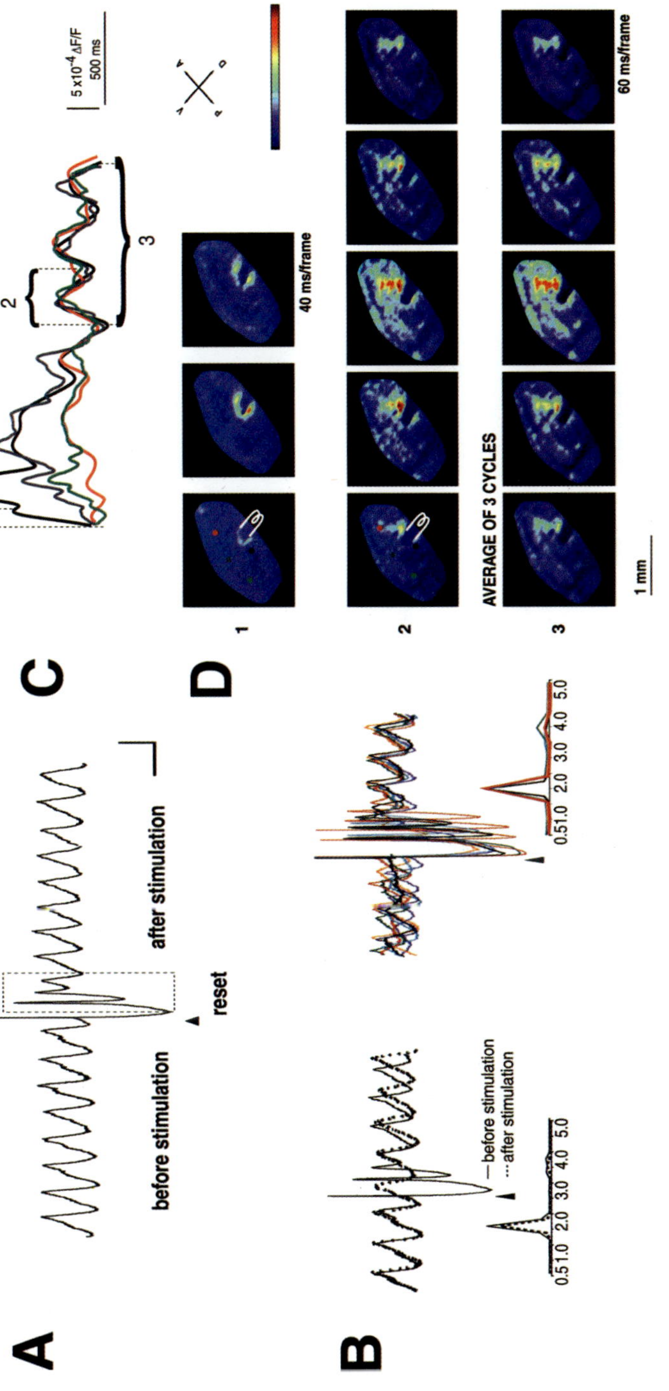

FIGURE 1. *See previous page for legend*

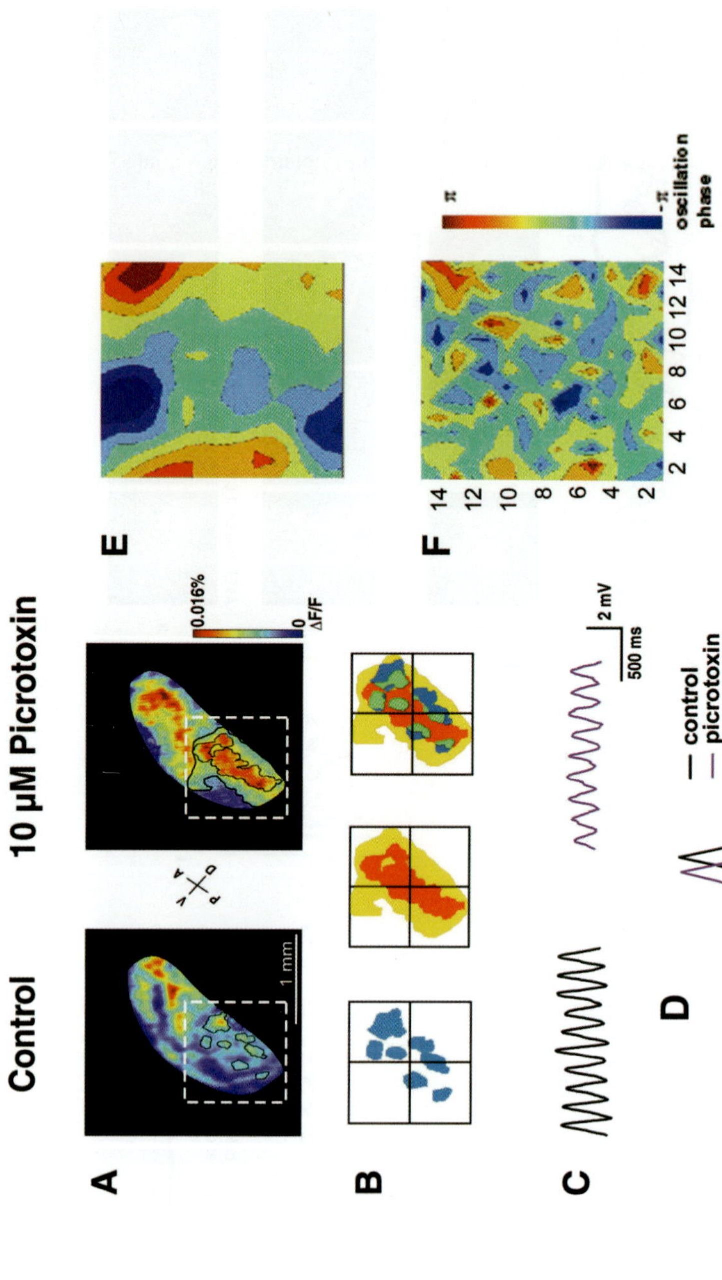

FIGURE 2. *See following page for legend*

change cluster size and location to control spatial organization. In addition, its weakly chaotic properties exhibit the necessary fluidity to control the number of degrees of freedom by rapidly modulating the phase relation between clusters.

Our results from computer simulation of single neurons[36] and a network of a considerable number of neurons[31] unequivocally support this view of IO functional dynamics. Based on this phenomenological approach, we succeeded in reproducing the main dynamic features of the IO neuron and in implementing variable electrotonic coupling among them. Furthermore, the simulated network of locally coupled neurons demonstrated that clustering had similar features to those observed in *in vitro* IO slices with voltage-sensitive dyes. Our main conclusion from this study is that the IO is a truly integrated system where the neuronal network and the macroscopic dynamic features (cluster permutation agility) are so well matched that their control properties almost match cellular time constants. This allows the system to effectively control innumerable movements without overloading brain function.

Neuron Model

This model is based on electrophysiological properties observed in slice IO recording.[21,35] Their main features are: (1) autonomous subthreshold activity and weakly chaotic oscillations (by "weakly chaotic" we mean periodic and close to regular oscillatory dynamics with sufficient variability to support a positive value of the biggest Lyapunov exponent);[35] (2) subthreshold oscillation and the presence of two levels of spike initiation; and (3) subthreshold oscillatory phase preservation following spike generation. On the basis of these properties we considered a model that mimics IO neuron dynamics, as just described. The model is composed of three autonomous interacting blocks responsible for the different functional modes observed experimentally.[36] The first block provides quasiharmonic oscillations; the other two blocks are responsible for spike generation at different thresholds.

FIGURE 2. (**A**) Frames of imaged ensemble neuronal oscillating clusters in control conditions (*left*) and after the addition of 20 µm of picrotoxin (*right*) are shown. Several representative clusters in the lower right part of the IO are outlined in *black*. Clusters were defined as areas with pixel values above the selected threshold level. After picrotoxin, an additional threshold level of higher value was chosen to delineate the highest areas of activity. (**B**) Cluster's contours shown within the *boxed area* (1 × 1 mm) are enlarged for the control condition (*right panel*) and after the addition of picrotoxin (*middle panel*). (*Left*) Areas shown in *right* and *middle panels* are superimposed. Control clusters are shown in *blue*; picrotoxin clusters of lower threshold are shown in *yellow,* and those of higher threshold are shaded in *red*. Overlapping regions of control clusters and picrotoxin clusters of lower threshold are shown in *green*. Note that picrotoxin significantly increased the size of clusters by activating additional regions and merging several small areas into larger activation areas. **(C)** Intracellular recording from an IO neuron showing spontaneous subthreshold oscillations before (*black*) and after (*purple*) the addition of 20 µM picrotoxin. **(D)** Power spectra of control (*in black*) and picrotoxin-modulated oscillations (*in purple*) are superimposed. The addition of picrotoxin reduced amplitude and frequency of subthreshold oscillations without changing the resting membrané potential of the cell. (**E, F**) Clustering in the modeled network as a function of coupling degree. (E) In the absence of inhibition, large clusters are observed and demonstrate large fluctuations in size (as in the case with picrotoxin). (F) In the presence of inhibition (reduced coupling), many small clusters are formed, but there is much less fluctuation in size. (Figure adapted from Leznik *et al.*[31])

As a particular implementation of the model we set:

$$
\left.
\begin{aligned}
&x' = y, && \textbf{(1A)}\\
&y' = (\gamma(1+\alpha u) - x^2)y - \omega^2(1+\beta u)x, && \\
&\varepsilon\omega' = g(w) - z - x, && \textbf{(1B)}\\
&z' = 0.5(w - I_2)(w^2 + 0.1), && \\
&\varepsilon u' = f(u) - v + hw, && \textbf{(1C)}\\
&v' = 0.05(u - I_1)(u^2 + 0.5) &&
\end{aligned}
\right\}\quad \textbf{(1)}
$$

where the variable u accounts for the membrane potential of the neuron and the parameters γ and ω account for the amplitude and frequency of quasiharmonic (subthreshold) oscillations. The coefficients α and β change the amplitude and frequency of oscillations depending on the membrane potential, u. The parameter h determines the strength of coupling between the two FitzHugh-Nagumo (FN) subsystems. $\varepsilon \ll 1$ defines the relaxation scale of the FN block. Note that although we made them identical, they may have two different time scales here. For our purpose we fixed the parameters $\varepsilon = 0.01$, $\gamma = 0.21$, $\omega^2 = 0.63$, $a = 1.8$, $b = 0.5$, and $h = 0.5$, and took I_1, I_2 as control parameters playing the (qualitative) role of a constant current stimulus that changes the membrane potential level.

System 1 consists of three interacting subsystems. The first subsystem, (1A) (variables x,y), represents a Van der Pol (VdP) oscillator, providing robust quasiharmonic oscillations. The other two subsystems, (1B) and (1C) (variables u,v and w,z), are based on an FN mathematical model and are responsible for high- and low-threshold-level spike generation, respectively. Nonlinear functions $f(u)$ and $g(u)$ are of cubic-shape form. For convenience, we approximate them in the following piecewise linear forms:

$$
f(u) = \begin{cases} -1.5u, & \text{if } u < a,\\ 0.2u - 1.7a, & \text{if } a < u < 4,\\ -1.6u - 1.7a + 9.6 & \text{if } u > 4, \end{cases}
$$

and **(2)**

$$
g(w) = \begin{cases} -2w, & \text{if } w < b,\\ 3w - 5b, & \text{if } b < w < 1,\\ -5w - 5b + 8 & \text{if } w > 1, \end{cases}
$$

where the parameters a and b define the high- and low-threshold values, respectively. The values of the coefficients in system 2 characterize the width, height, and refractory period of the pulses (spikes) generated from the two thresholds.

FIGURE 3. Reconstruction of attractor from a time series of membrane potential for a single neuron. The attractor has a structure close to a limit cycle with a regular periodic trajectory. Average Mutual Information (**A**) and False Nearest Neighbors (**B**) methods were used to reconstruct the attractor (**C**) and Lyapunov exponents (**D**). Exponent values (small positive, zero, and negative) indicate that the system displays low-dimensional chaotic dynamics. (Figure adapted from Makarenko and Llinás.[35])

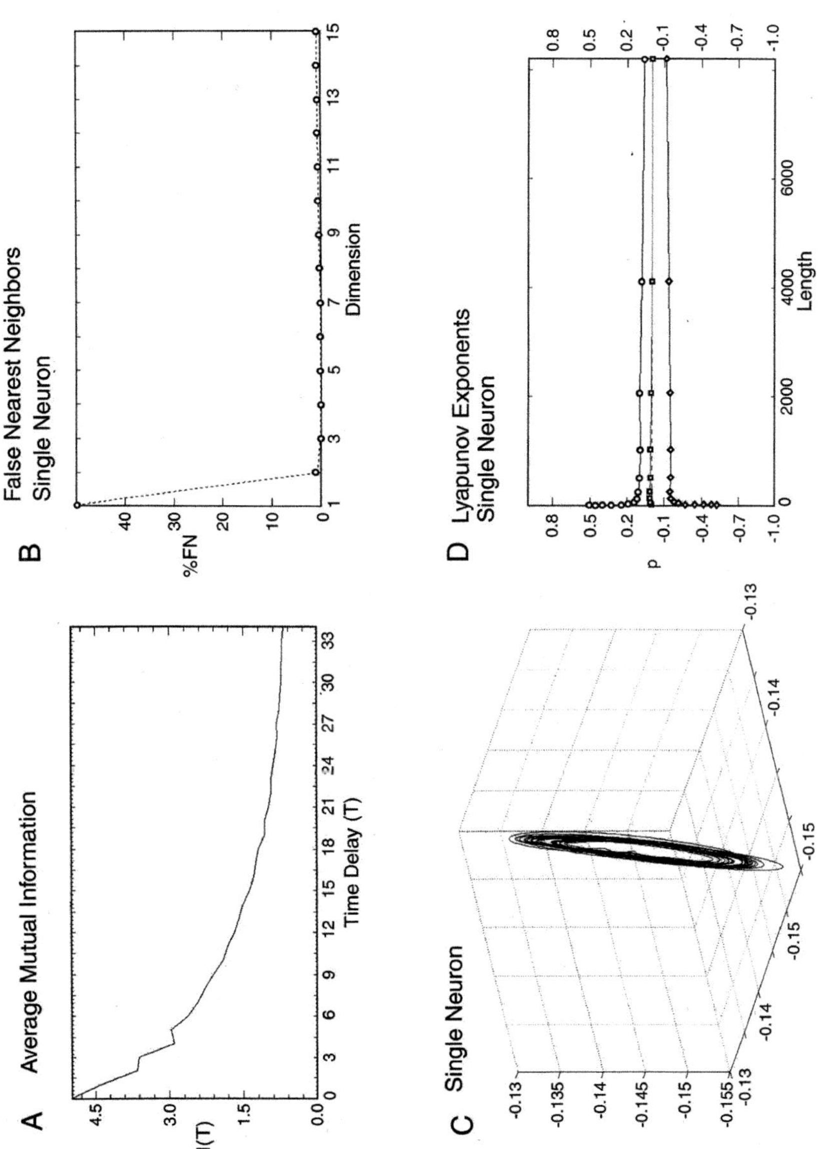

FIGURE 3. *See previous page for legend*

The model proved to be fully consistent with experimental data capable of reproducing all main features of IO neuron membrane potential dynamics (FIG. 2E and 2F).

Network Model

The motivation behind modeling neuronal activity clusters was to test the assumption that such clustering, and cluster size, are electrotonic-coupling dependent. Indeed, whereas this assumption may be considered obvious at first glance, it has been demonstrated that activity profiles other than clusters (such as spirals, waves, and concentric rings) can occur in coupled neuronal networks.[45]

Experimental data provided parameters such as membrane potential oscillation (from intracellular recordings) and cluster size and shape and their modulation by pharmacological intervention (from voltage-sensitive dye imaging[31]). These parameters defined the model type. The network properties were modeled by interconnecting a set of model neurons into a two-dimensional matrix with isotropic coupling properties. Such a model was easier to implement and to scale than were models that incorporated single cell cable and ionic conductance properties.[26,46]

In our model, single neurons are represented as distinct elements (points) in a rectangular 4-point connectivity lattice. Each element is endowed with quasiperiodic oscillatory properties defined by a set of equations having a periodic and a noise term as follows:

$$z' = z\,(i\,w_0 - g) + i\sqrt{2D}\,\xi\,(t), \qquad \textbf{(3)}$$

where the variable z characterizes neuron dynamics (with $x = \mathrm{Re}(z)$, that is, $z = x + iy$, prime implies differentiation), i is an imaginary unit, and x and y are real numbers), γ is a damping constant, w_0 is angular oscillation frequency in the absence of noise and damping ($w_0 = 2\pi * 10$ Hz), D is a parameter determining noise intensity scaling, and $\xi(t)$ is a noise term that has zero mean with a time correlation function given by

$$\langle \xi_{jk}(t)\xi_{lm}(t')\rangle = \delta_{jk}\delta_{lm}\delta(t - t') \qquad \textbf{(4)}$$

where the brackets imply average and δ denotes a function that is zero when its argument has a nonzero value. Relatively regular oscillations have variable amplitude with a sharp frequency peak in the region of 10 Hz, which is in good agreement with *in vivo* experimental data.[47]

CONCLUSIONS AND IMPLICATIONS

On the basis of the anatomy of the IO nucleus and widespread synchronicity of complex spikes in the cerebellum, it has been proposed that functionally the IO is organized in clusters of synchronously oscillating neurons that can generate rhythmic and coherent activity in the cerebellum.[43] In agreement with this proposal, we were able to visualize the IO oscillatory clusters and describe their spatiotemporal profiles *in vitro*.

Our optical imaging responses from the IO nucleus consisted of initial antidromic activation of the area around the stimulating electrode that was followed by stimulus reset ensemble IO oscillations. The optically recorded IO oscillations emanated from several spatially separated fluorescent clusters that had the same oscillation frequency and were phase coherent. Our imaging experiments with picrotoxin indicate that borders of these clusters were most likely determined by the level of effective electrotonic coupling among the IO cells and therefore by the state of the IO network.

Implication of Modeling Results

Our modeling results support the conclusion that regulation of electrotonic coupling in the IO by feedback inhibition from the cerebellar nuclei regulates IO cluster coupling, which controls their distribution and size.[11,15,31] That is, increased inhibition results in cluster size reduction and an increase in the number of independent clusters. Conversely, a decrease in inhibition results in increased cluster size and a simultaneous decrease in the number of independent clusters. These modeling results are in agreement with our imaging data and support our two main working hypotheses: first, the level of inhibition (i.e., coupling) is a main factor in regulating the average size of a cluster of synchronous activity; second, with the higher level of inhibition, the number of clusters increases. For statistical analysis we used the Markov Random Fields method described elsewhere,[27] which allows quantitative estimation of the level of spatial organization. Interestingly, cluster distribution is surprisingly stable, and counterintuitively fluctuations in cluster organization are inhibition dependent: the higher the level of inhibition, the lower the level of fluctuations. This is so both experimentally and as a modeling result. Furthermore, a high level of inhibition leads to a larger number of smaller clusters and thus to an increased number of possible functional states and degrees of freedom.

Indeed, as the degrees of freedom increase with an increased number of clusters, the probability of large fluctuations in cluster size decreases. In fact, the system becomes "less liquid" with a larger number of smaller clusters. That is, despite the large number of clusters, the system robustly preserves a particular level of spatial organization. More fundamentally, the results indicate the existence of a maximum number of possible functional states, implying that increased inhibitory feedback does not result in an ever-increasing control ability. Even at high levels of clustering the IO network demonstrates short-phase resetting time (~70 ms) without losing oscillatory robustness, thereby optimizing reset agility without sacrificing control.[35]

Functional Significance of Inferior Olive Oscillatory Clusters

Previous *in vivo* multielectrode experiments have shown that synchronized CSs are primarily observed among PCs located within the same parasagittal bands.[9,10,16,17] Groups of IO neurons that define such parasagittal bands consist of up to several hundred neurons.[17] Because this number is in the same order of magnitude as the calculated number of cells in our fluorescent clusters, we propose that our optical measurements might reflect the ensemble physiological properties formerly observed with multiple recordings at the PC level.[31] Hence, each IO cluster can control activity of a single parasagittal band in the cerebellum. Simultaneous activation of several IO clusters would result in coherent CS activity between different parasagittal bands, as observed during coordinated movements.

Ultimately, as further insight is gained into the oscillatory property of neurons and their relation to macroscopic motor execution and control, a general message becomes apparent: The dynamic properties of single neurons and the channel kinetics that generate them seem to hone the "impedance matching" from the molecular to the macroscopic to the point that, as in the IO, our motor timing is nothing other than the echo of our IO oscillation.

REFERENCES

1. RAMON Y CAJAL, S. 1888. Sobre las fibras nerviosas de la capa molecular del cerebelo. Rev. Trimestr. Histol. **2:** 33–41.
2. SZENTAGOTHAI, J. & K. RAJKOVITS. 1959. Ueber den Ursprung der Kletterfasern des kleinhirns. Z. Anat. Entwicklungsgeschichte. **121:** 130–141.
3. ECCLES, J.C., R. LLINÁS & K. SASAKI. 1966. The excitatory synaptic action of climbing fibres on the Purkinje cells of the cerebellum. J. Physiol. (Lond) **182:** 268–296.
4. THACH, W.T. 1968. Discharge of Purkinje and cerebellar nuclear neurons during rapidly alternating arm movements in the monkey. J. Neurophysiol. **31:** 785–797.
5. ARMSTRONG, D.M. & R.F. SCHILD. 1970. A quantitative study of the Purkinje cells in the cerebellum of the albino rat. J. Comp. Neurol. **139:** 449–456.
6. ELLER, T. & V. CHAN-PALAY. 1976. Afferents to the cerebellar lateral nucleus. Evidence from retrograde transport of horseradish peroxidase after pressure injections through micropipettes. J. Comp. Neurol. **166:** 285–301.
7. CRILL, W.E. 1970. Unitary multiple-spiked responses in cat inferior olive nucleus. J. Neurophysiol. **33:** 199–209.
8. LLINÁS, R. & Y. YAROM. 1981. Electrophysiology of mammalian inferior olivary neurones in vitro. Different types of voltage-dependent ionic conductances. J. Physiol. (Lond.) **315:** 549–567.
9. BELL, C.C. & T. KAWASAKI. 1972. Relations among climbing fiber responses of nearby Purkinje Cells. J. Neurophysiol. **35:** 155–169.
10. LLINÁS, R. & K. SASAKI. 1989. The functional organization of the olivo-cerebellar system as examined by multiple Purkinje cell recordings. Eur. J. Neurosci. **1:** 587–602.
11. LANG, E.J., I. SUGIHARA & R. LLINÁS. 1996. GABAergic modulation of complex spike activity by the cerebellar nucleoolivary pathway in rat. J. Neurophysiol. **76:** 255–275.
12. FUKUDA, M., T. YAMAMOTO & R. LLINÁS. 2001. The isochronic band hypothesis and climbing fiber regulation of motricity: an experimental study. Eur. J. Neurosci. **13:** 315–326.
13. LLINÁS, R., R. BAKER & C. SOTELO. 1974. Electrotonic coupling between neurons in cat inferior olive. J. Neurophysiol. **37:** 560–571.
14. SOTELO, C., R. LLINÁS & R. BAKER. 1974. Structural study of inferior olivary nucleus of the cat: morphological correlates of electrotonic coupling. J. Neurophysiol. **37:** 541–559.
15. LLINÁS, R. 1974. Eighteenth Bowditch Lecture. Motor aspects of cerebellar control. Physiologist **17:** 19–46.
16. LANG, E.J. *et al.* 1999. Patterns of spontaneous Purkinje cell complex spike activity in the awake rat. J. Neurosci. **19:** 2728–2739.
17. YAMAMOTO, T., M. FUKUDA & R. LLINÁS. 2001. Bilaterally synchronous complex spike Purkinje cell activity in the mammalian cerebellum. Eur. J. Neurosci. **13:** 327–339.
18. LLINÁS, R. & M. MUHLETHALER. 1988. Electrophysiology of guinea-pig cerebellar nuclear cells in the in vitro brain stem-cerebellar preparation. J. Physiol. (Lond.) **404:** 241–258.
19. WELSH, J.P. *et al.* 1995. Dynamic organization of motor control within the olivocerebellar system. Nature **374:** 453–457.

20. LLINÁS, R. & Y. YAROM. 1981. Properties and distribution of ionic conductances generating electroresponsiveness of mammalian inferior olivary neurones in vitro. J. Physiol. (Lond.) **315:** 569–584.
21. LLINÁS, R. & Y. YAROM. 1986. Oscillatory properties of guinea-pig inferior olivary neurones and their pharmacological modulation: an in vitro study. J. Physiol. (Lond.) **376:** 163–182.
22. BENARDO, L.S. & R.E. FOSTER. 1986. Oscillatory behavior in inferior olive neurons: mechanism, modulation, cell aggregates. Brain Res. Bull. **17:** 773–784.
23. BAL, T. & D.A. MCCORMICK. 1997. Synchronized oscillations in the inferior olive are controlled by the hyperpolarization-activated cation current I(h). J. Neurophysiol. **77:** 3145–3156.
24. LAMPL, I. & Y. YAROM. 1997. Subthreshold oscillations and resonant behavior: two manifestations of the same mechanism. Neuroscience **78:** 325–341.
25. HUTCHEON, B. & Y. YAROM. 2000. Resonance, oscillation and the intrinsic frequency preferences of neurons. Trends Neurosci. **23:** 216–222.
26. MANOR, Y. *et al.* 1997. Low-amplitude oscillations in the inferior olive: a model based on electrical coupling of neurons with heterogeneous channel densities. J. Neurophysiol. **77:** 2736–2752.
27. MAKARENKO, V. *et al.* 1997. A new approach to the analysis of multidimenshional neuronal activity: Markov Random Fields. Neural Networks **10:** 785–789.
28. BENNETT, M.V. 2000. Electrical synapses, a personal perspective (or history). Brain Res. Brain Res. Rev. **32:** 16–28.
29. DE ZEEUW, C.I. *et al.* 1996. Morphological correlates of bilateral synchrony in the rat cerebellar cortex. J. Neurosci. **16:** 3412–3426.
30. FREDETTE, B.J. & E. MUGNAINI. 1991. The GABAergic cerebello-olivary projection in the rat. Anat. Embryol. **184:** 225–243.
31. LEZNIK, E., V. MAKARENKO & R. LLINÁS. 2002. Electrotonically mediated oscillatory patterns in neuronal ensembles: an in vitro voltage-dependent dye-imaging study in the inferior olive. J. Neurosci. **22:** 2804–2815.
32. LANG, E.J. 2001. Organization of olivocerebellar activity in the absence of excitatory glutamatergic input. J. Neurosci. **21:** 1663–1675.
33. LANG, E.J. 2002. GABAergic and glutamatergic modulation of spontaneous and motor-cortex-evoked complex spike activity. J. Neurophysiol. **87:** 1993–2008.
34. EBNER, T.J. & G. CHEN. 1995. Use of voltage-sensitive dyes and optical recordings in the central nervous system. Prog. Neurobiol. **46:** 463–506.
35. MAKARENKO, V. & R. LLINÁS. 1998. Experimentally determined chaotic phase synchronization in a neuronal system. Proc. Natl. Acad. Sci. USA **95:** 15747–15752.
36. VELARDE, M.G. *et al.* 2002. Modeling inferior olive neuron dynamics. Neural Networks **15:** 5–10.
37. BLEASEL, A.F. & A.G. PETTIGREW. 1992. Development and properties of spontaneous oscillations of the membrane potential in inferior olivary neurons in the rat. Brain Res. Dev. Brain Res. **65:** 43–50.
38. SOTELO, C., T. GOTOW & M. WASSEF. 1986. Localization of glutamic-acid-decarboxylase-immunoreactive axon terminals in the inferior olive of the rat, with special emphasis on anatomical relations between GABAergic synapses and dendrodendritic gap junctions. J. Comp. Neurol. **252:** 32–50.
39. DE ZEEUW, C.I. *et al.* 1990. Mesodiencephalic and cerebellar terminals terminate upon the same dendritic spines in the glomeruli of the cat and rat inferior olive: an ultrastructural study using a combination of [3H]leucine and wheat germ agglutinin coupled horseradish peroxidase anterograde tracing. Neuroscience **34:** 645–655.
40. GRINVALD, A. *et al.* 1994. Cortical point-spread function and long-range lateral interactions revealed by real-time optical imaging of macaque monkey primary visual cortex. J. Neurosci. **14:** 2545–2568.
41. CUNNINGHAM, J.J. *et al.* 1999. Changes in the numbers of neurons and astrocytes during the postnatal development of the rat inferior olive. J. Comp. Neurol. **406:** 375–383.
42. DE ZEEUW, C.I. *et al.* 1998. Microcircuitry and function of the inferior olive. Trends Neurosci. **21:** 391–400.

43. WELSH, J.P. & R. LLINÁS. 1997. Some organizing principles for the control of movement based on olivocerebellar physiology. Progr. Brain Res. **114:** 449–461.
44. RUIGROK, T.J. & J. VOOGD. 1995. Cerebellar influence on olivary excitability in the cat. Eur. J. Neurosci. **7:** 679–693.
45. MAKARENKO, V.I. 1994. Neural network with embedded oscillators. Biol. Bull. **187:** 256–257.
46. SCHWEIGHOFER, N., K. DOYA & M. KAWATO. 1999. Electrophysiological properties of inferior olive neurons: a compartmental model. J. Neurophysiol. **82:** 804–817.
47. SASAKI, K., J.M. BOWER & R. LLINÁS. 1989. Multiple Purkinje cell recording in rodent cerebellar cortex. Eur. J. Neurosci. **1:** 572–586.

Historical Review of the Significance of the Cerebellum and the Role of Purkinje Cells in Motor Learning

MASAO ITO

Brain Science Institute, RIKEN, Wako, Saitama 351-0198, Japan

ABSTRACT: Classic studies of the cerebellum before the middle of the twentieth century established the structural entity of the cerebellum and characterized its function as enabling animals and humans to carry out smooth and accurate movements, even at a high speed and without visual feedback. In the 1960s, neuronal circuit structures of the cerebellum were analyzed in detail, which promoted computational approaches toward the study of neuronal network principles of the cerebellum. In the 1970s and 1980s, vestibulo-ocular reflex adaptation, adaptive locomotion, eye blink conditioning, and learning in hand/arm movement were established as effective experimental paradigms for investigating neural mechanisms of cerebellar functions. In the 1980s, long-term depression (LTD) was discovered and considered as a memory process in the cerebellum; in the 1990s, complex signal transduction processes underlying LTD were revealed. It was also in the 1980s that computational approaches were advanced for modeling control system functions of the cerebellum. Currently, there are two alternative models proposed for VOR adaptation. In this decade, we envisage new developments toward the fusion of knowledge of the cerebellum at molecular and cellular levels and those in systems and computation. Studies of LTD will play a key role in pursuing this direction.

KEYWORDS: neuronal circuit; long-term depression; vestibulo-ocular reflex; internal model

CLASSIC CONCEPTS OF THE CEREBELLUM

Two monographs compile classic data of cerebellar research obtained before the middle of the twentieth century. *The Anatomy of the Cerebellum* by Jansen and Brodal[1] established the structural entity of the cerebellum, lumping its diverse morphological variations among animal species. *The Physiology and Pathology of the Cerebellum* by Dow and Moruzzi[2] extracted functional features of the cerebellum from enormous amounts of data of classic lesion experiments on animals and clinical and pathological studies of patients with cerebellar diseases.

Dow and Moruzzi recorded in their monograph[2] the following events in the nineteenth century, when pioneers in cerebellar research observed symptoms arising

Address for correspondence: Masao Ito, Brain Science Institute, RIKEN, Wako, Saitama 351-0198, Japan. Voice: +81-48-462-1111 ext. 7542; fax: +81-48-467-9683.
masao@brain.riken.go.jp

Ann. N.Y. Acad. Sci. 978: 273–288 (2002).

from lesions in the cerebellum in various animal species. Roland discovered in 1809 that lesions in the cerebellum resulted in motor disturbances. Flourens in 1822 observed that animals with a lesioned cerebellum still moved spontaneously, but only clumsily without coordination. Luciani in 1891 observed in detail the recovery that occurred after partial or total resection of the cerebellum. From the turn of the nineteenth century up until the twentieth century, neurology progressed, and Babinski in 1902 recognized dysmetria as a unique characteristic sign of cerebellar diseases. Dysmetria can be tested by a simple examination such as the finger–nose test; patients with a cerebellar disorder failed to touch their nose with their index finger quickly and accurately with eyes closed. During World War I, Holmes reported that soldiers with a gunshot wound through the cerebellum exhibited a significantly slower onset of arm movement on the damaged side.

These observations altogether suggest that the cerebellum enables animals and humans to move accurately and smoothly even at a high speed and without visual feedback. It is difficult to believe that such subtle control of movements can be performed without learning. In fact, there were classic observations that suggested learning capabilities of the cerebellum. Flourens reported in 1842 that when he removed the superior one-half of the cerebellum in a young cock, he observed 15 days later that the equilibrium was totally re-established. However, when he removed the entire cerebellum from a hen, it did not recover even four months after the operation. In a dog, after full recovery from a partial lesion in the cerebellum, Luciani reported in 1891 a notable observation after placing a second lesion in an adjacent area of the cerebellum. A severe motor disturbance appeared as if the first and second lesions had been made simultaneously. This was interpreted to indicate that the second lesioned area was involved in the recovery from the first lesion. Moruzzi's group[3] thought that cerebellar tissues have an intrinsic learning capability or are required for recalling memory stored somewhere else, for example, the cerebral cortex.

NEURONAL CIRCUITS IN THE CEREBELLUM

The elaborate neuronal networks in the cerebellum have been well known since Cajal's[4] fine drawing of neuronal structures including Purkinje cells, basket cells, and climbing fibers (FIG. 1). Dow[5] correctly speculated that climbing fibers originate from the inferior olive, and Szentagothai and Rajkovits[6] demonstrated degenerating climbing fibers after lesioning the inferior olive. Based on the unique spatial pattern of wiring in the cerebellar cortex with basket-cell axons perpendicularly crossing parallel fiber bundles, Szentagothai[7] suggested that basket cells are inhibitory neurons. In the 1960s, Eccles *et al.*[8–10] in Canberra conducted a systematic study of synaptic connections in the cerebellar cortex. They demonstrated the excitatory nature of both mossy and climbing fibers as well as granule cells and their axons, namely, parallel fibers, and the inhibitory nature of basket, stellate, and Golgi cells.

Upon my return from Canberra to Tokyo in 1962, we intracellularly recorded from Deiters' neurons in the medulla oblongata and by chance found that stimulation of the cerebellar cortex induces inhibitory postsynaptic potentials with a monosynaptic delay.[11] We extended the work to cerebellar nuclei[12] and established that Purkinje cells exert a unique inhibitory action on their target neurons. We also dem-

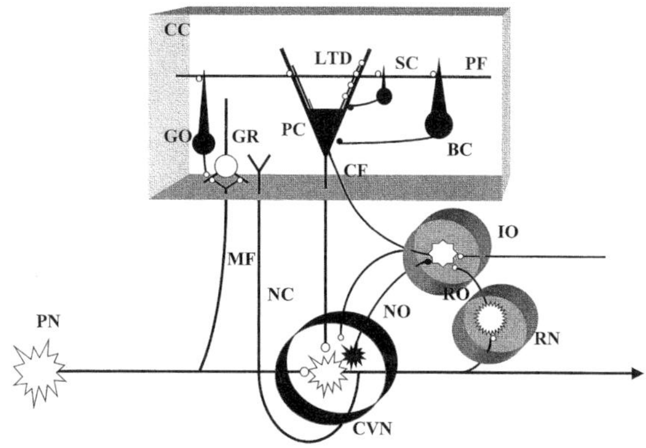

FIGURE 1. Neuronal circuit structure in the cerebellum. ABBREVIATIONS: CC, cerebellar cortex; CVN, cerebellar and vestibular nuclei; IO, inferior olive; PN, precerebellar nucleus; RN, red nucleus (parvocellular part); MF, mossy fiber; CF, climbing fiber; PC, Purkinje cell; GR, granule cell; PF, parallel fiber; BC, basket cell; SC, stellate cell; GO, Golgi cell. *Small hollow circle*, excitatory synapse. *Small filled circle*, inhibitory synapse.

onstrated that cerebellar afferent fibers ending in the cortex as either mossy or climbing fibers supply excitatory synapses via their collaterals to nuclear neurons.[13,14] Toyama *et al.*[15] showed that cerebellar nuclear neurons supply excitatory synapses to red nucleus neurons, and Uno *et al.*[16] demonstrated that nuclear neurons project excitatory axons to the ventrolateral nucleus of the thalamus. This work in the 1960s yielded a fine neuronal diagram of the entire cerebellum,[17] which was further elaborated later by a number of studies that revealed the nucleocortical mossy-fiber projection,[18,19] rubroolivary excitatory projection,[13,20] and nucleoolivary inhibitory projection[21–23] (FIG. 1).

The significant progress in the analyses of neuronal circuits of the cerebellum provoked worldwide interest in how such neuronal circuits operate and generate subtle functions of the cerebellum as indicated by classic studies. The international symposium organized by Francis O. Schmitt of the Neuroscience Research Program in Salishan Lodge on the Oregon Coast, Portland, in 1967 is of historical value. A large group of neuroscientists, mathematicians, information scientists, and bioengineers gathered there. In the group photo taken during the symposium, we find a number of authorities in the fields of cerebellar research at that time, such as J. Brookhart, R. Dow, J.C. Eccles, C.A. Fox, D. Mackay, R. Schneider, and J. Szentagothai, and also the younger generation of researchers, such as C. Bell, J. Hamori, R. Kado, R. Llinas, E. Mugnaini, O. Oscarsson, O. Pompeiano, K. Sasaki, and myself. Discussions held at the conference were exciting. I was particularly impressed by the lucid analysis by the late Professor D. MacKay of the possible functional roles of inhibitory neurons in the brain.

The Salishan Lodge conference did not yield fruit immediately, but a few years later the epochal theories on the cerebellar cortex by Marr[24] and Albus[25] were proposed. One of the major goals of neuroscience is to analyze the elaborate neuronal circuits present in the brain and to understand the principles of how these operate and generate subtle functions in the brain. In this respect, research on the cerebellum opened a gateway for neuroscience around 1970, and still remains a leading field in neuroscience.

LONG-TERM DEPRESSION (LTD)

A neuronal network, however elaborate, will not be able to construct a computer unless it contains memory elements. Hebb[26] proposed the concept of neuron assembly in which synaptic connections are modified depending on activities experienced. As a possible mechanism of modification, Hebb[26] proposed that the connection between two neurons firing synchronously is strengthened, and that between two neurons firing asynchronously is weakened. Because of this Hebbian type of synaptic plasticity, the neuron assembly will change its circuitry structure (corresponding to memory) and consequently modify its input-output relationship (corresponding to learning) depending on activities experienced. Brindley[27] pointed out the possibility that the convergence of parallel fibers and climbing fibers to Purkinje cells implies a Hebbian synapse. Climbing-fiber signals are so powerful that they inevitably excite Purkinje cells. Hence, when both parallel fibers and climbing fibers are activated synchronously, parallel fiber–Purkinje cell synapses are activated both presynaptically and postsynaptically, which is the condition that induces the Hebbian type of plasticity.

In an effort to test Hebb's concept of neuron assembly, Rosenblatt[28] constructed the Simple Perceptron that consists of three layers of neurons connected in one direction, that is, from the first sensory cell layer to the second association cell layer and to the third response cell layer. Connection from the first to the second layer is fixed, while that from the second to the third layer is modifiable according to the instruction from an outside teacher. The teacher increases the weight of connections at all junctions between the second and third layers when the Simple Perceptron responds to sensory stimuli correctly. The teacher decreases the weight at all second-to-third layer connections when the response is incorrect. When this training process is repeated, the Simple Perceptron improves its performance toward the success rate of 100%. Albus[25] assigned cells issuing mossy fibers to the first layer, granule cells to the second layer, and Purkinje cells to the third layer of the Simple Perceptron, and climbing fibers to the position of the outside teacher. Marr's[24] model is also compatible with the basic structure of the Simple Perceptron. While the outside teacher in the Simple Perceptron can either strengthen or weaken the second-to-third layer connections, it will be difficult for a climbing fiber to do this. Following Brindley's suggestion, Marr[24] assumed that climbing-fiber signals strengthen the parallel fiber–Purkinje cell synapses, while Albus[25] assumed that climbing-fiber signals weaken the parallel fiber–Purkinje cell synapses based on several reasons.

In the 1970s, although a reciprocal relationship in Purkinje cells between simple spikes representing activation through mossy fiber–parallel fiber pathways and complex spikes representing activation through climbing fibers suggests the applicability of Albus' assumption,[29,30] it was extremely difficult to demonstrate such synaptic plasticity more directly. This was because at that time experiments were conducted *in vivo*, and, therefore, stable intracellular recording for a sufficiently long period for demonstrating synaptic plasticity was impossible. The transmission across parallel fiber–Purkinje cell synapses was examined only by recording field potentials; however, because of their small size (ten times smaller than the field potentials recorded in the hippocampus), the potentials did not reliably reflect long-term modification of synaptic transmission. Around 1980, we in Tokyo examined parallel fiber–Purkinje cell transmission by measuring the firing index of Purkinje cells in response to near-threshold parallel fiber stimulation. After conjunctive stimulation of the parallel fibers and climbing fibers, the firing index persistently decreased, providing evidence of long-term depression (LTD).[31,32] LTD was also demonstrated by recording of field potentials using an electronic averaging instrument.[33] Later, LTD was shown to occur in cerebellar slices in which stable intracellular recording was possible.[34] Thus, LTD was established as a unique type of synaptic plasticity by the end of the 1980s.[35]

Since then, signal transduction processes underlying LTD have been extensively analyzed (see Refs. 36 and 37). Reduced forms of LTD induced in tissue-cultured Purkinje cells by glutamate application instead of parallel-fiber stimulation and by membrane-depolarization application instead of climbing-fiber stimulation were frequently used in these studies.[38] A complex flow chart of chemical signals involving more than 30 different molecules has thus been drawn[37] (FIG. 2). The chart has been analyzed using three strategies:[39] (1) distinction between mediators and modulators,[40] (2) identification of coincidence detectors,[41] and (3) dissection of neuronal circuits responsible for nonlinear characteristics of LTD such as threshold and frequency dependence.[42,43] This complexity appears to be a safety lock for protecting the LTD induction mechanism against the ever-changing chemical environments of Purkinje cells.

ADAPTIVE CONTROL OF THE VESTIBULO-OCULAR REFLEX (VOR)

For common cerebellar symptoms such as ataxia or dysmetria, it is presupposd that complex neuronal mechanisms are still difficult to analyze. Cerebellar research has been facilitated by the introduction in the 1970s–1980s of four major experimental paradigms for testing cerebellar function, namely, VOR adaptation (see below), eye-blink conditioning,[44,45] adaptive locomotion,[46,47] and learning in hand/arm movements.[30,48]

The VOR is one of the well-analyzed reflexes that is mediated by a three-neuron arc connecting primary vestibular afferents originating from vestibular labyrinths, vestibular relay neurons, and motoneurons for extraocular muscles[49] (FIG. 3). The VOR is driven by head movement and moves the eyes in the direction opposite to the head movement, automatically stabilizing vision relative to space. Around 1970, our group, including Fukuda and Highstein,[50] found that relay neurons of the VOR receive Purkinje cell inhibition signals directly from the flocculus. Maekawa and Simpson[51] also found that optic signals reach the flocculus via climbing fibers. Because the flocculus receives primary and secondary vestibular signals as mossy-fiber inputs, a complete set of afferent and efferent connections to and from the flocculus becomes available.

Since the VOR is a typical feedforward control system lacking negative feedback, I hypothesized that the flocculus replaces the feedback by its LTD-based adaptive mechanism to ensure precision of visual stability during movements.[52,53] At about

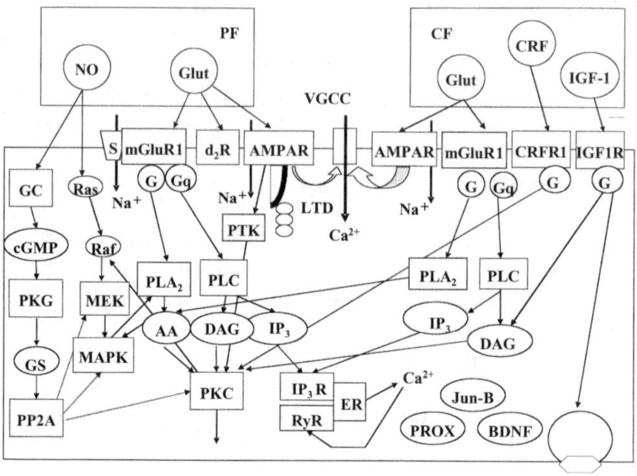

FIGURE 2. Pathway of signal transduction pathways following stimulation of a parallel fiber (PF) and a climbing fiber (CF). ABBREVIATIONS: NO, nitric oxide; Glut, glutamate; CRF, corticotropin-releasing hormone; IGF-1, insulin-like growth factor-1; AMPAR, AMPA receptor; MGluR1, type-1 metabotropic glutamate receptor; Δ2R, Δ2 glutamate receptor; CRFR1, type-1 CRF receptor; IGF1R, IGF-1 receptor; VGCC, voltage-gated calcium channel; S, slow EPSP generator associated with mGluR1; GC, guanylyl cyclase; CGMP, cyclic GMP; PKG, protein kinase G; GS, G-substrate; PP2A, protein phosphatase 2A; MAPK, mitogen-activated protein kinase; G, Gq, G′, G″, G-proteins; MEK, MAPK kinase; PLA2, phospholipase A2; PLC, phospholipase C; AA, arachidonic acid; DAG, diacylglycerol; IP_3, inositol trisphosphate; IP3R, IP3 receptor; RyR, ryanodin receptor; ER, endoplasmic reticulum; Jun-B, an immediate early gene expressed during LTD; PROX, a rapidly turned over protein that plays a role in the initial phase of LTD induction; BDNF, brain-derived neurotrophic factor. *Arrows* indicate signal flow. Broken arrow lines indicate inactivating effects. Modified from FIGURE 5 of Ito.[37]

the same time, Gonshor and Melvill-Jones[54] discovered that a marked adaptive change in the VOR occurs in altered visuovestibular environments. This VOR adaptation was soon confirmed to be eliminated by lesions in the flocculus.[55,56]

In the past three decades, the flocculus has been considered to play a crucial role in induction of VOR adaptation. However, the flocculus hypothesis that I proposed in 1982[57] has been challenged for the following reasons. First, it was questioned whether memory is stored within the flocculus (FIG. 3A) or elsewhere, for example, in the VOR relay neurons (FIG. 3B). In experiments where the VOR was adapted during visual–vestibular mismatch imposed for a relatively short period of 1–2 h, the adaptation once established was abolished following injection of lidocaine into the cerebellum of goldfish[58] or the flocculi of monkeys.[59] Therefore, it is now certain that at least during the first few hours of VOR adaptation, memory resides in the flocculus. Because, however, a residue of a long-adapted VOR gain after surgical or functional flocculectomy has been reported by several authors,[60–62] it is important to investigate neural events that would occur in and around the flocculus during prolonged VOR adaptation.

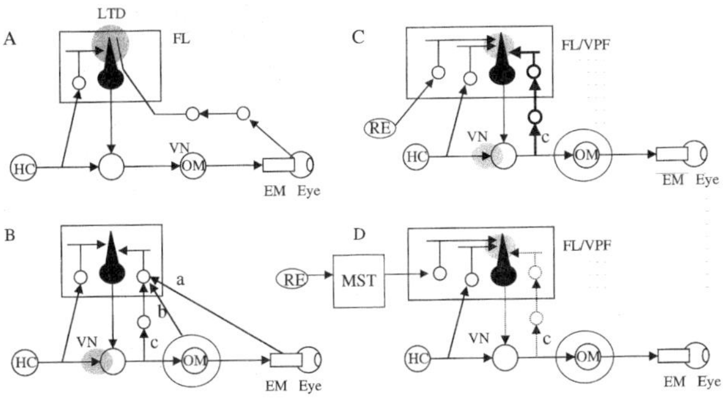

FIGURE 3. Various models of VOR-flocculus system. (**A**) Flocculus hypothesis.[57] ABBREVIATIONS: HC, horizontal canal; FL, flocculus; VN, VOR relay neurons in vestibular nuclei; OM, oculomotor motoneurons; EM, eye muscle. (**B**) Eye velocity pathways (**a,b,c**). VPF, ventral paraflocculus. (**C**) Gaze velocity theory.[74,75] RE, retina. (**D**) Parallel neural pathway hypothesis.[76] Possible memory sites are shaded.

Second, it was questioned whether the major mossy-fiber inputs to the flocculus are indeed vestibular signals, as proposed in the flocculus hypothesis (FIG. 3A), or eye velocity signals as proposed by Miles and Lisberger[63] (FIG. 3B). If they are vestibular signals, we expect that the VOR gain will change due to LTD-based modification of transmission of vestibular signals to flocculus Purkinje cells that modulate VOR relay neurons by inhibition. However, if they are eye velocity signals, we expect that a modification of transmission of vestibular signals to VOR relay neurons will be amplified by positive feedback of the eye velocity signals via the flocculus, resulting in a VOR gain change. In the flocculus hypothesis, the adaptive change in responses to head rotation of flocculus Purkinje cells[64–66] is due to LTD induction in these Purkinje cells. In the other hypothesis, by contrast, the adaptive change observed in the flocculus Purkinje cells is a mere consequence of a change in eye velocity caused by synaptic plasticity at the VOR relay neurons; no role is played by LTD in the flocculus Purkinje cells. Even though recent studies demonstrated a certain form of synaptic plasticity in cerebellar nuclei,[67] this synaptic plasticity is input-nonspecific, and hence it is difficult to assume that it plays a specific role as memory in cerebellar circuits.

There was a debate as to where the postulated eye velocity signals to the flocculus arise. The possibility that these eye velocity signals arise from eye muscles (FIG. 3Ba) is unlikely because deafferentation of the extraocular muscles in the rhesus monkey revealed no appreciable effect on the gain and phase of VOR.[68] Alternatively, eye velocity signals to the flocculus may arise from the pool of motoneurons innervating the extraocular muscles (FIG. 3Bc). Such eye velocity signals can be estimated only under a condition where there are no other eye-movement–related signals reaching the flocculus. Therefore, eye velocity signals have been estimated by measuring the difference in flocculus Purkinje cell discharges before and during visual suppression that nullifies eye velocity.[63] However, in visual suppression, the VOR is canceled by smooth pursuit eye movement to a head-fixed visual target, and hence smooth-pursuit–related signals would contaminate Purkinje cell discharges, as was pointed out previously.[69] On the other hand, measurements of flocculus Purkinje cell discharges in rabbits under various experimental conditions without relying on visual suppression suggest that only a modest portion (20%) of flocculus Purkinje cell responses to head rotation is related to eye velocity.[70–73]

A third possibility is that VOR relay neurons convert head velocity signals to eye velocity signals that are sent to the flocculus via certain brainstem neurons (FIG. 3, Cc) such as the midline paramedian tract neurons.[74] The crux of the above controversy thus lies in the question of whether the positive feedback that VOR relay neurons may receive via the flocculus is strong or weak. In his gaze velocity theory, Lisberger[75] assumed that the VOR, smooth pursuit, and visual suppression all proceed through a common mechanism equipped with a strong positive feedback via the cerebellum (FIG. 3C). This positive feedback explains how the smooth pursuit is maintained even when the driving retinal slip becomes zero during the steady state phase of smooth pursuit with gain 1 or during the temporary disappearance of the visual target. The gaze velocity theory assumes that this positive feedback plays a similar role in the VOR, and a computer simulation based on the gaze velocity theory reproduced the data collected in monkeys. In contrast to the flocculus hypothesis, the gaze velocity theory involves memory in both the cerebellar cortex and VOR relay neurons, and furthermore in the gaze velocity theory, memory in the cerebellar cortex needs to be opposite to LTD. In a recent system identification approach to the vertical VOR adaptation, Hirata and Highstein[74] analyzed their data on vertical VOR adaptation closely based on the gaze velocity theory and concluded that an adaptive VOR gain increase is due to the enhancement of transmission in the brainstem pathway, whereas an adaptive change also occurs in the flocculus pathway, but in the direction for reducing the VOR gain. More recently, however, Tabata *et al.*[76] formulated an alternative model in which the pursuit driving command is effectively generated in the medial superior temporal (MST) area of the cerebral cortex even when there is only a weak positive feedback via the cerebellum. This hypothesis assumes two parallel neural pathways consisting of the main pathway of the semicircular canals to the vestibular nucleus for VOR and the main pathway of the MST-dorsolateral pontine nuclei (DLPN)-flocculus/ventral paraflocculus to the vestibular nucleus for smooth pursuit (FIG. 3D). Tabata and Kawato[76] assigned the smooth pursuit command function to the MST-DLPN flocculus pathway, whereas Hirata and Highstein[74] regarded the function of this pathway merely to convey retinal slip signals. Tabata *et al.*'s[76] computer simulation using the parallel neural pathway hypothesis incorporating LTD as the sole memory mechanism reproduced VOR adaptation very well and explained most of the experimental data collected in monkeys.

The theoretical situation of the VOR adaptation is now clear and should be subject to rigorous experimental testings. We should be careful to observe the anatomical and functional distinction of the flocculus and ventral paraflocculus in monkeys that are often roughly lumped together in recent studies on VOR adaptation.[77] The primate flocculus receives major mossy-fiber inputs from the primary vestibular nerves, vestibular nuclei, nucleus reticularis tegmenti pontis, and central part of the mesencephalic reticular formation, whereas the primate ventral paraflocculus receives major mossy-fiber input from the pontine nucleus.[78] Purkinje cells in the flocculus project to the medial and ventrolateral parts of the medial vestibular nucleus, superior vestibular nucleus, and y group, whereas Purkinje cells in the ventral paraflocculus project to the caudoventral part of posterior interpositus and dentate nuclei, in addition to the medial and ventral parts of the medial vestibular nucleus, superior vestibular nucleus, and y group.[79] The VOR and optokinetic eye movement are reflexes, whereas the smooth pursuit eye movement is voluntary and the "ocular following" is intermediate. These functional subsystems may utilize Purkinje cells in both the flocculus and ventral paraflocculus as their common paths to VOR relay neurons, but whether they substantially overlap, as suggested in a recent ablation study,[80] or are differentially localized, remains open to future investigation.

SYSTEMS OPERATION OF THE CEREBELLUM

In the 1970s, a significant finding was also made on the compartmental structure of the cerebellum by Groenewegen *et al.*[81,82] One side of the cerebellum is divided into seven longitudinal bands (A, B, C1, C2, C3, D1, and D2), each having a specific connection to a portion of cerebellar or vestibular nuclei. The late Oscarsson[83] proposed a further smaller compartment of the cerebellar cortex, the microzone, as a functional unit of the cerebellar cortex. With emphasis on specific connections of a microzone with a subcortical structure, I proposed a corticonuclear microcomplex (referred to as microcomplex hereafter) as a modular unit of the cerebellum[72] (FIG. 4). A microcomplex contains a cortical microzone, a small group of nuclear neurons, and a small group of inferior olive neurons. Mossy-fiber afferents coming from precerebellar structures provide the major input signals to a microcomplex, and supply excitatory synapses via collaterals to the nuclear neuron group that generates output signals outgoing from the microcomplex. Mossy-fiber afferents, at the same time, pass to the microzone and supply excitatory synapses to granule cells, which, via parallel fibers, supply excitatory synapses to Purkinje cells and other inhibitory neurons. Purkinje cell axons supply inhibitory synapses to the nuclear cell group.

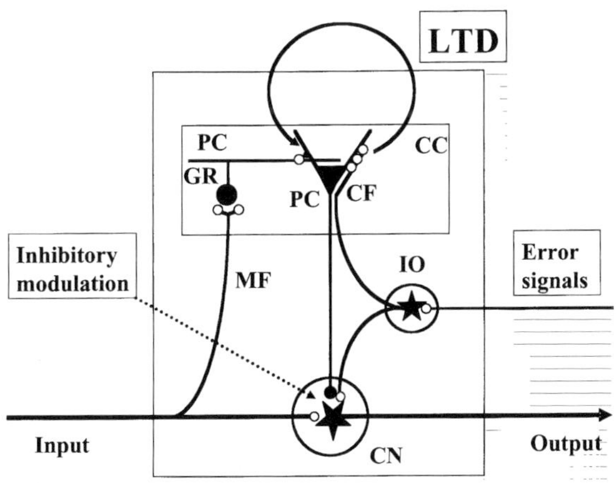

FIGURE 4. A microcomplex model of the cerebellum. Neuronal circuits shown in FIGURE 1 are simplified.

Here, the major signal flow across a microcomplex, from a mossy-fiber afferent to a group of nuclear neurons and their outgoing axons, is under the inhibitory regulation by the side path through the cerebellar cortex.

A microzone is equipped with other inputs, that is, climbing fibers originating from a small group of neurons in the inferior olive, which passes directly to Purkinje cells. A number of studies revealed that climbing fibers convey signals representing errors in the performance of an animal (see Ref. 37). Because LTD is induced by the coactivation of parallel fibers and a climbing fiber, and depresses the transmission from parallel fibers to Purkinje cells, I hypothesized that during repeated exercise, parallel-fiber inputs will be depressed as if they are being punished for their involvement in an unsuccessful operation of the microcomplex. The possible operational principle of a microcomplex is as follows. Initially, all parallel-fiber synapses on Purkinje cells may be functional so that Purkinje cells are fully activated by a mossy-fiber input and in turn inhibit nuclear neurons that otherwise might be activated by the mossy-fiber input. If the consequent silence of nuclear neurons results in errors, climbing-fiber signals are generated to depress the parallel fiber–Purkinje cell transmission, thereby releasing nuclear neurons from Purkinje cell inhibition and generating output signals from the microcomplex. This will lead to improvement of the overall system performance with reduced errors. Using this adaptive mechanism, microcomplexes attached to various reflex pathways would be able to adaptively modify their performance toward the minimization of errors. Microcomplexes will also adaptively regulate compound movements such as locomotion and saccadic eye movements.

The microcomplex model of the cerebellum suggests an interesting possibility that a microcomplex is able to copy input-output relationships of another system. If a microcomplex is fed with error signals representing the discrepancy of the outputs between the microcomplex and the system to be copied, the microcomplex will be progressively modified until it becomes a copy model of the system compared. This possibility, together with the anatomical finding that the paravermis of the cerebellum is included in a loop connection from the motor cortex back to the motor cortex, led me to hypothesize that a microcomplex in the paravermis forms an internal model mimicking a motor plant that is controlled by the motor cortex (FIG. 5). Using the internal feedback through this cerebellar model, the motor cortex would be able to perform a precise voluntary movement even without using the external sensory feedback. This hypothesis explains how we acquire skills by repeated exercises; that is, by means of the formation of an internal model. This hypothesis also explains dysmetria, a typical cerebellar symptom observed in patients with a cerebellar disease; such patients are unable to touch their own nose accurately with their fingertip with closed eyes, which may be due to the lack of an internal model in their cerebellum.

Kawato *et al.*[84] proposed another control system design in which the cerebellum provides an inverse model of the motor plant, which can act as a controller of the motor plant without any feedback (FIG. 6). In this scheme, a voluntary movement can be performed automatically by the cerebellum without conscious effort to activate the motor cortex. The combination of these two types of model-based control enables one to build a robot with superb motor learning capabilities.[85]

In the 1980s, Leiner *et al.*[86] proposed that the cerebellum is involved not only in movements but also in certain forms of mental activities, because the lateral hemisphere of the human cerebellum evolved in parallel with the cerebral association cor-

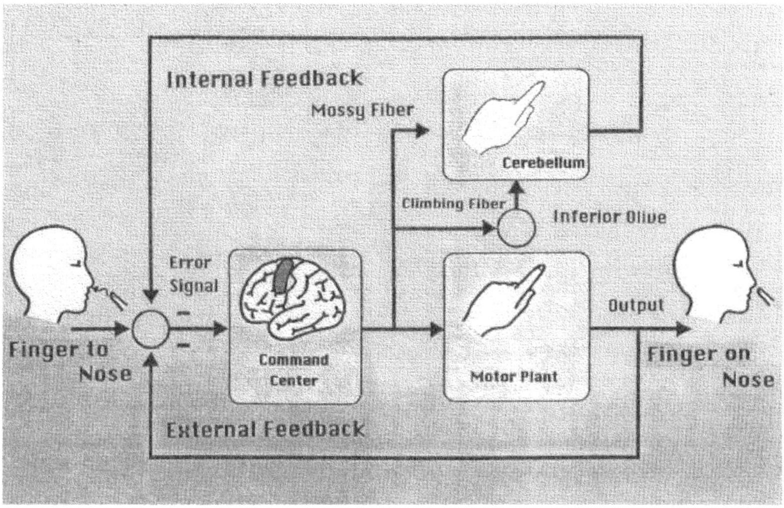

FIGURE 5. Block diagram showing the possible neuronal mechanism of the finger-nose test.

tex. Furthermore, I noticed that the control system operation can be identical for voluntary movement and thinking, either verbal or nonverbal, in the sense that while a voluntary movement manipulates a motor plant formed by muscles and bones, thinking manipulates images, concepts, or ideas encoded in the temporoparietal cortex.[87,88] The control system schemes (FIGS. 5 and 6) for movement should apply to thought processes for which the prefrontal cortex acts as a controller of images, concepts, or ideas generated in the temporoparietal cortex as control objects. While thought processes are repeated, a model of images, concepts, or ideas created in the association cortex will be copied by the cerebellum so that the thought process can be performed in a feedforward manner. One would be able to think without requiring conscious attention to the feedback of the outcome or even to the action of thinking itself.

PERSPECTIVES ON CEREBELLAR RESEARCH

During the past four decades, research on the cerebellum has proved exciting and productive for researchers. In particular, the wealth of knowledge of LTD and its signal transduction has enabled us to manipulate LTD induction either pharmacologically or genetically. It is now possible to examine the correlation between a motor learning behavior of an animal and the occurrence of LTD that may evidence their causal relationship. The generation of transgenic mice expressing a protein kinase C inhibitor that lacks VOR adaptation supports the flocculus hypothesis.[89] However, as reported recently, the PKC inhibitor–transfected mice still exhibit VOR adaptation when exposed to a visual–vestibular training paradigm for eight consecutive days.[90] This may be taken as evidence that LTD is required for short-term VOR ad-

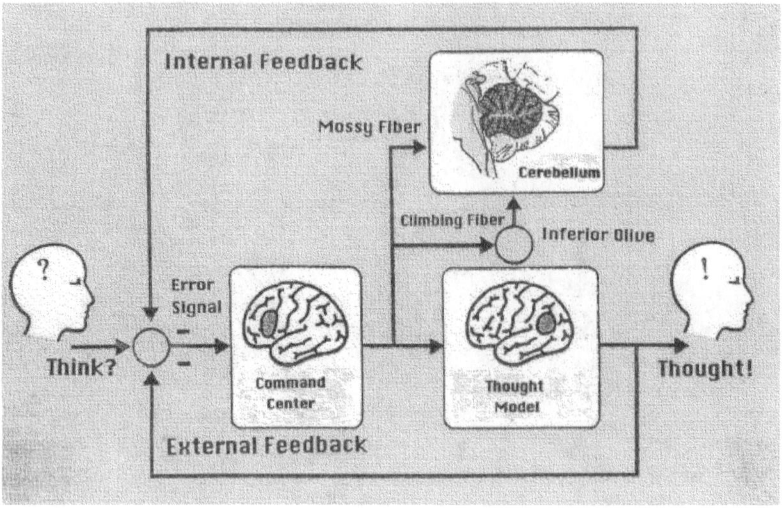

FIGURE 6. Block diagram showing the possible neuronal mechanism of thought.

aptation but not for long-term VOR adaptation or that LTD occurs even without PKC under a long-term training. Because LTD is supported by a complex signal transduction network, it may occur even in the absence of a component under a rigorous stimulation condition. In fact, nitric oxide or inositol trisphophate is indispensable for LTD induction in cerebellar slices, but is not required in induction of the reduced form of LTD in tissue-cultured Purkinje cells.[91,92] One needs to realize that presently we have no means of directly observing LTD in Purkinje cells in awake animals behaving under natural VOR-adaptation conditions. This inevitably makes it difficult to conduct a definitive testing of the above hypotheses. This shortcoming of such technologies is not specific to cerebellar research, but applies in general to the entire field of systems neuroscience. However, we remain hopeful in the light of the recent significant advances in visualization technologies for labeling molecular traces of cell activities and monitoring activities in living neurons.

REFERENCES

1. Jansen, J. & A. Brodal. Aspects of Cerebellar Anatomy. 1954. Forlagt Johan Grundt Tanum. Oslo.
2. Dow, R.S. & G. Moruzzi. 1958. The Physiology and Pathology of the Cerebellum. Univ. Minnesota, Minneapolis.: 580.
3. Batini, C., G. Moruzzi & O. Pompeiano. 1957. Cerebellar release phenomena. Arch. Ital. Biol. **95:** 71–95.
4. Cajal, R. y. 1955. Histologie du système nerveux de l'homme et des vertebrés. II. (original French edition, 1911, L. Azoulay, Ed.). Inst. Ramon y Cajal. Madrid.
5. Dow, R.S. 1942. The evolution and anatomy of the cerebellum. Biol. Rev. 17: 179–220.
6. Szentagothai, J. & K. Rajkovits. 1959. Ueber den Ursprung der Kletterfasern des Klainhirns. Z. Anat. Entwickl. Gesch. **121:** 130–141.

7. SZENTAGOTHAI, J. 1965. The use of degeneration methods in the investigation of short neuronal connexions. Prog. Brain Res. **14:** 1–32.
8. ECCLES, J.C., R. LLINAS & K. SASAKI. 1996. The inhibitory interneurones within the cerebellar cortex. Exp. Brain Res. **1:** 1–6.
9. ECCLES, J.C., R. LLINAS & K. SASAKI. 1996. Parallel fiber stimulation and the responses induced thereby in the Purkinje cells of the cerebellum. Exp. Brain Res. **1:** 17–39.
10. ECCLES, J.C., R. LLINAS & K. SASAKI. 1996. The excitatory synaptic action of climbing fibers on the Purkinje cells of the cerebellum. J. Physiol. (Lond.) **182:** 268–296.
11. ITO, M. & M. YOSHIDA. 1964. The cerebellar-evoked monosynaptic inhibition of Deiters' neurones. Experientia **20:** 515–516.
12. ITO, M., M. YOSHIDA & K. OBATA. 1964. Monosynaptic inhibition of the intracerebellar nuclei induced from the cerebellar cortex. Experientia **20:** 575–576.
13. ITO, M., K. OBATA & R. OCHI. 1966. The origin of cerebellar-evoked inhibition of Deiters' neurons. II. Temporal correlation between the trans-synaptic activation of Purkinje cells and the inhibition of Deiters' neurons. Exp. Brain Res. **2:** 350–364.
14. ITO, M., N. KAWAI, M. UDO & N. MANO. 1969. Axon reflex activation of Deiters' neurons from the cerebellar cortex through collaterals of the cerebellar afferents. Exp. Brain Res. **8:** 249–268.
15. TOYAMA, K., N. TSUKAHARA, K. KOSAKA & K. MATSUNAMI. 1970. Synaptic excitation of red nucleus neurons by fibres from interpositus nucleus. Exp. Brain Res. **11:** 187–198.
16. UNO, M., M. YOSHIDA & J. HIROTA. 1970. The mode of cerebellothalamic relay transmission investigated with intracellular recording from cells of the ventrolateral nucleus of cat's thalamus. Exp. Brain Res. **10:** 121–139.
17. ECCLES, J.C., M. ITO & J. SZENTAGOTHAI. 1967. The Cerebellum as a Neuronal Machine. Springer-Verlag. Berlin.
18. KOLSTEN, J., R. APPS & J.R. TROTT. 1995. A combined retrograde tracer and GABA immunocytochemical study of the projection from nucleus interpositus posterior to the posterior lobe C2 zone of the cat cerebellum. Eur. J. Neurosci. **7:** 926–933.
19. BATINI, C., C. COMPOINT, C. BUISSERET-DELMAS, *et al.* 1992. Cerebellar nuclei and the nucleocortical projections in the rat: retrograde tracing coupled to GABA and glutamate immunohistochemistry. J. Comp. Neurol. **315:** 74–84.
20. DE ZEEUW, C.I., A.M. VAN ALPEN, R.K. HAWKINS & T.J. RUIGROCK. 1997. Climbing fibre collaterals contact neurons in the cerebellar nuclei that provide a GABAergic feedback to the inferior olive. Neuroscience **80:** 981–986.
21. DE ZEEUW, C.I., J.C. HOLSTEGE, F. CALKOEN, *et al.* 1988. New combination of WGA-HRP anterograde tracing and GABA immunohistochemistry applied to afferents of the cat inferior olive at the ultrastructural level. Brain Res. 447: 369–375.
22. KIM, J.J., D.J. KRUPA & R.F. THOMPSON. 1998. Inhibitory cerebello-olivary projections and blocking effect in classical conditioning. Science **279:** 570–573.
23. HESSLOW, G. & M. IVARSSON. 1996. Inhibition of the inferior olive during conditioned responses in the decerebrate ferret. Exp. Brain Res. **110:** 36–46.
24. MARR, D. 1998. A theory of cerebellar cortex. J. Physiol. (Lond.) **202:** 437–470, 1969.
25. ALBUS, J.S. 1971. A theory of cerebellar function. Math. Biosci. **10:** 25–61.
26. HEBB, D.O. 1949. The Organization of Behavior. Wiley. New York.
27. BRINDLEY, G.S. 1964. The use made by the cerebellum of the information that it receives from sense organs. IBRO Bull. **3:** 80.
28. ROSENBLATT, F. 1962. Principles of Neurodynamics: Perceptron and the Theory of Brain Mechanisms. Spartan Books. Washington, D.C.
29. GHELARDUCCI, B., M. ITO & N. YAGI. 1975. Impulse discharge from flocculus Purkinje cells of alert rabbits during visual stimulation combined with horizontal head rotation. Brain Res. **87:** 66–72.
30. GILBERT, P.F.C. & W.T. THACH. 1977. Purkinje cell activity during motor learning. Brain Res. **128:** 309–328.
31. ITO, M., M. S AKURAI & P. TONGROACH. 1982. Climbing fibre induced depression of both mossy fibre responsiveness and glutamate sensitivity of cerebellar Purkinje cells. J. Physiol. (Lond.) **324:** 113–134.

32. EKEROT, C.-F. & M. KANO. 1985. Long-term depression of parallel fibre synapses following stimulation of climbing fibres. Brain Res. **342:** 357–360.
33. ITO, M. & M. KANO. 1982. Long-lasting depression of parallel fiber-Purkinje cell transmission induced by conjunctive stimulation of parallel fibers and climbing fibers in the cerebellar cortex. Neurosci. Lett. **33:** 253–258.
34. SAKURAI, M. 1987. Synaptic modification of parallel fiber-Purkinje cell transmission in in vitro guinea pig cerebellar slices.. J. Physiol. (Lond.) **394:** 463–480.
35. ITO, M. 1989. Long-term depression. Annu. Rev. Neurosci. **12:** 85–102.
36. DANIEL, H., C. LEVENES & F. CREPEL. 1998. Cellular mechanisms of cerebellar LTD. Trends Neurosci. **21:** 401–407.
37. ITO, M. 2001. Cerebellar long-term depression: characterization, signal transduction, and functional roles. Physiol. Rev. **81:** 1143–1195.
38. LINDEN, D.J., M.H. DICKINSON, M. SMEYNE & J.A. CONNOR. 1991. A long-term depression of AMPA currents in cultured cerebellar Purkinje neurons. Neuron **7:** 81–89.
39. ITO, M. 2002. The molecular organization of cerebelar long-term depression. Natl. Rev. Neurosci. **3:** 896–902.
40. SANES, J.R. & J.W. LICHTMAN. 1999. Can molecules explain long-term potentiation? Nat. Rev. Neurosci. **2:** 597–604.
41. BERRIDGE, M.J. 1996. Introduction. *In* Coincidence Detection in the Nervous System. A. Konnerth, R.Y. Tsien, K. Mikoshiba & J. Altman, Eds. Human Frontier Science Program: 9–13. Strasbourg, France.
42. BHALLA, U.S. & R. IYENGAR. 1999. Emergent properties of networks of biological signalling pathways. Science **283:** 381–387.
43. KURODA, S., N. SCHWEIGHOFER & M. KAWATO. 2001. Exploration of signal transduction pathways in cerebellar long-term depression by kinetic simulation. J. Neurosci. **21:** 5693–5702.
44. MCCORMICK, D.A. & R.F. THOMPSON. 1984. Cerebellum: essential involvement in the classically conditioned eyelid response. Science **223:** 296–299.
45. YEO, C.H., M.J. HARDIMAN & M. GLICKSTEIN. 1985a. Classical conditioning of the nictitating membrane response of the rabbit. I. Lesions of the cerebellar nuclei. Exp. Brain Res. **60:** 87–98.
46. ANDERSSON, G. & D.M. ARMSTRONG. 1987. Complex spikes in Purkinje cells in the lateral vermis (b zone) of the cat cerebellum during locomotion. J. Physiol. (Lond.) **385:** 107–134.
47. YANAGIHARA, D. & I. KONDO. 1996. Nitric oxide plays a key role in adaptive control of locomotion in cat. Proc. Natl. Acad. Sci. USA **93:** 13292–13297.
48. KITAZAWA, S., T. KIMURA & P.-B. YIN. 1998. Cerebellar complex spikes encode both destinations and errors in arm movements. Nature **392:** 494–497.
49. LORENTE DE NO, R. 1933. Vestibulo-ocular reflex arc. Arch. Neurol. Psychiatry **30:** 245–291.
50. FUKUDA, J., S.M. HIGHSTEIN & M. ITO. 1972. Cerebellar inhibitory control of the vestibulo-ocular reflex investigated in rabbit's IIIrd nucleus. Exp. Brain Res. **14:** 511–526.
51. MAEKAWA, K. & J.I. SIMPSON. 1973. Climbing fiber responses evoked in vestibulocerebellum of rabbit from visual pathway. J. Neurophysiol. **36:** 649–666.
52. ITO, M. 1970. Neurophysiological aspects of the cerebellar motor control system. Proceedings of the Fulton Society Symposium, New York. Int. J. Neurol. **7:** 126–176.
53. ITO, M. The control mechanisms of cerebellar motor system. 1974. *In* F.D. Schmitt, F.G. Worden (Eds), The Neuroscience, Third Study Program. F.D. Schmitt & F.G. Worden, Eds.: 293–303. MIT Press, Cambridge, MA.
54. GONSHOR, A. & G.M. MELVILL-JONES. 1974. Extreme vestibulo-ocular adaptation induced by prolonged optical reversal of vision. J. Physiol. (Lond.) **256:** 381–414.
55. ITO, M., N. SHIIDA, N. YAGI & M. YAMAMOTO. 1974. The cerebellar modification of rabbit's horizontal vestibulo-ocular reflex induced by sustained head rotation combined with visual stimulation. Proc. Jpn. Acad. **50:** 85–89.
56. ROBINSON, D.A. 1976. Adaptive gain control of vestibulo-ocular reflex by the cerebellum. J. Neurophysiol. **39:** 954–969.
57. ITO, M. 1982. Cerebellar control of the vestibulo-ocular reflex - around the floccculus hypothesis. Annu. Rev. Neurosci. **5:** 275–296.

58. McElligott, J.G., P. Beeton & J. Polk. 1998. Effect of cerebellar inactivation by lidocaine microdialysis on the vestibuloocular reflex in goldfish. J. Neurophysiol. **79:** 1286–1294.
59. Nagao, S. & H. Kitazawa. 2002. Effects of acute shutdown of the monkey flocculus on the memory trace for vestibulo-ocular reflex adaptation. Soc. Neurosci. Abstr. **28:** 766.13.
60. Pastor, A.M., R.R. De Cruz & R. Baker. 1994. Cerebellar role in adaptation of the goldfish vestibuloocular reflex. J. Neurophysiol. **72:** 1383–1394.
61. Partsail, A.M., Y. Zhanga & S.M. Highstein. 1995. Dorsal y group in squirrel monkey. II. Contribution of the cerebellar flocculus to neuronal responses in normal and adapted animals. J. Neurophysiol. **73:** 632–650.
62. Luebke, A.E. & D.A. Robinson. 1994. Gain changes of the cat's vestibulo-ocular reflex after flocculus inactivation. Exp. Brain Res. **98:** 379–390.
63. Miles, F.A. & S.G. Lisberger. 1981. Plasticity in the vestibuloocular reflex: a new hypothesis. Annu. Rev. Neurosci. **4:** 273–299.
64. Dufosse, M., M. Ito, P.J. Jastreboff & Y. Miyashita. 1978. A neuronal correlate in rabbit's cerebellum to adaptive modification of the vestibulo-ocular reflex. Brain Res. **150:** 611–616.
65. Nagao, S. 1989. Behavior of floccular Purkinje cells correlated with adaptation of vestibulo-ocular reflex in pigmented rabbits. Exp. Brain Res. **77:** 531–540.
66. Watanabe, E. 1985. Role of the primate flocculus in adaptation of the vestibulo-ocular reflex. Neurosci. Res. **3:** 20–38.
67. Aizenman, C.D. & D.J. Linden. 1999. Regulation of the rebound depolarization and spontaneous firing patterns of deep nuclear neurons in slices of rat cerebellum. J. Neurophysiol. **82:** 1697–1709.
68. Lewis, R.F., D.S. Zee, M.R. Hayman & R.J. Tamargo. 2001. Oculomotor function in the rhesus monkey after deafferentation of the extraocular muscles. Exp. Brain Res. **141:** 349–358.
69. Ito, M. 1984. The Cerebellum and Neural Control. Raven Press. New York.
70. Nago, S. 1990. Eye velocity is not the major factor that determines mossy fiber responses of rabbit floccular Purkinje cells to head and screen oscillation. Brain Res. **80:** 221–224.
71. Miyashita, Y. 1984. Eye-velocity responsiveness and its proprioceptive component in the floccular Purkinje cells of the alert pigment rabbits. Exp. Brain Res. **55:** 81–90.
72. Nago, S. 1991. Contribution of oculomotor signals to the behavior of rabbit floccular Purkinje cells during reflex eye movements. Neurosci. Res. **12:** 169–184.
73. Ito, M. 1998. Cerebellar learning in the vestibulo-ocular reflex. Trends Cogn. Sci. **2:** 313–321.
74. Hirata, Y. & S.M. Highstein. 2001. Acute adaptation of the vestibuloocular reflex: signal processing by floccular and ventral parafloccular Purkinje cells. J. Neurophysiol. **85:** 2267–2288.
75. Lisberger, S.G. 1994. Neural basis for motor learning in the vestibuloocular reflex of primates. III. Computational and behavioral analysis of the sites of learning. J. Neurophysiol. **72:** 974–998.
76. Tabata, H., K. Yamamoto & M. Kawato. 2002. Computational study on monkey VOR adaptation and smooth pursuit based on the parallel control-pathway theory. J. Neurophysiol. **87:** 2176–2189.
77. Gerrits, N.M. & J. Voogd. 1989. The topographical organization of climbing and mossy fiber afferents in the flocculus and the ventral paraflocculus in rabbit, cat, and monkey. Exp. Brain Res. Suppl. **17:** 26–29.
78. Nagao, S., T. Kitamura, M. Nakamura, *et al.* 1997. Differences of the primate flocculus and ventral paraflocculus in the mossy and climbing fiber input organization. J. Comp. Neurol. **382:** 480–498.
79. Nagao, S., T. Kitamura, M. Nakamura, *et al.* 1997. Location of efferent terminals of the primate flocculus and ventral paraflocculus revealed by anterograde axonal transport methods. Neurosci. Res. **27:** 257–269.
80. Rambold, H., A. Chruch, Y. Selig, *et al.* 2002. Partial ablation of the flocculus and ventral paraflocculus in monkeys causes linked deficit in smooth pursuit eye movements and adaptive modification of the VOR. J. Neurophysiol. **87:** 912–924.

81. GROENEWEGEN, H.J. & J. VOOGD. 1977. The parasagittal zonation within the olivocerebellar projection. I. Climbing fiber distribution in the vermis of cat cerebellum. J. Comp. Neurol. **174:** 417–488.
82. GROENEWEGEN, H.J., J. VOOGD & S.L. FREEDMAN. 1979. The parasagittal zonation within the olivocerebellar projection. II. Climbing fiber distribution in the intermediate and hemispheric parts of cat cerebellum. J. Comp. Neurol. **183:** 551–601.
83. OSCARSSON, O. 1976. Spatial distribution of climbing and mossy fibre inputs into the cerebellar cortex. *In* Afferent and Intrinsic Organization of Laminated Structures in the Brain. O. Creutzfeldt, Ed. Exp. Brain Res. Suppl. **1:** 34–42. Springer-Verlag. Berlin.
84. KAWATO, M., K. F URUKAWA & R. SUZUKI. 1987. A hierarchical neuronal network model for control and learning of voluntary movement. Biol. Cybern. **57:** 169–185.
85. KAWATO, M. 1990. Computational schemes and neural network models for formation and control of multijoint arm trajectory. *In* Neural Networks for Control. T. Miller, R.S. Sutton & P.J. Werbos, Eds.: 197–228. MIT Press. Cambridge, MA.
86. LEINER, H.C., A.L. LEINER & R.S. DOW. 1986. Does the cerebellum contribute to mental skill? Behav. Neurosci. **100:** 443–453.
87. ITO, M. 1993. Movement and thought: identical control mechanisms by the cerebellum. Trends Neurosci. **16:** 448–450.
88. ITO, M. 2000. Neural control of cognition and language. *In* Images, Language, Brain. Y. Miyashita & W. O'Neil, Eds.: 149–162. MIT Press. Cambridge, MA.
89. DE ZEEUW, C.I., C. HANSEL, F. BIAN, *et al.* 1998. Expression of a protein kinase C inhibitor in Purkinje cells blocks cerebellar LTD and adaptation of the vestibulo-ocular reflex. Neuron **20:** 495–508.
90. VAN ALPHEN A.M. & C.I. DE ZEEUW. 2002. Cerebellar LTD facilitates but is not essential for long-term adaptation of the vestibulo-ocular reflex. Eur. J. Neurosci. **16:** 486–490.
91. LINDEN, D.J., T.M. DAWSON & V.L. DAWSON. 1995. An evaluation of the nitric oxide/ cGMP/cGMP-dependent protein kinase cascade in the induction of cerebellar long-term depression in culture. J. Neurosci. **15:** 5098–5105.
92. NARASIMHAN, K., I.N. PESSAH & D.J. LINDEN. 1998. Inositol-1,4,5-trisphosphate receptor-mediated Ca mobilization is not required for cerebellar long-term depression in reduced preparations. J. Neurophysiol. **80:** 2963–2974.

Motor and Nonmotor Domains in the Monkey Dentate

RICHARD P. DUM,[b,c] CHEN LI,[c] AND PETER L. STRICK[a–d]

[a]*Pittsburgh Veterans Affairs Medical Center, Pittsburgh, Pennsylvania, USA*

[b]*Center for the Neural Basis of Cognition, and Departments of* [c]*Neurobiology, and* [d]*Psychiatry and Neurological Surgery, University of Pittsburgh School of Medicine, Pittsburgh, Pennsylvania 15261, USA*

ABSTRACT: Our concepts about the organization and functions of the cerebellum have changed substantially in the last 10 years. In recent studies, we used transneuronal virus tracing techniques to demonstrate that the output of the cerebellum of primates projects via the thalamus not only to its classical motor target, the primary motor cortex, but also to "nonmotor" cortical areas in the prefrontal and posterior parietal cortex. We found that dentate neurons projecting to different cortical areas originated from localized regions of the nucleus which we termed "output channels." To compare the locations of the output channels projecting to different cortical targets, we have created an unfolded map of the dentate. This unfolded map revealed that dentate output channels were segregated into spatially separate "motor" and "nonmotor" domains. The output channels in the motor domain exclusively targeted primary motor and premotor areas of the cerebral cortex. These channels were localized in the dorsal portion of the dentate. The output channels in the nonmotor domain projected to prefrontal and posterior parietal cortical areas. The nonmotor domain was confined to the ventral portion of the dentate. In recent studies, we defined a unique molecular marker, monoclonal antibody 8B3, which appears to differentially "recognize" these two domains. Taken together, our results suggest that dentate output is organized according to the functional capabilities of its cortical targets. This organization provides the dentate nucleus with the anatomical substrate to influence not only the control of movement, but also cognitive, higher-order executive and visuospatial functions.

KEYWORDS: dentate nucleus; cerebellum; motor; cognitive function; prefrontal cortex; posterior parietal cortex; thalamus

INTRODUCTION

The dentate nucleus is a major source of output from the cerebellum to the cerebral cortex. In the past, the dentate was thought to contain a single map of the body and project via the thalamus in a somatotopic fashion to the primary motor cortex

Address for correspondence: Dr. Peter L. Strick, Department of Neurobiology, University of Pittsburgh School of Medicine, W1640 Biomedical Science Tower, 200 Lothrop Street, Pittsburgh, PA 15261. Voice: 412-383-9961; fax: 412-383-9061.
strickp@pitt.edu

Ann. N.Y. Acad. Sci. 978: 289–301 (2002). © 2002 New York Academy of Sciences.

(M1), for example[1–2] (for review see Ref. 3). Our recent studies using transneuronal tracing with neurotropic virus have demonstrated that the output of the dentate projects to a more diverse set of cortical areas than previously recognized. These cortical areas include premotor, prefrontal, and posterior parietal areas of cortex.[4–8]

In this short review, we explore the topographic organization of the dentate projections to the cerebral cortex as revealed by the virus tracing method. Because the overall structure of the dentate is remarkably complex, we have created an unfolded map of the nucleus. By plotting the results of virus tracing studies on this map, we have been able to show that the dentate appears to be divided into spatially separate "motor" and "nonmotor" domains.[9] In the motor domain, dentate output projects in a somatotopic fashion to M1. Dentate projections to various premotor areas in the frontal lobe also originate from the motor domain.[6] In the nonmotor domain, dentate output projects to prefrontal and posterior parietal areas of cortex in a topographic manner. In addition, we have found that the monoclonal antibody 8B3 appears to differentially stain the motor and nonmotor domains of the dentate.[10] Our evidence suggests that the anatomical substrate exists for the output of the dentate to influence cognitive, higher-order executive and visuospatial functions as well as the control of skilled movement.

METHODS

The results described in this review are based on three procedures: (1) unfolding the dentate; (2) mapping dentate projections to the cerebral cortex; and (3) analyzing dentate staining with monoclonal antibody 8B3. Technical aspects of each procedure will be presented below.

Unfolded Map

To quickly and conveniently compare the location of neuronal labeling, antibody staining, or any potential topographically mapped feature, we created an unfolded map of the dentate.[9] Such maps have had considerable heuristic value when applied to other complex structures such as the inferior olive, cerebellar cortex, and cerebral cortex.[11–15] We believe that our unfolded map enables one to compare data from a large number of individual experiments in a common framework.

We first charted the outline of the dentate nucleus from every other serial 50-μm section throughout the nucleus. The dentate was divided into segments by placing a marker at the acute curves on each section outline. We drew a curved line halfway between the medial and lateral outlines of the nucleus (termed the "central" line, FIG. 1A). The segmental marks were projected onto the central line and the line was straightened. The unfolded lines from multiple sections were aligned on the border between the "c" segment (the hilus) and the "d" segment (horizontal ventral segment) of the dentate (FIG. 1A).

To map the results from virus tracing experiments, labeled neurons were "projected" onto the unfolded lines taken from sections through the dentate (for details, see Ref. 9). The number of labeled neurons was determined for every 200-μm interval along each unfolded line. The values for intervals from adjacent unfolded lines were summed to form 200-μm × 200-μm bins. The number of labeled neurons in a bin was indicated by a color code (FIG. 1C).

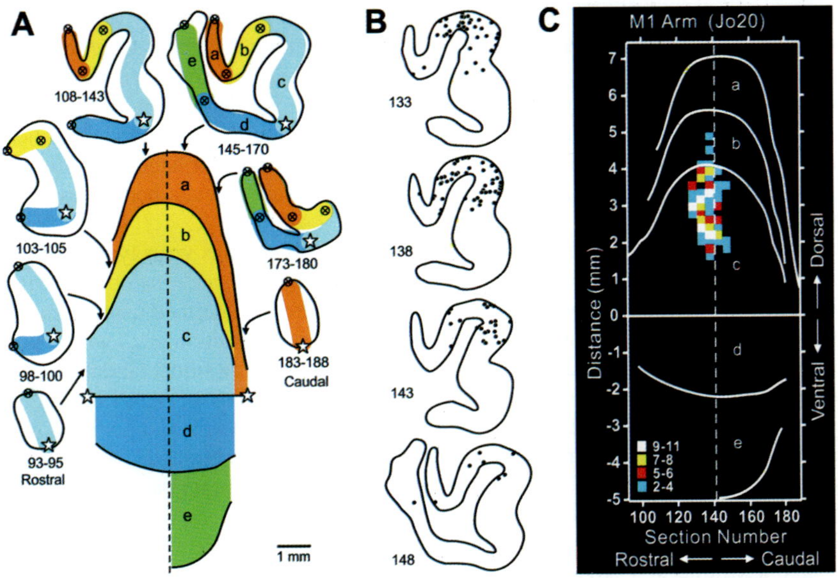

FIGURE 1. Unfolded map of the dentate nucleus. (**A**) This map of the dentate was created by unfolding serial coronal sections through the nucleus. The sections displayed around the map indicate the shape of the dentate at selected levels. The *central, colored lines* on these sections were used for unfolding. The color-coded segments of these lines correspond to the colored regions on the unfolded map. Each section is aligned to the junction of the "c" and "d" segments (*open star*). The rostro-caudal midpoint of the nucleus is indicated by a *vertical dashed line.* See METHODS for further explanation. (**B**) Cross-sections of the dentate nucleus showing the location of "second order" neurons (*dots*) labeled after transport of HSV1 from the arm representation of M1. **C:** The distribution and density of labeled neurons in the same animal are displayed on an unfolded map of the dentate. This map was reconstructed from every other coronal section. (From Dum and Strick.[9] Copyright © 2002 by the American Physiological Society.).

To map the results from staining with antibody 8B3, every fourth section through the dentate was captured using a high-resolution video camera connected to a digital image capture board in a PC.[16] We adjusted the overall intensity of each captured image to compensate for global differences in staining intensity between sections (Corel PhotoPaint®). We "thresholded" the image to remove intensities corresponding to background (unstained white matter). The intensity of staining in pixels 75 μm on either side of the central line was measured and then the line was unfolded as described above. The intensities along each line were smoothed with a sliding boxcar filter (7 pixels in width). Finally, staining intensity was contoured into four levels on the basis of deviations from the mean intensity.

Tracing with Virus

We used retrograde transneuronal transport of neurotropic viruses to identify the dentate regions that project via the thalamus to selected regions of motor, prefrontal, and posterior parietal cortex.[3,7,8] The methods used in these studies and the care provided to the experimental animals complied with the regulations detailed in the *Na-*

tional Institutes of Health Guide for the Care and Use of Laboratory Animals. All protocols were reviewed and approved by the Institutional Animal Care and Use committees. The biosafety precautions taken during these experiments conformed to or exceeded the Biosafety Level 2 (BSL-2) regulations detailed in *Biosafety in Microbiological and Biomedical Laboratories* (Health and Human Services publication CDC 93-8395). Complete descriptions of the procedures for handling virus and virus-infected animals, as well as technical details have been described previously.[3,7,8,17]

In brief: we injected the McIntyre-B strain of herpes simplex virus type I (HSV1) into specific regions of the cerebral cortex in cebus monkeys (*Cebus apella*). Most injection sites within motor areas of the cerebral cortex were defined using intracortical stimulation to generate a map of motor responses. Injection sites within prefrontal or posterior parietal areas were localized in relation to sulcal landmarks and verified by cytoarchitectonic analysis of the processed brain tissue. At the end of the survival period (4–5 days), each animal was deeply anesthetized and perfused with aldehyde fixatives. The cerebellum was frozen and serially sectioned (50 μm) in the coronal plane. To identify neurons infected with virus, we processed free-floating tissue sections according to the avidin–biotin peroxidase method (Vectastain; Vector Laboratories, Burlingame, CA) using a commercially available antibody to HSV1 (Dako, Carpinteria, CA; 1:2000 dilution).

Staining with 8B3

In prior studies we found that monoclonal antibody 8B3 differentially stains dorsal and ventral portions of the dentate nucleus in macaques.[10] In the present study we examined whether this antibody also differentially stains portions of the cebus monkey dentate. In addition, we performed a gradient density analysis of 8B3 staining to determine in a more quantitative manner the distribution of staining throughout the entire rostro-caudal extent of the nucleus. Our immunohistochemical procedures were modeled on those of Pimenta and colleagues.[10] In brief: we incubated free-floating sections through the dentate for 18–22 hours in monoclonal antibody 8B3 diluted 1:8000 in Blotto (4% non-fat dry milk in 0.1M phosphate buffered saline [PBS], pH 7.4) and 2% Triton X-100. After several washes in PBS, the sections were incubated for 2 hours in the secondary antibody, goat anti-mouse IgM-HRP conjugate, μ-chain-specific (Jackson Immunoresearch Labs, Inc., West Grove, PA), diluted 1:100 with Blotto, 2% Triton X-100, and 0.01% thimerosal. The antigen–antibody complexes were revealed by reaction in 3,3′-diaminobenzidine 0.05% in PBS and 0.01% H_2O_2.

RESULTS

Initially, we examined the somatotopy within the dentate projections to M1.[3] Unfolded maps of the dentate were used to provide a consistent frame of reference for comparing the distribution of labeled neurons that project to the different cortical areas. Injections of HSV1 into the arm representation of M1 ($n = 3$) labeled a compact cluster of neurons in the dorsal portion of the dentate at mid rostrocaudal levels (FIGS. 1B,C; 2B; 3A). Despite small variations in the placement of virus injections

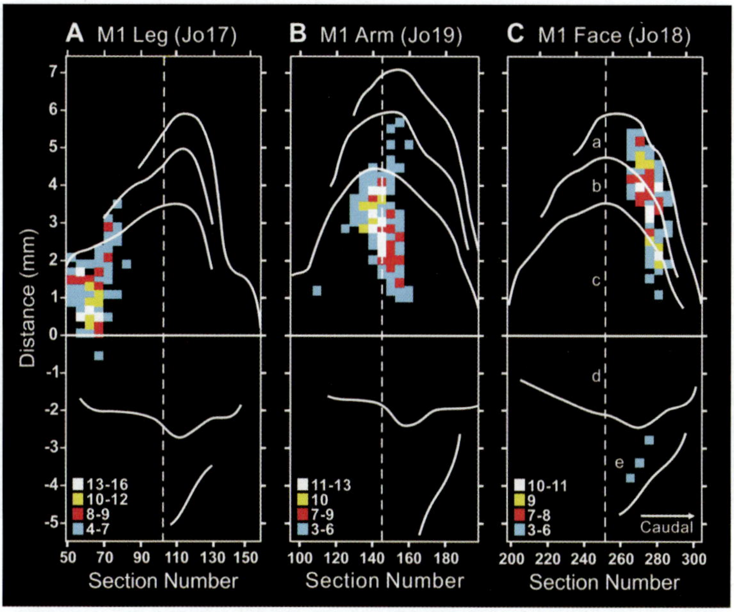

FIGURE 2. Somatotopic organization of dentate output channels to M1. Unfolded maps of the dentate illustrate the neurons labeled after HSV1 injections into the leg (**A**), arm (**B**), and face (**C**) representations of M1. Conventions and abbreviations as in FIGURE 1. The arm case shown in (**B**) is from a different animal than those illustrated in FIGURES 1 and 3A. (From Dum and Strick.[9] Copyright © 2002 by the American Physiological Society.)

within the arm representation of M1, the location of labeled neurons in the dorsal dentate was quite similar in all three animals (FIGS. 1C, 2B, 3A)(see Figure 11 in Hoover and Strick[3]).

Injections of HSV1 into the leg or the face representation of M1 also labeled dense clusters of neurons in a dorsal portion of the dentate. In each case, the location of labeled neurons was spatially separate from the location of neurons that projected to the arm representation of M1. Virus injections into the leg representation labeled neurons in the rostral pole of the dorsal portion of the dentate (Fig. 2A), whereas virus injections into the face representation labeled neurons at caudal levels of the dorsal dentate (FIG. 2C). We have used the term "output channels" to describe the topographically distinct origin of dentate projections to different cortical areas (for example, see Ref. 6). The rostral-to-caudal orientation of the dentate output channels to the leg, arm, and face representations in M1 is compatible with the rostral-to-caudal body map previously suggested for the dentate.[18–22]

The combined somatotopic map of dentate projections to M1 demonstrated in our studies occupied only a portion of the dorsal third of the nucleus.[3,9] This suggested that the large unlabeled portions of the dentate must provide projections to regions of the cerebral cortex that lie outside M1. To examine this issue, we injected virus into selected prefrontal and posterior parietal areas of the cortex. In every case, these injections resulted in transneuronal labeling of neurons that were concentrated ven-

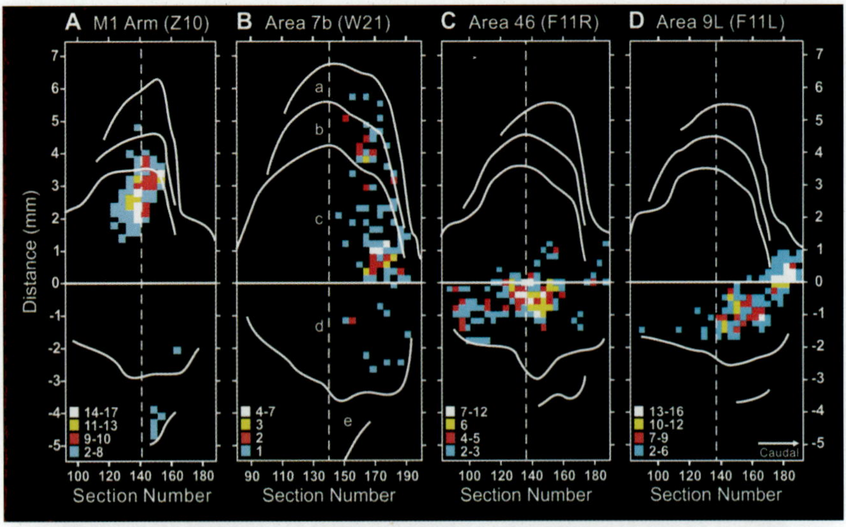

FIGURE 3. Dentate output channels to the prefrontal and posterior parietal cortex. Unfolded maps of the dentate illustrate the neurons labeled after HSV1 injections into *a*: the arm representation of M1 (**A**), area 7b (**B**), area 46 (**C**), and area 9L (**D**). Conventions and abbreviations as in FIGURE 1. The arm case shown in (**A**) is from a different animal than those illustrated in FIGURES 1 and 2B. (From Dum and Strick.[9] Copyright © by the American Physiological Society.)

tral to the regions labeled after virus injections into M1.[7,8] For the parietal lobe, dentate projections to a portion of area 7b located in the intraparietal sulcus were most concentrated ventral and caudal to the origin of projections to the arm representation of M1 (FIG. 3B). For the prefrontal cortex, the dentate projections to a part of area 46 located dorsal to the principal sulcus were most concentrated ventral to the output channel to the arm representation of M1 (FIG. 3C). Dentate projections to a part of area 9 located on the lateral surface of the hemisphere originated from yet another localized region within the caudal dentate (FIG. 3D). These results demonstrated that the dentate has substantial projections to nonmotor areas of the cortex. Furthermore, the projections to different nonmotor areas originate in a topographically organized manner from distinct regions within the ventral dentate (for example see Refs. 5–8).

We have summarized our findings by plotting the observations of individual experiments onto a single unfolded map of the dentate nucleus (FIG. 4A). On this map, we labeled the location of the greatest density of the neuronal labeling that defined the output channel to each cortical area examined. The summary map emphasizes that a sizeable portion of the dentate projects to parts of prefrontal and posterior parietal cortex, in addition to its classical target, M1. Furthermore, the dentate map reveals an important feature of the topographic organization of dentate output. The output channels to the prefrontal and posterior parietal areas of cortex are spatially segregated from those that target motor areas of the cortex. Thus, the dentate appears to be anatomically divided into separate domains that target "motor" and "nonmotor" areas of the cerebral cortex.

In recent studies we found that the monoclonal antibody, 8B3, which recognizes a proteoglycan localized to the cell surface, appeared to differentially stain the dorsal

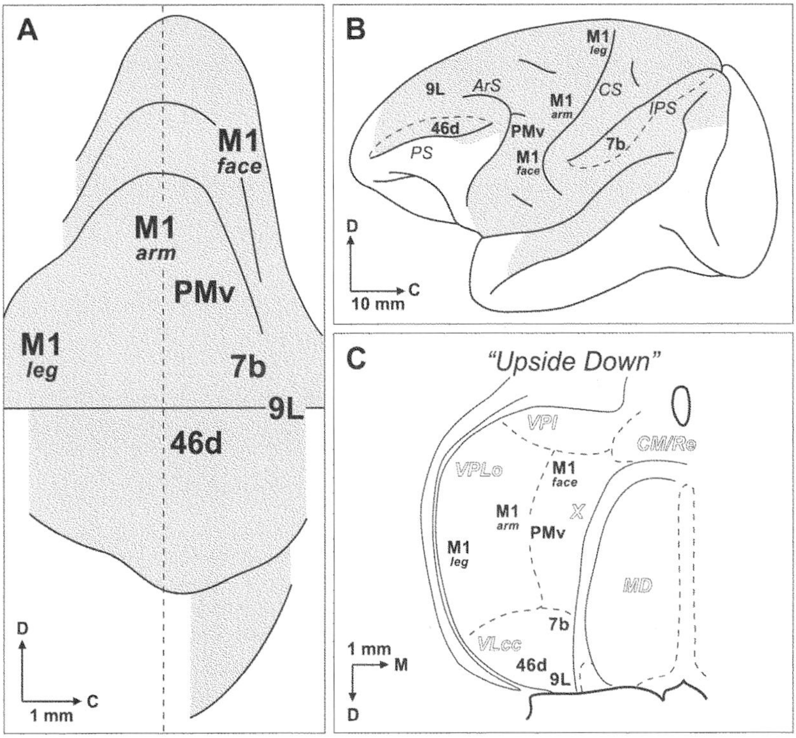

FIGURE 4. Topography within the cerebello-thalamocortical circuit. (**A**) Dentate output channels. *Labels* mark the location of the highest concentration of dentate projections to each cortical area studied. (**B**) Selected cortical targets of cerebello-thalamocortical circuits. Cortical *shading* designates regions of the lateral hemisphere that project to the cerebellum via the pons.[26–28] **C:** Origin of selected cortical projections from the ventrolateral thalamus. The cortical regions indicated receive input from regions of the ventrolateral thalamus that lie within the termination zone of cerebellar efferents. The thalamus has been displayed *upside down* to indicate the correspondence between its topography and that of the dentate. ArS, arcuate sulcus; CM/Re, nucleus centrum medianum/ nucleus reuniens; CS, central sulcus; IPS, intraparietal sulcus; MD, nucleus medialis dorsalis; PMv- ventral premotor area; PS, principal sulcus; VLcc, caudal portion of the nucleus ventralis lateralis, pars caudalis; VPI, nucleus ventralis inferior; VPLo, nucleus ventralis posterior lateralis, pars oralis; X, Area X. (From Dum and Strick.[9] Copyright © by the American Physiological Society.)

and ventral portions of the macaque dentate.[10] We examined whether 8B3 differentially stains the motor and nonmotor domains in the dentate nucleus of the cebus monkey. Reaction product resulting from immunostaining with 8B3 outlined the surface of cell bodies and proximal dendrites of a subpopulation of dentate neurons (FIG. 5). In addition, staining was present in a more diffuse and less clearly localized manner in the surrounding neuropil. Most of the dorsal portion of the dentate was stained at a light to moderate intensity. However, along the ventral limb of the nucleus, neuronal and neuropil labeling increased and exhibited a lateral to medial gradient of increasing staining intensity. Immunostaining for 8B3 reached a maximum

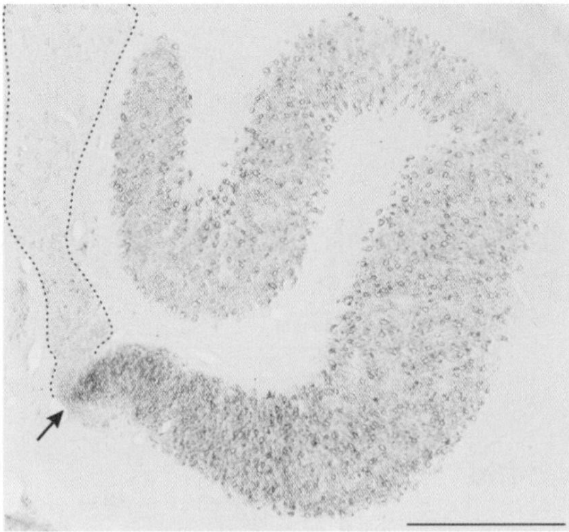

FIGURE 5. Photomicrograph of the dentate nucleus immunoreacted with the 8B3 antibody. Note the ring-like appearance of labeled neurons and the increase in intensity from the dorsal to the ventral portions of the dentate. *Arrow* indicates the "kink" in the profile of the dentate where the staining intensity decreases precipitously (see text). *Dotted line* outlines the medial ascending limb of the dentate that has very light staining. Calibration = 1 mm.

just before a sharp "kink" in the ventral dentate (*arrow,* FIG. 5). At this point, staining in the most medial portion of the nucleus precipitously declined to barely detectable levels (FIG. 5).

Unfolded maps of 8B3 labeling, together with gradient density analysis show that intense staining with the antibody was restricted to the ventral dentate (FIG. 6). This staining extended throughout the entire rostro-caudal extent of the nucleus. In contrast, weak to moderate staining was present throughout the entire dorsal dentate. The dorsal, lightly stained region represents approximately two-thirds of the entire dentate map.

DISCUSSION

Our studies have provided new insights into the topographic organization of dentate projections to the cerebral cortex. Our unfolded map of the dentate indicates that the nucleus is anatomically divided into separate domains that target "motor" and "nonmotor" areas of the cerebral cortex. The motor domain is located in the dorsal portion of the dentate map. It contains output channels that project to cortical areas concerned with the generation and control of movement. The neurons that project to the leg, arm, and face representations of M1 form a rostral to caudal somatotopy in the dorsal part of the nucleus (FIG. 4A; see also Ref. 3). In addition, the output channel to the arm representation of the ventral premotor area (PMv)[6] is located in the middle third of the dorsal portion of the dentate map, adjacent to the output channel

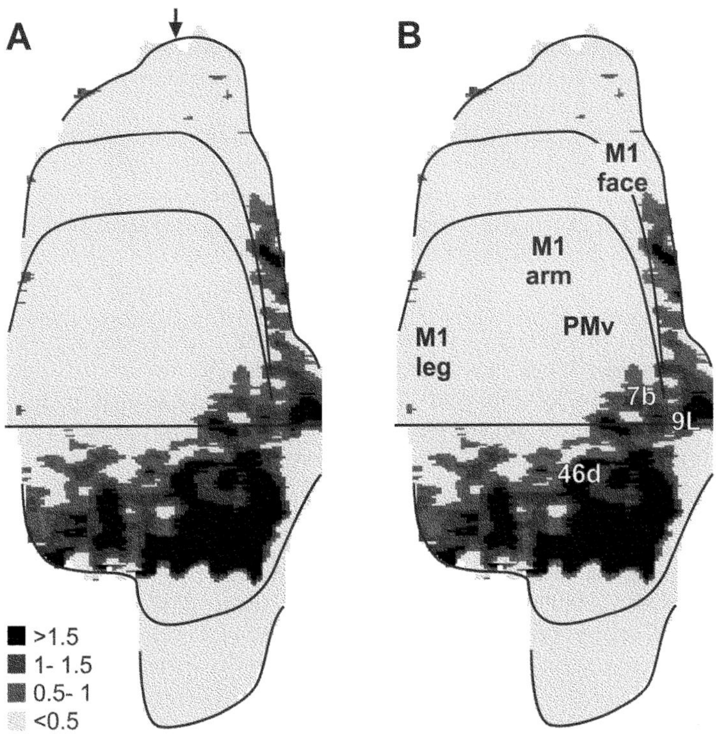

FIGURE 6. Unfolded map of the dentate after immunoreaction for the 8B3 antibody. The intensity of labeling was contoured into four grayscale levels. From the most intense staining (darkest) to the least intense staining (lightest) the four levels include: (1) intensities of >1.5 standard deviations (SDs) darker than the mean; (2) intensities 1.5–1.0 SDs darker than the mean; (3) intensities 1.0–0.5 SDs darker than the mean; and (4) the intensities 0.5–0 SDs darker than the mean as well as all intensities brighter than the mean. The *arrow* indicates the level of the section appearing in FIGURE 5.

to the arm representation of M1. Thus, the output channel directed at the arm representation in the PMv appears to align with the output channel that projected to the arm representation in M1. This observation raises the possibility that the dorsal motor domain of the dentate contains one integrated body map wherein each body segment (e.g., leg, arm, face) is represented as a mosaic of output channels to M1 and the various premotor areas. We plan to test this hypothesis in future studies.

In contrast, the nonmotor domain is located in the ventral portion of the dentate map. It contains output channels that project to cortical areas concerned with aspects of cognition and visuospatial function. Within the nonmotor domain, distinct output channels project to prefrontal and posterior parietal cortical areas. The output channels to these different areas display little evidence of overlap.

The division of the dentate into separate motor and nonmotor domains is strongly supported by the results of staining with the monoclonal antibody 8B3 (FIG. 6B). Dentate output channels that project to areas 9, 46, and 7b lie within the ventral region of the nucleus that has dense staining with 8B3. In contrast, dentate output channels that project to M1 and the PMv lie within the dorsal region of the nucleus that has light to moderate staining with 8B3. Thus, 8B3 appears to "recognize" the nonmotor domain of the nucleus.

Additional support for the separation of the dentate into motor and nonmotor domains comes from the results of neuron recording in awake trained monkeys. Neurons with activity that is well related to limb movements are concentrated in the dorsal part of the dentate (for example, see Refs. 23–25). In contrast, the particular features of behavior that activate neurons in the ventral part of the dentate have not been determined, but these neurons generally do not show changes in activity during simple motor tasks.

All of the dentate output channels that project to the cortex appear to be part of cerebro-cerebellar "loops." For each cerebral cortical area that we have examined, those that receive projections from the dentate also return projections to the cerebellum via efferents to the pontine nuclei (i.e., FIG. 4B, shaded region).[26–28] In contrast, cortical areas that do not appear to be the target of dentate output (e.g., the ventral portion of area 46, the lateral portion of area 12, and area TE in inferotemporal cortex [for discussion see Refs. 6 and 7]) fail to return projections to the cerebellum. These observations suggest that the interconnections between the cerebellum and the cerebral cortex can be characterized as an array of closed-loop circuits.[29] The cerebellum receives input from a number of higher-order nonmotor areas such as regions of anterior and posterior cingulate cortex.[26–28] If closed-loop circuits form the underlying unit of cerebro-cerebellar communication, then the higher-order cortical areas that project to the cerebellum are also likely to be the target of cerebellar output.

To understand how the output of the dentate maps upon the ventrolateral thalamus, we have displayed the origin of efferents to different cortical areas on a single section through the thalamus (FIG. 4C). Labels are placed at sites in the ventrolateral thalamus that receive cerebellar efferents and project to the cortical areas indicated. Although this map is a considerable simplification of the intricate three-dimensional nature of thalamocortical projections, it has heuristic value. The rostral-to-caudal topography in the dentate appears to become a lateral to medial topography in the thalamus. This is especially evident within the motor domain of the dentate where the rostral-leg to caudal-face somatotopy (Fig. 4A) is translated into a lateral-leg to medial-face somatotopy in the ventrolateral thalamus (FIG. 4C).[30–34] Similarly, the rostro-caudal topographic relationships between the output channels in the ventral, nonmotor domain appear to be preserved in the lateral to medial dimension within the VLcc of the thalamus (FIG. 4A, 4C, VLcc according to Holsapple *et al.*[35]).

The *dorsal to ventral* topography in the dentate is inverted to form a *ventral to dorsal* topography in the thalamus (note that the thalamus is displayed ***upside down*** to facilitate this comparison; FIG. 4C). As observed in the dentate, the origin of efferents to motor and nonmotor areas are located in distinct territories in the ventrolateral thalamus (FIG. 4A, 4C). Efferents from the motor domain in the dorsal portion of the dentate target the classic motor subdivisions in the ventrolateral thalamus (e.g., VPLo and area X according to Olszewski[36]). In contrast, efferents from the ventral, nonmotor domain terminate more dorsally in portions of VLcc (FIG. 4C). Thus, in both the dentate and VLcc, projections to widely dispersed areas of prefrontal and posterior parietal cortex have been collected together.

The dominant view of cerebellar function over the last century has been that it is concerned with the coordination and control of motor activity.[37] Consistent with that view, we found that the majority of the dentate provides efferents to traditional motor areas of the cerebral cortex. Nevertheless, our map shows that a significant portion of the output from the cerebellum is directed to nonmotor areas of the cerebral cortex.[3,6–8] As a consequence, it is abundantly clear that the anatomical substrate exists for cerebellar output to provide substantial influence over cognitive and visuospatial functions.[7,8,29] Indeed, there is evidence from human and monkey studies that ventral portions of dentate are activated during tasks that involve short-term working memory, rule-based learning and higher executive functions.[38–40] Since the proportion of dentate output that targets cortical regions involved in cognitive functions may be larger in humans than monkeys,[41,42] the dentate's influence over nonmotor functions may be enhanced in the human (for example, see Ref. 43).

In conclusion, our unfolded map of the dentate, like other flattened maps, provides an anatomical foundation for further investigation into the functional role of the dentate and cerebellum. The presence of separate motor and nonmotor domains within the monkey dentate suggests that similar domains may exist in the human. The fine mosaic of output channels combined with the complex three-dimensional shape of the dentate will make parsing out the contribution of individual portions of the nucleus quite difficult. Perhaps unfolding the human dentate and identifying its motor and nonmotor domains will lead to new insights into the topography of function that exists within this nucleus.

ACKNOWLEDGMENTS

We thank M. Page for the development of computer programs and W. Burnette, M. Evans, S. Fitzpatrick, K. Hughes and M. O'Malley-Davis for their expert technical support. We also thank Drs. D.I. Bernstein (Gamble Institute of Medical Research, Cincinnati, OH) and R.D. Dix (Jones Eye Institute, Little Rock, AR) for supplying HSV1 and Dr. A.F. Pimenta for supplying the 8B3 antibody and technical advice. This work was supported by the Veterans Affairs Medical Research Service and by US Public Health Service Grants NS24328 (PLS), MH56661 (PLS) and MH45156 (PLS).

REFERENCES

1. KEMP, J.M. & T.P.S. POWELL. 1971. The connexions of the striatum and globus pallidus: synthesis and speculation. Phil. Trans. R. Soc. Lond. B **262:** 441–457.
2. ASANUMA, C., W.T. THACH & E.G. JONES. 1983. Distribution of cerebellar terminations and their relation to other afferent terminations in the ventral lateral thalamic region of the monkey. Brain Res. Rev. **5:** 237–265.
3. HOOVER, J.E., & P.L. STRICK. 1999. The organization of cerebellar and basal ganglia outputs to primary motor cortex as revealed by retrograde transneuronal transport of herpes simplex virus type 1. J. Neurosci. **19:** 1446–1463.
4. LYNCH, J.C., J.E. HOOVER & P.L. STRICK. 1994. Input to the primate frontal eye field from the substantia nigra, superior colliculus, and dentate nucleus demonstrated by transneuronal transport. Exp. Brain Res. **100:** 181–186.
5. MIDDLETON, F.A, & P.L. STRICK. 1994. Anatomical evidence for cerebellar and basal ganglia involvement in higher cognitive function. Science **266:** 458–461.

6. MIDDLETON, F.A, & P.L. STRICK. 1996. New concepts regarding the organization of basal ganglia and cerebellar output. *In* Integrative and Molecular Approach to Brain Function. M. Ito & Y. Miyashita, Eds.: 253–269. Elsevier Science. Amsterdam.
7. MIDDLETON, F.A., & P.L. STRICK. 2001. Cerebellar "projections" to the prefrontal cortex of the primate. J. Neurosci. **21:** 700–712.
8. CLOWER, D.M., R.A. WEST, J.C. LYNCH & P.L. STRICK. 2001. The inferior parietal lobule is the target of output from the superior colliculus, hippocampus and cerebellum. J. Neurosci. **21:** 6283–6291.
9. DUM, R.P. & P.L. STRICK. 2002. An unfolded map of the cerebellar dentate nucleus and its projections to the cerebral cortex. J. Neurophysiol. (Nov. 6, 2002), 10.1152/jn.00626.2002.
10. PIMENTA, A.F., P.L. STRICK & P.R. LEVITT. 2001. A novel proteoglycan epitope is expressed in functionally discrete patterns in primate cortical and subcortical regions. J. Comp. Neurol. **430:** 369–388.
11. BRODAL, A. 1940. Experimentelle Untersuchungen über die olivocerebellare Lokalisation. Z. Gesamte Neurol. Psychiat. **169:** 1–153.
12. LARSELL, O. 1970. The comparative anatomy and histology of the cerebellum from monotremes through apes. The University of Minnesota Press. Minneapolis, MN.
13. VAN ESSEN, D.C. & J.H. MAUNSELL. 1980. Two-dimensional maps of the cerebral cortex. J. Comp. Neurol. **191:** 255–281.
14. DUM, R.P. & P.L. STRICK. 1991. The origin of corticospinal projections from the premotor areas in the frontal lobe. J. Neurosci. **11:** 667–689.
15. VAN ESSEN, D.C., J.W. LEWIS, H.A. DRURY, *et al.* 2001. Mapping visual cortex in monkeys and humans using surface-based atlases. Vision Res. **41:** 1359–1378.
16. DUM, R.P. & P.L. STRICK. 1996. Spinal cord terminations of the medial wall motor areas in macaque monkeys. J. Neurosci. **16:** 6513–6525.
17. STRICK P.L, & J.P. CARD. 1992. Transneuronal mapping of neural circuits with alpha herpesviruses. *In* Experimental Neuroanatomy: A Practical Approach, J.P. Bolam, Ed.: 81–101, Oxford Univerity Press. Oxford.
18. ALLEN, G.I., P.F. GILBERT & T.C. YIN. 1978. Convergence of cerebral inputs onto dentate neurons in monkey. Exp. Brain Res. **32:** 151–170.
19. STANTON, G.B. 1980. Topographical organization of ascending cerebellar projections from the dentate and interposed nuclei in *Macaca mulatta*: an anterograde degeneration study. J. Comp. Neurol. **190:** 699–731.
20. RISPAL-PADEL, L., F. CICIRATA & C. PONS. 1982. Cerebellar nuclear topography of simple and synergistic movements in the alert baboon (*Papio papio*). Exp. Brain. Res. **47:** 365–380.
21. ASANUMA, C., W.T. THACH & E.G. JONES. 1983. Anatomical evidence for segregated focal groupings of efferent cells and their terminal ramifications in the cerebellothalamic pathway of the monkey. Brain Res. Rev. **5:** 267–297.
22. THACH, W.T., J.G. PERRY, S.A. KANE & H.P. GOODKIN. 1993. Cerebellar nuclei: rapid alternating movement, motor somatotopy, and a mechanism for the control of muscle synergy. Rev. Neurol. (Paris) **149:** 607–628.
23. THACH, W.T. 1978. Correlation of neural discharge with pattern and force of muscular activity, joint position, and direction of intended next movement in motor cortex and cerebellum. J. Neurophysiol. **41:** 654–676.
24. WETTS, R., J.F. KALASKA & A.M. SMITH. 1985. Cerebellar nuclear cell activity during antagonist cocontraction and reciprocal inhibition of forearm muscles. J. Neurophysiol. **54:** 231–244.
25. VAN KAN, P.L.E., J.C. HOUK & A.L. GIBSON. 1993. Output organization of intermediate cerebellum of the monkey. J. Neurophysiol. **69:** 57–73.
26. BRODAL, P. 1978. The corticopontine projection in the rhesus monkey: origin and principles of organization. Brain **101:** 251–283.
27. GLICKSTEIN, M., J.G. MAY, 3D & B.E. MERCIER. 1985. Corticopontine projection in the macaque: the distribution of labeled cortical cells after large injections of horseradish peroxidase in the pontine nuclei. J. Comp. Neurol. **235:** 343–359.
28. SCHMAHMANN, J.D. 1998. Dysmetria of thought: clinical consequences of cerebellar dysfunction on cognition and affect. Trends Cognit. Sci. **2:** 362–371.

29. MIDDLETON, F.A., & P.L. STRICK. 1998. Cerebellar output: motor and cognitive channels. Trends Cognit. Sci. **2:** 348–354.
30. STRICK, P.L. 1976. Anatomical analysis of ventrolateral thalamic input to the primate motor cortex arm area. J. Neurophysiol. **39:** 1020–1031.
31. STRICK, P.L. 1976. Activity of ventrolateral thalamic neurons during movement. J. Neurophysiol. **39:** 1032–1044.
32. KIEVIT, J. & H.G. KUYPERS. 1977. Organization of the thalamo-cortical connexions to the frontal lobe in the rhesus monkey. Exp. Brain Res. **29:** 299–322.
33. MIYATA, M., & K. SASAKI. 1983. HRP studies on thalamocortical neurons related to the cerebellocerebral projection in the monkey. Brain Res. **274:** 213–224.
34. VITEK, J.L., J. ASHE, M.R. DELONG, & G.E. ALEXANDER. 1994. Physiologic properties and somatotopic organization of the primate motor thalamus. J. Neurophysiol. **71:** 1498–1513.
35. HOLSAPPLE, J.W., J.B. PRESTON & P.L. STRICK. 1991. The origin of thalamic inputs to the "hand" representation in the primary motor cortex. J. Neurosci. **11:** 2644–2654.
36. OSZEWSKI, J. 1952. The thalamus of the *Macaca mulatta*: an atlas for use with the stereotaxic instrument. S Karger AG. Basel.
37. BROOKS, V.B. & W.T. THACH. 1981. Cerebellar control of posture and movement. *In* Handbook of Physiology, Section 1: The Nervous System, Vol. 2, Motor Control. 877–946. Am. Physiol. Soc. Bethesda, MD.
38. MUSHIAKE, H. & P.L. STRICK. 1993. Preferential activity of dentate neurons during limb movements guided by vision. J. Neurophysiol. **70:** 2660–2664.
39. KIM, S.-G., K. UGURBIL & P.L. STRICK. 1994. Activation of cerebellar output nucleus during cognitive processing. Science **265:** 949–951.
40. JUEPTNER, M., C.D. FRITH, D.J. BROOKS, et al. 1997. Anatomy of motor learning. II. Subcortical structures and learning by trial and error. J. Neurophysiol. 77: 1325–1337.
41. LEINER, H.C., A.L. LEINER & R.S. DOW. 1991. The human cerebro-cerebellar system: its computing, cognitive, and language skills. Behav. Brain Res. **44:** 113–128.
42. MATANO, S. 2001. Brief communication: proportions of the ventral half of the cerebellar dentate nucleus in humans and great apes. Am. J. Phys. Anthropol. **114:** 163–165.
43. SCHMAHMANN, J.D. & J.C. SHERMAN. 1998. The cerebellar cognitive affective syndrome. Brain **121:** 561–579.

The Cerebellum and Event Timing

RICHARD B. IVRY,[a,b] REBECCA M. SPENCER,[c] HOWARD N. ZELAZNIK,[c] AND JÖRN DIEDRICHSEN[c]

[a]*Department of Psychology, University of California, Berkeley, California, USA*

[b]*Veterans Administration Medical Center, Martinez, California, USA*

[c]*Department of Health and Kinesiology, Purdue University Integrative Program in Neuroscience, West Lafayette, Indiana, USA*

ABSTRACT: Damage to the cerebellum disrupts performance on a range of tasks that require precise timing including the production of skilled movements, eyeblink conditioning, and perceptual tasks such as duration discrimination. We hypothesize that such tasks involve event timing, a form of representation in which the temporal goals are explicitly represented. For example, during finger tapping, the goal to produce evenly paced intervals invokes an explicit temporal representation of the time between successive contact points with the tapping surface. In contrast, timing can be an emergent property in other actions, reflecting temporal consistencies that arise through the control of other movement parameters. One example is continuous circle drawing, a task in which temporal consistency can be achieved by maintaining a constant angular velocity or minimizing higher-order derivatives (e.g., jerk). Temporal consistency on event and emergent timing tasks is not correlated and patients with cerebellar damage show no increase in temporal variability during continuous circle drawing. While the cerebellum likely contributes to performance of a wide range of skilled behaviors, it appears to be especially important when the tasks entail event timing.

KEYWORDS: cerebellum; motor control; cognition; event timing; emergent timing

INTRODUCTION

We address two issues concerning the role of the cerebellum in motor control and, more generally, cognition. First, we review the hypothesis that the cerebellum can be characterized as an internal timing system. This hypothesis offers a functional account of how the cerebellum contributes to sensorimotor coordination, consistent with the clinical literature that has stressed the importance of this subcortical structure in the regulation of the temporal aspects of movement,[1,2] and suggests a general framework for how to conceptualize the operation of specific neural regions. Historically, behavioral neurology has tended to organize neural regions in terms of task domains. In this view, the cerebellum is considered part of the motor system, a prominent subcortical structure that can influence the output of both the extrapyramidal

Address for correspondence: Richard B. Ivry, University of California, Berkeley, 3210 Tolman Hall #1650, Berkeley, CA 94720. Voice: 510-642-0135; fax: 510-642-5293.
ivry@socrates.berkeley.edu

Ann. N.Y. Acad. Sci. 978: 302–317 (2002). © 2002 New York Academy of Sciences.

and pyramidal descending pathways. An alternative to this task-based division has its roots in the information processing approach that is central to cognitive psychology and, more recently, cognitive neuroscience. The emphasis of this approach has been to describe mental operations in terms of the representation and transformation of information. These computations are likely to be invoked across task domains, and recruited whenever their specialized operations are required.

Second, we turn to more recent studies from our laboratories that specify constraints on the functional domain of the cerebellum. Temporal regularities may be manifest in a wide range of behaviors. However, these regularities do not mandate an explicit representation of these temporal phenomena and, correspondingly, that timing is directly controlled. Indeed, we will argue for a fundamental distinction between tasks in which timing is explicitly represented and those in which timing is emergent. We review evidence consistent with this hypothesis and show that the cerebellum is especially critical for tasks that involve explicit temporal representations, situations that we describe as involving event timing.

CEREBELLAR TIMING IN MOVEMENT, LEARNING, AND PERCEPTION

The cerebellar timing hypothesis is intended to provide a broad characterization of the computational basis of cerebellar function. Precise timing is important for the production of skilled movements but is not the only computation required for skilled movement; thus damage to many other brain regions can disrupt skilled behavior. John Hore and colleagues (this volume), in a series of elegant studies of ball throwing, asked how the multiple actions across different joints can be efficiently and flexibly coordinated. Their results show that the timing of the opening of the fingers, required to release the gripped ball, is crucial for spatial accuracy. Moreover, this timing appears to be preprogrammed (i.e., open-loop), at least in skilled throwers. The lack of development of this temporal control can be seen in the poorly aimed throws of healthy individuals when using their non-dominant hand.[3] A breakdown in the temporal relationship between the forward movement of the limb and the ball release at the wrist is a cardinal feature of cerebellar damage.[4]

To provide a more direct test of the cerebellar timing hypothesis, we have generally opted to employ tasks that would, at least on the surface, appear to place a high demand on temporal control. Many of these studies use a repetitive tapping task in which the trial begins with the presentation of an auditory metronome indicating the target interval (FIG. 1A). Participants tap in synchrony with the tones for a short period of time and then are required to continue tapping when the tones end, trying to maintain the specified rate. Internal timing control is assessed by the overall variability of performance during this continuation phase, as well as by the application of a model that uses the covariance function of the time series data to provide estimates of two component sources of variability, one associated with motor implementation processes and a second associated with central processes including an internal clock (see Ref. 5). Patients with cerebellar lesions are impaired on this simple tapping task.[6] Interestingly, the increased implementation variability is associated with lesions of the medial cerebellum; in contrast, patients with lesions of the lateral cerebellar hemispheres exhibit a prominent increase in central variability.[7,8] This

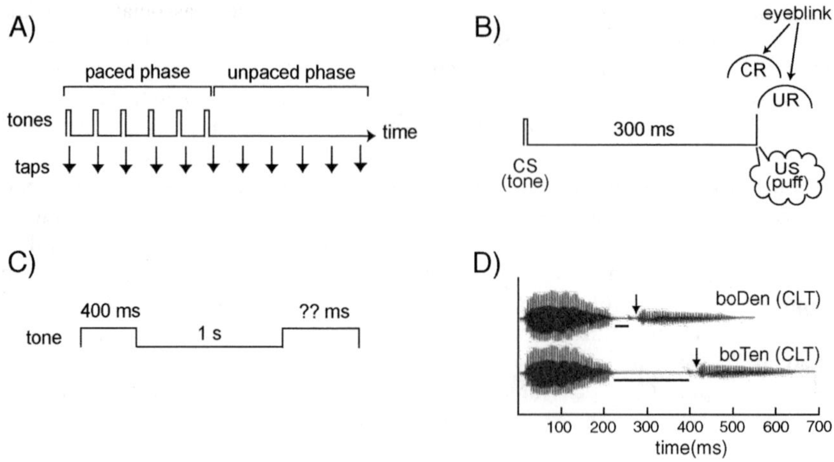

FIGURE 1. Schematic depiction of event structure in four tasks associated with timing functions of the cerebellum. (**A**) Repetitive tapping task. Participants are asked to produce series of evenly spaced responses. The primary dependent measure is the variability of intervals during unpaced phase. (**B**) Eyeblink conditioning. After repeated pairing of CS (conditioned stimulus) and US (unconditioned stimulus), animal learns to produce CR (conditioned respose) in anticipation of the US. (**C**) Duration discrimination. Participants judge whether comparison interval is shorter or longer than standard interval. A difference threshold is determined as measure of temporal acuity. (**D**) Speech perception. Perception of voiced or voiceless medial consonant is a function of duration of the silent interval between the two syllables [CLT (closure time)]. (Adapted from Ackermann *et al.*[17]).

dissociation is in accord with the anatomical projections of the cerebellum, given that the output of medial regions primarily targets descending motor pathways while the output from the hemispheric region influences the prefrontal, premotor, and motor cortex via thalamocortico projections.[9–11]

The timing functions of the cerebellum are not limited to motor control. A substantial literature has emerged over the past two decades implicating the cerebellum in simple forms of sensorimotor learning. For example, in eyeblink conditioning (FIG. 1B), an animal can learn to blink in anticipation of an aversive stimulus (e.g., an airpuff to the eyelid) following repeated pairings with a neutral stimulus (e.g., a tone). Such learning is only adaptive if the animal is able to learn the precise timing between the conditioned and unconditioned stimuli. Thus, the fundamental association is not only between two stimulus events, but also the precise temporal relationship of these events. Lesions of the cerebellum produce profound impairments on such forms of conditioning in a range of species, including rabbits, rats, turtles, and humans. While the exact locus of plasticity within the cerebellum remains contentious (see Yeo, this volume), the cerebellar cortex appears to play a critical role in regulating the fine timing of the conditioned response.[12] Detailed and highly constrained computational models of this task have been developed by a number of theorists (e.g., Refs. 13 and 14). Interestingly, these models have incorporated a range of mechanisms to allow for real-time representations, a feature that is typically absent in models of other neural learning systems.

Perceptual tasks that require precise timing also have been associated with the cerebellum (FIG. 1C and D). Patients with cerebellar lesions are impaired when asked to judge the duration of intervals in the range of 400 to 4 s.[6,15,16] These same patients do not exhibit comparable deficits when judging other properties of the same stimuli such as their intensity.

A more radical demonstration of the cerebellar role in temporal processing comes from the domain of speech perception. Most speech contrasts are signaled by a variety of cues. For example, the words "BOP" and "POP" can be distinguished on the basis of redundant cues: spectral differences exist in that the acoustic information is more dispersed at release for the /p/ sound. Temporal differences also exist in that energy at the fundamental frequency, generated by the periodic oscillations of the vocal cords, is delayed for the /p/. However, these sources of information can be isolated. For example, the temporal cues for "LABEL" and "LADLE" are essentially identical, with the distinction between the two sounds conveyed by the acoustic changes created by the way in which the airflow is occluded prior to the onset of the medial consonant. In contrast, words such as "BIDDEN" and "BITTEN" can be distinguished solely on the basis of a temporal cue, the period of silence at the time of occlusion. Patients with bilateral cerebellar degeneration are unable to discriminate between speech contrasts that involve purely temporal cues (FIG. 1D).[17] A recent functional imaging study has found that such sounds produce increased activation in only the cerebellum and inferior left frontal gyrus compared to speech perception conditions in which similar word contrasts required the analysis of spectral cues.[18] Perception of the latter sounds was associated with greater activation in the auditory regions of the superior temporal lobe. Thus, even a task as specialized as speech perception appears to engage a distributed network of processors. Presumably, these processors are invoked when their specialized computations, such as timing, are required.

While our emphasis has been on the timing functions of the cerebellum, it is important to note that other neural regions have been associated with the operation of an internal clock. These include the basal ganglia and regions of the frontal cortex, especially in the right hemisphere (reviewed in Ref. 19). It is possible that an amodal property such as time may be represented in a distributed or equipotential manner. For example, temporal codes may be derived from relatively slow physiological processes that operate across the brain.[20] At present, discrepant results in both the neuropsychological and neuroimaging literatures fail to provide a clear picture on this issue. At least some of the confusion here surely reflects the fact that even tasks as simple as repetitive tapping or duration perception involve many computations, and our analytic skills for isolating specific operations remain limited.

EVENT VS. EMERGENT TIMING

It is also possible that tasks may vary in terms of their requirements for a precise temporal representation or that the form in which temporal information is represented may vary as a function of task requirements. The Ackermann *et al.*[17] study emphasizes this point. The cerebellar patients performed normally in categorizing speech sounds that could be identified on the basis of either spectral or temporal cues. However, when only a temporal cue was available, their performance fell to near chance, reflecting the stringent demands placed on an internal timing system in this condition. Although there exist temporal regularities in the former condition, the patients were able to make the discriminations, presumably relying on computations involving the non-temporal cues.

Even when a task exhibits temporal regularities, it cannot be concluded a priori that these regularities will be represented in a direct manner. Consider the situation of waiting for your morning toast. Assuming that the heating element operates in a reliable manner, the amount of time required for the toast to reach the right level of brownness is relatively invariant from day to day. One could anticipate this moment by measuring the passage of time. Whether the cerebellar timing system is involved in such measurements is an open question; to date there have been few studies on the role of this structure in intervals spanning up to a minute. It is also possible that anticipating the opening of the toaster is not based on the ticking of an internal timer. From habit you might know that the toast will be ready just as soon as you take the butter out of the refrigerator and retrieve a plate and knife from the cupboard. Your behavior demonstrates a temporal consistency, yet there is no need to posit that a dedicated timing system is regulating all of these actions. Rather, the timing of these events is what we will call "emergent." Timing in this case may be highly subject to the vagaries of attention and arousal (e.g., Refs. 21 and 22). For example, if you are distracted by a particularly interesting story on the radio, you may be surprised to see the toast ready earlier than expected, perhaps inferring that the toaster was operating at a faster speed than normal.

Researchers of motor control have also stressed that temporal properties of movements can arise in an emergent manner, that is, without positing the explicit control of timing information.[23,24] For example, during bimanual circle drawing, temporal variability is much greater when the two hands follow parallel trajectories compared to when they follow mirror-symmetric trajectories.[25,26] This difference is not attributed to different control processes regulating the timing of the two types of movements. Rather, the increased timing variability is hypothesized to result from interactions created by the less compatible spatial requirements in the parallel condition. The temporal perturbations here, although regular and predictable, are emergent properties.

The idea that temporal properties differ across movement tasks is made clear in a series of studies reported by Zelaznik and colleagues.[27] Previous correlational studies had shown that the consistency of an internal timing system could account for a significant proportion of the variance between individuals in their performance on a range of motor and non-motor tasks.[28] Indeed, this correlational work had laid the groundwork for the studies of motor and perceptual timing in cerebellar patients. However, Zelaznik and colleagues failed to find such a dependency across a set of movement tasks. Across all of the tasks, participants were required to produce repetitive movements by tapping or drawing circles in different conditions. Each type of movement was performed at two different rates. A correlation matrix was constructed, using the normalized standard deviation as the primary indicator of temporal variability. Surprisingly, this matrix revealed that correlations were high across rates within a given task (e.g., tapping at 550 ms correlated with tapping at 800 ms), but were consistently low between tasks, even when rate was held constant (e.g., tapping and circle drawing failed to correlate when both were performed at 800 ms). Furthermore, correlations between tapping and circle drawing were low when both tasks were performed at the subjects' preferred rates.[29]

A number of differences exist between the movement tasks that might account for these null results. First, the movements involve very different sets of joints: while tapping involved flexion/extension of the index finger, circle drawing required movements about the shoulder and elbow. However, Keele *et al.*[28] had reported significant correlations between finger and foot tapping. Second, the spatial demands associated with the tasks were quite different. For example, tapping involves motion along a single axis and the only spatial requirement is that the finger touch the table in each cycle. In contrast, circle drawing requires that the participants maintain a particular spatial trajectory throughout the movement cycle. Third, the tasks appear

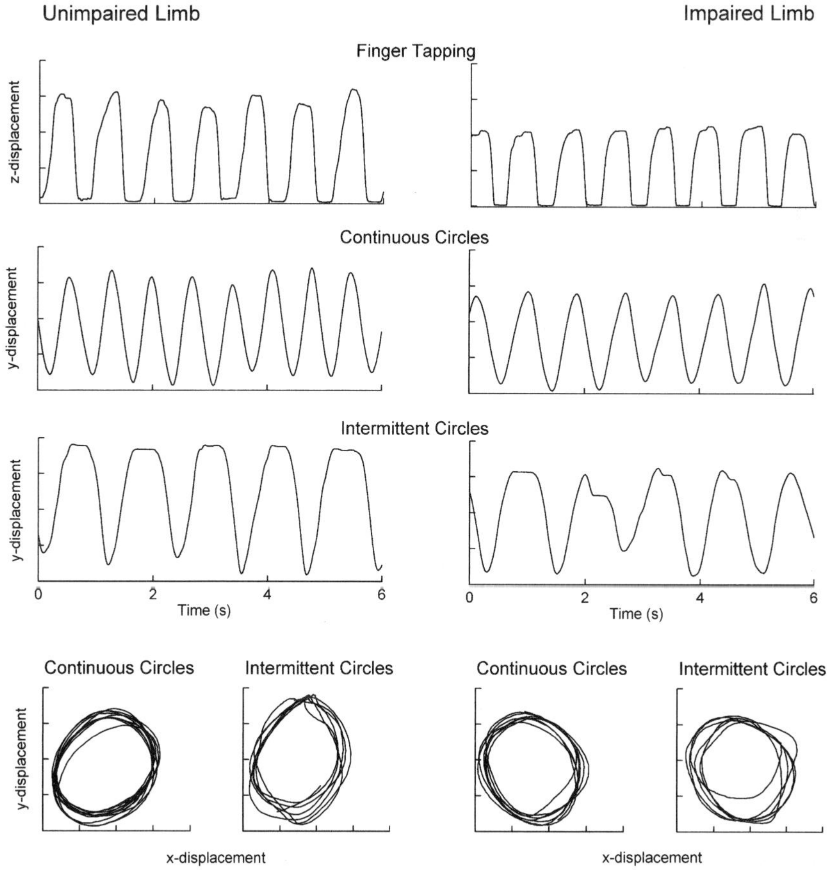

FIGURE 2. Kinematic traces from the performance of a patient with a unilateral cerebellar lesion on three repetitive movement tasks—finger tapping, continuous circle drawing, and intermittent circle drawing. Participants are instructed to insert a pause between each cycle in the intermittent drawing task. Note that a pause is introduced spontaneously during finger tapping at the point of maximum extension (*top of trace*). The pause at the point of maximum flexion indicates time during which the finger was in contact with the table surface. (Adapted from Spencer *et al.*[40])

to differ in terms of how they demarcate successive cycles. During finger tapping, each movement is separated from the next one by a small pause at the point of maximal extension. Thus, the series of movements have a discrete aspect to them, with certain points such as the onset of each downstroke or the contact with the table being especially salient (FIG. 2). The drawing tasks used by Robertson *et al.*[27] are of a more continuous nature. Perhaps such differences are also reflected in the control operations. For example, models of tapping have embodied discrete notions such as singular points at which central timing commands are directed to the periphery, with much of the operation being open-loop.[30–33] In contrast, models designed to account for temporal regularities during the drawing of curved shapes have focused on more continuous variables such as angular velocity (e.g., Refs. 34 and 35).

We designed a study to compare these hypotheses.[37] Two of the tasks had been included in the earlier studies, finger tapping and continuous circle drawing. The third task, intermittent circle drawing was designed to be a hybrid. For this task, the participants were instructed to divide each cycle into two parts, a movement phase in which they were to produce a complete circle, and a pause phase in which they were to remain at one position. The pitch of the pacing signal alternated, and the participants were instructed to begin each movement phase with the high-pitched tone and end this phase with the onset of the low-pitched tone, thus attempting to alternate between a 400-ms movement and a 400-ms pause. Spatial and temporal measures were obtained for all three tasks, with the focus of our analysis on the variability of cycle duration.

The intermittent and continuous circle drawing tasks involve similar joint movements and spatial demands. If either of these factors were critical in terms of the underlying sources of temporal variability, we should observe a high interindividual correlation between these two tasks, and performance on neither task would correlate with tapping. On the other hand, if the key variable centers on whether the tasks are conceptualized as composed of continuous movements or a series of discrete events (e.g., tap or circle onset), then performance on the tapping and intermittent circle drawing tasks would be correlated. The results clearly favored the latter hypothesis. The correlation between the tapping and intermittent circle tasks was .50. In comparison, the correlations between the continuous circle task and the other two tasks were below .30. Performance on the two circling tasks did correlate with one another on a measure of spatial consistency. Nonetheless, the surprising result was that on the measure of temporal variability, these two circle drawing tasks were only weakly related whereas a strong correlation was found between temporal variability on tapping and intermittent circle drawing.

In the experiment described above, the target cycle duration for all three tasks had been 800 ms. Given that the movement phase of the intermittent circling task was completed in approximately half of this time, the movements were performed at a considerably higher velocity than in the continuous circling task. A second experiment was conducted to roughly equate movement velocity across these tasks.[37] The cycle duration was reduced to 400 ms in the continuous circling condition. It remained at 800 ms in the intermittent condition. In addition, we included a perception task, duration discrimination, to measure non-motor temporal acuity.

The results closely replicated that of the first experiment. Significant correlations were again observed between measures of temporal variability on the tapping and intermittent circle drawing tasks. Neither task correlated with temporal variability on the continuous circling task. Even more impressive, both temporal consistency during tapping and intermittent circling were correlated with temporal acuity on the perception task (values of .39 and .56, respectively). The correlation between the perception task and continuous circle drawing was −.15.

In sum, the results clearly point to a shared timing process involved in finger tapping, multijoint intermittent circle drawing, and time perception. While the positive correlations between these three tasks are impressive, it is equally important to note that this shared process is not generic. Temporal variability on the continuous circle drawing task was not related to these three tasks. We hypothesize that an explicit representation of temporal information is the commonality between the tapping, intermittent circle drawing, and duration discrimination tasks. We refer to this as event timing, to capture the idea that the actions or judgments are made with reference to this explicit temporal representation. In tapping, the representation may be of the desired time between successive downstrokes or taps.[38] Similarly, for participants to consistently maintain a target cycle duration in the intermittent drawing task, they must account for both the movement and pause time. As such, an efficient representation of the task goal would be to include an explicit representation of the time between the onsets of successive cycles. We also posit that judging the duration of an empty interval requires that the interval be explicitly represented and compared with a representation of an internalized standard interval.

In contrast to tasks that entail event timing, we suggest that temporal regularities can also occur without explicit temporal representations. We refer to these temporal properties as emergent or secondary to some other form of control. At present, we can only speculate as to what this type of control might be. For continuous circle drawing, one hypothesis would build on the idea that optimal performance occurs when angular velocity is held constant.[35] A related idea would be that control here would be based on operations that minimize discontinuities in the velocity profile (or higher-order derivatives such as jerk; see Ref. 39). This is not to ignore the fact that the participants are easily able to produce the circles to match the target duration defined by the metronome. We propose that at the beginning of each trial an explicit temporal representation is derived based on the intervals constructed by the beats of the metronome. However, this representation is then transformed into some other control parameter that allows the circles to be completed within the target duration. In this manner, timing during continuous circle drawing is emergent in that the goal is represented in terms of a target velocity or some other secondary form of control rather than explicitly defined temporal events. This form of control would, of course, fail to work for intermittent circling because, by definition, the task requires periods of zero velocity. In a similar manner, the salient points within the tapping cycle appear to favor an event-based representation rather than a representation that reflects some sort of continuous control signal.

EVENT TIMING AND THE CEREBELLUM

The correlational studies reviewed above provide compelling behavioral evidence that non-identical control processes are involved in the timing of repetitive movements that are either performed in a continuous or segmented manner. We have also investigated this dissociation in a series of neuropsychological studies.[40] In our

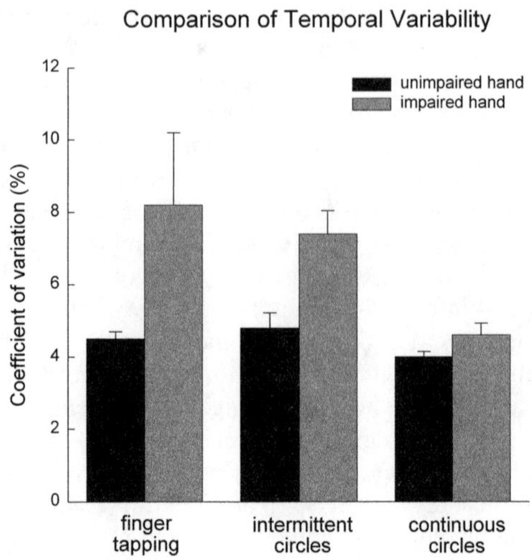

FIGURE 3. Temporal variability on the three movement tasks for a group of patients with unilateral cerebellar lesions. To compensate for differences in movement rate, the results are expressed in terms of the coefficient of variation (standard deviation of the intervals divided by the mean interval). (Adapted from Spencer *et al.*[40])

first study, patients with focal lesions of the cerebellum were tested on three tasks: finger tapping, continuous circle drawing, and intermittent circle drawing. The target cycle duration was fixed at 800 ms with the instructions in the intermittent circling task emphasizing that this interval should be divided into a movement phase and a pause phase. The patients performed each task twice, once with the ipsilesional, impaired limb and once with the contralesional, normal limb. In this manner, each person served as his or her own control. The patients were tested at least one year after their stroke or tumor resection surgery, at a point where their cerebellar symptoms were stable. In fact, some of the patients showed minimal sign of cerebellar dysfunction at the time of testing.

As our primary dependent variable, we used a normalized measure of temporal variability, dividing the standard deviation of the cycle durations by the mean cycle duration. Given that timing variability increases with cycle duration,[32,41] this measure allowed us to compare variability scores across tasks and hands even if the actual produced cycle durations were unequal. Replicating earlier findings,[7,8] the patients were more variable when tapping with their ipsilesional hand. A similar deficit was also observed on the intermittent circle drawing task. However, no difference was found on the continuous circle drawing task (FIG. 3).

We introduced two changes in a second experiment to further explore this dissociation. First, in addition to the patients with unilateral cerebellar lesions, we also tested patients with bilateral cerebellar degeneration and age-matched control participants. Second, the tapping and intermittent circling tasks were modified such that the intervals within a trial were produced one at a time with a random interval separating successive productions rather than as a continuous series. During the pacing phase, two sounds were presented, defining the target interval for either two taps or one circling cycle. During the continuation phase, a single tone indicated when the participants should produce the next response, trying to match the target duration for a single cycle. The results replicated what had been observed in the first experiment. The unilateral patients were selectively impaired on the single cycle circling and tapping tasks when using their ipsilesional limb. Similarly, the bilateral patients were more variable than the controls on these two tasks. Importantly, both patient groups performed comparable to the controls on the continuous circling drawing task.

The lack of a deficit on the continuous circle drawing task is quite striking. While the correlational work had suggested that this task involved different control processes, a priori, it seemed reasonable to expect that cerebellar dysfunction would also be manifest during circle drawing with the ipsilesional arm. Indeed, there were a number of reasons to expect that this task would place greater demands on the cerebellum than finger tapping. First, the average movement speed is much greater during circle drawing given that the circumference of the circle is about five times greater than the amplitude of tapping. Second, the circle drawing condition required more visuomotor control given the participants were provided with a visual template for this task. Third, circle drawing requires coordination across the shoulder and elbow joints. Such multijoint movements will generate interactional torques that are likely minimal during finger tapping. All of these variables—movement speed, visuomotor control, multijoint coordination, and interactional torques—have been shown to be sensitive to cerebellar dysfunction.[42] Yet the results show that the patients exhibited no impairment when the circles were produced in a continuous manner.

We believe that the distinction between event and emergent timing provides the most parsimonious account of these results. In this view, the cerebellum is essential for those tasks that involve event timing. It provides the explicit representations that specify the timing of these events: when the next tap should be produced or when the next movement cycle should begin during intermittent circle drawing. The null findings with respect to the continuous circling task suggest that the cerebellum is not a critical component of the control processes for tasks in which timing is emergent. At present, we have little to say about the neural mechanisms that control timing during continuous circle drawing. As proposed above, optimal performance here can be achieved by maintaining a constant angular velocity. From this, we would infer that the cerebellum plays, at most, a secondary role in the control of the relevant parameters for these continuous movements. We suspect that cortical mechanisms might be primary. However, it remains to be seen whether the reverse pattern of results would be obtained in studies with a different patient group: namely, a selective impairment on the continuous circle drawing task.

TEMPORAL COORDINATION DURING CONTINUOUS AND DISCRETE MOVEMENTS

The performance of callosotomy patients during bimanual movements provides converging evidence of a dissociation between subcortical and cortical systems in the temporal coordination of tasks that involve discrete movements and those that involve continuous movements. A number of studies have shown that the two limbs remain temporally coupled during such movements.[43–45] However, we recently tested three callosotomy patients on a version of the continuous circle drawing task, under conditions in which the participants were asked to move at their preferred speed or at maximal speed.[46] A second variable was based on the direction of the two hands. The hands either followed symmetric trajectories (e.g., one clockwise rotation, the other counterclockwise) or asymmetric trajectories (e.g., both clockwise). Control participants have been shown to exhibit less temporal stability, expressed as an increase in variability of cycling time and relative phase between the hands when the hands perform asymmetric movements, especially when performing at maximal speed.[26] This instability has been modeled as resulting from the interactions of coupled spatiotemporal oscillators. We found that the callosotomy patients did not show this difference between symmetric and asymmetric circling. More importantly, the patients exhibited a striking lack of temporal coupling during all circling conditions. Indeed, the two hands frequently moved at completely different frequencies. This suggests that in normal individuals the interaction between the two hands during continuous circle drawing involves intercortical processing mediated by callosal fibers.

The previous reports of persistent temporal coupling in callosotomy patients had all involved movements containing a discrete event, such as the contact with an object[44,45,47] or the on- and offset of movements while drawing line segments.[43] To directly test whether the inclusion of such discrete events is critical, we tested the patients on two modified "tapping" tasks.[46] In both, the movements were restricted to flexion and extension of the index fingers, and these movements were made in the air (e.g., no tactile feedback from contacting a surface). In one condition, the participants were instructed to make the movements continuously, with smooth transitions between flexion and extension. In the other condition, they were instructed to insert a pause prior to each flexion onset. This subtle manipulation produced a dramatic change in performance. Whereas the degree of temporal coupling was similar for control participants under both conditions, the callosotomy patients exhibited a threefold increase in the standard deviation of the phase differences in the continuous movement condition. Similar to what we had observed in circle drawing, the two hands exhibited extended epochs of temporal uncoupling (FIG. 4).

When considered in combination with the studies of cerebellar patients, a consistent story begins to emerge. The temporal control of continuous, rhythmic movements does not invoke the cerebellar timing system. The temporal regularities observed under such conditions are emergent; there is no explicit representation of timing signals once the initial temporal goal has been transformed into a different control parameter. Our working hypothesis is that the control of such movements is cortically mediated, perhaps through the operation of an oscillatory process that regulates smooth changes in spatial variables over time. During bimanual movements, these oscillators become coupled. We assume that these interactions involve cortically based representations and that interhemispheric communication is primarily dependent on the corpus callosum. Such a mechanism can also account for spatial compatibility effects that are observed during bimanual movements.[48]

In contrast, the temporal control of a repetitive series of discrete movements requires an explicit temporal representation. Psychologically, this representation specifies when particular events should occur or have occurred. Timing remains an explicit part of the representation of the task goal. We hypothesize that the cerebellum is specialized for providing this explicit form of temporal representation. In finger tapping, the cerebellum signals when the next response should occur. In other

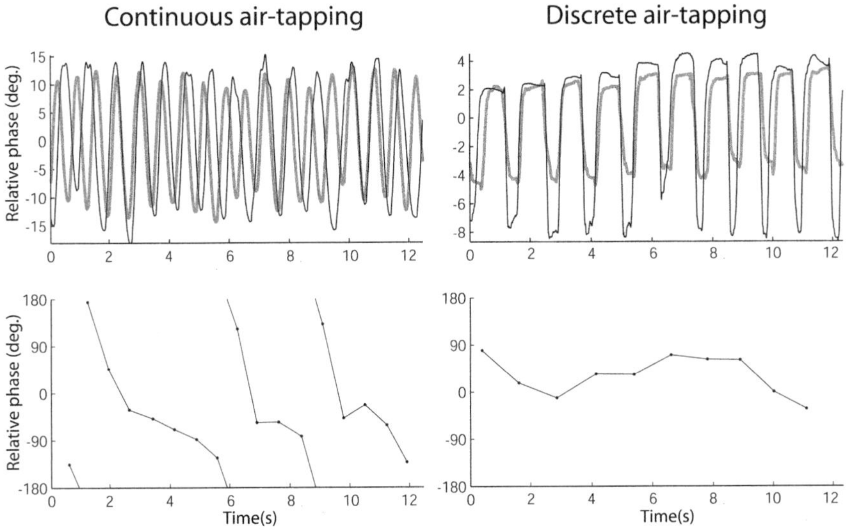

FIGURE 4. Kinematic traces produced by a callosotomy patient during two bimanual finger movement tasks. In the discrete task, the instructions emphasize that a brief pause should be made prior to each flexion phase. In the continuous task, the instructions emphasize that the movements should be made as smoothly as possible. The patient's movements are temporally uncoupled in the continuous condition but remain coupled in the discrete condition. (Adapted from Kennerley *et al.* [46])

work, we have proposed that independent temporal control signals for each finger are generated within the cerebellum during bimanual tapping, but that these signals are constrained to act through a common, subcortical gate. This gating process introduces a nonlinearity in the control process, retaining the event structure of the representation of these movements.[31]

SUMMARY

Our previous work has emphasized the contribution of the cerebellum in tasks that require a precise representation of temporal information. The cerebellar event-timing hypothesis specifies an important boundary condition for this functional account of the cerebellum, underscoring the observation that while temporal regularities may be observed in a wide range of behaviors, the use of an explicit representation of temporal information may be more limited.[49] In our work to date, we have compared two classes of repetitive movements, those which are performed in a continuous manner and those in which the successive actions are marked by some discrete event or boundary such as a finger tap or pause. We also believe the notion of event structure applies to other temporal processing tasks that have been linked to the cerebellum. For example, eyeblink conditioning requires the association of two salient events and the interval that separates them. Similarly, speech perception tasks that are dependent on the cerebellum are ones in which the only cue is the duration of a silent interval that separates two articulatory events.[17]

An obvious question is, how does one specify tasks that entail an event structure? The hypothesis would risk becoming circular if the defining criterion were based solely on determining whether or not the task presents difficulties for patients with cerebellar lesions. One way to address this concern is by looking for converging evidence such as in our studies with callosotomy patients, or with other methodologies. Computationally, one can ask whether the underlying representations embody the notion of an event structure. For example, Semjen and Ivry[50] have argued that limitations in bimanual tapping should be attributed to the manner in which the temporal goals are represented rather than to the dynamics of coupled oscillators. In support of this view, they observed that identical rhythmic distortions are found for uni- and bimanual tapping, reflecting a common event-based representation.

The event timing hypothesis is also likely to be relevant in evaluating the current debate about the relationship of timing and prediction. Representing when the next tap should occur or when an airpuff will be experienced are, of course, forms of prediction. It has been proposed that the fundamental property of cerebellar function is its operation as a predictive device. In sensorimotor control, this idea is captured by the formalism of internal, forward models.[51,52] More generally, this idea has been offered to account for putative cerebellar contributions to higher-level cognition such as language, problem solving, and attention.[53,54] For example, the cerebellum might contribute to problem solving by supporting internal simulations that test out possible solutions to see if the end result matches the goal. While the empirical evaluation of such ideas is limited at present, the event timing hypothesis offers a conservative starting point for assessing the role of the cerebellum in cognition. Prediction is ubiquitous. However, there is a difference between anticipating something in a generic sense and specifying when that event will occur.[55] We expect that, similar to motor control, the functional domain of the cerebellum in cognition will be limited to situations that require real-time predictions.

ACKNOWLEDGMENTS

This work was supported by NIH grants NS 30256, NS 1778, and DC 02527, and the National Science Foundation grant ECS-9873474.

REFERENCES

1. Hallett, M., B. Shahani & R. Young. 1975. EMG analysis of patients with cerebellar deficits. J. Neurol. Neurosurg. Psychiatry **38:** 1163–1169.
2. Holmes, G. 1939. The cerebellum of man. Brain **62:** 1–30.
3. Hore, J., S. Watts, D. Tweed & B. Miller. 1996. Overarm throws with the nondominant arm: kinematics of accuracy. J. Neurophysiol. **76:** 3693–3704.
4. Timman, D., S. Watts & J. Hore. 1999. Failure of cerebellar patients to time finger opening precisely causes ball high-low inaccuracy in overarm throws. J. Neurophysiol. **2:** 103–114.
5. Vorberg, D. & A. Wing. 1996. Modeling variability and dependence in timing. *In* Handbook of Perception and Action, Vol. 2: Motor Skills. H. Heuer & S.W. Keele, Eds.: 181–262. Academic Press. London.

6. IVRY, R.B. & S.W. KEELE. 1989. Timing functions of the cerebellum. J. Cognit. Neurosci. **1:** 136–152.
7. IVRY, R.B., S.W. KEELE & H.C. DIENER. 1988. Dissociation of the lateral and medial cerebellum in movement timing and movement execution. Exp. Brain Res. **73:** 167–180.
8. FRANZ, E.A., R.B. IVRY & L.L. HELMUTH. 1996. Reduced timing variability in patients with unilateral cerebellar lesions during bimanual movements. J. Cognit. Neurosci. **8:** 107–118.
9. ASANUMA, C., W.T. THACH & E. JONES. 1983. Anatomical evidence for segregated focal groupings of efferent cells and their ramifications in the cerebellothalamic pathway of the monkey. Brain Res. Rev. **5:** 267–298.
10. ASANUMA, C., W.T. THACH & E. JONES. 1983. Brainstem and spinal projections of the deep cerebellar nuclei in the monkey with observations on the brainstem projections of the dorsal column nuclei. Brain Res. Rev. **5:** 299–322.
11. MIDDLETON, F.A. & P.L. STRICK. 1997. Cerebellar output channels. Int. Rev. Neurobiol. **41:** 31–60.
12. PERRETT, S., B. RUIZ & M. MAUK. 1993. Cerebellar cortex lesions disrupt learning–dependent timing of conditioned eyelid responses. J. Neurosci. **13:** 1708–1718.
13. FIALA, J.C., S. GROSSBERG & D. BULLOCK. 1996. Metabotropic glutamate receptor activation in cerebellar Purkinje cells as substrate for adaptive timing of the classically conditioned eye-blink response. J. Neurosci. **16:** 3760–3774.
14. MEDINA, J.F., K.S. GARCIA, W.L. NORES, *et al.* 2000. Timing mechanisms in the cerebellum: testing predictions of a large-scale computer simulation. J. Neurosci. **15:** 5516–5525.
15. MANGELS, J.A., R.B. IVRY & N. SHIMIZU. 1998. Dissociable contributions of the prefrontal and neocerebellar cortex to time perception. Cognit. Brain Res. **7:** 15–39.
16. NICHELLI, P., D. ALWAY & J. GRAFMAN. 1996. Perceptual timing in cerebellar degeneration. Neuropsychologia **34**(9)**:** 863–871.
17. ACKERMANN, H., S. GRAEBER, I. HERTRICH & I. DAUM. 1997. Categorical speech perception in cerebellar disorders. Brain Lang. **60:** 323–331.
18. MATHIAK, K. *et al.* 2002. fMRI of cerebellar speech perception. J. Cognit. Neurosci. **14:** 902–912.
19. MANGELS, J.A. & R.B. IVRY. 2001. Time perception. *In* Handbook of Cognitive Neuropsychology: What Deficits Reveal about the Mind. B. Repp, Ed.: 467–493. Psychology Press. Philadelphia.
20. BUONOMANO, D.V. 2000. Decoding temporal information: a model based on short-term synaptic plasticity. J. Neurosci. **20:** 1129–1141.
21. BLOCK, R.A. & D. ZAKAY. 1996. Models of psychological time revisited. *In* Time and Mind. H. Helfrich, Ed.: 171–195. Hogrefe and Huber Publishers. Bern.
22. ORNSTEIN, R.E. 1969. On the Experience of Time. Penguin Books. Middlesex, England.
23. MICHAELS, C.F. & C. CARELLO. 1981. Direct Perception. Prentice Hall. Englewood Cliffs, NJ.
24. TURVEY, M.T. 1977. Preliminaries to a theory of action with reference to vision. *In* Perceiving, Acting, and Knowing. R. Bransford, Ed.: 211–265. Lawrence Erlbaum. Hillsdale, NJ.
25. CARSON, R.G., J. THOMAS, J.J. SUMMERS, *et al.* 1997. The dynamics of bimanual circle drawing. Q. J. Exp. Psychol. Sect. A Hum. Exp. Psychol. **50:** 664–683.
26. SEMJEN, A., J.J. SUMMERS & D. CATTAERT. 1995. Hand coordination in bimanual circle drawing. J. Exp. Psychol. Hum. Percept. Perform. **21:** 1139–1157.
27. ROBERTSON, S.D., H.N. ZELAZNIK, D. LANTERO, *et al.* 1999. Correlations for timing consistency among tapping and drawing tasks: evidence against a single timing process for motor control. J. Exp. Psychol. Hum. Percept. Perform. **25:** 1316–1330.
28. KEELE, S., R. POKORNY, D. CORCOS & R.B. IVRY. 1985. Do perception and motor production share common timing mechanisms? Acta Psychologia **60:** 173–193.
29. ZELAZNIK, H.N., R.M. SPENCER & J.G. DOFFIN. 2000. Temporal precision in tapping and circle drawing movements at preferred rates is not correlated: further evidence against timing as a general purpose ability. J. Mot. Behav. **32:** 193–199.

30. HELMUTH, L. & R. IVRY. 1996. When two hands are better than one: reduced timing variability during bimanual movements. J. Exp. Psychol. Hum. Percept. Perform. **22:** 278–293.
31. IVRY, R.B. & T.R. RICHARDSON. 2002. Temporal control and coordination: the multiple timer model. Brain Cog. **48:** 117–132.
32. WING, A. 1980. The long and short of timing in response sequences. *In* Tutorials in Motor Behavior. G. Stelmach & J. Requin, Eds.: 469–486. North Holland Publishing Co. New York.
33. WING, A. & A.B. KRISTOFFERSON. 1973. Response delays and the timing of discrete motor responses. Percept. Psychophys. **14** (1): 5–12.
34. VIVIANI, P. & C. TERZUOLO. 1980. Space-time invariance in learned motor skills. *In* Tutorials in Motor Behavior. G.E.S.J. Requin, Ed.: 525–533. North Holland Publishing Company. Amsterdam.
35. VIVIANI, P. & C. TERZUOLO. 1982. Trajectory determines movement dynamics. Neuroscience **7:** 431–437.
36. BASTIAN, A.J., T.A. MARTIN, J.G. KEATING & W.T. THACH. 1996. Cerebellar ataxia: abnormal control of interaction torques across multiple joints. J. Neurophysiol. **76:** 492–509.
37. ZELAZNIK, H., R. SPENCER & R.B. IVRY. 2002. Dissociation of explicit and implicit timing processes in repetitive tapping and drawing movements. J. Exp. Psychol. Hum. Percept. Perform. **28:** 575–588.
38. BILLON, M., A. SEMJEN & G.E. STELMACH. 1996. The timing effects of accent production in periodic finger-tapping sequences. J. Mot. Behav. **28:** 198–210.
39. HOGAN, N. & T. FLASH. 1987. Moving gracefully: quantitative theories of motor coordination. Trends Neurosci. **10:** 170–174.
40. SPENCER, R.M., R.B. IVRY & H.N. ZELAZNIK. Temporal variability on discrete and continuous motor tasks in patients with cerebellar lesions. In preparation.
41. IVRY, R.B. & R.E. HAZELTINE. 1995. The perception and production of temporal intervals across a range of durations: evidence for a common timing mechanism. J. Exp. Psychol. Hum. Percept. Perform. **21:** 1–12.
42. BASTIAN, A.J., T.A. MARTIN, J.G. KEATING & W.T. THACH. 1996. Cerebellar ataxia: abnormal control of interaction torques across multiple joints. J. Neurophysiol. **76:** 492–509.
43. FRANZ, E., J. ELIASSEN, R.B. IVRY & M. GAZZANIGA. 1996. Dissociation of spatial and temporal coupling in the bimanual movements of callosotomy patients. Psychol. Sci. **7:** 306–310.
44. IVRY, R.B. & E. HAZELTINE. 1999. Subcortical locus of temporal coupling in the bimanual movements of a callosotomy patient. Hum. Mov. Sci. **18:** 345–375.
45. TULLER, B. & J.A. KELSO. 1989. Environmentally-specified patterns of movement coordination in normal and split-brain subjects. Exp. Brain Res. **75:** 306–316.
46. KENNERLEY, S., J. DIEDRICHSEN, A. SEMJEN & R.B. IVRY. 2002. Callosotomy patients exhibit temporal and spatial uncoupling during continuous bimanual movements. Nature Neurosci. **5:** 376–381.
47. ELIASSEN, J.C., K. BAYNES & M.S. GAZZANIGA. 2000. Anterior and posterior callosal contributions to simulataneous bimanual movements of the hand and fingers. Brain **123:** 2501–2511.
48. DIEDRICHSEN, J., E. HAZELTINE, S. KENNERLEY & R.B. IVRY. 2001. Moving to directly cued locations abolishes spatial interference during bimanual actions. Psychol. Sci. **12:** 493–498.
49. ZEILER, M.D. 1998. On sundials, springs, and atoms. Behav. Processes **44:** 89–99.
50. SEMJEN, A. & R.B. IVRY. 2001. The coupled oscillator model of between-hand coordination in alternate-hand tapping: a reappraisal. J. Exp. Psychol. Human Percept. Perform. **27:** 251–265.
51. WOLPERT, D.M., Z. GHAHRAMANI & M.I. JORDAN. 1995. An internal model for sensorimotor integration. Science **269:** 1880–1882.
52. WOLPERT, D.M., R.C. MIALL & M. KAWATO. 1998. Internal models in the cerebellum. Trends Cognit. Sci. **2:** 313–321.

53. Block, R.A. & D. Zakay. 1996. Models of psychological time revisited. *In* Time and Mind. H. Helfrich, Ed.: 171–195. Hogrefe and Huber Publishers. Bern.
54. Bower, J.M. 1997. Control of sensory data acquisition. Int. Rev. Neurobiol. **41:** 489–513.
55. Courchesne, E. & G. Allen. 1997. Prediction and preparation, fundamental functions of the cerebellum. Learn. Mem. **4:** 1–35.
56. Ivry, R.B. 2000. Exploring the role of the cerebellum in sensory anticipation and timing: commentary on Tesche and Karhu. Hum. Brain Mapping **9:** 115–118.

Role of the Posterolateral Cerebellum in Language

ANDREA L. GEBHART, STEVEN E. PETERSEN, AND W.T. THACH

Washington University School of Medicine, Department of Anatomy and Neurobiology, St. Louis, Missouri 63110, USA

ABSTRACT: Historically, scientists have believed that the cerebellum controls only movement. However, recent evidence from neuroimaging and human lesion studies suggests that the right posterolateral cerebellar hemisphere is involved, independently of movement, in helping an individual to generate verbs for given nouns. We sought to elucidate the key factors contributing to the verb generation deficits of subjects with right posterolateral cerebellar damage and thus to better understand the specific contributions of the posterolateral cerebellum to language. We compared the performance of subjects with focal left-sided posterolateral cerebellar lesions, those with focal right-sided posterolateral cerebellar lesions, and neurologically normal pilot control subjects on an antonym generation task, noun (category member) generation task, verb selection task, and lexical decision task. Preliminary results show that subjects with right cerebellar lesions are impaired relative to other subjects only on the antonym generation task. The results provide evidence that the right cerebellar language deficit is not due solely to deficits in "mental movement" coupled to a verb and that internal generation of a word seems to be a key factor in eliciting a deficit. In addition, a semantic processing demand may be necessary but insufficient to elicit a right cerebellar language deficit. The results support the theory that the right posterolateral cerebellar hemisphere assists the left cerebral hemisphere in helping an individual learn to generate specific types of spoken, two-word associations. The full nature of this process awaits further investigation.

KEYWORDS: cognition; cerebellum; language; learning; verbal ability; lesion

INTRODUCTION

The cerebellum has classically been implicated in motor control and coordination. Hallmark clinical signs of cerebellar damage include ataxia of the upper and lower extremities, disturbances in gait, abnormal eye movements, and dysarthric speech.[1,2] Such motor symptoms are most commonly observed in patients who have damage to the *superior* and *medial* regions of the cerebellum. By contrast, individuals who have damage to the *posterolateral* regions of the cerebellum typically have fewer and less severe motor symptoms.[3]

Address for correspondence: W.T. Thach, Department of Anatomy and Neurobiology, Washington University School of Medicine, 660 S. Euclid Ave., Campus Box 8108, St. Louis, MO 63110. Voice: 314-362-3538; fax: 314-747-4370.
thachw@pcg.wustl.edu

Ann. N.Y. Acad. Sci. 978: 318–333 (2002). © 2002 New York Academy of Sciences.

The specific functions of the posterolateral cerebellum are largely unknown. However, recent evidence supports a role for the posterolateral cerebellum in higher cognitive function. Neuroanatomically, the cerebellum is ideally situated to contribute to cognition, particularly to language. Recent work reveals extensive contralateral projections between the cerebellum and nearly all regions of the cerebral cortex, including the prefrontal cortex, in monkeys.[4–6]

Neuroimaging and human lesion studies have provided evidence that the right posterolateral cerebellum is involved in language, independent of movement. In a positron emission tomography (PET) study, Petersen and colleagues[7,8] unexpectedly found that when healthy adult subjects generated aloud an appropriate verb for each of several visually presented nouns, they showed strong right lateral cerebellar activation concurrent with left prefrontal cortical activation. This activation occurred during verb generation but *not* during noun reading. This research group later reported that with less than 15 minutes of practice on the same set of nouns, subjects increased their accuracy of responses and reduced their reaction (voice onset) times. As subjects improved their performance on this task, right lateral cerebellar and left prefrontal activations diminished concurrent with increasing activation of the insula, so that the pattern of brain activation was nearly identical to that during simple noun reading.[9] This study suggested that the right lateral cerebellum and left prefrontal cortex are important in practice-related skill learning and in the automation of performance on a language task. However, the *specific contributions* of the posterolateral cerebellum to verbal learning are not well understood.

In a separate, single-case study, a lawyer with right-sided posterolateral cerebellar damage (who had a WAIS-R Verbal IQ above the 98th percentile) showed a number of deficits on the verb generation task relative to neurologically normal control subjects, in that he made more inappropriate responses, lacked insight into his errors, and showed no tendency to correct his errors with practice on the same set of nouns.[10] In a follow-up study, a group of three subjects with right cerebellar damage showed these same deficits on the verb generation task relative to both control subjects and those with left cerebellar damage.[11]

Nonetheless, the foregoing studies do not pinpoint the specifics of the posterolateral cerebellar contribution to language or verbal learning. In the present study, we addressed this open issue. Specifically, we designed and implemented a series of experiments to investigate which underlying factors may have contributed to the right cerebellar subjects' deficits on the verb generation task. We compared the performance of subjects with left-sided posterolateral cerebellar lesions, those with right-sided posterolateral cerebellar lesions, and neurologically normal control subjects on follow-up experiments to the verb generation task.

First, we investigated whether the right cerebellar language deficit is restricted to the domain in which subjects are required to produce *verbs*. Consistent with classic theories of cerebellar function, right cerebellar subjects may have had deficits in implicit actions associated with the production of verbs in particular (i.e., difficulty in making "mental movements"). Neuroimaging studies have demonstrated that the cerebellum is activated when individuals are mentally imagining movement (without actually moving).[12–15] Hence, cerebellar damage could have impaired "mental movement," such as that which might occur when a subject sees the stimulus "bike" and needs to think of an appropriate verb such as "ride." It is possible that a deficit in such mental movement ability impaired the performance of the subjects with right cerebellar damage on the verb generation task. In addition, it is possible that subjects with right cerebellar damage would be impaired on word generation tasks that *do not* necessitate the consideration or production of verbs (and do not involve mental movement).

To investigate this possibility, we tested subjects on an *antonym generation task* and *noun (category member) generation task*, both of which retained the requirement for internal generation of semantically based word association responses but eliminated the necessity of an *action component* of these responses. We hypothesized that subjects with right cerebellar damage would show deficits on these tasks relative to the other subject groups, thereby providing evidence that their language deficit *is not* restricted to the domain in which they must generate *verbs*.

Secondly, we assessed the relative importance of the *internal generation requirement* in eliciting a language deficit in subjects with right cerebellar damage. We investigated whether the deficit was due more to difficulty in *internally* thinking of and then saying an appropriate verb for each given noun than to global difficulty in recognizing appropriate noun-verb relationships. In the *verb selection task*, rather than being required to think of and say an appropriate verb for each given noun (as in the verb generation task), subjects saw a series of nouns, each of which had a choice of verbs below it. Subjects were instructed to select the one verb choice that was appropriate for each noun and say it into the microphone. This eliminated the requirement for subjects to *internally* generate verbs, but it retained the requirement to recognize appropriate noun-verb relationships. We hypothesized that subjects with right cerebellar damage would be impaired on this task relative to other subjects, but not to the same extent as they were on the verb generation task.

Finally, to reduce speech and to eliminate the requirement for semantic association responses, we tested subjects on a lexical decision task in which they had to decide (via a key-press response) if items were words or non-words. We predicted that subjects with right cerebellar damage would retain the ability to accurately make such judgments.

METHODS

Subjects

Three subject groups participated in the experiments: six neurologically normal pilot control subjects, two subjects with left-sided posterolateral cerebellar damage, and three subjects with right-sided posterolateral cerebellar damage. All lesions were verified by both a radiologist's report and the visual inspection of an MRI or CT scan by one of us (W.T. Thach). All lesions were confined to the cerebellum.

The pilot control subjects were aged 62–77 years, with an education ranging from 12–17 years. Refer to TABLE 1 for the cerebellar subject demographics. All subjects were native English speakers who had normal or corrected-to-normal vision. No subjects had dysarthric speech. A future publication will detail the results of the completed study in which more cerebellar subjects will be tested, and a neurologically normal control subject will be matched to each cerebellar subject (on the basis of age, gender, handedness, and education) and tested.

Psychometric Tests

To verify that any deficits shown by the cerebellar subjects on the experimental tasks were not attributable to a low level of standardized verbal intelligence and/or memory, all cerebellar subjects were tested on the verbal subtests of the Wechsler Adult Intelligence Scale-Revised (WAIS-R[16]) as well as the Logical Memory and Verbal Paired Associates subtests of the Wechsler Memory Scale-III (WMS-III[17]). The verbal intelligence quotients of individuals who were over the age of 74 at the time of testing were computed using the MAYO WAIS-R population norms, because regular WAIS-R norms do not exist for individuals in this age bracket.[18] Any subject who had a verbal IQ that was classified as borderline or lower (according to the WAIS-R or MAYO WAIS-R population norms) was to be excluded from the study; however, no subjects needed to be excluded on this basis.

Experimental Setup and Overview

Subjects were seated in a chair that was located approximately 45 cm horizontally from a computer screen, which displayed the stimuli. Each experimental task was programmed using custom PsyScope software for the Macintosh computer. All stimuli were presented in white Geneva font against a black background at the center of the screen.

Each subject completed four experiments in the following fixed order: noun generation, verb selection, antonym generation, and lexical decision. For the noun generation, verb selection, and antonym generation tasks, the presentation of each stimulus was synchronized to a beep (inaudible to the subject). For these experiments, subjects spoke into a microphone, and a voice-activated box relayed reaction times in milliseconds (i.e., the time between stimulus presentation and the onset of the subject's verbal response) to the computer. Reaction times were recorded by the computer and stored in data files. All verbal responses were recorded through an amplifier onto an audiotape. Response times that the computer failed to record accurately (e.g., those due to the computer program registering an extraneous sound as the subject's response) were digitized using Sound Edit Pro 16 software (Macromedia, Inc.). The experimenter scored each response trial as "appropriate," "inappropriate," or "no response." For the lexical decision task, subjects made key-press responses on a PsyScope button box that were recorded by the computer. The key-press response selections and reaction times were stored in data files.

In each task, each stimulus appeared on the screen either until the subject responded or for 4750 ms, whichever occurred first. When subjects responded before the end of the time limit on the antonym generation, noun generation, and verb selection tasks, the experimenter pressed the spacebar to advance to the next stimulus. When subjects responded on the lexical decision task before the end of the time limit, the computer program automatically advanced to the next stimulus. The interstimulus interval in all tasks was 250 ms, during which the subjects fixated on a crosshair at the center of the computer screen.

Subjects were required to successfully complete several practice items and were corrected on any errors before each experimental task to verify that they understood the instructions.

Based on our pilot data analysis, we decided that no changes in the structure of the experimental tasks or the stimuli were necessary before testing the subjects with

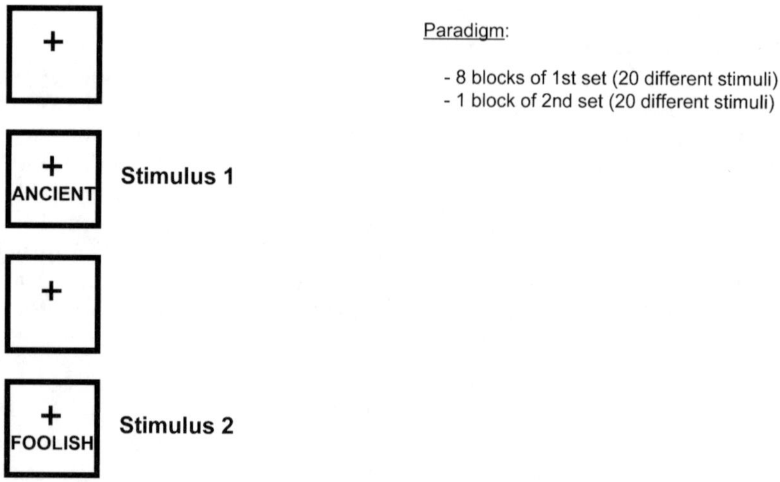

FIGURE 1. Antonym generation task paradigm. This shows examples of two stimuli (adjectives) to which subjects were instructed to respond as accurately and quickly as possible, with a word opposite in meaning. See text for further details.

cerebellar damage. Therefore, the pilot control subjects and cerebellar subjects performed the identical tasks and saw the identical stimuli. Pilot control subject data are presented in this paper as a mean across the subject group, whereas cerebellar subject data are presented individually.

Antonym Generation Task

Subjects were presented with 8 successive blocks of one set of 20 adjectives. The 20 adjectives were randomized in each block and presented one at a time. In a ninth block, subjects saw a new set of 20 adjectives presented one at a time. Subjects were instructed to say a word that was opposite in meaning to each stimulus (FIG. 1); for example, a correct response to the stimulus "ancient" would be "modern." Subjects were instructed to respond with the first correct opposite that came to mind, as quickly as possible. In addition, they were told that they would see some of the same words repeatedly across blocks, but that this was *not* an exercise in creativity. Subjects were told that it was fine to give the same responses to the same items repeatedly. Stimuli were presented in 48-point font and had a median Kucera-Francis written frequency of 41 occurrences per million words (range: 3–660).[19]

Data were analyzed for subjects' improvement in accuracy and reaction time across Blocks 1 to 8. In addition, data were analyzed for the difference in each subject's median reaction time between Blocks 8 and 9.

For each subject, a median reaction time for appropriate responses was computed for Block 1. A "normalized median reaction time" was then computed for each of Blocks 2 through 9. Each normalized median reaction time was equal to the median raw reaction time for appropriate responses for that block divided by the median raw reaction time for appropriate responses for Block 1. This method allowed us to eval-

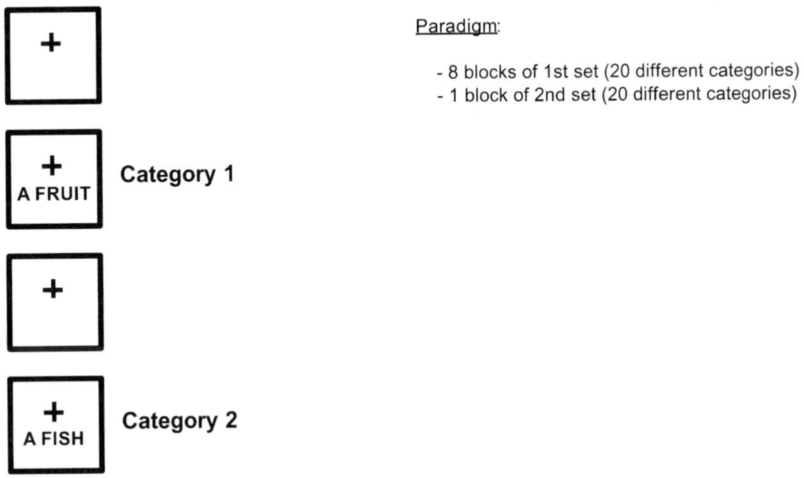

FIGURE 2. Noun generation task paradigm. This shows examples of two stimuli (categories) to which subjects were instructed to respond, as accurately and quickly as possible, with a member of each category. See text for further details.

uate the extent to which each subject reduced his or her reaction time across blocks as well as to standardize the comparison across subjects. The reaction times across blocks for each experimental task were normalized in this manner. Furthermore, in each experimental learning paradigm, the mean of the median reaction times for Blocks 1 to 4 was compared to that of Blocks 5 to 8 as a measure of each subject's practice-related learning.

Pilot control subjects were not tested on the antonym generation task, as the task was added to the experimental battery after the piloting phase had ended and just before the testing of the cerebellar subjects. In the future, a matched control subject will be tested for each cerebellar subject on this task.

Noun Generation Task

Subjects were presented with 8 successive blocks of one set of 20 categories (for example, "a fruit"). The 20 categories were randomized in each block and presented one at a time. Subjects then saw a ninth block of a new set of 20 categories presented one at a time. Subjects were instructed to say an appropriate member of each category (FIG. 2). For example, a correct response to the category "a fruit" would be "apple." Subjects were instructed to respond to each category as quickly as possible with the first correct category member that came to mind. In addition, subjects were told that they would see some of the same categories repeatedly across blocks, but that this was *not* an exercise in creativity. Subjects were told that it was fine to give the same responses to the same categories repeatedly. Categories were presented in 46-point font and were part of the Battig and Montague category norms.[20]

Data were analyzed as in the antonym generation task.

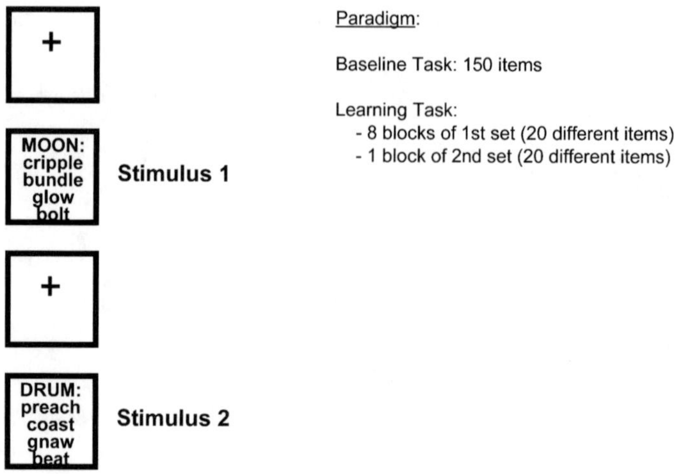

FIGURE 3. Verb selection task paradigm. This shows examples of two stimulus items to which subjects were instructed to respond, as accurately and quickly as possible, with the appropriate verb choice. See text for further details.

Verb Selection Task

First, subjects saw 150 baseline items presented one at a time. Each item was a noun presented with four choices of verbs below it, only one of which was an appropriate verb for the given noun (FIG. 3). Subjects were instructed to say, as accurately and quickly as possible, the one verb (out of the four choices) that was appropriate for each noun. For example, the correct response to the stimulus "moon" in FIGURE 3 is "glow." The purpose of this baseline part of the task was to provide a large corpus of items over which to compute subjects' accuracy and median reaction time.

Next, subjects completed a learning paradigm with 9 blocks of stimuli, all of which were different from the stimuli in the baseline part of the task. Subjects were presented with 8 successive blocks of one set of 50 items. The 50 items were randomized in each block and presented one at a time. Subjects then saw a ninth block of a new set of 50 items presented one at a time. Subjects were instructed to say, as accurately and quickly as possible, the one verb (out of the four choices) that was appropriate for each noun, as they had in the baseline part of the task. In addition, they were told that they would see some of the same items repeatedly and that they should respond as accurately and quickly as possible.

Stimulus nouns were presented in 48-point font, and the verb choices were presented in 32-point font. The median Kucera-Francis written frequency of the nouns was 14 occurrences per million words (range: 1–464). The median Kucera-Francis written frequency of the verb choices was 6 occurrences per million words (range: 1–772).

Data were analyzed for subjects' accuracy, reaction times, and improvement in accuracy and reaction time with practice across Blocks 1 to 8. In addition, data were analyzed for each subject's difference in normalized median reaction time between Blocks 8 and 9.

Lexical Decision Task

First, subjects saw 150 baseline stimuli presented one at a time. For each item, subjects were instructed to press one key on a PsyScope button box if it was an English word and another key if it was not an English word, as accurately and quickly as possible. Subjects used two fingers (of their choice) on their dominant hand for the key-press responses. One finger rested on one key, and the other finger rested on the other key. Subjects were told that the "word" stimuli would be relatively common English words. Therefore, if they did not recognize an item as a word, they were told to respond that it was not a word. Of the 150 stimuli, half were words (nouns) and half were non-words (FIG. 4). The non-words were equally divided among letter strings (e.g., "lnhrt"), phonologically acceptable non-words that did *not* sound like real words (e.g., "slig"), and pseudohomophones (phonologically acceptable non-words that sounded like real words, such as "brane"). The purpose of this baseline

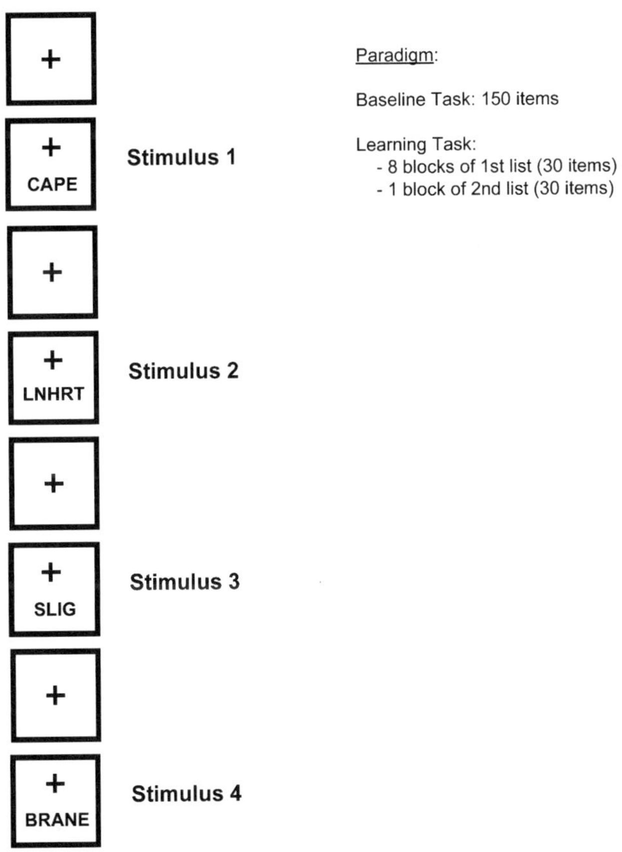

FIGURE 4. Lexical decision task paradigm. This shows examples of four stimuli to which subjects were instructed to make word/non-word decisions (via key-press response on a PsyScope button box), as accuratcly and quickly as possible. See text for further details.

part of the task was to provide a large corpus of items over which to compute subjects' accuracy and median reaction time.

Next, subjects were presented with 8 successive blocks of one set of 30 items, all of which were different from the items in the baseline part of the task. The 30 items were randomized in each block and presented one at a time. Subjects then saw a ninth block of a new set of 30 items presented one at a time. As in the baseline part of the task, subjects were instructed to make word/non-word decisions by pressing the appropriate key on a PsyScope button box. In each block, half of the stimuli were words (nouns) and half were non-words, as in the baseline part of the task. Subjects were told that they would see some of the same stimuli repeatedly and that they should respond as accurately and quickly as possible.

Stimuli were presented in 48-point font. The median Kucera-Francis written frequency of the "word" stimuli was 9 occurrences per million words (range: 1–298).

Data were analyzed for subjects' accuracy, reaction times, and improvement in accuracy and reaction time with practice across Blocks 1 to 8. Also, data were analyzed for each subject's difference in median reaction time between Blocks 8 and 9.

RESULTS

Psychometric Tests

All cerebellar subjects had a WAIS-R Verbal IQ that was classified as at least "average" according to the WAIS-R or MAYO WAIS-R population norms (TABLE 1). Moreover, subjects with cerebellar damage did not show deficits on the Logical

TABLE 1. Cerebellar subject demographics

Subject	Age (yr)	Gender	Education (yr)	Diagnosis	Time interval: lesion onset to testing	Verbal WAIS-R score and classification
LC1	94	M	10	Left PICA territory infarct	3.9 yr	101[a] (average)
LC2	77	M	12	Left PICA territory infarct	3.7 yr	95 (average)
RC1	80	F	12	Right PICA territory hemorrhage	1.5 yr	111[a] (high average)
RC2	64	F	17	Right PICA territory damage[b]	2 mo?[b]	120 (superior)
RC3	47	F	12	Right PICA territory infarct	2.4 yr	92 (average)

PICA, posterior inferior cerebellar artery. "Age" in years reflects subjects' ages at the time of testing on the four experimental tasks. All subjects except for RC2 were tested between 2 and 3 years earlier on the WAIS-R than on the four experimental tasks.

[a]Score computed according to MAYO WAIS-R population norms, which were used for individuals who were above age 74 at the time of testing.

[b]It was unclear to this patient's physicians whether the cerebellar damage had resulted from a recent infarct or had been present since birth.

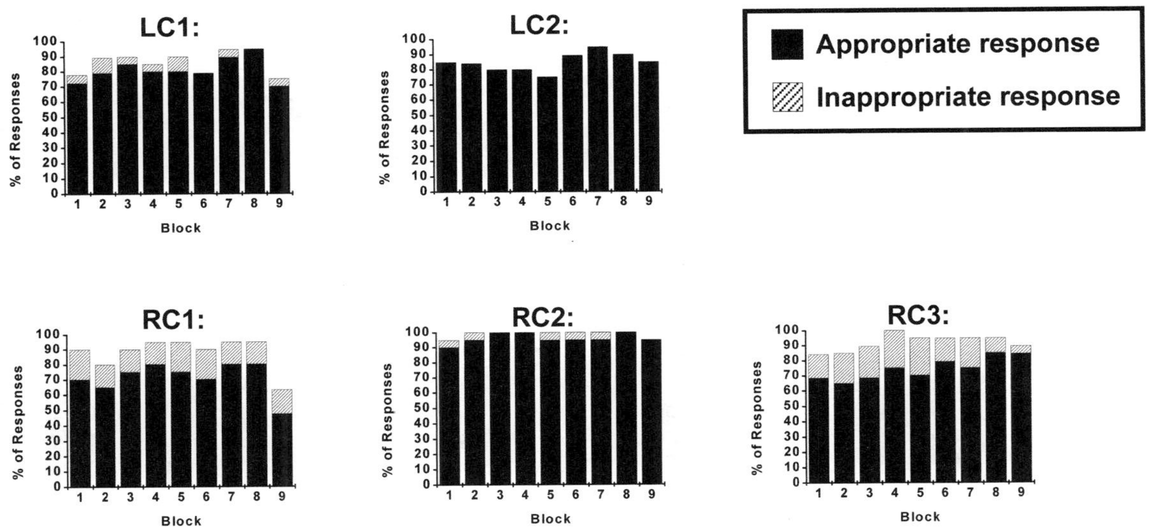

FIGURE 5. Antonym generation task: accuracy. Each subject's percentage of appropriate responses for each block is plotted in *black*, and the percentage of inappropriate responses for each block is plotted in a *hatched pattern*. As illustrated, RC1 and RC3 showed fewer accurate responses than did the other subjects.

Memory or the Verbal Paired Associates subtests of the WMS-III. This provides evidence that any deficits shown by the cerebellar subjects on the experimental tasks were not due to a low level of standardized verbal intelligence and/or memory.

Antonym Generation Task

On the antonym generation task, the right cerebellar subject group showed less accuracy and more perseveration in responses than the left cerebellar subject group. The right cerebellar subjects showed a higher percentage of inappropriate responses than the left cerebellar subjects (FIG. 5). Across Blocks 1 to 9, the right cerebellar subject group showed a mean of 12.37% (±SE 4.78) inappropriate responses, and the left cerebellar subject group showed a mean of 2.58% (±SE 2.58) inappropriate responses. Compared to the left cerebellar subjects, the right cerebellar subjects gave more responses that were semantically associated with the presented stimuli, but not antonyms. For example, both RC1 and RC3 repeatedly responded "old" to the stimulus "ancient." RC3 gave a number of perseverative responses. For example, to the stimulus "stale," she repeatedly responded "old" (a semantic associate); to the stimulus "fancy," she responded "cheap" (which was one of the stimulus items that she saw repeatedly); and to the stimulus "lazy," she responded "smart" (which was a correct response to the stimulus "foolish," which she had given to that stimulus item repeatedly).

As illustrated in FIGURE 6, RC1 and RC2 showed little improvement in reaction time for appropriate responses with practice across Blocks 1 to 8. However, RC3 did

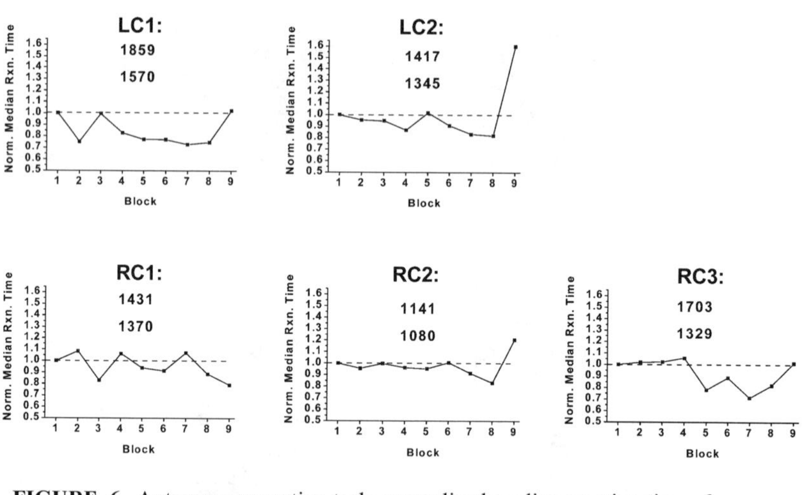

FIGURE 6. Antonym generation task: normalized median reaction times for appropriate responses across Blocks 1 to 9. The *top number* (underneath each subject label) is the mean of the median raw reaction times (in ms) for appropriate responses across Blocks 1 to 4. The *bottom number* is the mean of the median raw reaction times (in ms) for appropriate responses across Blocks 5 to 8. As illustrated, RC1 and RC2 showed less reduction in normalized median reaction time with practice than did the other subjects. However, all subjects except RC1 had an increase in normalized median reaction time between Blocks 8 and 9.

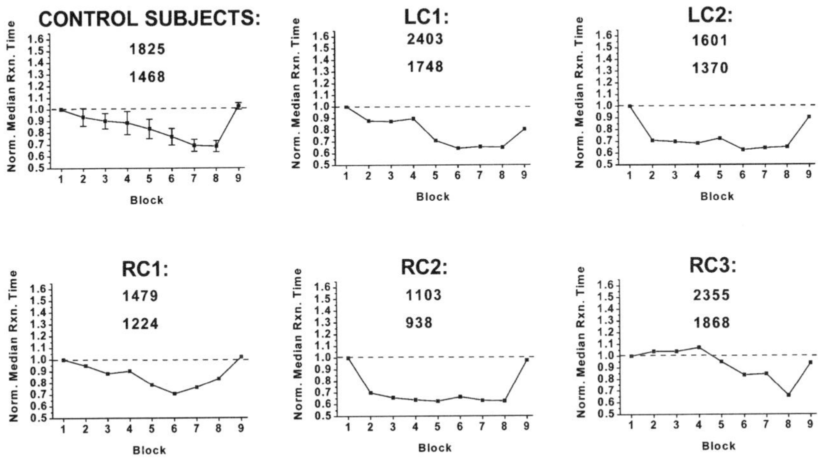

FIGURE 7. Noun generation task: normalized median reaction times for appropriate responses across Blocks 1 to 9. The *top number* (underneath each subject label) is the mean of the median raw reaction times (in ms) for appropriate responses across Blocks 1 to 4. The *bottom number* is the mean of the median raw reaction times (in ms) across Blocks 5 to 8 for appropriate responses. As illustrated, all subjects reduced their normalized median reaction times across Blocks 1 to 8 and increased their normalized median reaction times between Blocks 8 and 9.

improve her reaction time across blocks. Overall, the results suggest that right cerebellar subjects are not able to automate their antonym generation responses to the same extent as left cerebellar subjects. However, all subjects (except for RC1) increased their normalized median reaction time from Block 8 to 9, suggesting that they did learn to automate their responses across Blocks 1 to 8 to some extent.

Noun Generation Task

All subjects performed accurately on the noun generation task. Individual subjects' mean accuracy for Blocks 1 to 9 ranged from 82.28–100.00%. Each cerebellar subject had a mean accuracy of 87.08% or higher. Across Blocks 1 to 9, mean (±SE) accuracy of each subject group was as follows: control subject group: 95.45% (±2.68); left cerebellar subject group: 90.75% (±3.67); right cerebellar subject group: 97.40% (±1.31). As shown in FIGURE 7, all subjects dramatically reduced their normalized median reaction times for appropriate responses across Blocks 1 to 8 and increased their normalized median reaction times for the new set of categories in Block 9. Thus, the subject groups showed no striking differences in their accuracy or improvement with practice on this task.

Verb Selection Task

All subjects performed accurately on the verb selection task. Individual subjects' mean accuracy for Blocks 1 to 9 ranged from 86.26–99.09%, with a comparable lev-

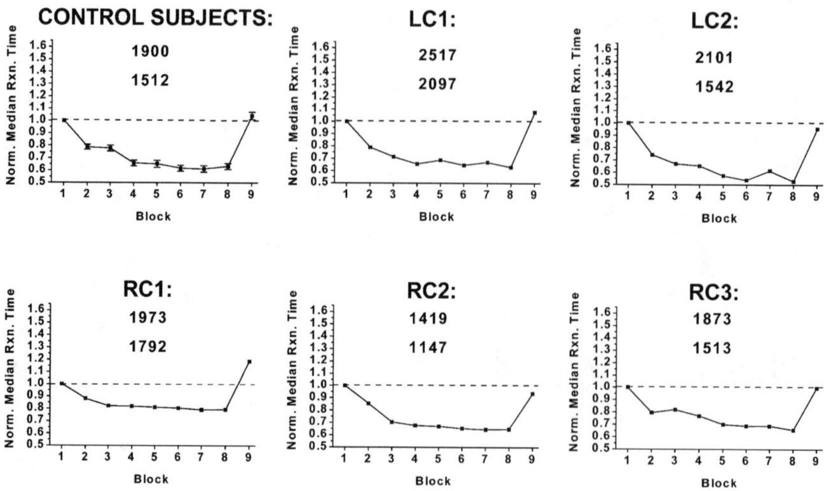

FIGURE 8. Verb selection task: normalized median reaction times for appropriate responses across Blocks 1 to 9. The *top number* (underneath each subject label) is the mean of the median raw reaction times (in ms) for appropriate responses across Blocks 1 to 4. The *bottom number* is the mean of the median raw reaction times (in ms) for appropriate responses across Blocks 5 to 8. As shown, all subjects reduced their normalized median reaction times across Blocks 1 to 8 and increased their normalized median reaction times between Blocks 8 and 9.

el of accuracy on the 150-item baseline part of the task. Each subject with cerebellar damage had a mean accuracy of 86.26% or higher. The mean (±SE) accuracy for each subject group across Blocks 1 to 9 was as follows: control subject group: 94.71% (±1.58); left cerebellar subject group: 90.09% (±3.83); and right cerebellar subject group: 95.45% (±1.73). As illustrated in FIGURE 8, all subjects significantly reduced their normalized median reaction times for appropriate responses across Blocks 1 to 8 and increased their normalized median reaction times for the new set of items in Block 9.

Lexical Decision Task

On the lexical decision task, most subjects performed very accurately. Individual subjects' mean accuracy for Blocks 1 to 9 ranged from 73.59–100.00%, with a comparable level of accuracy on the 150-item baseline part of the task. Each subject with cerebellar damage had a mean accuracy of 95.56% or higher. The mean (±SE) accuracy for each subject group across Blocks 1 to 9 was as follows: control subject group: 93.00% (±4.10); left cerebellar subject group: 96.30% (±0.74); and right cerebellar subject group: 98.27% (±0.89). Subjects did not tend to reduce their normalized median reaction times with practice on this task or to increase their normalized median reaction times for the new set of items in Block 9 (FIG. 9). Thus, the subject groups did not show striking differences in their performance on this task.

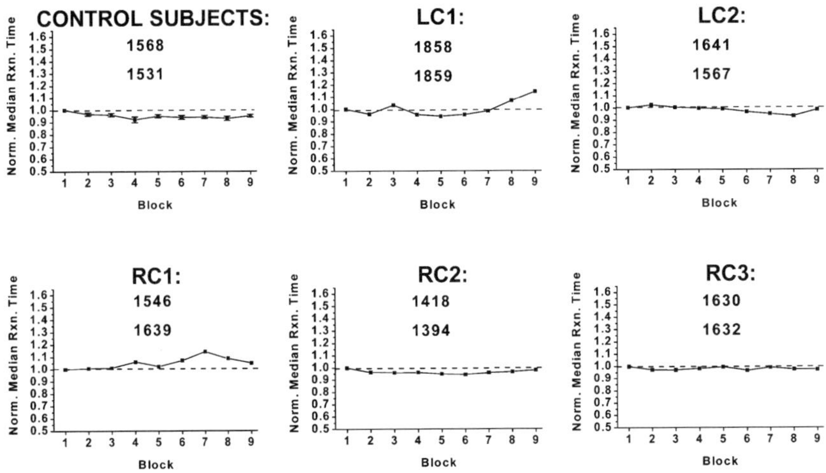

FIGURE 9. Lexical decision task: normalized median reaction times across Blocks 1 to 9. The *top number* (underneath each subject label) is the mean of the median raw reaction times across Blocks 1 to 4. The *bottom number* is the mean of the median raw reaction times across Blocks 5 to 8. As shown, subjects did *not* reduce their normalized median reaction times across Blocks 1 to 8 nor did they substantially increase their normalized median reaction times between Blocks 8 and 9 (with the exception of LC1).

DISCUSSION

In the current study, we investigated the specific contributions of the right posterolateral cerebellum to language. Specifically, we sought to determine the key factors underlying the right cerebellar subjects' deficits on the verb generation task. We compared the performance of subjects with focal left-sided posterolateral cerebellar lesions, those with focal right-sided posterolateral cerebellar lesions, and neurologically normal control subjects on a battery of follow-up tests to the verb generation task.

On the antonym generation task, the right cerebellar subject group showed less accuracy and less reduction in median reaction time with practice than the left cerebellar subject group. On the noun generation, verb selection, and lexical decision tasks, all subject groups performed similarly.

This study provides evidence that the cerebellum is involved in language independent of movement. The results suggest that right cerebellar subjects have difficulty internally generating specific types of word association responses. The deficits of the right cerebellar subjects on the antonym generation task provide evidence that the right cerebellar language deficit is not due solely to overt or mental movement coupled to a verb. Thus, results from the present study argue against the explanation that "mental movement" is necessary to elicit a right cerebellar language deficit.

The absence of a right cerebellar deficit on the noun generation task suggests that a semantic processing demand may be necessary, but not sufficient, to elicit a language deficit in right cerebellar subjects. The right posterolateral cerebellum may be more involved in associative semantics than in categorical semantics. Specifically, the right posterolateral cerebellum may be important in helping an individual to generate constrained semantic association responses. The range of appropriate responses in the antonym generation task (on which the right cerebellar subjects showed deficits) was much more highly constrained than that in the noun generation task. That is, for example, in the antonym generation task, there were fewer appropriate antonyms to the stimulus “ancient” than there were appropriate exemplars of a category such as “a fruit” in the noun generation task. Likewise, the range of appropriate responses in the verb generation task (on which the right cerebellar subjects performed poorly) was much more highly constrained than in the noun generation task. Thus, the element of constraint in semantic association responses may be a key factor in eliciting a right cerebellar language deficit.

Right cerebellar subjects showed no deficits on experimental tasks in which they did not need to internally generate semantic word association responses. The striking dissociation between right cerebellar subjects’ impaired performance on the verb generation task and their excellent performance on the verb selection task suggests that they are able to recognize appropriate noun-verb relationships. Thus, right cerebellar subjects have the ability to accurately make semantic associations when they are presented with verb choices and do not need to internally think of and generate such associations. In addition, they were able to significantly reduce their normalized median reaction times across Blocks 1 to 8 when they were not required to internally generate verbs. In right cerebellar subjects, internal generation of a word may be a key element in eliciting a language deficit, as seen in the verb generation task. However, as suggested by the results of the noun generation task, internal generation may be necessary, but is not sufficient, to elicit a deficit.

The absence of a right cerebellar deficit on the lexical decision task provides evidence that right cerebellar subjects are able to distinguish words from non-words. Subjects may not have reduced their normalized median reaction time with practice on this task (FIG. 9) because it likely requires a lower level of processing than the other tasks (which had required subjects to make semantic association responses). The absence of a right cerebellar deficit further supports the theory that the *internal generation requirement* of a word association response may be a key element in eliciting a right cerebellar language deficit. In addition, active speech and the requirement to generate a *semantic association* may be necessary to elicit a language deficit.

We conclude that the right posterolateral cerebellum is involved in strictly cognitive aspects of language. The results provide preliminary evidence that the right posterolateral cerebellum may assist the left cerebral cortex in helping an individual to learn to internally generate specific types of constrained, spoken two-word associations. In addition, the right posterolateral cerebellum may help an individual to automate word associations during speech. The right posterolateral cerebellum does not seem to be involved in the ability to recognize appropriate noun-verb relationships or in the ability to distinguish words from non-words. The role of the right posterolateral cerebellum in language seems to contrast with the theorized role of the left posterolateral cerebellum in spatial operations.[21,22] Future investigation will further reveal the specific contributions of the right posterolateral cerebellum to language.

REFERENCES

1. HOLMES, G. 1939. The cerebellum of man. Brain **62:** 1–30.
2. GILMAN, S., J. BLOEDEL & R. LECHTENBERG. 1981. The symptoms and signs of cerebellar disease. *In* Disorders of the Cerebellum. S. Gilman, J. Bloedel & R. Lechtenberg, Eds. :189–221. Philadelphia. F.A. Davis.
3. TOHGI, H., S. TAKAHASHI, K. CHIBA, *et al.* 1993. Cerebellar infarction. Clinical and neuroimaging analysis in 293 patients. The Tohoku Cerebellar Infarction Study Group. Stroke **24:** 1697–1701.
4. MIDDLETON, F. & P. STRICK. 1994. Anatomical evidence for cerebellar and basal ganglia involvement in higher cognitive function. Science (Washington, DC) **266:** 458–461.
5. SCHMAHMANN, J. & D. PANDYA. 1997. Anatomic organization of the basilar pontine projections from prefrontal cortices in rhesus monkey. J. Neurosci. **17:** 438–458.
6. MIDDLETON, F. & P. STRICK. 1998. Cerebellar output: motor and cognitive channels. Trends Cognit. Sci. **2:** 348–354.
7. PETERSEN, S., P. FOX, M. POSNER, *et al.* 1988. Positron emission tomographic studies of the cortical anatomy of single-word processing. Nature (Lond.) **331:** 585–589.
8. PETERSEN, S., P. FOX, M. POSNER, *et al.* 1989. Positron emission tomographic studies of the processing of single words. J. Cognit. Neurosci. **1:** 153–170.
9. RAICHLE, M., J. FIEZ, T. VIDEEN, *et al.* 1994. Practice-related changes in human brain functional anatomy during nonmotor learning. Cereb. Cortex **4:** 8–26.
10. FIEZ, J., S. PETERSEN, M. CHENEY, *et al.* 1992. Impaired non-motor learning and error detection associated with cerebellar damage: A single case study. Brain **115:** 155–178.
11. GEBHART, A., S. PETERSEN & W.T. THACH. 2000. Evidence for functional lateralization of language operations to the right posterior cerebellum. Soc. Neurosci. Abstr. **26:** 1246.
12. DECETY, J., H. SJOHOLM, E. RYDING, *et al.* 1990. The cerebellum participates in mental activity: tomographic measurements of regional cerebral blood flow. Brain Res. **535:** 313–317.
13. RYDING, E., J. DECETY, H. SJOHOLM, *et al.* 1993. Motor imagery activates the cerebellum regionally. A SPECT rCBF study with 99mTc-HMPAO. Cognit. Brain Res. **1:** 94–99.
14. DECETY, J., D. PERANI, M. JEANNEROD, *et al.* 1994. Mapping motor representations with positron emission tomography. Nature (Lond.) **371:** 600–602.
15. LOTZE, M., P. MONTOYA, M. ERB, *et al.* 1999. Activation of cortical and cerebellar motor areas during executed and imagined hand movements: an fMRI study. J. Cognit. Neurosci. **11:** 491–501.
16. WECHSLER, D. 1981. Wechsler Adult Intelligence Scale-Revised. Manual. Psychological Corp. New York.
17. WECHSLER, D. 1997. Wechsler Memory Scale-III. Manual. Psychological Corp. San Antonio.
18. IVNIK, R., J. MALEC, G. SMITH, *et al.* 1992. Mayo's older Americans normative studies: WAIS-R norms for ages 56 to 97. Clin. Neuropsychologist **6:** 1–30.
19. KUCERA, H. & W. FRANCIS. 1967. Computational Analysis of Present-Day American English. Brown University Press. Providence.
20. BATTIG, W. & W. MONTAGUE. 1969. Category norms for verbal items in 56 categories: a replication and extension of the Connecticut category norms. J. Exp. Psychol. Monogr. **80:** 1–46.
21. VAN MIER, H., L. TEMPEL, J. PERLMUTTER, *et al.* 1998. Changes in brain activity during motor learning measured with PET: effects of hand of performance and practice. J. Neurophysiol. (Bethesda) **80:** 2177–2199.
22. GEBHART, A. & W.T. THACH. 1999. Cerebellar subjects show cognitive impairments on a novel maze learning task. Soc. Neurosci. Abstr. **25:** 650.

Role of the Cerebellum in Motor Cognition

HANNEKE I. VAN MIER[a] AND STEVE E. PETERSEN[b]

[a]*Neurocognition, Faculty of Psychology, University Maastricht, Maastricht, 63110 The Netherlands*

[b]*Departments of Neurology and Neurological Surgery, Radiology, Anatomy and Neurobiology, Psychology, Washington University School of Medicine, St. Louis, Missouri, USA*

ABSTRACT: Cerebellar data from five experiments using different groups of subjects performing the same motor learning task are presented. Positron emission tomography (PET) as well as functional magnetic resonance imaging (fMRI) was used to study changes in cerebellar activations as an effect of learning. Cerebellar brain activations obtained during the performance of a new motor task were compared to activations during the performance of the same task after as well as during practice. To account for changes in velocity and somatosensory processing as an effect of practice, two control conditions were included. Behavioral data showed that as an effect of practice performance speed as well as accuracy increased in all five experiments and groups. Neuroimaging data from adults as well as children showed differential changes in brain activations in different cerebellar areas. In all experiments an area in the left lateral cerebellum showed practice-related decreases, which were most likely related to a decrease in errors. In two experiments a highly significant correlation was found between the decrease in errors and the decrease in left cerebellar activation. An area in the right lateral cerebellum and one in the ipsilateral anterior vermis showed activations that seemed related to the level of capacity at which the subjects were performing and might refer to timing-related aspects of the task.

KEYWORDS: motor learning; neuroimaging; PET; fMRI; timing; adults; children

INTRODUCTION

It is now commonly accepted that the cerebellum is involved in processes beyond that of motor coordination and motor control. Studies involving cerebellar patients have suggested a role for the cerebellum in cognition in both adults[1–4] and children.[5,6] Anatomical studies[7–9] have shown that the cerebellum projects to several prefrontal areas via the thalamus, linking the cerebellum to areas known to be involved in cognitive processes. However, studies using functional neuroimaging techniques have shown beyond doubt that the cerebellum is involved in a wide range of

Address for correspondence: Dr. Hanneke van Mier, Maastricht University, Psychology, Neurocognition, P.O. Box 616, 6200 MD Maastricht, 63110 The Netherlands. Voice: 043-3884010; fax: 043-3884125.

h.vanmier@psychology.unimaas.nl

Ann. N.Y. Acad. Sci. 978: 334–353 (2002). © 2002 New York Academy of Sciences.

cognitive processes. In an impressive overview including 275 neuroimaging studies to explore brain areas involved in cognitive functions, Cabeza and Nyberg[10] showed that the cerebellum is engaged in a variety of cognitive tasks. Furthermore, findings of cerebellar activation in neuroimaging studies addressing motor imagery[11–13] and mental navigation[14] also suggest a role for the cerebellum beyond that of motor control and motor execution.

As for the role of the cerebellum in motor cognition, several hypotheses have been advanced. The cerebellum is thought to be involved in motor learning[15–17] (for a review of neuroimaging studies showing cerebellar activations in motor learning see van Mier[18]), in the coordination of multijoint movements,[19–22] in the coordination between eye and hand,[23] in error detection and correction,[24] as well as in the control of motor timing.[25–27] Data from anatomical, lesion, electrophysiological, and neuroimaging studies have shown evidence for each hypothesis. However, which theory or combination of theories will be able to describe the apparently diverse role of the cerebellum[28] is not clear at the moment. Furthermore, it is not unlikely that different parts of the cerebellum might be involved in different aspects of motor control and learning.

The changes in brain activations observed during and after the learning of a sequential motor task are described. In five experiments, five different groups of subjects had to learn a sequence of movement directions by trial and error in order to traverse a maze design as quickly and accurately as possible. Subjects were instructed to move a pen through a cutout design with their eyes closed. Although the whole brain was scanned, this paper focuses only on activations observed in cerebellar areas. Brain activations were measured using two different brain imaging techniques, positron emission tomography (PET) and functional magnetic resonance imaging (fMRI). The first technique was used in experiments 1 to 3, wherease the latter was applied in experiments 4 and 5. In all experiments but one, subjects performed the task with the right hand; in experiment 2, subjects traced the mazes with the left hand. Adults (experiments 1 to 4) as well as children (experiment 5) were scanned. Results from these experiments clearly showed the involvement of cerebellar areas in procedural motor learning.

METHODS

Subjects

Cerebellar activations were measured in a total of 36 subjects. In experiments 1 to 3, eight subjects were tested in each experiment (five females and three males, mean age 26 in experiment 1; five females and three males, mean age 23 in experiment 2; and four females and four males, mean age 24 in experiment 3). As for experiments 4 and 5, six subjects were tested in each experiment (four females and two males, mean age 23 years in experiment 4; and four girls and two boys, aged 7, 8, 9, 10, 11, and 12 years, in experiment 5).

All subjects were tested for handedness on the Edinburgh handedness inventory[29] and were shown to be strongly right handed. None of the subjects showed any motor problems or had a history of neurological disorders. The Human Studies Committee and for the PET experiments also the Radioactive Drug Research Committee of

Washington University approved the study. Informed consent conforming to the guidelines and procedures set forth by these committees was obtained from all subjects (and/or one of the parents/caretakers of the children in experiment 5). All subjects were paid for participation.

MATERIAL

Data regarding performance on a square design and two maze designs (FIG. 1) are reported in this paper. The square consisted of four segments, whereas the maze designs consisted of eight line segments that were connected by intersections. Selecting a wrong turn at an intersection led to a dead end. The length of each dead end path was 0.5 cm. Total (correct) path length of the mazes and squares was 24 cm. All designs formed closed loops, that is, the end-point of one loop was the starting point of the next loop. Subjects were instructed to move continuously with a pen through the designs with their eyes closed. Mazes and square were reverse-embossed patterns, with cutout paths for the pen to traverse. Depth of each path was 0.15 cm in the experiments including adult subjects and 0.30 cm in the experiment with the children. The latter was done to prevent the children from jumping out of the designs. In experiments 1 to 3, two maze versions were used, versions A and B, with version B being a 90-degree rotation of version A. One version was traced during naive and practiced performance, the other version during novel performance, with the version of maze presentation counterbalanced across subjects (FIG. 1).

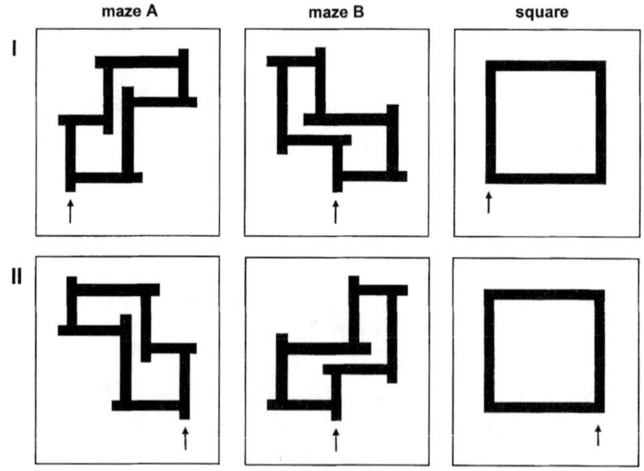

FIGURE 1. Designs used in the experiments. *Arrows* indicate the starting position for each design. I. Designs used in experiments 1, 3, 4, and 5, in which subjects traced the designs with the right hand in a clockwise direction. II. Designs used in experiment 2 in which subjects performed the task with the left hand in a counterclockwise direction.

Task Description

As stated before, tracing of the designs was done with the eyes closed. Subjects were not allowed to inspect the maze designs visually or tactilely. The experimenter guided the hand of the subject that held the pen into the proper starting position of the maze. Subjects were instructed that the first movement was upwards, that they had to move continuously through the designs, and that they should try to avoid retracing a path. For most conditions the instruction to the subjects was to trace the designs as quickly and accurately as possible. To explain the task and the two-choice principle at intersections, an example maze was shown to the subjects before they were placed in the scanner. In experiments 1, 3, 4, and 5, subjects were instructed to perform the tracing with the right hand in a clockwise direction; in experiment 2, tracing had to be done with the left hand in a counterclockwise direction. The pen was kept on the writing tablet or writing table during each scan.

In the PET experiments, subjects traced the designs on a Calcomp 2500 digitizer using a specially designed pen,[30] both of which were connected to a Gateway 2000 PC in order to record the X- and Y-coordinates of the pen movements. Recording was done at a frequency of 100 Hz and a precision of 1 mm. The digitizer was placed on an adjustable table at the right or left side of the subject (see also Ref. 31). Subjects traced the designs continuously for 1 minute during the experimental conditions and were verbally instructed when to start and to stop tracing.

In the fMRI experiments, subjects used a fiberoptic light pen and traced the designs on a Plexiglas table. Pen and table were specifically designed for the motor learning experiments. The table was positioned over the subject's upper/middle body and fitted in slots on both sides of the scanner bed, assuring stability. Height and angle of the tabletop could be adjusted, assuring a comfortable position for each subject. The maze and square designs were cutout designs in Plexiglas blocks, which fitted in special slots in the table. Only one design was in the table slot during a particular run. When needed, the experimenter changed the design between runs. Subjects traced the designs continuously for 30 seconds during experimental blocks and were verbally instructed when to start and to stop tracing. Subjects received the instructions through headphones that they were wearing in order to dampen the loud noises generated by the scanner. Because the tabletop was positioned under an angle and was translucent (as were the maze and square designs), the trajectory of the moving light dot (produced by the light pen) was visible. To make detection of the light dot possible, all lights in the scanner room were turned off. The light dot was recorded on high quality videotape with a videocamera positioned outside the scanner room in front of the window between control and scanner rooms. With special software designed at the Washington University Neuroimaging Laboratory the X- and Y-coordinates of the videotaped light dot were recorded with a frequency of 60 Hz and a precision of 1 mm. No metal parts were used for the tracing setup in the scanner room, thus minimizing the conduction of electrical noise into the scanner bore and room and possibly degradation of the MR signal.

TASK PROCEDURE

Although in some experiments additional conditions were performed, in this report behavioral and imaging data are presented while subjects performed under the following conditions:

initial unpracticed maze tracing (**NAIVE**);
first training trial on the same maze (**TRAIN1**);
second training trial on the same maze (**TRAIN2**);
third training trial on the same maze (**TRAIN3**);
fourth training trial on the same maze (**TRAIN4**);
after 5 to10 minutes of practice on the same maze (**PRAC**);
following practice, but tracing a novel maze (**NOVEL**);
tracing a square at high speed (**SQ FAST**);
tracing the same square at low speed (**SQ SLOW**); and
holding the pen in their hand without moving it (**REST**).

(For a more detailed description of the conditions, see Ref. 31). TABLE 1 presents an overview of the conditions that were used in each experiment. As can be seen, the naïve, practiced, and square fast conditions were used in all experiments. In the PET experiments, subjects traced the designs continuously for 1 minute; in the fMRI experiments, tracing was done continuously in blocks of 30 seconds.

The square design was included in the experiments as a control condition, because it is motorically familiar and quickly "learned."[32] Great differences in movement speed between the naive and practiced maze conditions as well as in somatosensory feedback were expected. To make sure that differences in brain activations were not mainly related to changes in speed and somatosensory processes, subjects were instructed to trace the square as fast as possible (square fast condition) and to perform the square tracing slowly and to change direction only after bumping into the sides at the end of each segment (square slow condition). Regarding speed and somatosensory feedback, the slow square tracing would be comparable to naive maze tracing, while practiced maze tracing would be comparable to fast square tracing. As for the children, they had practiced the square already outside the scanner and during a practice session in the mockup scanner and were therefore also very familiar with the design.

In experiments 1, 2, and 5 in which subjects were not scanned during training trials, subjects practiced the same maze as they had traced in the naive condition for about 10 minutes during the interval between scans or runs. Resting periods were included during this practice interval to prevent subjects from slowing down as an effect of fatigue (TABLE 1).

Imaging Procedure

In both PET and fMRI experiments, subjects were positioned in the scanner in a supine position on an adjustable scanner table. An individually molded thermoplastic mask was securely and closely fit over the subject's face to minimize head movement during the experiments. To also minimize movement of the upper arm and shoulder, the arm that was used to perform the tasks was fixed with straps at the upper arm (PET experiments) or was supported by padding (fMRI experiments).

PET Imaging

The standard PET scanning activation method developed at Washington University[33–38] was used in the three PET experiments using a Siemens-CTI 953B

TABLE 1. Conditions during which subjects were scanned (indicated by X) for the five experiment

	Conditions								
Exp.	Naive	Train1	Train2	Train3	Train4	Prac	Novel	Sq.fast	Sq.slow
1	X					X	X	X	X
2	X					X	X	X	X
3	X	X	X	X	X	X	X	X	
4	X	X	X	X	X	X		X	X
5	X					X		X	

NOTE: Trainx, training x; Prac = practiced, Sq. fast = square fast; Sq. slow = square slow.

PET scanner that collected data in three-dimensional mode for 31 axial planes with center-to-center slice separation of 3.38 mm (see also Ref. 31). Subjects were positioned in the scanner so that the lower two thirds of the brain was scanned, thereby insuring coverage of the entire cerebellum.

A venous catheter was inserted in either the left (experiments 1 and 3) or the right arm (experiment 2) to administer water labeled with the positron-emitting oxygen isotope ^{15}O acting as a blood-flow tracer, which was given as a bolus of 8–10 mL of saline containing 15 mCi. Blood flow was measured as an indirect marker of neuronal activation. As soon as the bolus had entered the brain, a 40-second scan was started. Between scans, an interval of 10 minutes was inserted to allow radioactivity levels to return to baseline. During each experimental scan, subjects traced the designs for a period of 1 minute, starting 10 seonds before and finishing 10 seconds after each scan.

FMRI Imaging

MRI scans were obtained on a Siemens 1.5-Tesla Vision System (Erlangen, Germany) using the blood oxygenation level-dependent (BOLD) contrast method as an indirect marker of neuronal activation. A high resolution sagittal MP-RAGE T1-weighted sequence was used to acquire structural images with a voxel size of $1.25 \times 1 \times 1$ mm. Functional images were collected in runs with the use of an asymmetric spin-echo-planar imaging (EPI) sequence sensitive to BOLD contrast[39,40] (TR = 2,400 ms, T2 * = 50 ms, $\alpha = 90^o$). During each functional run, 118 sets of 16 contiguous, axial images were acquired, 8 mm thick, resulting in a 3.75×3.75 mm in-plane resolution. These images were acquired parallel to the anteroposterior commissure plane, offering coverage of the whole brain at a high signal-to-noise ratio.[41] Each functional run was examined for motion artifacts and automatically corrected for movement within and across runs.[42]

Each run consisted of 5 off blocks (rest), during which the subject was holding the pen in the design without moving it, and 4 on blocks (tracing), during which they traced the design. Each block lasted 30 seconds. Each run started and ended with a rest block and lasted approximately 4.5 seconds. Between runs there was an interval of approximately 5 minutes during which the experimenter could change the designs when needed.

Before the scanner was turned on, subjects were given complete instructions on the task. During each run, verbal instructions to either “start tracing” or “stop tracing” were given to the subjects at the beginning and end of each experimental block through headphones. In experiment 4 in the square conditions, the instruction was either to “start tracing fast” or “start tracing slow” in those instances in which subjects had to trace the square at high or low speed.

In experiment 5, each child underwent a training session in a special mock-up scanner 1–6 days before the actual fMRI scanning session, during which they could get accustomed to the scanner environment and scanner noises and practiced the procedure of stopping and starting while tracing the square with their eyes closed. No maze designs were used in the mock-up training sessions.

DATA ANALYSIS

Behavioral Performance

In the PET experiments, data records for all conditions were low-pass filtered at 10 Hz and the drawing trajectories were displayed on a computer screen. The beginning and end of the 1-minute tracing movement was determined by means of an interactive computer program.

In the fMRI experiments, the drawing trajectories were displayed on a computer screen using OASIS 6.0,[43] and the beginning and end of each 30-second tracing period was determined. In all experiments, mean velocity as well as the number of errors per loop was calculated for each tracing period. Tracing a dead-end path or retracing (part of) a segment was counted as an error.[32] Repeated measurement ANOVA’s were performed to study effects of condition for velocity and errors (superANOVA, version 1.1, Abacus Concepts). Probabilities were adjusted using the Greenhouse-Geisser epsilon.

Imaging Experiments

PET images were reconstructed with filtered back projection. A Butterworth 0.5 filter with an order of 5 was used, producing a transaxial resolution of 14 mm full width at half maximum. Linear normalization was applied to the smoothed images. Furthermore, due to sampling characteristics of the PET scanner resulting in much noisier uppermost and lowermost slices than those in the center, 7 slices from the top and bottom were excluded from further analysis.

The fMRI images were realigned and normalized to a common mode. The fMRI data were analyzed using an implementation of the general linear model. For the analysis in experiment 4, data of two blocks of 30 seconds were combined. Therefore, run 1 included the naive and first training condition, run 2 the second and third training condition, whereas run 3 contained the fourth training and the practiced condition. In the square fast and square slow conditions, only data from the first two blocks were used for the analysis. In experiment 5, four blocks were combined in each condition.

For each subject, images were grouped into reference–control pairs with the rest condition being subtracted from one of the experimental conditions (in the PET experiments) or the off blocks being subtracted from the on blocks (in the fMRI experiments). No excessive movement artifacts were found in any of the subtracted image pairs. All subtraction images were transformed into stereotaxic space defined in Talairach and Tournoux's atlas[44] based on a lateral skull X-ray (PET experiments) or MPRAGE (fMRI experiments). In each experiment, images acquired during identical conditions were averaged across subjects to improve the signal-to-noise ratio. In the next step a composite, summed difference image was created for each experiment that included all the experimental conditions or on-blocks minus their rest condition or off-blocks. A maximum-detection algorithm was used to identify positive local maxima in the summed image that recorded the maxima by location in stereotaxic coordinates and by magnitude. For the cerebellar regions reported in this paper, for each subtraction the magnitude was calculated per region and per subject. Repeated measurement ANOVA's were performed on all conditions for the magnitudes in each cerebellar region. Probabilities were adjusted using the Greenhouse-Geisser epsilon.

RESULTS

Behavioral Data

As can be seen in TABLE 2 and FIGURE 2, in all experiments the speed at which the mazes were traced increased significantly with practice. It is also clear that all subjects performed the fast square tracing at high speed while tracing at a slow speed in the square slow condition. In all five experiments a significant effect of condition was found for mean velocity (p <0.001). Furthermore, the velocity during practiced and square fast tracing was significantly higher (p <0.01) than during naive, novel, and square slow tracing. The only exception was found in experiment 5. The speed with which the children traced the practiced maze was not significantly faster than the speed during naive tracing. Tracing during square fast, however, was. FIGURE 2 and TABLE 2 show also that in all experiments, practice resulted in more accurate performance. Practiced maze tracing showed a significant decrease in the number of errors compared to naive or novel maze tracing. Also for errors the effect of condition was significant (p <0.01). None of the adults made any errors in the square conditions. This was in contrast to the children who made some errors in the square fast condition. However, it was only in a few instances that they retraced a segment in the square (TABLE 2 and FIG. 2).

Neuroimaging Data

Three cerebellar areas showed conditional effects in all experiments. An area in the left lateral cerebellum showed activations that were related to practice, with high activations found in naive and novel conditions, whereas activations were significantly less after practice and in any of the simple over-learned square conditions (FIG. 3, upper graph). That this change in activation was not related to changes in velocity can be seen in FIGURE 3, upper graph, when comparing left cerebellar activations in both square conditions. Although the velocity was almost four times as high in the square fast condition, left cerebellar activations were more or less the same in both square conditions. In all five experiments a significant effect of condi-

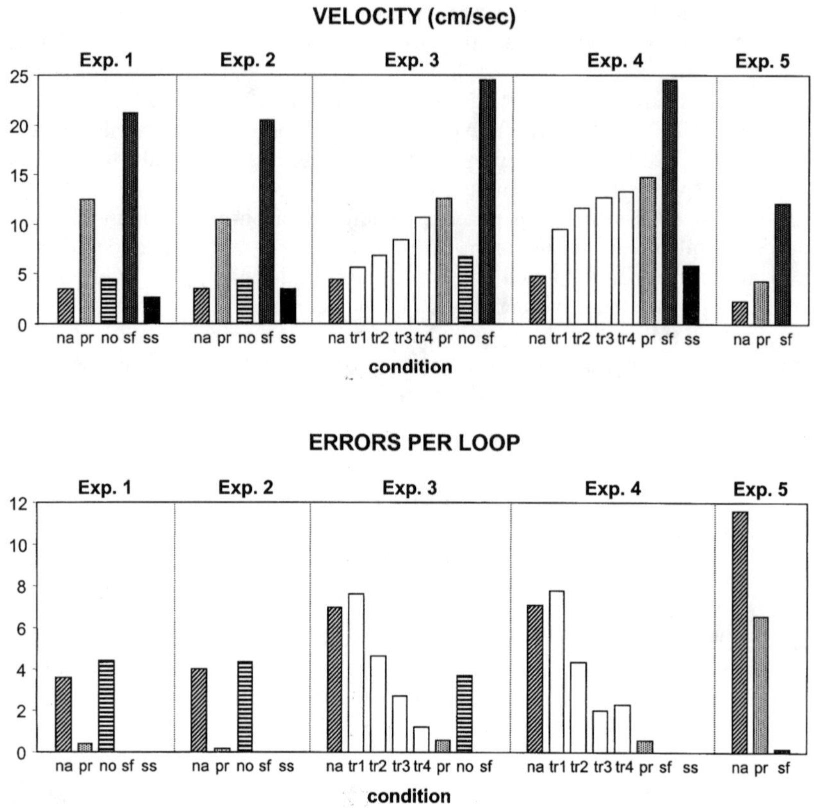

FIGURE 2. Mean velocity (*upper graph*) and number of errors per loop (*lower graph*) for each experiment as a function of condition.

tion was found on the magnitudes ($p < 0.05$). Skilled performance activated the left cerebellum significantly less than did unskilled performance. In experiments 3 and 4 in which subjects were also scanned during practice, the decrease in activation in the left cerebellum over the six practice conditions (naive, training 1-4, and practiced tracing) correlated significantly ($p < 0.01$) with the decrease in errors in these conditions, with a correlation of 0.92 in experiment 3 and 0.83 in experiment 4 (FIGS. 2 [lower graph] and 3 [upper graph], the first six columns in the graphs of experiments 3 and 4).

An area in the right lateral cerebellum showed an effect that might be related to timing requirements. As can be seen in FIGURE 3, middle graph, in those conditions in which subjects received the instruction to perform the task as quickly as possible, in the naive, novel, training, practiced, and square fast conditions, this area in the right cerebellum was highly activated. Only under the condition in which subjects did not receive the instruction to perform at high speed, the square slow condition, right cerebellum, was significantly less activated (FIG. 3, middle graph, black bars).

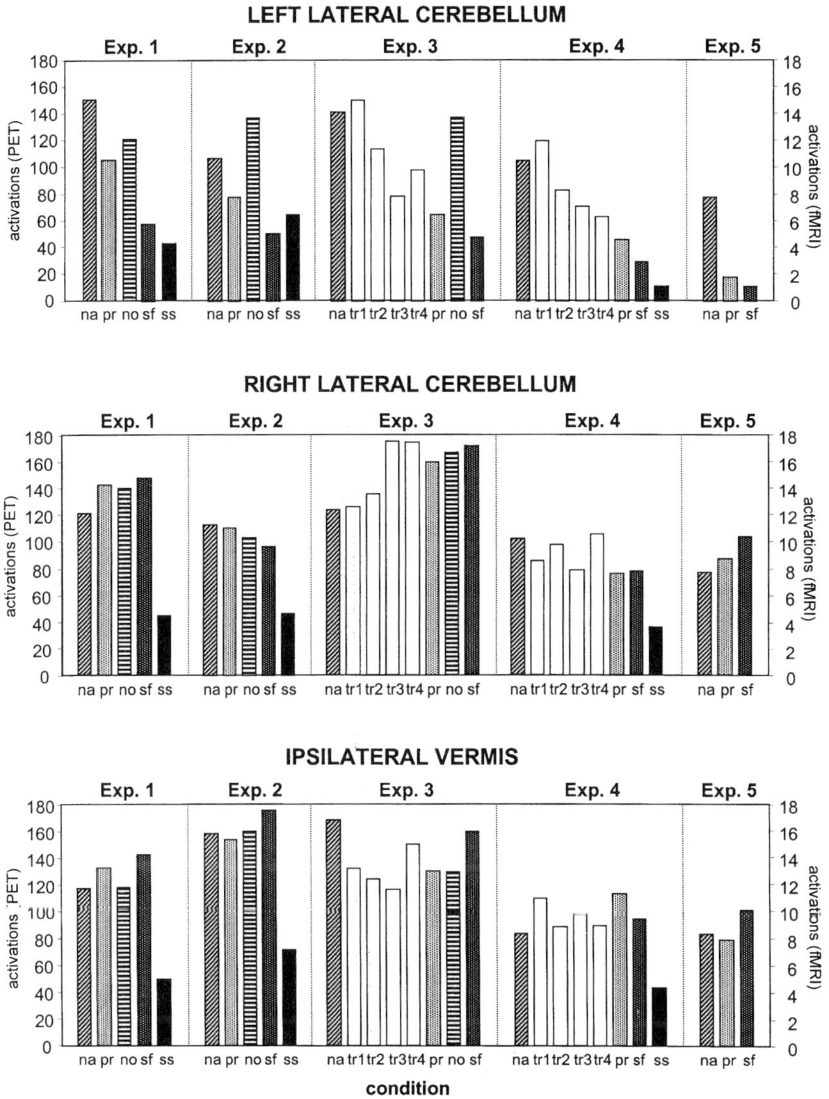

FIGURE 3. Activations in left lateral cerebellum (*upper graph*), right lateral cerebellum (*middle graph*), and ipsilateral cerebellar vermis (*lower graph*) for each experiment as a function of condition. (For coordinates, see TABLE 2.)

As expected, only in the experiments that included the square slow condition (experiments 1, 2, and 4) was a significant effect of condition measured (p <0.05).

A second cerebellar area that showed the same "timing" effects as those found in the right lateral cerebellum was observed in the anterior vermis, ipsilateral to the hand of performance (FIG. 3, lower graph). Also in this area, all conditions resulted

TABLE 2. Values of dependent variables (velocity and errors) and magnitudes in cerebellar areas per experiment and condition

		Conditions								
	Coordinates	Na	Tr1	Tr2	Tr3	Tr4	Pr	No	Sf	Ss
Velocity										
Exp. 1		3.5					12.5	4.4	21.1	2.7
Exp. 2		3.5					10.5	4.3	20.4	3.5
Exp. 3		4.4	5.6	6.8	8.4	10.7	12.6	6.7	24.5	
Exp. 4		4.8	9.5	11.6	12.7	13.4	14.8		24.5	5.9
Exp. 5		2.3					4.3		12.1	
Errors										
Exp. 1		3.6					0.4	4.4	0	0
Exp. 2		4.0					0.2	4.3	0	0
Exp. 3		7.0	7.6	4.6	2.7	1.2	0.6	3.7	0	
Exp. 4		7.1	7.8	4.3	2.0	2.3	0.6		0	0
Exp. 5		11.6					6.6		0.2	
Left lateral cerebellum										
Exp. 1	–31, –45, –30	151					106	121	58	42
Exp. 2	–31, –45, –30	107					78	138	50	64
Exp. 3	–33, –55, –32	142	151	114	79	99	65	138	48	
Exp. 4	–29, –57, –34	10.6	12.0	8.3	7.1	6.3	4.6		2.9	1.1
Exp. 5	–33, –61, –26	7.8					1.8		1.1	
Right lateral cerebellum										
Exp. 1	21, –53, –16	121					143	140	148	45
Exp. 2	25, –57, –14		119				117	109	102	49
Exp. 3	25, –51, –22	124	126	136	176	175	160	167	172	
Exp. 4	19, –53, –24	10.2	8.6	9.8	7.9	10.6	7.6		7.8	3.6
Exp. 5	27, –49, –24	7.7					8.7		10.4	
Ipsilateral vermis										
Exp. 1	11, –55, –8	124					140	125	143	49
Exp. 2	–15, –59, –8	158					154	160	176	71
Exp. 3	3, –57, –14	169	132	125	117	151	131	130	160	
Exp. 4	5, –65, –20	8.3	11.0	8.8	9.8	8.9	11.3		9.4	4.3
Exp. 5	7, –55, –8	8.3					7.9		10.1	

NOTE: Coordinates are given in Talairach[44] format in the order x, y, and z, where x is the lateral displacement from the midline (- for left hemisphere); y is the anteroposterior displacement relative to the anterior commissure (- for positions posterior to this); and z is the vertical position relative to the AC-PC line (- if below this line). Exp. X, Experiment X; Na, naïve; TrX., training X; Pr, practiced, No, novel; Sf, square fast; Ss, square slow.

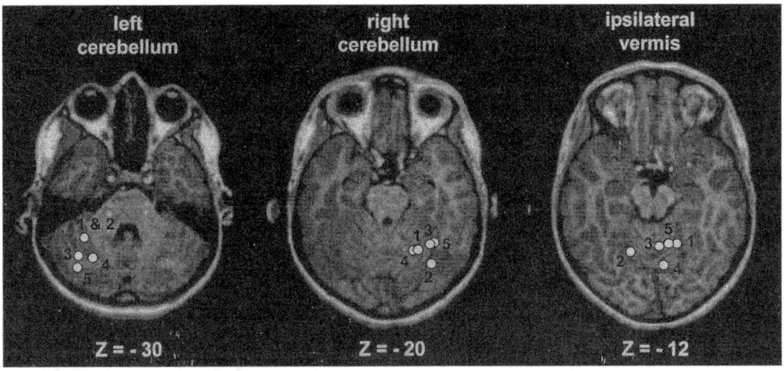

FIGURE 4. Cerebellar activations of all five experiments overlaid on horizontal anatomical images, shown in Talairach space with the z-coordinate label. *Left* in the images corresponds to left and *top* to frontal.

in high activations, except again in the square slow condition. Again, a significant conditional effect was only found in the experiments that included the square slow condition (p <0.05).

As can be seen in TABLE 2 and FIGure 4, activations in the aforementioned cerebellar areas were observed in almost the same location in all five experiments. An ANOVA comparing the x, y, and z coordinates in all five experiments showed that they could not be significantly distinguished from each other ($p = 0.46$ for the left cerebellum; $p = 0.26$ for the right cerebellum; and $p = 0.45$ for cerebellar vermis. (For analysis in vermal coordinates, all x coordinates were assumed to be located in the same hemisphere).

Note that the only cerebellar area that showed a lateralized activation was the vermis. This area was activated ipsilateral to the hand with which the task was performed. Left and right lateral cerebellum showed activations that were independent of the hand used and most likely reflect processing at a more abstract level (FIGS. 3 and 4).

DISCUSSION

The results of the five experiments described in this paper clearly show that the cerebellum is involved in sequential motor learning. Performing the same task, in all five experiments the same cerebellar areas were activated. However, these areas showed differential conditional effects. Left lateral cerebellum showed effects that were related to practice, with lower activations during practiced and over-learned performance. Right lateral cerebellum and ipsilateral vermis were activated to almost the same extent in all maze conditions and square fast tracing, with only significantly less activation observed during square slow tracing. These cerebellar changes were measured with PET as well as fMRI and were found in adults as well as in children, stressing the robustness of these findings. These findings are in agreement with the results reported in an fMRI study by Cui *et al.*[45] in which subjects performed a delayed sequential finger movement task with the right hand. The authors report activation in lateral cerebellum in both hemispheres. As in our experiments, subjects in this study were instructed to perform the sequence, which was presented in advance of an imperative signal, as quickly as possible, which might suggest involvement of the right cerebellum. The sequence of the required movements changed on each trial, resulting in the production of rather unpracticed sequences. This might suggest involvement of the left cerebellum.

To trace the maze accurately, skillfully, and smoothly, errors need to be detected and corrected, the correct sequence of the segments has to be learned, and the appropriate velocities and movement timings need to be produced. Therefore, error detection, sequential, spatial, and timing aspects (as well as other types of information, e.g., visualization, somatosensory feedback, memory, and muscle agonist-antagonist pairing) all may play a role. Results from the experiments that we report in this paper suggest that the cerebellar involvement in motor learning in the maze task is most likely related to errors and timing aspects.

Learning-Related Effects in the Left Lateral Cerebellum

There is little doubt that the cerebellum is involved in motor learning, even the simplest types of motor learning. Increased activation in the left cerebellum during early learning in the naive and novel conditions and decreased activation in this area when the task was learned or when an over-learned square was traced clearly indicate the role of the cerebellum in procedural learning. This role is also emphasized by studies involving cerebellar patients that reported impaired motor learning in these patients.[46–48] Molinari and colleagues[49] have shown that patients with left cerebellar lesions were more impaired on a procedural motor learning task than were patients with right cerebellar lesions, supporting the findings of left cerebellar involvement in our motor learning task. Furthermore, Molinari's patients were impaired on the task during both right- and left-hand performance. This finding is consistent with our data showing left cerebellar activation with both right- and left-hand performance during naive and novel maze tracing. When Gebhart and Thach[50] tested left cerebellar patients using our maze and square designs, they observed impaired learning on the maze task compared to that of controls. Patients showed less improvement and more errors. On the other hand, repeated tracing of the square design showed a learning effect in the patients that was comparable to that of the controls.

Evidence that the cerebellum plays a role in motor learning also comes from animal studies. Lalonde and Botez[51] found that animals with cerebellar damage were impaired on maze learning. Rats who had observed 200 Morris water maze trails performed by companion rats before undergoing hemicerebellectomy displayed no spatial deficits and normal exploration abilities comparable to those of control rats. This contrasts with the performance of rats who had undergone hemicerebellectomy before observation training.[52] These rats showed a complete lack of learning.

Involvement of the cerebellum during motor learning tasks was also shown in several other imaging studies. Although some studies showed increased cerebellar activation as an effect of practice,[53–55] other studies showed a decrease in cerebellar activations following learning and practice of a motor task.[56–60] Furthermore, in an experiment in which subjects were tested on a perceptual maze with little pre-training, activations were also observed in the left cerebellum.[61]

The high correlation between decreases in left cerebellar activations and decreases in errors over the course of practice in experiments 3 and 4 suggests that activation in the left cerebellum might be related to error detection and correction. Increased cerebellar activity related to performance errors during naive performance or early on during learning has been observed in other imaging studies.[23,56,58,62] The patients with left cerebellar damage tested by Gebhart and Thach[50] on the maze task showed only a moderate decrease in errors, quite different from the control subjects who showed a decrease comparable to that of subjects in our experiments 3 and 4. Fiez *et al.*[2] described a right cerebellar patient who hardly showed any decrease in errors on a verb generation task, a task that is known to engage the right cerebellum.[63] Furthermore, several theoretical models suggest involvement of the cerebellum in error-based learning.[24,64,65]

Our data as well as the aforementioned studies suggest a specialized role for the cerebellum in motor learning. As Thach states in his 1998 paper,[66] the "cognitive" cerebellum might be involved in movement planning, learing new movements, and keeping track of movement errors.

Timing-Related Effects in Ipsilateral Vermis and Right Lateral Cerebellum

The finding that the ipsilateral vermis and right lateral cerebellum were activated to more or less the same extent in all maze conditions as well as the square fast condition suggests that these areas are somehow or other involed in performance under speed instructions. In all these conditions the instruction was to perform as quickly as possible. This in contrast to the instruction during the square slow condition, where subjects were explicitly instructed to perform the task slowly and only change direction when a forced stop was experienced when the subject bumped into one of the sides. In this condition no acceleration and deceleration pattern had to be learned or used in order to move fast and smoothly through the design. In the other conditions, however, a precise timing of accelerations and decelerations is needed to produce the appropriate velocity to trace the designs as quickly as possible and to avoid bumping into the side. It is likely that this velocity-based acceleration/deceleration timing pattern might be controlled by the ipsilateral vermis and right lateral cerebellum. Research findings indeed suggest involvement of the cerebellum in the control of movement velocity.[67] It also has been shown that patients with cerebellar lesions have impairments regarding movement velocity and acceleration.[68,69] Furthermore, research with cerebellar patients also has demonstrated that these patients have deficits in velocity perception.[70,71]

In order to carry out well-learned skilled movements, especially when they are performed quickly and more or less amtomatically, it seems crucial that these movements need to be organized with correct timing. It is hypothesized that the cerebellum performs this task by sending trigger signals to the motor cortex.[72] Research with cerebellar patients also suggests a role for the cerebellum in the timing of coordinated movements.[73] Deficits in the temporal coordination of internally generated multijoint movement sequences have been observed in cerebellar patients.[74,75]

Results from brain imaging studies have linked the control of motor timing to lateral and vermal cerebellum. Reproducing rhythms of increasing temporal complexity that were presented separately in auditory and visual modalities resulted in PET activation of both lateral cerebellar cortex and cerebellar vermis.[27] In another imaging study, subjects were imaged using fMRI during synchronization and continuation of a paced finger-tapping task.[76] In this study, tapping with the right finger resulted in activations in right lateral cerebellum and vermis. Lateral cerebellar activations also were observed in another fMRI study in which subjects performed a memory-timed finger movement task.[77]

Although both lateral and vermal cerebellum seem to be involved in motor timing, it seems unlikely that they are involved in the same timing processes. Animal and human lesion studies suggest a differential contribution of medial and lateral cerebellum in motor timing. A study with rats showed differential effects of midline and lateral cerebellar lesions. While midline lesions resulted in deficits in motor coordination, lesions in lateral areas showed deficits in spatial orientation.[78] However, in this study lesions were bilateral, and lateral lesions included the cerebellar hemispheres and dentate nucleus. A follow-up study with rats, in which the dentate nucleus was selectively lesioned,[79] showed the same deficit in spatial orientation.

The fact that in our experiments the vermis was activated ipsilateral to the hand that was used to perform the tasks (right vermis in experiments 1, 3, 4, and 5, and left vermis in experiment 2) suggests that this vermal activation might be related to timing aspects involved in motor execution. A study[80] involving patients with cerebellar lesions in either the medial zone of the cerebellum or the lateral cerebellum reported that all patients showed an increased variability in a repetitive tapping task; however, this variablity was attributable to different processes. While patients with lateral lesions showed a deficit in central timing processes as to determine when a response should be made, patients with lesions in the medial area showed a variance related to motor processes, most likely showing an impairment in the implementation and execution of the timed response. Results of a more recent study[81] regarding lateral and vermal cerebellar involvement in temporal processing support the data by Ivry *et al.*[80] mentioned earlier. When comparing cerebellar vermis-lesioned and cerebellar hemisphere-lesioned rats on a temporal discrimination task,[82] it was observed that only the rats with lesions in the cerebellar hemisphere showed deficits in perceptual timing. Neuroimaging data with respect to cerebellar activations also provided results supporting differential roles for vermis and lateral cerebellum in motor timing.[83,84] In a PET study in which subjects performed a visuomanual tracking task by tracking a moving target at different rates, it was found that activation in ipsilateral anterior vermis correlated with the rate of movement, implicating this area in the control of low-level kinematic or dynamical parameters of movement execution.[83] An fMRI study in which subjects perceived visual and auditory rhythms but no motor response was made during scanning, showed that only lateral cerebellum was activated, no activation in medial areas was found,[84] again suggesting that activations in the vermis are most likely related to motor execution.

Movement timing is one of the processes involved in motor control. More and more evidence supports the view that the cerebellum plays an important role in this process. Research data also suggest that different cerebellar areas control different aspects of motor timing. Although the exact role of different cerebellar regions in motor timing is not known, it seems that the vermis is most likely involved in lower processes of motor timing, like the execution of timed movements, while the lateral cerebellum appears to play a role in the cognitive aspects of motor timing.

Comparisons between PET and fMRI

As for comparisons between the two imaging techniques, PET and fMRI, several studies that compared both techniques directly, either in different subjects focusing on a specific cortical area[85,86] or in the same subjects focusing on the whole brain,[87] have shown good colocalization of the activations obtained with both techniques. Our data regarding cerebellar areas confirm these results and are consistent with the right lateral cerebellar activation observed in subjects who performed self-paced right index finger movements during PET and fMRI scanning.[87] We can add to these findings that there is also good colocalization between cerebellar activations in children and adults.

CONCLUSION

The results from our experiments might be helpful in understanding the role of the cerebellum in motor learning especially regarding the cognitive aspects involved in motor learning. The differential activations in different cerebellar areas support the hypothesis that the cerebellum is most likely involved in several functions[88] that might be controlled by different cerebellar areas. Each area seems to be specialized for a specific aspect of motor learning.

REFERENCES

1. WALLESCH, C.W. & A. HORN. 1990. Long-term effects of cerebellar pathology on cognitive functions. Brain Cogn. **14:** 19–25.
2. FIEZ, J.A., S.E. PETERSEN, M.K. CHENEY & M.E. RAICHLE. 1992. Impaired non-motor learning and error detection associated with cerebellar damage. A single case study. Brain **115:** 155–178.
3. SILVERI, M.C., A.M. DI BETTA, V. FILIPPINI, *et al.* 1998. Verbal short-term store-rehearsal system and the cerebellum. Evidence from a patient with a right cerebellar lesion. Brain **121:** 2175–2187.
4. TIMMANN, D., J. DREPPER, M. MASCHKE, *et al.* 2002. Motor deficits cannot explain impaired cognitive associative learning in cerebellar patients. Neuropsychologia **40:** 788–800.
5. LEVISOHN, L., A. CRONIN-GOLOMB & J.D. SCHMAHMANN. 2000. Neuropsychological consequences of cerebellar tumour resection in children: cerebellar cognitive affective syndrome in a paediatric population. Brain **123:** 1041–1050.
6. RIVA, D. & C GIORGI. 2000. The cerebellum contributes to higher functions during development: evidence from a series of children surgically treated for posterior fossa tumours. Brain **123:** 1051–1061.
7. MIDDLETON, F.A. & P.L. STRICK. 1994. Anatomical evidence for cerebellar and basal ganglia involvement in higher cognitive function. Science **266:** 458–461.
8. MIDDLETON, F.A. & P.L. STRICK. 1998. Cerebellar output: motor and cognitive channels. Trends Cogn. Sci. **2:** 348–354.
9. MIDDLETON, F.A. & P.L. STRICK. 2001. Cerebellar projections to the prefrontal cortex of the primate. J. Neurosci. **21:** 700–712.
10. CABEZA, R. & L. NYBERG. 2000. Imaging cognition II: An empirical review of 275 PET and fMRI studies. J. Cogn. Neurosci. **12:** 1–47.
11. DECETY, J., H. SJOHOLM, E. RYDING, *et al.* 1990. The cerebellum participates in mental activity: Tomographic measurement of regional cerebral blood flow. Brain Res. **535:** 313–317.

12. RYDING, E., J. DECETY, H. SJOHOLM, *et al.* 1993. Motor imagery activates the cerebellum regionally. A SPECT rCBF study with 99mTc-HMPAO. Cogn. Brain Res. **1:** 94–99.
13. LOTZE, M., P. MONTOYA, M. ERB, *et al.* 1999. Activation of cortical and cerebellar motor areas during executed and imagined hand movements: an fMRI study. J. Cogn. Neurosci. **11:** 491–501.
14. INO, T., Y. INOUE, M. KAGE, *et al.* 2002. Mental navigation in humans is processed in the anterior bank of the parieto-occipital sulcus. Neurosci. Lett. **322:** 182–186.
15. MARR, D. 1969. A theory of cerebellar cortex. J. Physiol. **202:** 437–470.
16. ALBUS, J.S. 1971. A theory of cerebellar function. Math. Biosci. **10:** 22–61.
17. RAYMOND, J.L., S.G. LISBERGER & M.D. MAUK. 1996. The cerebellum: a neuronal learning machine? Science **272:** 1126–1131.
18. VAN MIER, H. 2000. "Human learning." *In* Brain Mapping: The Systems. A.W. Toga & J.C. Mazziotta, Eds. : 605–621. Academic Press. San Diego.
19. THACH, W.T., H.P. GOODKIN & J.G. KEATING. 1992. The cerebellum and the adaptive coordination of movement. Annu. Rev. Neurosci. **15:** 403–442.
20. GOODKIN, H.P., J.G. KEATING, T.A. MARTIN & W.T. THACH. 1993. Preserved simple and impaired compound movement after infarction in the territory of the superior cerebellar artery. Can. J. Neurol. Sci. **20:** S93–104.
21. BASTIAN, A.J., T.A. MARTIN, J.G. KEATING & W.T. THACH. 1996. Cerebellar ataxia: abnormal control of interaction torques across multiple joints. J. Neurophysiol. **76:** 492–509.
22. TOPKA, H., J. KONCZAK, K. SCHNEIDER, *et al.* 1998. Multijoint arm movements in cerebellar ataxia: abnormal control of movement dynamics. Exp. Brain Res. **119:** 493–503.
23. MIALL, R.C., H. IMAMIZU & S. MIYAUCHI. 2000. Activation of the cerebellum in coordinated eye and hand tracking movements: an fMRI study. Exp. Brain Res. **135:** 22–33.
24. KAWATO, M. & H. GOMI. 1992. A computational model of four regions of the cerebellum based on feedback-error learning. Biol. Cybern. **68:** 95–103.
25. IVRY, R.B. & S.W. KEELE. 1989. Timing functions of the cerebellum. J. Cogn. Neurosci. **1:** 136–152.
26. THACH, W.T. 1996. On the specific role of the cerebellum in motor learning and cognition: clues from PET activation and lesion studies in man. Behav. Brain Sci. **19:** 411–431.
27. PENHUME, V.B., R.J. ZATORRE & A.C. EVANS. 1998. Cerebellar contributions to motor timing: a PET study of auditory and visual rhythm reproduction. J. Cogn. Neurosci. **10:** 752–765.
28. MAUK, M.D., J.F. MEDINA, W.L. NORES & T. OHYAMA. 2000. Cerebellar function: coordination, learning or timing? Curr. Biol. **10:** R522–525.
29. RACZKOWSKI, D., J.W. KALAT & R. NEBES. 1974. Reliability and validity of some handedness questionnaire items. Neuropsychologia **6:** 43–47.
30. MAARSE, F.J., H.J.J. JANSSEN & F. DEXEL. 1988. A special pen for an XY-tablet. In Computers. *In* Psychology; Methods, Instrumentation, and Psychodiagnostics. F.J. Maarse, L.J.M. Mulder, W.P.B. Sjouw & A.E. Akkerman, Eds. :133–139. Swets and Zeitlinger. Amsterdam.
31. VAN MIER, H., L.W. TEMPEL, J.S. PERLMUTTER, *et al.* 1998. Changes in brain activity during motor learning measured with PET; effects of hand of performance and practice. J. Neurophysiol. **80:** 2177–2200.
32. VAN MIER, H., W. HULSTIJN & S.E. PETERSEN. 1993. Changes in motor planning during the acquisition of movement patterns in a continuous task. Acta Psychol. **82:** 291–312.
33. FOX, P.T., J.S. PERLMUTTER & M.E. RAICHLE. 1985. A stereotactic method of anatomical localization for positron emission tomography. J. Comput. Assisted Tomogr. **9:** 141–153.
34. FOX, P.T., M.A. MINTUN, E.M. REIMAN & M.E. RAICHLE. 1988. Enhanced detection of focal brain responses using intersubject averaging and change-distribution analysis of subtracted pet images. J. Cereb. Blood Flow Metab. **8:** 642–653.

35. FOX, P.T. & M.A. MINTUN. 1989. Noninvasive functional brain mapping by change-distribution analysis of averaged pet images of $H_2{}^{15}O$ tissue activity. J. Nucl. Med. **30:** 141–149.
36. HERSCOVITCH, P., J. MARKHAM & M.E. RAICHLE. 1983. Brain blood flow measured with $H_2{}^{15}O$. I. Theory and error analysis. J. Nucl. Med. **24:** 782–789.
37. MINTUN, M.A., P.T. FOX & M.E. RAICHLE. 1989. A highly accurate method of localizing regions of neuronal activation in the human brain with positron emission tomography. J. Cereb. Blood Flow Metab. **9:** 96–103.
38. RAICHLE, M.E., W.R.W. MARTIN, P. HERSCOVITCH, *et al.* 1983. Brain blood flow measured with $H_2{}^{15}O$. II. Implementation and validation. J. Nucl. Med. **24:** 790–798.
39. KWONG, K.K., J.W. BELLIVEAU, D.A. CHESLER, *et al.* 1992. Dynamic magnetic resonance imaging of human brain activity during primary sensory stimulation. Proc. Natl. Acad. Sci. USA **89:** 5675–5679.
40. OGAWA, S., D.W. TANK, R. MENON, *et al.* 1992. Intrinsic signal changes accompanying sensory stimulation: Functional brain mapping with magnetic resonance imaging. Proc. Natl. Acad. Sci. USA **89:** 5951–5955.
41. CONTURO, T.E., R.C. MCKINSTRY, E. AKBUDAK, *et al.* 1996. Sensitivity optimization and experimental design in functional magnetic resonance imaging. Neurosci. Abstr. **22:** 7.
42. SNYDER, A.Z. 1996. Difference image vs. ratio image error function forms in PET-PET realignment. *In* Quantification of Brain Function Using PET. R. Myers, V. Cunningham, D. Bailey & T. Jones, Eds. :131–137. Academic Press. San Diego.
43. DE JONG, W.P., W. HULSTIJN, B.J.M. KOSTERMAN & B.C.M. SMITS-ENGELSMAN. 1996. OASIS software and its application in experimental handwriting research. *In* Handwriting and Drawing Research: Basic and Applied Issues. M.L. Simner, C.G. Leedham & A.J.W.M. Thomassen, Eds. IOS. Amsterdam.
44. TALAIRACH, J. & P. TOURNOUX. 1988. Co-planar Stereotaxic Atlas of the Human Brain. Thieme. New York.
45. CUI, S.Z., E.Z. LI, Y.F. ZANG, *et al.* 2000. Both sides of human cerebellum involved in preparation and execution of sequential movements. Neuroreport **11:** 3849–3853.
46. SANES, J.N., B. DIMITROV & M. HALLETT. 1990. Motor learning in patients with cerebellar dysfunction. Brain **113:** 103–120.
47. PASCUAL-LEONE, A, J. GRAFMAN, K. CLARK, *et al.* 1993. Procedural learning in Parkinson's disease and cerebellar degeneration. Ann. Neurol. **34:** 594–602.
48. LAFORCE, R., JR. & J. DOYON. 2001. Distinct contribution of the striatum and cerebellum to motor learning. Brain Cogn. **45:** 189–211.
49. MOLINARI, M., M.G. LEGGIO, A. SOLIDA, *et al.* 1997. Cerebellum and procedural learning. Evidence from focal cerebellar lesions. Brain **120:** 1753–1762.
50. GEBHART, A.L. & W.T. THACH. 1999. Cerebellar subjects show cognitive impairments on a novel maze learning task. Neurosci. Abstr. **25:** 650.
51. LALONDE, R. & M.I. BOTEZ. 1990. The cerebellum and learning processes in animals. Brain Res. Rev. **15:** 325–332.
52. LEGGIO, M.G., M. MOLINARI, P. NERI, *et al.* 2000. Representation of actions in rats: the role of cerebellum in learning spatial performances by observation. Proc. Natl. Acad. Sci. USA **97:** 2320–2325.
53. Doyon, J., A.M. Owen, M. Petrides, *et al.* 1996. Functional anatomy of visuomotor skill learning in human subjects examined with positron emission tomography. Eur. J. Neurosci. **8:** 637–648.
54. GRAFTON, S.T., R.P. WOODS & M. TYSZKA. 1994. Functional imaging of procedural motor learning: relating cerebral blood flow with individual subject performance. Human Brain Mapping **1:** 221–234.
55. SEITZ, R.J., A.G. CANAVAN, L. YAGUEZ, *et al.* 1994. Successive roles of the cerebellum and premotor cortices in trajectorial learning. Neuroreport **20:** 2541–2544.
56. JENKINS, I.H., D.J. BROOKS, P.D. NIXON, *et al.* 1994. Motor sequence learning: a study with positron emission tomography. J. Neurosci. **14:** 3775–3790.
57. FRISTON, K.J., C.D. FRITH, R.E. PASSINGHAM, *et al.* 1992. Motor practice and neurophysiological adaptation in the cerebellum: a positron tomography study. Proc. Roy. Soc. Lond.-Series B. Biol. Sci. **248:** 223–228.

58. FLAMENT, D., J.M. ELLERMANN, S-G. KIM, *et al.* 1996. Functional magnetic resonance imaging of cerebellar activation during the learning of a visuomotor dissociation task. Human Brain Mapping **4:** 210–226.
59. TONI, I., M. KRAMS, R. TURNER & R.E PASSINGHAM. 1998. The time course of changes during motor sequence learning: a whole-brain fMRI study. Neuroimage **8:** 50–61.
60. JUEPTNER, M., K.M. STEPHAN, C.D. FRITH, *et al.* 1997. Anatomy of motor learning. I. Frontal cortex and attention to action. J. Neurophysiol. **77:** 1313–1324.
61. GHATAN, P.H., J.-C. HSEIH, A. WIRSEN-MEURLING, *et al.* 1995. Brain activation induced by the perceptual maze test: A PET study of cognitive performance. Neuroimage **2:** 112–124.
62. IMAMIZU, H., S. MIYAUCHI, T. TAMADA, *et al.* 2000. Human cerebellar activity reflecting an acquired internal model of a new tool. Nature **403:** 192–195.
63. RAICHLE, M.E., J.A. FIEZ, T.O. VIDEEN, *et al.* 1994. Practice-related changes in human brain functional anatomy during nonmotor learning. Cereb. Cortex **4:** 8–26.
64. DOYA, K. 1999. What are the computations of the cerebellum, the basal ganglia and the cerebral cortex? Neural Networks **12:** 961–974.
65. HOUK, J.C., J.T. BUCKINGHAM & A.G. BARTO. 1996. Models of the cerebellum and motor learning. Behav. Brain Sci. **19:** 368–383.
66. THACH, W.T. 1998. What is the role of the cerebellum in motor learning and cognition? Trends Cogn. Sci. **9:** 331–337.
67. EBNER, T.J. 1998. A role for the cerebellum in the control of limb movement velocity. Curr. Opin. Neurobiol. **8:** 762–769.
68. DIENER, H.C. & J. DICHGANS. 1992. Pathophysiology of cerebellar ataxia. Mov. Disord. **7:** 95–109.
69. HORE, J., B. WILD & H.C. DIENER. 1991. Cerebellar dysmetria at the elbow, wrist, and fingers. J. Neurophysiol. **65:** 563–571.
70. IVRY, R. & H.C. DIENER. 1991. Impaired velocity perception in patients with lesions of the cerebellum. J. Cogn. Neurosci. **3:** 355–366.
71. NAWROT, M. & M. RIZZO. 1998. Chronic motion perception deficits from midline cerebellar lesions in human. Vision Res. **38:** 2219–2224.
72. HIKOSAKA, O., H. NAKAHARA, M.K. RAND, *et al.* 1999. Parallel neural networks for learning sequential procedures. Trends Neurosci. **22:** 464–471.
73. IVRY, R. 1993. Cerebellar involvement in the explicit representation of temporal information. Ann. N.Y. Acad. Sci. **682:** 214–230.
74. FELLOWS, S.J., J. ERNST, M. SCHWARZ, *et al.* 2001. Precision grip deficits in cerebellar disorders in man. Clin. Neurophysiol. **112:** 1793–1802.
75. TIMMANN, D., S. WATTS & J. HORE. 1999. Failure of cerebellar patients to time finger opening precisely causes ball high-low inaccuracy in overarm throws. J. Neurophysiol. **82:** 103–114.
76. JANCKE, L., R. LOOSE, K. LUTZ, *et al.* 2000. Cortical activations during paced finger-tapping applying visual and auditory pacing stimuli. Cogn. Brain Res. **10:** 51–66.
77. KAWASHIMA, R., J. OKUDA, A. UMETSU, *et al.* 2000. Human cerebellum plays an important role in memory-timed finger movement: an fMRI study. J. Neurophysiol. **83:** 1079–87.
78. JOYAL, C.C., C. MEYER, G. JACQUART, *et al.* 1996. Effects of midline and lateral cerebellar lesions on motor coordination and spatial orientation. Brain Res. **739:** 1–11.
79. JOYAL, C.C., C. STRAZIELLE & R. LALONDE. 2001. Effects of dentate nucleus lesions on spatial and postural sensorimotor learning in rats. Behav. Brain Res. **122:** 131–137.
80. IVRY, R.B., S.W. KEELE & H.C. DIENER. 1988. Dissociation of the lateral and medial cerebellum in movement timing and movement execution. Exp. Brain Res. **73:** 167–180.
81. MALAPANI, C., B. DUBOIS, G. RANCUREL & J. GIBBON. 1998. Cerebellar dysfunctions of temporal processing in the seconds range in humans. Neuroreport **9:** 3907–391.
82. BREUKELAAR, J.W. & J.C. DALRYMPLE–ALFORD. 1999. Effects of lesions to the cerebellar vermis and hemispheres on timing and counting in rats. Behav. Neurosci. **113:** 78–90.
83. TURNER, R.S., S.T. GRAFTON, J.R. VOTAW, *et al.* 1998. Motor subcircuits mediating the control of movement velocity: a PET study. J. Neurophysiol. **80:** 2162–2176.
84. SCHUBOTZ, R.I., A.D. FRIEDERICI & D.Y. VON CRAMON. 2000. Time perception and motor timing: a common cortical and subcortical basis revealed by fMRI. Neuroimage **11:** 1–12.

85. Kraut, M.A., S. Marenco, B.J. Soher, *et al.* 1995. Comparison of functional MR and (H_2O)-O-15 positron emission tomography in stimulation of the primary visual cortex. Am. J. Neuroradiol. **16:** 2101–2107.
86. Ramsey, N.F., B.S. Kirkby, T. Vangelderen, *et al.* 1996. Functional mapping of human sensorimotor cortex with 3D BOLD fMRI correlates highly with (H_2O)-O-15 PET rCBF. J. Cereb. Blood Flow Metab. **16:** 755–764.
87. Joliot, M., D. Papathanassiou, E. Mellet, *et al.* 1999. FMRI and PET of self-paced finger movement: comparison of intersubject stereotaxic averaged data. Neuroimage **10:** 430–447.
88. Mauk, M.D., J.F. Medina, W.L. Nores, *et al.* 2000. Cerebellar function: coordination, learning, or timing? Curr. Biol. **10:** R522–525.

Cognitive Processing by the Cerebellum

Summary of Discussion

S. PETERSEN, *Rapporteur*

Pursuant to Raichle's introduction to this session, Bower commented that functional imaging in humans has provoked two enormous changes in our thinking about the cerebellum. The first is the intense activation that accompanies many task performances that are explicitly and implicitly *not* motor. Based on clinical lesion studies of the late 1800s and early 1900s, the belief was that the exclusive domain of the cerebellum was in control of posture and movement. The second is that in learning situations with repeated task performances, the cerebellum is activated *during* the learning, but not after the learning is acquired. This would seem "to go 180 degrees against" the popular idea that a key role of the cerebellum in motor learning is to copy the instructions from cerebral cortex, store the memory, and subsequently control the performance unconsciously and automatically. Ito also puzzled as to why the cerebellar activation was so great at the beginning of motor learning and less at the end. He did point out, however, that in studies by Kawato, there was still some cerebellar activity associated with mental learned performance, albeit in a very much smaller area. Raichle responded that the relation between the dynamics of blood flow, oxygen extraction, and neural activity is still not understood and might underlie some of these observations. Thach replied to Bower's first comment that lesion studies are now suggesting cognitive deficits after cerebellar lesions, but it remains a puzzle as to why they are not as obvious as cerebellar motor deficits. As to the second comment, he suggested that cerebellar "activation" during learning could possibly reflect predominantly metabolic activity and blood flow preferentially associated with the massive synaptic activity of the climbing fiber, which is argued to be preferentially active during the leaning process and not, or less so, during retention and storage.

Bower then went on to ask whether any study has shown that cerebellar activation scales with task difficulty. Ivry replied that cerebellar activation could reflect the overt or covert or mental rehearsal of a motor component of the task. For example, in the verb generate task, presentation of the noun "apple" could cause the subject variously to consider the verbs "picking," "peeling," or "typing" and the actions that would go with them. The more potential actions, the more the cerebellar activation. Van Mier then replied that lateralization of activation would argue *against* such a movement-based notion: that maze learning activates the left cerebellar hemisphere (and the right cerebral hemisphere) independent of which hand is used and independent of the cerebellar activation that is tied to the use of the ipsilateral hand.

Cohen challenged the conclusion that right cerebellar lesions give significant deficits in language. He pointed out that the data of Gebhart *et al.* showed that one right cerebellar subject had a language deficit much less severe than that of the other subjects and scarcely deviant from her normal control. Gebhart replied that the differ-

Ann. N.Y. Acad. Sci. 978: 354–356 (2002). © 2002 New York Academy of Sciences.

ence on the language tests between the right and the left cerebellar groups showed considerable variation, and the one subject was indeed minimally impaired. Indeed, there had been some question as to whether the right cerebellar lesion found on MRI by her physician had been a congenital lesion rather than a fresh infarct and whether it might have been compensated over time. Cohen suggested that it remains to be seen if right cerebellar lesions produce clinically significant language deficits. Gebhart agreed that more subjects with equivalent lesions need to be studied. Nevertheless, several of the patients with right cerebellar lesions (including the one studied by Fiez *et al.*) were noted by immediate family and caregivers to have speech problems after their stroke: they could not immediately give names of relatives, could not think of the correct word or gave wrong names or wrong words, and were usually unaware of their errors when made. These deficits would be similar in kind to some aspects of an expressive dysphasia resulting from a left cerebral lesion, although less severe in degree.

Cohen asked if there had been imaging studies in other mental activities such as the mental rehearsals of athletes performed in the absence of overt motor and EMG activity. Furthermore, could cerebellar activation in such rehearsal be used to *prevent* the unwanted, but usually associated, motor activity? Van Mier replied that lateralization of the cerebellar activation (maze learning, left cerebellum; language learning, right cerebellum) would weigh against so simple a mechanism concerned with a specific body part. Van Mier and Petersen both mentioned that there indeed have been imaging studies in which the cerebellum was activated during motor planning and rehearsal without an accompanying EMG (e.g., studies of Decety and Ingvar).

Ivry presented work showing that patients with cerebellar lesions make greater errors during discontinous movements at "starts" and "stops" than during continuous movements, consistent with the possibility that a principal control function of the cerebellum is timing significant events. Someone asked about deficits that might result from cerebellar lesions in sensory discrimination tasks, such as visual acuity in the peripheral visual fields, and perception of location in space. Would such a finding weigh against a pivotal role of the cerebellum in perceptual and motor timing? Ivry replied that in his view the disturbances in sensory discriminatory performances occurred only in cerebellar degenerative disease and therefore might reflect malfunction of some other region involved in parallel or cerebral diaschesis.

Horak asked whether Ivry's proposal of preferential control of timing was consistent with Ito and others' view of the cerebellum as an internal model of motor and mental performance. Ivry replied that the two concepts had many features in common. Petersen replied to both Cohen's and Horak's comments that indeed spoken language performance in particular and mental performance in general appeared less accurate and smooth after a cerebellar lesion, consistent with the Ito model.

Llinas asked Strick whether in his studies the mossy fiber pontocerebellar projection went to deep nuclei as well as to cerebellar cortex, and, if so, to what relative extent. Strick replied that the projection went to deep nuclei as well as to granule and Golgi cells, but that the projection to the granule cells was greater. Highstein asked Strick about projections to the inferior olive from the deep cerebellar nuclei, and Strick also confirmed this projection if time was allowed for virus to pass from deep nuclei to the next stage.

Welsh then questioned Strick about the association of autism with both hypoplasia of the cerebellar vermis *and* abnormal frontal lobe function. Specifically, are there preferential connections between the cerebellar vermis and the frontal lobe? Strick replied first that in autism there is actually more widespread hypoplasia of the cerebellum than is popularly believed, including Crus II lateral cerebellar cortex, which *is* disturbed preferentially.

Strick was asked about the meaning of phylogenetic differences in the cerebro-ponto-cerebellar projection ranging from entire cerebrum in "lower" animals to only parts of the cerebrum in "higher" animals and man. Specifically, he noted a lack of cerebral projection from area 17 and the "ventral" visual pathway and wondered if that suggests a cerebellar cognitive role more in spatial processing and orientation than in sensory discriminations. Strick replied that it did seem that the cerebrocerebellar projection (including prefrontal regions) had more to do with the anatomy of the "dorsal visual pathway" and was different from the basal ganglia, where there were instead two-way connections with the TE of the temporal lobe not shared with the cerebellum. Petersen replied that right cerebellar activation seemed specific to language production (and left cerebellar to spatial activity) and that these activities were two-way connected through corresponding activities in the contralateral cerebrum. Llinas wondered if there was a *left* cerebellar contribution to the presumed *right* cerebral cortical role in language *prosody*. Ivry and Raichle commented that this was pretty much an open question and that the cerebellum had been "overlooked" in studies of prosody.

Thach wondered if some, as yet unrecognized, control variable might be operating in the right cerebellum and left cerebrum that might be more fundamental than either "language" and "eye–limb timing" but that might underlie them both. Might there be some other control variable underlying cerebellar and cerebral roles in spatial operations? In any event, it would seem likely that the cerebrum and the cerebellum do not do the same thing for language, spatial operations, or timing. The question rests in what each might specifically contribute to complement each other.

Plasticity and Dynamics in the Olivocerebellar System

Introduction to Presentations

C.I. DE ZEEUW

This VIth series of papers deals with recent advances in the field of olivocerebellar research on motor learning and motor timing. The series starts off with a chapter by Schmolesky *et al.*, which attempts to solve the puzzle of the composition of the complex spike by dissecting it into several pieces that are formed by various conductances in the different compartments of the Purkinje cell. Special emphasis is put on dendritic calcium spike components, which may act as a key player in several forms of synaptic plasticity such as long-term depression of the parallel fiber-Purkinje cell synapse (PF-LTD) and possibly even at the climbing fiber-Purkinje cell synapse (CF-LTD). The investigators explain how this latter form of plasticity raises new possibilities, which may make cerebellar synaptic plasticity an even more interesting model to study mechanisms of learning and memory.

The fact that the complex spike is not as invariable as assumed does not exclude the possibility that it may also serve as the beating instrument of a timing device that controls motor timing. Llinas, Welsh, Lang, and others have provided evidence that synchronous complex spike activities can be correlated with the onset of particular types of movements.[1,2] It is their tenet that chemical synaptic inputs to dendritic spines in the inferior olive that are coupled by gap junctions orchestrate dynamic patterns of synchronous complex spike activities and thereby the timing of corresponding motor domains. (For a review see De Zeeuw *et al.*[3]) Surprisingly, the experiments by Kistler *et al.* described in the second chapter of this series do not provide supportive evidence for this hypothesis. They investigated the motor behavior and complex spike synchrony of a knockout of Connexin36, which so far is the only gap junction protein known to be involved in the formation of electrical junctions between olivary neurons. Even though these null-mutants do not appear to have functional olivary gap junctions (see also Long *et al.*[4]), they were not ataxic, they showed normal performance on the accelerating rotorod, and they had a fairly regular walking pattern. Moreover, they showed normal compensatory eye movements during sinusoidal visual and/or vestibular stimulation, and they even showed comparable tremors after application of harmaline, which is known to specifically turn olivary neurons into an oscillatory state at a high frequency. The Connexin36 null-mutants only showed a small delay in a velocity-step response to optokinetic stimulation. However, this deficit probably reflects a delay in the retina, because visually evoked potentials in these animals have a similar delay, whereas the same optokinetic paradigm does not show an increased latency in *lurcher* mice, which is a spontaneous mutant in which the entire cerebellar output is missing due to the lack of Purkinje

Ann. N.Y. Acad. Sci. 978: 357–358 (2002). © 2002 New York Academy of Sciences.

cells. (See also Van Alphen *et al.*, this issue.) Equally surprising was the finding that the complex spike synchrony in the mutant did not differ from that in the wild type. Thus, data obtained with the Connexin36 null-mutant do not support the hypothesis that electrotonic coupling by gap junctions underlies synchronization of olivary spike activity or that these gap junctions are essential for normal motor performance.

In contrast, the third chapter by Hirano *et al.* shows that Golgi cells are essential for a normal motor performance. In their elegant study, they specifically eliminated Golgi cells and abolished GABA and synaptic inhibition in the granular layer by injecting exotoxin coupled to anti-IL2Rα antibody in the cerebellum of a transgenic mouse in which Golgi cells express the human interleukin-2 receptor α subunit (IL2Rα). Their findings suggest that elimination of the GABAergic inhibitory output of Golgi cells initially causes severe ataxia due to overexcitation of the granule cells and that motor control gradually and partly recovers due to a compensatory downregulation of NMDA receptors in granule cells. In addition, they showed that the inhibitory interneurons in the cerebellar cortex are not only necessary for proper motor performance, but also the subject of a rather unique form of homosynaptic plasticity. They show that the rebound potentiation of the inhibitory synapses onto Purkinje cells, which occurs after depolarization of the Purkinje cell, is suppressed when the presynaptic stellate cell or basket cell is activated during depolarization of the Purkinje cell and that this suppression is mediated by a cAMP-dependent protein kinase pathway.

The fourth chapter by Van Alphen *et al.* shows what happens when all forms of plasticity in the cerebellar cortex, such as PF-LTD and CF-LTD, described in Schmolesky's chapter, or rebound potentiation and its suppression, described in Hirano's chapter, are absent due to removal of the postsynaptic target, the Purkinje cell. They show that in *lurcher* mice not only motor performance but also all forms of motor learning, that is, both short-term adaptation and long-term adaptation are disturbed. Thus, even though some plastic mechanisms may take place downstream of the cerebellar cortex, they probably still require an intact output of this structure. Interestingly, the motor performance of the *lurcher* appears to be affected in that the amplitude and time course of the movement, rather than the latency of the movement, are impaired. Thus, the role of the cerebellum in motor control may be more prominent in guiding ongoing movements than in initiating them. Finally, as described in the other chapters of this series and as mentioned during the discussion of the meeting, Van Alphen and colleagues address the potential compensatory mechanisms that may occur in mutant mice.

REFERENCES

1. De Zeeuw, C.I., J.I. Simpson, C.C. Hoogenraad, *et al.* 1998. Microcircuitry and function of the inferior olive. Trends Neurosci. **21:** 391–400.
2. Lang, E.J. 2002. GABAergic and glutamatergic modulation of spontaneous and motor-cortex-evoked complex spike activity. J. Neurophysiol. **87:** 1993–2008.
3. Long, M.A., M.R. Deans, S.L. Patrick, *et al.* 2001. Connexin36-dependent electrical coupling synchronizes subthreshold rhythmic activity of inferior olivary neurons. *In* Soc. Neurosci. Abstr. 70.13.
4. Welsh, J.P., E.J. Lang, I. Sugihara & R. Llinás. 1995. Dynamic organization of motor control within the olivocerebellar system. Nature **374:** 453–457.

The Making of a Complex Spike: Ionic Composition and Plasticity

MATTHEW T. SCHMOLESKY, JOHN T. WEBER, CHRIS I. DE ZEEUW, AND CHRISTIAN HANSEL

Department of Neuroscience, Erasmus Medical Center, 3000 DR Rotterdam, The Netherlands

ABSTRACT: Climbing fiber (CF) activation evokes a large all-or-nothing electrical response in Purkinje cells (PCs), the complex spike. It has been suggested that the role of CFs (and thus complex spikes) is that of a "teacher" in simple learning paradigms such as associative eyeblink conditioning. An alternative hypothesis describes the olivocerebellar system as part of a timing device and denies a role of the CF input in learning. To date, neither of these hypotheses nor others can definitively be verified or discounted. Similarly, the complex spike evades a clear understanding when it comes to the cellular events underlying complex spike generation. What is known, however, is that complex spikes are associated with large dendritic calcium signals that are required for the induction of long-term depression (LTD) at the parallel fiber (PF)-PC synapse. PF-LTD is a form of long-term synaptic plasticity that has been suggested to underlie certain forms of cerebellar motor learning. In contrast to the PF input, the CF input has been considered invariant. Our recent discovery of LTD at the CF input shows that complex spikes are less static than previously assumed. In addition to depression of CF-evoked excitatory postsynaptic currents, long-lasting, selective reduction of slow complex spike components could be observed after brief CF tetanization. To understand the functional implications of CF-LTD, it is crucial to know the types of currents constituting the specific complex spike components. Here we review the "anatomy" of the complex spike as well as our observations of activity-dependent complex spike waveform modifications. In addition, we discuss which properties CF-LTD might add to the circuitry of the cerebellar cortex.

KEYWORDS: cerebellum; complex spike; Purkinje cell; synaptic plasticity

INTRODUCTION

The climbing fiber (CF) provides one of two types of excitatory input that cerebellar Purkinje cells (PCs) receive. CF activity is assumed to convey information related to the unconditioned stimulus in reflex conditioning pathways, such as the airpuff signal in associative eyeblink conditioning. In Marr-Albus-Ito models of cerebellar motor learning,[1–3] the CF thus functions as a "teacher" that activates PCs

Address for correspondence: Christian Hansel, Department of Neuroscience, Erasmus Medical Center, P.O. Box 1738, 3000 DR Rotterdam, The Netherlands. Voice: +31-10-4087404; fax: +31-10-4089459.
hansel@anat.fgg.eur.nl

Ann. N.Y. Acad. Sci. 978: 359–390 (2002).

when an error signal, such as an airpuff, occurs. Learning is assumed to take place at the other excitatory input to PCs, the parallel fibers (PFs), which convey information related to the conditioned stimulus. (For a review see Refs. 4 and 5.) The CF input has also been described as a timing device,[6] based on the synchronous oscillatory activity in the inferior olive, the structure from which CFs originate.[7] Another possibility is that CFs simply participate in sensorimotor integration by providing an additional relay pathway parallel to the mossy fiber/granule cell/parallel fiber system. (For a review see Refs. 8 and 9.)

At the cellular level, the CF-PC synapse is unique. In the adult mammalian brain, each PC receives input from only one CF. However, this input is extremely powerful, as it provides ~1,500 synaptic contact sites that are characterized by a high release probability.[10–12] These CF synapses primarily contact spines on the proximal part of the PC dendritic tree. (For a review see Ref. 13.) CF stimulation results in a large all-or-nothing electrical event in PCs[14] that was termed the "complex spike."[15] Extensive studies on the complex spike from the 1960s on have led to the assumption that they are composed of somatic sodium conductances and dendritic calcium conductances. Although this picture largely holds true to this day, the advent of patch-clamp technology and the development of more selective and potent pharmacologic agents have provided greater knowledge of the spectrum of ligand- and voltage-gated conductances present in PCs, which could be involved in the generation of complex spikes.

A key towards understanding CF function might be found in its role in synaptic plasticity. The discovery of various forms of synaptic plasticity at excitatory and inhibitory inputs to PCs has changed our perception of how these neurons integrate synaptic information. (For a review see Ref. 5.) Long-term changes in synaptic strength can be either positive (long-term potentiation, LTP) or negative (long-term depression, LTD) and are generally assumed to provide the cellular substrate for learning and memory. The type of synaptic plasticity in the cerebellar cortex that has received the most attention is LTD at PF-PC synapses.[16] PF-LTD is dependent on simultaneous activation of PFs and CFs (corresponding to the concept of "associativity" in learning paradigms), but the CF input itself has been considered invariant. CF activity is required for PF-LTD induction, as it contributes a large dendritic Ca signal that is needed to trigger the induction process.[17] Similarly, CF-evoked Ca signals are required for the induction of "rebound potentiation," a form of LTP of inhibitory inputs to PCs,[18] as well as several forms of short-term plasticity at synapses contacting PCs. (For a review see Ref. 5.)

In contrast to previous assumptions, we recently discovered that the CF-PC synapse also shows a form of synaptic plasticity. Brief tetanization of the CF resulted in long-term depression of CF-evoked excitatory postsynaptic currents (EPSCs) recorded in voltage-clamp mode.[19] This tetanization also resulted in a modification of the complex spike measured in current-clamp mode. Whereas the fast, initial Na spike was largely unaffected, a moderate but reliable reduction occurred in slow complex spike components, which presumably reflect dendritic Ca conductances. This selective modification of the complex spike waveform led us to reexamine the ionic conductances that are believed to underlie the multiple components of a complex spike. Here, we attempt to bring both sides of complex spike electrophysiology together by summarizing our current knowledge of the ionic basis of complex spikes as well as describing how the complex spike waveform can be modified.

ELECTRICAL EVENTS UPON CLIMBING FIBER ACTIVATION

Excitatory postsynaptic potentials (EPSPs) that trigger complex spike generation when the spike threshold is reached were first described in detail in anesthetized cats by Llinas and Nicholson[20] who studied the reversal properties of CF-EPSPs. A characterization of CF-evoked synaptic events at the receptor level was made possible by the development of selective glutamate receptor antagonists as well as the application of patch-clamp technology to brain slices. Performing whole-cell patch-clamp recordings in voltage-clamp mode from PCs of rat cerebellar slices, it was shown that CF-EPSCs preserve the all-or-nothing character of complex spikes (see below) and that CF-EPSCs undergo paired-pulse depression (PPD) upon repeated stimulation. Furthermore, it was shown that CF-EPSCs can be blocked by CNQX, a selective antagonist of the AMPA/kainate subtype of glutamate receptors, but not by APV, a selective antagonist of the NMDA subtype of glutamate receptors.[21–24] This pharmacological profile as well as recent studies using specific antibodies and quantitative immunogold analyses suggest that glutamate is the CF transmitter. (For a review see Ref. 25; see also Refs. 26 and 27 for reviews on the previously discussed CF transmitter candidates aspartate and homocysteic acid, respectively.)

The first detailed electrophysiological characterization of responses triggered by CF activity in cerebellar PCs was provided by Eccles and colleagues.[14,28] These investigators recorded extracellularly and intracellularly from PCs of anesthetized cats *in vivo*. The extracellular recordings revealed a strong all-or-nothing excitatory response after CF stimulation, which consisted of several spike components of diminishing amplitude. Intracellular recordings showed a response that consisted of a fast, initial spike followed by several slower spikelets of smaller amplitude that emerged out of a slow depolarization plateau (FIG. 1).[14,28] The complex nature of this electrical response suggested that several, possibly independent sites of activation of voltage-dependent conductances might be involved. The hypothesis that some complex spike components were initiated in PC dendrites was supported by the finding that variations in somatic current injection affected synaptically evoked simple spikes to a greater degree than complex spikes.[29] (See also Ref. 30.) This finding suggested that relative to the simple spike initiation site, at least some components of the complex spike were initiated further away from the soma. Additional evidence for the participation of PC dendrites in spike firing was provided by intradendritic recordings from PCs of anesthetized alligators.[31]

How are complex spikes generated in PCs and what is their ionic basis? A direct demonstration of ionic conductances specifically involved in complex spike generation is complicated by the fact that pharmacological blockers of specific voltage-dependent conductances also have an impact on synaptic release. For this reason, the available information is mostly indirect in nature. The most complete description of ionic conductances involved in PC electroresponsiveness has been provided by Llinas and Sugimori.[32,33] Using intrasomatic and intradendritic sharp microelectrode recordings from PCs of guinea pig cerebellar slices, these investigators described several types of Na, Ca, and K conductances in PCs. Direct activation of PCs intrasomatically or intradendritically by current injection elicited two types of spike firing, fast somatic spikes and/or depolarizing spike bursts. Through amplitude comparisons of these spike components at different recording sites and through the use of blockers of voltage-dependent Na or Ca conductances, the investigators were able

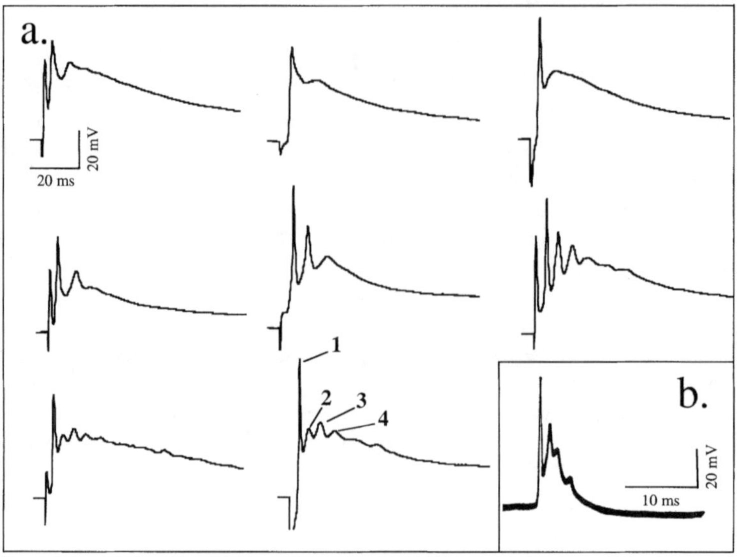

FIGURE 1. (**a**) Whole-cell patch-clamp recordings of complex spikes from eight Purkinje cells (PCs) in rat cerebellar slices *in vitro*. In all cases, the waveform displays a single fast Na spike (spike 1). The number of subsequent slow spikes (spikes 2–4) varied between cells. Sometimes the second complex spike component was simply a depolarization plateau or "hump" (*top row*). In four of the eight cases, stimulation also activated the PC axon, eliciting an antidromic spike seen preceding the complex spike itself (e.g., *top left*). (**b**) Sharp microelectrode recording of a complex spike from a PC in a cat *in vivo*. Apart from differences in kinetics (note *scale bars*), the waveform and amplitudes of complex spikes recorded in the brain slice and in the intact brain are remarkably similar. Panel **b** is modified from Eccles *et al.*[28] Used with permission from The Physiological Society.

to describe the fast spikes as somatically activated Na conductances and the depolarizing spike bursts as dendritically activated Ca conductances. The same distribution of conductances has been found in PCs of the turtle cerebellum.[34] Two important observations link these data to the complex spike: (a) fast spikes evoked by antidromic stimulation were always smaller in the dendrites than in the soma and (b) the depolarizing spike burst resembled the response observed *in vivo* following CF activation.[32,33] While the first observation suggests that fast Na spikes are generally elicited in the soma or the axon hillock and only passively spread into the dendrites, the latter suggests that the late complex spike components are, as the depolarizing spike burst, generated by dendritic Ca conductances. (For a discussion see Ref. 33.) In addition to the Na and Ca conductances just mentioned, Llinas and Sugimori described a non-inactivating Na conductance in the soma. It is important to note that the complex spike waveform can vary from cell to cell and from trial to trial, possibly involving different relative conductances at different times. FIGURE 1a shows examples of complex spike waveforms obtained from our own whole-cell patch-clamp recordings from PCs in rat cerebellar slices. Although the number of smaller spikelets varied typically between one and three and, on rare occasions, even four to five, it is obvious that the complex spike waveform recorded under these *in vitro* conditions strongly resembles those measured *in vivo.* (For comparison, see FIG. 1b.) Thus, the slice preparation is adequately suited to study complex spike components.

Na, Ca, and K conductances are all important for both complex spike waveform generation and synaptic plasticity. The large Ca influx following CF activity is certainly the dominant cellular event associated with a complex spike and can trigger a variety of intracellular cascades, ranging from those inducing various types of synaptic plasticity to those promoting cell death. The K conductances present in PCs deserve attention as well. They mediate the resting membrane potential, repolarization and afterhyperpolarization for example, and thus have a significant impact on cell excitability and spike activity patterns. Finally, Na conductances in PCs are receiving increasing attention, as they determine a major part of the electroresponsiveness of the soma and thus provide the crucial link between electrical input collected in the dendrites and the output initiated in the axon hillock. While this notion holds true for almost all types of neurons, Na conductances in PCs are particularly interesting because of a unique current promoting repetitive burst firing in the soma. In the following sections, we describe these various types of ion conductances in PCs and their involvement in the generation of complex spikes. Most of the experiments described were performed using PCs from rodent cerebellar slices.

SODIUM CONDUCTANCES

Llinas and colleagues described inactivating and non-inactivating tetrodotoxin (TTX)-sensitive Na currents in PC somata that generate fast spikes and a slow plateau potential, respectively.[32,35] A TTX-sensitive, biphasic Na current was also found when depolarization step protocols were applied to outside-out patches from somatic membranes of PCs from organotypic cultures.[36] The current amplitudes measured in single-channel recordings during the early and late phase of the Na current were comparable, indicating that both Na current phases could be generated by the same type of Na channel.

Another TTX-sensitive Na current was described in dissociated PCs, a preparation that preserves the soma and stump of the proximal dendrite of PCs. This slow current could be elicited upon repolarization after a depolarization to positive potentials and was termed the "resurgent Na current."[37–39] Single-channel recordings and biophysical studies showed that the same Na channel can mediate both fast and resurgent current components. In fact, the resurgent current appears to be a component of the sodium current that results from inactivated sodium channels recovering through an open state. The reopening of channels upon repolarization from positive potentials has two important consequences for high frequency or burst firing. First, it produces a depolarization that can promote additional short-latency spikes. Second, the reopening is, by definition, a recovery from inactivation, which can occur in these channels even in the absence of strong hyperpolarization.[39] Because a single action potential can inactivate Na channels by more than 90%,[37] recovery is necessary to restore an availability of Na channels that is sufficient to produce bursts of action potentials.

Various types of Na channel α subunits have been found in PCs, including Nav1.1, Nav1.2, and Nav1.6 (equivalent to rat brain I, rat brain II, and Scn8a; see

Refs. 40–42; see also Ref. 43 for nomenclature). Voltage-dependent Na channels consist of an α subunit, forming the channel core, that can be associated with accessory β subunits. The resurgent Na current, the non-inactivating Na current, and, to a lesser degree, the fast Na current were reduced in Nav1.6 mutant mice, indicating that this subunit generates most of the slow Na current components.[44] These mice also showed strongly reduced spontaneous and evoked burst firing in PCs, an observation that not only highlights the importance of slow, subthreshold Na currents in general burst firing, but also indicates that these currents might participate in slow complex spike components.

A key result of the studies conducted by Llinas and Sugimori was that Na spikes were generated in the soma, but not the dendrites of PCs, and that dendritic spike activity resulted from the activation of Ca conductances.[32,33] This spatial distribution of Na influx into PCs was confirmed by simultaneous electrophysiological and Na imaging experiments using the fluorescent Na indicator SBFI.[45,46] In these experiments, fast spikes, elicited by current injections into the soma or the dendrites, were always associated with Na signals in the soma and the axon, but not in the dendrites of PCs. Dendritic Na signals were observed after CF stimulation, but are presumably caused by Na influx through ligand-gated channels, as direct dendritic current injection did not result in similar Na signals.

The question of where sodium action potentials are initiated and how they spread within PCs was addressed using simultaneous somatodendritic as well as somatoaxonal double-patch recordings.[47] Whether the sodium spikes were evoked by somatic or by dendritic current injections, the spike occurred first in the soma and then in the dendrites, and the dendritic spike amplitude decreased drastically with the increasing distance from the soma. Somatoaxonal recordings demonstrated more specifically that the spikes were always initiated in the axon or axon hillock. This finding is in contrast to results published earlier,[48] which showed that sodium action potentials can occur in PC dendrites. This latter study, however, was conducted after axotomy using somatic voltage-clamp recordings, presenting recording conditions that reduce the probability of somatoaxonal spike initiation. Moreover, using outside-out patches, it was clearly demonstrated that PC Na channel density declines with increasing distance from the soma.[47] Although this observation explains the low level of Na influx into PC dendrites even after direct current injection,[46] it was suggested that a more significant reason for the failure of Na spikes in PCs to backpropagate into the dendrites lies in the high degree of dendritic branching in PCs.[49]

Somatodendritic double-patch recordings were also used to monitor electric responses associated with complex spike firing in PCs.[47] Similar to the results obtained with direct somatic or dendritic current injection, the action potential was first recorded in the soma and then in the dendrite (FIG. 2). This observation directly demonstrates that the initial, fast component of the complex spike is also initiated not in the dendrite, but at a site closer to the soma.

The slow complex spike components have been described as resulting from the activation of dendritic Ca conductances.[33] An alternative view that has been expressed is based on the finding that complex spike-like events can be evoked in the soma by intrasomatic depolarization or anode break stimulation.[46] Based on this observation, the investigators suggested that, at least at the somatic level, the slow complex spike components are predominantly mediated by Na conductances and that a high probability to fire intrinsic regenerative potentials might be a general feature of

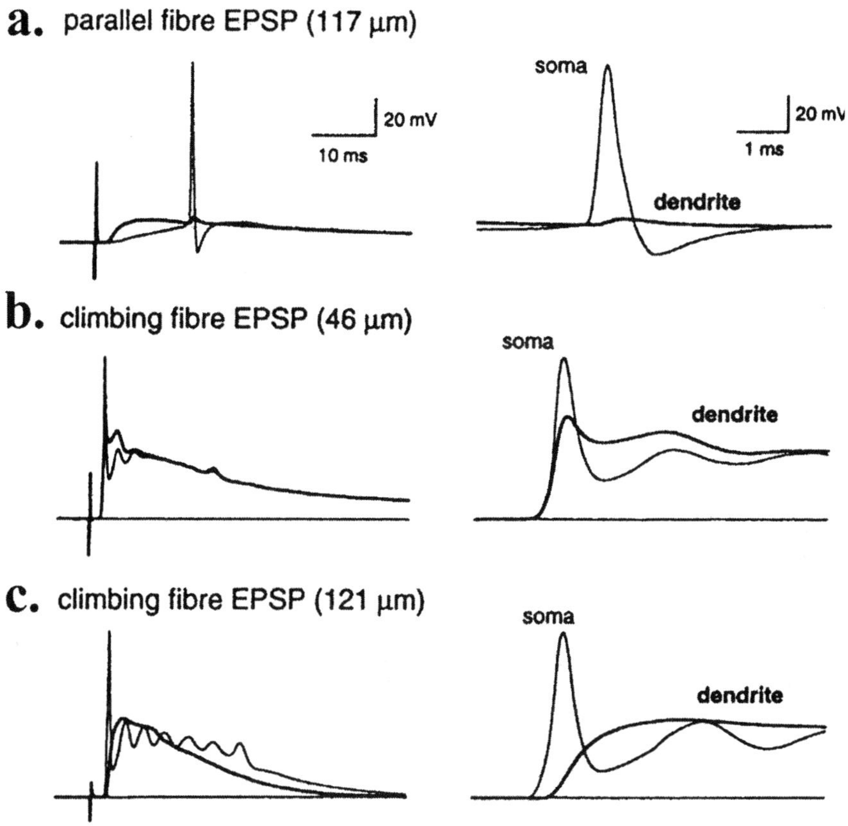

FIGURE 2. Action potential initiation by synaptic potentials. (**a-c**) (*Left)* Simultaneous somatic and dendritic whole-cell recordings from Purkinje cells (PCs) in rat cerebellar slices showing action potentials evoked by a parallel fiber (PF) excitatory postsynaptic potential (EPSP) (**a**) or climbing fiber (CF) EPSPs (**b, c**) at different distances from the soma. (*Right*) The same responses shown at higher time resolution. The *thicker traces* show the response recorded in the dendrite. This figure is modified from Stuart and Häusser.[47] Used with permission from Elsevier Science.

PC somas. This view is certainly strengthened by the characterization of the resurgent Na current in PCs,[37,38] which contributes to Na-dependent burst firing in the soma. Interestingly, in somatodendritic double-patch recordings by Stuart and Häusser,[47] a higher number of spikelets can be seen in the soma than in the dendrites during complex spike firing (FIG. 2). Thus, Ca spikes in the dendrites may solely be reflected as a depolarizing plateau in the soma, and the somatically recorded spikelets may be locally generated.

When interpreting these data it must be kept in mind that a tendency for repetitive spike firing in PC somata in no way precludes the possibility that Ca spikes also occur in the dendrites. The previous results of Llinas and Sugimori[33] convincingly demonstrate dendritic Ca spike activity. Moreover, it was shown that spike firing

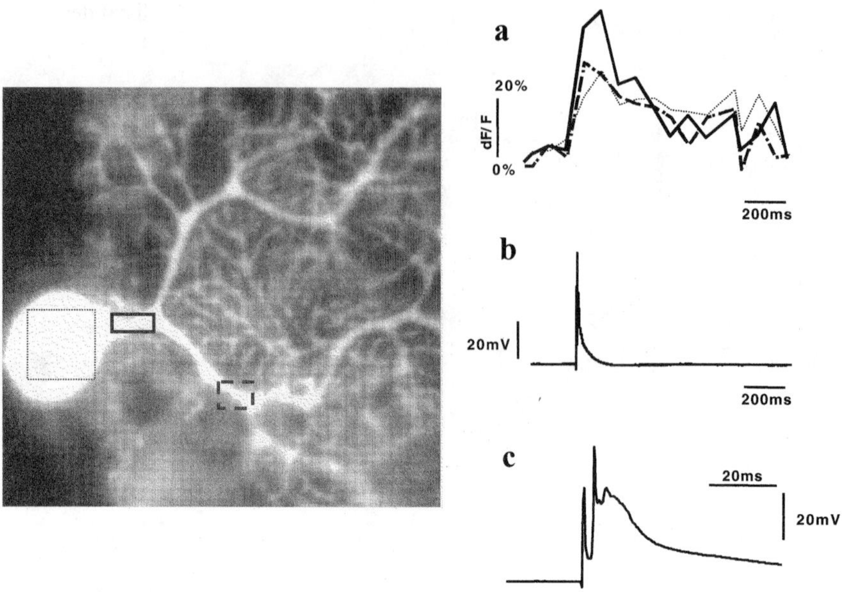

FIGURE 3. Climbing fiber (CF)-evoked Ca transients in a Purkinje cell (PC) from a rat cerebellar slice. The cell was loaded with the fluorescent Ca indicator dye Oregon Green BAPTA-2 (200 μM) and visualized using a cooled CCD camera. (**a**) Ca transients evoked by a single CF stimulus at the somatic and dendritic locations indicated in the image on the *left*. (**b**) Corresponding complex spike evoked by the same CF stimulus. (**c**) The same complex spike as in **b** at a higher temporal resolution.

evoked by intrasomatic current injection was reduced when external Ca was substituted by Co.[44] As this burst firing depended on the activation of Scn8a sodium currents, it was hypothesized that both voltage-dependent Ca currents as well as Scn8a Na currents contribute to somatic burst firing. Currently, the question as to what degree the spikelets that are somatically recorded during the complex spike faithfully reflect dendritic spike activity remains open.

CALCIUM CONDUCTANCES

One of the most significant cellular events following CF activation is a large, widespread Ca transient in PC dendrites. CF-evoked Ca transients were initially monitored using simultaneous electrophysiological and optical recordings with either photodiode arrays or charge-coupled device (CCD) cameras and Ca indicator dyes such as Arsenazo III[50] or fura-2.[17,51–53] CF activation results in Ca transients throughout the cell that typically reach higher amplitudes in the dendrite than in the soma (FIG. 3). Measurements of the spatial distribution of CF-evoked Ca transients have shown that their amplitudes were relatively constant in the proximal dendrites, but considerably more variable in the distal dendrites where they could be either larger or smaller than in the proximal dendrites.[52] Interestingly, the distal dendritic Ca signals could be enhanced by pairing the CF activation with the somatic injection of depolarizing current pulses, indicating that the spatial spread could depend on the ongoing electric activity in the dendrite. Large dendritic Ca signals could also be evoked by intrasomatic current injection alone.[54] These Ca signals could reach higher amplitudes in fine branches of the distal dendrites than in the proximal dendrites or the soma, often resembling CF-evoked Ca transients in their amplitude and spatial distribution.[52] This finding indicates that the major source of CF-evoked Ca signals are voltage-dependent Ca channels. Similar imaging studies indicate that intrasomatic injection of depolarizing currents leads initially to Ca plateau potentials in the fine branches of the distal dendrites and subsequently to Ca spikes in the main branches.[55,56] However, the recordings by Lev-Ram *et al.*[54] indicate that Ca spikes can occur in individual fine branches of the distal dendrites alone. The development of confocal and 2-photon laser scanning microscopy has enabled imaging studies at the level of PC dendritic spines. Studies at the PF-PC synapse suggested that the smallest unit of dendritic integration and Ca signaling may be the individual dendritic spine or the spine plus the adjacent dendritic shaft.[57,58] It was also shown that simultaneous CF and PF activation leads to supralinear Ca signals in spines contacted by PF inputs.[59] However, CF evoked Ca signals in spines contacted by CF terminals have not yet been characterized. In addition, it remains to be described where along the dendritic axis and under what conditions CF stimulation alone can evoke Ca transients in PF spines.

Which Ca sources contribute to the large Ca transients evoked by CF activity? For this consideration we must focus on the participation of various types of voltage-dependent Ca channels as these are directly involved in the electrogenic processes underlying the formation of the complex spike. It has to be kept in mind, however, that other sources, such as Ca release from internal stores, contribute to the Ca signal as well. (For a review see Ref. 60.)

Voltage-dependent Ca channels are hetero-oligomeric proteins composed of a channel-forming subunit (α1) and several other subunits that are not strictly required for channel formation, but rather have regulatory or auxiliary functions. The α1 subunits are derived from a number of genes (α1 A-I; for a review see Ref. 61). The various α1 subunits can be assigned to different Ca channel types that have been characterized on the basis of electrophysiological and pharmacological properties. The low voltage-activated T-type channel can be formed by α1G, α1H, and α1I subunits, whereas the intermediate voltage-activated R-type channel is presumably formed by the α1E subunit. There are different types of high voltage-activated Ca channels, namely, P/Q-type (α1A), N-type (α1B), and L-type channels (α1C, α1D, and α1S). Based on immunocytochemical studies of α1 subunits, a general pattern of expression in different neuronal compartments could be established for the various Ca channel types. (For a review see Ref. 61.) N-type channels are predominantly expressed in dendrites.[62] P/Q-type channels have been found in the dendrites and somata of PCs.[63,64] Similarly, L-type channels are expressed in the soma as well as the proximal dendrites.[65] R-type channels are also mainly expressed in the soma and dendrites of PCs, but they show a more distal dendritic localization pattern than do the L-type channels.[66]

How do these different Ca channel types participate in the generation of a complex spike upon CF activation? The experimental evidence is once more indirect, as most of the available data have been obtained by analyzing spike patterns evoked by intracellular current injections. Using this technique, it was shown by Llinas and Sugimori[33] that Ca-mediated spikes can be generated in PC dendrites. Due to its abundant expression in the dendrites, it was suggested that the P-type channel might be involved in the generation of dendritic Ca spikes.[67,68] (For a review see Ref. 69.) Llinas and colleagues directly demonstrated such an involvement of P-type channels in Ca spike generation using funnel web spider toxin, a selective blocker of P-type channels.[67] Additional support for the assumption that P-type channels mediate dendritic Ca spike firing came from a study that systematically analyzed the involvement of the different types of Ca channels in dendritic Ca spike generation using a combination of patch-clamp recordings and fluorescent Ca measurements with a cooled CCD camera.[70] Dendritic Ca spikes were evoked by intrasomatic current injection, and the effect of selective Ca channel blockers on the dendritic as well as somatic Ca transients was monitored. These Ca spikes were abolished by blocking P-type channels (ω-Agatoxin-IVA), but not by blocking L-type (nifedipine) or N-type channels (ω-conotoxin-GVIA). T-type channels were blocked by low concentrations of Ni^{2+}, resulting in a relative delay in spike onset. These data demonstrate that Ca spikes in PC dendrites are mainly generated by activation of P/Q-type channels and that the low voltage-activated T-type channels might have a role in spike boosting. Interestingly, in agreement with previous somatic recordings of P-type currents,[68] ω-Agatoxin-IVA application also abolished the somatic Ca transients.

A significant, but slightly different role for P-type and T-type channels in dendritic Ca spike generation was suggested by recordings from PCs in cerebellar slice culture.[71] In this study, it was shown that P-type channels do not directly generate Ca spikes, but rather contribute a Ca-dependent plateau component that can be observed when K channels are blocked by TEA. The Ca spikes, in contrast, were attributed to the activation of low voltage-activated Ca currents. In combination with results obtained from *in situ* hybridization experiments that demonstrated the presence of α1G subunits in the Purkinje cell and molecular layer,[72] these studies indicate the presence of functional T-type Ca channels in PC dendrites. However, these findings are in contrast to previous suggestions that no low voltage-activated Ca currents are present[63,68] or that they are present only in a certain percentage of PCs.[73] This discrepancy might be explained by a possible decrease in functional T-type channel density during development as, in contrast to the aforementioned studies, the presence of low voltage-activated Ca currents sensitive to Ni^{2+} and amiloride was clearly demonstrated in PCs from rat cerebellar slice cultures.[74,75] These organotypic cultures were obtained from slices prepared from newborn rats (thus representing an early developmental stage) and were used for electrophysiological recordings after 2 weeks.[76]

Currently, we have no available data that directly show which types of voltage-dependent Ca channels participate in the generation of the CF-evoked complex spike. The studies just summarized analyzed the involvement of different Ca channels in dendritic Ca spike firing that can be elicited by current injection. Under these conditions, there is a strong indication for a dependence on P/Q-type channels and/or T-type channels. CF activation and somatic or dendritic current injection lead to different forms of dendritic excitation. Whereas the former involves synaptic events that locally lead to the activation of voltage-dependent Ca conductances, the latter leads to a general spread of depolarization into the dendrites. Thus, the location and type of activated Ca channels need not be the same. Despite this limitation, it still seems fair to assume that of all types of voltage-dependent Ca channels present, P/Q-type and T-type channels are the most likely to participate in the generation of the dendritic Ca spike activity that occurs during the complex spike. Furthermore, any type of voltage-dependent Ca channel in the soma, such as P-type channels, are likely to be involved in the somatic electrogenesis that occurs during the complex spike.

POTASSIUM CONDUCTANCES

Potassium channels can be divided into four broad groups based on genetic, structural, physiological, and pharmacological characterizations: (1) voltage-activated potassium (Kv) channels; (2) calcium-activated potassium (K_{ca}) channels (big, intermediate, and small conductance channels known as BK, IK, and SK, respectively); (3) inward rectifying (K_{ir}) potassium channels; and (4) two-pore domain (K_{2P}) or "leak" potassium channels. (For a review see Refs. 77–81.) A vast array of functionally discrete potassium channels are formed because of the variety of potassium subunits known to exist (>35 in the Kv family alone), posttranslational modification of these subunits, the potential for subunit heteromerization, and functional alteration of the channel via assembly with auxiliary subunits (e.g., Kvβ subunits).[79]

VOLTAGE-ACTIVATED POTASSIUM (KV) CHANNELS

Voltage-activated potassium (Kv) channels play a role in action potential firing threshold, frequency, burst duration, repolarization, and afterhyperpolarization. Each of these features can be seen in PC recordings (FIG. 4), and different Kv channels have been implicated in PCs to serve them.

The classical A-type K current (I_A) operates in the subthreshold range of action potential generation and activates rapidly at relatively low voltage thresholds. This current also inactivates quickly in an essentially V-independent manner. It is known that Kv4 proteins are the key subunits underlying I_A[82] and that at least one Kv1 subunit, Kv1.4, is also capable of forming transient A-type channels.[83] *In situ* hybridization reveals an intense signal in PCs for Kv4.3, a weaker signal for Kv4.1, and no signal at all for Kv4.2.[83] Multiple studies have also shown that Kv1.1, 1.2, β1, and β2 and, to a lesser extent, Kv1.4 and 1.6 are present in the cerebellum.[84–93] More specifically, intense staining has been found in the PC somata for Kv1.1,[94] Kv1.5,[95] and Kvβ1.[96] Thus, the Kv4 and/or Kv1 subunits are likely to underlie the transient A-type currents that have been recorded in the PC somata.[53,87] Because A-type channels are active in a voltage range below action potential threshold, where many other channels are closed, the modification of these channels could dramatically affect PC spike output.

In contrast to somatic A-type currents, PC dendrites express a macroscopic current with slower inactivation kinetics.[53] For example, somatic current injection protocols result in a series of calcium action potentials of dendritic origin that show a characteristic delay.[32–34,53,54,70] The application of 4-aminopyridine reduces this delay.[70] While A-type channels are 4-aminopyridine sensitive, so too are the slow-inactivating "delayed" or D-type K currents first described in hippocampal cells.[97,98]

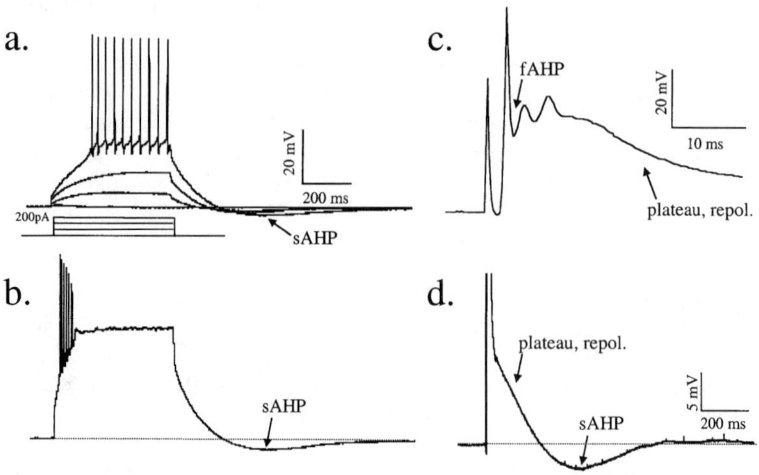

FIGURE 4. Somatic current injection (**a, b**) and climbing fiber (CF) synaptic activation (**c, d**) of Purkinje cells (PCs) in cerebellar slices can be used to study K currents. The traces shown in **a** and **b** are from a typical PC that displays a slow afterhyperpolarization (sAHP) following current offset, a threshold for action potential generation (200 pA in this case), and (**b**) reduced spike count at higher intensities (500 pA) where Na channels remain inactivated. Note also the fast afterhyperpolarization (fAHP) between each AP and the lack of spike frequency adaptation. (**c, d**) Repolarization kinetics, fAHP and sAHP each contribute to the complex spike waveform as well and reflect the various K currents activated by CF stimulation.

Thus, it was suggested that the dendritic K currents initially ascribed to A-type channels are in large part actually subserved by D-type channels,[99] and models based on this premise replicate many experimental findings.[100] (For an alternative view see Refs. 101 and 102.) D-type K currents are most likely subserved by Kv1 subunits,[98] and it is known that at least Kv1.5 is highly expressed in the dendritic compartments of PCs.[95] The 4-aminopyridine–sensitive D-type and/or A-type K currents in the PC dendrites may be particularly important in regulating dendritic calcium spike firing.[53,103]

The Kv2.1 and 2.2 channels form another set of delayed rectifiers. Both channels are found in PCs. As with other neuronal types, both Kv2.1 and 2.2 are found in the PC soma, but whereas Kv2.2 is found throughout the dendritic tree, Kv2.1 is restricted to the proximal dendrites.[104,105]

The Kv3 class of K channels are typified by a high voltage for activation (>–10 mV) and fast activation and deactivation kinetics.[106] Thus, these delayed rectifier channels are ideal for contributing to action potential repolarization and fast afterhyperpolarization (FIG. 4). As Rudy *et al.*[106] note, although these fast current properties increase the rate of recovery of Na channel inactivation, they do not lengthen the refractory period of the neuron and, therefore, should confer upon a neuron the ability to fire at high frequencies. PCs express both Kv3.3 and 3.4 but do not appear to express Kv3.2.[107–109] The Kv3.1 channel may be selectively expressed in PCs of the flocculus and paraflocculus,[109] but this issue is not yet resolved.[110] Raman and Bean[111] recorded somatic K currents that are likely to reflect Kv3 channel activity: they had high activation threshold (no current until >–30 mV), submillisecond deactivation kinetics, high TEA sensitivity, and they were the currents most likely to be activated during high frequency firing, as indicated by voltage-clamp with action-potential waveforms. In agreement with these data, Southan and Robertson[112] also isolated a PC somatic outward current that best fits a Kv3 channel profile. Finally, Martina *et al.*[113] recorded a fast activating and inactivating current in the distal PC dendrite, which they attribute to Kv3.4 channels. Immunohistological staining supported this correlation, showing intense preferential labeling for Kv3.4 in the distal PC dendrites and preferential localization of Kv3.3 to the PC somata and proximal dendrites.[113] Thus, different Kv3 channels appear to regulate AP repolarization and fAHP in the PC somatodendritic compartments.

The ether-a-gogo (eag or ERG)-related Kvα subunits make up one of the three major families of the Kv superfamily[79] and ERG K currents are expressed in PCs. Most recently, Sacco *et al.*[114] reported that blockade of ERG K currents in the cerebellar slice preparation using the selective antagonist WAY-123,398 increased the frequency of action potential firing evoked by somatic depolarization and suppressed spike frequency adaptation. Of particular interest to the current discussion, ERG antagonism also resulted in prolongation of the complex spike and even addition of another spikelet to the complex spike waveform.

Finally, in one of the most complete investigations of K conductances in the PC to date, Gruol and colleagues[115] observed six K channel types in organotypic cerebellar cultures using cell-attached single-channel recordings under symmetrical K conditions. The three highest conducting channel types were also observed under physiological K conditions and were activated by membrane depolarization. The three lowest conductance channels were active at resting or hyperpolarized potentials. All of the somatic and dendritic patches studied contained K channels and usually contained more than one type. Unfortunately, apart from one notable exception (see BK channel section below), none of these K channel types have yet been further characterized using, for example, specific channel antagonists.

CALCIUM-SENSITIVE POTASSIUM (K_{CA}) CHANNELS

Calcium-activated potassium channels may contribute to action potential repolarization, spike adaptation, and slow AHP (sAHP) (FIG. 4). The three main types of K_{ca} channels are the big, small, and intermediate conductance channels, referred to most commonly as BK, SK, and IK, respectively. The current discussion focuses solely on the first two channel types as little is known about IK conductances in neurons.

The big conductance potassium (BK) channels found in multiple types of neurons have low calcium sensitivity (0.6–10 μM $[Ca^{2+}]_i$), significant voltage sensitivity, and high single-channel conductance (100–400 pS). By contrast, BK channels in PCs are reported to have higher calcium sensitivity, on the order of 10–250 nM $[Ca^{2+}]_i$.[116] Because of their fast activation kinetics and steep voltage dependence, BK channels are activated during the upstroke of an action potential, but rapidly deactivate as the membrane potential nears resting levels. Thus, BK channels in diverse neurons contribute to AP repolarization and fast AHP, but have little role in the slow AHP that often follows AP firing.[81,117]

In situ hybridization shows coexpression of the BKα and BKβ subunits in Purkinje neurons.[118] The high-conductance, Ca^{2+}-activated K protein *Slo* is found in the PC somata and proximal dendrite.[119] Patch-clamp recordings from PCs have revealed BK-like currents in both the somatic and dendritic regions.[115,116] In correspondence, highly TEA-sensitive Ca-dependent K currents were recorded in Purkinje cell somata and appear to participate strongly in repolarization of individual action potentials, as indicated by voltage-clamp with action-potential waveforms.[111] There have been reports of colocalization of BK channels and voltage-dependent Ca channels.[78] If such colocalization occurs in the PC dendritic region, then BK channels would be particularly capable of controlling Ca influx through voltage-dependent Ca channels.

The small conductance potassium (SK) channels are typified by high calcium sensitivity (100–600 nM $[Ca^{2+}]_i$), weak or nonexistent voltage sensitivity, and low single-channel conductance (5–20pS). The rise in intracellular calcium following action potential firing leads to slow activation of SK channels, which remain active during the long $[Ca^{2+}]_i$ decay, on the order of 100 ms to seconds. This SK activity generates the sAHP seen in many neurons.[77,78,81,120,121]

Two forms of sAHP can be recognized. The first is termed I_{AHP}, $gK_{ca,1}$, or *m*-I_{AHP}. This current is sensitive to the bee venom toxin apamin, activates rapidly after a single action potential, shows a single-channel conductance of 4-14pS, and decays with an approximate time constant of 150 ms.[78,121] The second is termed sI_{AHP}, slow-I_{AHP}, $gK_{ca,2}$, or sAHP. Unfortunately, it has also been labeled I_{AHP} in hippocampal pyramidal cells.[81] This current is apamin insensitive, activates slowly (0.5–1 second to peak), shows a single-channel conductance of approximately 18 pS, and decays with a time constant of about 1.5 seconds.[78,121] Only three SK subunits are presently known to exist. SK2 subunits are highly apamin sensitive, SK1 subunits are not sensitive at all, and SK3 subunits show intermediate sensitivity[122,123] (but see Ref. 124). Thus, it appears likely that SK2 and SK3 subunits form I_{AHP} channels, whereas SK1 subunits underlie sI_{AHP}.[81] PCs demonstrate low levels of SK2 mRNA and very low to nonexistent levels of SK1 mRNA.[125] Neither SK3 mRNA nor protein have been found in PCs.[125,126] The properties of SK channels suggest that they may play a dominant role in the slow AHP seen in PCs, but this possibility awaits experimental verification.

INWARD RECTIFYING POTASSIUM (K_{IR}) CHANNELS

K_{ir} channels have a role in setting the resting membrane potential and determining input resistance, which greatly affects the response of the neuron to synaptic inputs. In this way, K_{ir} channels have a large impact on synaptic integration. The K_{ir} family contains K channels that are active most of the time and channels whose activity is acutely sensitive to transmitters and the internal metabolic state. (See Refs. from Ref. 127.) For example, $K_{ir}2.1$ (IRK1) channels are constitutively open and demonstrate an open probability of near 1.0, whereas $K_{ir}3$ (GIRK) channels activate in the presence of the $\beta\gamma$ subunits of G proteins mobilized by metabotropic glutamate receptors (mGluRs).[127] One type of K_{ir} channel that is particularly abundant in the

PC layer is K_{ir} 7.1.[128] The hyperpolarization-activated (I_h) current is also an inward-rectifying current but is attributed to HCN (hyperpolarization-activated, cyclic nucleotide–sensitive, cation-nonselective) channels. Physiological studies have reported hyperpolarization-induced membrane potential "sag" in the PC dendrites and the soma, suggesting that I_h currents are present in both compartments.[111,129,130]

It is worth noting that Kv channels and K_{ir} channels belong to two distantly related families. Kv channels are intrinsically sensitive to voltage due to V-sensors built into the protein. By contrast, K_{ir} channels lack V-sensors but show V-dependence because these channels are blocked from the inside by cations such as Mg and polyamines that are normally present in all cells. Thus, K ions flow through K_{ir} channels more readily into the cell than out of the cell, that is, they show inward rectification.[127]

TWO-PORE DOMAIN K (K_{2P}) CHANNELS

Two-pore domain K channels are unique in that they have two K selective pores, whereas other K channels have one. These channels are constitutively active at resting membrane potentials and are therefore termed "leak" channels. Similar to K_{ir} channels, they play a major role in setting the resting membrane potential and membrane input resistance. However, these channels show little voltage dependence, being similarly active at all voltages. Instead, they are sensitive to temperature, pH, oxygen tension, osmolarity, and membrane stretch. A recent study examined seven of the eight K_{2P} channels in the central nervous system and found that six are significantly expressed in the cerebellar granule cell layer. By contrast, PCs expressed only one of the seven K_{2P} channels tested, the tandem of P domains in weak inward rectifier K^+ channel type 1 (TWIK-1) channel.[131]

THE MAKING OF A COMPLEX SPIKE

Cerebellar Purkinje cells are one of the electrophysiologically most thoroughly characterized types of neurons in the mammalian brain. However, despite decades of research aimed at understanding the ionic conductances present in PCs, our knowledge of processes underlying discrete discharge patterns in these cells remains very limited. This results partially from the aforementioned experimental limitations. Most data were obtained rather indirectly by characterizing discharge patterns that were evoked by current injection. These data have, nevertheless, provided the knowledge about the different types of ion channels present in PCs that is required if we are to understand how any spike activity pattern, in this case the complex spike, is electrically generated. In this section, we attempt to integrate the experimental data described in the previous sections and discuss what the major electrical events during the complex spike might be. It must be noted, however, that this description is, by necessity, somewhat speculative in nature.

Upon CF activity, glutamate is released into the synaptic cleft and binds to postsynaptic AMPA receptors and mGluRs (participation of the latter is discussed below). The opening of the ligand-gated ion channels leads to local EPSPs. As the CF synapses contact large parts of the primary, mostly proximal dendritic tree of PCs and there is a high release probability at every CF terminal, CF activity results in widespread depolarization of the dendrite that reaches the threshold for activation of voltage-dependent Ca conductances. Dendritic Ca spike activity upon CF stimulation was convincingly demonstrated using intradendritic recording techniques.[33] It is likely that several types of voltage-dependent Ca channels participate in the Ca spikes, but the available experimental data clearly emphasize the importance of P/Q-type channels[67] as well as T-type channels.[70–72] The resulting depolarization spreads towards the soma. The fast Na spike that can be recorded as the first complex spike component is most likely generated in the axon hillhock[47] by fast, inactivating Na conductances. The following, slower complex spike components, as they are recorded in the soma, are somewhat more difficult to interpret. Most likely, non-inactivating Na conductances,[32] resurgent Na currents,[37] and Ca conductances are involved in the generation of the depolarization plateau as well as the spikelets riding on top of it. It should be noted that the non-inactivating Na currents might correspond largely to the resurgent Na currents in their reopening state. Support for this idea comes from the ~80% reduction of steady-state current in Scn8a (Nav1.6) null mice.[39]

In the soma and the dendrite, both the depolarization and the Ca entry will activate discrete sets of K conductances. It has been shown that CF activation of an individual PC results in an increase in $[K^+]_o$,[132] and it was assumed that K conductances mediate the afterhyperpolarization following repetitive spike firing in PCs. (See, e.g., Ref. 36.) Because of the large variety of K channels present in PCs, it is very difficult to make more than a general prediction about which ones are activated predominantly during the complex spike. A- and D-type currents are localized in the soma and in the dendrites, respectively. These currents will undoubtedly shape the activity pattern both above and below the spike firing threshold. The Ca-dependent BK and SK conductances are likely to have a strong impact on the complex spike waveform as well. BK conductances have been characterized in the soma and dendrites of PCs. Because of their fast inactivation kinetics, BK conductances likely contribute to the fast repolarization following spike activity. On the basis of kinetics, the slow afterhyperpolarization following complex spikes most likely results from slow SK conductances, although this possibility requires experimental verification. Generally, TEA-sensitive, Ca-dependent currents account for half the K current that flows during spiking.[111]

One approach that can be helpful in putting the pieces of the puzzle together is the use of realistic computer models to mimic physiologically observed spike activity patterns. Promising models have been reported that integrate various experimentally described conductances.[133–135] It can be expected that in the future such models can be instrumental in studying synaptically evoked spike patterns such as the complex spike.

The remaining uncertainties regarding the "anatomy" of the complex spike not only concern the conductances involved, but also the question of how the electrical activity is transferred from one cellular compartment to the next. It is known that the somatic Na spikes do not backpropagate into the dendrites.[47] It remains unclear, however, to what degree the slow complex spike components recorded in the soma (the one to three spikelets riding on top of the depolarization plateau) faithfully reflect Ca spike activity in the dendrites. This is a result of the seemingly high probability of repetitive spike firing in the soma. We also have only limited information on the axonal spike pattern that results from a complex spike. Initial data indicate that every spikelet can activate an all-or-nothing Na spike in the axon, but this observation was not made for every complex spike.[136] We also do not have much information at hand regarding spatial characteristics of the dendritic Ca spikes. Are these localized events in separate branches of the dendritic tree? If so, do they remain discrete signals as they propagate towards the soma or do they merge with others into a spike pattern spreading over the proximal dendritic tree?

ASPECTS OF DENDRITIC INTEGRATION

The degree to which different PC compartments are electrically connected is determined by several factors, including the passive electrotonic structure, the distribution of the various forms of ion channels as well as the dendritic architecture. These factors will not only influence how strongly somatic and dendritic signals are interdependent, but also how locally restricted Ca spikes can be within the dendrite. It has been pointed out that the PC dendritic tree, with its extensive degree of branching, favors the propagation of steady-state potentials, but strongly attenuates fast voltage transients.[49,137] The electrical compactness of PCs has been demonstrated using voltage-sensitive dyes *in vitro*.[138,139] Thus, depolarization plateaus, such as the dendritic Ca-dependent plateau characterized by Llinas and Sugimori,[33] might well spread into the distal dendrites, possibly activating further Ca channels. By contrast, fast spikes, such as the fast Na spike, will be heavily attenuated as a result of the high degree of branching. This effect, rather than the low density of voltage-dependent Na channels in the dendrites, might be the predominant cause for the lack of backpropagating action potentials.[49] In a similar way, it can be expected that the dendritic branch points will limit the spread of dendritic Ca spikes, favoring local regenerative events.[137] Thus, the architecture of the dendritic tree has a major impact on the interaction of the different cell compartments during a widespread electrical event, such as the complex spike.

MODIFIABILITY AND PLASTICITY

Since its first detailed description by Eccles and colleagues,[14,28,140] the cerebellar complex spike has gained a reputation as being invariant. This widespread assumption results from the role of the CF in coding for the unconditioned stimulus in reflex conditioning pathways, the high probability of release at CF terminals, the all-or-nothing character of the postsynaptic responses, and the large postsynaptic depolarization that accompanies complex spikes. As will be described, experimental evidence indeed confirms that the degree of synaptic reliability at the CF-PC synapse is very high. Under a variety of experimental conditions it was not possible to suppress the occurrence of the complex spike. However, a different perspective of the "invariance" of the complex spike emerges when it comes to modifications of the complex spike waveform. In the following, we discuss transient alterations of the complex spike waveform that result from simultaneous inhibition, changes in synaptic release, modification of K conductances, or neuromodulation. In addition, we describe a sustained alteration of the complex spike waveform that results from long-term plasticity at the CF-PC synapse, and we discuss the possible functional implications of this alteration.

TRANSIENT MODIFICATIONS

A synaptic factor that might be crucial for this consideration is inhibition. It has been shown that the activity of inhibitory interneurons can dramatically change the spike-firing pattern of PCs.[130] The shunting effect caused by the activation of $GABA_A$ receptors[138] can be expected to have an impact on the complex spike waveform. Interestingly, using combined electrical recording and Ca imaging techniques, it has been shown that inhibitory inputs do not necessarily modify simultaneously evoked complex spikes at the somatic level. However, responses to CF stimulation recorded dendritically with a sharp microelectrode were reduced and the evoked dendritic Ca signals could be attenuated by up to 90%.[141] These findings are in line with earlier, but less detailed observations by Eccles and colleagues[140] who described reductions in the late complex spike components upon simultaneous inhibition.

Activity at excitatory synapses can modify the PC spike pattern as well. A hallmark of the CF-PC synapses is the high release probability at the synaptic terminals that results in PPD upon repeated stimulation with an interpulse-interval of up to 5 seconds.[21,22] The postsynaptically recorded PPD is a consequence of a reduction in transmitter release, and this effect could be expected to transiently modify the complex spike. Surprisingly, it was demonstrated that complex spiking can follow stimulation frequencies of up to 100 Hz.[140] Thus, PPD does not seem to reduce the EPSC amplitude below threshold. However, at interpulse intervals of up to 1 second the second complex spike has shown a reduced number of spikelets.[11] Thus, it is possible that the complex spike waveform constantly changes as a function of discharge frequency.

At the postsynaptic side, the activation of mGluRs might be relevant in this context. The activation of mGluR has been shown to elicit a transient inward current followed by an outward current,[142–144] indicating that mGluRs, in addition to AMPARs, are participating in electrogenesis resulting from synaptic activity. The activation of mGluR 1 receptors has been shown to transiently increase PC excitability through activation of phospholipase C.[145] As protein kinase C activity can attenuate K conductances in PCs,[146] it is possible that mGluR 1 activation changes PC excitability through a protein kinase C-mediated reduction in K conductances (as discussed in Ref. 145). Local dendritic K conductances almost certainly influence dendritic complex spike components. Using intracellular recordings in combination with imaging techniques, it was observed that K currents can modify the amplitude of dendritic Ca spikes as well as the spatial spread of Ca spike-associated Ca increases in the dendrites.[53] This finding could explain the variability in the spatial spread of CF-evoked Ca signals that was described earlier.[52] This effect of K conductances is a fine example of the possible level of complexity of the processes regulating the complex spike.

In a recent study, the effect of diverse neuromodulators on the CF-evoked Ca transients was tested. Somewhat surprisingly, neither carbachol, serotonin, noradrenaline, nor dopamine had any significant effect,[147] while the 5-HT_{1A} agonist buspirone caused only a small attenuation. By contrast, corticotropin-releasing factor (CRF) causes a drastic reduction in the complex spike AHP.[148,149] This finding is particularly interesting since the CF itself may be the source of the corticotropin-releasing factor release.[150]

The data just summarized clearly demonstrate that complex spike-related events, such as the Ca spikes and the spatial spread of associated Ca signals, can be modulated by processes such as inhibition or activation of voltage-gated K conductances.[53,141] Moreover, the AHP following complex spikes (which belongs to the complex spike waveform) is subject to modifications as well.[148,149] In this context, it is important to note that synaptically evoked complex spikes occur on top of a high spontaneous spike-firing frequency in PCs of up to 150 Hz.[130] Thus, the complex spikes will be shaped in dependence on the given activation state of the diverse voltage-gated conductances. In turn, the complex spikes will interrupt and possibly modify the ongoing electrical activity as well.

Obviously, there are many more examples of how voltage-dependent ion channels in PCs can be modified, and any such alteration has the theoretical potential to alter the complex spike waveform. Although a detailed description of all these modifications is beyond the scope of this paper, the aforementioned examples indicate the existing plasticity range.

Whereas we have discussed several transient modifications of the complex spike (mostly at the dendritic level), we will describe in the following the long-term changes of the complex spike waveform that can be recorded somatically.

LONG-TERM PLASTICITY

For this consideration, we must return to the phenomenon of synaptic plasticity. Changes in synaptic efficacy are widely assumed to provide a cellular basis for learning and memory. The type of synaptic plasticity in the cerebellum that has received the most attention is PF-LTD.[16] There are several requirements for PF-LTD induction. It is assumed that CF activity, which provides massive excitation of the dendritic tree, contributes a large Ca signal to PF-LTD induction.[17] PF-LTD has also been observed after PF stimulation alone,[151,152] but the intense stimulation paradigms used seem to be outside the physiological range. It can be reasoned that there is a Ca threshold amplitude for PF-LTD induction that can, under physiological conditions, only be reached when co-activity of the CF input generates supralinear Ca signal amplitudes at the PF input sites.[59] Such Ca amplitude-related thresholds have been demonstrated for both LTD and LTP induction at excitatory inputs to hippocampal and neocortical pyramidal cells using microfluorometric Ca imaging techniques and photolytic caged Ca release.[153,154] In addition to the dendritic Ca transient, PF-LTD has a requirement for the activation of AMPARs and mGluR1 receptors at the PF input and subsequent activation of protein kinase C. Furthermore, activation of a nitric oxide/cGMP/protein kinase G pathway has also been implicated in PF-LTD induction. (For a review see Refs. 4, 5, and 155.)

Our interest to search for forms of synaptic plasticity at the CF-PC synapse stemmed from the observation that the molecular machinery required for PF-LTD induction is present at the CF input as well. Both the AMPA and mGluR1 receptors are present at the CF-PC synapse[156–158] as are P/Q-type voltage-dependent Ca

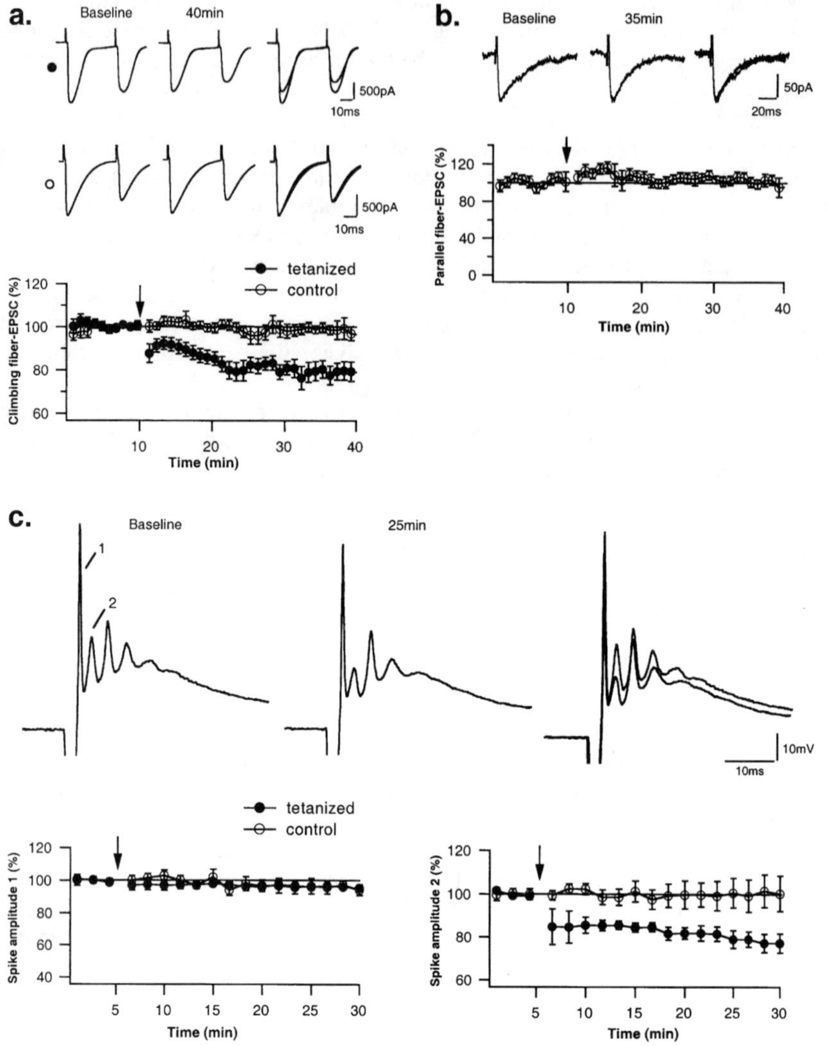

FIGURE 5. Climbing fiber (CF) long-term depression. (**a**) 5-Hz CF tetanization resulted in a depression of CF-evoked excitatory postsynaptic currents (EPSCs; $n = 15$). Under control conditions, no change in EPSC amplitude was observed ($n = 5$). Each test stimulus consisted of a pulse pair applied with an interval of 50 ms. Traces are averages of 2–5 sweeps. In the time-course graph, each data point represents the average of three successive responses evoked at 0.05 Hz. Only the amplitude of the first EPSC is shown. (**b**) CF tetanization did not change the amplitudes of parallel fiber (PF) EPSCs as measured in a subset of the tetanized cells ($n = 9$) shown in **a**. (**c**) CF stimulation produced an alteration in the complex spike waveform (current-clamp mode). Representative single traces are shown before and 25 minutes after tetanization. The time graphs show the changes in the amplitudes of the first two spike components under tetanization ($n = 5$) and control conditions ($n = 4$). Each data point represents the average of five successive responses evoked at 0.05 Hz. This figure is modified from Hansel and Linden.[19] Used with permission from Elsevier Science.

channels[66] and protein kinase C.[159,160] Thus, we monitored CF responses using whole-cell patch-clamp recordings from PCs in rat cerebellar slices. In contrast to the previous assumption that the CF-PC synapse is invariant, we observed LTD of CF-evoked EPSCs (recorded in voltage-clamp mode) after 5 Hz CF tetanization for 30 seconds (FIG. 5a).[19] This stimulation frequency is in the upper range of olivary firing *in vivo*.[8,161] The duration of tetanization is certainly longer than what is typically observed *in vivo*, but CF-LTD was recently also obtained using repetitive application of 5 pulses at 5 Hz with 10-second interburst intervals (D.J. Linden, personal communication), a protocol that more closely resembles *in vivo* conditions. CF-LTD was input specific, as it did not spread to neighboring PF synapses (FIG. 5b) and could be saturated on repeated tetanization. A postsynaptic locus of expression was indicated by the observation that CF-LTD was not associated with changes in the PPD ratio. As outlined above, the molecular machinery required for PF-LTD induction can also be found at the CF-PC synapse. Indeed, we observed that elements of the PF-LTD induction cascade are needed to trigger CF-LTD as well. CF-LTD could be blocked by loading PCs with the Ca chelator BAPTA or by bath application of the group I mGluR antagonist AIDA or the protein kinase C inhibitor chelerythrine.[19]

The characterization of CF-LTD and its mechanisms was initially restricted to the level of EPSCs, as we were interested in forms of true synaptic plasticity. Thus, we intended to isolate the electrical synaptic events from the subsequent activation of voltage-dependent conductances occurring during the complex spike by holding the cells in voltage-clamp mode. (For a detailed discussion see Ref. 19.) To test whether the naturally occurring PC response to CF activity, the complex spike, could be affected by the same LTD-inducing tetanization, we performed long-term recordings of complex spikes in current-clamp mode. Interestingly, under these conditions the 5-Hz tetanization did not result in spike failures, but rather in a specific modification of the complex spike waveform that was observed as long as the recording could be maintained. Whereas the amplitude of the fast Na spike was unaffected, there was a selective reduction in the following slower spikelets that was most reliably observed in the second spike component (FIG. 5c). This is an important finding, as it shows that CF-LTD does not just modify the EPSC amplitude within the suprathreshold range. Instead, CF-LTD is associated with a moderate, but selective modification of the second component (occasionally later components as well) of the complex spike waveform. These slow spikes are most likely produced by a somatic resurgent Na current and somatic Ca currents following initial Ca spike activity in the dendrites (see above). Given the uncertainties regarding the identity of the slow complex spike components and especially the unresolved question of whether or not the somatically recorded spikelets faithfully reflect Ca spike activity in the dendrites, it is currently impossible to say which type of current is reduced. This aspect illustrates nicely how crucial it is to know the different conductances present in PCs as at least the available information provides a reasonable starting point to experimentally address this question and focus on a few candidate conductances. Currently, we do not know whether the observed modification of the complex spike waveform is entirely caused by the EPSC amplitude reduction or whether a more direct activity-dependent modification of voltage-dependent conductances takes place. For now, we will use the term "CF-LTD" to generally refer to both the effect at the level of the EPSCs and that at the spike level.

What is the functional significance of CF-LTD and the associated reduction in slow complex spike components? One possibility is that CF-LTD is associated with long-term changes in dendritic Ca signaling. Such an effect would be highly significant for cerebellar network function as dendritic Ca transients are required for the induction of various forms of synaptic plasticity in the cerebellar cortex, including PF-LTD. (For a review see Ref. 5.) One important consequence could be a shift in the probability for PF-LTD induction. A reduction in the amplitude of CF-evoked Ca transients could be significant in another context as well. High amplitude Ca signals can reach neurotoxic levels when the frequency of complex spikes is transiently or persistently high.[162] Thus, CF-LTD could provide a neuroprotective mechanism after periods of elevated complex spike frequencies. It should be pointed out that such a protective function and the possible heterosynaptic effect on, for example, PF-LTD need not be mutually exclusive.

CF-LTD could also be associated with long-term changes in PC potassium conductances. In our laboratory, we have started to characterize the possible effects of CF-LTD on the slow AHP following complex spikes. Preliminary data indicate that, in fact, the tetanus-induced complex spike modification can be associated with a reduction in the amplitude of the AHP (FIG. 6a and b). Changes in the AHP could be crucial for at least two reasons. First, either a spontaneous or stimulus-driven complex spike leads to a pause in ongoing high frequency simple spike activity (FIG. 6c and d).[8] One standing hypothesis is that this "complex spike pause" is due in part to the same calcium-activated K currents that generate the complex spike AHP. Initially, a broad comparison of the complex spike pause duration (usually <100 ms) and the AHP timing (150–600 ms *in vitro*, FIG. 6) suggests a discrepancy. However, this timing mismatch is largely resolved when the focus is limited to pauses recorded in unanesthetized, *in vivo* preparations (10–30 ms; FIG. 6)[163–168] and the AHP recorded *in vivo* (10–50 ms).[28,164,169] If CF-LTD does consistently attenuate the AHP, it could, therefore, also have a significant impact upon PC output. Second, learning-associated reductions in K-mediated hyperpolarizations have been reported in multiple regions of the brain,[170] including the cerebellum where AHP reductions in PCs of rabbits have been observed following classical conditioning.[171] Thus, a possible change in the AHP amplitude accompanying CF-LTD is interesting in the context of PC output and cerebellar mediated learning, and we are currently analyzing this effect in further detail.

Aside from the possible modification of the AHP amplitude, CF-LTD could have an even more direct impact on the PC output. It has been demonstrated that complex spikes can lead to an axonal spike pattern in which each of the slow complex spike components may be associated with an all-or-nothing spike in the axon.[136] It is possible that a CF-LTD–related reduction in one or more of the slow spikelets occurring during the complex spike leads to a failure to initiate the respective spike in the axon hillock and thus a change in the axonal spike pattern. This scenario is illustrated in FIGURE 7. Furthermore, action potentials in PC axons have been shown to be accompanied by Ca entry into the axon.[172] In the axon hillock and the initial part of the axon, this Ca signal might very well be enhanced by complex spike activity in the soma and thus might be changed after CF-LTD. Also in this scenario, CF-LTD could change the axonal spike pattern as the axonal spike frequency was demonstrated to vary in dependence of these Ca signals.[172]

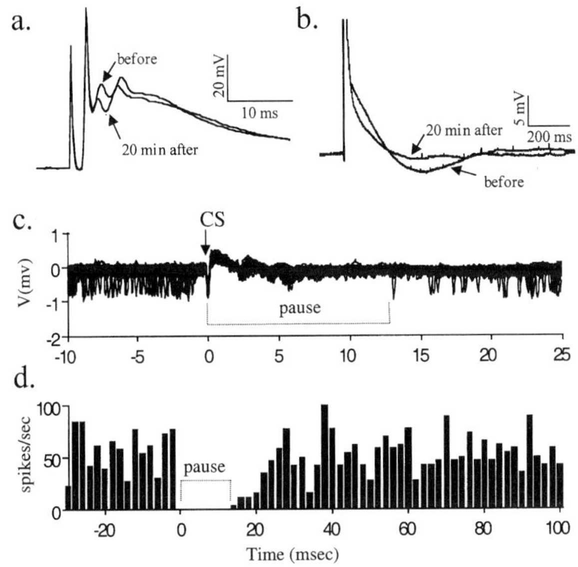

FIGURE 6. (**a, b**) Climbing fiber (CF) tetanization-induced modifications of the complex spike waveform can be associated with a reduction in the slow afterhyperpolarization (sAHP). (**a**) CF tetanization at 5 Hz for 30 seconds resulted in the attenuation of the second and, in this case, the third spikelet of the complex spike. (**b**) Examination of the same complex spike recordings at a longer time base revealed that the sAHP following the complex spike was also significantly reduced after CF tetanization. Note also that the initial rate of repolarization may be enhanced. (**c, d**) Extracellular recordings of spontaneous activity from a single Purkinje cell (PC) in an awake mouse reveal a high rate of simple spike firing that is interrupted by a single complex spike. The simple spike activity before and after a complex spike is presented by overlaying 130 traces locked to complex spike generation (**c**). The peri-response time histogram (2-ms bin width) generated from these data reveal an absolute pause of ~13 ms (**d**). FIGURES 6c and d were kindly provided by A. v. Alphen.

Here, we have outlined just a few of the possible functional implications of CF-LTD. Certainly, these possibilities have to be experimentally addressed to evaluate the role of CF-LTD in modifying cerebellar network function. One aspect should already be highlighted. As much as the CF-PC synapse is a unique type of synapse in the brain, CF-LTD is a unique type of synaptic plasticity. LTP and LTD at most other types of synapses serve to locally change the synaptic input weight, ultimately resulting in a change in the postsynaptic spike-firing probability. CF-LTD, however, might not only have an impact on the spike firing activity in PC somata or their axons, but also most likely has a significant influence on dendritic Ca signaling and dendritic integration.

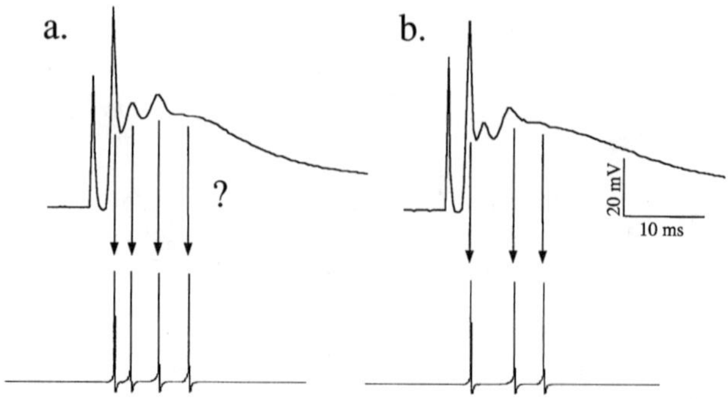

FIGURE 7. Does climbing fiber long-term depression (CF-LTD) change the electrical Purkinje cell (PC) output? (**a**) Each spike (or spikelet) component of the complex spike might activate an axonal all-or-nothing spike. (**b**) Selective attenuation of one spikelet could result in alteration of the axonal spike pattern. Traces in **a** and **b** were obtained from electrical recordings in rat cerebellar slices. However, somatic and axonal activity patterns were not obtained from the same cells and are solely used in this figure for illustrative reasons.

SUMMARY

The "anatomy" of the complex spike in PCs is described in an attempt to solve the puzzle of conductances in different neuronal compartments that constitute the complex spike. A hallmark of the cerebellar complex spike is the repetitive and consistent pattern of spikelets following the initial fast spike. This burst of spikelets most likely has separate dendritic as well as somatic components, namely, local dendritic Ca spikes, on the one hand, and Na spikes intermingled with Ca currents in the soma, on the other. Dendritic spike activity resulting from the backpropagation of fast Na spikes evoked in the axon hillock has recently obtained a great deal of attention in the context of synaptic plasticity.[173] While backpropagating spikes are typically observed in hippocampal or neocortical pyramidal cells, PCs do not have Na spike activity in the dendrites. However, the complex spike provides an analogous example of an electrical signal associated with a Ca transient that covers large parts of the dendritic tree. In this case, the Ca signal evoked by the complex spike makes this regenerative electrical event a key player in multiple forms of synaptic plasticity, such as in the case of PF-LTD. The observation of plasticity of the complex spike raises an impressive array of functional considerations that makes cerebellar synaptic plasticity an even more interesting model to study mechanisms of learning and memory.

ACKNOWLEDGMENTS

This work was supported by an EUR fellowship (M.T.S.), Hersenstichting Nederland grant (M.T.S., C.I.D.Z., and C.H.), an EEC grant (C.I.D.Z.), NWO-MW (J.T.W. and C.I.D.Z.), NWO-ALW 810-37-003 (C.H.), and the Royal Dutch Academy of Sciences (C.H.). We thank D.J. Linden and I.M. Raman for comments on the manuscript.

REFERENCES

1. MARR, D. 1969. A theory of cerebellar cortex. J. Physiol. **202:** 437–470.
2. ALBUS, J.S. 1971. A theory of cerebellar function. Math. Biosci. **10:** 25–61.
3. ITO, M. 1984. The Cerebellum and Neural Control. Raven Press. New York.
4. DANIEL, H., C. LEVENES & F. CRÉPEL. 1998. Cellular mechanisms of cerebellar LTD. Trends Neurosci. **21:** 401–407.
5. HANSEL, C., D.J. LINDEN & E. D'ANGELO. 2001. Beyond parallel fiber LTD: the diversity of synaptic and non-synaptic plasticity in the cerebellum. Nature Neurosci. **4:** 467–475.
6. WELSH, J.P., E.J. LANG, I. SUGIHARA & R. LLINAS. 1995. Dynamic organization of motor control within the olivocerebellar system. Nature **374:** 453–457.
7. SZENTAGOTHAI, J. & K. RAJKOVITS. 1959. Über den Ursprung der Kletterfasern des Kleinhirns. Anat. Entwicklungsgesch. **121:** 130–141.
8. SIMPSON, J.I., D.R. WYLIE & C.I. DE ZEEUW. 1996. On climbing fiber signals and their consequence(s). Behav. Brain Sci. **19:** 384–398.
9. DE ZEEUW, C.I., J.I. SIMPSON, C.C. HOOGENRAAD, *et al.* 1998. Microcircuitry and function of the inferior olive. Trends Neurosci. **21:** 391–400.
10. DITTMAN, J.S. & W.G. REGEHR. 1998. Calcium dependence and recovery kinetics of presynaptic depression at the climbing fiber to Purkinje cell synapse. J. Neurosci. **18:** 6147–6162.
11. HASHIMOTO, K. & M. KANO. 1998. Presynaptic origin of paired-pulse depression at climbing-fiber Purkinje cell synapses in the rat cerebellum. J. Physiol. **506:** 391–405.
12. SILVER, R.A., A. MOMIYAMA & S.G. CULL-CANDY. 1998. Locus of frequency-dependent depression identified with multiple-probability fluctuation analysis at rat climbing fiber-Purkinje cell synapses. J. Physiol. **510:** 881–902.
13. STRATA, P. & F. ROSSI. 1998. Plasticity of the olivocerebellar pathway. Trends Neurosci. **21:** 407–413.
14. ECCLES, J.C., R. LLINAS & K. SASAKI. 1964. Excitation of cerebellar Purkinje cells by the climbing fibers. Nature **203:** 245–246.
15. THACH, W.T. 1967. Somatosensory receptive fields of single units in cat cerebellar cortex. J. Neurophysiol. **30:** 675–696.
16. ITO, M., M. SAKURAI & P. TONGROACH. 1982. Climbing fiber induced depression of both mossy fiber responsiveness and glutamate sensitivity of cerebellar Purkinje cells. J. Physiol. **324:** 113–134.
17. KONNERTH, A., J. DREESSEN & G.J. AUGUSTINE. 1992. Brief dendritic calcium signals initiate long-lasting synaptic depression in cerebellar Purkinje cells. Proc. Natl. Acad. Sci. USA **89:** 7051–7055.
18. KANO, M., U. REXHAUSEN, J. DREESSEN & A. KONNERTH. 1992. Synaptic excitation produces a long-lasting rebound potentiation of inhibitory synaptic signals in cerebellar Purkinje cells. Nature **356:** 601–604.
19. HANSEL, C. & D.J. LINDEN. 2000. Long-term depression of the cerebellar climbing fiber-Purkinje neuron synapse. Neuron **26:** 473–482.
20. LLINAS, R. & C. NICHOLSON. 1976. Reversal properties of climbing fiber potential in cat Purkinje cells: an example of a distributed synapse. J. Neurophysiol. **39:** 311–323.
21. KONNERTH, A., I. LLANO & C.M. ARMSTRONG. 1990. Synaptic currents in cerebellar Purkinje cells. Proc. Natl. Acad. Sci. USA **87:** 2662–2665.
22. PERKEL, D.J., S. HESTRIN, P. SAH & R.A. NICOLL. 1990. Excitatory synaptic currents in Purkinje cells. Proc. R. Soc. Lond. B Biol. Sci. **241:** 116–121.
23. LLANO, I., A. MARTY, C.M. ARMSTRONG & A. KONNERTH. 1991. Synaptic- and agonist-induced excitatory currents of Purkinje cells in rat cerebellar slices. J. Physiol. **434:** 183–213.
24. HÄUSSER, M. & A. ROTH. 1997. Dendritic and somatic glutamate receptor channels in rat cerebellar Purkinje cells. J. Physiol. **501.1:** 77–95.
25. OTTERSEN, O.P., & F. WALBERG. 2002. Neurotransmitters in the cerebellum. *In* The Cerebellum and Its Disorders. M.-U. Monto & M. Pondolfo, Eds. :38–48. Cambridge University Press. Cambridge.

26. CUÉNOD, M., K.-Q. DO, F. VOLLENWEIDER, *et al.* 1989. The puzzle of the transmitters in the climbing fibers. Exp. Brain Res. **17:** 161–176.
27. DO, K.-Q., P. GRANDES, C. HANSEL, *et al.* 1992. Sulphur containing excitatory amino acids: release, activity and localization. Mol. Neuropharmacol. **2:** 39–42.
28. ECCLES, J.C., R. LLINAS & K. SASAKI. 1966. The excitatory synaptic action of climbing fibers on the Purkinje cells of the cerebellum. J. Physiol. **182:** 268–296.
29. MARTINEZ, F.E., W.E. CRILL & T.T. KENNEDY. 1971. Electrogenesis of cerebellar Purkinje cell responses in cats. J. Neurophysiol. **34:** 348–356.
30. FUJITA, Y. 1968. Activity of dendrites of single Purkinje cells and its relationship to so-called inactivation response in rabbit cerebellum. J. Neurophysiol. **31:** 131–141.
31. LLINAS, R. & C. NICHOLSON. 1971. Electrophysiological properties of dendrites and somata in alligator Purkinje cells. J. Neurophysiol. **34:** 532–551.
32. LLINAS, R. & M. SUGIMORI. 1980a. Electrophysiological properties of *in vitro* Purkinje cell somata in mammalian cerebellar slices. J. Physiol. **305:** 171–195.
33. LLINAS, R. & M. SUGIMORI. 1980b. Electrophysiological properties of *in vitro* Purkinje cell dendrites in mammalian cerebellar slices. J. Physiol. **305:** 197–213.
34. HOUNSGAARD, J. & J. MIDTGAARD. 1988. Intrinsic determinants of firing pattern in Purkinje cells of the turtle cerebellum in vitro. J. Physiol. **402:** 731–749.
35. KAY, A.R., M. SUGIMORI & R. LLINAS. 1998. Kinetic and stochastic properties of a persistant sodium current in mature guinea pig cerebellar Purkinje cells. J. Neurophysiol. **80:** 1167–1179.
36. GÄHWILER, B.H. & I. LLANO. 1989. Sodium and potassium conductances in somatic membranes of rat Purkinje cells from organotypic cerebellar cultures. J. Physiol. **417:** 105–122.
37. RAMAN, I.M. & B.P. BEAN. 1997. Resurgent sodium current and action potential formation in dissociated cerebellar Purkinje neurons. J. Neurosci. **17:** 4517–4526.
38. RAMAN, I.M. & B.P. BEAN. 1999. Properties of sodium currents and action potential firing in isolated cerebellar Purkinje neurons. Ann. N.Y. Acad. Sci. **868:** 93–96.
39. RAMAN, I.M. & B.P. BEAN. 2001. Inactivation and recovery of sodium currents in cerebellar Purkinje neurons: evidence for two mechanisms. Biophysiol. J. **80:** 729–737.
40. WESTENBROEK, R.E., D.K. MERRICK & W.A. CATTERALL. 1989. Differential subcellular localization of the RI and RII Na channel subtypes in central neurons. Neuron **3:** 695–704.
41. VEGA-SAENZ DE MIERA, E., B. RUDY, M. SUGIMORI & R. LLINAS. 1997. Molecular characterization of the sodium channel subunits expressed in mammalian cerebellar Purkinje cells. Proc. Natl. Acad. Sci. USA **94:** 7059–7064.
42. GONG, B., K.J. RHODES, Z. BEKELE-ARCURI & J.S. TRIMMER. 1999. Type I and II Na channel α-subunit polypeptides exhibit distinct spatial and temporal patterning, and association with auxiliary subunits in rat brain. J. Comp. Neurol. **412:** 342–352.
43. GOLDIN, A.L. 1999. Diversity of mammalian voltage-gated sodium channels. Ann. N.Y. Acad. Sci. **868:** 38–50.
44. RAMAN, I.M., L.K. SPRUNGER, M.H. MEISLER & B.P. BEAN. 1997. Altered subthreshold sodium currents and disrupted firing patterns in Purkinje neurons of Scn8a mutant mice. Neuron **19:** 881–891.
45. LASSER-ROSS, N. & W.N. ROSS. 1992. Imaging voltage and synaptically activated sodium transients in cerebellar Purkinje cells. Proc. R. Soc. Lond. B **247:** 35–39.
46. CALLAWAY, J.C. & W.N. ROSS. 1997. Spatial distribution of synaptically activated sodium concentration changes in cerebellar Purkinje neurons. J. Neurophysiol. **77:** 145–152.
47. STUART, G. & M. HÄUSSER. 1994. Initiation and spread of sodium action potentials in cerebellar Purkinje cells. Neuron **13:** 703–712.
48. REGEHR, W.G., A. KONNERTH & C.M. ARMSTRONG. 1992. Sodium action potentials in the dendrites of cerebellar Purkinje cells. Proc. Natl. Acad. Sci. USA **89:** 5492–5496.
49. VETTER, P., A. ROTH & M. HÄUSSER. 2001. Propagation of action potentials in dendrites depends on dendritic morphology. J. Neurophysiol. **85:** 926–937.
50. ROSS W.N. & R. WERMAN. 1987. Mapping calcium transients in the dendrites of Purkinje cells from the guinea-pig cerebellum in vitro. J. Physiol. **389:** 319–336.

51. KNÖPFEL, T., I. VRANESIC, C. STAUB & B.H. GÄHWILER. 1990. Climbing fiber responses in olivo-cerebellar slice cultures. II. Dynamics of cytosolic calcium in Purkinje cells. Eur. J. Neurosci. **3:** 343–348.
52. MIYAKAWA, H., V. LEV-RAM, N. LASSER-ROSS & W.N. ROSS. 1992. Calcium transients evoked by climbing fiber and parallel fiber synaptic inputs in guinea-pig cerebellar Purkinje neurons. J. Neurophysiol. **68:** 1178–1189.
53. MIDTGAARD, J., N. LASSER-ROSS & W.N. ROSS. 1993. Spatial distribution of Ca influx in turtle Purkinje cell dendrites in vitro: role of a transient outward current. J. Neurophysiol. **70:** 2455–2469.
54. LEV-RAM, V., H. MIYAKAWA, N. LASSER-ROSS & W.N. ROSS. 1992. Calcium transients in cerebellar Purkinje neurons evoked by intracellular stimulation. J. Neurophysiol. **68:** 1167–1177.
55. TANK, D.W., M. SUGIMORI, J.A. CONNOR & R. LLINAS. 1988. Spatially resolved calcium dynamics of mammalian Purkinje cells in cerebellar slice. Science **242:** 773–777.
56. SUGIMORI, M. & R. LLINAS. 1990. Real-time imaging of calcium influx in mammalian cerebellar Purkinje cells in vitro. Proc. Natl. Acad. Sci. USA **87:** 5084–5088.
57. DENK, W., M. SUGIMORI & R. LLINAS. 1995. Two types of calcium response limited to single spines in cerebellar Purkinje cells. Proc. Natl. Acad. Sci. USA **92:** 8279–8282.
58. EILERS, J., G.J. AUGUSTINE & A. KONNERTH. 1995. Subthreshold synaptic Ca signalling in fine dendrites and spines of cerebellar Purkinje neurons. Nature **373:** 155–158.
59. WANG, S.S.-H., W. DENK & M. HÄUSSER. 2000. Coincidence detection in single dendritic spines mediated by calcium release. Nature Neurosci. **3:** 1266–1273.
60. ROSE, C.R. & A. KONNERTH. 2001. Stores not just for storage: intracellular calcium release and synaptic plasticity. Neuron **31:** 519–522.
61. MORENO DAVILA, H. 1999. Molecular and functional diversity of voltage-gated calcium channels. Ann. N.Y. Acad. Sci. **868:** 102–117.
62. WESTENBROEK, R.E., J.W. HELL, C. WARNER, *et al.* 1992. Biochemical properties and subcellular distribution of an N-type calcium channel alpha 1 subunit. Neuron **9:** 1099–1115.
63. USOWICZ, M.M., M. SUGIMORI, B. CHERKSEY & R. LLINAS. 1992. P-type calcium channels in the somata and dendrites of adult cerebellar Purkinje cells. Neuron **9:** 1185–1199.
64. WESTENBROEK, R.E., T. SAKURAI, E.M. ELLIOTT, *et al.* 1995. Immunochemical identification and subcellular distribution of the α1A subunits of brain calcium channels. J. Neurosci. **15:** 6403–6418.
65. AHLIJANIAN, M.K., R.E. WESTENBROEK & W.A. CATTERALL. 1990. Subunit structure and localization of dihydropyridine sensitive calcium channels in mammalian brain, spinal cord and retina. Neuron **4:** 819–832.
66. YOKOYAMA, C.T., R.E. WESTENBROEK, J.W. HELL, *et al.* 1995. Biochemical properties and subcellular distribution of the neuronal class E calcium channel alpha 1 subunit. J. Neurosci. **15:** 6419–6432.
67. LLINAS, R., M. SUGIMORI, J.W. LIN & B. CHERSKY. 1989. Blocking and isolation of a calcium channel from neurons in mammalians and cephalopods utilizing a toxin fraction FTX from funnel web spider poison. Proc. Natl. Acad. Sci. USA **86:** 1689–1693.
68. MINTZ, I.M., M.E. ADAMS & B.P. BEAN. 1992. P-type calcium channels in rat central and peripheral neurons. Neuron **9:** 85–95.
69. LLINAS, R., M. SUGIMORI, D.E. HILLMAN & B. CHERSKY. 1992. Distribution and functional significance of the P-type, voltage-dependent Ca channels in the mammalian central nervous system. Trends Neurosci. **15:** 351–355.
70. WATANABE, S., H. TAKAGI, T. MIYASHO, *et al.* 1998. Differential roles of two types of voltage-gated Ca channels in the dendrites of rat cerebellar Purkinje neurons. Brain Res. **791:** 43–55.
71. POUILLE, F., P. CAVELIER, T. DESPLANTEZ, *et al.* 2000. Dendro-somatic distribution of calcium-mediated electrogenesis in Purkinje cells from rat cerebellar slice cultures. J. Physiol. **527.2:** 265–282.

72. TALLEY, E.M., L.L. CRIBBS, J.-H. LEE, *et al.* 1999. Differential distribution of three members of a gene family encoding low voltage-activated (T-type) calcium channels. J. Neurosci. **19:** 1895–1911.
73. REGAN, L.J. 1991. Voltage-dependent calcium currents in Purkinje cells from rat cerebellar vermis. J. Neurosci. **11:** 2259–2269.
74. MOUGINOT, D., J.-L. BOSSU & B.H. GÄHWILER. 1997. Low-threshold Ca currents in dendritic recordings from PCs in rat cerebellar slice cultures. J. Neurosci. **17:** 160–170.
75. BOSSU, J.L., L. FAGNI & A. FELTZ. 1989. Voltage-activated calcium channels in rat Purkinje cells maintained in culture. Pflügers Arch. **414:** 92–94.
76. GÄHWILER, B.H. 1981. Organotypic monolayer cultures of nervous tissue. J. Neurosci. Methods **4:** 329–342.
77. SAH, P. 1996. Ca(2+)-activated K^+ currents in neurones: types, physiological roles and modulation. Trends Neurosci. **19:** 150–154.
78. VERGARA, C., R. LATORRE, N.V. MARRION & J.P. ADELMAN. 1998. Calcium-activated potassium channels. Curr. Opin. Neurobiol. **8:** 321–329.
79. PONGS, O. 1999. Voltage-gated potassium channels: from hyperexcitability to excitement. FEBS Lett. **452:** 31–35.
80. COETZEE, W.A., Y. AMARILLO, J. CHIU, *et al.* 1999. Molecular diversity of K^+ channels. Ann. N.Y. Acad. Sci. **868:** 233–285.
81. SAH, P. & P. DAVIES. 2000. Calcium-activated potassium currents in mammalian neurons. Clin. Exp. Pharmacol. Physiol. **27:** 657–663.
82. BALDWIN, T.J., M.L. TSAUR, G.A. LOPEZ, *et al.* 1991. Characterization of a mammalian cDNA for an inactivating voltage-sensitive K^+ channel. Neuron **7:** 471–483.
83. SERODIO, P. & B. RUDY. 1998. Differential expression of Kv4 K^+ channel subunits mediating subthreshold transient K^+ (A-type) currents in rat brain. J. Neurophysiol. **79:** 1081–1091.
84. MCNAMARA, N.M., Z.M. MUNIZ, G.P. WILKIN & J.O. DOLLY. 1993. Prominent location of a K^+ channel containing the alpha subunit Kv 1.2 in the basket cell nerve terminals of rat cerebellum. Neuroscience **57:** 1039–1045.
85. MCNAMARA, N.M., S. AVERILL, G.P. WILKIN, *et al.* 1996. Ultrastructural localization of a voltage-gated K+ channel alpha subunit (KV 1.2) in the rat cerebellum. Eur. J. Neurosci. **8:** 688–699.
86. SHENG, M., Y.J. LIAO, Y.N. JAN & L.Y. JAN. 1993. Presynaptic A-current based on heteromultimeric K+ channels detected in vivo. Nature **365:** 72–75.
87. WANG, Y., J.C. STRAHLENDORF & H.K. STRAHLENDORF. 1991. A transient voltage-dependent outward potassium current in mammalian cerebellar Purkinje cells. Brain Res. **567:** 153–158.
88. WANG, Y., J.C. STRAHLENDORF & H.K. STRAHLENDORF. 1992. Serotonin reduces a voltage-dependent transient outward potassium current and enhances excitability of cerebellar Purkinje cells. Brain Res. **571:** 345–349.
89. SCOTT, V.E., Z.M. MUNIZ, S. SEWING, *et al.* 1994. Antibodies specific for distinct Kv subunits unveil a heteroologomeric basis for subtypes of alpha-dendrotoxin-sensitive K+ channels in bovine brain. Biochemistry **33:** 1617–1623.
90. RHODES, K.J., S.A. KEILBAUGH, N.X. BARREZUETA, *et al.* 1995. Association and colocalization of K+ channel alpha- and beta-subunit polypeptides in rat brain. J. Neurosci. **15:** 5360–5371.
91. RHODES, K.J., M.M. MONAGHAN, N.X. BARREZUETA, *et al.* 1996. Voltage-gated K+ channel beta subunits: expression and distribution of Kv beta 1 and Kv beta 2 in adult rat brain. J. Neurosci. **16:** 4846–4860.
92. RHODES, K.J., B.W. STRASSLE, M.M. MONAGHAN, *et al.* 1997. Association and colocalization of the Kvbeta1 and Kvbeta2 beta-subunits with Kv1 alpha-subunits in mammalian brain K+ channel complexes. J. Neurosci. **17:** 8246–8258.
93. VEH, R.W., R. LICHTINGHAGEN, S. SEWING, *et al.* 1995. Immunohistochemical localization of five members of the Kv1 channel subunits: contrasting subcellular locations and neuron-specific co-localizations in rat brain. Eur. J. Neurosci. **7:** 2189–2205.
94. LAUBE, G., J. ROPER, J.C. PITT, *et al.* 1996. Ultrastructural localization of Shaker-related potassium channel subunits and synapse-associated protein 90 to septate-like junctions in rat cerebellar Pinceaux. Brain Res. Mol. Brain Res. **42:** 51–61.

95. CHUNG, Y.H., C. SHIN, M.J. KIM, *et al.* 2001. Immunohistochemical study on the distribution of six members of the Kv1 channel subunits in the rat cerebellum. Brain Res. **895:** 173–177.
96. BUTLER, D.M., J.K. ONO, T. CHANG, *et al.* 1998. Mouse brain potassium channel beta1 subunit mRNA: cloning and distribution during development. J. Neurobiol. **34:** 135–150.
97. STORM, J.F. 1988. Temporal integration by a slowly inactivating K+ current in hippocampal neurons. Nature **336:** 379–381.
98. GOLDING, N.L., H.Y. JUNG, T. MICKUS & N. SPRUSTON. 1999. Dendritic calcium spike initiation and repolarization are controlled by distinct potassium channel subtypes in CA1 pyramidal neurons. J. Neurosci. **19:** 8789–8798.
99. MAGEE, J.C. 1999. Voltage-gated ion channels in dendrites. *In* Dendrites. G. Stuart, N. Spruston & M. Häusser, Eds. :139–160. Oxford University Press. New York.
100. MIYASHO, T., H. TAKAGI, H. SUZUKI, *et al.* 2001. Low-threshold potassium channels and a low-threshold calcium channel regulate Ca spike firing in the dendrites of cerebellar Purkinje neurons: a modeling study. Brain Res. **891:** 106–115.
101. ETZION, Y. & Y. GROSSMAN. 1998. Potassium currents modulation of calcium spike firing in dendrites of cerebellar Purkinje cells. Exp. Brain Res. **122:** 283–294.
102. ETZION, Y. & Y. GROSSMAN. 2001. Highly 4-aminopyridine sensitive delayed rectifier current modulates the excitability of guinea pig cerebellar Purkinje cells. Exp. Brain Res. **139:** 419–425.
103. HOUNSGAARD, J. & J. MIDTGAARD. 1989. Synaptic control of excitability in turtle cerebellar Purkinje cells. J. Physiol. **409:** 157–170.
104. HWANG, P.M., A.M. CUNNINGHAM, Y.W. PENG & S.H. SNYDER. 1993. CDRK and DRK1 K+ channels have contrasting localizations in sensory systems. Neuroscience **55:** 613–620.
105. LIM, S.T., D.E. ANTONUCCI, R.H. SCANNEVIN & J.S. TRIMMER. 2000. A novel targeting signal for proximal clustering of the Kv2.1 K+ channel in hippocampal neurons. Neuron **25:** 385–397.
106. RUDY, B., A. CHOW, D. LAU, *et al.* 1999. Contributions of Kv3 channels to neuronal excitability. Ann. N.Y. Acad. Sci. **868:** 304–343.
107. GOLDMAN-WOHL, D.S., E. CHAN, D. BAIRD & N. HEINTZ. 1994. Kv3.3b: a novel Shaw type potassium channel expressed in terminally differentiated cerebellar Purkinje cells and deep cerebellar nuclei. J. Neurosci. **14:** 511–522.
108. VEGA-SAENZ DE MIERA, E., M. WEISER, C. KENTROSE, *et al.* 1994. *Shaw*-related K^+ channels in mammals. *In* Handbook of Membrane Channels: Molecular and Cellular Physiology. C. Peracchia, Ed. :41–78. Academic Press. San Diego.
109. WEISER, M., E. VEGA-SAENZ DE MIERA, C. KENTROS, *et al.* 1994. Differential expression of Shaw-related K+ channels in the rat central nervous system. J. Neurosci. **14:** 949–972.
110. SEKIRNJAK, C., M.E. MARTONE, M. WEISER, *et al.* 1997. Subcellular localization of the K+ channel subunit Kv3.1b in selected rat CNS neurons. Brain Res. **766:** 173–187.
111. RAMAN, I.M. & B.P. BEAN. 1999a. Ionic currents underlying spontaneous action potentials in isolated cerebellar Purkinje neurons. J. Neurosci. **19:** 1663–1674.
112. SOUTHAN, A.P. & B. ROBERTSON. 2000. Electrophysiological characterization of voltage-gated K(+) currents in cerebellar basket and purkinje cells: Kv1 and Kv3 channel subfamilies are present in basket cell nerve terminals. J. Neurosci. **20:** 114–122.
113. MARTINA, M., G.L. YAO & B.P. BEAN. 2001 Molecular and functional characterization of voltage-gated potassium channels in the somata and dendrites of cerebellar Purkinje cells. Soc. Neurosci. Abstr. **382.11**.
114. SACCO, T., E. WANKE & F. TEMPIA. 2001. ERG Potassium current in mouse cerebellar Purkinje cells. Soc. Neurosci. Abstr. **272.23**.
115. GRUOL, D.L., T. JACQUIN & A.J. YOOL. 1991. Single-channel K+ currents recorded from the somatic and dendritic regions of cerebellar Purkinje neurons in culture. J. Neurosci. **11:** 1002–1015.
116. JACQUIN, T.D. & D.L. GRUOL. 1999. Ca regulation of a large conductance K+ channel in cultured rat cerebellar Purkinje neurons. Eur. J. Neurosci. **11:** 735–739.

117. JOHNSTON, D., D.A. HOFFMAN, J.C. MAGEE, *et al.* 2000. Dendritic potassium channels in hippocampal pyramidal neurons. J. Physiol. **525:** 75–81.
118. CHANG, C.P., S.I. DWORETZKY, J. WANG & M.E. GOLDSTEIN. 1997. Differential expression of the alpha and beta subunits of the large-conductance calcium-activated potassium channel: implication for channel diversity. Brain Res. Mol. Brain Res. **45:** 33–40.
119. KNAUS, H.G., C. SCHWARZER, R.O. KOCH, *et al.* 1996. Distribution of high-conductance Ca(2+)-activated K+ channels in rat brain: targeting to axons and nerve terminals. J. Neurosci. **16:** 955–963.
120. LANCASTER, B. & R.S. ZUCKER. 1994. Photolytic manipulation of Ca^{2+} and the time course of slow, Ca(2+)-activated K+ current in rat hippocampal neurones. J. Physiol. **475:** 229–239.
121. BOND, C.T., J. MAYLIE & J.P. ADELMAN. 1999. Small-conductance calcium-activated potassium channels. Ann. N.Y. Acad. Sci. **868:** 370–378.
122. KÖHLER, M., B. HIRSCHBERG, C.T. BOND, *et al.* 1996. Small-conductance, calcium-activated potassium channels from mammalian brain. Science **273:** 1709–1714.
123. ISHII, T.M., J. MAYLIE & J.P. ADELMAN. 1997. Determinants of apamin and d-tubocurarine block in SK potassium channels. J. Biol. Chem. **272:** 23195–23200.
124. GRUNNET, M., B.S. JENSEN, S.P. OLESEN & D.A. KLAERKE. 2001. Apamin interacts with all subtypes of cloned small-conductance Ca-activated K+ channels. Pflügers Arch. **441:** 544–550.
125. STOCKER, M. & P. PEDARZANI. 2000. Differential distribution of three Ca(2+)-activated K(+) channel subunits, SK1, SK2, and SK3, in the adult rat central nervous system. Mol. Cell Neurosci. **15:** 476–493.
126. TACCONI, S., R. CARLETTI, B. BUNNEMANN, *et al.* 2001. Distribution of the messenger RNA for the small conductance calcium-activated potassium channel SK3 in the adult rat brain and correlation with immunoreactivity. Neuroscience **102:** 209–215.
127. YI, B.A., D.L. MINOR, JR., Y.F. LIN, JR., *et al.* 2001. Controlling potassium channel activities: Interplay between the membrane and intracellular factors. Proc. Natl. Acad. Sci. USA **98:** 11016–11023.
128. KRAPIVINSKY, G., I. MEDINA, L. ENG, JR., *et al.* 1998. A novel inward rectifier K^+ channel with unique pore properties. Neuron **20:** 995–1005.
129. CRÉPEL, F. & J. PENIT-SORIA. 1986. Inward rectification and low threshold calcium conductance in rat cerebellar Purkinje cells. An in vitro study. J. Physiol. **372:** 1–23.
130. HÄUSSER, M. & B.A. CLARK. 1997. Tonic synaptic inhibition modulates neuronal output pattern and spatiotemporal synaptic integration. Neuron **19:** 665–678.
131. TALLEY, E.M., G. SOLORZANO, Q. LEI, *et al.* 2001. CNS distribution of members of the two-pore-domain (KCNK) potassium channel family. J. Neurosci. **21:** 7491–7505.
132. TEN BRUGGENCATE, G., C. NICHOLSON & H. STÖCKLE. 1976. Climbing fiber evoked potassium release in cat cerebellum. Pflügers Arch. **367:** 107–109.
133. DE SCHUTTER, E. & J.M. BOWER. 1994. An active model of the cerebellar Purkinje cell I. Stimulation of current clamps in slice. J. Neurophysiol. **71:** 375–400.
134. DE SCHUTTER, E. & J.M. BOWER. 1994. An active model of the cerebellar Purkinje cell II. Stimulation of synaptic inputs. J. Neurophysiol. **71:** 401–419.
135. Jaeger, D., E. De Schutter & J.M. Bower. 1997. The role of synaptic and voltage-gated currents in the control of Purkinje cell spiking: a modelling study. J. Neurosci. **17:** 91–106.
136. ITO, M. & J.I. SIMPSON. 1971. Discharges in Purkinje cell axons during climbing fiber activation. Brain Res. **31:** 215–219.
137. ROTH, A. & M. HÄUSSER. 2001. Compartmental models of rat cerebellar Purkinje cells based on simultaneous somatic and dendritic patch-clamp recordings. J. Physiol. **535.2:** 445–472.
138. STAUB, C., E. DE SCHUTTER & T. KNÖPFEL. 1994. Voltage-imaging and simulation of effects of voltage- and agonist-activated conductances on soma-dendritic voltage coupling in cerebellar Purkinje cells. J. Comp. Neurosci. **1:** 301–311.
139. BORST, A., D. HECK & M. THOMANN. 1997. Voltage signals of individual Purkinje cell dendrites in rat cerebellar slices. Neurosci. Lett. **238:** 29–32.

140. ECCLES, J.C., R. LLINAS, K. SASAKI & P.E. VOORHOEVE. 1966. Interaction experiments on the responses evoked in Purkinje cells by climbing fibers. J. Physiol. **182:** 297–315.
141. CALLAWAY, J.C., N. LASSER-ROSS & W.N. ROSS. 1995. IPSPs strongly inhibit climbing fiber-activated $[Ca^{2+}]_i$ increases in the dendrites of cerebellar Purkinje neurons. J. Neurosci. **15:** 2777–2787.
142. STAUB, C., I. VRANESIC & T. KNÖPFEL. 1992. Responses to metabotropic glutamate receptor activation in cerebellar Purkinje cells: induction of an inward current. Eur. J. Neurosci. **4:** 832–839.
143. BATCHELOR, A.M. & J. GARTHWAITE. 1993. Novel synaptic potentials in cerebellar Purkinje cells: probable mediation by metabotropic glutamate receptors. Neuropharmacology **32:** 11–20.
144. VRANESIC, I., C. STAUB & T. KNÖPFEL. 1993. Activation of metabotropic glutamate receptors induces an outward current which is potentiated by methylxanthines in rat cerebellar Purkinje cells. Neurosci. Res. **16:** 209–215.
145. NETZEBAND, J.G., K.L. PARSONS, D.D. SWEENEY & D.L. GRUOL. 1997. Metabotropic glutamate receptor agonists alter neuronal excitability and Ca levels via the phospholipase C transduction pathway in cultured Purkinje neurons. J. Neurophysiol. **78:** 63–75.
146. LINDEN, D.J., M. SMEYNE, S.C. SUN & J.A. CONNOR. 1992. An electrophysiological correlate of protein kinase C isozyme distribution in cultured cerebellar neurons. J. Neurosci. **12:** 3601–3608.
147. PISANI, A. & W.N. ROSS. 1999. Weak effect of neuromodulators on climbing fiber-activated $[Ca^{2+}]_i$ increases in rat cerebellar Purkinje cells. Brain Res. **831:** 113–118.
148. FOX, E.A. & D.L. GRUOL. 1993. Corticotropin-releasing factor suppresses the after-hyperpolarization in cerebellar Purkinje neurons. Neurosci. Lett. **149:** 103–107.
149. MIYATA, M., D. OKADA, K. HASHIMOTO, *et al.* 1999. Corticiotropin-releasing factor plays a permissive role in cerebellar long-term depression. Neuron **22:** 763–775.
150. PALKOVITS, M., C. LERNANTH, T. GORCS & W.S. YOUNG, III. 1987. Corticotropin-releasing factor in the olicocerebellar tract of rats: demonstration by light- and electron-microscopic immunohistochemistry and in situ hybridisation histochemistry. Proc. Natl. Acad. Sci. USA **84:** 3911–3915.
151. HARTELL, N.A. 1996. Strong activation of parallel fibers produces localized calcium transients and a form of LTD that spreads to distant synapses. Neuron **16:** 601–610.
152. EILERS, J., H. TAKECHI, E.A. FINCH, *et al.* 1997. Local dendritic Ca signalling induces cerebellar long-term depression. Learn. Mem. **4:** 159–168.
153. HANSEL, C., A. ARTOLA & W. SINGER. 1997. Relation between dendritic Ca levels and the polarity of synaptic long-term modifications in rat visual cortex neurons. Eur. J. Neurosci. **9:** 2309–2322.
154. YANG, S.N., Y.G. TANG & R.S. ZUCKER. 1999. Selective induction of LTP and LTD by postsynaptic $[Ca^{2+}]_i$ elevation. J. Neurophysiol. **81:** 781–787.
155. ITO, M. 2001. Cerebellar long-term depression: characterization, signal transduction, and functional roles. Physiol. Rev. **81:** 1143–1194.
156. BAUDE, A., E. MOLNÁR, D. LATAWIEC, *et al.* 1994. Synaptic and nonsynaptic localization of the GluR1 subunit of the AMPA-type excitatory amino acid receptor in the rat cerebellum. J. Neurosci. **14:** 2830–2843.
157. NUSSER, Z., E. MULVIHILL, P. STREIT & P. SOMOGYI. 1994. Subsynaptic segregation of metabotropic and ionotropic glutamate receptors as revealed by immunogold localization. Neuroscience **61:** 421–427.
158. LANDSEND, A.S., M. AMIRY-MOGHADDAM, A. MATSUBARA, *et al.* 1997. Differential localization of δ glutamate receptors in the rat cerebellum: coexpression with AMPA receptors in parallel fiber-spine synapses and absence from climbing fiber-spine synapses. J. Neurosci. **17:** 834–842.
159. ASE, K., N. SAITO, M.S. SHEARMAN, *et al.* 1988. Distinct cellular expression of beta I- and beta II-subspecies of protein kinase C in rat cerebellum. J. Neurosci. **8:** 3850–3856.
160. HASHIMOTO, T., K. ASE, S. SAWAMURA, *et al.* 1988. Postnatal development of a brain specific subspecies of protein kinase C in rat. J. Neurosci. **8:** 1678–1683.

161. LANG, E.J., I. SUGIHARA, J.P. WELSH & R. LLINAS. 1999. Patterns of spontaneous Purkinje cell complex spike activity in the awake rat. J. Neurosci. **19:** 2728–2739.
162. O'HEARN, E. & M.E. MOLLIVER. 1997. The olivocerebellar projection mediates ibogaine-induced degeneration of Purkinje cells: a model of indirect, trans-synaptic excitotoxicity. J. Neurosci. **15:** 8828–8841.
163. GRANIT, R. & C.G. PHILLIPS. 1956. Excitatory and inhibitory processes acting upon individual Purkinje cells of the cerebellum in cats. J. Physiol. **133:** 520–547.
164. BLOEDEL, J.R. & W.J. ROBERTS. 1971. Action of climbing fibers in cerebellar cortex of the cat. J. Neurophys. **34:** 17–31.
165. MANO, N., I. KANAZAWA & K. YAMAMOTO. 1986. Complex spike activity of cerebellar Purkinje cells related to wrist tracking movement in monkey. J. Neurophysiol. **56:** 137–158.
166. SATO, Y., A. MIURA, H. FUSHIKI & T. KAWASAKI. 1992. J. Neurophysiol. **68:** 2051–2062.
167. SCHIFFMAN, S.N., G. CHERON, A. LOHOF, *et al.* 1999. Impaired motor coordination and Purkinje cell excitability in mice lacking calretinin. Proc. Natl. Acad. Sci. USA **96:** 5257–5262.
168. GOOSSENS, J., H. DANIEL, A. RANCILLAC, *et al.* 2001. Expression of protein kinase C inhibitor blocks cerebellar long-term depression without affecting Purkinje cell excitability in alert mice. J. Neurosci. **21:** 5813–5823.
169. MURPHY, J.T. & N.H. SABAH. 1970. The inhibitory effect of climbing fiber activation on cerebellar Purkinje cells. Brain Res. **19:** 486–490.
170. GIESE, K.P., M. PETERS & J. VERNON. 2001. Modulation of excitability as a learning and memory mechanism: a molecular genetic perspective. Physiol. Behav. **73:** 803–810.
171. SCHREURS, B.G., P.A. GUSEV, D. TOMSIC, *et al.* 1998. Intracellular correlates of acquisition and long-term memory of classical conditioning in Purkinje cell dendrites in slices of rabbit cerebellar lobule HVI. J. Neurosci. **18:** 5498–5507.
172. CALLEWAERT, G., J. EILERS & A. KONNERTH. 1996. Axonal calcium entry during fast 'sodium' action potentials in rat cerebellar Purkinje neurones. J. Physiol. **495.3:** 641–647.
173. LINDEN, D.J. 1999. The return of the spike: postsynaptic action potentials and the induction of LTP and LTD. Neuron **22:** 661–666.

Analysis of Cx36 Knockout Does Not Support Tenet That Olivary Gap Junctions Are Required for Complex Spike Synchronization and Normal Motor Performance

W. M. KISTLER,[a,d] M. T. G. DE JEU,[a,d] Y. ELGERSMA,[a] R. S. VAN DER GIESSEN,[a] R. HENSBROEK,[a] C. LUO,[a] S. K. E. KOEKKOEK,[a] C. C. HOOGENRAAD,[a] F. P. T. HAMERS,[b] M. GUELDENAGEL,[c] G. SOHL,[c] K. WILLECKE,[c] AND C. I. DE ZEEUW[a]

[a]*Department of Neuroscience, Erasmus MC, University Medical Center Rotterdam, 3000DR Rotterdam, The Netherlands*

[b]*Department of Medical Pharmacology and Anatomy, Rudolf Magnus Institute for Neurosciences, University Medical Center Utrecht, 3584CG Utrecht, The Netherlands*

[c]*Institute of Genetics, Division of Molecular Genetics, University of Bonn, 53117 Bonn, Germany*

ABSTRACT: Electrotonic coupling by gap junctions between neurons in the inferior olive has been claimed to underly complex spike (CS) synchrony of Purkinje cells in the cerebellar cortex and thereby to play a role in the coordination of movements. Here, we investigated the motor performance of mice that lack connexin36 (Cx36), which appears necessary for functional olivary gap junctions. Cx36 null-mutants are not ataxic, they show a normal performance on the accelerating rotorod, and they have a regular walking pattern. In addition, they show normal compensatory eye movements during sinusoidal visual and/or vestibular stimulation. To find out whether the normal motor performance in mutants reflects normal CS activity or some compensatory mechanism downstream of the cerebellar cortex, we determined the CS firing rate, climbing-fiber pause, and degree of CS synchrony. None of these parameters in the mutants differed from those in wildtype littermates. Finally, we investigated whether the role of coupling becomes apparent under challenging conditions, such as during application of the tremorgenic drug harmaline, which specifically turns olivary neurons into an oscillatory state at a high frequency. In both the mutants and wildtypes this application induced tremors of a similar duration with similar peak frequencies and amplitudes. Thus surprisingly, the present data does not support the notion that electrotonic coupling by gap junctions underlies synchronization of olivary spike activity and that these gap junctions are essential for normal motor performance.

Address for correspondence: C.I. De Zeeuw, Department of Neuroscience, Erasmus MC, University Medical Center Rotterdam, Postbox 1738 3000DR Rotterdam, the Netherlands. Use the following when sending by Fedex: Department of Neuroscience, Room Ee1202, Erasmus MC, University Medical Center Rotterdam, Dr. Molewaterplein 50, 3015 GE Rotterdam, the Netherlands. Voice: +31(10)4087299; fax: +31(10)4089459.

dezeeuw@anat.fgg.eur.nl

[d]Both authors contributed equally to this paper.

Ann. N.Y. Acad. Sci. 978: 391–404 (2002).

Keywords: gap junction; connexin36; inferior olive; cerebellum; complex spike synchrony; motor behavior; multielectrode recording; cross-correlation; harmaline; tremor; mouse

INTRODUCTION

Despite the fact that the microcircuitry of the olivocerebellar system is known in considerable detail, its function is still a matter of fervent debate. The function that is classically assigned to the cerebellum is the coordination of motor behavior—a conclusion that is widely supported by clinical observations.[32] A design principle that could explain convincingly how cerebellar function emerges from its morphological and electrophysiological peculiarities, however, is still elusive, despite the many attempts that have been made to link "form and function."[2,4,16,19,33] Recently, attention has been paid particularly to the inferior olivary nucleus, which is an extraordinary structure in several aspects. First, the inferior olive (IO) is the sole source of cerebellar climbing fibers that form highly efficient synaptic contacts with their respective Purkinje cell. A single climbing-fiber spike suffices to trigger a postsynaptic action potential followed by a burst of spikelets, the so-called complex spike response.[11] Second, olivary neurons fire at a frequency of only 1 Hz, but they exhibit a high degree of temporal order. Slice recordings reveal synchronous subthreshold activities,[10,21,28] and Purkinje cells within a common rostro-caudal strip tend to fire complex spikes synchronously.[3,36]

This synchronization of complex spikes, which can occur in a millisecond time scale, has been suggested to result from dense coupling by fast electrical synapses between olivary dendrites.[5–7,22,23,26,27,30] Moreover, since the dendrodendritic gap junctions are surrounded by strategically positioned chemical synaptic inputs[7] and GABA antagonists can influence the complex spike synchrony,[23,27] it has been reasoned that regulation of dynamic coupling in the olive may play an important role in the timing and coordination of motor behavior.[25,26,41]

Recently, global knockout mice of gap junction protein connexin36 (Cx36) have been created, and the inferior olive of these mutants turns out to be devoid of functional neuronal gap junctions.[31] Hence, $Cx36^{-/-}$ mice provide an exciting substrate to test the hypotheses on the role of gap junctions in complex spike synchrony and motor coordination. In the present paper we address these hypotheses using both behavioral tests and *in vivo* electrophysiology. Our findings raise the possibility that the major current ideas on the role of olivary coupling gap junctional coupling may have to be reevaluated.

MATERIALS AND METHODS

Cx36-deficient and wildtype mice were generated and characterized as described by Gueldenagel *et al.*[13] Tests were performed on adult animals only. All animal procedures described were carried out under animal care protocols approved by the Erasmus University Rotterdam.

Walking Pattern Analysis

Walking patterns were recorded and analyzed using the Catwalk system described previously.[15] In short, mice were directed over a clear, Plexiglas runway (60 cm long, 5 cm wide). Light was passed through the Plexiglas floor using a fluorescent light bulb, and wherever the floor was touched, light was deflected. This deflected light was recorded by a camera allowing a spatiotemporal analysis of walking patterns. The following parameters were calculated: step regularity (percentage of regular step patterns), step length, base of support (the width between paws), and paw contact area (the paw area contacted at the moment of maximum paw-floor contact).

Rotorod Test

More complicated motor behaviors and motor learning were assessed with the accelerating rotorod test. The rotorod consists of a smooth plastic roller (8 cm diameter; 14 cm long) flanked by two large round plates (30 cm diameter) to prevent animals from escaping. On day 1, mice were placed on the stationary roller, and the time was measured until they fell off again (2 minutes max.). The next day, the test was repeated with the rod rotating at a constant velocity of 2 rpm. On days 3–6, the mice were initially put on the rod spinning at a constant velocity (2 rpm). After 10 s the rotational velocity was gradually increased in 2 minutes to 12 rpm until the mice fell off. Each test was repeated three times per day per mouse.

Eye Movement Recording

Horizontal eye movements were recorded using the magnetic search coil technique described previously.[8,39] In short, the mouse head was restrained during _recording by an acrylic pedestal fixed to the scull. After induction of halothane/nitrous oxide anesthesia, the scalp was incised, and four stainless steel screws (M 1×1.5 mm) were implanted in the calvarium and then encased in dental acrylic. Additional acrylic was used to fix two M2 nuts to this mass.

Under general (halothane/nitrous oxide) and local (Novesine, 0.4%; BournonvillePharma B. V., the Netherlands) anesthesia, a scleral search coil (1 mm outside diameter, 60 turns of copper wire, 1.0 mg) was implanted in the conjunctiva on the temporal side of the eyeball and fixed to the sclera with two sutures (10/0 nylon, Ethicon). The leads of the coil were carefully tunnelled subconjunctivally around the inferior margin of the eye and then out of the medial superior aspect of the orbit to a miniature coaxial MMCX connector that was attached to the acrylic head pedestal. After the mouse recovered from surgery (3 days), compensatory eye movements were recorded using a magnetic measurement system (Skalar Medical). Visual and vestibular stimuli were delivered with the use of a drum with a black-and-white strip pattern and a turntable. Stimulus parameters and data acquisition was controlled by a PC via a Power1401 interface (Cambridge Electronics Design, Cambridge, UK) and Spike2 for Windows. The performance of the optokinetic (OKR) and the vestibulo-ocular reflex (VOR) was assessed on four consecutive days by using sinusoidal stimuli at various frequencies and amplitudes. The response latency of the OKR was measured by using velocity steps from –4°/s to +4°/s in intervals with random durations of 1–2 seconds. Each recording session lasted for about one hour. Eye-movement data was analyzed offline with custom-made C programs and Mathematica (Wolfram Research).

Electrophysiology

The mouse head was restrained during recording in a stereotaxic instrument by an acrylic pedestal that was built one day before the recording, as described above. To prepare the animal for recording it was anesthetized with ketamine (50 mg/kg, i.p.) and xylazine (8 mg/kg, i.p.). Additional doses were delivered every 60 min or as needed to maintain anesthesia. Neck muscles, occipital bone, and dura were removed to expose the dorsal surface of the cerebellum, which was then covered by a layer of agar (2% Agar Agar in 0.9% NaCl) to prevent dehydration of the tissue.

Extracellular recordings of complex spike (CS) and simple spike (SS) activity were obtained with glass microelectrodes with a tip diameter of about 2 µm (~2 MΩ) containing 2 M saline. Electrodes were positioned independently with two precision micromanipulators (TSE Systems, Germany). Particular care was taken to position the electrodes close to each other (rostrocaudal tip distance 0–200 µm). As soon as a stable recording was obtained with one of the electrodes, the other electrode was used to explore the neighborhood mediolaterally in steps of 100 µm (mediolateral tip distance < 200 µm). Spike trains were recorded for about 20 min per cell pair. Single units of Purkinje cells could easily be identified by the presence of complex spikes followed by a brief pause in simple spike activity. Signals where preamplified (×50 ultralow noise differential amplifier AI 402, Axon Instruments, USA), band-pass filtered in the range of 300 Hz–4 kHz (CyberAmp 380, Axon Instruments, USA), digitalized with a sampling rate of 10 kHz and 16 bit resolution (Power1401, Cambridge Electronic Design, Cambridge UK), and stored on computer hard disk for off-line analysis.

Spike Sorting and Analysis

Complex spikes were identified by their characteristic wave form. In a first step, all spikes were extracted from the recording files using a simple threshold criterion. The form of the spikes (0.5 ms before and 5.0 ms after the first downstroke) was analyzed by principal component analysis (PCA) using custom-made C-programs and Mathematica (Wolfram Research, USA). Projecting the high-dimensional set of data points to the first few principal components usually produced two well-separated clouds for simple and complex spikes.

Trains of complex spikes were discretized using bin widths of Δt equal to 1, 2, 5, or 10 milliseconds. The degree of synchrony between two complex spike trains was quantified by the zero-time difference cross-correlation coefficient,

$$C_{1,2}(0) = (n^2\bar{S}_1\bar{S}_2)^{-1/2}\left[\sum_{i=1}^{n} S_1(i\Delta t)S_2(i\Delta t) - n\bar{S}_1\bar{S}_2\right], \quad (1)$$

with $S_k(i\Delta t) = 1$; if time bin i contains the onset of a complex spike of neuron k, $k =$ 1, 2, or else $S_k(i\Delta t) = 0$. Furthermore, n is the total number of time bins and

$$\bar{S}_k = n^{-1} \sum_{n=1}^{n} S_k(i\Delta t) .$$

Tremor Measurements and Analysis

Harmaline-induced tremors were studied in 10 Cx36 wildtype (WT), 10 Cx36 knockout (KO), and 3 Lurcher mice (ages 71–152 days). Animals were shortly anesthetized with isoflurane (1.5–2%) and equipped with a harness (Instech Laboratories, Plymouth Meeting, PA) that contained three accelerometer chips (VTI Hamlin, Vantaa, Finland). After putting on the harness, harmaline (Sigma, St. Louis, USA) was injected intraperitoneally (50 mg/kg body weight). The accelerometer chips were oriented in the anterior-posterior (x axis), medial-lateral (y axis), and dorsal-ventral direction (z axis). Signals were acquired using a CyberAmp (Axon Instruments, USA) and relayed by a Power1401 interface (CED, Cambridge, UK) to a personal computer equipped with Spike2 (CED, Cambridge, UK). Signals were recorded at a sample rate of 200 Hz.

The acceleration traces of all three axes (1200 s) were analyzed by using a Hann windowed 1024-point Fast Fourier Transformation. Every second a power spectrum was generated by analysing the last 5.12 seconds. Episodes of tremor were detected by the peaks in the power spectra at the 5–15 Hz frequency band. These episodes of tremor were then used to generate a global power spectrum. The central peak of this spectrum was fitted to a Gaussian function.

RESULTS

We have addressed the role of electrotonic coupling of IO neurons in mice at both the behavioral and electrophysiological level. We compare results obtained in wildtype animals with those in mice that had a homozygous deletion of connexin36 and have been reported to be devoid of functional neuronal gap junctions.[31]

Motor Behavior

To our great surprise Cx36$^{-/-}$ mice showed no obvious signs of motor impairment or ataxia. In order to obtain quantitative data, we carefully analyzed the walking pattern of Cx36$^{-/-}$ and wildtype mice using the "catwalk" technique, described in the METHODS section. Parameters such as step regularity, step length, base of support, and paw contact area did not exhibit a significant difference (FIG. 1A–D), while the base of support of the hind limbs, but not of the forelimbs, was significantly smaller in the Cx36$^{-/-}$ mice (p value = 0.014, two-sided Student t test). This difference, however, turned out not to be significant in an ANOVA test that also accounted for variations in body weight. As a positive control, we also tested a weight-matched group of lurcher mice. These mice had, for example, a significantly ($p < 0.004$) reduced step regularity, signifying their general ataxia (data not shown). To see whether Cx36$^{-/-}$ mice are impaired in a motor task that is more challenging than walking on a smooth surface, we tested the Cx36$^{-/-}$ mice on the accelerating rotorod. Again, no significant difference in the performance of Cx36$^{-/-}$ and the wildtype control group could be detected (FIG. 1E).

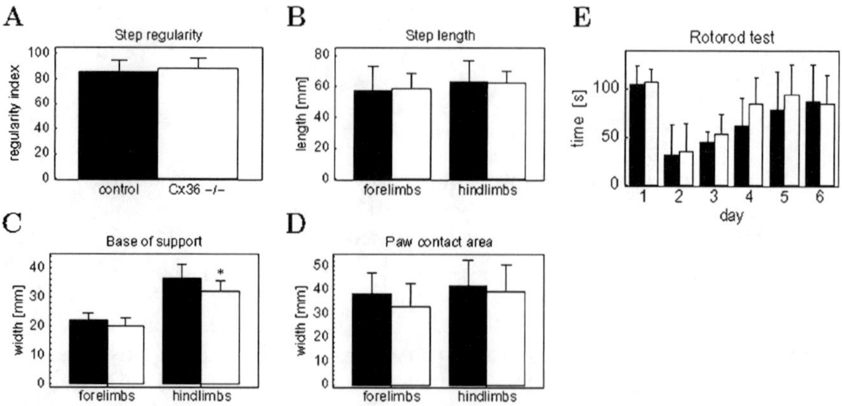

FIGURE 1. Motor behavior. (**A–D**) Walking pattern analysis of Cx36$^{-/-}$ (n = 13, *open bars*) and wildtype (n = 12, *filled bars*) mice. None of the measured parameters revealed a significant difference between the two groups except the base of support of the hind limbs (see RESULTS). (**E**) Rotorod test. The *bars* give the time the mice managed to stay on the stationary (day 1), slowly rotating (day 2), or accelerating rod (days 3–6). No difference in the performance of Cx36$^{-/-}$ (n = 7, *open bars*) and wildtype mice (n = 7, *filled bars*) could be detected. Error bars correspond to 1 SD.

The olivocerebellar system is known to be critically involved in the control of compensatory eye movements, such as OKR and VOR.[17] To check whether the absence of olivary gap junctions had any effect on the performance of these reflexes, we measured amplitude and phase lag of the eye movements elicited by sinusoidal visual and vestibular stimuli. We used a broad range of different stimulation frequencies and amplitudes, but none of these paradigms revealed a significant difference between Cx36$^{-/-}$ mice and the wildtype control group (FIG. 2A–D).

The absence of the connexin protein not only affects olivary neurons but also gap junctions in the retina.[14,35] The impaired coupling of AII amacrine cells and bipolar cells causes deficiencies in early vision.[13] To test whether these deficiencies show up at the behavioral level, we measured the latency of the OKR to velocity steps of the visual stimulation. In wildtype animals ($n = 6$) the latency amounts to 82.1 ms ± 4.0 ms (mean ± SEM), whereas in Cx36$^{-/-}$ ($n = 7$), we found slightly, but significantly, delayed responses with a latency of 93.8 ms ± 2.3 ms ($p < 0.033$, two-sided Student t test) (FIG. 2E). Thus, even though the optokinetic signals are delayed, probably due to retinal dysfunction, the control of the vestibulocerebellum over the compensatory eye movements appears intact.

Complex Spike Activity

To find out whether a lack of coupling in the inferior olive affects the level of complex spike synchrony, we recorded activities of Purkinje cell pairs by means of extracellular recordings with two independently manipulable glass electrodes.

We recorded a total of 94 pairs of nearby Purkinje cells that were located in the same parasagittal plane in either the Vermis or Crus IIa. About half of these record-

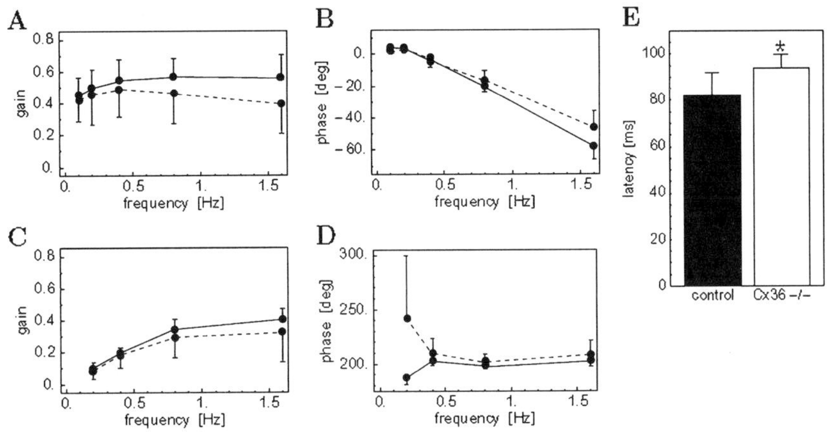

FIGURE 2. Compensatory eye movements. (**A–D**) Gain and phase lag for OKR (**A** and **B**) and VOR (**C** and **D**) performance of Cx36$^{-/-}$ ($n = 5$, *solid line*) and wildtype mice ($n = 4$, *dashed line*) at various stimulus frequencies. *Error bars* correspond to 1 SD. There is no significant difference in the performance of both groups. (**E**) Latency of the OKR response to velocity steps of the visual stimulus. The response of Cx36$^{-/-}$ mice ($n = 7$, *empty bar*) was significantly ($p < 0.033$, two-sided Student t test) delayed with respect to the wildtype control group ($n = 6$, *solid bar*). *Error bars* correspond to 1 SD.

ings ($n = 50$ pairs) showed a sufficiently high signal-to-noise ratio so that not only the complex spikes but also the simple spikes could be reliably discerned as single units. Apart from two extreme cases (see below), the complex spikes of the cell pairs in both Cx36$^{-/-}$ and wildtype mice were only loosely synchronized; the corresponding peaks in their cross-correlograms had a width of tens of milliseconds (FIG. 3).

To quantify the degree of synchrony we calculated the cross-correlation coefficient at time difference zero using bin widths of 1, 2, 5, and 10 ms for each cell pair. The mean cross-correlation coefficients in Cx36$^{-/-}$ mice were 0.0027 ± 0.0027, 0.0047 ± 0.0046, 0.0107 ± 0.0118, 0.0214 ± 0.0231 for bin widths of 1, 2, 5, and 10 ms, respectively (mean ± SD), while in the wildtype control group the corresponding coefficients equalled 0.0018 ± 0.0023, 0.0032 ± 0.0031, 0.0081 ± 0.0072, and 0.0162 ± 0.0154 (FIG. 3). The degree of synchrony was not significantly different between knockout and wildtype animals whatever bin width was used (two-sided Student t test, $p > 0.20$) (FIG. 3D).

In only two cases both cells fired complex spikes in an almost perfectly synchronized fashion. In these cases (both Cx36$^{-/-}$) all complex spikes were accompanied by a complex spike in the other cell, and the onsets of both complex spikes occurred within the same millisecond (FIG. 4). The cross-correlation coefficients of these two pairs, $C_{1,2}(0)$, which equalled 0.85, differed by several orders of magnitude from those of the group of loosely synchronized cell pairs described above, and the data were therefore not pooled for analysis. Since the tips of the recording electrodes were relatively close, sometimes at a distance of about 160 μm in the rostro-caudal direction, we had to exclude the possibility that the highly synchronized activities were, in fact, derived from the same single cell. Therefore, we cross-correlated the

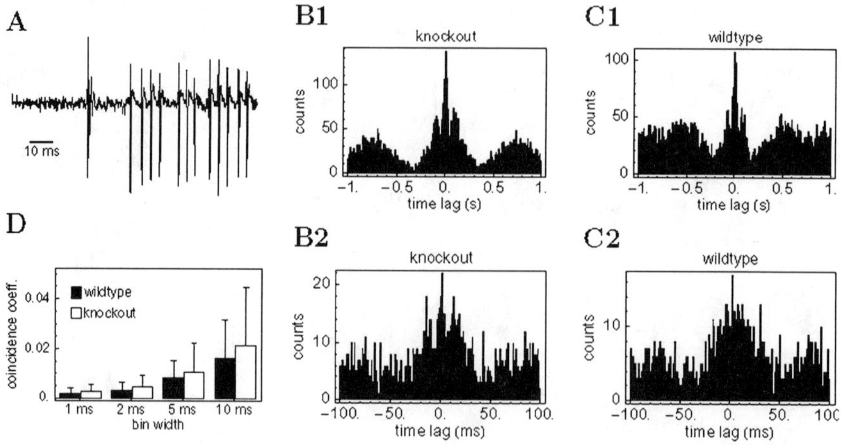

FIGURE 3. Complex spike synchrony in Cx36 knockout and wildtype mice. (**A**) Extracellular recording of Purkinje cell activity. Complex and simple spikes can clearly be discerned by their wave form. (**B**) Typical cross-correlograms of complex spike activity in Cx36 knockout compiled for 10-ms bins (**B1**) and 1-ms bins (**B2**). (**C**) Cross-correlograms of complex spike activity for wildtype mice compiled for 10-ms bins (**Cl**) and 1-ms bins (**C2**). (**D**) Mean coincidence coefficients for Cx36 knockout ($n = 21$ cell pairs, *empty bars*) and wildtype ($n = 29$ cell pairs, *solid bars*) mice compiled at different bin widths. *Error bars* correspond to 1 SD. The degree of synchrony in knockout and wildtype is not significantly different whatever the bin width.

simple spike activities of these cell pairs, and we examined the zero-bin peaks. Since these simple spike correlograms were virtually flat and both cells showed a clean climbing-fiber pause (FIGS. 4B and C), we concluded that the activities were indeed obtained from two distinct Purkinje cells.

In addition to the complex spike correlations of two Purkinje cells, we also checked single cell parameters, such as the average complex spike firing rate (Cx36$^{-/-}$: 0.95 ± 0.48 Hz; control: 1.03 ± 0.45 Hz), mean interspike interval (Cx36$^{-/-}$: 1.19 ± 0.6 s; control: 1.10 ± 0.5 s), coefficient of variation (Cx36$^{-/-}$: 0.74 ± 0.39; control: 0.66 ± 0.45), and the duration of the complex spike pause (Cx36$^{-/-}$: 23.3 ± 7.6 ms; control: 27.0 ± 8.5 ms; all numbers mean ± SD). None of these parameters revealed a significant difference between Cx36$^{-/-}$ and controls.

Harmaline-Induced Tremor

Because we were unable to detect any effect due to the absence of functional olivary gap junctions, either on motor performance or on complex spike activity, we wanted to investigate whether the role of the coupling becomes apparent under more challenging conditions. We therefore applied the tremorgenic drug harmaline, which specifically turns olivary neurons into an oscillatory state at a high frequency.[29] If the degree of spike synchrony in the IO was relatively reduced in the Cx36 knockout under this artificial condition, we would expect to see a reduced residual tremor.

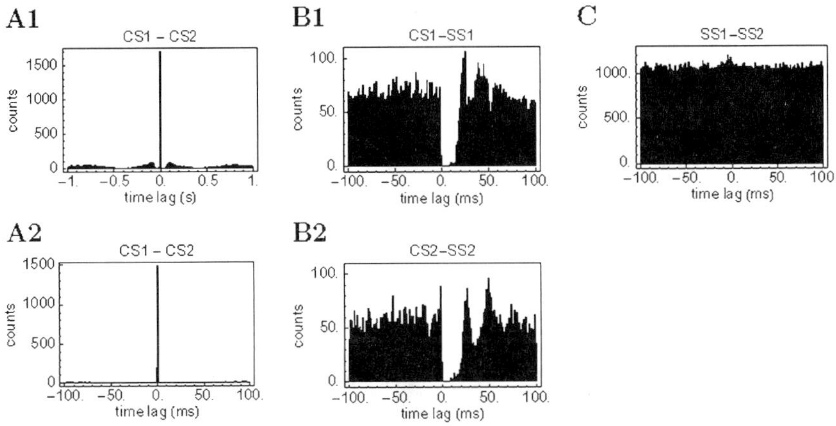

FIGURE 4. Ultraprecisely synchronized complex spikes. (**A**) Complex spike cross-correlograms with 10-ms (**A1**) and 1-ms (**A2**) bin width. The width of the central peak in **A2** is less than 1 ms. (**B**) Complex-spike triggered histogram of simple spikes for cell 1 (**B1**) and cell 2 (**B2**). Each complex spike is followed by a clear pause in the simple spike activity, indicating that we are recording from a single unit. (**C**) The simple spike cross-correlogram is virtually flat, which implies that the recordings have been obtained from two different Purkinje cells.

The tremorgenic effect of harmaline was studied in wildtype, $Cx36^{-/-}$, and, as a control, in lurcher mice. Harmaline induced pronounced tremors in the $Cx36^{-/-}$ and wildtype mice, but not in the lurcher mice (FIG. 5). Tremor frequency, amplitude, and spectral distribution were not significantly different between $Cx36^{-/-}$ and wildtype mice (two-sided Student t test, all $p > 0.1$; see FIG. 5); the average tremor frequency amounted to 8.1 ± 0.4 Hz for $Cx36^{-/-}$ mice and 8.5 ± 0.4 Hz for wildtype mice. The lack of harmaline-induced tremors in the lurcher mice has already been reported by Milner *et al.*[34] and was used as a negative control.

DISCUSSION

Given the large amount of attention that gap junctions have attracted throughout the last decades and the importance that has been assigned to olivary electrotonic coupling for complex spike synchronization and motor coordination, it was surprising to see that $Cx36^{-/-}$ mice, which appear to lack functional neuronal gap junctions in their inferior olive,[31] exhibit no obvious motor impairments. The astounding absence of an overt behavioral phenotype in $Cx36^{-/-}$ mice raises some questions about the functional role of olivary gap junctions. The animals that we used were global knockouts, and it is thus possible that compensatory mechanisms occlude the effect of missing gap junctions. Possibly, other unknown gap junction proteins or particular conductances may be up- or downregulated to change the kinetics of the olivary neurons, but it is difficult to envision potential mechanisms other than gap

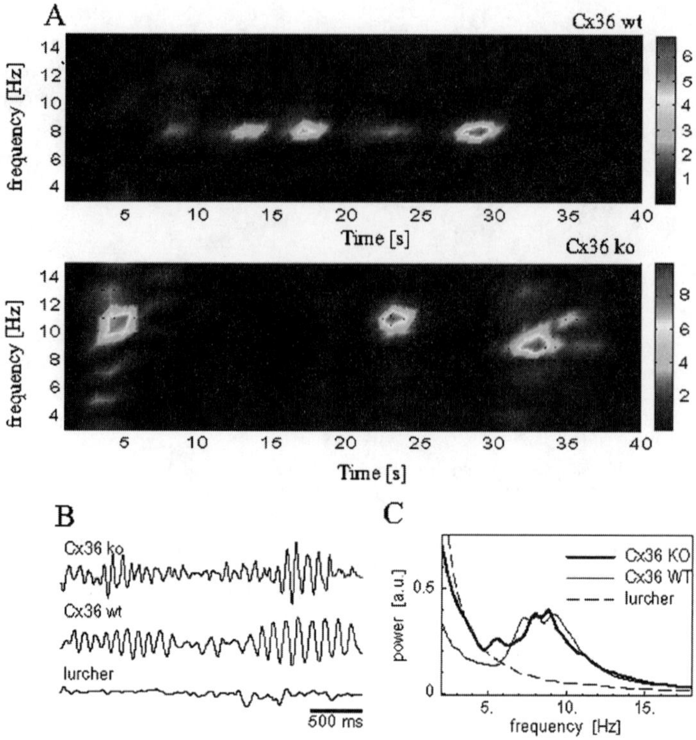

FIGURE 5. Harmaline-induced tremors in Cx36$^{-/-}$, wildtype, and lurcher mice. (**A**) Time-tremor frequency representation of a Cx36$^{-/-}$ and wildtype mouse injected with harmaline. The amplitude of the power spectra is coded in color so that [red (in on-line version only)] corresponds to the higher amplitude. (**B**) Acceleration recordings of a Cx36$^{-/-}$, wildtype, and lurcher mouse injected with harmaline. (**C**) Averaged power spectra of harmaline-induced tremors of Cx36$^{-/-}$ (n = 10), wildtype (n = 10), and lurcher (n = 3) mice.

junctions that can account for fast electrical coupling leading to synchronization of action potentials. Therefore, it remains possible that the primary function of olivary gap junctions is *not* the synchronization of olivary firing patterns. This option would be consistent with the observations that the level of synchrony between complex spikes is not significantly higher than that of the simple spikes of the same cell pairs in the alert behaving animal,[5] that the electrical conductivity of Cx36 gap junctions is exceptionally weak,[38] and that the coupling coefficient of olivary cells is usually well below 5 percent.[9]

We have directly addressed the role of electrotonic coupling of inferior olivary neurons by means of dual-electrode recordings of complex spike activities in coupling-deficient mice. We recorded extracellularly from nearby Purkinje cells located in the same parasagittal plane and found that the degree of complex spike synchrony in $Cx36^{-/-}$ mice is *not* different from wildtype control animals. We have also shown that there are two different types of synchrony. Most of the cell pairs were only loosely synchronized; cross-correlograms had broad central peaks, and not all spikes in one cell were accompanied by a spike in the other cell. Additionally, there were two pairs that exhibited perfect synchrony, in the sense that each spike was accompanied by a spike in the other cell with a temporal precision better than 1 millisecond.

Both forms of synchrony can easily be accounted for by mechanisms that do not rely on electrotonic coupling. A loose synchronization on a time scale of several tens of milliseconds is most likely due to common synaptic input or to locking phenomena within the reverberating loop formed by olivary neurons, Purkinje cells, and the deep cerebellar nuclei.[12,19,20] Perfectly synchronized complex spikes with a one-to-one relationship in the firing times, on the other hand, are a signature of two Purkinje cells receiving input from the very same olivary neuron via climbing-fiber collaterals.[3] This interpretation is also consistent with the results of Sugihara *et al.*[40] who found that collaterals from a given olivary neuron are restricted to a narrow parasagittal strip of the cerebellar cortex and that climbing-fiber collaterals can contact Purkinje cells in the same folium.

Our results on the degree of complex spike synchrony in (wildtype) mice confirm results obtained previously with a similar recording technique in guinea pigs.[3] These results are, however, in conflict with multielectrode recordings of the rat[22–25] and dual-electrode recordings of the rabbit,[5,42] which showed higher cross-correlation coefficients at time zero. The reasons for this discrepancy elude us. They could be either technical, such as the induction of artifactual synchrony due to capacitive coupling of multiple electrode arrays,[43] or they could reflect species differences. For example, recordings of CS activities in cat and monkey have also shown timing properties that diverge from those reported for rat and rabbit;[1,18] or, the synchrony levels in the mouse may be, in contrast to those in the rabbit, too low to be detected with dual-electrode recording techniques.

The fact that the absence of functional neuronal gap junctions does not affect the degree of complex spike synchrony is further corroborated by experiments, where we compared the amplitude and frequency of harmaline-induced tremors in Cx36 knockouts and wildtypes. Harmaline specifically increases the resting activity of olivary neurons and thus leads to a pronounced tremor at about 7–12 Hz, that is, at the frequency of the intrinsic oscillatory properties of IO neurons.[29] If the olivary neurons were no longer synchronized in Cx36 knockout, we would expect to see a reduced residual tremor. However, the amplitude and distribution of frequencies were comparable in both groups.

Altogether, the conclusions that are to be drawn are twofold. First, the submillisecond synchronization of complex spikes is the result of collateral innervation of two Purkinje cells with the same IO neuron. In particular, this form of synchrony is not due to gap junctional coupling and thus not subject to a dynamical modulation by GABAergic terminals in the IO. Second, gap junctions may not be essential for the observed loose synchronization of complex spikes, because this form of synchrony is not affected in mice that lack functional gap junctions. Moreover, common synaptic input to IO neurons provides a more likely mechanism for the generation of synchrony on a 10-ms timescale than fast electrotonic coupling. Thus, the present data do not support the idea that olivary gap junctions synchronize complex spike activity on a millisecond timescale.

Even though we have been unable to correlate the level of complex spike synchrony in the vestibulocerebellum of the alert rabbit to a particular stimulus paradigm or eye movement behavior,[5] the millisecond synchronization of complex spikes and a dynamic uncoupling of olivary neurons by activation of GABAergic synapses at the site of the gap junctions has been suggested to play an important role in the precise coordination and timing of movements.[25,26,41] We have tested motor behavior on different levels, including compensatory eye movements, locomotion, and complex motor tasks. The only difference between $Cx36^{-/-}$ and wildtype mice we could detect was in the latency of the OKR to velocity steps in the visual stimulus. This difference, however, is most likely due to an impairment of signal transmission in the retina[13] and not to the absence of gap junctions in the inferior olive, because the onset of climbing fiber activation following velocity step stimulation is not sufficiently early to explain the latency of the eye movements.[37] Other eye-movement parameters, such as gain and phase-lag of the vestibulo-ocular reflex and the OKR, were also normal in the $Cx36^{-/-}$ mutant. The results from all the other behavioral tests we performed were very much in the same line. The walking pattern analysis demonstrated that $Cx36^{-/-}$ mice do not show any sign of ataxia. Even performance of complex motor tasks as tested in the accelerating rotorod test is not impaired in $Cx36^{-/-}$ mice. In short, the results of our experiments indicate that mice that have no functional olivary gap junctions hardly show any behavioral deficiencies.

At this point, the natural question is: What is the function of (olivary) gap junctions? In the inferior olive, gap junctions *and* synapses are located at dendritic spines with exceptionally long and thin spine necks.[7] Spines from several neurons and GABAergic as well as glutamatergic terminals are packed together in glomeruli shielded by several layers of glia. Assuming that the electrical resistance of spine neck and gap junctions are of the same order of magnitude (Ref. 38, and Kistler and De Zeeuw, unpublished), the coupling between spine heads must be much more pronounced than the coupling between somata. Therefore, we propose that olivary gap junctions link spines within the glomerulus to a functional unit. Active ion channels in the spine heads can boost the synaptic input and create local "micro spikes" that readily propagate through gap junctions to neighboring spines. Synaptic input to only one or two spine(s) in a given glomerulus can thus trigger unitary responses in all the other spines within the glomerulus, and ultimately lead to synaptic potentials at the somata of all "postglomerular" neurons. In this sense, gap junctions synchronize synaptic input to olivary cells rather than spike output. The appealing aspect of this hypothesis is that it opens new vistas on potential compensatory mechanisms for the absence of Cx36. Synchronization of synaptic input, for example, can also be achieved by an altered pattern of synaptic innervation in the glomeruli. Moreover, it raises the possibility that the dynamics in the olivary glomeruli ultimately regulate the firing patterns of olivary neurons over time in absolute terms, without necessarily affecting the synchrony.

ACKNOWLEDGMENTS

This work has been supported by NWO, EEC, and HFSP. We thank Eddie Dalm and Hans van der Burg for technical assistance.

REFERENCES

1. AGGELOPULOS, N.C., C. DUKE & S.A. EDGLEY. 1995. Non-uniform conduction time in the olivo-cerebellar pathway in the anaesthetized cat. J. Physiol. (Lond.) **486:** 763–768.
2. ALBUS, J.S. 1971. A theory of cerebellar function. Math. Biosci. **10:** 25–61.
3. BELL, C.C. & T. KAWASAKI. 1972. Relations among climbing fiber responses of nearby Purkinje cells. J. Neurophysiol. **35:** 155–169.
4. BRAITENBERG, V. 1967. Is the cerebellar cortex a biological clock in the millisecond range? *In* Progress in Brain Research. C.A. Fox & R.S. Snider, Eds.: **25:** 334–346. Elsevier Publishing Company. Amsterdam.
5. DE ZEEUW, C.I., S.K.E. KOEKKOEK, D.R.W. WYLIE & J.I. SIMPSON. 1997. Association between dendritic lamellar bodies and complex spike synchrony in the olivocerebellar system. J. Neurophysiol. **77:** 1747–1758.
6. DE ZEEUW, C.I., E.J. LANG, I. SUGIHARA, *et al.* 1996. Morphological correlates of bilateral synchrony in the rat cerebellar cortex. J. Neurosci. **16:** 3412–3426.
7. DE ZEEUW, C.I., J.I. SIMPSON, C.C. HOOGENRAAD, *et al.* 1998. Microcircuitry and function of the inferior olive. Trends Neurosci. **21:** 391–400.
8. DE ZEEUW, C.I., C. HANSEL, F. BIAN, *et al.* 1998. Expression of a protein kinase C inhibitor in Purkinje cells blocks cerebellar LTD and adaptation of the vestibulo-ocular reflex. Neuron **20:** 495–508.
9. DEVOR, A. & Y. YAROM. 2002. Electrotonic coupling in the inferior olivary nucleus revealed by simultaneous double patch recordings. J. Neurophysiol. **87:** 3048–3058.
10. DEVOR, A. & Y. YAROM. 2002. Generation and propagation of subthreshold waves in a network of inferior olivary neurons. J. Neurophysiol. **87:** 3059-3069.
11. ECCLES, J.C., R. LLINAS & K. SASAKI. 1966. The excitatory synaptic action of climbing fibres on the purinje cells of the cerebellum. J. Physiol. **182:** 268–96.
12. GERSTNER, W., J.L. VAN HEMMEN & J.D. COWAN. 1996. What matters in neuronal locking? Neural Comput. **8:** 1653–1676.
13. GUELDENAGEL, M., J. AMMERMUELLER, A. FEUGENSPAN, *et al.* 2001. Visual transmission deficits in mice with targeted disruption of the gap junction gene connexin36. J. Neurosci. **21:** 6036-6044.
14. GUELDENAGEL, M., G. SOHL, A. PLUM, *et al.* 2000. Expression patterns of connexin genes in mouse retina. J. Comp. Neurol. **425:** 193-201.
15. HAMERS, F.P., A.J. LANKHORST, T.J. VAN LAAR, *et al.* 2001. Automated quantitative gait analysis during overground locomotion in the rat: its application to spinal cord contusion and transection injuries. J. Neurotraurna. **18:** 187-201.
16. HOUK, J.C., J.T. BUCKINGHAM & A.G. BARTO. 1996. Models of the cerebellum and motor learning. Behav. Brain Sci. **19:** 368-383.
17. ITO, M. 1984. The Cerebellum and Neural Control. Raven Press. New York.
18. KEATING, J.G. & W.T. THACH. 1995. Nonclock behaviour of inferior olive neurons: interspike interval of Purkinje cell complex spike discharge in the awake behaving monkey is random. J. Neurophysiol. **73:** 1329-1340.
19. KISTLER, W.M. & J.L. VAN HEMMEN. 1999. Delayed reverberation through time windows as a key to cerebellar function. Biol. Cybern. **81:** 373-380.
20. KISTLER, W.M., J.L. VAN HEMMEN & C.I. DE ZEEUW. 2000. Time window control: A model for cerebellar function based on synchronization, reverberation, and time slicing. Progr. Brain Res. **124:** 275-297.
21. LAMPL, I. & Y. YAROM. 1993. Subthreshold oscillations of the membrane potential: A functional synchronizing and timing device. J. Neurophysiol. **70:** 2181-2186.
22. LANG, E.J. 2001. Organization of olivocerebellar activity in the absence of excitatory glutamatergic input. J. Neurosci. **21:** 1663-1675.
23. LANG, E.J., I. SUGIHARA & R. LLINÁS. 1996. GABAergic modulation of complex spike activity by the cerebellar nucleoolivary pathway in rat. J. Neurophysiol. **76:** 255-275.
24. LANG, E.J., I. SUGIHARA & R. LLINÁS. 1997. Differential roles of apamin- and charybdotoxin-sensitive K^+ conductances in the generation of inferior olive rhythmicity in vivo. J. Neurosci. **17:** 2825-2838.
25. LANG, E.J., I. SUGIHARA, J.P. WELSH & R. LLINÁS. 1999. Patterns of spontaneous Purkinje cell complex spike activity in the awake rat. J. Neurosci. **19:** 2728–2739.

26. LLINÁS, R. 1991. The noncontinuous nature of movement execution. *In* Motor Control: Concepts and Issues. D.R. Humphrey & H.J. Reund, Eds.: 223–242. Wiley. New York.
27. LLINÁS, R. & K. SASAKI. 1989. The functional organization of the olivo-cerebellar system as examined by multiple Purkinje cell recordings. Eur. J. Neurosci. **1:** 587–602.
28. LLINÁS, R. & Y. YAROM. 1986. Oscillatory properties of guinea-pig inferior olivary neurones and their pharmacological modulation: An *in vitro* study. J. Physiol. (Lond.) **376:** 163–182.
29. LLINÁS, R. & R.A. VOLKIND. 1973. The olivo-cerebellar system: functional properties as revealed by harmaline-induced tremor. Exp. Brain Res. **18:** 69–87.
30. LLINÁS, R.R., R. BAKER & C. SOTELO. 1974. Electrotonic coupling between neurons in cat inferior olive. J. Neurophysiol. **37:** 560–571.
31. LONG, M.A., M.R. DEANS, S.L. PATRICK, *et al.* 2001. Connexin36-dependent electrical coupling synchronizes subthreshold rhythmic activity of inferior olivary neurons. Soc. Neurosci. Abstr. 70.13.
32. LUCIANI, L. 1891. Il cervelletto: nuovi studi di fisiologia normale e patologica. Le Monnier. Florenz.
33. MARR, D. 1969. A theory of cerebellar cortex. J. Physiol. (Lond.) **202:** 437–470.
34. MILNER, T.E., G. CADORET, L. LESSARD & A.M. SMITH. 1995. EMG analysis of harmaline-induced tremor in normal and three strains of mutant mice with Purkinje cell degeneration and the role of the inferior olive. J. Neurophysiol. **73:** 2568–2577.
35. RASH, J.E., W.A. STAINES, T. YASUMURA, *et al.* 2000. Immunogold evidence that neuronal gap junctions in adult brain and spinal cord contain connexin-36 but not connexin-32 or connexin-43. Proc. Natl. Acad. Sci. USA **97:** 7573–7578.
36. SASAKI, K., J.M. BOWER & R. LLINÁS. 1989. Multiple Purkinje cell recording in rodent cerebellar cortex. Eur. J. Neurosci. **1:** 572–586.
37. SIMPSON, J.I., D.R. WYLIE & C.I. DE ZEEUW. 1996. On climbing fiber signals and their consequence(s). Behav. Brain Sci. **19:** 384–398.
38. SRINIVAS, M., R. ROZENTAL, T. KOJIMA, *et al.* 1999. Functional properties of channels formed by the neuronal gap junction protein connexin36. J. Neurosci. **19:** 9848–9855.
39. STAHL, J.S., A.M. VAN ALPHEN & C.I. DE ZEEUW. 2000. A comparison of video and magnetic search coil recordings of mouse eye movements. J. Neurosci. Methods **99:** 101–110.
40. SUGIHARA, I., H.S. WU & Y. SHINODA. 2001. The entire trajectories of single olivocerebellar axons in the cerebellar cortex and their contribution to cerebellar compartmentalization. J. Neurosci. **21:** 7715–7723.
41. WELSH, J.P., E.J. LANG, I. SUGIHARA & R. LLINÁS. 1995. Dynamic organization of motor control within the olivocerebellar system. Nature **374:** 453–457.
42. WYLIE, D.R., C.I. DE ZEEUW & J.I. SIMPSON. 1995. Temporal relations of the complex spike activity of Purkinje cell pairs in the vestibulocerebellum of rabbits. J. Neurosci. **15:** 2875–2887.
43. ZHU, Z., K. LIN & T. KASAMATSU. 2002. Artifactual synchrony via capacitance coupling in multi-electrode recording from cat striate cortex. J. Neurosci. Methods **115:** 45–53.

Roles of Inhibitory Interneurons in the Cerebellar Cortex

TOMOO HIRANO,[a,b] DAI WATANABE,[c] SHIN-YA KAWAGUCHI,[a,b] IRA PASTAN,[d] AND SHIGETADA NAKANISHI[c]

[a]*Department of Biophysics, Graduate School of Science, Kyoto University, Sakyo-ku, Kyoto 606-8502, Japan*

[b]*CREST, Japan Science and Technology Corporation, Kawaguchi, Saitama 332-0012, Japan*

[c]*Department of Biological Sciences, Faculty of Medicine and Department of Molecular and System Biology, Graduate School of Biostudies, Kyoto University, Sakyo-ku, Kyoto 606-8501, Japan*

[d]*Laboratory of Molecular Biology, Division of Basic Science, National Cancer Institute, National Institute of Health, Bethesda, Maryland 20892, USA*

ABSTRACT: The roles of inhibitory interneurons in the cerebellar cortex were investigated. First, Golgi cells were specifically eliminated in transgenic mice in which Golgi cells expressed human interleukin-2 receptor α subunit (IL2Rα). Injection of exotoxin coupled to anti-IL2Rα antibody in the cerebellum of the transgenic mouse eliminated Golgi cells and abolished GABA and synaptic inhibition in the granular layer. After elimination of Golgi cells, acute severe ataxia and subsequent mild motor discoordination were observed. In the latter chronic phase, NMDA receptor–mediated synaptic response was reduced in granule cells. Our findings indicate that elimination of GABAergic inhibition in the granular layer caused overexcitation of granule cells resulting in severe ataxia, and then NMDA receptors in granule cells were downregulated, compensating for the reduction of GABAergic inhibition and improving motor control. In the second part, we report on the regulation mechanism of synaptic plasticity at inhibitory synapses on Purkinje cells (PCs). Inhibitory synaptic transmission on a PC is potentiated after repetitive PC depolarization. This synaptic plasticity (rebound potentiation, RP) was suppressed when a presynaptic neuron was activated during the PC depolarization. This synaptic regulation is unique in the sense that the homosynaptic activity suppresses the induction of synaptic plasticity. The mechanism of how presynaptic activity suppresses RP was examined. GABA released from the presynaptic terminal activated not only $GABA_A$ receptor but also $GABA_B$ receptor. The latter was coupled to Gi/o proteins, which downregulated adenylyl cyclase reducing cAMP and inactivated cAMP-dependent protein kinase (PKA). Downregulation of PKA suppressed RP induction.

KEYWORDS: cerebellum; interneuron; Golgi cell; transgenic mouse; GABA; synaptic plasticity

Address for correspondence: Tomoo Hirano, Department of Biophysics, Graduate School of Science, Kyoto University, Sakyo-ku, Kyoto 606-8502, Japan. Voice: 81-75-753-4237; fax: 81-75-753-4229.
thirano@nb.biophys.kyoto-u.ac.jp

Ann. N.Y. Acad. Sci. 978: 405–412 (2002). © 2002 New York Academy of Sciences.

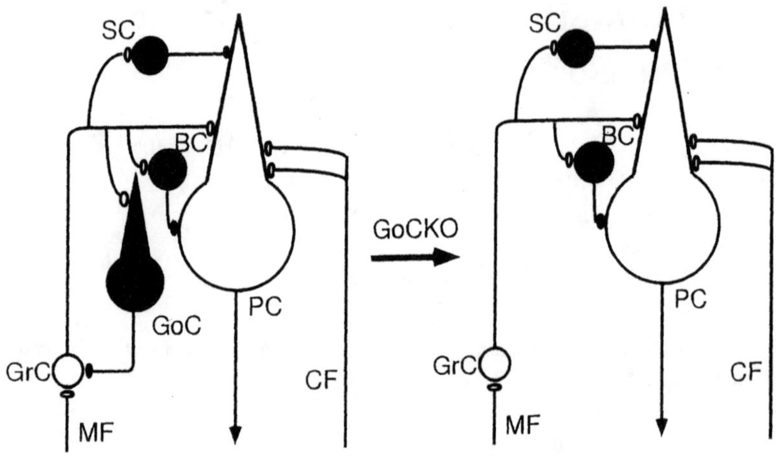

FIGURE 1. Cerebellar cortical circuits and ablation of Golgi cells. ABBREVIATIONS: GrC, granule cell; GoC, Golgi cell; SC, stellate cell; BC, basket cell; PC, Purkinje cell; MF, mossy fiber; CF, climbing fiber.

BACKGROUND

Major neurons in the cerebellar cortex are Purkinje cells (PCs), granule cells, stellate cells, basket cells, and Golgi cells (FIG. 1).[1,2] PCs are sole neurons sending output from the cerebellar cortex. Granule cells are excitatory neurons relaying mossy-fiber inputs to PCs. The other three neurons are inhibitory interneurons whose roles are less clearly understood. Cell bodies of Golgi cells are located in the granular layer. Golgi cells receive inputs from granule cells and inhibit granule cells, thus constituting a negative feedback circuit. Basket and stellate cells receive excitatory inputs from granule cells and climbing fibers and inhibit PCs. Branching patterns of stellate and basket cells suggest that they are implicated in the lateral inhibition.[1, 2]

One way to investigate roles of a particular type of neuron is to specifically rule them out. In the first part of this paper, we describe the effects of Golgi cell ablation in transgenic mice (FIG. 1).[3] The cerebellar cortex has been implicated in motor learning, and it has been suggested that the long-term depression at synapses between parallel fibers and PCs is a cellular mechanism of motor learning.[4,5] The long-term depression is the long-lasting decrease in the excitatory transmission efficacy that is induced by conjunctive activation of parallel fibers and a climbing fiber. Synaptic plasticity was also found at inhibitory synapses on a PC. Repetitive depolarization of a PC enhanced the inhibitory synaptic transmission for a long time.[6] This phenomenon is called rebound potentiation (RP), and it might also contribute to motor learning.[7,8] In the second part, the regulation mechanism of RP is described.[9] Such studies eventually contribute to clarification of the roles of stellate and basket cells.

GOLGI CELL KNOCKOUT

Generation of Transgenic Mice and Selective Elimination of Golgi Cells

Transgenic mice in which IL2Rα is specifically expressed in Golgi cells in the cerebellar cortex were generated.[3] The transgene consisted of the mouse mGluR2 promoter, the human IL2Rα cDNA, and the sequence containing the polyadenylation signals. mGluR2 is a subtype of metabotropic glutamate receptors preferentially expressed in Golgi cells in the cerebellar cortex.[10] Several independent transgenic lines were generated. The transgenic mice showed a common expression pattern of IL2Rα mRNA, which was similar to that of mGluR2 mRNA. Thus, human IL2Rα mRNA was selectively expressed in Golgi cells in the cerebellar network.

To selectively eliminate Golgi cells expressing human IL2Rα, anti-Tac(FV)-PE38 was administered intrathecally to adult transgenic mice.[3] Anti-Tac(FV)-PE38 is a recombinant protein composed of the variable heavy and light chains of a monoclonal antibody against human IL2Rα and fused to a catalytic domain of *Pseudomonas* exotoxin (PE38).[11,12] Anti-Tac(FV)-PE38 is internalized by cells bearing human IL2Rα and kills them. Thus, administration of the immunotoxin to the cerebellum of transgenic mice specifically eliminates Golgi cells.

Morphological and Physiological Consequences of Golgi Cell Elimination

The immunotoxin injection to the transgenic mouse ablated mGluR2-immunoreactive Golgi cells without affecting an overall layer structure of the cerebellar cortex and the morphology and numbers of other neurons such as Purkinje, granule, basket, and stellate cells.[3] We next examined the amount of inhibitory transmitter GABA in the cerebellar cortex using GABA antibody. GABA immunoreactivity was observed in the molecular, Purkinje, and granular layers of wild-type mice. In immunotoxin-treated transgenic mice, GABA immunoreactivity in the granular layer was almost or completely lost with little change in the molecular and Purkinje cell layers. This loss of GABA immunoreactivity correlated well with the disappearance of mGluR2-positive Golgi cells.

To investigate the physiological effects of Golgi cell elimination on the spatiotemporal pattern of excitation propagation in the cerebellar cortex, optical recording of membrane potential with a voltage-sensitive dye RH482 was applied to cerebellar slices.[3] White matter stimulation in the slice preparation evoked horizontal spread of excitation in the granular layer and then vertical propagation to the molecular layer followed. Most of these signals were blocked by an AMPA receptor antagonist CNQX (6-cyano-7-nitroquinoxaline-2,3-dione). To clarify the roles of GABA-mediated inhibition in the excitation propagation, we examined the effects of $GABA_A$ receptor antagonist, (–)-bicuculline methochloride. In the wild-type mice, bicuculline slowed the decay of optical signal in the granular layer. In contrast, no obvious effect of bicuculline on the spatiotemporal pattern of optical signal was observed in the granular layer of immunotoxin-treated transgenic mice. These results indicate that GABA-mediated inhibition is diminished by Golgi cell elimination in the granular layer. Bicuculline increased the spatial spread of optical signal in the molecular layers of both wild-type mice and transgenic mice treated with the immunotoxin, indicating that GABAergic inhibition derived from other interneurons remains intact.

Golgi Cell Elimination Induced Unexpected Downregulation of NMDA Receptor Function in Granule Cells

The optical recording unexpectedly revealed that the optical signal in the granular layer of immunotoxin-treated transgenic slice decayed faster than that of the bicuculline-treated slice prepared from wild-type mice.[3] These results indicate that Golgi cell elimination diminishes a slow excitatory postsynaptic component that is sensitive to GABA-mediated inhibition in granule cells.

NMDA receptors are highly expressed in granule cells and contribute to the slow component of excitatory synaptic potential.[13,14] Therefore, we considered that NMDA receptors mediate the slow component decreased by Golgi cell elimination. To test this possibility, we applied a NMDA receptor antagonist D-APV (D-2-amino-5-phosphonovalerate) to the cerebellar slices. D-APV decreased the prolonged depolarization sensitive to GABA inhibition in the wild-type slices; however, in slices prepared from immunotoxin-treated transgenic mice, D-APV did not affect the time course of depolarization in the granular layer. These results suggest that Golgi cell elimination attenuates functional NMDA receptors in granule cells (FIG. 2).

To confirm this conclusion, the whole-cell voltage-clamp technique was applied to granule cells, and excitatory postsynaptic currents (EPSCs) evoked by mossy-fiber stimulation were recorded.[3] The EPSC consists of a fast component mediated by AMPA receptors and a slow component mediated by NMDA receptors. The ratio of the components (slow/fast) was 0.48 ± 0.11 in the wild-type slices and 0.15 ± 0.03 in the transgenic slices treated with the immunotoxin. Thus, the NMDA receptor-mediated synaptic component was decreased in the granule cells in the cerebellar slices prepared from the transgenic mice treated with the immunotoxin.

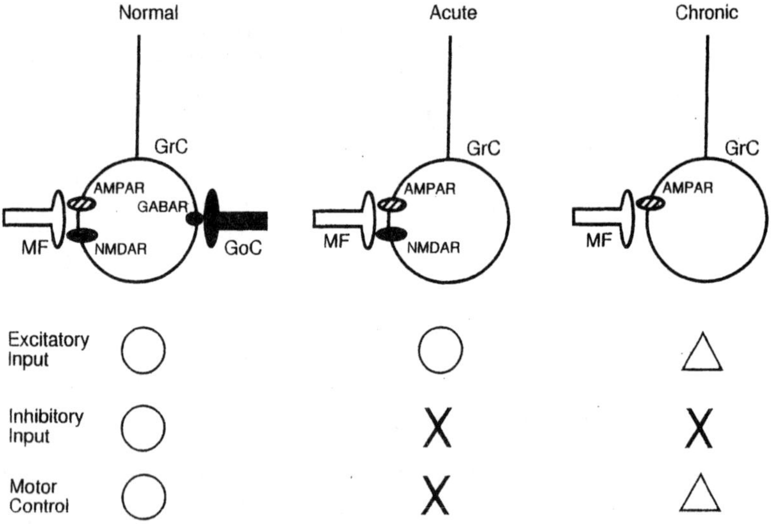

FIGURE 2. Effects of Golgi cell elimination. Ablation of Golgi cells results in acute severe ataxia followed by chronic mild motor discoordination. ABBREVIAITONS: GrC, granule cell; GoC, Golgi cell; MF, mossy fiber; AMPAR, AMPA receptor; NMDAR, NMDA receptor; GABAR, GABA receptor.

Motor Coordination Failure by Golgi Cell Elimination

The immunotoxin injection caused severe ataxia after 4–5 days; then the mice gradually recovered in a week.[3] The time course of motor disturbance was quantified by the rotating rod test. When the rod turned at 10 rpm, the transgenic mice treated with the immunotoxin showed increasing difficulty balancing on the rod 4–5 days after toxin injection. Thereafter, they gradually recovered and eventually managed to stay on the rod after 5 days. However, when the rod was turned at 20 rpm, the immunotoxin-treated animal could not balance on the rod even 17 days after the injection. The wild-type mice did not show any difficulty staying on the rod rotated at 20 rpm. Histological examination showed that Golgi cells never recovered after toxin injection. Thus, the Golgi cell ablation caused biphasic motor coordination failures, the early phase with severe ataxia and the late chronic phase with mild motor discoordination.

SUMMARY OF EFFECTS OF GOLGI CELL ELIMINATION AND DISCUSSION

The effects of Golgi cell elimination in the transgenic mice has been described in the first part of this paper. Golgi cell elimination depleted GABA and abolished the synaptic inhibition in the granular layer and caused short-lasting severe ataxia and chronic mild motor discoordination. How then is motor control recovered after Golgi cell elimination? We propose that downregulation of NMDA receptor function in granule cells contributes to this recovery. We think that severe ataxia soon after the Golgi cell elimination (FIG. 2) was caused by overexcitation of granule cells due to the absence of GABAergic synaptic inhibition. Then, NMDA receptor–mediated slow excitatory synaptic component started to decrease, reducing excitatory synaptic drive and thus partly compensating for the lack of synaptic inhibition. Therefore, in the chronic phase, excitatory and inhibitory drives were adaptively balanced at a low level, which enabled better control of motion. Our findings suggest that synaptic integration involving the inhibitory (GABA) and excitatory (NMDA) receptors is critical for compound motor coordination.

In this study we have applied the immunotoxin-mediated cell targeting (IMCT) technique to examine the roles of Golgi cells in the cerebellar cortex. This technique can be applied to any neurons in the central nervous system that possess specific promoter sequences. Indeed, the same transgenic mice were used to ablate striatal cholinergic neurons and starburst amacrine cells in the retina.[15,16] Thus, the IMCT technique is a promising method in the analysis to determine the roles of specific neurons and the regulatory mechanisms of synaptic transmission in the neural network.

REGULATION OF SYNAPTIC PLASTICITY IN THE MOLECULAR LAYER

Rebound Potentiation and Suppression by Presynaptic Activity

Synaptic plasticity has been considered as a cellular mechanism for learning and memory.[17,18] In the cerebellum, the excitatory synaptic transmission at synapses between parallel fibers and a Purkinje cell undergoes long-term depression (LTD) after conjunctive activation of parallel fibers and a climbing fiber.[4,5] The LTD has been a candidate mechanism for motor learning. Inhibitory synapses on a Purkinje cell also show synaptic plasticity. Depolarization of a Purkinje neuron, such as that evoked by activation of a climbing fiber, potentiates the inhibitory GABAergic synaptic transmission on a Purkinje cell.[6] This phenomenon is called rebound potentiation (RP).

Recently, a novel regulatory mechanism of RP was reported.[9] Activation of the presynaptic neuron during conditioning depolarization of a PC suppressed the induction of RP. This regulation is unique because presynaptic activity usually induces (but does not suppress) synaptic plasticity. The physiological significance of RP might be to prevent overexcitation of PCs and keep the firing frequency of action potential to a certain range. Because of the heterosynaptic induction nature of RP, RP itself is not a synapse-specific mechanism. However, suppression of RP could be synapse-specific, because it takes place only at the synapse where the presynaptic fiber was activated.

Involvement of GABA$_B$ Receptors, Gi/o, and PKA Activity in the Suppression of RP

The suppression mechanism of RP was studied in a cerebellar culture.[9] The RP is accompanied by the increase in postsynaptic GABA$_A$ receptor responsiveness (FIG. 3). Thus, the RP was monitored by the amplitude of GABA$_A$ receptor-mediated current in a cultured PC. The GABA response was potentiated by depolarization of the PC and lasted for more than 30 minutes. In this condition, presynaptic activation could be mimicked by GABA application. GABA application during the conditioning depolarization suppressed the potentiation of GABA$_A$ receptor responsiveness.

The suppressive effect of GABA to RP was mediated by GABA$_B$ receptors (FIG. 3). GABA application could be replaced by application of a GABA$_B$ receptor agonist, baclofen, in the suppression of RP. It was also shown that the suppressive effect of GABA on RP was blocked by a GABA$_B$ receptor antagonist, SCH50911.

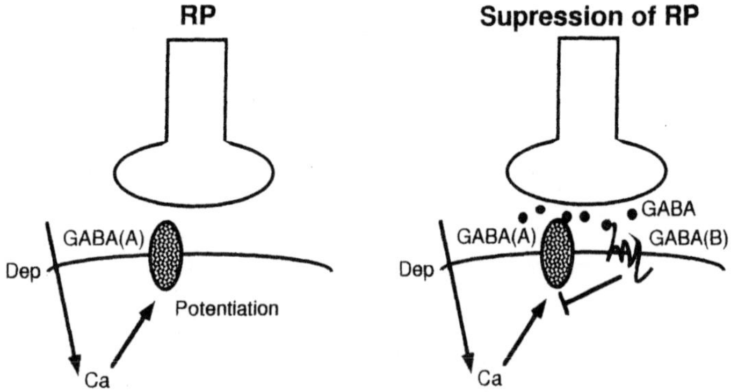

FIGURE 3. Rebound potentiation (RP) and suppression. Depolarization (Dep) of a Purkinje cell induces RP through the increase in intracellular Ca^{2+} (Ca). RP is expressed as the increase in responsiveness of GABA$_A$ receptors. Presynaptic activation during postsynaptic depolarization suppresses RP induction through postsynaptic GABA$_B$ receptors.

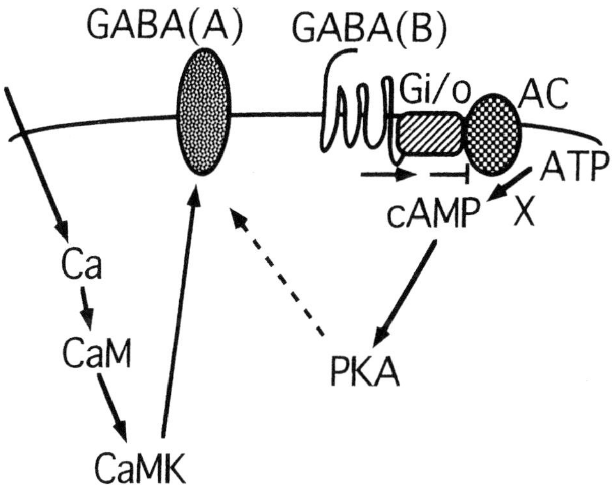

FIGURE 4. Signaling cascade for suppression of RP. Downregulation of PKA activity through $GABA_B$ receptors and Gi/o proteins, etc., suppresses RP. ABBREVIATIONS: CaM, calmodulin; AC, adenylyl cyclase.

In some neurons, $GABA_B$ receptors are known to couple to Gi/o proteins and downregulate adenylyl cyclase, resulting in reduction of cAMP and active PKA (FIG. 4). We investigated whether this pathway works in a PC in the suppression of RP. Treatment of cultured neurons with pertussis toxin, an inhibitor of Gi/o proteins, blocked the suppression of RP. Application of Sp-cAMP, an activator of PKA, also blocked the suppression of RP. On the other hand, a PKA inhibitor, KT5720, suppressed the induction of RP. These results suggest that $GABA_B$ receptor, Gi/o protein, adenylyl cyclase, cAMP, and PKA are implicated in the suppression of RP (FIG. 4).

Overall Signaling Cascade Regulating RP

A previous study showed that CaMKII is necessary for RP induction.[19] We therefore analyzed functional interaction of CaMKII and PKA during RP induction. In the hippocampus, CaMKII activity is counteracted by the activity of protein phosphatase 1 (PP1),[20] and the activity of PP1 is regulated by inhibitor-1 (I-1). I-1 is a native inhibitor protein that binds to PP1 and inhibits it when phosphoryrated by PKA. I-1 is also known to be dephosphorylated by calcineurin. Thus, PKA activity can influence CaMKII activity via I-1 and PP1. We examined whether a similar signaling cascade works to regulate RP induction, and we have obtained the data that PP1, calcineurin, and DARPP-32 are involved in RP regulation.[21] DARPP-32 is a molecule analogous to I-1 which is expressed in PCs.[22,23] Thus, we suggest that downregulation of PKA activity facilitates dephosphorylation of DARPP-32 by calcineurin, which releases PP1 from inhibition. PP1 in turn opposes the CaMKII activity and suppresses RP induction.

REFERENCES

1. ITO, M. 1984. The Cerebellum and Neural Control. Raven Press. New York. p580.
2. LLINAS, R.R. & K.D. WALTON. 1998. Cerebellum. *In* The Synaptic Organization of the Brain, 4th edit. G.M. Shepherd, Ed.: 255–288. Oxford University Press. New York.
3. WATANABE, D., H. INOKAWA, K. HASHIMOTO, *et al.* 1998. Ablation of cerebellar Golgi cells disrupts synaptic integration involving GABA inhibition and NMDA receptor activation in motor coordination. Cell **95:** 17–27.
4. ITO, M. 1989. Long-term depression. Annu. Rev. Neurosci. **12:** 85–102.
5. LINDEN, D.J. & J.A. CONNOR. 1995. Long-term synaptic depression. Annu. Rev. Neurosci. **18:** 319–357.
6. KANO, M., U. REXHAUSEN, J. DREESSEN & A. KONNERTH. 1992. Synaptic excitation produces a long-lasting rebound potentiation of inhibitory synaptic signals in cerebellar Purkinje cells. Nature **356:** 601–604.
7. KANO, M. 1995. Plasticity of inhibitory synapses in the brain: a possible memory mechanism that has been overlooked. Neurosci. Res. **21:** 177–182.
8. HANSEL, C., D.J. LINDEN & E. D'ANGELO. 2001. Beyond parallel fiber LTD: the diversity of synaptic and non-synaptic plasticity in the cerebellum. Nature Neurosci. **4:** 467–475.
9. KAWAGUCHI, S. & T. HIRANO. 2000. Suppression of inhibitory synaptic plasticity by presynaptic activity through postsynaptic $GABA_B$ receptors in a Purkinje neuron. Neuron **27:** 339–347.
10. OHISHI, H., R. OGAWA-MEGURO, R. SHIGEMOTO, *et al.* 1994. Immunohistochemical localization of metabotropic glutamate receptors, mGluR2 and mGluR3, in rat cerebellar cortex. Neuron **13:** 55–66.
11. PASTAN, I., M.C. WILLINGHAM & D.J. FITZGERALD. 1988. Immunotoxins. Cell **47:** 641–648.
12. KOBAYASHI, K., S. MORITA, H. SAWADA, *et al.* 1995. Immunotoxin-mediated conditional disruption of specific neurons in transgenic mice. Proc. Natl. Acad. Sci. USA **92:** 1132–1136.
13. D'ANGELO, E., G. DE FILIPPI, P. ROSSI & V. TAGLIETTI. 1995. Synaptic excitation of individual rat cerebellar granule cells in situ: evidence for the role of NMDA receptors. J. Physiol. (Lond.) **484:** 397–413.
14. KADOTANI, H., T. HIRANO, M. MASUGI, *et al.* 1996. Motor discoordination results from combined gene disruption of the NMDA receptor NR2A and NR2C subunits, but not from single disruption of the NR2A or NR2C subunit. J. Neurosci. **16:** 7859–7867.
15. KANEKO, S., T. HIDAKA, D. WATANABE, *et al.* 2000. Synaptic integration mediated by striatal cholinergic interneurons in basal ganglia function. Science **289:** 633–637.
16. YOSHIDA, K., D. WATANABE, H. ISHIKANE, *et al.* 2001. A key role of starburst amacrine cells originating retinal directional selectivity and optokinetic eye movement. Neuron **30:** 771–780.
17. BLISS, T.V. & G.L. COLLINGRIDGE. 1993. A synaptic model of memory: long-term potentiation in the hippocampus. Nature **361:** 31–39.
18. MALENKA, R.C. 1994. Synaptic plasticity in the hippocampus: LTP and LTD. Cell **78:** 535–538.
19. KANO, M., M. KANO, K. FUKUNAGA & A. KONNERTH. 1996. Ca^{2+}-induced rebound potentiation of gamma-aminobutyric acid-mediated currents requires activation of Ca^{2+}/calmodulin-dependent kinase II. Proc. Natl. Acad. Sci. USA **93:** 13351–13356.
20. MULKEY, R.M., S. ENDO, S. SHENOLIKAR & R.C. MALENKA. 1994. Involvement of a calcineurin/inhibitor-1 phosphatase cascade in hippocampal long-term depression. Nature **369:** 486–488.
21. KAWAGUCHI, S. & T. HIRANO. 2002. Signaling cascade regulating long-term potentiation of GABAA receptor responsiveness in cerebellar Purkinje neurons. J. Neurosci. **22:** 3969–3976.
22. HEMMINGS, H.C., JR., P. GREENGARD, H.Y. TUNG & P. COHEN. 1984. DARPP-32, a dopamine-regulated neuronal phosphoprotein, is a potent inhibitor of protein phosphatase-1. Nature **310:** 503–505.
23. SCHALLING, M., M. DJURFELDT, T. HOKFELT, *et al.* 1990. Distribution and cellular localization of DARPP-32 mRNA in rat brain. Brain Res. Mol. Brain Res. **7:** 139–149.

Motor Performance and Motor Learning in Lurcher Mice

A.M. VAN ALPHEN, T. SCHEPERS, C. LUO, AND C.I. DE ZEEUW

Department of Neuroscience, Erasmus University Rotterdam, Dr. Molewaterplein 50, 3000 DR Rotterdam, the Netherlands

ABSTRACT: In adult Lurcher mice virtually all cerebellar Purkinje cells have degenerated as a direct consequence of mutant gene action, providing a natural model for studying the effect of cerebellar cortical lesions on the generation of compensatory eye movements. Lurcher mice possess both optokinetic (OKR) and vestibular (VOR) compensatory reflexes. However, clear differences were observed in control of the OKR consisting of a large reduction in gain and a moderate increase in phase lag. Minor differences were also observed in the VOR in that gain and phase lead of the reflex were both increased in Lurcher animals. Subjecting Lurcher animals to eight days of visuovestibular training tested the assumption that increased VOR gain reflected an adaptive mechanism within remaining brainstem oculomotor pathways to compensate for the reduced OKR. Contrary to control animals, Lurcher animals were unable to modify either VOR or OKR in the course of training and therefore confirmed that an intact cerebellum is indispensable for the implementation of adaptive modifications to the oculomotor system.

KEYWORDS: Lurcher mouse; OKR; VOR; visuovestibular adaptation

INTRODUCTION

Oculomotor reflexes like the optokinetic reflex (OKR) and the vestibulo-ocular reflex (VOR) reduce slip of visual images across the retina by generating eye movements that are compensatory to head motion or movement of the visual surround Both reflexes are mediated by a neural network that is largely located in the brain stem[1–3] and supported by a parallel pathway through the cerebellum. The cerebellar side loop has been deemed particularly necessary for maintaining accuracy of compensatory ocular reflexes,[4,5] a process referred to as motor learning.

With the discovery of Purkinje cell–specific gene promotors, it has become possible to switch individual genes on or off at will, allowing us to study cerebellar information processing and motor learning in a controlled way at the level of individual proteins.[6,7] As a reference against which to compare such specific molecular modifications, it seems prudent to quantify in mice the worst intervention possible, that is, complete ablation of the cerebellar cortex. Rather than attempting

Address for correspondence: C.I. De Zeeuw, Dept. of Neuroscience, Erasmus University Rotterdam, Dr. Molewaterplein 50, 3000 DR Rotterdam, The Netherlands. Voice: +31-10-4087299; fax: +31-10-4089459.
dezeeuw@anat.fgg.eur.nl

Ann. N.Y. Acad. Sci. 978: 413–424 (2002). © 2002 New York Academy of Sciences.

extensive surgical procedures, we chose to use Lurcher mice, a strain of mutant mice that completely lacks cerebellar Purkinje cells, making their deficit equivalent to a complete cerebellar cortical lesion. The underlying pathology of Purkinje cell death has been traced to a mutation of δ2-glutamate receptor (GluRδ2) resulting in a gain of this receptor's function that triggers apoptotic cell death of Purkinje cells in these animals.[8–10]

The Lurcher strain provides us with a model for cerebellar cortical lesions, which is highly reproducible between individual animals and allows us to judge how ocular reflexes are controlled on a subcortical level. Previous studies have attempted to describe the oculomotor performance of Lurcher mice, but they failed to be quantitatively accurate due to the use of a large scleral search coil that partially impeded eye movements.[11,12] We therefore expand upon these previous studies by using a less invasive "mini"-search coil technique to investigate Lurcher oculomotor behavior over a wider range of stimulus parameters and by assessing the extent to which motor plasticity still occurs in these Purkinje cell– deficient animals.

METHODS

The eye movement performance of 15 Lurcher mutant mice and 13 normal control littermates was tested. The procedures for implanting a head fixation pedestal and a mini-search coil were identical to those previously described for this laboratory.[13] Animals were allowed to recover for at least three days prior to the first recording session. The local ethical committee of the Erasmus University Rotterdam approved animal care and experimental protocols.

The optokinetic reflex (OKR) and vestibulo-ocular reflex (VOR) and visually enhanced VOR (VVOR) were measured. Two different approaches were used to test the performance of the optokinetic system. First, the frequency response of the OKR was tested by sinusoidally rotating a vertically striped (width of bars 4°) optokinetic drum around the animal at frequencies of 0.1, 0.2, 0.4, 0.8, and 1.6 Hz while peak stimulus velocity was kept constant at 8°/s. Second, the effect of peak velocity on gain of the OKR was evaluated by varying peak stimulus velocity between 2, 4, 8, 16, and 32°/s at 0.4 Hz stimulus frequency. The VOR was elicited by sinusoidal, whole-body rotation in the dark. Sinusoidal stimulation was performed at 0.1, 0.2, 0.4, 0.8, and 1.6 Hz and 10° stimulus amplitude. The VVOR was tested using the VOR stimulus parameters while the animal viewed the illuminated, earth-fixed drum.

Gain of the eye movement (eye velocity/head velocity) and phase of eye movement with respect to stimulus movement were calculated by fitting a sine wave to the average response using least-square optimization. When eye movement lagged stimulus movement, phase was expressed with a negative sign. Phase relations of VOR were shifted by 180°, making the phase angle zero for perfectly compensatory responses. All values reported are mean and standard error, weighting each mouse equally. All signal analysis was done with MATLAB (Mathworks Inc., Natick, MA).

A VOR reversal paradigm was used to test VOR adaptation. In this paradigm the animal had to learn to reverse the natural direction of the VOR. Reversal training lasted for eight days and began by rotating the optokinetic drum in phase, that is, 0° phase difference, with table rotation at 5° amplitude. In the following five days, amplitude of the optokinetic drum was increased by 1°/day until it was 10° on day 6. At this point the optokinetic drum was rotating in phase with table rotation but at twice the amplitude. Amplitude of the optokinetic drum remained 10° from day 6 on. Every day the VOR was checked before training to monitor progression of VOR change.

Differences between Lurcher and control animals in gain or phase of OKR, VOR, and VVOR were tested for statistical significance using a two-way, repeated measures ANOVA. The effect of visuovestibular training on OKR and VOR was tested for each mouse separately using a one-way, repeated measures ANOVA. Statistical analysis was performed, using a commercial software package called SAS-6.12 (SAS Institute Inc., Cary, NC).

RESULTS

Motor Performance

Lurcher mice were easily distinguished from control littermates by their ataxic behavior, which was visible as a wobbly, lurching gate. Oculomotor control of Lurcher mice was considerably less accurate than that of normal control animals, as reflected by a large decrease in OKR gain across all tested peak velocities (FIG. 1A,

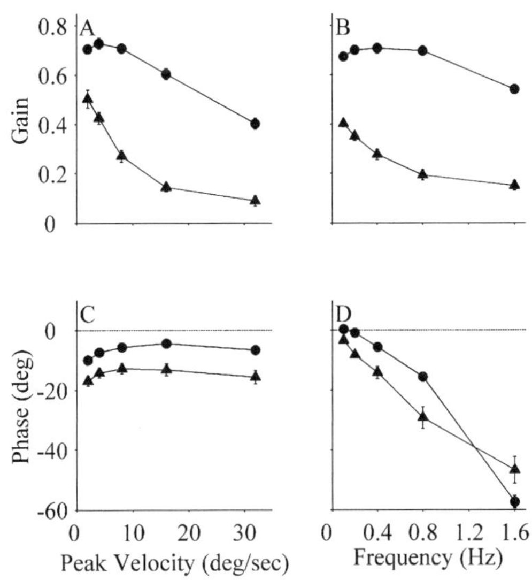

FIGURE 1. The OKR of Lurcher (*triangles*) and control (*circles*) animals is presented in BODE plots relating gain and phase of the response to peak stimulus velocity (**A** and **C**) and stimulus frequency (**B** and **D**). Lurcher OKR gain was found to be smaller than that of control animals across all stimulus conditions. In addition to a reduced OKR gain, Lurcher animals also showed increased phase lag of eye velocity with respect to stimulus velocity. Phase lag at 1.6 Hz was not significantly different from that of control animals ($p = 0.14$). The error bars indicate 1 SEM.

p <0.001). While OKR gain of normal animals remained fairly constant at 0.7 for stimulus velocities smaller than 8°/s and decreased for higher peak velocities, that of Lurcher animals was already considerably smaller at 2°/s and declined immediately when peak stimulus velocity was increased. The increased dependence of Lurcher OKR on peak velocity was reflected in the two-way ANOVA by a significant interaction (p <0.001) between the factors: animal type and peak stimulus velocity. Performance of the OKR was also tested against stimulus frequency while peak stimulus velocity was kept constant at 8°/s. OKR gain in Lurcher animals again showed a significant depression over the entire range of tested frequencies when compared to control animals (FIG. 1B, p <0.001). Again, interaction between animal type and stimulus frequency proved significant (p <0.001) in the analysis, reflecting the fact that OKR gain remained relatively constant up to 0.8 Hz in control animals, but showed no such stability in Lurcher animals. The reduced OKR gain of Lurcher animals was accompanied by an increase in phase lag of the eye movement with respect to stimulus movement. Eye movements of Lurcher mice generally lagged stimulus movement more than those of control animals (FIG. 1D, $p = 0.03$). Phase lag was significantly different at all tested frequencies (p <0.01, two-tailed t test with Bonferroni correction) except 1.6 Hz. The stronger dependence of OKR gain on frequency and the increased phase lag indicate that the absence of cerebellar cortical output in Lurcher animals has changed the low-pass filter characteristics of the OKR in the closed-loop situation by an apparent increase of the overall time constant.

The VOR of Lurcher animals was similar to that of control animals in that VOR gain increased with frequency, measuring 0.07 ± 0.008 and 0.59 ± 0.02 at 0.1 Hz and 1.6 Hz, respectively (FIG. 2A). The relation of gain with frequency was complemented by a reduction in phase lead for increasing stimulus frequencies. At 0.1 Hz phase lead measured 110 ± 6.5° and decreased to 12 ± 1.2° at 1.6 Hz (FIG. 2C). In normal animals, VOR gain increased from 0.02 ± 0.009 to 0.55 ± 0.02, and VOR phase decreased from 75 ± 9.5° to 7.8 ± 0.7° at 0.1 Hz and 1.6 Hz, respectively. Both VOR gain and phase curves were significantly different between Lurcher and control animals (p <0.01). A post hoc t test with Bonferroni correction for multiple comparisons pinpointed the differences between Lurcher and control animals at 0.1, 0.2, and 0.4 Hz. At these three frequencies, VOR gain of Lurcher mice was significantly larger (p <0.05) than that of control animals, while phase lead of the VOR in Lurcher mice was significantly larger than that of control animals (p <0.05) at 0.1 and 0.2 Hz. Because both phase lead and gain of the Lurcher VOR increased, the difference between control and Lurcher animals cannot be attributed to a different overall VOR time constant between Lurcher and control animals. In fact, a decrease in the dominant time constant of the VOR could be expected in Lurcher mutants, but such a decrease would result in a concomitant decrease in VOR gain at low frequencies. From the present results we witness the exact opposite: VOR gain is relatively increased in Lurcher animals at low-stimulus frequencies.

One of the functions assigned to the flocculus is that it provides the site of direct interaction between OKR and VOR,[4] in such a way that the OKR would supplement VOR to provide a stable, combined response. In control animals, the VVOR indeed proved stable across the range of tested frequencies. Gain was 0.74 ± 0.02 at 0.1 Hz and 0.67 ± 0.03 at 1.6 Hz, while phase measured 2.5 ± 0.4 and 6.1 ± 2.1 at the same frequencies (FIG. 2B and D). The VVOR response of Lurcher mice was more variable, gain first declined from 0.51 ± 0.03 at 0.1 Hz to 0.44 ± 0.02 at 0.2 Hz and then

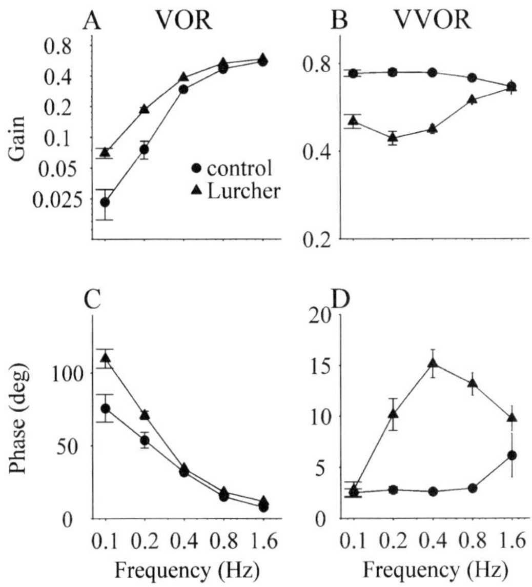

FIGURE 2. BODE plots for Lurcher (*triangles*) and control (*circles*) animals are shown of the VOR (**A** and **C**) and VVOR (**B** and **D**). VOR gain increased with increasing stimulus frequency for both Lurcher and control animals. The interaction of VOR and OKR provided a stable VVOR in control animals across the tested frequency range. VVOR gain and phase were more variable in Lurcher animals, reflecting the inability of their severely impaired OKR system to complement VOR.

increased again to 0.66 ± 0.03 at 1.6 Hz. VVOR phase followed a similar change in Lurcher mice, measuring 3 ± 0.8° and 10 ± 1.2° at 0.1 and 1.6 Hz, respectively, and reaching a peak of 15 ± 1.4° at 0.4 Hz. The variable response of Lurcher mice can be reasonably well explained from a linear addition of VOR and OKR. Because VVOR is a combination of both vestibular and optokinetic inputs, the response at high-stimulus frequency is largely driven by the VOR, while that at low-stimulus frequency will be relatively more dependent on the OKR. The OKR of Lurcher animals did not extend far enough into the higher-frequency range to complement VOR and achieve a stable VVOR across all tested frequencies. The validity of explaining differences in VVOR between Lurcher and controls from a defect in Lurcher OKR may be illustrated by a linear reconstruction of the VVOR at 0.4 Hz using separately measured OKR and VOR responses.

By vectorially subtracting VOR of Lurcher mice from the VVOR stimulus, the residual vector that drives the optokinetic system can be estimated. At 0.4 Hz this estimate comes to a residual optokinetic stimulus of 18°/s at 162° phase angle with ipsilateral head velocity. FIGURE 1A can now be used as a reference table that maps the response of the OKR at 0.4 Hz for different peak stimulus velocities. OKR gain and phase at 18°/s can be read from the curve at 0.15 and –14°, with respect to stimulus movement. Converting these values to the VVOR coordinate system yields an eye velocity of 2.7°/s at a phase lag of –32° with ipsilateral head velocity. Vectorially adding the estimated eye velocity vector for optokinetic component with vestibular component yields a VVOR gain of 0.44 and a phase lead of 21° with ipsilateral head velocity. These values are close enough to the real values (0.48 and 15°, respectively) to assume that deficits in Lurcher VVOR are largely caused by a normal integration of an impaired OKR and VOR.

Motor Learning

Visuovestibular adaptation was tested in 4 Lurcher mice and 5 control animals. Both Lurcher and control animals displayed a VVOR in response to the training stimulus that was different from their normal VOR at the same vestibular stimulus. The difference between VVOR and VOR indicates that visual information was used to guide the eye movements during training. Control animals responded to the training stimulus by slowly decreasing VOR gain at 0.6 Hz from 0.30 ± 0.03 on the first to 0.13 ± 0.02 on the sixth day and then slightly increasing VOR gain again to 0.16 ± 0.03 on day 8 (FIG. 3A). Changes in VOR gain were accompanied by a slow rise in phase of the eye movement with respect to turntable movement, measuring 30 ± 5°, 111 ± 26° and 115 ± 31° on the same days (FIG. 3B). No changes were observed in Lurcher mice. VOR gain measured 0.37 ± 0.1 before and 0.42 ± 0.03 after train-

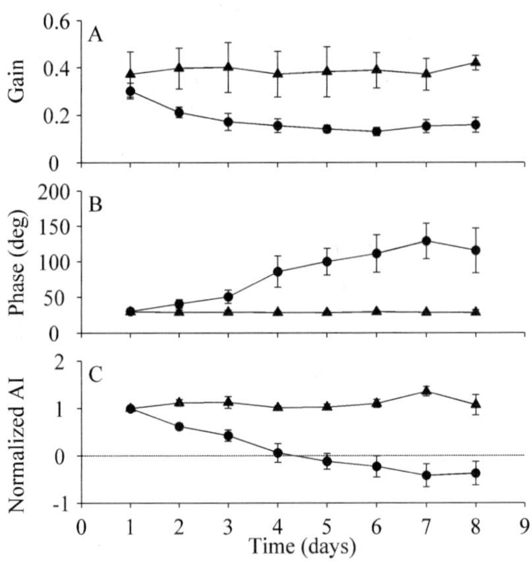

FIGURE 3. Change in properties of the VOR at 0.6 Hz in response to visuovestibular training is plotted against time. (**A**) Absolute VOR gain for Lurcher (*triangles*) and control animals (*circles*) is presented against time. VOR in control animals is reduced with training while that of Lurcher animals does not change. Error bars indicate SEM. (**B**) Phase of the eye movement with respect to table movement is plotted in the same way as gain. Control animals show a pronounced advance in phase lead with training reflecting a change in direction of the VOR. (**C**) Normalized adaptation index is shown to outline two distinct features. First, VOR reverses direction as phase lead is larger than 90° and the adaptation index changes sign. Second, normalizing the data of each animal to their initial performance clearly shows that the Lurcher animals do not adapt their VOR.

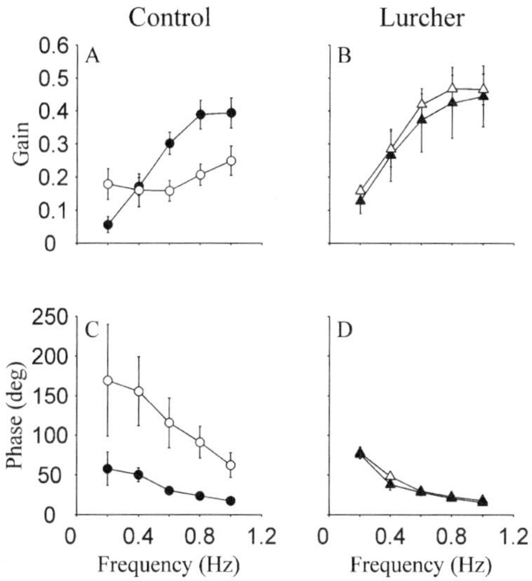

FIGURE 4. Change in properties of the VOR in control (**A** and **C**) and Lurcher (**B** and **D**) animals is presented in the form of BODE plots. *Closed symbols* represent data before and *open symbols* after eight days of visuovestibular training. Error bars indicate SEM.

ing, while phase of the eye movement with respect to table movement was 30 ± 2° before and 28 ± 4° after training. In control animals, phase lead changed so much that direction of the VOR actually reversed from out of phase with table rotation to in phase. In order to accentuate the actual reversal of direction we calculated an adaptation index (AI) according to the equation: $AI = G^{*}\cos(\theta)$. In this equation, G equals VOR gain, and θ is the phase angle of the eye movement with respect to table movement. As phase lead of the eye movement with respect to table movement increased to values larger than 90°, AI became negative, indicating that VOR had reversed direction. FIGURE 3E shows AI, normalized to its initial value, for all animals at the training frequency (0.6 Hz). As expected, control animals differed significantly from Lurcher animals ($p = 0.002$, two-way ANOVA with repeated measures), in that control animals showed a steady decrease of AI, while Lurcher animals showed no change in AI throughout the entire eight days of training. In addition to the training frequency, VOR was also tested for four other frequencies centered about the training frequency (FIG. 4). Although training was performed exclusively at 0.6 Hz, changes in VOR of control animals were not limited to the training frequency but encompassed the entire tested spectrum. AI changed significantly at all frequencies in control animals ($p < 0.001$ at all frequencies), while in Lurcher animals it did not change at any of the tested frequencies ($p > 0.05$ at all frequencies).

After visuovestibular training, behavioral changes in control animals were not limited to the VOR but also included the OKR. Every day the OKR was tested at the same amplitude (5°) and frequencies as the VOR with the addition of 0.1 Hz. The experimental protocol to monitor progression of learning in the OKR therefore differed from that described earlier in that peak stimulus velocity increased as frequen-

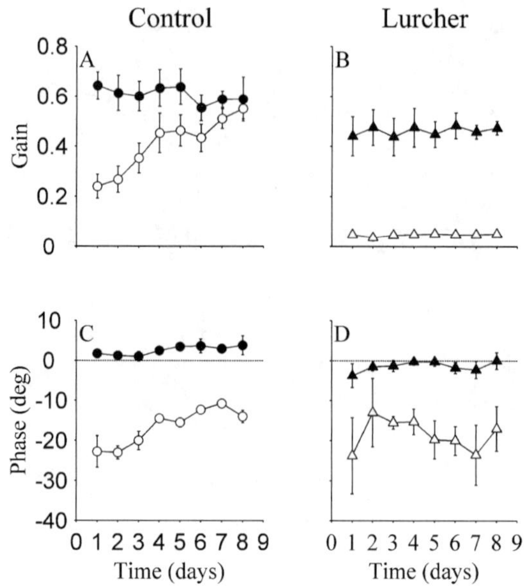

FIGURE 5. The progression of OKR changes in time is presented for control (**A** and **C**) and Lurcher (**B** and **D**) animals. Only data taken at 0.1 Hz (*closed symbols*) and 1 Hz (*open symbols*) are shown because those data represent the smallest and largest effect of training on OKR, respectively.

cy was increased. During visuovestibular training, OKR gain progressively increased and phase lag decreased in control animals (FIG. 5A and C). At 1 Hz (31°/s peak stimulus velocity), OKR gain increased from 0.24 ± 0.05 to 0.55 ± 0.04 ($p = 0.007$), and phase lag decreased from –24 ± 9° to –14 ± 2° ($p = 0.006$). In Lurcher animals, neither OKR gain nor phase changed during training (FIG. 5B and D). Adaptive changes of the OKR in control animals were limited to higher-stimulus frequencies (FIG. 6A and C). Changes in gain proved significant at 0.6, 0.8, and 1 Hz ($p = 0.02$, $p < 0.001$, and $p < 0.001$), while changes in phase were significant at all frequencies except 0.1 Hz ($p = 0.03$ at 0.2 Hz and $p < 0.001$ for all other frequencies). In Lurcher mice, the OKR remained unchanged at all tested frequencies (FIG. 6B and D). An intact cerebellar cortex is therefore required both for adaptation of the VOR as well as for adaptation of the OKR.

DISCUSSION

The oculomotor performance of Lurcher mice is largely dominated by an impaired OKR. Closed-loop OKR gain is severely reduced, and phase lag of the eye movement with respect to stimulus movement is increased compared to control animals. Lurcher OKR, like that of control animals, is clearly nonlinear, rendering accuracy of the OKR not only dependent on stimulus frequency but also on stimulus

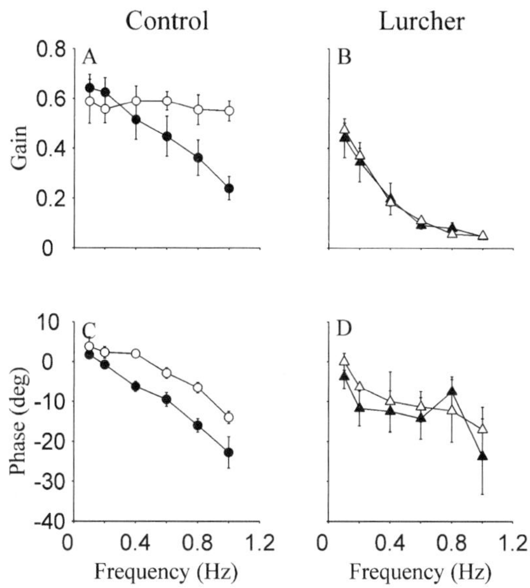

FIGURE 6. Changes in OKR due to visuovestibular training are represented in the form of BODE plots for control (**A** and **C**) and Lurcher (**B** and **D**) animals. *Closed symbols* represent data before and *open symbols* after eight days of training. BODE plots of the OKR before training differ from those shown in FIGURE 1 because a constant stimulus amplitude was used to assess the OKR during training. Because of the constancy of stimulus amplitude, peak stimulus velocity increased as frequency was increased. Stimulus frequencies of 0.4, 0.8, and 1.2 correspond to peak stimulus velocities of 12.6, 25.1, and 37.7°/s, respectively.

velocity. This nonlinear relation of the OKR appeared more pronounced in Lurcher animals as reflected by the stronger dependence of OKR gain on peak stimulus velocity. Control animals showed a stable OKR gain for sinusoidal stimuli with a peak velocity smaller than 9°/s, while Lurcher mice did not exhibit any such plateau response. In relation to frequency, the OKR (peak velocity 8°/s) both in Lurcher and control animals revealed typical low-pass characteristics of declining gain and increased phase lag with increasing stimulus frequency. However, in addition to a severely reduced OKR gain, phase lag was also larger in Lurcher mice than in control animals. The Lurcher OKR represents the function of remaining extracerebellar pathways, which are known to exist at the level of the brain stem.[14–16]

Changes in the OKR of Lurcher mice may be explained by considering the OKR, in its reduced form, to be a low-pass system within a negative feedback loop. The internal filter element can be expressed as the Laplacian transfer function $g_{int}/(s\tau+1)$ in which g_{int} represents the internal gain, s denotes complex frequency, and τ represents the time constant. Placing such an element inside a negative feedback loop results in a systems performance $E'/D' = (g_{int}/g_{int}+1)/(s(\tau/1+g_{int})+1)$. A reduction in the internal gain of the system would therefore result in a reduction of the closed-loop OKR gain and an increase in its time constant. The result of such a deficit was present in Lurcher animals as reduced OKR gain and increased phase lag with respect to control animals. In addition to explaining the increased phase lag, this interpretation of a reduced internal gain could also explain the more pronounced dependence of Lurcher OKR on stimulus velocity, because a reduction of the internal gain would emphasize the effect of a saturation element in the system. Saturation is known to occur in the optokinetic system at the level of the accessory optic system involving the detection of retinal slip.[17,18] The nature of the deficits observed in Lurcher mice hint at the possibility that cerebellar output is necessary in mice to partially compensate for the nonlinear nature of the OKR by boosting internal gain of the system.

Assigning a role to the cerebellum in gain control is definitely not new.[4,19] The cerebellum has been assigned a short-term role in oculomotor control concerned with directly fine-tuning VOR based on optokinetic information,[4] and also a long-term role in which vestibular and optokinetic interaction would shape dynamics of the underlying VOR.[4,5,20] In this respect it was striking that in Lurcher mice no severe impairments emerged in either VOR alone or in the ability to combine VOR and OKR. VVOR gain was reduced in Lurcher mice, but this impairment is attributable to the obvious OKR deficit. Lack of a functional cerebellum did not appear to prevent normal integration of optokinetic and vestibular signals into a VVOR.

VOR of Lurcher mice was similar to that of control animals and only differed significantly at 0.1, 0.2, and 0.4 Hz. Contrary to expectation,[14,21,22] at these low frequencies Lurcher mice performed better, that is, they had a higher VOR gain than normal animals. An increase in phase lead of the eye movement with respect to stimulus movement was also observed. Increased phase lead of the VOR has been described before in Lurcher[23] and PCD[24] mice and can be attributed to a reduced function of the velocity to position neural integrator,[25] which depends on cerebellar output.[26] However, a reduction in performance of the neural integrator would invariably lead to a reduction in VOR gain at low-stimulus frequencies. Rather than a reduction, we observed an increased VOR gain in Lurcher animals, which is comparable to that reported in previous studies of Purkinje cell–deficient mutants.[23,24] It is unclear whether increased VOR gain in cerebellar mutant mice is a primary effect of the cerebellar lesion or a secondary compensatory mechanism that allows the animals to achieve a reasonable performance of the VVOR despite an impaired OKR. Lesion studies provide no unequivocal answer to this question because lesions of the flocculus have been reported to cause both a decrease[14,21,22,27] or an increase[5,28] of VOR gain, and some studies report no effect at all.[2,22] The same ambivalent results have been reported for lesions of the nodulus, ascribing both an increasing[24,27] and decreasing[29] influence on angular VOR gain to the lesion. Rather than adding to this list of lesion studies, we asked whether Lurcher animals might have retained some adaptive capacity in the remaining, extracerebellar circuitry that would allow them to adapt VOR gain.

The extent to which Lurcher and control animals were able to adapt their VOR was investigated by subjecting both groups to a vision-reversal paradigm. In this paradigm, control animals responded by an initial drop of VOR gain followed by an increase in phase lead. Changes in VOR were paralleled by an increased OKR gain and reduced phase lag of the OKR. Lurcher animals proved incapable of adapting their VOR or OKR, demonstrating once more the necessity of an intact cerebellum to express these forms of plasticity.[5,20,22,27] Though cerebellar involvement in adaptation of the VOR and OKR is evident, long-term depression of the parallel fiber to Purkinje cell synapse[4] does not appear to be an absolute prerequisite to express the adaptive behavior because L7-PKCi transgenic animals, in which LTD was selectively disrupted, did show attenuated adaptation when subjected to the same adaptation protocol.[30] The present data differ from previous adaptation studies[27,31–33] in that control mice did not show frequency-specific adaptation of either VOR or OKR. The different findings are possibly caused by differences in experimental protocols. Particularly the fact that we extended training to eight consecutive days may have prevented us from seeing any frequency-specific changes.

REFERENCES

1. SZENTÁGOTHAI, J. 1950. The elementary vestibulo-ocular reflex arc. J. Neurophysiol. **13:** 395–407.
2. BARMACK, N.H. & V.E. PETTOROSSI. 1985. Effects of unilateral lesions of the flocculus on optokinetic and vestibuloocular reflexes of the rabbit. J. Neurophysiol. **53:** 481–496.
3. ZEE, D.S., A. YAMAZAKI, P.H. BUTLER & G. GUCER. 1981. Effects of ablation of flocculus and paraflocculus of eye movements in primate. J. Neurophysiol. **46:** 878–899.
4. ITO, M. 1982. Cerebellar control of the vestibulo-ocular reflex—around the flocculus hypothesis. Annu. Rev. Neurosci. **5:** 275–296.
5. ROBINSON, D.A. 1976. Adaptive gain control of vestibuloocular reflex by the cerebellum. J. Neurophysiol. **39:** 954–969.
6. KITAYAMA, K., M. ABE, T. KAKIZAKI, *et al.* 2001. Purkinje cell-specific and inducible gene recombination system generated from C57BL/6 mouse ES cells. Biochem. Biophys. Res. Commun. **281:** 1134–1140.
7. BARSKI, J.J., K. DETHLEFFSEN & M. MEYER. 2000. Cre recombinase expression in cerebellar Purkinje cells. Genesis **28:** 93–98.
8. WULLNER, U., P.A. LOSCHMANN, M. WELLER & T. KLOCKGETHER. 1995. Apoptotic cell death in the cerebellum of mutant weaver and lurcher mice. Neurosci. Lett. **200:** 109–112.
9. DOUGHTY, M.L., P.L. DE JAGER, S.J. KORSMEYER & N. HEINTZ. 2000. Neurodegeneration in Lurcher mice occurs via multiple cell death pathways. J. Neurosci. **20:** 3687–3694.
10. ZUO, J., P.L. DE JAGER, K.A. TAKAHASHI, *et al.* 1997. Neurodegeneration in Lurcher mice caused by mutation in delta2 glutamate receptor gene. Nature **388:** 769–773.
11. DE ZEEUW, C.I., A.M. VAN ALPHEN, S.K. KOEKKOEK, *et al.* 1998. Recording eye movements in mice: a new approach to investigate the molecular basis of cerebellar control of motor learning and motor timing. Otolaryngol. Head Neck Surg. **119:** 193–203.
12. KOEKKOEK, S.K., A.M. V. ALPHEN, J. V.D. BURG, *et al.* 1997. Gain adaptation and phase dynamics of compensatory eye movements in mice. Genes Funct. **1:** 175–190.
13. VAN ALPHEN, A.M., J.S. STAHL & C.I. DE ZEEUW. 2001. The dynamic characteristics of the mouse horizontal vestibulo-ocular and optokinetic response. Brain Res. **890:** 296–305.
14. KELLER, E.L. & W. PRECHT. 1978. Persistence of visual response in vestibular nucleus neurons in cerebellectomized cat. Exp. Brain Res. **32:** 591–594.
15. HOLSTEGE, G. & H. COLLEWIJN. 1982. The efferent connections of the nucleus of the optic tract and the superior colliculus in the rabbit. J. Comp. Neurol. **209:** 139–175.
16. PRECHT, W. & P. STRATA. 1980. On the pathway mediating optokinetic responses in vestibular nuclear neurons. Neuroscience **5:** 777–787.
17. MAEKAWA, K. & J.I. SIMPSON. 1973. Climbing fiber responses evoked in vestibulo-cerebellum of rabbit from visual system. J. Neurophysiol. **36:** 649–666.
18. MAEKAWA, K. & J.I. SIMPSON. 1972. Climbing fiber activation of Purkinje cells in the flocculus by impulses transferred through the visual pathway. Brain Res. **39:** 245–251.
19. NAGAO, S., N. YOSHIOKA, T. HENSCH, *et al.* 1991. The role of cerebellar flocculus in adaptive gain control of ocular reflexes. Acta Otolaryngol. Suppl. **481:** 234–236.

20. LISBERGER, S.G., F.A. MILES & D.S. ZEE. 1984. Signals used to compute errors in monkey vestibuloocular reflex: possible role of flocculus. J. Neurophysiol. **52:** 1140–1153.
21. VAN NEERVEN, J., O. POMPEIANO & H. COLLEWIJN. 1989. Depression of the vestibulo-ocular and optokinetic responses by intrafloccular microinjection of GABA-A and GABA-B agonists in the rabbit. Arch. Ital. Biol. **127:** 243–263.
22. ITO, M., P.J. JASTREBOFF & Y. MIYASHITA.1982. Specific effects of unilateral lesions in the flocculus upon eye movements in albino rabbits. Exp. Brain Res. **45:** 233–242.
23. STAHL, J.S. 2002. Calcium channelopathy mutants and their role in ocular motor research. Ann. N. Y. Acad. Sci. **956:** 64–74.
24. KILLIAN, J.E. & J.F. BAKER. 2002. Horizontal vestibuloocular reflex (VOR) head velocity estimation in Purkinje cell degeneration (pcd/pcd) mutant mice. J. Neurophysiol. **87:** 1159–1164.
25. SKAVENSKI, A.A. & D.A. ROBINSON. 1973. Role of abducens neurons in vestibuloocular reflex. J. Neurophysiol. **36:** 724–738.
26. ROBINSON, D.A. 1974. The effect of cerebellectomy on the cat's bestibulo-ocular integrator. Brain Res. **71:** 195–207.
27. NAGAO, S. 1983. Effects of vestibulocerebellar lesions upon dynamic characteristics and adaptation of vestibulo-ocular and optokinetic responses in pigmented rabbits. Exp. Brain Res. **53:** 36–46.
28. PARTSALIS, A.M., Y. ZHANG & S.M. HIGHSTEIN. 1995. Dorsal Y group in the squirrel monkey. II. Contribution of the cerebellar flocculus to neuronal responses in normal and adapted animals. J. Neurophysiol. **73:** 632–650.
29. BARMACK, N.H., P. ERRICO, A. FERRARESI, *et al.* 2002. Cerebellar nodulectomy impairs spatial memory of vestibular and optokinetic stimulation in rabbits. J. Neurophysiol. **87:** 962–975.
30. VAN ALPHEN, A.M. & C.I. DE ZEEUW. 2002. Cerebellar LTD facilitates but is not essential for long-term adaptation of the vestibulo-ocular reflex. Eur. J. Neurosci. **16:** 486–490.
31. COLLEWIJN, H. & A.F. GROOTENDORST. 1978. Adaptation of the rabbit's vestibulo-ocular reflex to modified visual input: importance of stimulus conditions. Arch. Ital. Biol. **116:** 273–280.
32. SCHAIRER, J.O. & M.V. BENNETT. 1986. Changes in gain of the vestibulo-ocular reflex induced by combined visual and vestibular stimulation in goldfish. Brain Res. **373:** 164–176.
33. DE ZEEUW, C.I., C. HANSEL, F. BIAN, *et al.* 1998. Expression of a protein kinase C inhibitor in Purkinje cells blocks cerebellar LTD and adaptation of the vestibulo-ocular reflex. Neuron **20:** 495–508.

Trying to Understand the Cerebellum Well Enough to Build One

TATSUYA OHYAMA, JAVIER F. MEDINA, WILLIAM L. NORES, AND MICHAEL D. MAUK

Department of Neurobiology and Anatomy and Keck Center for the Neurobiology of Learning and Memory, University of Texas Medical School at Houston, Houston, Texas 77030, USA

ABSTRACT: The development of an increasingly detailed computer simulation of the cerebellum is briefly described. Specific and relatively direct evaluation of the performance of this simulation is made possible by the straightforward way in which pavlovian eyelid conditioning engages the cerebellum. Inputs to the simulation are based on recordings of mossy fiber and climbing fiber responses to the stimuli used in eyelid conditioning, and the output of the simulation can be evaluated with respect to the extensively characterized behavioral properties of eyelid conditioning. Because construction of the simulation has been guided by a strong aversion to errors of commission, both failures and successes of the simulation have proven informative. The behavior of the simulation related to the inhibitory nucleo-olivary feedback connection and spontaneous activity of climbing fibers is described. A prediction of the simulation concerning extinction is confirmed by experiment.

KEYWORDS: cerebellum; extinction; eyelid conditioning; inferior olive; motor learning; Pavlov; rabbit; simulation

INTRODUCTION

The diversity of talks at a conference in St. Louis illustrates the variety of behaviors and neural processes to which the cerebellum is believed to contribute. Despite this variety there appears to be considerable uniformity of synaptic organization throughout all regions of the cerebellum.[1,2] For this reason it may be more productive to think of the cerebellum as having one function: it receives inputs, processes them according to the computation its synaptic organization supports, and sends the results to other brain regions. From this single computation, diversity of function arises from differences in what inputs encode and from differences in downstream targets. This view highlights the potential value of approaches characterizing the nature of the computation accomplished by the cerebellum.

Address for correspondence: Tatsuya Ohyama, Ph.D., Department of Neurobiology and Anatomy, University of Texas Medical School, 6431 Fannin Street MSB 7.628, Houston, TX 77030. Voice: 713-500-5598; fax: 713-500-0621.
Tatsuya.Ohyama@uth.tmc.edu

Ann. N.Y. Acad. Sci. 978: 425–438 (2002). © 2002 New York Academy of Sciences.

That such approaches are even possible is a tribute to the seminal work of Eccles, Ito, Szentàgothai, and many others who have provided a remarkably detailed characterization of the synaptic organization and physiology of the cerebellum.[1,2] With this advantage it should, in principle, be possible to accomplish for the cerebellum what Hodgkin and Huxley accomplished for the action potential.[3] It should be possible to understand the components of the cerebellum well enough to build one. Here we describe efforts to build a representation of the cerebellum in the computer, the ways in which the properties of pavlovian eyelid conditioning facilitate evaluation of cerebellar simulations, and the insights gained from a recent success of the simulation.

APPROACH TO AND SCOPE OF THE SIMULATIONS

During the development of the simulation we were mindful of the important distinctions between errors of omission and errors of commission.[4] Errors of omission refer to the unavoidable limitation that no simulation can contain the complete set of biological details at all levels. While errors of omission can cause a simulation to fail, this is simply part of the process of asking what set of details is required to produce certain capabilities. By contrast, errors of commission refer to the addition of unwarranted assumptions for which there is no independent evidence and whose inclusion is motivated primarily to allow a simulation to succeed in explaining only the phenomenon that the assumption was intended to explain. Thus, we suggest that the best we can do is to recognize errors of omission, while avoiding errors of commission at all cost. The latter is especially important since otherwise (1) the simulations cannot fail, (2) successes are only as informative as the hypothetical features that are added, and (3) the process is not a fair test of whether enough is known about the parts to explain the whole. By avoiding errors of commission, simulations can be informative with both their failures and their successes.

Building a computer simulation of the cerebellum is possible owing to the wealth of knowledge available for the anatomy and synaptic physiology of cerebellar neurons.[5,6] In our simulation, each cell is represented as a leaky integrate-and-fire neuron whose membrane potential is determined by both leak and synaptic conductances taken from available physiologic data. (See Ref. 6 for details.) The cells are interconnected with divergence and convergence ratios that closely conform to anatomic data, and connectivity in the cerebellar cortex is represented to respect the known two-dimensional geometry. The simulation incorporates plasticity at the granule cell-to-Purkinje cell (gr-Pkj) synapse.[7] As in early cerebellar models of motor learning,[2,8] the simulation incorporates long-term depression (LTD) at this synapse, as supported by a wealth of data.[2,9–15] In contrast to an early version of the simulation that included only LTD,[16] the current simulation also incorporates long-term potentiation (LTP) at the gr-Pkj synapse. Based on recent data,[17] the LTD and LTP in the simulation are mutually reversing. In the following sections, we describe how the incorporation of this bidirectional plasticity, together with the inhibitory nucleo-olivary connection, has greatly improved the performance of the simulation over that of its predecessor. In particular, we focus on a specific prediction that follows from the simulation regarding the extinction of conditioned eyelid responses (see below).

TESTING SIMULATIONS WITH PAVLOVIAN EYELID CONDITIONING

The properties of pavlovian eyelid conditioning provide us with a unique opportunity to test a simulation with realistic inputs, to compare outputs of the simulation to behavior, and to maintain close relations between simulations and experiments. In eyelid conditioning, rabbits are given repeated presentations of a neutral stimulus such as a tone paired with electrical stimulation of the periorbital muscles (or an air-puff to the eye), much in the way Pavlov originally conditioned his dogs to salivate to stimuli predicting food.[18] Initially, the tone or "conditioned stimulus" (CS) has no behavioral effect, whereas the stimulation or "unconditioned stimulus" (US) evokes a reflexive closure of the eyelids. With repeated pairings of the CS and US, the CS acquires the ability to evoke a conditioned eyelid response. If after acquisition the CS is repeatedly presented alone, the CS undergoes extinction, that is, it loses the ability to evoke conditioned responding. The wealth of parametric behavioral data on this form of associative learning[19] has made it a popular system for the neural analysis of learning. In particular, the learned timing of these responses[20–23] has proven to be especially informative about the cerebellar mechanisms that may be involved (FIG. 1).[5,6,24–29]

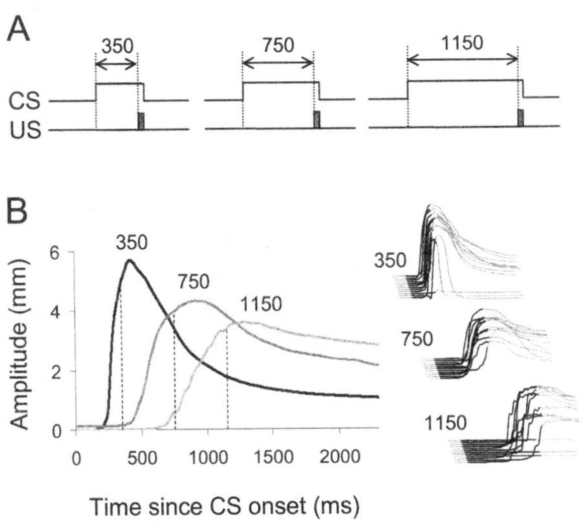

FIGURE 1. (**A**) Three conditioning protocols in which the conditioned stimulus (CS) duration (400, 800, and 1200 ms) and the interval between CS and unconditioned stimulus (US) onsets (350, 750, and 1,150 ms) are respectively varied. (**B**) Sample conditioned responses to auditory CSs from individual rabbits respectively trained on the protocols shown in **A**. *(Left)* Averages across 12 CS alone trials for each rabbit on the seventh day of training. *(Right)* Individual sweeps from which the average data were obtained. Note the sensitivity of the timing of the conditioned response to variation in the CS-US interval. (See Refs. 5, 6, and 24–29 for the role of the cerebellar cortex in response timing.) *Dark lines* indicate the presence of the CS.

For simulations, the key advantage of eyelid conditioning stems from evidence indicating the relatively direct relations between (1) the stimuli used in eyelid conditioning and afferent inputs to the cerebellum, and (2) output from the anterior interpositus nucleus of the cerebellum and the expression of the conditioned responses. These relations make it possible to provide inputs to the simulation and to test the outputs against the well-characterized behavioral properties of eyelid conditioning. Inputs to the simulation are taken from published peri-stimulus histograms of tone-elicited mossy fiber activity[30] and US-elicited climbing fiber responses.[31] Evidence from recording, stimulation, and lesion studies indicates that activation of certain cells in the anterior interpositus nucleus is responsible for driving the expression of the responses.[32,33] The output of the simulation can therefore be evaluated with respect to the behavioral properties of eyelid conditioning. Thus, our goal is to understand how the cerebellum takes a particular pattern of mossy and climbing fiber inputs and transforms them into the specific pattern of nucleus cell output that generates the conditioned eyelid response (FIG. 2).

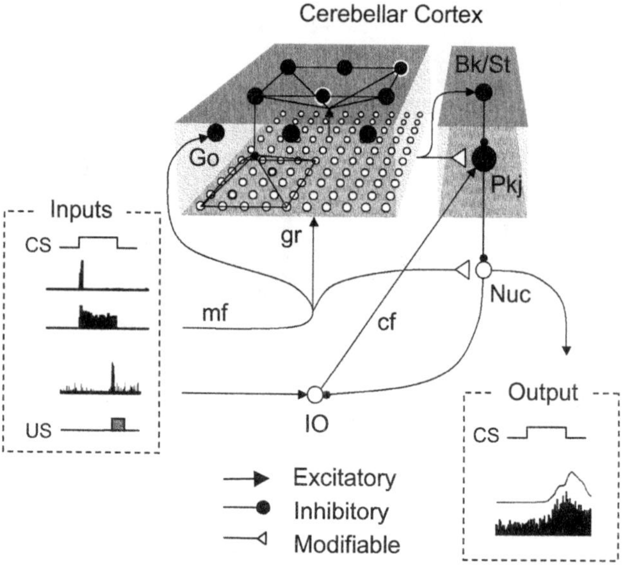

FIGURE 2. Basic eyelid circuitry of the cerebello-olivary system and the way in which it is engaged by pavlovian conditioning. The conditioned (CS) and unconditioned stimuli (US), respectively, activate mossy (mf) and climbing fibers (cf). Data for mossy and climbing fibers are taken from Aitkin and Boyd[30] and Sears and Steinmetz,[31] respectively. The granule cell-to-Purkinje cell (gr-Pkj) synapses in the cerebellar cortex are assumed to undergo plastic changes during conditioning. (See Refs. 5, 6, 26, and 27 for plasticity at a second site in the cerebellar nucleus.) As a result, deep nucleus cells acquire the ability to produce output that drives brain stem motor neurons. Data for interpositus nucleus cells are taken from McCormick and Thompson.[32] Bk/St, basket and stellate cells; gr, granule cell; Go, Golgi cell; IO, inferior olive; Nuc, cerebellar nucleus cell; Pkj, Purkinje cell.

As we argued previously,[34] LTD at the gr-Pkj synapses activated by the CS satisfies both the necessity of convergence and the capacity for expression criteria, which cannot be violated if LTD mediates the expression of conditioned eyelid responses. Inputs from the mossy and climbing fibers (respectively from the CS and US) converge at these synapses.[34–39] Since Purkinje cells are known to tonically inhibit cerebellar nucleus cells[2,40] and since Purkinje cell activity inhibits the expression of conditioned responses (FIG. 2),[41] LTD at the gr-Pkj synapses activated by a CS would promote changes in Purkinje cell activity capable of mediating response expression. Given a sufficient amount of LTD, the resulting decrease in Purkinje cell activity in response to the CS could release the deep nucleus from tonic Purkinje cell inhibition and generate a conditioned response. Conversely, increasing the strength of gr-Pkj synapses via LTP could increase Purkinje cell activity, thereby increasing the tonic inhibitory input to the nucleus cell. Such a process could account for why conditioned responses undergo extinction. Our simulations permit specific and detailed tests of the feasibility of these as yet unproven ideas.

Finally, the straightforward ways in which eyelid conditioning engages the cerebellum permit a close correspondence between experiments and simulations. Since we can train a simulation with the same patterns of inputs with which we can train rabbits, predictions from the simulations can be empirically tested directly.

RULES FOR PLASTICITY

The rules that govern the induction of synaptic plasticity are fundamentally important assumptions for a simulation of the cerebellum, for they determine the exact computations that the simulation will accomplish in response to the inputs with which it is provided. Fortunately, there now exists a large body of evidence on various forms of synaptic plasticity in the cerebellum and their rules for induction. (See Ref. 42 for a review.)

Consistent with early models of cerebellar motor learning, evidence is convincing that gr-Pkj synapses undergo LTD when presynaptic activity occurs in close temporal proximity to a climbing fiber spike.[2,9–15] The exact temporal relationship required for the induction of LTD is not yet well established. Various studies have suggested that LTD induction occurs when parallel fiber stimulation either precedes,[15] is simultaneous with,[2,9,11,13,14] or follows[10,12] climbing fiber stimulation. We have generally found that the ability of the simulation to learn properly timed responses and to show extinction of the responses is largely unaffected by the precise temporal assumptions regarding LTD induction. Based on indirect evidence from VOR studies in monkey,[43] LTD in the current simulations is induced when activation of the gr-Pkj synapse falls within a 100-ms window centered 100 ms prior to the climbing fiber input. However, the results presented here do not depend on the precise temporal assumptions for LTD induction. LTP at the gr-Pkj synapses has also been characterized, and the evidence suggests that gr-Pkj synapses are strengthened when presynaptic activity occurs in the absence of climbing fiber input.[11,13,17,44,45] We therefore incorporated an activity-dependent LTP rule at the gr-Pkj synapse, in which LTP occurs when a granule cell is active outside of the climbing fiber window.

Recent evidence suggests that LTD and LTP at gr-Pkj synapses induced by the presence or absence of climbing fiber inputs, respectively, are mutually reversing.[17,46] In the cerebellum-like structure of mormyrid fish, it was found that the synapses corresponding to the gr-Pkj synapses in the mammalian cerebellum undergo presynaptic forms of mutually reversible LTP and LTD.[46] More recently, it was shown in rat cerebellar slices that there are two forms of cerebellar LTP, one of which is presynaptic and the other of which is postsynaptic.[17] The latter form of LTP has now been shown to be mutually reversing with LTD,[17] a process itself that is considered postsynaptic.[47] These data are fully consistent with the reversible LTP/LTD rule in the simulation. The simulation also incorporates plasticity in the cerebellar nucleus at the mossy fiber to nucleus synapses. Because this plasticity is not essential to the results that follow, we omit discussion. (See Ref. 48 for details.)

A BRIEF HISTORY OF FAILURES

The current simulations display many of the main behavioral properties of eyelid conditioning and have produced mechanistic predictions that were recently subjected to empirical test.[6,49,50] These are, however, relatively recent events. For many years the simulations failed to produce even the most basic aspects of eyelid conditioning. Here we briefly describe aspects of these failures, particularly of the transition from failures to successes, from which a degree of insight can be gleaned.

Due to limitations imposed by computer power and by lack of empirical information, our early simulations involved only the cerebellar cortex and were designed to address whether Purkinje cells could learn appropriately timed pauses in activity (and hence appropriately timed nucleus output) given eyelid conditioning-like inputs.[16] These simulations contained only a rule for LTD at the gr-Pkj synapses, in part because LTP at these synapses was less understood than LTD and in part because the focus was on acquisition (as opposed to extinction) of conditioned responses. The simulations were comprised of mossy fiber inputs, a layer of granule cells, a layer of Golgi cells, and a single Purkinje cell whose granule cell synapses could undergo LTD when a climbing fiber input was activated (as when the US was presented, for example). Thus, there was only unidirectional plasticity at the gr-Pkj synapses. Two other major features were missing from the early simulations: (1) a representation of the deep nuclei and its inhibitory connection to the inferior olive and (2) spontaneous climbing fiber activity. These were all errors of omission, because it had already been shown that LTP could be induced at the gr-Pkj synapse,[11,13,44] recording studies had found that climbing fibers were spontaneously active at 1 Hz,[31] and anatomic and physiologic studies supported the existence of the inhibitory nucleo-olivary projection.[51–54] Despite their capacity to generate a fine-grained temporal code in the cerebellar cortex that, in principle, might permit the proper timing of conditioned responses, these early simulations failed to learn when spontaneous climbing fiber activity was introduced between trials.

The problem posed by spontaneous climbing fiber activity was related to the eventual saturation of all synapses at their minimum strengths.[55] If climbing fibers were spontaneously active, then assuming only LTD, gr-Pkj synapses would continue to decrease in strength every time presynaptic activity coincided with a spontaneous climbing fiber spike. Because granule cells are also spontaneously active,[56] all gr-Pkj synapses exhibiting non-zero firing rates would be expected to decrease in strength and eventually saturate (albeit at different rates depending on their spontaneous firing rates) at their minimum strength, even without external input to mossy or climbing fibers. Such considerations reveal that a model assuming only LTD would fail to explain the stability of motor memories.[57] Indeed, this problem of saturation was the primary reason why our early simulations failed to learn properly.

The early failure of the simulation suggested that merely showing that cerebellar LTD occurs during learning would not be a sufficient explanation of eyelid conditioning. In the next section, we describe how a more detailed representation of the simulation can learn under realistic conditions in which climbing fibers are spontaneously active at their observed rate of 1 Hz. The success shows that far from posing a problem, spontaneous climbing fiber activity is central to the cerebellum's ability to learn and retain motor memories. As will become evident, the cerebellar features whose addition brought the simulation to a critical mass highlights a more systems-oriented view.

NUCLEO-OLIVARY FEEDBACK AND EXTINCTION

The failures just described clearly indicated that a more detailed simulation was needed even to explain the most basic learning features of the cerebellum. As noted above, there were three clear errors of omission in the early simulation: an LTP process at the gr-Pkj synapse,[11,13,44] spontaneous climbing fiber activity,[31,57] and a nucleus cell representation with its inhibitory connection to the inferior olive.[51–54] The most recent version of our simulation sets the leak conductance of the inferior olive such that action potentials are generated spontaneously at their natural firing rate of 1 Hz.[31] The inclusion of these features into the simulation marked a crucial turning point. By incorporating these crucial missing pieces into the simulation, we were able to account for various features of eyelid conditioning, including the timing of conditioned responses,[6] the rapidity with which reacquisition occurs to an extinguished CS,[46] and the ability of conditioned responses to extinguish.[58] These successes have led to novel and testable predictions that were recently subjected to empirical test.[6,49,50] Here, we focus on a specific prediction of the simulation that we recently tested involving the role of the inhibitory nucleo-olivary feedback loop in extinction.

Understanding the role of nucleo-olivary feedback in extinction requires an appreciation of the self-regulatory process accomplished by nucleo-olivary inhibition.[55] To begin with, the simulation suggests that the spontaneous activity of climbing fibers is self-regulated by the cerebellum to a special level. FIGURE 3 shows results from simulations that were initiated with starting synaptic weights that made the simulation badly out of equilibrium. Initially, the strengths of gr-Pkj synapses are at a minimum, and thus there is no Purkinje cell activity (FIG. 3A, left). Nucleus cells are therefore disinhibited (FIG. 3A, center), and the climbing fibers are silent due to the strong nucleo-olivary inhibition (FIG. 3A, right). As time elapses, however, the system eventually settles into an equilibrium state in which climbing fiber activity stabilizes at 1 Hz (FIG. 3A, right). The same equilibrium state is attained if the gr-Pkj synapses are initialized at strengths higher than those required for equilibrium (FIG. 3B). Thus, the spontaneous activity of climbing fibers is naturally and necessarily driven to the level where, on average, there are no net changes in the strength of gr-Pkj synapses and no net changes in the level of output of the deep nucleus neurons.[55]

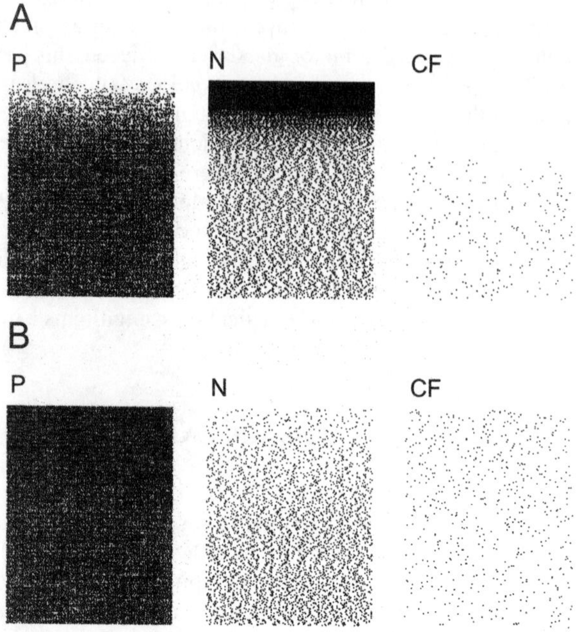

FIGURE 3. Simulated rasterplots of Purkinje cell (P), deep nucleus cell (N), and climbing fiber (CF) activity. The simulation runs from *top to bottom* and *left to right.* (**A**) Behavior of simulation initiated with synaptic weights that were minimal for gr-Pkj synapses. (**B**) Behavior of simulation initiated with synaptic weights that were maximal for gr-Pkj synapses. In both simulations, the synapses are eventually adjusted and stabilized to a level in which CF activity settles at its equilibrium level of 1 Hz.

It was largely the absence of this self-regulating equilibrium that precluded the proper behavior of our early simulations. The current simulation solves the aforementioned problem of synaptic saturation by accomplishing this self-regulation. Recall that with only LTD, spontaneous climbing fiber activity eventually drives spontaneously active gr-Pkj synapses to a minimum. By allowing LTD to be reversed by LTP, however, the synapses are freed from being pinned down to a minimum strength, because any LTD that occurs by chance will be reversed by LTP upon spontaneous occurrences of presynaptic granule cell activity in the absence of climbing fiber input. The result is that despite moment-to-moment changes in synaptic strength, each gr-Pkj synapse stabilizes on average at an intermediate strength determined by the spontaneous firing rate of the presynaptic granule cell.[51] By contrast, if the LTP and LTD are not mutually reversing, then the saturation problem remains unsolved, because spontaneous activity in both granule cells and climbing fibers eventually drives each of the processes to saturate to its minimum (for LTD) or maximum (for LTP), that is, the strengths of the gr-Pkj synapses spontaneously drift until they are either all as strong or all as weak as they can be. Eventually, the capacity for synaptic modification would permanently be lost for all gr-Pkj synapses, a condition that would preclude any new learning or the retention of previous learning.

The preceding analysis suggests that the spontaneous climbing fiber activity is fundamental for the capacity to retain learning and that learning in the cerebellum can only be understood in terms of the way training disrupts this equilibrium.[58] Because the equilibrium level of spontaneous climbing fiber activity precludes net changes in input and output, the training procedures involved in acquisition and extinction of conditioned responses must systematically disrupt this equilibrium to produce the observed changes in responding to the CS. How this would happen during acquisition is relatively intuitive. Normally, the gr-Pkj synapses activated by the CS would encounter a climbing fiber input at the equilibrium probability, but presentation of the US in conjunction with the CS ensures that the probability of a climbing fiber input is much higher in the latter situation. Accordingly, there is an accumulation of net LTD at these synapses as training proceeds. These changes are maintained because the modified synapses experience the equilibrium level of climbing fiber activity between training trials.

An important consequence of the inhibitory nucleo-olivary connection is that as the simulation acquires conditioned responses (and thus learns an increased output), the inhibition of climbing fibers by the cerebellum increases. In this way the simulation parallels the important empirical observation that the ability of the US to elicit a climbing fiber response diminishes as conditioned responses increase.[31,59,60] FIGURE 4 shows results from simulations that were trained from an equilibrium state. During repeated presentations of the CS alone (shown in gray), there is no change in the responses of Purkinje cells, nucleus cells, or climbing fibers (FIG. 4, light gray boxes). When the simulation is subsequently trained with repeated pairings of the CS and US, Purkinje cells gradually learn to pause and disinhibit deep nucleus cells in response to the CS (FIG. 4, thin dark gray box). As the nucleus output increases, the climbing fiber response evoked by the US becomes increasingly suppressed (FIG. 4, thick dark gray box).

The simulation makes specific and testable predictions about the mechanisms that operate during extinction of conditioned responses. As shown in FIGURE 4 (thin black box), if the CS is repeatedly presented alone after acquisition, Purkinje cell output increases and nucleus cell output decreases in response to the CS across trials. This extinction process is mediated by a restrengthening via LTP of those gr-Pkj synapses that had originally undergone net LTD during acquisition. What, however, is the signal that drives this process? Due to nucleo-olivary inhibition, climbing fiber activity is suppressed below equilibrium during extinction near the time the US had occurred during acquisition (FIG. 4, thick black box). In the simulation, it is this transient suppression of climbing fiber activity below the normal spontaneous level that drives the induction of LTP at gr-Pkj synapses consistently active in the absence of a climbing fiber input.[58] The simulation therefore predicts that if the inhibitory feedback from the deep nucleus to the inferior olive is blocked, climbing fiber activity should not be transiently suppressed, and hence extinction should be prevented.

We recently found strong evidence in support of this prediction.[50] Rabbits with a cannula implanted in the inferior olive were trained in a standard delay conditioning protocol for five sessions, during which a tone CS was paired with periorbital stimulation. After acquisition, they were given three separate extinction sessions (each preceded by two training sessions to establish baseline responding) in which the CS was repeatedly presented alone. A separate treatment occurred for each extinction session: a control treatment with no infusion, a control treatment with continuous in-

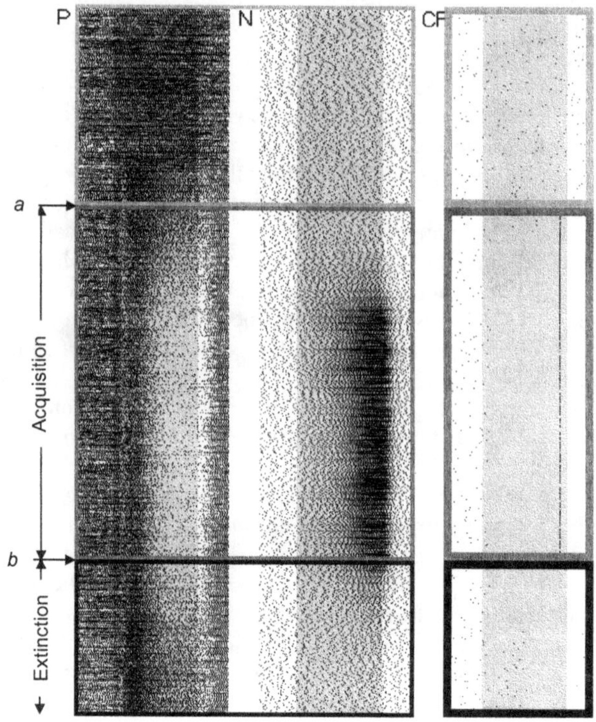

FIGURE 4. Simulated rasterplots of Purkinje cell (P), deep nucleus cell (N), and climbing fiber (CF) activity. Each sweep represents a segment of one trial cycle comprised of a 1-second CS (*shaded area*) and a 5-second intertrial interval. Trials run from *top to bottom* and *left to right*. The simulation begins at equilibrium and is initially given repeated presentations of the CS alone. No changes in activity are observed in response to the CS (*thin light gray box*), and the CF fires at its equilibrium rate in both the CS and the intertrial periods (*thick light gray box*). Acquisition training in which the CS and US are paired begins at **a** and continues until **b**. P activity gradually becomes suppressed and N activity becomes increased in response to the CS (*thin dark gray box*). CF responses evoked by the US occur 50 ms prior to the termination of the CS (*thick dark gray box*). Starting at **b**, the CS is extinguished. P activity increases and N activity decreases to the CS with repeated extinction trials (*thin black box*). Note the marked suppression of CF activity below its spontaneous firing rate near the time at which the US occurred during acquisition (*thick black box*).

fusion of artificial cerebrospinal fluid (ACSF), or a continuous infusion of the GABAA antagonist picrotoxin to block nucleo-olivary inhibition. Consistent with the prediction, extinction failed to occur when nucleo-olivary feedback was blocked (FIG. 5A, right), whereas extinction proceeded normally during the control treatments (FIG. 5A, left).

A second prediction was then tested: blocking excitatory transmission in the inferior olive should block the US signal and hence produce extinction, despite continued pairings of the CS and US. Consistent with this prediction, when the subjects

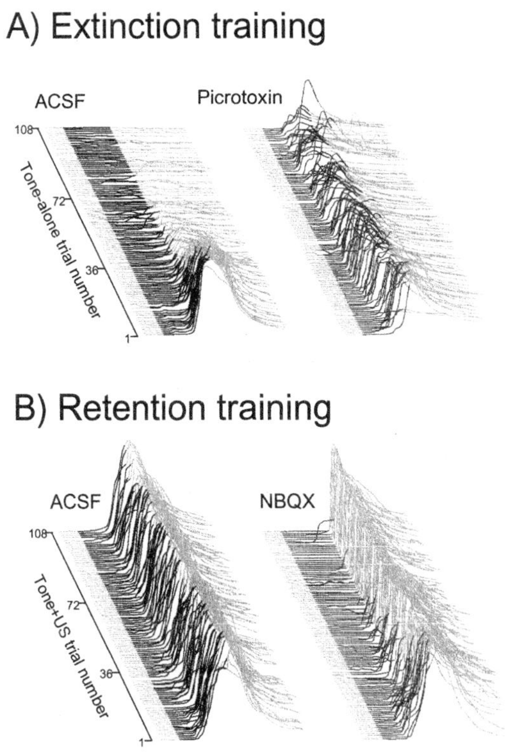

FIGURE 5. Trial-by-trial sweeps for different treatment sessions averaged across four rabbits given cannula implants in the inferior olive. (**A**) Average sweeps during extinction sessions in which either artificial cerebrospinal fluid (ACSF; *left*) or the GABAA antagonist picrotoxin (*right*) was continuously infused into the inferior olive. Extinction is blocked by picrotoxin. (**B**) Average sweeps during paired training (retention) sessions in which either ACSF (*left*) or the AMPA receptor antagonist NBQX (*right*) was infused continuously into the inferior olive. NBQX produced extinction. The concentration for both GABAA and NBQX was 150 μM; the rate of infusion was 0.1 μM min^{-1}. *Dark lines* indicate the presence of the CS.

were subsequently retrained and then tested with continuous infusions of the AMPA antagonist NBQX that began just before a paired training session, the conditioned responses extinguished (FIG. 5B, right). Extinction did not occur during sessions with ACSF infusions (FIG. 5B, left). A second control treatment showed that this extinction-like effect was due not merely to a time-dependent effect of the drug on response expression, but to an experience-dependent effect; when the session was begun 40 minutes after the start of the infusion, extinction still required a similar number of trials. These results confirmed the predictions of the simulation regarding extinction and suggest that extinction of conditioned eyelid responses is driven by the transient suppression of climbing fiber activity below equilibrium.

CONCLUSIONS

We have described a computer simulation of the cerebellum and the way in which it has been tested using Pavlovian eyelid conditioning. The simulation is based on integrate-and-fire neurons whose properties are based on available physiologic data and are wired in accord with the known synaptic organization of the cerebellum. The inputs to the cerebellum are based on recordings of mossy and climbing fiber responses to the stimuli used in eyelid conditioning. The output from the cerebellum can be evaluated with respect to the behavioral output of eyelid conditioning, because recording and stimulation studies suggest that the deep nucleus neurons of the anterior interpositus directly drive the conditioned response. The simulated gr-Pkj synapses in the cerebellar cortex undergo bidirectional changes in strength. Early failures of the simulation were due to the omission of a bidirectional plasticity rule at gr-Pkj synapses and the nucleo-olivary feedback connection. Recent tests of successful simulations incorporating both features suggest that the transient suppression of climbing fiber activity below equilibrium via nucleo-olivary inhibition is the signal that drives extinction.

The insights we have gained into the mechanisms of cerebellar computation may shed light on how the cerebellum operates in other domains.[26] If cerebellar function is considered in terms of how the cerebellum transforms its inputs into outputs, attempts to apply the simulation to data from VOR adaptation, saccade adaptation, and smooth pursuit, all of which are known to be affected by cerebellar lesions,[26,61] should prove highly informative. So long as the inputs to and outputs from the cerebellum are known in these preparations, the simulation can be applied to investigate whether and how its computational principles solve the required input-output transformations. Hopefully, we will then discover how a single computation allows the cerebellum to perform diverse functions.

REFERENCES

1. ECCLES, J.C., M. ITO & J. SZENTÀGOTHAI. 1967. The Cerebellum as a Neuronal Machine. Springer. Berlin.
2. ITO, M. 1984. The Cerebellum and Motor Control. Raven Press. New York.
3. HODGKIN, A.L. & A.F. HUXLEY. 1952. A quantitative description of membrane current and its application to conduction and excitation in nerve. J. Physiol. **117:** 500–544.
4. MAUK, M.D. 2000. The potential effectiveness of simulations versus phenomenological models. Nat. Neurosci. **3:** 649–651.
5. MEDINA, J.F. & M.D. MAUK. 2000. Computer simulation of cerebellar information processing. Nat. Neurosci. **3** (Suppl.)**:** 1205–1211.
6. MEDINA, J.F. *et al.* 2000. Timing mechanisms in the cerebellum: testing predictions of a large-scale computer simulation. J. Neurosci. **20:** 5516–5525.
7. MARR, D. 1969. A theory of cerebellar cortex. J. Physiol. **202:** 437–470.
8. ALBUS, J.S. 1971. A theory of cerebellar function. Math. Biosci. **10:** 25–61.
9. ITO, M. & M. KANO. 1982. Long-lasting depression of parallel fiber-Purkinje cell transmission induced by conjunctive stimulation of parallel fibers and climbing fibers in the cerebellar cortex. Neurosci. Lett. **33:** 253–258.
10. EKEROT, C.F. & M. KANO. 1985. Long-term depression of parallel fibre synapses following stimulation of climbing fibres. Brain Res. **342:** 357–360.
11. SAKURAI, M. 1987. Synaptic modification of parallel fiber-Purkinje cell transmission in in vitro guinea-pig cerebellar slices. J. Physiol. **394:** 463–480.
12. EKEROT, C.F. & M. KANO. 1989. Stimulation parameters influencing climbing fibre induced long-term depression of parallel fibre synapses. Neurosci Res. **6:** 264–268.

13. HIRANO, T. 1990. Depression and potentiation of the synaptic transmission between a granule cell and a Purkinje cell in rat cerebellar culture. Neurosci. Lett. **119:** 141–144.
14. LINDEN, D.J. *et al.* 1991. A long-term depression of AMPA currents in cultured cerebellar Purkinje neurons. Neuron **7:** 81–99.
15. CHEN, C. & R.F. THOMPSON. 1995. Temporal specificity of long-term depression in parallel fiber--Purkinje synapses in rat cerebellar slice. Learn. Mem. **2:** 185–198.
16. BUONOMANO, D. & M.D. MAUK. 1994. Neural network model of the cerebellum: temporal discrimination and the timing of motor responses. Neural Comp. **6:** 38–55.
17. LEV-RAM, V.S. *et al.* 2001. Two distinct forms of cerebellar long-term potentiation. Soc. Neurosci. Abstr. **27:** 682.
18. PAVLOV, I.P. 1927. Conditioned Reflexes. Oxford University Press. London.
19. GORMEZANO, I., E.J. KEHOE & B.S. MARSHALL. 1983. Twenty years of classical conditioning research with the rabbit. *In* Progress in Psychobiology and Physiological Psychology. J.M. Sprague & A.N. Epstein, Eds.: 197–275. Academic Press. New York.
20. SCHNEIDERMAN, N. & I. GORMEZANO. 1964. Conditioning of the nictitating membrane of the rabbit as a function of the CS-US interval. J. Comp. Physiol. Psychol. **2:** 188–195.
21. FREY, P.W. & L.E. ROSS. 1968. Classical conditioning of the rabbit eyelid response as a function of the interstimulus interval. J. Comp. Physiol. Psychol. **65:** 246–250.
22. SMITH, M.C. 1968. The CS-US interval and US intensity in classical conditioning of the rabbit's nictitating membrane response. J. Comp. Physiol. Psychol. **66:** 679–687.
23. MAUK, M.D. & B.P. RUIZ. 1992. Learning-dependent timing of Pavlovian eyelid responses: differential conditioning using multiple interstimulus intervals. Behav. Neurosci. **106:** 666–681.
24. PERRETT, S.P., B.P. RUIZ & M.D. MAUK. 1993. Cerebellar cortex lesions disrupt learning-dependent timing of conditioned eyelid responses. J. Neurosci. **13:** 1708–1718.
25. PERRETT, S.P. & M.D. MAUK. 1995. Extinction of conditioned eyelid responses requires the anterior lobe of cerebellar cortex. J. Neurosci. **15:** 2074–2080.
26. RAYMOND, J.L., S.G. LISBERGER & M.D. MAUK. 1996. The cerebellum: a neuronal learning machine? Science **272:** 1126–1131.
27. MAUK, M.D. & N.H. DONEGAN. 1997. A model of Pavlovian eyelid conditioning based on the synaptic organization of the cerebellum. Learn. Mem. **4:** 130–158.
28. GARCIA, K.M. & M.D. MAUK. 1998. Pharmacological analysis of cerebellar contributions to the timing and expression of conditioned eyelid responses. Neuropharmacology **37:** 471–480.
29. GARCIA, K.S., P.M. STEELE & M.D. MAUK. 1999. Cerebellar cortex lesions prevent acquisition of conditioned eyelid responses. J. Neurosci. **19:** 10940–10947.
30. AITKIN, L.M. & J. BOYD. 1978. Acoustic input to the lateral pontine nuclei. Hear. Res. **1:** 67–77.
31. SEARS, L.L. & J.E. STEINMETZ. 1991. Dorsal accessory inferior olive activity diminishes during acquisition of the rabbit classically conditioned eyelid response. Brain Res. **545:** 114–122.
32. MCCORMICK, D.A. & R.F. THOMPSON. 1984. Cerebellum: essential involvement in the classically conditioned eyelid response. Science **223:** 296–299.
33. BERTHIER, N.E. & J.W. MOORE. 1990. Activity of deep cerebellar nuclear cells during classical conditioning of nictitating membrane extension in rabbits. Exp. Brain Res. **83:** 44–54.
34. MAUK, M.D. *et al.* 1998. Does cerebellar LTD mediate motor learning? Toward a resolution without a smoking gun. Neuron **20:** 359–362.
35. MAUK, M.D., J.E. STEINMETZ & R.F. THOMPSON. 1986. Classical conditioning using stimulation of the inferior olive as the unconditioned stimulus. Proc. Natl. Acad. Sci. USA **83:** 5349–5353.
36. STEINMETZ, J.E. *et al.* 1987. Initial localization of the acoustic conditioned stimulus projection system to the cerebellum essential for classical eyelid conditioning. Proc. Natl. Acad. Sci. USA **84:** 3531–3535.
37. STEINMETZ, J.E., D.G. LAVOND & R.F. THOMPSON. 1989. Classical conditioning in rabbits using pontine nucleus stimulation as a conditioned stimulus and inferior olive stimulation as an unconditioned stimulus. Synapse **3:** 225–233.

38. STEINMETZ, J.E. 1990. Classical nictitating membrane conditioning in rabbits with varying interstimulus intervals and direct activation of cerebellar mossy fibers as the CS. Behav. Brain Res. **38:** 97–108.
39. GOULD, T.J., L.L. SEARS & J.E. STEINMETZ. 1993. Possible CS and US pathways for rabbit classical eyelid conditioning: electrophysiological evidence for projections from the pontine nuclei and inferior olive to cerebellar cortex and nuclei. Behav. Neural Biol. **60:** 172–185.
40. ITO, M. 1970. Neurophysiological aspects of the cerebellar motor control system. J. Int. Neurol. **2:** 162–176.
41. HESSLOW, G. 1994. Inhibition of classically conditioned eyeblink responses by stimulation of cerebellar cortex in the decerebrate cat. J. Physiol. **476:** 245–256.
42. HANSEL, C., D.J. LINDEN & E. D'ANGELO. 2001. Beyond parallel fiber LTD: the diversity of synaptic and non-synaptic plasticity in the cerebellum. Nat. Neurosci. **4:** 467–475.
43. RAYMOND, J.L. & S.G. LISBERGER. 1998. Neural learning rules for the vestibulo-ocular reflex. J. Neurosci. **18:** 9112–9129.
44. SHIBUKI, K. & D. OKADA. 1992. Cerebellar long-term potentiation under suppressed post-synaptic Ca^2+ activity. Neuroreport **3:** 231–234.
45. SALIN, P.A., R.C. MALENKA & R.A. NICOLL. 1996. Cyclic AMP mediates a presynaptic form of LTP at cerebellar parallel fiber synapses. Neuron **16:** 797–803.
46. HAN, V.Z., K. GRANT & C.C. BELL. 2000. Reversible associative depression and nonassociative potentiation at a parallel fiber synapse. Neuron **27:** 611–622.
47. WANG, Y.T. & D.J. LINDEN. 2000. Expression of cerebellar long-term depression requires postsynaptic clathrin-mediated endocytosis. Neuron **25:** 635–647.
48. MEDINA, J.F. & M.D. MAUK. 1999. Simulations of cerebellar motor learning: computational analysis of plasticity at the mossy fiber to deep nucleus synapse. J. Neurosci. **19:** 7140–7151.
49. MEDINA, J.F., K.S. GARCIA & M.D. MAUK. 2001. A mechanism for savings in the cerebellum. J. Neurosci. **21:** 4081–4089.
50. MEDINA, J.F., W.L. NORES & M.D. MAUK. 2001. Inhibition of climbing fibers is necessary for the extinction of conditioned eyelid responses. Nature. **416:** 330–333.
51. ANGAUT, P. & C. SOTELO. 1987. The dentato-olivary projection in the rat as a presumptive GABAergic link in the olivo-cerebello-olivary loop. An ultrastructural study. Neurosci Lett. **83:** 227–231.
52. ANDERSSON, G., M. GARWICZ & G. HESSLOW. 1988. Evidence for a GABA-mediated cerebellar inhibition of the inferior olive in the cat. Exp. Brain Res. **72:** 450–456.
53. DE ZEEUW, C.I. *et al.* 1989. Ultrastructural study of the GABAergic, cerebellar, and mesodiencephalic innervation of the cat medial accessory olive: anterograde tracing combined with immunocytochemistry. J. Comp. Neurol. **284:** 12–35.
54. FREDETTE, B.J. & E. MUGNAINI. 1991. The GABAergic cerebello-olivary projection in the rat. Anat. Embryol. (Berl.). **184:** 225–243.
55. KENYON, G.T., J.F. MEDINA & M.D. MAUK. 1998. A mathematical model of the cerebellar-olivary system I: self-regulating equilibrium of climbing fiber activity. J. Comput. Neurosci. **5:** 17–33.
56. ECCLES, J.C., R. LLINAS & K. SASAKI. 1966. The mossy fiber-granule cell relay of the cerebellum and its inhibitory control by Golgi cells. Exp. Brain Res. **1:** 82–101.
57. LLINAS, R., E.J. LANG & J.P. WELSH. 1997. The cerebellum, LTD, and memory: alternative views. Learn. Mem. **3:** 445–455.
58. KENYON, G.T., J.F. MEDINA & M.D. MAUK. 1998. A mathematical model of the cerebellar-olivary system II: motor adaptation through systematic disruption of climbing fiber equilibrium. J. Comput. Neurosci. **5:** 71–90.
59. HESSLOW, G. & M. IVARSSON. 1996. Inhibition of the inferior olive during conditioned responses in the decerebrate ferret. Exp. Brain Res. **110:** 36–46.
60. KIM, J.J., D.J. KRUPA & R.F. THOMPSON. 1998. Inhibitory cerebello-olivary projections and blocking effect in classical conditioning. Science **279:** 570–573.
61. ROBINSON, F.R. & A.F. FUCHS. 2001. The role of the cerebellum in voluntary eye movements. Annu. Rev. Neurosci. **24:** 981–1004.

Cerebellar Plasticity and the Ocular Following Response

KENJI YAMAMOTO,[a,b] YASUSHI KOBAYASHI,[c] AYA TAKEMURA,[a,d] KENJI KAWANO,[a,d] AND MITSUO KAWATO[e]

[a]*Neuroscience Research Institute, National Institute of Advanced Industrial Science and Technology, Central 2, 1-1-1, Umezono, Tsukuba, Ibaraki 305-8568, Japan*

[b]*Japan Society for the Promotion of Science, Ibaraki 305-856, Japan*

[c] *Osaka University Graduate School of Frontier Biosciences, Laboratories for Neuroscience, Osaka 560-8531, Japan*

[d]*CREST, Japan Science and Technology Corporation*

[e]*ATR Human Information Science Laboratory, Kyoto 619-0288, Japan*

ABSTRACT: We constructed a realistic simulation model to elucidate whether the characteristics of the cerebellar synaptic plasticity reported *in vitro* guide the acquisition and adaptation of the ocular following response (OFR). The model reconstructed the firing frequency of the inputs of granule cell axons (GCA), inhibitory cells (IC), and climbing fibers (CF) to cerebellar Purkinje cells for the OFR, to simulate the reported cerebellar plasticity, including long-term depression, long-term potentiation, and rebound potentiation. When the model used the same visual inputs as reported for monkeys, it successfully simulated the real characteristics of simple spikes in Purkinje cells of adult monkeys and adaptation of gain and direction. The success of our simulation relied on the temporal relationship of the synaptic weight changes when CF inputs preceded GCA and IC inputs, corresponding to the relationship reported by Chen and Thompson and reanalysis of the data of Karachot *et al.* The success of our simulation strongly suggests that acquisition and adaptation of the OFR arise from cerebellar plasticity.

KEYWORDS: cerebellum; ocular following; LTD; MST; adaptation; simulation

INTRODUCTION

The ocular following response (OFR) is a reflex eye movement that stabilizes the gaze. It is induced with short latencies by motion of a large-field visual stimulus.[1,2] FIGURE 1. shows the neural parts reported to control the OFR. Simple (SSs) and complex (CSs) spikes of Purkinje cells (P-cells) in the cerebellar ventral paraflocculus (VPFL) are modulated during the OFR.[3–6] The movement of the retinal image con-

Address for correspondence: Kenji Yamamoto, Neuroscience Research Institute, National Institute of Advanced Industrial Science and Technology, Central 2, 1-1-1, Umezono, Tsukuba, Ibaraki 305-8568, Japan. Voice: +81-298-61-5848; fax: +81-298-61-5849.
k.yamamoto@aist.go.jp

Ann. N.Y. Acad. Sci. 978: 439–454 (2002). © 2002 New York Academy of Sciences.

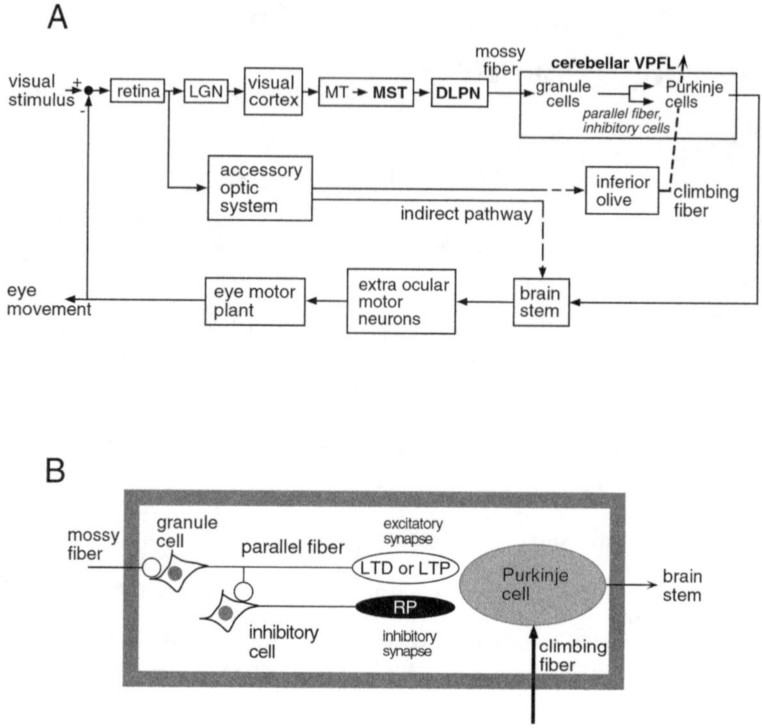

FIGURE 1. (**A**) The neural structures producing the OFR. MST: medial superior temporal area. DLPN: dorsolateral pontine nucleus. VPFL: ventral paraflocculus. (**B**) A diagram of the cerebellar cortex. LTD: long-term depression. LTP: long-term potentiation. RP: rebound potentiation.

structs SSs in the VPFL via neural activity in the visual cortex, medial-superior temporal area (MST),[7] dorsolateral pontine nucleus (DLPN),[8] mossy fibers,[9] and granule-cell axons (GCAs) and inhibitory cells (ICs) (i.e., stellate cells and basket cells) in the VPFL. The movement of the retinal image also generates CSs in the VPFL via the accessory optic tract,[10] inferior olivary nucleus, and climbing fiber (CF).

The direction, speed, and temporal characteristics of cell firing differ markedly between the mossy-fiber inputs and the SS firing outputs of the VPFL[2,11–13] (FIG. 2). First, the preferred direction changes from a uniform distribution in the MST,[7] DLPN,[8] and mossy fibers[9] (FIG. 2, first row, left and central columns), to the horizontal and vertical axes only in SSs[4,6] (FIG. 2, first row, right column). Second, the wide distribution of preferred velocities in the MST[7] and DLPN[8] changes to a narrow distribution in SSs in the VPFL[4,6] (FIG. 2, second row). Third, the wide distribution of waveforms in the VPFL inputs changes to a narrow distribution in VPFL outputs[11] (FIG. 2, third row).

The OFR undergoes two types of adaptation: gain adaptation after repeated "speed-step" stimuli and directional adaptation after repeated "direction-step" stimuli.[14] The speed-step stimuli contain a "step-up stimulus" and a "step-down stimu-

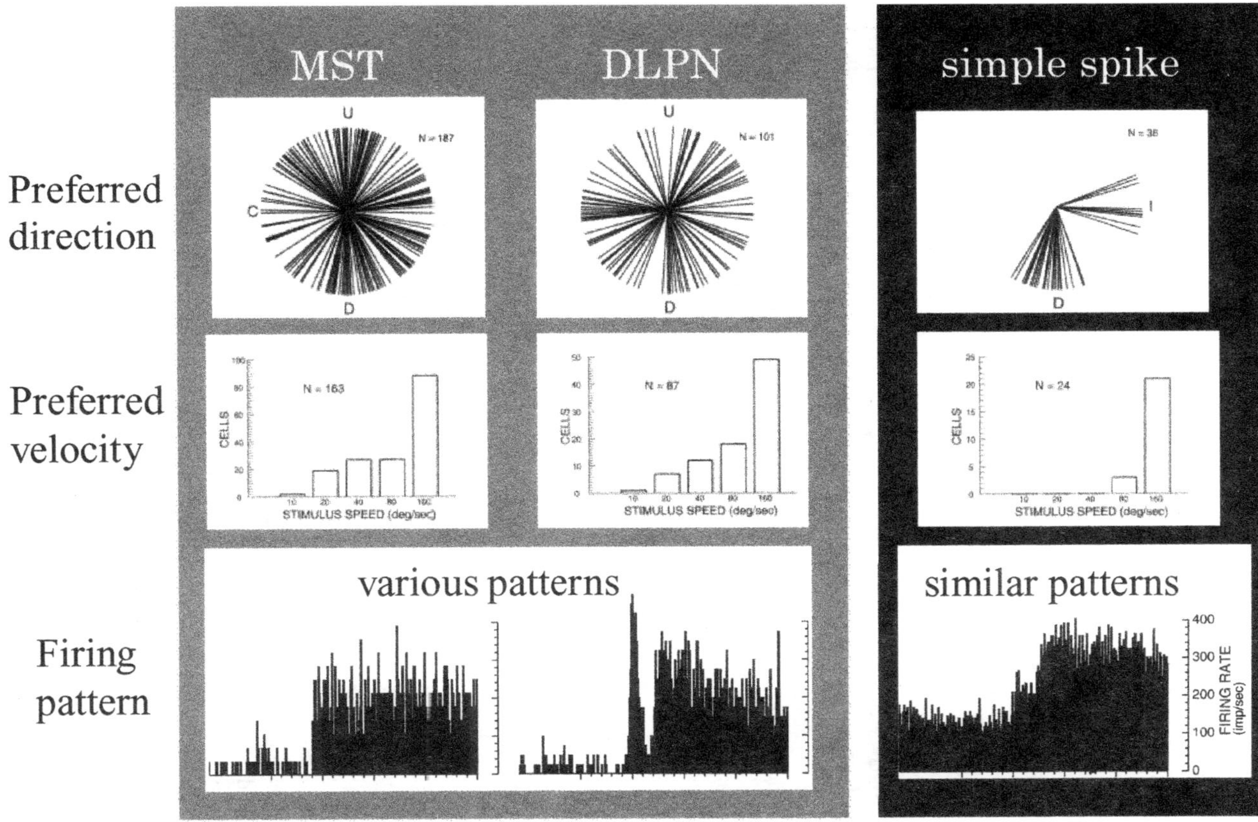

FIGURE 2. Characteristics of neural firing of the MST (*left column*), DLPN (*middle column*), and simple spikes of the VPFL (*right column*) for the preferred direction (*1st row*), preferred velocity (*2nd row*), and temporal pattern of firing frequency (*3rd row*) (Kawano *et al.*[13])

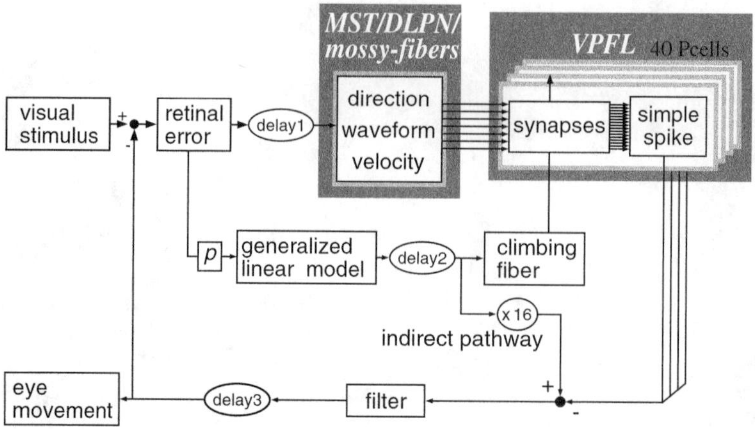

FIGURE 3. A diagram of the model. The model contains 1080 MST cells and 40 P-cells. MST firing is computed from the retinal error, using preferred directions ("direction"), temporal patterns ("waveform"), and preferred velocities ("velocity"). The granule cell axons and inhibitory cells carry the MST firing to the 40 P-cells via excitatory and inhibitory synapses with individual weights ("synapses"). The weighted summation of these inputs becomes SS firing ("simple spike"). The SS outputs from the VPFL are added to the output from the indirect pathway before the filter to become the eye movement. The retinal error is the visual stimulus motion minus the eye movement. *p* denotes a Laplace transformation operator.

lus". For the step-up stimulus, the monkeys were given a ramp stimulus slower than 100°/s for the first 150 ms, and the velocity of the stimulus was changed to 100°/s for the next 150 ms. For the step-down stimulus, the monkeys were given a ramp stimulus slower than 100°/s for the first 150 ms, and the velocity of the stimulus was changed to 0°/s for the next 150 ms. After repeating the step-up trials, the OFR gain of the monkey, which was defined as the maximum eye velocity divided by the stimulus velocity, increased adaptively for the test ramp stimuli at the initial speed. After repeated step-down trials, the OFR gain decreased for the 100-ms test ramp. In the direction-step trials, the direction of the stimulus was changed 90° counterclockwise 150 ms after the onset of the ramp stimulus. After repeated direction-step trials, the OFR direction was changed to the counterclockwise direction for the 100-ms test ramp.

It is widely accepted that the cerebellum plays an important role in motor learning and adaptation. We constructed a detailed network model to examine whether the characteristics of cerebellar plasticity reported *in vitro* are the origin of the characteristics of the SSs and adaptations of the OFR reported *in vivo* (FIG. 3).[15]

MODELING THE OFR SYSTEM

Modeling Mossy-Fiber Inputs

The temporal firing patterns of MST cells were constructed in order to model the reported characteristics as 1080 models (12×3×30) characterized by 12 different

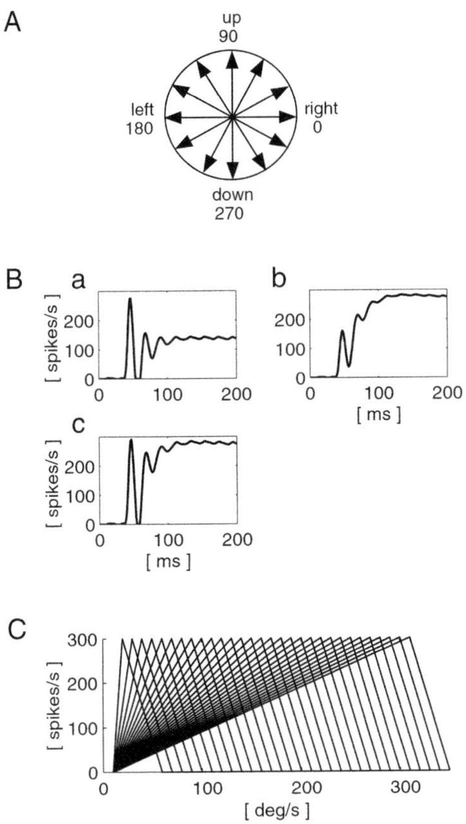

FIGURE 4. Directional (**A**), temporal (**B**), and velocity (**C**) tuning characteristics of model MST firing. (**A**) Modeling the direction-selectivity of MST cells. Twelve groups of cells, one every 30° possess preferred directions. (**B**) Modeling the temporal firing patterns of MST cells. The temporal patterns were modeled by summing the acceleration and velocity of the retinal error weighted using three ratios, such that the maximum of the phasic and tonic components of the firing exhibited the three ratios 2:1 (**a**), 1:2 (**b**), and 1:1 (**c**). (**C**) Modeling the selectivity of MST cells on the stimulus velocity. Thirty curves show saturated firing as a function of stimulus velocity for 30 cells with preferred velocities distributed from 10 to 300°/s every 10°/s from left to right. The abscissa denotes the stimulus velocity. The ordinate denotes the maximum possible firing-frequency output from each MST cell.

preferred directions (from 0 to 360° in every 30°), 3 different temporal wave forms, and 30 different preferred velocities (10–300°/s) (FIG. 4). Physiologically, no remarkable difference has been reported in the firing characteristics of MST cells, DLPN cells, and mossy fibers.[8–9] Accordingly, we used MST cell firing as the model for both DLPN and mossy-fiber cell firing. In addition, no recording has been attempted from GCAs or ICs during the OFR. Therefore, our model assumed that GCA firing is the same as mossy-fiber firing. IC firing was assumed to be the same as GCA firing. Therefore, the MST-DLPN-mossy-fiber pathway contained 1080 GCAs and 1080 ICs.

Modeling Climbing-Fiber Input

We constructed models of CF inputs that reproduced the reported temporal patterns and preferred directions of CSs in monkeys. The temporal patterns of CF inputs were reconstructed using the generalized linear model of the velocity of retinal error on the basis of the report by Kobayashi *et al.*[6] We prepared 10 different temporal patterns based on the CSs of 10 Purkinje cells recorded by Kobayashi *et al.*[6] The preferred directions of CF inputs were set to be upward or contralateral, as in adult monkeys.[6]

Modeling Purkinje Cells

Each P-cell model had one CF input and 1080×2×40 synaptic inputs from GCAs and ICs. We prepared 40 (2×2×10) P-cell models characterized by the difference in the two hemispheres (left and right) and differences in the innervated climbing fibers (two preferred directions [upward or contralateral] and 10 temporal patterns). We called the P-cells connected to CFs with an upward preferred direction vertical P-cells (v-P-cells), and those connected to CFs with a contralateral preferred direction horizontal P-cells (h-P-cells).

Modeling Cerebellar Plasticity

Cerebellar synaptic plasticity includes long-term depression (LTD) and long-term potentiation (LTP) of excitatory GCA synapses, and rebound potentiation (RP) of inhibitory synapses from ICs (FIG. 1B, TABLE 1). LTD was induced by the conjunctive stimulation of GCAs and CFs.[16,17] LTP was induced by stimulating GCA in the absence of CF stimulation.[18,19] RP was induced by conjunctive activation of IC and P-cells, resulted in increases in the synaptic efficacy of IC synapses.[20,21]

In the experiments using slice preparations, the amount of change in the synaptic weights was found to depend on the time interval between CF and GCA stimulation. We call the interval the temporal window of the change. Chen and Thompson[22] reported that maximal LTD was induced when GCA inputs were excited 250 ms before CF activation. To simulate the temporal window, we used a Gaussian temporal window, which had its center 200 ms before the CF input and a standard deviation (SD) of 50 ms, on the basis of data of Chen and Thompson.[22] Since there have been no reports of temporal windows for LTP and RP, we used the same temporal window for both in our simulations.

TABLE 1. Reported changes in synaptic transmission efficacy in the cerebellar cortex

	Climbing fiber	
	Fire	Silent
Parallel fiber fires	LTD[a]	LTP[b]
Inhibitory cell fires	RP[c]	no change

[a]Long-term depression.[16,17]
[b]Long-term potentiation.[18,19]
[c]Rebound potentiation.[20,21]

The following equation represents the change in the synaptic weight of GCA inputs in LTD and LTP during each stimulus presentation, on the basis of the reports on the cerebellar plasticity shown in TABLE 1 and its temporal window:

$$\frac{d\omega_{GCA}(t)}{dt} = -1/n_{LTD}GCA(t)\int_0^{350} G(t-t')\cdot 1(CF(t')-CFsp)dt + 1/n_{LTP}GCA(t)\int_0^{350} G(t-t')\cdot 1(CFsp-CF(t'))dt' - 1/\tau(\omega_{GCA}(t)-\omega_0) \quad (1)$$

Here,

$$1(x) = \begin{cases} 0, \text{ for } x < 0 \\ x, \text{ for } x \geq 0 \end{cases} \quad (2)$$

In Equation (**1**), the first term is the change caused by LTD. The increase of the firing frequency of GCA inputs with the increase of the firing frequency of CF input ("firing–firing" relationship) shown in TABLE 1 constructs the LTD. The increase at time t' of CF firing frequency on its spontaneous frequency, $1(CF(t')\text{-}CFsp)$, affects the GCA inputs that modulate it at time t, GCA(t), within the temporal window, $G(t\text{-} t')$, and the product constructs LTD, which decreases the synaptic weights of GCAs and P-cells, $\omega_{GCA}(t)$. $1/n_{LTD}$ denotes the coefficient for change caused by LTD.

The second term represents the change caused by LTP. The increase of the firing frequency of GCA inputs with the decrease of the firing frequency of CF input ("firing–silence" relationship) shown in TABLE 1 constructs LTP. LTP increases ω_{GCA} in proportion to the product of the GCA inputs and the CF decrease from its spontaneous level. $1/n_{LTP}$ is the LTP coefficient. The third term represents the exponential self-decay effect on the initial weight, ω_0.

The following equation represents the change in the synaptic weight of IC inputs in RP:

$$\frac{d\omega_{IC}(t)}{dt} = -1/n_{RP}IC(t)\int_0^{350} G(t-t')\cdot 1(CF(t')-CFsp)dt - 1/\tau(\omega_{IC}(t)-\omega_0) \quad (3)$$

In Equation (**3**), the first term is due to RP. The firing–firing relationship in TABLE 1 constructs RP. RP decreases the synaptic weight between IC and P-cells at time t, $\omega_{IC}(t)$, in proportion to the product of the IC inputs and the CF increase from its spontaneous level. $1/n_{RP}$ is the RP coefficient. The second term represents the self-decay. Equation RP does not contain a term corresponding to the second term in Equation (**1**), because in the absence of CFs, synaptic weights are not affected by IC stimulation (TABLE 1).[21]

These equations represent the temporal changes in the synaptic weights with GCA(t), IC(t), and CF(t) simulated in the system.

Modeling the Structure Downstream from the Cerebellum

The neural structure between P-cells and eye movements was modeled using low band pass filters that represent the dynamical characters of the system, based on our previous studies.[23,24] The h-P-cells were connected to a filter that moves the eyes horizontally. The v-P-cells were connected to a filter that moves the eyes vertically (FIG. 3).

Initial Settings

The firing frequencies of SSs were constructed by summing all the inputs from the GCAs and ICs. In the initial state, the synaptic weights of GCAs and ICs were set randomly with positive values for GCAs and negative values for ICs. As mentioned above, the firing frequencies of ICs were constructed using the same values as for GCAs. Consequently, the balanced positive and negative inputs did not modulate the firing frequency of SSs much in the initial state, although the firing frequencies of GCAs, ICs, and CF could. We prepared this initial state as the inborn state.

The diameters of the small circles in FIGURE 5A represent the synaptic weights averaged over GCAs (*open circles*) and ICs (*solid circles*) with the same preferred direction on P-cells. The positions of the circles indicate the preferred direction of the synapses. Circles had the same diameters for GCAs and ICs in each preferred direction in the initial state.

FIGURE 3 shows the entire system. The visual inputs to the system and the eye movement produce the retinal error, which modulates the GCA, IC, and CF inputs to P-cells. These inputs change the synaptic weight using the equations. Consequently, we can simulate the changes in SSs and eye movements derived from changes in the synaptic weight.

RESULTS

In the initial state, no test ramp stimuli modulated SSs (FIG. 5B-a), and the eyes did not follow the visual stimuli (FIG. 5B-b).

Simulation of OFR Acquisition

First, we re-input the velocity-step visual stimuli to the system to determine whether cerebellar plasticity can construct the same characters as the SSs in adult monkeys. This resulted in SSs (FIG. 5B-c) of P-cells that moved the eyes to follow the visual stimuli (FIG. 5B-d).

FIGURE 5C shows the changes in the synaptic weights of v-P-cells. With a downward visual stimulus (*left column*), the firing frequencies of GCAs and ICs with downward preferred directions increased (*1st row*). Simultaneously, the firing frequencies of CSs of v-P-cells decreased below their spontaneous level (*3rd row*) because their preferred directions were upward. The firing–silence relationship shown in TABLE 1 and Equation (**1**) constructed LTP in v-P-cells at the synapses with pre-

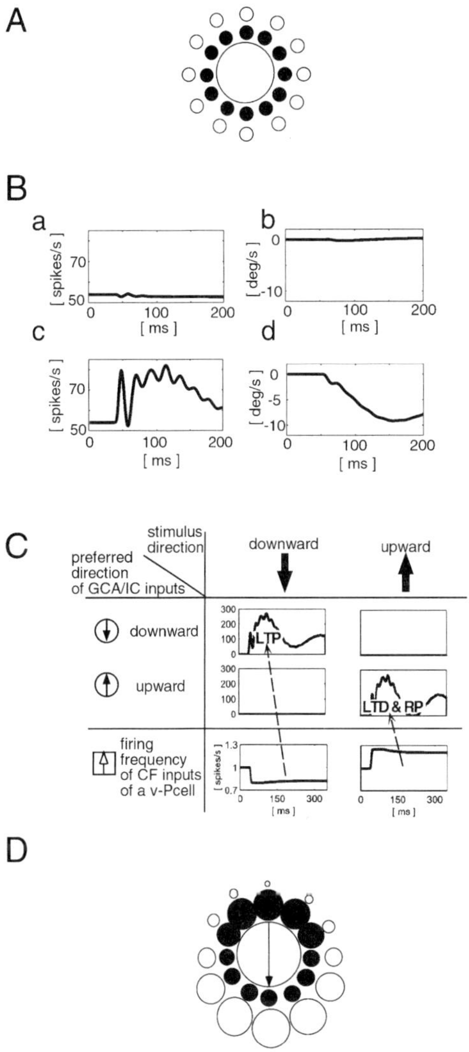

FIGURE 5. (**A**) Visualized presentation of synaptic weights on P-cells at the initial state. The *large circles* denote P-cells. The diameters of the *open* and *solid small circles* denote the averaged synaptic weights of GCA excitatory synapses and inhibitory IC synapses. The positions of the *small circles* indicate the preferred directions of these synaptic inputs. (**B**) The SS firing frequency pattern (**a**, **c**) and the temporal pattern of eye velocity (**b**, **d**) with a downward 10°/s test ramp in the initial and adult states, respectively. (**C**) Simulated GCA, IC, and CF firing frequency patterns and synaptic plasticity mechanisms in the initial state model during downward and upward stimuli. The *1st and 2nd rows* show the temporal patterns of some GCAs and ICs with downward and upward preferred directions, respectively. The *3rd row* shows the temporal patterns of CF inputs for v-P-cells. The *left* and *right columns* show temporal firing frequency patterns with downward and upward stimuli, respectively. The *two broken-line arrows* indicate that the designated synaptic plasticity is induced by the interaction between the GCA/IC inputs and CF inputs. (**D**) Visualized presentation of the synaptic weights simulated on v-P-cells in the adult state, formatted as in FIG. 5A. The *arrows* within the large circle (v-P-cells) show the acquired mean preferred direction (downward) of SSs.

ferred downward directions (the *broken-line arrow*). With an upward stimulus (*right column*), the firing frequencies of the GCAs and ICs with an upward preferred direction increased (*2nd row*). Simultaneously, the firing frequencies of the CSs of v-P-cells increased above their spontaneous level (*3rd row*). The firing–firing relationship in TABLE 1 and Equations (**1**) and (**2**) constructed LTD and RP in v-P-cells at the synapses with preferred upward direction.

As a result, the balance between positive and negative disappeared at the synapses for each preferred direction. For downward-preferring synapses on v-P-cells, the weights of excitatory inputs exceeded negative ones because of the LTP (FIG. 5D,

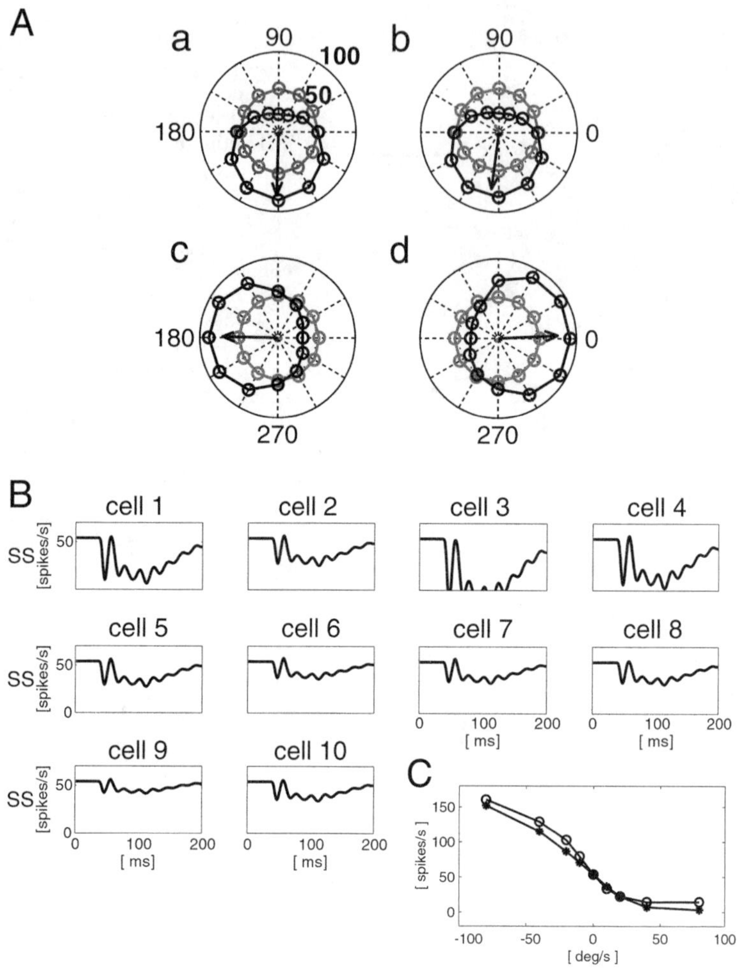

FIGURE 6. (**A**) Direction-dependent SS activation and preferred directions of SSs before (*gray circles*) and after (*black circles* and *arrows*) repeated ramp trials. **a**, **b**, **c**, and **d** are for the averages of 10 left v-P-cells, 10 right v-P-cells, 10 left h-P-cells, and 10 right h-P-cells, respectively. The circles denote the SS frequency averaged over the time interval between 50 and 150 ms after the stimulus motion onset for stimuli in 12 different directions. When the maximum modulation was 100 spikes/s, a plotted point appears on the *outer solid circle*. The *arrows* denote the average preferred directions for each of the four groups of P-cells. (**B**) Temporal firing frequency patterns of the SSs of the 10 v-P-cell models on the *left side* with an upward 10°/s ramp stimulus after repeated ramp trials. (**C**) Comparison of the results of the simulations (solid circles) and experiments (open circles) (6) on the dependence of SS mean firing rates on the stimulus velocity.

lower larger open and *smaller solid circles*). For the upward-preferring synapses, the weights of negative inputs exceeded positive ones because of the LTD and RP (FIG. 5D, *upper smaller open* and *larger solid circles*). Consequently, v-P-cells started to increase their firing frequencies with downward stimuli and to decrease them with upward stimuli, indicating that their preferred direction had become downward. Similarly, the preferred direction of h-P-cells became ipsilateral.

The simulated preferred directions of SSs were downward for all 20 v-P-cells and ipsilateral for all 20 h-P-cells (FIG. 6A). The learned patterns of all SSs were similar over all P-cells (FIG. 6B), and the simulated SSs preferred high speed (FIG. 6C). These simulated preferred directions, waveforms, and velocities corresponded to those of SSs reported in adult monkeys, shown in FIGURE 2. We call this state the "adult state."

Simulation of Gain Adaptation

Next, we simulated gain adaptation. After simulating OFR acquisition, the step-up (upward and rightward) and step-down (downward and leftward) stimuli were applied to the adult state model. The Gaussian temporal window for the changes was centered on 200 ms before the CF modulation. Thus, the window produced the largest changes on the GCAs and ICs that fired during the first 150 ms preceding the changes in the firing frequency of CSs during the last 150 ms. For the downward step-down stimulus (*left column* in FIG. 7A), the GCAs and ICs with a downward-preferred direction increased their firing frequency for the first 150 ms (*1st row*), while the CS firing frequency (*3rd row*) of v-P-cells increased above its spontaneous level for the next 150 ms, because the downward eye velocity surpassed the reduced stimulus velocity and generated an upward retinal error. Thus, the time-lagged firing–firing relationship shown in TABLE 1 and Equations (**1**) and (**2**) constructed LTD and RP on the synapses of GCAs and ICs that had downward preferred directions. Consequently, the LTD and RP reduced the increase in SS firing frequency of v-P-cells for a downward stimulus, and this decreased the downward OFR gain (FIG. 7B).

For the upward step-up stimulus (right column in FIG. 7A), the GCAs and ICs with upward preferred directions increased their firing frequencies for the first 150 ms (*2nd row*), while the CS firing frequency of v-P-cells (*3rd row*) increased markedly over the spontaneous level for the next 150 ms. The firing–firing relationship shown in TABLE 1 and Equations (**1**) and (**2**) constructed LTD and RP on the synapses of GCAs and ICs with upward preferred directions. As a consequence, the LTD and RP further decreased the SS firing frequency of v-P-cells, causing the upward stimulus and resulting in the increase in upward OFR gain (FIG. 7C). Similarly, the OFR gain for a leftward stimulus was reduced by repetition of the leftward step-down stimuli, and the OFR gain for rightward stimulus was increased with repetition of rightward step-up stimuli.

Simulation of Directional Adaptation

Next, we simulated directional adaptation. After simulating OFR acquisition, direction-step stimuli were presented to the adult state model. Left-down, up-left, right-up, and down-right direction step stimuli were presented to the adult state mod-

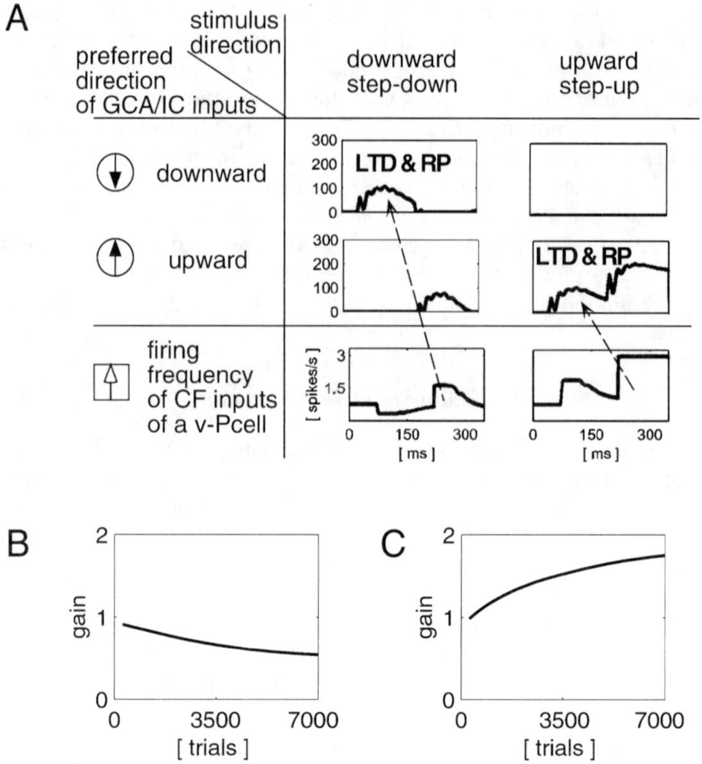

FIGURE 7. (**A**) Simulated GCA, IC, and CF firing frequency patterns and synaptic plasticity mechanisms in the "adult state" model during 40°/s speed-step stimuli. The format of this FIGURE is similar to that of FIG. 5C. The *1st* and *2nd rows* show the temporal patterns of GCAs and ICs with preferred downward and upward directions, respectively, and a preferred velocity of 100°/s. The left and *right columns* show temporal firing frequency patterns with downward step-down and upward step-up stimuli, respectively. (**B**) Changes in OFR gain observed for a 10°/s downward test ramp. (**C**) Changes in OFR gain observed for a 10°/s upward test ramp.

el at 10, 20, 30, 40, 60, 80, and 100 °/s. In the left-down trial, a leftward ramp stimulus was presented for the first 150 ms, and then the direction was changed to downward for the next 150 ms. Similarly, in the up-left, right-up, and down-right trials, the respective first ramp stimulus was given for the first 150 ms, and then the next ramp stimulus was given for the next 150 ms.

For the left-down step stimulus (FIG. 8A), the GCAs and ICs with a leftward preferred direction increased their firing frequency for the first 150 ms (*2nd row*), while the CS firing frequency of v-P-cells (5th row) fell below its spontaneous level for the last 150 ms. The time-lagged firing–silence relationship shown in TABLE 1 and Equation (**1**) constructed LTP on the synapses of GCAs and ICs with leftward preferred directions. Consequently, the v-P-cells started to modulate their firing frequency

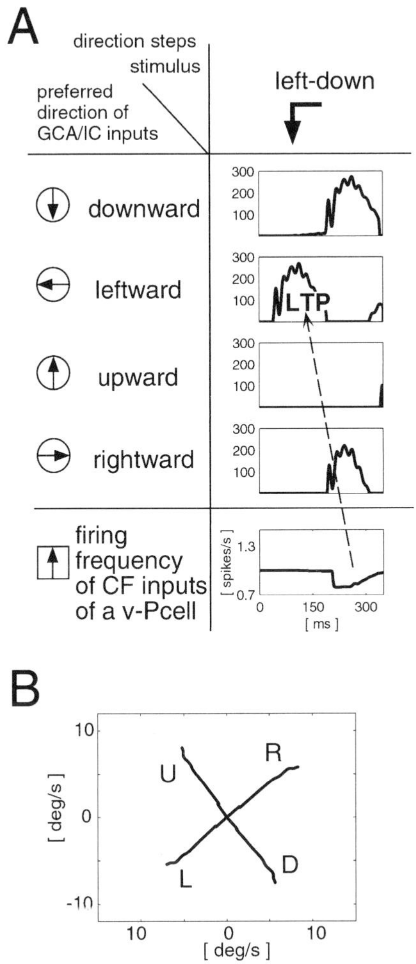

FIGURE 8. (**A**) Simulated firing frequency patterns and synaptic plasticity induced in the "adult state" model for left-down direction-step stimuli. The format is similar to that in FIG. 5C. The *1st, 2nd, 3rd,* and *4th rows* show the temporal patterns of GCA/IC inputs with downward, leftward, upward, and rightward preferred directions, respectively, and CF inputs for a v-P-cell are shown in the *5th row*. The *column* shows the temporal firing frequency patterns for the left-down step. The *broken-line arrow* indicates that the designated synaptic plasticity is induced by the interaction between the GCA/IC inputs and CF inputs. (**B**) A two-dimensional plot of eye velocity showing the trajectories evoked by 10°/s ramp stimuli after repeated direction-step stimuli. The plots labeled R, L, U, and D are for eye velocity for right, left, up, and down test ramp stimuli, respectively. Each plot starts at zero velocity at the center.

with leftward stimuli. The firing of v-P-cells with leftward stimuli constructed a left-downward OFR with leftward visual stimuli (FIG. 8B, line "L"). Similarly, the OFR direction for upward, rightward, and downward visual stimuli became up-leftward (FIG. 8B, line "U"), right-upward (line "R"), and down-rightward (line "D").

DISCUSSION

On the basis of previous behavioral and physiological studies of the OFR, a large-scale network model for two-dimensional OFR was developed and connected with a model of cerebellar plasticity. It successfully simulated OFR acquisition after birth. Specifically, we found that the drastic transition in neural representation from the population codes in the MST to firing rate codes in the VPFL regarding preferred directions, velocities, and temporal firing waveforms[2,11,12] were produced at GCA and IC synapses on P-cells. Second, the adaptation of OFR gains induced by speed-steps was successfully reproduced. Third, the adaptation of OFR kinematic characteristics induced by direction-step stimuli was successfully reproduced. The retinal error–MST–DLPN–mossy-fiber pathway involved visual coordinates, while the cerebellum–brain stem–eye movements pathway involved movement coordinates. In our simulation, cerebellar plasticity succeeded in changing the axis between visual and motor coordinates to produce the directional adaptation. The success of our simulations suggests that acquisition and adaptation of the OFR *in vivo* result from cerebellar plasticity involving the characteristics reported *in vitro*.

The temporal window we used for the changes is based on the report of Chen and Thompson.[22] The temporal window produced the largest changes in the synaptic weights of GCAs and ICs that fired before CSs. As a result, late modulation of CSs in speed-step or direction-step stimuli affected the synaptic weights of the inputs with early modulation. In another report on the temporal window, Karachot *et al.* [25] reported that LTD occurred almost uniformly when the GCA inputs followed the CF input within a 2-s interval. We also attempted simulations using a temporal window spread uniformly between 0 and 2 s after the CF input, based on their report. In that situation, we could not simulate the gain or direction adaptations because the temporal window did not affect the synaptic weights of the inputs that had a modulatory effect before CF modulation.[15] However, our reanalysis of the data of Karachot *et al.*, which considered the cycle of stimulation, found that their data suggested that LTD was effectively induced when CF was delayed by about 100 ms (see FIG. 11 in Ref. 15). We were able to reproduce all of the acquisition and adaptation simulation results using a Gaussian-function temporal window with its peak 100 ms before the CF inputs and a standard deviation of 50 milliseconds.[15]

The cerebellar motor learning hypothesis[12,16,17] requires that CF inputs carry error information regarding the movements executed. This error information is available only after the movements have been executed and only after the GCA and IC inputs have been given to P-cells to issue motor commands. Both our reanalysis of the data of Karachot *et al.*[25] and our simulation results strongly suggest that the width of the temporal window should be restricted to before CF modulation, so that LTD is useful for motor learning.

Kawaguchi and Hirano[26] recently reported that conjunctive stimulation of CF and ICs did not construct RP, countering the reports of Kano *et al.*[22,21] Besides RP, Jorntell and Ekerot[27] more recently reported on additional synaptic plasticity between GCAs and ICs. The conjunctive stimulation of CF and GCAs increased the relative synaptic weights of GCAs and ICs. The increase in the synaptic weights of the inputs into ICs can work similarly to RP and produce an increase in the outputs of ICs. Thus, the model described in this manuscript should successfully simulate the OFR, even if the plasticity reported by Jorntell and Ekerot is used in place of the RP.

ACKNOWLEDGMENTS

We thank Drs. S. Highstein and T. Thach for inviting us to contribute to this issue.

REFERENCES

1. MILES, F.A., K. KAWANO & L.M. OPTICAN. 1986. Short-latency ocular following responses of monkey. 1. Dependence of temporospatial properties of visual input. J. Neurophysiol. **56:** 1321–1354.
2. KAWANO, K. 1999. Ocular tracking: behavior and neurophysiology. Curr. Opin. Neurobiol. **9:** 467–473.
3. SHIDARA, M., K. KAWANO, H. GOMI, *et al.* 1993. Inverse-dynamics encoding of eye movements of Purkinje cells in the cerebellum. Nature **365:** 50–52.
4. SHIDARA, K. & K. KAWANO. 1993. Role of Purkinje cells in the ventral paraflocculus in short-latency ocular following responses. Exp. Brain Res. **93:** 185–195.
5. GOMI, H., M. SHIDARA, A. TAKEMURA, *et al.* 1998. Temporal firing patterns of Purkinje cells in the cerebellar ventral paraflocculus during ocular following responses in monkeys. 1. Simple spike. J. Neurophysiol. **80:** 818–831.
6. KOBAYASHI, Y., K. KAWANO, A. TAKEMURA, *et al.* 1998. Temporal firing patterns of Purkinje cells in the cerebellar ventral paraflocculus during ocular following responses in monkeys. 2. Complex spikes. J. Neurophysiol. **80:** 832–848.
7. KAWANO, K., M. SHIDARA, Y. WATANABE, *et al.* 1994. Neural activity in cortical area of MST of alert monkey during ocular following responses. J. Neurophysiol. **71:** 2305–2324.
8. KAWANO, K., M. SHIDARA & S. YAMANE. 1992. Neural activity in dorsolateral pontine nucleus of alert monkey during ocular following responses. J. Neurophysiol. **67:** 680–703.
9. KAWANO, K. & M. SHIDARA. 1993. The role of the ventral paraflocculus in ocular following in the monkey. *In* Role of the Cerebellum and Basal Ganglia in Voluntary Movement: 195–202. Elsevier. New York.
10. INOUE, Y., A. TAKEMURA, K. KAWANO, *et al.* 2000. Role of the pretectal nucleus of the optic tract in short-latency ocular following responses in monkeys. Exp. Brain Res. **131:** 269–281.
11. TAKEMURA, A., Y. INOUE, H. GOMI, *et al.* 2001. Change in neuronal firing patterns in the process of motor command generation for the ocular following response. J. Neurophysiol. **86:** 1750–1763.
12. KAWATO, M. 1999. Internal models for motor control and trajectory planning. Curr. Opin. Neurobiol. **9:** 718–727.
13. KAWANO, K., A. TAKEMURA, Y. INOUE, *et al.* 1996. Visual inputs to cerebellar ventral paraflocculus during ocular following responses. Prog. Brain Res. **112:** 415–422.
14. MILES, F.A. & K. KAWANO. 1986. Short-latency ocular following responses of monkey. 3. Plasticity. J. Neurophysiol. **56:** 1381–1396.
15. YAMAMOTO, K., Y. KOBAYASHI, A. TAKEMURA, *et al.* 2002. Computational studies on acquisition and adaptation of ocular following responses based on cerebellar plasticity. J. Neurophysiol. **87:** 1554–1571.
16. ITO, M., M. SAKURAI & P. TONGROACH. 1982. Climbing fibre induced depression of both mossy fibre responsiveness and glutamate sensitivity of cerebellar Purkinje cells. J. Physiol. (Lond.) **324:** 113–134.
17. ITO, M. & M. KANO. 1982. Long-lasting depression of parallel fiber-Purkinje cell transmission induced by conjunctive stimulation of parallel fibers and climbing fibers in the cerebellar cortex. Neurosci. Lett. **33:** 253–258.
18. SAKURAI, M. 1987. Synaptic modification of parallel fibre-Purkinje cell transmission in *in-vitro* guinea-pig cerebellar slices. J. Physiol. (Lond.) **394:** 463–480.
19. HIRANO, T. 1990. Depression and potentiation of the synaptic transmission between a granule cell and a Purkinje cell in rat cerebellar culture. Neurosci. Lett. **119:** 141–144.
20. KANO, M., U. REXHAUSEN, J. DREESSEN, *et al.* 1992. Synaptic excitation produces a long-lasting rebound potentiation of inhibitory synaptic signals in cerebellar Purkinje cells. Nature **356:** 601–604.
21. KANO, M. 1996. Long-lasting potentiation of GABAergic inhibitory synaptic transmission in cerebellar Purkinje cells: Its properties and possible mechanisms. Behav. Brain Sci. **19:** 354–361.
22. CHEN, C. & R.F. THOMPSON. 1995. Temporal specificity of long-term depression in parallel fiber-Purkinje synapses in rat cerebellar slice. Learn. Mem. **2:** 185–198.

23. YAMAMOTO, K., Y. KOBAYASHI, A. TAKEMURA, *et al.* 1997. A mathematical model that reproduces vertical ocular following responses from visual stimuli by reproducing the simple spike firing frequency of Purkinje cells in the cerebellum. Neurosci. Res. **29:** 161–169.
24. YAMAMOTO, K., Y. KOBAYASHI, A. TAKEMURA, *et al.* 2000. A mathematical analysis of the characteristics of the system connecting the cerebellar ventral paraflocculus and extraoculomotor nucleus of alert monkeys during upward ocular following responses. Neurosci. Res. **38:** 425–435.
25. KARACHOT, L., R.T. KADO & M. ITO. 1994. Stimulus parameters for induction of long-term depression in *in-vitro* rat Purkinje cells. Neurosci. Res. **21:** 161–168.
26. KAWAGUCHI, S. & T. HIRANO. 2000. Suppression of inhibitory synaptic potentiation by presynaptic activity through postsynaptic GABA(B) receptors in a Purkinje neuron. Neuron **27:** 339–347.
27. JORNTELL, H. & C.F. EKEROT. 2002. Reciprocal bidirectional plasticity of parallel fiber receptive fields in cerebellar Purkinje cells and their afferent interneurons. Neuron **34:** 797–806.

Modeling Cerebellar Flocculus and Paraflocculus Involvement in Complex Predictive Smooth Eye Pursuit in Monkeys

RONALD E. KETTNER, MINAH SUH,[a] DITA DAVIS, AND HOI-CHUNG LEUNG[b]

Department of Physiology and the Neuroscience Institute, Northwestern University Medical School, Chicago, Illinois 60611, USA

[a]*Department of Physiology/Neuroscience, New York University School of Medicine, New York, New York 10016, USA*

[b]*Department of Psychology, State University of New York, Stony Brook, New York 11794, USA*

ABSTRACT: **The role of flocculus and paraflocculus neurons in the cerebellar control of predictive eye movements was examined using two modeling techniques. The first study characterized the dependence of individual Purkinje-cell firing patterns on oculomotor output, visual input, and response timing using multilinear regression techniques. Interestingly, no dependence on visual input was detected. Purkinje cell firing was explained by sensitivities to eye position and eye velocity alone. However, complex responses occurred when sensitivity vectors pointed in different directions. For example, some neurons showed a preference for circular pursuit in a particular rotation direction. Responses also tended to lead the eye during predictable pursuit and to lag during unpredictable, visually driven pursuit. This suggests that flocculus and paraflocculus neurons played a stronger role during predictive pursuit than visually driven pursuit. A second modeling study demonstrated how the flocculus/paraflocculus system might generate predictive pursuit. A biologically realistic neural network was simulated based on the known anatomy and physiology of this cerebellar system. It included mossy and climbing fibers with realistic responses, Purkinje cells acting on well-characterized brain-stem circuits, and granule, Golgi, basket, and stellate cells with appropriate connections. The network was able to learn new pursuit trajectories based on long-term alterations in synaptic connectivity at parallel-to-Purkinje synapses. Interestingly, this model was able to generate predictive pursuit without visual input based only on eye-motion input. Thus, both models provide complementary evidence for the generation of nonvisual predictive control by flocculus and paraflocculus neurons.**

KEYWORDS: **cerebellum; modeling; smooth pursuit; eye movements; monkey; oculomotor; flocculus**

Address for correspondence: Ronald E. Kettner, Department of Physiology and the Neuroscience Institute, Northwestern University Medical School, Chicago, IL 60611. Voice: 312-503-0456; fax: 312-503-5101.
r-kettner@nwu.edu

Ann. N.Y. Acad. Sci. 978: 455–467 (2002).

INTRODUCTION

From the earliest anatomical studies, it was clear that the cerebellum has enormous computational power. In terms of cell count alone, it contains more neurons than does any other brain structure including the neocortex. Divergence-convergence ratios also suggest impressive information-processing capacity. Granule-cell interneurons are 300 times more plentiful than mossy-fiber inputs and 3,000 times more numerous than Purkinje-cell outputs. This complexity is amplified by a large number of synaptic connections along this input-output pathway. Each mossy fiber makes synaptic contact with 1,200 granule cells, and each Purkinje cell receives 200,000 parallel-fiber synapses from granule cells. Importantly, the majority of these synapses are modifiable, so that cerebellar function can be optimized for a myriad of environmental demands. Golgi, stellate, and basket cells also have extensive connections that enhance the computational power of the cerebellar system.

How then is one to study such a complex and powerful brain system? A behavior is needed that is computationally demanding enough to require the cerebellum, while remaining amenable to quantitative measurement and analysis. The behavior selected for our initial studies was complex predictive eye pursuit. This class of behaviors exhibits nonlinearities and a dependence on spatial context that cannot be generated by a simple neural circuit constructed from a handful of neurons. Furthermore, the timing accuracy of predictive pursuit suggested a system like the cerebellum that was tightly linked to both motor output as well as efferent-copy feedback about that motor output. The cerebellum also receives short latency information from other sources of nonvisual afferent input that could be useful in the generation of nonvisual predictive control. The flocculus/paraflocculus region was selected for our initial studies of neural processing because it has been linked to the control of smooth pursuit eye movements and has been well characterized both anatomically and physiologically. Finally, oculomotor behaviors provided several advantages for studies of complex motor control. They were accurately measured using search-coil techniques, and the relative simplicity of the oculomotor plant allowed a better focus on complex predictive processing.

This report reviews our recent progress in using mathematical models to better understand the neural basis of predictive control during smooth pursuit by cerebellar flocculus and paraflocculus neurons in the monkey.[1–7]

BEHAVIORAL STUDIES

To study visual control while minimizing the influence of predictive control, pursuit was first studied during unpredictable changes in the direction of ongoing pursuit.[1] More specifically, rhesus monkeys performed in a perturbation task that required pursuit of a target that moved along a circle trajectory for several cycles before it executed an abrupt right-angle change in direction along a circle meridian. The time and location of the right-angle perturbation were varied so that the monkey could not predict its occurrence. The results are shown in FIGURE 1A. The eye accurately followed the circular trajectory before the right-angle perturbation. It then continued along the expected circular trajectory for ~90 ms before it began to turn towards the altered target motion. Thus, as expected, the system was able to respond

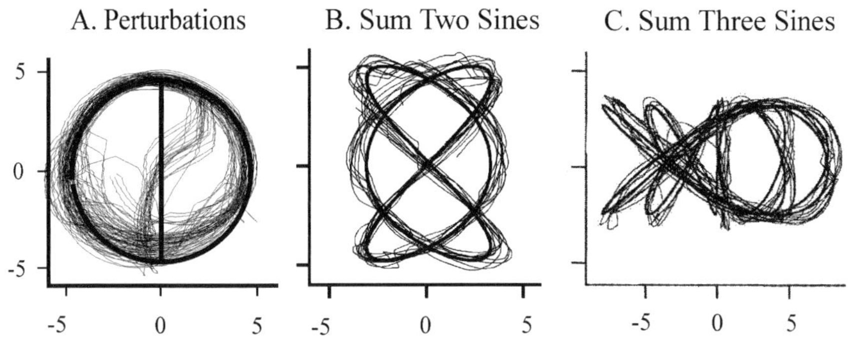

FIGURE 1. Overshoot errors result from the 90-ms delay in visual processing during unpredictable changes in target direction (**A**). By contrast, pursuit timing is highly accurate during predictable sum-of-two-sines (**B**) and sum-of-three-sines (**C**) trajectories. Here *thin lines* indicate pursuit behavior, *thick lines* indicate target motion, and *scales* indicate eye position in degrees.

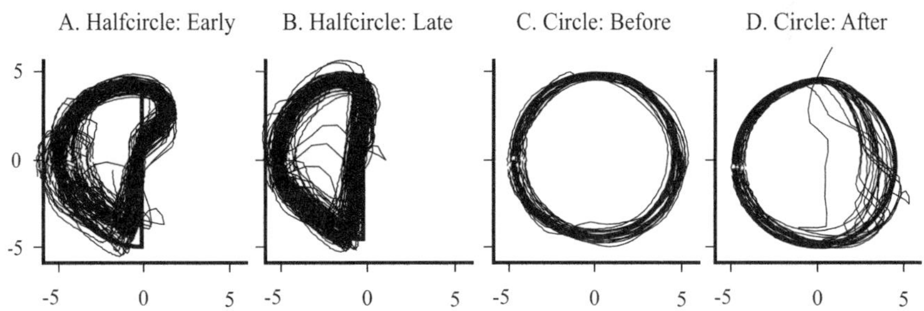

FIGURE 2. Training improves pursuit accuracy in the half-circle task. Overshoot during the first block of training (**A**) is gradually reduced until it is nearly absent during the final block of training (**B**). This training produced clear aftereffects that were evident in comparisons of circle pursuit before (**C**) and after (**D**) training. Here *thin lines* indicate pursuit behavior, *thick lines* indicate target motion, and *scales* indicate eye position in degrees.

to unexpected changes in target motion. However, there was considerable delay in executing these changes in direction. The result was a large overshoot error that moved the target image well away from the fovea and sharply reduced visual acuity after the perturbation. Although this paradigm focuses on delays during ongoing pursuit, similar delays have been reported during the initiation of pursuit.[8,9]

In light of these results, it was surprising that the same monkeys showed average delays between eye and target of only 5 ms during a circle task that required pursuit along sinusoids in four orientations and circular pursuit in counterclockwise (CCW) and clockwise (CW) directions (FIG. 2C). Thus, the smooth pursuit system predicted slow changes in target direction before visual information about these changes had been processed with a 90-ms delay. Simple feedback controllers driven directly by image slip on the retina would produce larger lags closer to the 90-ms delays observed after unexpected changes in target motion. Other investigators have also described accurate pursuit of sinusoidal, circular, and other simple periodic motions.[1,10–15]

We therefore set out to see if predictive control was possible along trajectories that were more complex. It turned out that with extended practice the monkeys were able to track more complex trajectories with only very small lags. FIGURE 1B and C shows typical pursuit performance along two-dimensional trajectories created by summing sinusoids of different frequencies on horizontal and vertical axes. Although complex, these sum-of-sines trajectories had relatively short repeat periods that allowed predictive control. As for the circle trajectories, performance in these sum-of-sines tasks was excellent. The average delay was 4 ms for a variety of sum-of-two-sines trajectories that were studied. In fact, a slight average lead of 9 ms was observed for several sum-of-three-sines trajectories. Both averages were much smaller than the 90-ms delay expected for controllers based on visual error feedback. Previous studies of sum-of-sines pursuit in humans have also suggested the presence of predictive control. This was true even though these studies used trajectories with longer repeat periods that were not designed to be highly predictable. Some have also reported small phase lags for high-frequency components and phase leads for low-frequency components.[16–20]

The existence of predictive performance suggests that regularities in a repeated trajectory are used to improve future performance. This can only occur if the pursuit controller can store information in memory about a particular trajectory for later use. Furthermore, predictive performance should improve with training. This idea was tested by seeing if the pursuit system could learn the right-angle perturbation illustrated in FIGURE 1A by making it more predictable during what was called the half-circle task. Here, monkeys pursued a repeated half-circle trajectory for thousands of cycles. Initially, the eye overshot the target as previously observed for unexpected right-angle perturbations during circular pursuit (FIG. 2A). However, after several hundred repetitions of the half-circle trajectory, performance improved until the

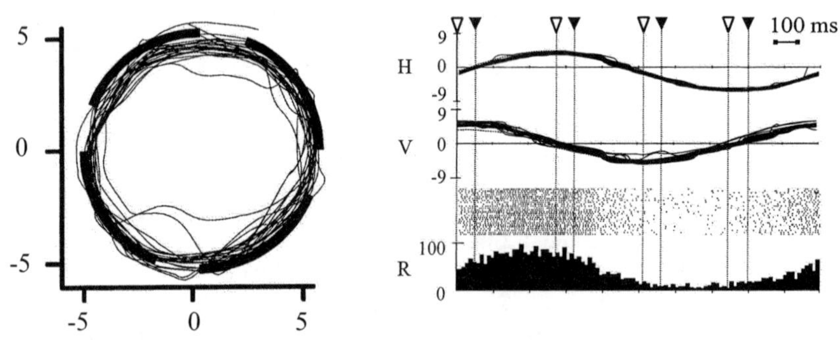

FIGURE 3. At the *left*, there is no alteration in pursuit behavior (*thin traces*) during four gaps in a circular target motion (*thick lines*). At the *right*, horizontal (H) and vertical (V) component traces also indicate a smooth continuation of smooth eye motion relative to target offset (*open triangles*) and onset (*closed triangles*) times. *Below*, raster and average histogram displays of Purkinje cell activity indicate no response at the four gap periods during 20 repetitions of the gap trajectory.

right angle was executed with little or no error (FIG. 2B). In addition, a comparison of circle tracking before (FIG. 2C) and after (FIG. 2D) half-circle training indicated clear aftereffect errors during the previously well-executed circle trajectory. These aftereffects remained for several hundred cycles until accurate circle pursuit was again observed. The fact that accurate half-circle pursuit required hundreds of cycles and that aftereffects remained for hundreds of cycles indicates that permanent alterations in pursuit processing occur when one switches trajectories. These findings are compatible with similar slow patterns of learning during paradigms that study the initiation of pursuit.[21]

Predictive control was also examined by turning off the target for 100 ms at up, down, right, and left locations during a circular target motion in what was termed the gap task. Interestingly, the monkeys showed no difference in pursuit lag for circular trajectories with (FIG. 3) and without (FIG. 2C) gaps. We had expected some decline in performance when the target went off, albeit at a delay of ~90 ms, but no such decline was observed. The continuation of pursuit during short target gaps suggests some form of predictive control during the extension of an ongoing behavior in the absence of visual drive. Similar continuations during short target gaps have been observed during straight-line trajectories[22] that place smaller computational burdens on the predictive machinery. Of course, there is ultimately a decline in pursuit performance for gaps of longer duration.[23]

CEREBELLAR SINGLE-NEURON RECORDING STUDIES

Thus far, the use of predictive control has been demonstrated at the behavioral level. More difficult is an understanding of how the brain generates this predictive control. The cerebellar flocculus/paraflocculus was selected for initial study because lesions of this region produce severe declines in pursuit[24] and responses from single neurons are highly correlated with pursuit behavior.[25–29] Our intent was to contrast single-neuron responses during predictive versus visually driven pursuit in terms of both sensory and motor influence and response latency.

Although our objectives were easily stated, distinctions between sensory and motor influences can be quite subtle during ongoing pursuit. One needs to know if modulations in a neuron's response are due to links with motor output systems or instead result from variations in visual input that occur at the same time. The estimation of neural response latencies is further complicated by the need to understand the kinematics of a neuron's response. For example, neurons related to eye velocity will fire earlier than neurons related to eye position during the same sinusoidal motion. In addition, a neuron's response timing is influenced by visual inputs that can vary in magnitude during pursuit along a changing trajectory. Without quantitative knowledge of a particular neuron's relation to the kinematics of eye and visual motion variables, estimates of response latency are difficult, if not impossible.

We addressed these concerns by developing models that explicitly characterized the firing rate of each neuron in terms to eye motion and visual motion influences. This potentially complicated relation was modeled using the following multilinear equation:

$$R(t) = \beta + \rho_m \cdot P_m(t+\tau_m) + \nu_m \cdot V_m(t+\tau_m) + \alpha_m \cdot A_m(t+\tau_m) + \rho_v \cdot P_v(t+\tau_v) + \nu_v \cdot V_v(t+\tau_v) + \alpha_v \cdot A_v(t+\tau_v) \quad (1)$$

Here, R is the neural firing rate, *t* is the time, β is the baseline firing rate, ρ, ν, and α are sensitivity vectors, P, V and A are position, velocity, and acceleration vectors, τ measures the time difference between neural firing and eye or visual image motion, and the subscripts m and v distinguish variables related to eye motion and visual image motion on the retina, respectively. This vector equation can also be expressed in terms of horizontal (x) and vertical (y) components by expanding each dot product operation so that, for example, $\rho_m \cdot P_m(t + \tau_m) = \rho_{mx} P_{mx}(t + \tau_m) + \rho_{my} P_{my}(t + \tau_m)$. However, vector notation was preferred because it was more compact and provided a more appropriate representation for the two-dimensional vector variables that might influence firing rate.

This equation was fit to data collected in the perturbation, sum-of-sines, and gap tasks using standard multivariate regression techniques. Importantly, each regression fit was based on a number of different variations of each task so that the influences of individual variables could be separated. For example, the complete sum-of-sines task consisted of four complex trajectories as well as the individual sinusoidal components used to create them. In addition, regression fits were based on raw data streams instead of data averaged across repetitions of a trajectory. This further reduced the correlation between visual and eye motion influences. However, saccadic epochs were deleted so that smooth control could be isolated from the processing of saccades.

Our first major finding was that Purkinje cell responses were not related to visual input. This was most apparent in raster and histogram displays of Purkinje cell firing during the gap task. Here the target was turned off for 100 ms at four points along a circle trajectory without any obvious effect on Purkinje cell firing (FIG. 3). This result was confirmed by regression analyses. Partial R^2 values for visual-position, visual-velocity, and visual-acceleration sensitivities were small, and their deletion from *Equation 1* produced no change in goodness of fit for all three tasks studied. By contrast, deletion of the eye-position, eye-velocity, and eye-acceleration terms produced large declines in goodness of fit. Although a lack of visual drive makes some sense for highly predictable circular and sum-of-sines target motions that were tracked with high accuracy and minimal retinal slip, this result was more perplexing for the perturbation task that produced large amounts of slip. These results suggest that pursuit drive generated by retinal-image motion is supplied by other brain regions.

The second major finding was that while eye-motion sensitivity vectors could account for the firing patterns of individual Purkinje cells, these vectors did not always point in the same direction. As a result, many Purkinje cells showed relatively complex response patterns. Some had different directional preferences during CW versus CCW pursuit. Other showed preferences for pursuit towards a central location, whereas others preferred motion away from center. A striking preference for CCW motion is illustrated in FIGURE 4. This specificity resulted from a rightward eye-position vector and an upward eye-velocity vector with similar magnitudes but with preferred directions at right angles to each other. As a result, peak firing profiles due to position and velocity components reinforced during CCW motion and cancelled during CW motion. This type of complexity is interesting, because it is not what one expects from a neuron that has exclusive control of a single muscle. Differences in eye-position and eye-velocity preferred direction were common. More than half the neurons studied had sensitivity vectors that differed by more than 90 degrees.

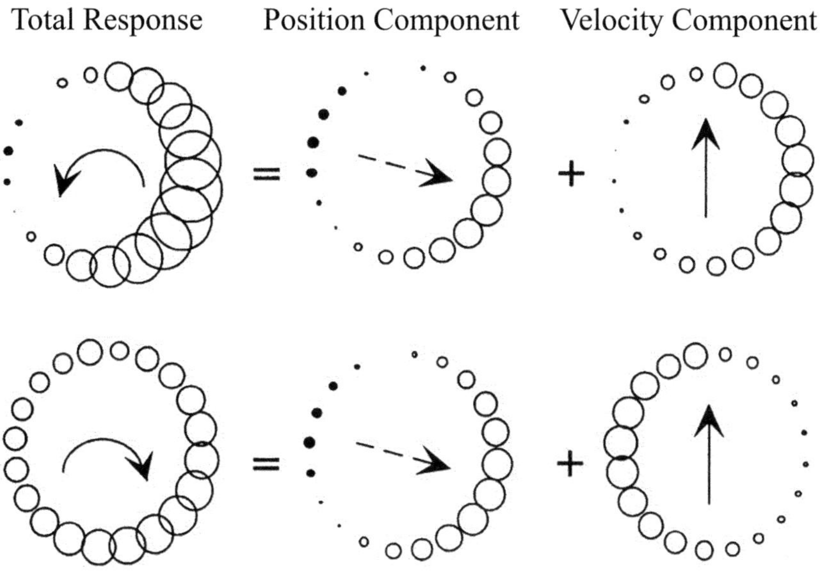

FIGURE 4. A neuron with a preference for counterclockwise (CCW) rotation was explained by eye-position and eye-velocity sensitivities that pointed in different directions. Here a rightward position preference and an upward velocity preference created component responses that reinforced during CCW pursuit and canceled during clockwise (CW) pursuit. The areas of individual circles are proportional to the firing rate at locations along each trajectory.

The third major finding was a tendency for Purkinje cell firing to lead pursuit eye movements during tasks that required predictive pursuit and to lag pursuit during tasks requiring visually driven pursuit. This result was based on estimates of τ in *Equation 1* for the perturbation, sum-of-sines, and gap tasks. It is clear from FIGURE 5 that the temporal structure of the neural responses in these tasks was well fit by the model. Indeed, the complexity of these responses provided necessary temporal constraints for good estimates of response timing. On average, Purkinje cell responses led eye motion by 12 ms during sum-of-sines pursuit, by 6 ms during circle pursuit with gaps, and by only 2 ms during unpredictable right-angle perturbations. Given the 9-ms transmission delay between flocculus stimulation and eye motion,[30,31] this suggests that the flocculus/paraflocculus drove pursuit during the predictable sum-of-sines target motions, but played a follow-up role during visually driven pursuit in the perturbation task. Performance during the gap task suggested an intermediate mixture of predictive and visual control.

Thus, the flocculus/paraflocculus region of the cerebellum appears to be actively involved in predictive control during pursuit and acts in conjunction with other brain regions that generate visual control signals. During well-learned, predictable trajectories, flocculus/paraflocculus predictive signals drive eye motion, whereas other brain areas provide relatively weaker visual control signals. During unpredictable

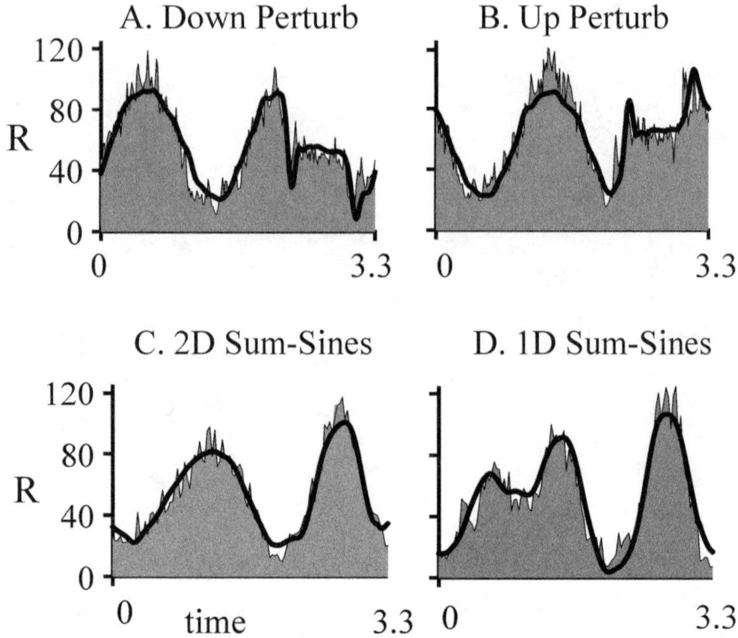

FIGURE 5. Neural response histograms (in *gray*) during pursuit are well explained by *Equation 1* (*thick lines*) in terms of eye-motion sensitivities. Examples illustrate good fits during abrupt changes in horizontal motion for downward (**A**) and upward (**B**) perturbations from circle pursuit as well as for complex two-dimensional (**C**) and one-dimensional (**D**) sum-of-sines motions.

pursuit, visually driven signals from other brain regions dominate the control of pursuit, whereas flocculus/paraflocculus plays a relatively weaker and delayed role.

COMPUTER MODELING STUDIES

Thus far, we have demonstrated the existence of complex predictive control during smooth pursuit and its possible locus in the flocculus/paraflocculus. It is then natural to ask how this cerebellar region might generate predictive control. This is a difficult question. It cannot be answered at the single neuron level. Even if one had complete knowledge of the firing patterns of every flocculus/paraflocculus neuron involved in predictive control as well as a complete understanding of each neuron's processing capabilities at the biochemical level, one would still need to know how this neural population acts at the systems level to generate oculomotor output.

To better understand how a population of flocculus/paraflocculus neurons might generate complex predictive smooth pursuit, we used neural network models[2,3] based on the known anatomy and physiology of the flocculus/paraflocculus system. The architecture of our current model[3] is illustrated in FIGURE 6. This model is noteworthy in being the only model that has been able to provide biologically realistic

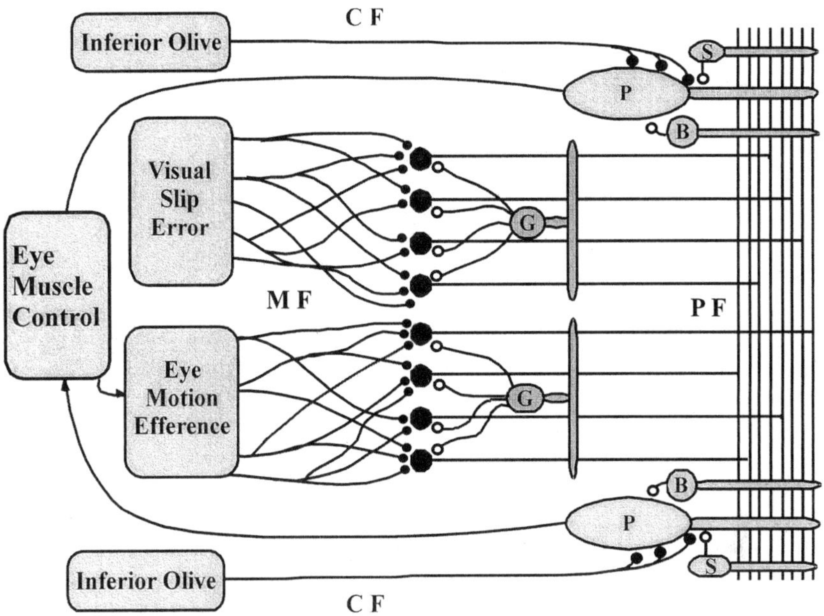

FIGURE 6. Architecture of the cerebellar model. Visual and eye-motion mossy fibers (MF) excite granule cells (*solid circles*) with axons that ascend and then bifurcate as parallel fibers (PF). These axons excite Purkinje (P), stellate (S), basket (B), and Golgi (G) cells that have inhibitory actions on their targets. Climbing fibers (CF) from the inferior olive excite Purkinje cells.

mechanisms for generating all of the predictive behaviors just described. (See examples in FIGURE 7.) The model also mimics the slow time course of half-circle learning and the development of aftereffects when the monkey is returned to circle pursuit. Slow learning times and trajectory specificity are explained by a training algorithm that learns a specific trajectory by making slow alterations in synaptic connectivity weights.

The model also accounts for several puzzling features of flocculus/paraflocculus physiology and anatomy. It explains the dominance of eye-motion mossy-fiber input (~80%) and the relative paucity of visual mossy-fiber input (~5%) in a structure involved in the pursuit of visual targets. In fact, the model generates good predictive control when visual mossy-fiber input is eliminated. Only occasional corrections from a rudimentary saccade generator are needed to synchronize the eye with a visual target. In addition, the model accounts for the strong Golgi-cell inhibition observed in many cerebellar studies. This inhibition is so strong that it must often be blocked in slice preparations before other modes of synaptic processing can be studied. However, in the model this inhibition provides an enhancement of input patterns that is essential for model performance.

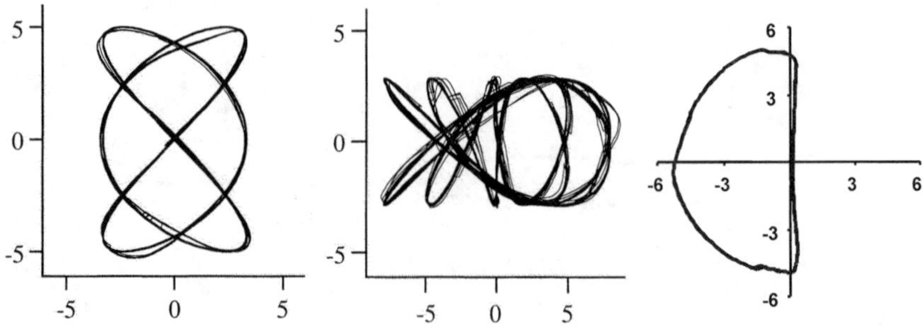

FIGURE 7. Performance by the model was excellent for sum-of-two-sines (**A**), sum-of-three-sines (**B**), and half-circle (**C**) trajectories.

The model learns new trajectories by modifying parallel-fiber-to-Purkinje-cell synapses using a computer algorithm[32,33] that is motivated by empirical studies of long-term plasticity.[34,35] Simply stated, synaptic changes occur when parallel-fiber synapses are active shortly before a complex spike is generated by climbing fiber input from the inferior olive. This simple rule eventually leads to accurate pursuit performance. More specifically, climbing-fiber pulses are triggered by visual errors after a 100-ms delay.[36–39] Parallel-fiber activation arises from granule cells that receive mossy-fiber excitation and Golgi-cell inhibition. Finally, the best learning is obtained when a trace is used to match parallel-fiber and climbing-fiber activations. In combination, these results suggest that long-term plasticity should be optimal when parallel-fiber activation precedes climbing-fiber activation, a result that has received some experimental support.[40,41] A similar algorithm based on a continuous climbing-fiber error signal was used in a previous model.[2] Interestingly, training stability improved when climbing fiber activation was made less frequent by introducing a 500-ms refractory period. This offers a possible rationale for the very low occurrence of complex spikes observed in biological systems.

Eye-motion mossy-fiber input to the model is based on empirical studies[26,27] that describe a linear increase in firing rate with progressive eye movement in an "on" direction after a threshold eye position is exceeded. Importantly, thresholds vary across mossy fibers. This means that different subpopulations of mossy fibers are activated as the eye moves throughout two-dimensional space. Golgi-cell inhibition within synaptic glomeruli further fractionates these activation patterns, so that distinct subpopulations of parallel fibers are active at each point along a trajectory. The model generates predictive control by linking these subpopulations to specific Purkinje output cells via appropriate parallel-to-Purkinje synaptic connection weights. For example, during half-circle learning the subpopulation of parallel fibers activated as the eye moves towards the top of the half-circle (FIG. 2B) is trained to activate Purkinje cells that drive the eye downward. This allows the predictive initiation of downward eye motion before the target begins to move downward and well before visual information about this change in trajectory is processed visually. Of course, this strategy only works when a right-angle change in target motion consistently occurs at a particular point on the half-circle motion. It will produce an inappropriate prediction of downward motion during circle pursuit. In other words, aftereffect errors can occur during trajectories that require a different pattern of predictive control, and this is what is observed behaviorally.

DISCUSSION

Sophisticated modeling techniques were used to demonstrate how the cerebellar flocculus and paraflocculus might generate complex predictive control to improve the accuracy of smooth pursuit eye movements. These results have implications for other motor systems that utilize cerebellar control. Similar mechanisms could explain the exquisite motor control that is observed during skilled performance by athletes and musicians. The resulting predictive control would allow changes in body motion to be initiated at the proper time and in the proper cadence and rhythm. This would allow better performance than possible using external auditory and visual cues processed with large delays by sluggish neural systems. This type of predictive control is hypothesized to use internal signals that allow prediction without external sensory input. For the oculomotor system, these signals arise from neurons that either directly or indirectly drive eye muscles. Brain systems controlling other movements could improve performance using their own sources of efference-copy information in addition to proprioceptive signals derived from somatosensory, joint, and muscle-spindle afferents.

More specifically, we have suggested that the flocculus/paraflocculus region of the cerebellum generates nonvisual predictive control during smooth pursuit eye movements. Its activity is highly correlated with eye motion and only weakly correlated with visual image motion. In addition, neural activity in this brain area leads predictable target motion but lags unpredictable target motion. Our modeling efforts indicate how the flocculus/paraflocculus might generate predictive control using biologically realistic simulations of this neural system. Specific subpopulations of its parallel-fiber input become active as the eye moves along a well-practiced trajectory, and these subpopulations generate appropriate motor outputs via their connectivity strengths with Purkinje-cell output neurons. Appropriate parallel-fiber-to-Purkinje-cell synaptic strengths are learned using an algorithm based on experimental reports of how long-term plasticity develops. A prediction of this model is that relatively arbitrary trajectories can be acquired with extended practice and that this learning interferes with previously learned behaviors. This prediction was confirmed by behavioral studies of pursuit learning that demonstrate dramatic improvements in half-circle pursuit with extended training and the existence of aftereffects during a previously well-executed circle trajectory.

ACKNOWLEDGMENTS

This work was supported by grants from the National Institutes of Health (R01 EY-12984, P50 MH-48185, and T32 DC-00015) and the National Science Foundation (IBN-9723846).

REFERENCES

1. LEUNG, H.-C. & R.E. KETTNER. 1997. Predictive smooth pursuit of complex two-dimensional trajectories demonstrated by perturbation responses in monkeys. Vision Res. **37:** 1347–1354.
2. KETTNER, R.E., S. MAHAMUD, H.-C. LEUNG, *et al.* 1997. Prediction of complex two-dimensional trajectories by a cerebellar model of smooth pursuit eye movement. J. Neurophysiol. **77:** 2115–2130.
3. KETTNER, R.E. & M. SUH. 1998. Biological realism improves the performance of a flocculus/paraflocculus model of predictive smooth pursuit eye movements. Soc. Neurosci. Abstr. **24:** 1410.
4. LEUNG, H.-C., M. SUH & R.E. KETTNER. 2000. Cerebellar flocculus and paraflocculus purkinje cell activity during circular pursuit in monkey. J. Neurophysiol. **83:** 13–30.
5. SUH, M., H.-C. LEUNG & R.E. KETTNER. 2000. Cerebellar flocculus and ventral paraflocculus purkinje cell activity during predictive and visually driven pursuit in monkey. J. Neurophysiol. **84:** 1835–1850.
6. KETTNER, R.E., M. SUH & D. DAVIS. 2000. Smooth eye movements improve with training and predictability along half-circle trajectories in monkey. Soc. Neurosci. Abstr. **26:** 1330.
7. SUH, M., D. DAVIS & R.E. KETTNER. 2000. Smooth eye movements improve with training during random target gaps in monkey. Soc. Neurosci. Abstr. **26:** 1716.
8. FUCHS, A.F. 1967. Saccadic and smooth pursuit eye movements in the monkey. J. Physiol. (Lond.) **191:** 609–631.
9. LISBERGER, S.G. & L.E. WESTBROOK. 1985. Properties of visual inputs that initiate horizontal smooth pursuit eye movements in monkeys. J. Neurosci. **5:** 1662–1673.
10. WESTHEIMER, G. 1954. EYE movement response to a horizontally moving visual stimulus. Arch. Ophthalmol. **52:** 932–941.
11. STARK, L., G. VOSSIUS & L.R. YOUNG. 1962. Predictive control of eye tracking movements. IRE Trans. Human Factors Electron. HFE **3:** 52–57.
12. KOWLER, E. & R.M. STEINMAN. 1979. The effect of expectations on slow oculomotor control - I. Periodic target steps. Vision Res. **19:** 619–632.
13. LISBERGER, S.G., C. EVINGER, G.W. JOHANSON & A.F. FUCHS. 1981. Relationship between acceleration and retinal image velocity during foveal smooth pursuit in man and monkey. J. Neurophysiol. **46:** 229–249.
14. BAHILL, A.T. & J.D. MCDONALD. 1983. Smooth pursuit eye movements in response to predictable target motions. Vision Res. **23:** 1573–1583.
15. BARNES, G.R. & P.T. ASSELMAN. 1991. The mechanisms of prediction in human smooth pursuit eye movements. J. Physiol. (Lond.) **439:** 439–461.
16. DALLOS, P.J. & R.W. JONES. 1963. Learning behavior of the eye fixation control system. IEEE Trans. Autom. Control **8:** 218–227.
17. COLLEWIJN, H. & E.P. TAMMINGA. 1984. Human smooth and saccadic eye movements during voluntary pursuit of different target motions on different backgrounds. J. Physiol. (Lond.) **351:** 217–250.
18. YASUI, S. & L.R. YOUNG. 1984. On the predictive control of foveal eye tracking and slow phases of optokinetic and vestibular nystagmus. J. Physiol. (Lond.) **347:** 17–33.
19. BARNES, G.R., S.F. DONNELLY & R.D. EASON. 1987. Predictive velocity estimation in the pursuit reflex response to pseudo-random and step displacement stimuli in man. J. Physiol. (Lond.) **389:** 111–136.
20. KETTNER, R.E., H.-C. LEUNG & B.W. PETERSON. 1996. Predictive smooth pursuit of complex two-dimensional trajectories in monkey: component interactions. Exp. Brain Res. **108:** 221–235.
21. KAHLON, M. & S.G. LISBERGER. 1996. Coordinate system for learning in the smooth pursuit eye movements of monkeys. J. Neurosci. **16:** 7270–7283.
22. NEWSOME, W.T., R.H. WURTZ & H. KOMATSU. 1988. Relation of cortical areas MT and MST to pursuit eye movements. II. Differentiation of retinal from extraretinal inputs. J. Neurophysiol. **60:** 604–620.
23. BECKER, W. & A.F. FUCHS. 1985. Prediction in the oculomotor system: smooth pursuit during transient disappearance of a visual target. Exp. Brain Res. **57:** 562–575.

24. ZEE, D.S., Y. ATSUMI, P.H. BUTLER & G. GÜÇER. 1987. Effects of ablation of flocculus and paraflocculus on eye movements in primate. J. Neurophysiol. **46:** 878–899.
25. LISBERGER, S.G. & A.F. FUCHS. 1978. Role of primate flocculus during rapid behavioral modification of vestibuloocular reflex. I. Purkinje cell activity during visually guided horizontal smooth–pursuit eye movements and passive head rotation. J. Neurophysiol. **41:** 733–763.
26. LISBERGER, S.G. & A.F. FUCHS. 1978. Role of primate flocculus during rapid behavioral modification of vestibuloocular reflex. II. Mossy fiber firing patterns during horizontal head rotation and eye movement. J. Neurophysiol. **41:** 764–777.
27. MILES, F.A., J.H. FULLER, D.J. BRAITMAN & B.M. DOW. 1980. Long-term adaptive changes in primate vestibuloocular reflex. III. Electrophysiological observations in flocculus of normal monkeys. J. Neurophysiol. **43:** 1437–1476.
28. NODA, H. & D.A. SUZUKI. 1979. The role of the flocculus of the monkey in fixation and smooth pursuit eye movements. J. Physiol. (Lond.) **294:** 335–48.
29. STONE, L.S. & S.G. LISBERGER. 1990. Visual responses of Purkinje cells in the cerebellar flocculus during smooth pursuit eye movements in monkeys I. Simple spikes. J. Neurophysiol. **63:** 1241–1261.
30. LISBERGER, S.G. & T.A. PAVELKO. 1988. Brain stem neurons in modified pathways for motor learning in the primate vestibulo-ocular reflex. Science **242:** 771–773.
31. SHIDARA, M. & K. KAWANO. 1993. Role of Purkinje cells in the ventral paraflocculus in short-latency ocular following responses. Exp. Brain Res. **93:** 185–195.
32. KLOPF, A.H. 1972. Brain Function and Adaptive Systems: A Heterostatic Theory. Air Force Cambridge Research Laboratories (AFCRL-72-0164). Bedford, MA.
33. SUTTON, R.S. & A.G. BARTO. 1981. Toward a modern theory of adaptive networks: expectation and prediction. Psychol. Rev. **88:** 135–170.
34. ITO, M. 2001. Cerebellar long-term depression: characterization, signal transduction, and functional roles. Physiol. Rev. **81:**1143–1195.
35. LINDEN, D.J. & J.A. CONNOR. 1995. Long-term synaptic depression. Ann. Rev. Neurosci. **18:** 319–357.
36. FUSHIKI, H., Y. SATO, A. MIURA & T. KAWASAKI. 1994. Climbing fiber responses of Purkinje cells to retinal image movement in cat cerebellar flocculus. J. Neurophysiol. **71:** 1336–1350.
37. KOBAYASHI, Y., K. KAWANO, A. TAKEMURA, *et al.* 1998. Temporal firing patterns of purkinje cells in the cerebellar ventral paraflocculus during ocular following responses in monkeys. II. Complex spikes. J. Neurophysiol. **80:** 832–848.
38. SIMPSON, J.I. & K.E. ALLEY. 1974. Visual climbing fiber input to rabbit vestibulo-cerebellum: a source of direction-specific information. Brain Res. **82:** 302–308.
39. STONE, L.S. & S.G. LISBERGER. 1990. Visual responses of Purkinje cells in the cerebellar flocculus during smooth pursuit eye movements in monkeys II. Complex spikes. J. Neurophysiol. **63:** 1262–1275.
40. SCHREURS, B.G, M.M. OH & D.L. ALKON. 1996. Pairing-specific long-term depression of Purkinje cell excitatory postsynaptic potentials results from a classical conditioning procedure in the rabbit cerebellar slice. J. Neurophysiol. **75:** 1051–1060.
41. CHEN, C. & R.F. THOMPSON. 1995. Temporal specificity of long-term depression in parallel fiber-Purkinje synapses in rat cerebellar slice. Learn. Mem. **2:** 185–198.

Surface-Based Atlases of Cerebellar Cortex in the Human, Macaque, and Mouse

DAVID C. VAN ESSEN

Department of Anatomy and Neurobiology, Washington University School of Medicine, St. Louis, Missouri 63110, USA

ABSTRACT: This study describes surface reconstructions and associated flat maps that represent the highly convoluted shape of cerebellar cortex in three species: human, macaque, and mouse. The reconstructions were based on high-resolution structural MRI data obtained from other laboratories. The surface areas determined for the fiducial reconstructions are about 600 cm^2 for the human, 60 cm^2 for the macaque, and 0.8 cm^2 for the mouse. As expected from the ribbon-like pattern of cerebellar folding, the cerebellar flat maps are elongated along the axis parallel to the midline. However, the degree of elongation varies markedly across species. The macaque flat map is many times longer than its mean width, whereas the mouse flat map is only slightly elongated and the human map is intermediate in its aspect ratio. These cerebellar atlases, along with associated software for visualization and for mapping experimental data onto the atlas, are freely available to the neuroscience community (see http:/brainmap.wustl.edu).

KEYWORDS: cerebellar cortex; atlas; humans; macaques; mice

INTRODUCTION

The most striking gross morphologic feature of the cerebellum is its highly convoluted cortical mantle. The convolutions allow a large cortical surface area to be contained within a compact cerebellar volume, but they present a major obstacle to experimentalists interested in analyzing diverse aspects of cerebellar structure and function.

A valuable approach to dealing with a convoluted cortex is to generate surface reconstructions that capture the shape of the convolutions and can be inflated and flattened to facilitate visualization of buried cortex. This strategy has been applied very successfully to the cerebral cortex, especially in humans and macaques.[1–3] Extending the same strategy to the cerebellar cortex has proven more challenging, because it is much thinner (about one third as thick as cerebral cortex) and its convolutions are more complex. In many species, including human and macaques, the convolutions span several scales, from large cerebellar lobules to medium-sized lamellae, to

Address for correspondence: Department of Anatomy and Neurobiology, Washington University School of Medicine, 660 S. Euclid Ave., Box 8108, St. Louis, MO 63110. Voice: 314-326-7043; fax; 314-747-3436.
vanessen@v1.wustl.edu

Ann. N.Y. Acad. Sci. 978: 468–479 (2002). © 2002 New York Academy of Sciences.

fine-grained folia. To reconstruct the surface, it is necessary to obtain structural data in which the cerebellar folds are accurately visualized and are in register across neighboring slices.

Presented here are the first surface reconstructions that represent the overall pattern of convolutions in human, macaque, and mouse cerebellum. These reconstructions, based on high-resolution structural magnetic resonance imaging (MRI) data, faithfully represent the cerebellar lobules and lamellae, and they capture many but not all of the finer-grained folia. They differ from previously reported cerebellar surface reconstructions that represented the shape of the external cerebellar hull but not the complexity of its convolutions.[4,5] In this chapter, reconstructions are used for two main purposes: to examine the intrinsic geometry of the cerebellar sheet and to illustrate the potential of surface-based cerebellar atlases for visualizing many types of experimental information.

A distinctive aspect of cerebellar surface geometry is that the folds tend to run parallel to one another, as in an accordion or a piece of ribbon candy. One might therefore expect that the unfolded cerebellar ribbon would be highly elongated. Indeed, the cerebellar flat maps that Sultan and Braitenberg[6] generated from a variety of species using an indirect, semiquantitative mapping procedure were all highly elongated. On the other hand, the geometry is complicated by extensive branching and fusion of sulci in the cerebellar hemispheres. As shown here, computationally generated cerebellar flat maps vary greatly in shape: the macaque map is highly elongated, but the mouse map is not, and the human map is intermediate.

Surface-based atlases are useful for viewing many types of spatially localized experimental data that are brought from individuals onto a common atlas framework. This is well established for the cerebral cortex, where current atlases include, for example, a diversity of areal partitioning schemes, information about anatomic connectivity, and a variety of neuroimaging data.[3,7,8] The cerebellar surface reconstructions illustrated here, along with associated surface visualization and analysis software, are freely available, and they provide an attractive substrate for representing a large and rapidly growing body of information about the cerebellar cortex.

METHODS

Surface reconstructions were based on high resolution structural MRI data sets (T1-weighted) obtained for mouse, macaque, and human. The mouse data set (case map015, provided by A. Toga and R. Jacobs) was scanned at 56 μm^3 voxel size; the macaque data set (case F99UA1, provided by N. Logothetis) was scanned at 0.5 mm^3 voxel size. The human data set (case colin_avg20, provided by A. Toga) was from an individual subject scanned repeatedly at 1 mm^3 voxel size; the individual scans were registered to a common stereotaxic space, averaged, and resampled to 0.5 mm^3 voxel size. This is the same case described in a study by Holmes *et al.*[9] and in the cerebellar atlas of Schmahmann *et al.*[5] In the human case the contrast between gray and white matter was significantly lower in ventral than in dorsal portions of the cerebellum, perhaps because of regional differences in the accuracy of the registration algorithm.

For each species, the whole brain MRI volume was cropped to include just the cerebellum. It was then resampled so that the cerebellar cortex was about two or three voxels thick on average. The resampled volumes had voxel dimensions of 84 m^3 for the mouse, 0.25 mm^3 for macaque, and 0.5 mm^3 for human. A modification of the SureFit segmentation method, originally developed for cerebral cortex,[1] was applied to the structural MRI volumes to generate a binary volume (segmentation) whose boundary represents the shape of the cerebellar cortex, including its convolutions. The general SureFit strategy involves generating probabilistic maps related to characteristics of gray matter, white matter, and cerebrospinal fluid (CSF) that are discernible in the image data. Most of cerebellar white matter (except for the central core) is very thin and cannot be robustly identified on the basis of intensity cues alone. A more reliable cue is that thin slabs of white matter are adjoined on either side by slabs of gray matter. Consequently, the image intensity gradient in such regions should be pointed towards the white matter on two opposing sides and should have an appropriate intermediate magnitude (based on the average intensity difference between gray and white matter and the average cortical thickness). An algorithm sensitive to such patterns at any arbitrary three-dimensional orientation was used to generate a probabilistic map of cerebellar white matter. An analogous process was used to generate a probabilistic map for the outer cortical boundary (the pial surface). This was based on the cue that in regions of CSF sandwiched between opposing banks of gray matter, the intensity gradient should be pointed away from the CSF and should have an appropriate magnitude (based on the average intensity difference between gray matter and CSF and on cortical thickness). The maps of thin white matter and of the outer boundary maps were combined to form a probabilistic map of position along the radial axis of cortical gray matter. The radial position map was then thresholded to yield an initial segmentation whose boundary on average ran approximately midway through the cortical gray matter. To the extent that this is achieved, each unit area on the associated surface reconstruction represents a comparable amount of cortex, irrespective of whether it is at the tip or the base of a lobule.[10] The initial segmentations generated with this method contained numerous topological errors, which generally reflect ambiguities associated with low contrast and noise in the original image data. Extensive error correction using an interactive editing tool in SureFit was used to achieve a topologically correct segmentation for each species. For each corrected cerebellar segmentation, a surface reconstruction whose nodes lie on the boundary of the segmented volume was generated in SureFit. This slightly blocky "raw" surface was smoothed to generate a fiducial surface that constitutes the best available estimate of cortical shape.

Cortical flattening was achieved using Caret surface visualization and analysis software.[1,11] For the mouse, cerebellar flattening involved straightforward application of the automated Caret multiresolution method for flattening and distortion reduction.[12] For the macaque and human, flattening was applied piecemeal to restricted subregions of the cerebellar surface. This was necessary because of the inherent difficulty in flattening a highly irregular cerebellar surface that has a far higher degree of intrinsic (Gaussian) curvature than does the cerebral cortex (akin to the fingers on a rubber glove rather than the folds of a crumpled sheet). Once the individual subregions were satisfactorily flattened, they were combined into a composite whole, and the map was locally smoothed in the vicinity of the sutures.

RESULTS

The results are presented in a similar format for each species, including images of the structural MRI data and the segmented cerebellar volume plus surface reconstructions in multiple configurations and viewing perspectives. The mouse cerebellum is illustrated first, as its relatively simple geometry provides a useful baseline for comparisons with the macaque and human.

Mouse Cerebellum

In the structural MRI volume used for the mouse atlas, the cerebellar gray and white matter can readily be discriminated, as shown in a sagittal slice (FIG. 1, top left). The adjacent panel (FIG. 1, top center) shows the segmented cerebellar cortex, whose boundary runs consistently through the cortical gray matter. Comparison with the Nissl-stained cerebellar sections from a mouse brain atlas[13] indicates that the full pattern of cerebellar folds is visible in the structural MRI images and is preserved in the segmented volume. The cerebellar lobules, identified with the aid of this atlas, are labeled on the segmented slice.

The remaining panels of FIGURE 1 show different configurations and viewing perspectives of the mouse cerebellar surface reconstruction. Fiducial, inflated, and spherical configurations are shown from posterior and lateral perspectives in the left and upper right parts of the figure, respectively. The fiducial surface reveals the shape of the intact cortex. In the inflated surface the folds are largely smoothed out, but the overall cerebellar shape is preserved. The spherical map provides a useful substrate for surface-based coordinates and surface-based registration (see Discussion). The flat map, which shows the entire cerebellar surface in a single view, is displayed in the lower right part of FIGURE 1. The location of the cuts made to reduce distortions on the flat map are best seen on the spherical and inflated maps. All of the surfaces are shaded to show surface folding (computed on the fiducial configuration), with light creases signifying outward folds along lobule crests and dark creases signifying inward (sulcal) folds.

The total cerebellar surface area for the fiducial configuration is 80 mm^2, which is about half the combined cortical surface area of the left and right cerebral hemispheres reconstructed from the same mouse atlas (174 mm^2). Sultan and Braitenberg[6] reported a much larger cerebellar surface area (190 mm^2 average for three mice), but it is difficult to evaluate whether systematic biases affected their indirect measurement strategy.

Overall, the cerebellar flat map is about as wide as it is long, making its shape quite different from the approximately threefold elongated map for the schematically unfolded mouse cerebellum illustrated by Sultan and Braitenberg.[6] The vermis, which occupies 43% of the fiducial surface area, forms an elongated strip along the vertical axis. The cerebellar hemispheres, on the left and right sides of the map, predominantly involve expansions of lobules VI–VIII and occupy 51% of the fiducial surface. The adjoining paraflocculus and flocculus (6% of the fiducial surface) are extensions from the copula (lobule VII) and paramedian lobule (lobule VIII).

As with maps of cerebral cortex, there are substantial distortions in the representation of cortical surface area on the cerebellar flat map and spherical map (particu-

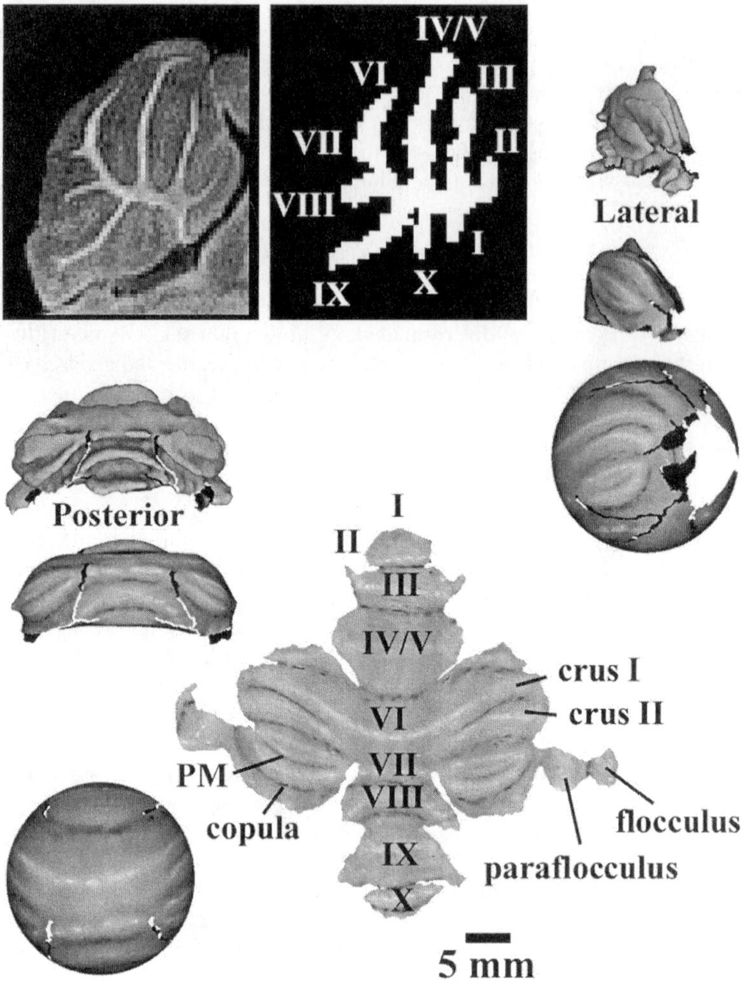

FIGURE 1. A surface-based atlas of mouse cerebellar cortex. *Upper left* and *upper center* panels show the structural MRI and segmentation (with lobules identified) from a sagittal section through the vermis. The remaining panels show cerebellar surface reconstructions, with posterior views of the fiducial, inflated, and spherical surface shown on the *left*, lateral views of the same three configurations on the *right*, and a flat map in the *lower center.*

larly on the spherical map because it is geometrically more constrained). Although it is important to know that these distortions exist, they are not a major impediment to quantitative analyses, because the surface area of any region of interest on a flat map or spherical map can readily be computed for the corresponding region on the fiducial surface (as illustrated in the preceding paragraph). Having a topologically correct spherical surface allows location to be specified by spherical coordinates of

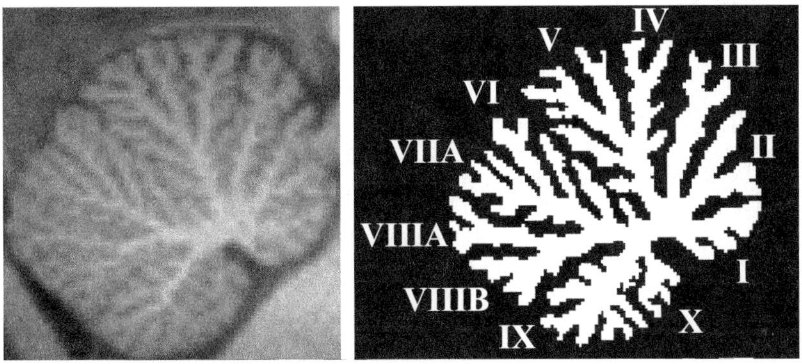

FIGURE 2. Sagittal sections through cerebellar vermis of the macaque monkey. *(Left)* Structural MRI; *(right)* cortical segmentation.

latitude and longitude,[14,15] in which nearby points on the cerebellar surface always have similar coordinates.

Macaque Cerebellum

The vastly greater complexity of cerebellar folding in the macaque than in the mouse is obvious in a midline sagittal view of the structural MRI (FIG. 2, left). I assessed the fidelity of the structural MRI by making comparisons to sagittal sections of adult macaque cerebellum shown in tracings by Larsell and Jansen[16] (their Figs. 176–177). A comparable number of distinct folia (~90–100) can be discerned in sagittal slices from both data sets. This implies that the structural MRI data used here reveals nearly all of the cerebellar convolutions, at least along the midline. In the segmented volume (FIG. 2, right), all of the lobules and major lamellae are successfully represented, as are a majority of the individual folia. However, about a third of the folia are not explicitly captured in the segmentation, owing to their small size combined with image noise and low contrast. In more lateral regions (not shown in FIGURE 1), the detailed structure of the flocculus was difficult to discern, and it was not included in the segmentation.

FIGURE 3 shows the macaque cerebellar surface reconstruction, with fiducial, inflated, and spherical configurations displayed from posterior views on the left and lateral views on the right. The total surface area of the fiducial configuration is 61 cm^2, which is 25% of total cerebral cortex for the left and right hemispheres (248 cm^2, respectively) and is close to the value of 81 cm^2 for the macaque reported by Sultan and Braitenberg.[5] It is almost 80-fold greater than the surface area of the mouse atlas cerebellum.

Compared to the mouse, not only are the convolutions of macaque cerebellar cortex deeper and more numerous, but also the overall pattern in the hemispheric regions is more complex. About 65% of the fiducial surface area is associated with the two hemispheres versus 33% for the vermis and 2.4% for the portion of the paraflocculus included in the segmentation. The hemispheric portions of lobules VII and VIII have a disproportionately large surface area, and they form a ribbon that curls

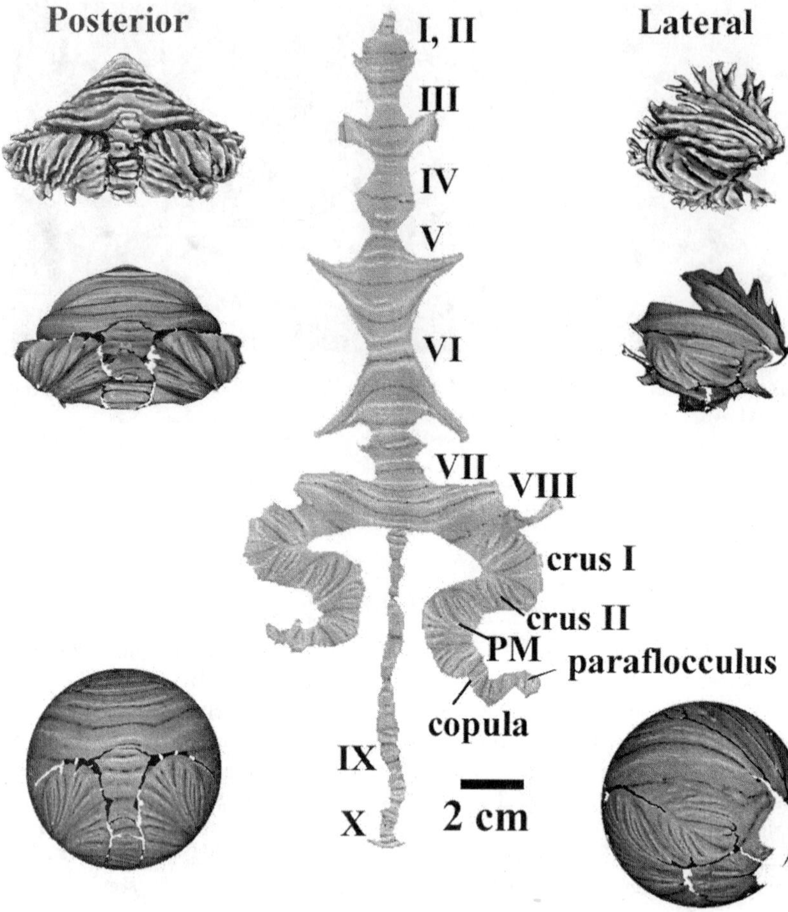

FIGURE 3. Surface representations of the macaque cerebellar cortex. (*Left*) Posterior views of the fiducial, inflated, and spherical surface. (*Right*) Lateral views of the same three configurations. (*Center*) Cerebellar flat map.

from lateral back towards the midline, ventral to the shelf formed by lobule VI. Along the boundary between the posterior vermis and lobule VIII, the cortex is not physically continuous. Instead, there are natural terminations, evident in stained sections from a macaque atlas,[17] although they are not directly discernible in the structural MRI data. The cuts used to reduce distortions in the flat map were made along the estimated trajectory of these natural cortical boundaries. To avoid severe distortions, it was necessary to make many additional cuts that interrupted genuinely contiguous cortical regions, as can be seen in the inflated and spherical maps.

Because the macaque vermis is relatively narrow (4–6 mm) but has deep and complex folds (cf. FIG. 2), it forms a highly elongated strip that runs vertically on the flat map. In the upper part of the map, the hemispheric portions of lobules I–VI appear as stubby flanges attached to the vermis, reflecting the fact that lobules I–VI become progressively shallower and simpler in more lateral portions of the fiducial surface.

In contrast, the hemispheric portions of lobules VII and VIII form prominent "legs" whose elongated but bowed shape largely reflects the ribbon-like configuration of this region in the fiducial surface.

Human Cerebellum

In a sagittal view, the structural MRI of the human cerebellar vermis (FIG. 4, left) shows about the same apparent degree of complexity as shown in the corresponding slice through the macaque cerebellum (cf. FIG. 2) in terms of the number of lobules and lamellae. However, the overall image quality is somewhat lower for the human MRI data set compared to the macaque, particularly in the ventral cerebellum (see Methods). In reality, the human vermis is substantially more convoluted than that in the macaque, as sagittal section tracings [18,19] reveal about twice as many small folia visible on each of the lobules as in the macaque. Because of the lower image quality, the segmented human cerebellum (FIG. 4, right) certainly missed most of these smaller folia, but it successfully represents all lobules, most or all lamellae, and many individual folia. The lobules were identified using published atlases.[5,18]

FIGURE 5 shows the human cerebellar surface reconstruction, with fiducial, inflated, and spherical configurations shown from posterior views on the left and lateral views on the right. The total surface area of the fiducial configuration is 540 cm^2. This is about ninefold greater than the macaque cerebellum and is 28% of the total cerebral cortex for the left and right hemispheres combined (1,900 cm^2). This is certainly a substantial underestimate because, as just noted, many cerebellar folia were missed in the segmentation. The larger estimate of 1,128 cm^2 for human cerebellum reported by Sultan and Braitenberg[6] is likely to be more accurate. Not surprisingly, the spherical map is severely compressed in some regions, expanded in others, and anisotropically stretched in others. However, the fact that it preserves the topology

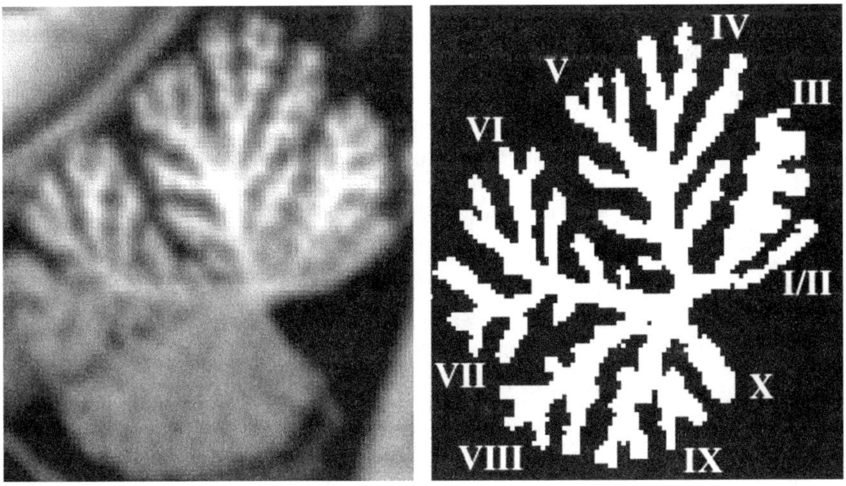

FIGURE 4. Sagittal sections through the human cerebellar vermis. (*Left*) Structural MRI; (*right*) cortical segmentation.

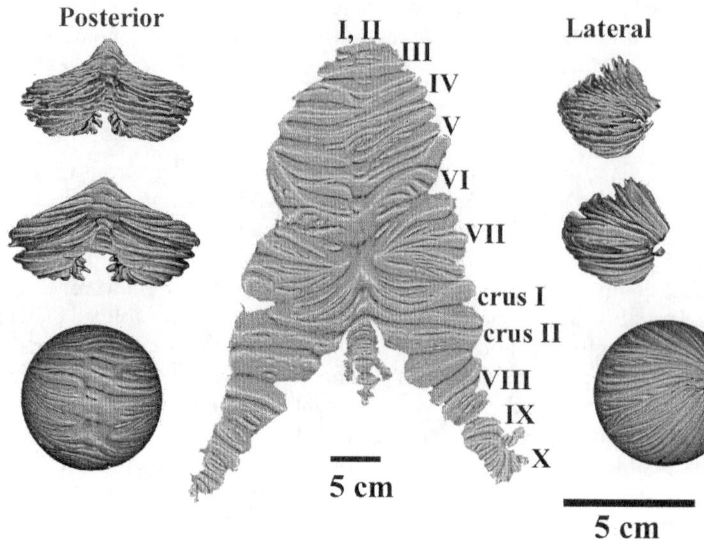

FIGURE 5. Surface representations of the human cerebellar cortex. (*Left*) Posterior views of the fiducial, inflated, and spherical surface. (*Right*) Lateral views of the same three configurations. (*Center*) Cerebellar flat map.

of the cortical sheet gives it utility as a substrate for surface-based coordinates and surface-based registration.

In comparing human and macaque cerebellum, the most striking difference is the greater elaboration of hemispheric cortex in all lobules, with an estimated 88% of human cerebellar cortex identified as hemispheric in the fiducial configuration. Also, the representation of hemispheric portions of all lobules is more balanced than in the macaque. In the more anterior lobules (I–III), this arises because the hemispheric regions are relatively wide and retain about the same number and depth of lamellae over most of their width. For the more posterior lobules, the expansion reflects convolutions that are markedly deeper and more numerous. These characteristics result in a human cerebellar flat map that is relatively wide throughout and is not nearly as elongated as in the macaque or as in the schematic unfolding shown by Sultan and Braitenberg.[6]

DISCUSSION

Cerebellar Geometry and Development

The distinctive surface geometry of cerebellar cortex reflects several key aspects of cerebellar morphogenesis. The cerebellum receives its full complement of Purkinje cells very early in development, before extensive convolutions have formed. The immature Purkinje cells are initially stacked many cells deep, especially in the hemispheric regions of the cerebellar plate.[19] The vastly more numerous granule cells proliferate extensively as they migrate tangentially along the external granule cell layer. The granule cells extend parallel fiber axons before diving down to the granule cell layer, thereby establishing an axis that is correlated with the pattern of folding and anisotropic growth in surface area.[20] An attractive hypothesis is that mechanical tension along parallel fibers is responsible for both the anisotropic growth (preferential expansion along the axis orthogonal to parallel fibers) and the axis of folding parallel to that of the parallel fibers.[21]

Cerebral cortex has a very different surface geometry that reflects its very different pattern of morphogenesis. For cerebral cortex, a thin, smooth, balloon-like cortical sheet forms early in development as a result of cell proliferation in the ventricular zone and migration along the radial axis. The increases in cerebral cortical surface area arise mainly from cell growth rather than differential migration. The formation of cortical folds is probably caused by mechanical tension along long-distance corticocortical connections as they are established.[21] Because the convolutions arise, in effect, from crumpling of a quasispherical cortical balloon, the distortions are relatively modest during the reverse process of smoothing out the convoluted adult cerebral cortex.

Surface-Based Cerebellar Atlases

Brain atlases provide natural gateways for navigating and visualizing an immense range of neuroscientific data.[8,11,22] For cortical structures, surface-based atlases are especially valuable, because they provide flexible visualization options, plus they can be used for registering data across individuals and across species, and as a substrate for surface-based coordinates that respect the topology of the cortical sheet.[3,7,8,15]

The maps of cerebellar cortex presented here are intended for general use as surface-based atlases. The atlas data sets for all three species are freely available, along with associated surface visualization software.[26] The structural volume data are also available and can be viewed with freely available volume visualization software.[23,24] Data that are currently available on the atlas (not shown here because of space constraints) include representations of cerebellar geography (individual lobules identified and painted different colors) plus exemplar functional MRI data on the human cerebellar maps.

A variety of currently available tools for data mapping should allow the amount of experimental data mapped to cerebellar atlases to increase rapidly. These include options for registering data from individual sections or from surface reconstructions of part or all of the cortex. An additional option is to map individual neuroimaging activation foci or entire fMRI volumes that are represented in stereotaxic space.

Interspecies Comparisons

The major cerebellar lobules can consistently be identified in mouse, macaque, and human, and they emerge from a common developmental pattern.[16,18] Presumably there are important functional correspondences between lobules of the same name, but the degree of correspondence is not well established in detail. Making such comparisons entails dealing with major species differences in folding patterns and in the relative sizes of different lobules and their internal subdivisions. There is also the prospect that the basic topology of cerebellar organization may differ markedly across species. The availability of spherical maps for the cerebellum should facilitate interspecies comparisons because it allows mappings that preserve the topology of the cortical sheet, analogous to those that have proven useful for interspecies comparisons of cerebral cortex.[25,26]

ACKNOWLEDGMENTS

This work was supported by a Human Brain Project/Neuroinformatics research grant funded jointly by the National Institute of Mental Health, National Science Foundation, National Cancer Institute, National Library of Medicine, and the National Aeronautics and Space Administration (R01 MH60974-06). I thank J. Harwell and D. Hanlon for software development and T. Thach for advice.

REFERENCES

1. Van Essen, D.C., J. Dickson J. Harwell, *et al.* 2001. An integrated software system for surface-based analyses of cerebral cortex. J. Am. Med. Inform. Assoc. **8:** 443–459.
2. Dale, A.M., B. Fischl & M.I. Sereno. 1999. Cortical surface-based analysis. I. Segmentation and surface reconstruction. NeuroImage **9:** 179–194.
3. Van Essen, D.C. 2002. Organization of visual areas in macaque and human cerebral cortex. *In* The Visual Neurosciences. L. Chalupa & J.S. Werner. MIT Press. In press.
4. Hurdal, M.K., P.L. Bowers, K. Stephenson, *et al.* 1999. Quasi-conformally flat mapping the human cerebellum, *In* Medical Image Computing and Computer-Assisted Intervention-MICCAI'99, Vol. 1679. Lecture Notes in Computer Science. C. Taylor & A. Colcheste, Eds. :279–286. Springer. Berlin.
5. Schmahmann, J.D., J. Doyon, D. McDonald, *et al.* 1999. Three-dimensional MRI atlas of the human cerebellum in proportional stereotaxic space. NeuroImage **10:** 233–260.
6. Sultan, F. & V. Braitenberg. 1993. Shapes and sizes of different mammalian cerebella. A study in quantitative comparative neuroanatomy. J. Hirnforsch. **34:** 79–92.
7. Van Essen, D.C. & H.A. Drury. 1997. Structural and functional analyses of human cerebral cortex using a surface-based atlas. J. Neurosci. **17:** 7079–7102.
8. Van Essen, D.C. 2002. Windows on the brain: the emerging role of atlases and databases in neuroscience. Curr. Opin. Neurobiol. **12:** 574–579.
9. Holmes, C.J., R. Hoge, I. Collins, *et al.* 1998. Enhancement of MR images using registration for signal averaging. J. Comput. Assist. Tomogr. **22:** 324–333.
10. Van Essen, D.C. & J.H.R. Maunsell. 1980. Two-dimensional maps of the cerebral cortex. J. Comp. Neurol. **191:** 255–281.
11. Van Essen, D.C., H.A. Drury, J. Harwell & D. Hanlon. 2002. CARET User's Guide and Tutorial. Part II. Surface Manipulation, Flattening, and Surface-Based Registration. Accessible at http://brainmap.wustl.edu/caret/pdf/caret_user_guide_part2.pdf
12. Drury, H.A., D.C. Van Essen, C.H. Anderson, *et al.* 1996. Computerized mappings of the cerebral cortex. A multiresolution flattening method and a surface-based coordinate system. J. Cogn. Neurosci. **8:** 1–28.
13. Paxinos, G. & K.B.J. Franklin. 2000. The Mouse Brain in Stereotaxic Coordinates, 2nd ed. Academic Press. New York.
14. Fischl, B., M.I. Sereno & A.M. Dale. 1999. Cortical surface-based analysis. I. Inflation, flattening, and a surface-based coordinate system. NeuroImage **9:** 195–207.
15. Fischl, B., M.I. Sereno, R.B. Tootell & A.M. Dale. 1999. High-resolution intersubject averaging and a coordinate system for the cortical surface. Hum. Brain Mapping **8:** 272–284.

16. LARSELL, O. & J. JANSEN. 1970. The Comparative Anatomy and Histology of the Cerebellum from Monotremes through Apes. University of Minnesota Press. Minneapolis.
17. PAXINOS, G., X.-F. HUANT & A.W. TOGA. 2000. The Rhesus Monkey Brain in Stereotaxic Coordinates. Academic Press. New York.
18. LARSELL, O. & J. JANSEN. 1972. The Comparative Anatomy and Histology of the Cerebellum. The Human Cerebellum, Cerebellar Connections, and Cerebellar Cortex. Uniersity of Minnesota Press. Minneapolis.
19. RAKIC, P. & R.L. SIDMAN. 1970. Histogenesis of cortical layers in human cerebellum, particularly the lamina dissecans. J. Comp. Neurol. **139:** 473–500.
20. SIDMAN, R.L. & P. RAKIC. 1973. Neuronal migration, with special reference to developing human brain: a review. Brain Res. **62:** 1–35.
21. VAN ESSEN, D.C. 1997. A tension-based theory of morphogenesis and compact wiring in the central nervous sytem. Nature **385:** 313–318.
22. TOGA, A.W. & P.M. THOMPSON. 2002. New approaches in brain morphometry. Am. J. Geriatr. Psychiatry **10:** 13–23.
23. VAN ESSEN, D.C., H.A. DRURY, D. HANLON & J. HARWELL. 2002. User's Guide to SureFit Cortical Segmentation and Surface Reconstruction. Accessible at http://brainmap.wustl.edu/SureFit/SureFitUsersGuide.pdf
24. COX, R.W. AFNI: Introduction, Concepts, and Principles. URL: http://afni.nimh.nih.gov/afni/edu
25. VAN ESSEN, D.C., H.A. DRURY, S. JOSHI & M.I. MILLER. 1998. Functional and structural mapping of human cerebral cortex: solutions are in the surfaces. Proc. Natl. Acad. Sci. USA **95:** 788–795.
26. VAN ESSEN, D.C., H.A. DRURY, J. HARWELL & D. HANLON. 2002. CARET User's Guide and Tutorial. Part I. Surface-Based Atlases of Cerebral and Cerebellar Cortex in Human, Macaque, and Mouse. Accessible at http://brainmap.wustl.edu/caret/pdf/caret_user_guide_part1.pdf

Plasticity of the Vertical VOR

A System Identification Approach to Localizing the Adaptive Sites

YUTAKA HIRATA[a] AND STEPHEN M. HIGHSTEIN[b]

[a]*Department of Electronic Engineering, Chubu University College of Engineering, Aichi 487-8501, Japan*

[b]*Department of Otolaryngology, Washington University School of Medicine, St. Louis, Missouri 63110, USA*

ABSTRACT: The vestibulo-ocular reflex (VOR) stabilizes the visual image on the retina during head movement by counter-rotating the eyes in the head. The VOR is under adaptive control and has been extensively used to investigate sensorimotor transformations and motor learning. The cerebellar flocculus (FL) is intimately involved in VOR motor control and learning, because its sole output, the Purkinje cell firing pattern, modulates during visual-vestibular interaction paradigms that induce motor learning, and flocculectomy impairs the ability to modify the VOR. However, the role of the FL in VOR motor learning and the possibility of another neuronal site responsible for learning have been controversial. Currently, we performed single unit recordings of FL Purkinje cells during adaptation of the vertical (V) VOR in squirrel monkeys and used a system identification approach to localize the adaptable neuronal sites responsible for VVOR motor learning. We demonstrated that there are multiple adaptive sites: one upstream or in the FL, another in the non-FL pathway; change in the former site is in the wrong direction to cause the observed VVOR gain change, whereas that in the latter site is in the correct direction. The possibility that FL Purkinje cells convey an error signal to their target neurons that can be used to aid learning is discussed.

KEYWORDS: plasticity; vertical VOR; vestibulo-ocular reflex

INTRODUCTION

The vestibulo-ocular reflex (VOR) stabilizes the visual image on the retina during head motion by counter-rotating the eyes in the head. The VOR is an adaptive control system that self-calibrates motor commands to achieve compensatory eye movements. The cerebellar flocculus (FL) is necessary for VOR adaptation, as flocculectomy precludes the adaptability of the VOR, yet its possible role or roles in adaptive VOR control have long been controversial. The FL receives multimodal sensory

Address for correspondence: Dr. Yutaka Hirata, Department of Electronic Engineering, Chubu University College of Engineering, 1200 Matsumoto-cho, Kasugai, Aichi 487-8501, Japan. Voice: +81-568-51-9476; fax: +81-568-51-1478.
yutaka@isc.chubu.ac.jp

Ann. N.Y. Acad. Sci. 978: 480–495 (2002).

inputs and transforms them into motor-related signals that will, in turn, be fed back to the FL after signal processing in vestibular pre-motor neurons. The complexity of this circuitry prevents the specification of causality among the changes in the neuronal activities observed within the circuitry and the assignment of a role to the FL in the process of the adaptation and performance of the VOR.

A system identification approach may be suitable for solving this problem. In a system identification approach, a mathematical model of the system is constructed and model parameters are determined so that the model output best fits the real system output. If the model can reproduce system behaviors, not only ones used to estimate its parameters but also ones that were not, it is likely that the model extracted the system characteristics within the dynamic range of the stimulus employed. Therefore, if we can construct a model of the VOR system in which each candidate adaptive site is explicitly described and the system is identified before and after VOR adaptation, it may be possible to pinpoint a site(s) that changed its system characteristics in parallel with VOR adaptation. That is, the neuronal site(s) that changed its system characteristics in parallel with VOR adaptation is the one responsible for the adaptation.

We currently employed a system identification approach to localize the site or sites responsible for vertical (V) VOR adaptation in squirrel monkeys and evaluated the FL role in VVOR control and motor learning. We show that there are multiple adaptive sites for VVOR adaptation. The possibility that FL Purkinje cell firing patterns can be used as an error signal to train brain-stem neurons is discussed.

METHODS

Experiment

Two adult squirrel monkeys were used. For head fixation, a stainless steel bolt was secured to the occiput. This bolt fit into a receptacle on the primate chair. A Teflon-coated stainless steel eye coil was implanted under the conjunctiva for eye movement recording with the scleral search coil method. A stainless steel recording chamber aimed at the FL was implanted and secured to the skull. All these surgical procedures and experiments were approved by the Animal Welfare and Use Committee of Washington University.

Animals were seated in a primate chair with their heads fixed in the center of a magnetic field generated by two sets of field coils driven in quadrature (FIG. 1). The primate chair was placed in the center of a white cylindrical screen 1 m in diameter (extending 36 cm above and 50 cm below the animal's head) on which black random dots were projected. This is the optokinetic stimulus (OKS). The right side of the animal was placed down so that rotation of the OKS about an earth vertical axis produced upward or downward nystagmus. The axis of the chair and the OKS rotations were aligned with the animal's interaural axis. Horizontal and vertical eye position, OKS velocity, and chair velocity were continuously digitized at a sampling frequency of 200 Hz with the use of a CED 1401 interface (Cambridge Electronic Design) for display and storage using the Spike2 program. The output of a threshold detector, signaling the occurrence of action potentials, was sampled with a resolution of 0.01 ms and stored in the same way.

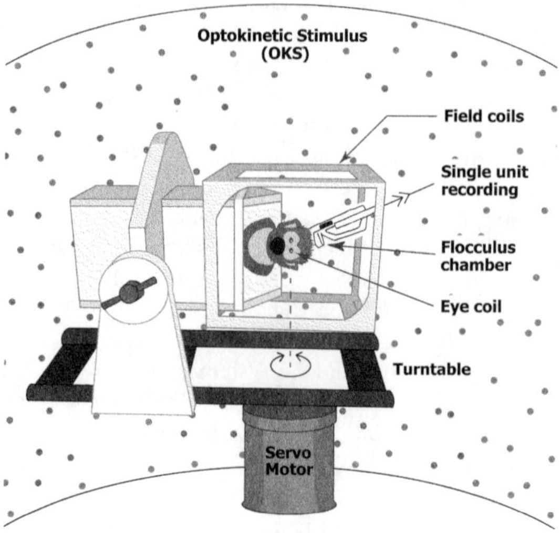

FIGURE 1. Experimental setup. The animal sitting in the primate chair with its head fixed was tilted 90 degrees with its right side down. The chair was fixed on the turntable, so that the center of the animal's semicircular canals was aligned to the center of the rotational axis of the turntable. Black random dots pattern (OKS) was projected on a white cylindrical screen surrounding the monkey at the same center as the rotational axis of the turntable. The screen is 1 m in diameter, 86 cm in width, and extends 36 cm above and 50 cm below the animal's midline.

Flocculus and ventral parafloccular vertical zone Purkinje cells were recorded at various VOR gains acutely modified by visual-vestibular mismatch stimuli (VORs or VORe at 0.5 Hz; see below). The Purkinje cells were identified by the occurrence of complex spikes and by their characteristic simple spike discharge patterns during vertical visual-vestibular interaction paradigms. Once a single unit was isolated, OKR, VOR in dark (VORd), VOR suppression (VORs), and VOR enhancement (VORe) paradigms were applied for 1 minute each. In VORs and VORe paradigms, the OKS moves in-phase and 180 degrees out of phase with head rotation, respectively. The maximum amplitude of head rotation and OKS were limited to 40°/s to assure experimentation within the linear range of the system. We used sum-of-sines (SOS) stimuli that consisted of five frequencies (0.05, 0.27, 0.5, 1.27, and 2.5 Hz) to identify the VVOR/OKR system in a wide frequency range. We also applied conventional single frequency stimuli (0.1, 0.5, and 2.5 Hz).

Data Handling

Saccades and postsaccadic drifts, if any, were desaccaded from eye movement traces by a semiautomatic desaccade program written in Matlab. Data in the same periods of time were eliminated from head, OKS, and Purkinje cell instantaneous firing rate. The VOR gain was calculated as the slope (a) of the regression, $x(t - \phi) = a\,h(t) + b$, where $x(t)$ and $h(t)$ are eye velocity and head velocity, respectively, and a, 180ϕ, and b are the VOR gain, phase at 0.5 Hz and eye velocity offset, respectively.

For performance check of the model, the desaccaded eye movement and Purkinje cell firing data were averaged over more than 14 samples measured when VOR gains were between 0.8 and 1.0 (normal VOR gain) to evaluate the general (average) performance of the model. For other analyses, each sample was used without averaging.

Model Description

Structure of the model is the same as that in a previous study[1] in which stimuli consisting of a single frequency (0.5 Hz) were applied. In the present study, we tested the performance of the model over the wider frequency range just mentioned. The model (FIG. 2B) was configured around the FL based upon the known anatomy and physiology (FIG. 2A). The entire vertical (V) VOR/OKR system was divided into a pathway that includes FL (FL pathway) and one that does not (nonFL pathway). The FL pathway was further divided into one upstream to FL including FL itself (preFL&FL pathway) and one downstream to FL (postFL pathway). The preFL&FL pathway was further divided into pathways representing the three different input modalities to FL (preFL&FL vestibular, visual, and efference copy pathways). In the same manner, the nonFL pathway was divided into nonFL vestibular and nonFL visual pathways. The system equations and transfer function of each subsystem are as follows:

$$F(s) = H(s)\, G\ vestib\ preFL\&FL\ (s) + R(s)\, G\ visual\ preFL\&FL\ (s) + E(s)\, G\ ecopy\ preFL\&FL\ (s), \quad (1)$$

$$X(s) = H(s)\, G\ vestib\ nonFL\ (s) + R(s)\, G\ visual\ nonFL\ (s) + F(s)\, G\ postFL\ (s), \quad (2)$$

where s denotes the Laplace operator, $F(s)$, $H(s)$, $R(s)$, $E(s)$, and $X(s)$ denote Laplace transforms of FL Purkinje cell firing pattern, head position, retinal slip position, and efference copy signal (eye velocity) and eye position, respectively. Each subsystem is described as the following equations[1] for determination of the equations):

preFL&FL vestibular pathway:

$$G\ vestib\ preFL\&FL\ (s) = (\alpha_h s^2 + \beta_h s) \exp(-\tau_h s) \quad (3)$$

preFL&FL visual pathway:

$$G\ visual\ preFL\&FL\ (s) = (\alpha_r s^2 + \beta_r s) \exp(-\tau_r s) \quad (4)$$

preFL&FL efference copy pathway:

$$G\ ecopy\ preFL\&FL\ (s) = \beta_e \exp(-\tau_e s) \quad (5)$$

postFL pathway:

$$G\ postFL\ (s) = \exp(-\tau_x s)/(a_x s^2 + b_x s + c) \quad (6)$$

nonFL vestibular pathway:

$$G\ \overset{vestib}{nonFL}\ (s) = (a_h s^2 + b_h s)\exp(-(\tau_x + \tau_h)s)/(a_x s^2 + b_x s + c) \quad (7)$$

nonFL visual pathway:

$$G\ \overset{visual}{nonFL}\ (s) = (a_r s^2 + b_r s)\exp(-(\tau_x + \tau_r)s)/(a_x s^2 + b_x s + c) \quad (8)$$

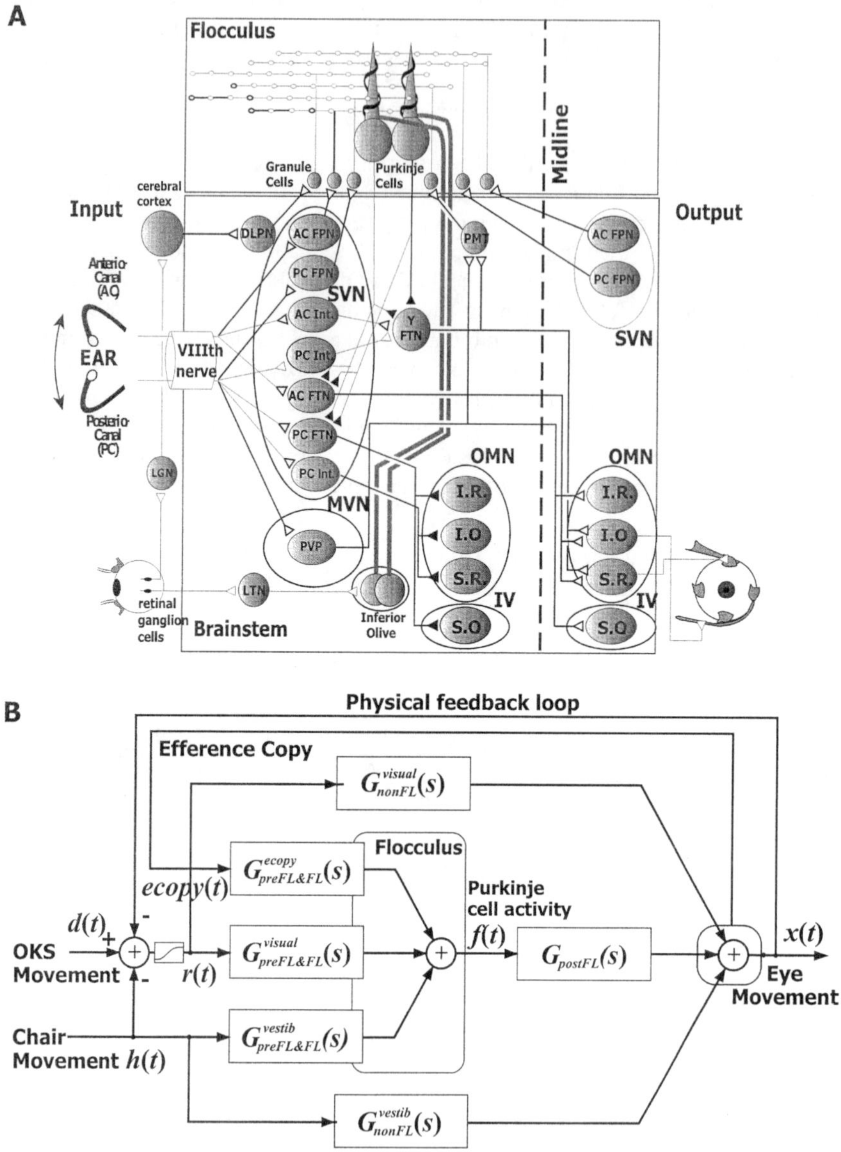

FIGURE 2. Vertical VOR and OKR neuronal circuit (**A**) and model (**B**). (**A**) FPN, floccular projecting neurons; FTN, floccular target neurons; SVN, superior vestibular nuclei; Y, dorsal Y group; Int., the interneurons in SVN projecting to Y; P, floccular Purkinje cell; g, granular cell; PVP, position vestibular pause neurons; MVN, medial vestibular nuclei; MT, middle temporal visual area; MST, medial superior temporal area; DLPN, dorsolateral pontine nuclei; OMN·IV, extraocular motor neurons; PMT, paramedian tract; LTN, lateral terminal nucleus of the accessory optic system. (**B**) *G ecopy preFL&FL(s)*, *G vestib*

The model was described in the Matlab Simulink environment for parameter estimation and simulations.

Parameter Estimation and Model Validation

To test the validity of the model over a wide frequency range, experimental data of the SOS paradigms were used. Model parameters were estimated by fitting model output (eye velocity and Purkinje cell firing rate) to the SOS experimental data averaged over more than 14 samples in response to the same stimuli. The experimental data recorded from animals whose VOR gains were in a normal range (0.8–1.0) were used to test the model performance. A nonlinear least square method (BFGS method) implemented in Matlab was used to find an optimal parameter set that minimizes the squared sum of the residual (experimental data minus model output). To avoid multicolinearity among input signals (efference copy and head velocity), first, parameters included in subsystems that are activated by visual stimuli (*G visual preFL&FL* (s), *G ecopy preFL&FL* (s), *G visual nonFL* (s), and *G postFL* (s)) were determined by using the VF paradigm. Then, in the second step, by using the VORd paradigm, those included in subsystems that are activated by vestibular stimuli (*G vestib preFL&FL* (s) and *G vestib nonFL* (s)) were determined. In the second step, parameters included in *G ecopy preFL&FL* (s) and *GpostFL* (s) that are activated during both VF and VORd paradigms were fixed at values estimated in the first step. Once all parameters were determined, eye movements and Purkinje cell firing patterns during VORs and VORe were simulated to test if the model can predict experimental data that were not used to determine the parameters.

After confirming the validity of the model at the normal VOR gain in a wide frequency range, the model parameters were estimated by using experimental data of each Purkinje cell recorded at a specific VOR gain in response to stimuli at 0.5 Hz. Length of the experimental data employed was 1 minute and data were not averaged (For more details of the parameter estimation from each Purkinje cell, see Hirata and Highstein.[1]) System gain and phase of each subsystem at 0.5 Hz at each VOR gain were then calculated from the estimated model parameters using Equations 3 to 8.

RESULTS

A total of 53 FL and ventral paraflocculus Purkinje cells were well isolated at 78 VOR gains between 0.4 and 1.5 from two monkeys. Among them, the SOS para-

---

preFL&FL(s), and *G visual preFL&FL*(s) are pre-floccular/floccular subsystems, each of which represents a transfer function of pre-floccular/floccular efference copy pathway, vestibular pathway, and visual pathway, respectively. *G postFL*(s) represents a transfer function of postfloccular pathway, which transfers the Purkinje cell activity to a part of the eye movement. *G visual nonFL*(s) and *G vestib nonFL*(s) represent transfer functions of nonfloccular visual and vestibular pathways, respectively. Corresponding neuronal circuit to *G visual nonFL*(s) is not shown in **A**. $h(t)$, head position; $d(t)$, OKS position; $r(t)$, retinal slip position; $f(t)$, Purkinje cell simple spike firing rate; $ecopy(t)$, efference copy signal (eye velocity); $x(t)$, eye position.

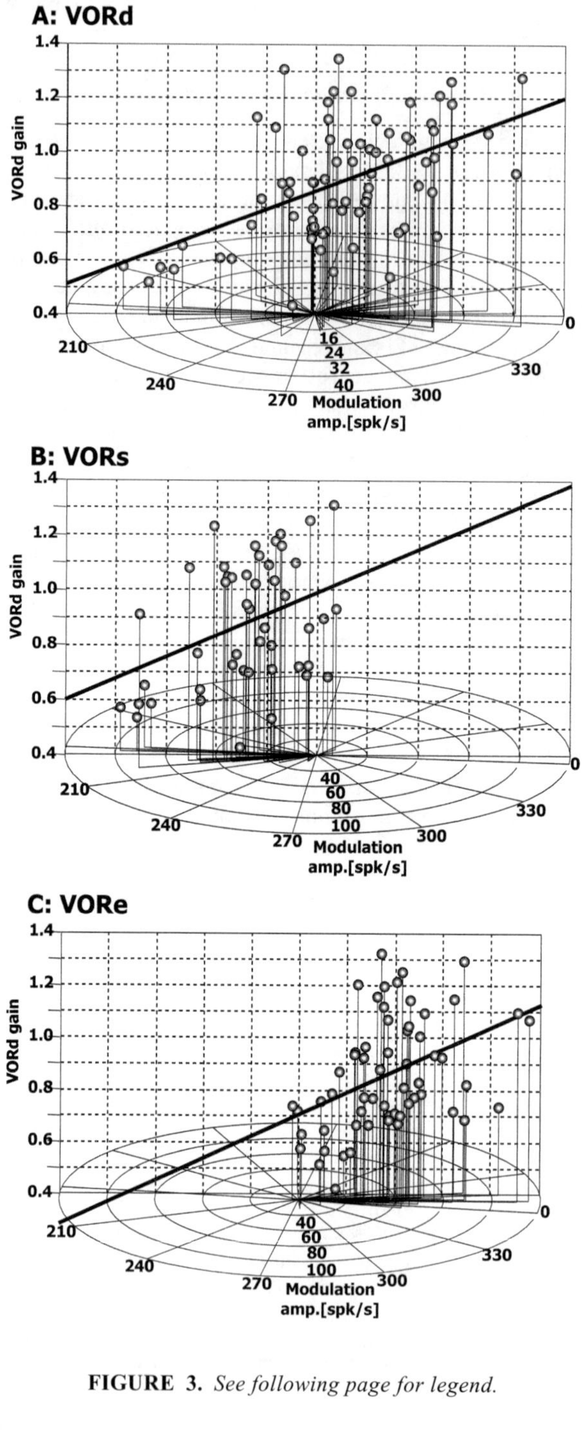

FIGURE 3. *See following page for legend.*

digms were applied to 28 cells recorded at the normal VOR gains (0.8–1.0). VOR gain means vertical VOR gain in what follows, unless otherwise specified.

Flocculus Purkinje Cell Firing Modulation

FIGURE 3 illustrates VOR gain versus amplitude and phase of Purkinje cell firing modulation during VORd (A), VORs (B), and VORe (C) paradigms in a three-dimensional polar coordination. The z axis indicates VOR gain at which a Purkinje cell was recorded, and the radius and angle denote amplitude and phase (regarding head velocity) of the Purkinje cell firing modulation. Each black circle represents data from each cell recorded at a VOR gain. Note that the Purkinje cell modulation amplitudes during VORd are much smaller than those during VORs and VORe. The heavy black line in each panel is the regression line that corresponds to the first principle component. The slopes of the regression line against the VOR gain in the three paradigms are significantly different from 0, indicating that Purkinje cells' firing modulation changed in parallel with VOR gain in these paradigms. During the VORs paradigm that leads VOR gain to decrease, amplitude of the Purkinje cell modulation out of phase with head velocity increased when VOR gain decreased. On the other hand, during the VORe paradigm that causes an increase in VOR gain, the Purkinje cell modulation amplitude in phase with head velocity increased when VOR gain increased. The in-phase and out-of-phase Purkinje cell modulations with head velocity during VORe and VORs paradigm are causal in generation of the suppressed and enhanced VOR, respectively. Direction of the change in the VORd paradigm is causal to induce the observed VOR adaptation. Namely, the increment in Purkinje cell modulation in-phase with head velocity observed at higher VOR gains can cause an increment of 180 degrees out-of-phase modulation in their target neurons that, in turn, increases the amplitude of eye velocity out of phase with head velocity. However, these changes in FL Purkinje cell modulation do not necessarily mean that FL is the adaptive site. There are several possibilities that could cause the changes in FL Purkinje cell firing modulation: (1) change in *G vestib preFL&FL* (*s*), (2) change in *G postFL* (*s*), (3) change in *G ecopy preFL&FL* (*s*), (4) change in *G vestib nonFL* (*s*), or (5) any combinations of these. The following system identification approach makes it possible to be more specific about the choice among these possibilities.

Model Validation

Before applying the model to the system identification problem, we confirmed its validity by comparing its behavior with experimental data. FIGURE 4 compares model behaviors with experimental eye velocity (upper panel) and FL Purkinje cell firing pattern (lower panel) during the OKR (A) and VORd paradigms (B) with SOS. The experimental data are averages of more than 14 samples. Model parameters were ad-

FIGURE 3. Three-dimensional polar diagram of Purkinje cell firing pattern during VORd (**A**), VORs (**B**), and VORe (**C**) recorded at different VOR gains. Each plot represents each cell. The *x-y* plane represents amplitude (radius) and phase (angle) of the Purkinje cell firing modulation in a polar coordination. The *z* axis represents VOR gains at which the cells were recorded. The *heavy black line* is a regression line (1st principle component), which is in the plane (*dotted line*) perpendicular to the polar plane.

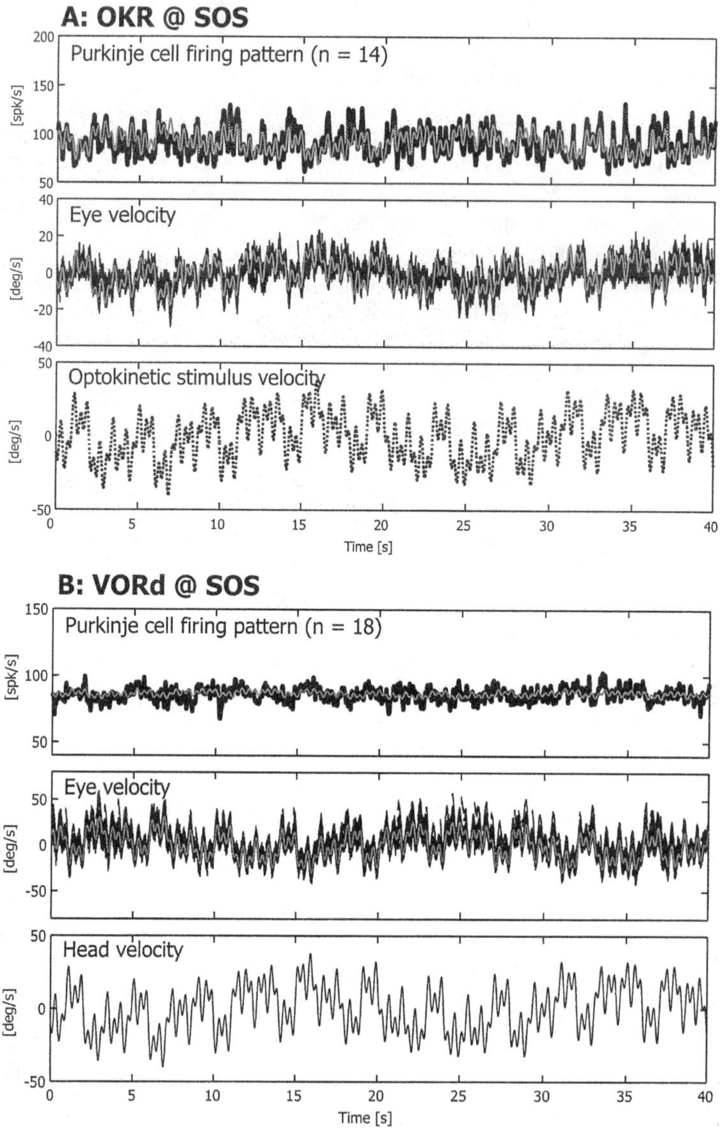

FIGURE 4. Model performance check using OKR (**A**) and VORd paradigm (**B**) with the sum-of-sines (SOS) stimulus (*bottom panels*). Model parameters were adjusted so that model outputs (*gray traces*) fit to experimental Purkinje cell firing pattern (*top panel, black trace*) and eye velocity (*middle panel, black trace*) simultaneously. Experimental data are averages of *n* samples (cells) recorded under the same condition.

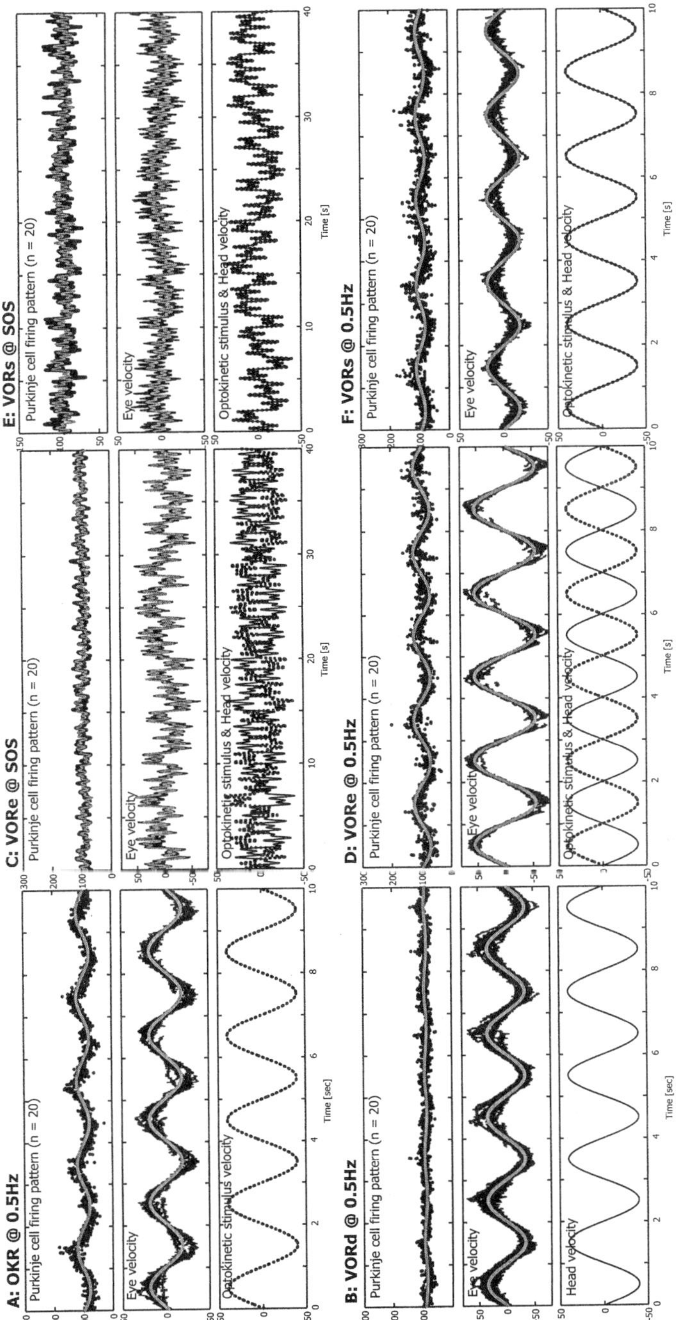

FIGURE 5. Model performance check using OKR at 0.5 Hz (**A**), VORd at 0.5 Hz (**B**), VORe at 0.5 Hz, and SOS (**C,D**), and VORs at 0.5 Hz and SOS (**E,F**). Formats are the same as those in FIGURE 4 except that all experimental data are superimposed rather than showing their average. The same model with the parameters estimated in FIGURE 4 was used to simulate all the paradigms.

justed to fit model outputs to the experimental eye velocity and Purkinje cell firing pattern. The OKR paradigm was used to determine parameters in *G ecopy preFL&FL(s)*, *G visual preFL&FL (s)*, *G visual nonFL (s)*, and *G postFL (s)*, whereas the VORd paradigm was used to determine those in *G vestib preFL&FL(s)* and *G vestib nonFL(s)*. Note that the model outputs fit well the experimental data in both A and B, indicating that the model is capable of reproducing the average characteristics of these experimental data in response to a separate visual (OKR) or vestibular stimulus (VORd).

FIGURE 5 illustrates results of model prediction of experimental eye velocity (upper panel) and FL Purkinje cell firing pattern (lower panel) recorded during the OKR and VORd at 0.5 Hz, VORs and VORe of SOS, and those at 0.5 Hz. More than 20 samples are superimposed in each trace to show variability. Note that the model predictions fit the experimental eye velocities and Purkinje cell firing patterns very well in all paradigms, indicating that the model can predict average eye velocities and Purkinje cell firing patterns during these visual-vestibular interaction paradigms that were not used in the parameter estimation. The results in FIGURES 4 and 5 assure that the model is valid within the frequency (0.05–2.5 Hz) and amplitude range (~40°/s) currently employed.

Identified Characteristics of Each Subsystem

We applied this model to data of each Purkinje cell recorded at each VOR gain and calculated the characteristics of subsystems (gain and phase) at each VOR gain using the estimated model parameters.

FIGURE 6 illustrates VOR gain (*z* axis) versus gain (radius) and phase (angle) characteristics of each subsystem at 0.5 Hz in a three-dimensional polar coordination identified by the system identification approach. Each black circle represents an estimate from each Purkinje cell recorded at a particular VOR gain. The heavy black line is a regression line (the first principle component) representing the average characteristics of the system in relation to VOR gain. It is in the plane formed by dotted lines that is perpendicular to the polar plane and parallel to the paper. It was found that the system characteristics of the preFL&FL vestibular pathway *G vestib preFL&FL(s)* and nonFL vestibular pathway *G vestib nonFL(s)* changed in parallel with VOR gain (slopes of the regression lines are significantly different from 0; $p < 0.0075$ [$F_0 = 7.8635$], $p < 0.000011$ [$F_0 = 24.5431$], respectively), whereas the postFL pathway G_{postFL}, preFL&FL efference copy pathway *G ecopy preFL&FL(s)*, visual pathway *G visual preFL&FL(s)*, and nonFL visual pathway *G visual nonFL(s)* did not show significant changes ($p > 0.804$ [$F_0 = 0.0621$], $p > 0.4380$ [$F_0 = 0.6125$], $p > 0.2026$ [F0 = 1.6723], $p > 0.273$ [$F_0 = 1.2290$], respectively). These results suggest

---

FIGURE 6. Three-dimensional polar representation of VOR gain versus system characteristics at 0.5 Hz of each subsystem. Each *dot* represents an estimate from each Purkinje cell. Length of the *thin black line* under each dot represents gain of the VOR at which the cell was recorded. The intersection of the base of each thin vertical line with the polar plane indicates the estimated gain (radius) and phase (angle) of the system. The *heavy black line* is a regression line (1st principle component), which is in the plane (*dotted line*) perpendicular to the polar plane. *Asterisks* indicate that the slope of the regression line is statistically different from 0 ($p < 0.0075$).

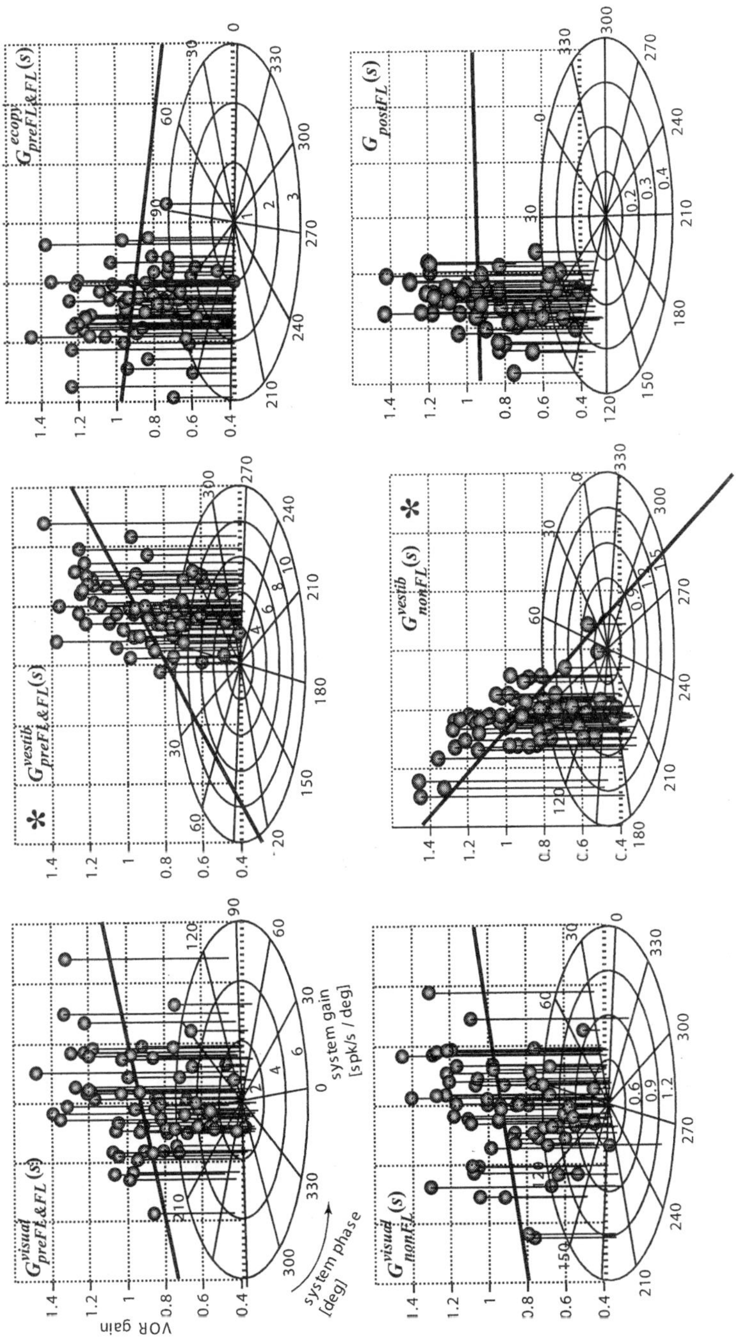

FIGURE 6. *See previous page for legend.*

that there are multiple sites of adaptation for VVOR motor learning: One is in the preFL&FL vestibular pathway *G vestib preFL&FL*(*s*) and the other is in the nonFL vestibular pathway *G vestib nonFL*(*s*).

DISCUSSION

System Identification Approach

A system identification approach to study biological systems was pioneered by N. Wiener and was first applied by Stark and his colleagues[2,3] to the human pupillary light reflex system. A number of studies have applied the technique to various biological systems since then. The VOR system is one of the most well understood biological systems in terms of its anatomy and physiology. In fact, a number of mathematical models of the VOR system have been constructed.[4–11] However, no attempt was made to apply the system identification technique and identify model characteristics before and after VOR adaptation to localize the neuronal sites responsible for the motor learning. One reason for this may be the difficulties in determining the parameters of an anatomically plausible model by using actual experimental data. To identify model parameters describing the characteristics of each of the candidates of the adaptive sites, in principle, we need to know the input and output signals of the candidates simultaneously. In the case of the VOR system, this requires simultaneous single or multiunit recordings from multiple neuronal sites. In the present study, we designed a method that makes it possible to identify all model parameters from visual and vestibular sensory input signals, motor output signals, and FL Purkinje cell firing patterns by configuring the model structure appropriately. Due to the model configuration, it was not possible to pinpoint a cell type or a specific type of synaptic connection responsible for VOR adaptation, but it was possible to localize the adaptive sites in either preFL, postFL, or nonFL pathways.

Neuronal Sites Responsible for Vertical Vestibulo-ocular Reflex Adaptation

The present system identification approach revealed multiple neuronal sites responsible for VVOR adaptation: at least one in the FL or upstream from the FL (*G vestib preFL&FL*(*s*)) and another in a non-FL vestibular pathway (*G vestib nonFL*(*s*)). Our approach is not biased in that unlike most model-based studies in which the adaptable elements were presumed and the model parameters were determined heuristically, we did not make any presumption on the adaptability of any subsystem in the model and all of the subsystems were free to change their system characteristics. Included in the preFL&FL vestibular pathway *G vestib preFL&FL*(*s*) are the vestibular end organs (anterior and posterior semicircular canals), floccular projecting neurons (FPNs), and floccular cortical neurons. Miles *et al.*[12] have shown that activities of the vestibular end organ did not change following VOR adaptation. FPNs express pure vestibular sensitivity at 0.5 Hz and do not seem to receive any type of error signals that can teach them if the current VOR performance is not adequate.[13] Therefore, it is unlikely that the vestibular end organs and FPNs are adaptive elements. The remaining candidates in the pathway are FL cortical neurons. Several types of plasticity found in the cerebellar cortex including LTD, LTP, and rebound potentiation could be underlying mechanisms responsible for the changes in the system characteristics of the *G vestib preFL&FL*(*s*). Included in the nonFL vestibular pathway are the vestibular end organs, the position-vestibular-pause neurons (PVPs) in medial vestibular nucleus (MVN), floccular target neurons (FTNs) and interneurons in the superior vestibular nucleus (SVN), Y group neurons, oculomotor motor neurons, and the oculomotor muscle plant. Because OKR eye movements did not change after VOR gain change, it is unlikely that the motor neurons and muscle plant changed their characteristics. Lisberger *et al.*[14] demonstrated that PVP activity did not change significantly after horizontal VOR adaptation. As for the inter neurons in the SVN, the same discussion is possible as for the FPNs. Therefore, the remaining candidates are Y group neurons and FTNs. Several lines of experiment have demonstrated that these neurons change their firing patterns after VOR adaptation,[15,16] supporting the current result.

Role of Flocculus

A proposed role of FL Purkinje cells is the encoding of a dynamic component of motor commands to achieve the desired eye movements by counteracting the viscous and inertial components of the oculomotor plant.[17,18] The theory has been proved to work in the case of the ocular following response (OFR). In other words, Purkinje cell firing patterns during the OFR can be reconstructed by a linear combination of eye position, velocity, and acceleration, an inverse dynamic model. If this is true, the same inverse dynamic model should work for other paradigms such as VOR and visual-vestibular interactions. We have shown that the same type of inverse dynamic model can reconstruct Purkinje cell firing patterns during these paradigms, but it requires a different set of parameters for each paradigm and a different set of parameters at different VOR gains.[19,20] This result suggests that FL Purkinje cells encode something else in the process of the VOR adaptation as predicted by the feedback error learning scheme.[21]

For Y group neurons and FTNs to learn, an error signal must be available. A proposed theory is that such a signal is transferred from FL Purkinje cells.[7,22,23] To test this theory, we analyzed Purkinje cell firing modulation during VORs and VORe paradigms that induce low and high VOR gain adaptation, respectively, to see if Purkinje cell activity conveys information useful for Y group neurons and FTNs to learn. An error signal must contain at least information about which direction VOR gain should change and, preferably, how much more VOR gain should change in the given visual-vestibular mismatch paradigm. We demonstrated that the direction of the Purkinje cell modulation is opposite in VORs and VORe paradigms that lead VOR gains in the opposite directions, and further, that the modulation amplitudes changed in parallel with the VOR gain changes. Therefore, Purkinje cell activity might possibly be an error signal that teaches target neurons which direction and how much they should change to achieve appropriate VOR gains in a given visual-vestibular mismatch environment.

ACKNOWLEDGMENT

We thank Mr. Ichiro Takeuchi for his help in model simulations, Dr. Pablo Blazquez for his discussions on this material, and Pat Keller for her animal care. This work was supported by National Institutes of Health Research Grants EY-05433 (S.M.H.), grants from the Ministry of Education, Culture, Sports, Science and Technology, Japan (Y.H.), and Japan Science and Technology Corporation overseas research fellowship (Y.H.).

REFERENCES

1. Hirata, Y. & S.M. Highstein. 2001. Acute adaptation of the vestibule ocular reflex: signal processing by floccular and ventral parafloccular Purkinje cells. J. Neurophysiol. **85:** 2267–2288.
2. Stark, L. & P.M. Sherman. 1957. A servoanalytic study of consensual pupil reflex to light. J. Neurophysiol. **20:**17–26.
3. Stark, L. 1984. The pupil as a paradigm for neurological control systems. IEEE Trans. Biomed. Eng. **31:** 919–924.
4. Robinson, D.A. 1981. The use of control systems analysis in the neurophysiology of eye movements. Annu. Rev. Neurosci. **4:** 463–503.
5. Galiana, H.L. & J.S. Outerbridge. 1984. A bilateral model for central neural pathways in the vestibulo-ocular reflex. J. Neurophysiol. **51:** 210–241.
6. Gomi, H. & M. Kawato. 1992. Adaptive feedback control models of the vestibulocerebellum and spinocerebellum. Biol. Cybern. **68:** 105–114.
7. Miles, F.A. & S.G. Lisberger. 1981. Plasticity in the vestibulo-ocular reflex: a new hypothesis. Annu. Rev. Neurosci. **4:** 273–299.
8. Lisberger, S.G. & T.J. Sejnowski. 1992. Motor learning in a recurrent network model based on the vestibulo-ocular reflex. Nature **360:** 159–161.
9. Lisberger, S.G. 1994. Neural basis for motor learning in the vestibuloocular reflex of primates. III. Computational and behavioral analysis of the sites of learning. J. Neurophysiol. **72:** 974-998.
10. Quinn, K.J., A.J. Didier, J.F. Baker & B.W. Peterson. 1998. Modeling learning in brain stem and cerebellar sites responsible for VOR plasticity. Brain Res. Bull. **46:** 333–346.
11. Tabata, H., K. Yamamoto & M. Kawato. 2002. Computational study on monkey VOR adaptation and smooth pursuit based on the parallel control-pathway theory. J. Neurophysiol. **87:** 2176–2189.
12. Miles, F.A. & D.J. Braitman. 1980. Long-term adaptive changes in primate vestibuloocular reflex. II. Electrophysiological observations on semicircular canal primary afferents. J. Neurophysiol. **43:** 1426–1436.
13. Zhang, Y., A.M. Partsalis & S.M. Highstein. 1993. Properties of superior vestibular nucleus neurons projecting to the cerebellar floccular in the squirrel monkey. J. Neurophysiol. **69:** 642–645.
14. Lisberger, S.G., T.A. Pavelko & D.M. Broussard. 1994. Neural basis for motor learning in the vestibulo-ocular reflex of primates. I. Changes in the responses of brain stem neurons. J. Neurophysiol. **72:** 928–953.
15. Lisberger, S.G., T.A. Pavelko, H.M. Bronte-stewart & L.S. Stone. 1994. Neural basis for motor learning in the vestibuloocular reflex of primates. II. Changes in the responses of horizontal gaze velocity Purkinje cells in the cerebellar floccular and ventral parafloccular. J. Neurophysiol. **72:** 954–973.
16. Partsalis, A.M., Y. Zhang & S.M. Highstein. 1995. Dorsal Y group in the squirrel monkey. I. Neuronal responses during rapid and long-term modifications of the vertical VOR. J. Neurophysiol. **73:** 615–631.
17. Shidara, M., K. Kawano, H. Gomi & M. Kawato. 1993. Inverse dynamics model eye movement control by Purkinje cells in the cerebellum. Nature **365:** 50–52.
18. Gomi, H., M. Shidara, A. Takemura, *et al.* 1998. Temporal firing patterns of Purkinje cells in the cerebellar ventral paraflocculus during ocular following responses in monkeys. I. Simple spikes. J. Neurophysiol. **80:** 818–831.
19. Hirata, Y. & S.M. Highstein. 2001. Evaluation of possible error signals used in vertical vestibuloocular reflex adaptation in squirrel monkeys. Soc. Neurosci. Abstr. **27:** CD-ROM.

20. HIRATA, Y., P.M. BLAZQUEZ, Y. TASAKI & S.M. HIGHSTEIN. 2002. Analysis of cerebellar flocculus Purkinje cell firing patterns during visual-vestibular interactions before and in the process of VOR motor learning. Proc. IEEE EMBS. In press.
21. KAWATO, M. & H. GOMI. 1992. The cerebellum and VOR/OKR learning models. Trends Neurosci. **15:** 445–453.
22. PARTSALIS, A.M., Y. ZHANG & S.M. HIGHSTEIN. 1995. Dorsal Y group in the squirrel monkey. II. Contribution of the cerebellar floccular to neuronal responses in normal and adapted animals. J. Neurophysiol. **73:** 632–649.
23. HIGHSTEIN, S.M., A. PARTSALIS & R. ARIKAN. 1997. Role of Y-group of the vestibular nuclei and flocculus of the cerebellum in motor learning of the vertical vestibulo-ocular reflex. *In* Progress in Brain Research. C.I. de Zeeuw, P. Strata & J. Voogd, Eds. :383–397. Elsevier Science. Amsterdam.

Models of Cerebellar Function

Discussion

STEPHEN G. LISBERGER AND LANCE M. OPTICAN, *Rapporteurs*

STEVE LISBERGER (*University of California at San Francisco*): As I see it, we have had four different talks that use modeling for four different reasons. Each of the talks was quite detailed and I wouldn't be surprised if some of the audience had missed a few points. I certainly missed some of the fine points, and so I thought I would start this by asking the speakers to answer a more general question and tell us (1) why they made their model and what they were trying to learn, and (2) not how their model succeeded, but rather how it failed, because it is from the failures of our models that we learn. This may not be unanimously the case and I would be happy to answer this question too, but I don't think that's my place here. Let us start with that line of thinking and see whether it provokes some questions from the audience to clarify things.

MIKE MAUK (*University of Texas Medical School at Houston, Houston, TX*): I do modeling because it helps me think about things. So I'd say that first and foremost I really do believe that since the brain processes information, the level at which we need to understand it is at the level of information processing. Trying to build computer versions of our circuits is the most direct way to address that.

LISBERGER: So how does your model fail?

MAUK: In so many ways that I can't name them! But this is where Medina's work with me was so brilliant. In a nutshell, we went from a computer simulation that couldn't do eyelid conditioning to one that could. When we stopped doing what was intuitive to me, which was to try to keep the model focused on the things that we thought might be important for acquisition, it started to work. Medina said "Let's put it all in," including things that might be important for extinction, and it began to work. A theme that has emerged since is that even when we are training the animal, extinction processes play an important role in things like timing and learning and maintaining the response so that things don't saturate. I guess the idea was that I had intuitions about what could be excluded and they were all wrong.

RON KETTNER (*Northwestern University, Chicago, IL*): I would agree that ideally we have to explain everything. One of the things that I was most focused on is that in the pursuit literature you have a phenomenon that develops in two or three cycles. We were seeing a fairly low learning rate, and actually it turned out that the monkeys were also learning slowly, at least for some tasks, so another concern that I have concerns the initiation of pursuit that was brought up. Obviously what is happening is that lots of systems are interacting to generate the motor output, and I am focused on one. I don't want to exclude the others, but I would like to work in the unipolar brush cells—one holy grail is to try to understand what the glomeruli are doing. We think

Ann. N.Y. Acad. Sci. 978: 496–502 (2002).

we have some idea of what this is. A key problem is the notion that the inferior olive is a poor training signal, and so we are trying to think of ways in which we can understand why it is so slow. Since we have some ideas there, it is the tip of the iceberg.

LISBERGER: If he were here, Bower would probably say that it seemed a little arbitrary to put the Golgi cells in, but not the other stuff.

KETTNER: I agree. My feeling is that it really keeps you honest if you have an idea about how some mechanism might work. The first step is to just try to see if you can get things to work at all. Then you test and go back and forth. So my philosophy is to add only things that I need.

LISBERGER: I don't want to get too much into specifics, but how would your model have failed if the Golgi cells weren't there?

KETTNER: It just doesn't work. In order to get these very complicated patterns you need to have some sort of fairly simple mechanism. You just have the receptive fields, and if you don't have the Golgi cells, you don't have finite receptive fields, because the mossy-fiber activation is very broad.

LISBERGER: I think you are saying that to get your model to work you needed that sparse coding, and that this was a way to get it. I'm just trying to get you to say that in a way that's clearly understood.

KETTNER: An approach to take is that if you want to be devil's advocate on the other side, you pick and choose from the myriad mechanisms. You try not to be too selective, and you try to account for each of the major findings. I'll admit right off that I don't have all the channels and all the plasticities at all the different synapses. I am working with Medani and Slater to try to improve my knowledge of these cellular mechanisms, but we clearly do pick and choose.

LISBERGER: Dr. Yamamoto, do you want to take on the same question: What are you trying to accomplish with your model and how does it fail?

KENJI YAMAMOTO (*Neuroscience Research Institute, Ibaraki, Japan*): I am a physiologist and only use the results of physiological experiments. Sometimes we cannot say what is real. To connect those results I use a model. If I only use the results of experimental data, and generally if there is no mistake about the experimental data, I believe the model can be right.

LISBERGER: To summarize before I go to Dr. Hirata, I know Mauk's model has failed many ways and that he has conducted a lot of experiments to test its predictions. I think Yamamoto's data predicted that if you have a monkey go through directional learning in ocular following, the direction tuning of the Purkinje cells should change. Am I correct that that is a prediction of your model?

YAMAMOTO: Yes.

LISBERGER: So, if your model is right, then that should happen. That would be a beautiful demonstration and to my mind it would carry the modeling to the next level, and show the real strength of your model, which is predicting something that you didn't already know. I wanted to comment that in order to understand where this kind of thing can go, and I have no doubt, since you are working with Kawano, that that is the kind of experiment that Kawano would ultimately do. Finally, Dr. Hirata, your modeling is for a slightly different purpose, but it makes the answer to the question a little more difficult.

YUTAKA HIRATA (*Chubu University College of Engineering, Aichi, Japan*): Basically, we made a model to pinpoint where the VOR adaptation occurs. This kind of approach has been used in other biological systems, so we just applied that method to a problem that has been controversial in this field for a long time. I think so far our model works fine.

JOSÉ DELGADO-GARCÍA (*Universitdad Pablo de Olavide, Seville, Spain*): I have a question for Dr. Mauk. I liked your representation of what is going on in the cerebellum during classic conditioning, but this is, let's say, only half of the truth. Obviously you have Purkinje cells that decreased their firing during the CR, and you also have a nuclear cell that increased firing during the CR. But besides that you have the same amount of Purkinje cells that increased firing, and also at the level of the interpositus you have nuclear cells that decreased firing. This is all because to close the lid you need to activate the orbicularis ocular muscle, but also you have to release the levator palpebrae, which is holding the position of the lid open during alertness. The second question is related to the philosophy of the model. When you use delay conditioning with tone and air puff, the CS is supposed to go through the pontine nuclei and the US is going through the inferior olive, but when you use trace conditioning using air-puff conditioning, you get exactly the same kind of conditioned response. It is difficult to understand the pathways that are taking the CS and US; if both are supposed to be going through the inferior olive, I can't understand how a system working with one stimulus is going to deal with two stimuli that are going to appear and disappear in less than half a second.

MAUK: I'll answer the second question first, because our simulation can't do trace conditioning. If we train it with a trace protocol, it doesn't learn. The literature says that unlike delay conditioning, trace conditioning is sensitive to lesions of the hippocampus. The canonical idea is that the reason you need the hippocampus for trace conditioning is that it provides mossy-fiber inputs during the trace interval so that the cerebellum can use its normal rules to learn, but it's an idea that has more followers than evidence.

DELGADO-GARCÍA: As I said, the cerebellum is firing both for trace and delay condition in exactly the same way, and papers published this year say that pyramidal cells in the hippocampus fire exactly the same way during trace and late conditioning. In both cases the hippocampal cell starts firing after the beginning of the CR, so they are not involved in any kind of motor response, but are involved in the cognitive process.

MAUK: I agree that the evidence from hippocampus recordings support simple ideas of how it "helps" the cerebellum, and trace conditioning doesn't follow very well. The cerebellum by itself can't do trace conditioning, our simulation can't do trace conditioning, and that is about as good as I can do right now.

LISBERGER: Would you explain the difference between trace and delay conditioning?

MAUK: Delay conditioning is thought to be simpler. The tone comes on, and the reinforcer comes on while the tone is on. In trace conditioning the tone comes on, then it goes off, and there's a so-called trace interval, a silent interval, and then the shock comes on. The key difference it seems is that cerebellar lesions abolish the responses for both of those kinds of learning, but hippocampal lesions have any effect only for trace conditioning.

CURTIS BELL (*Oregon Health & Sciences University, Beaverton, OR*): I think everyone had plasticity at the parallel fiber to Purkinje-cell synapse. Have any of you explored different learning rules at that synapse to see the efficacy of different learning rules in terms of the output in the behavior of your model?

MAUK: We've looked at one subtle variant. We began by using the LTD rule with the old conjunctive stimulation rule, which is that parallel fiber and climbing fiber together would make LTD and that learns fine, except the simulations learn inappropriately in a way that the rabbit doesn't when the tone shock interval is very short. Raymond and Lisberger came up with evidence that climbing fibers modify parallel-fiber synapses that were active about 100 ms ago. When we added that, it didn't change any of the already successful properties of the simulation, but it took away the bad behavior.

CHRISTOPHER YEO (*University College London, London, UK*): Dr. Mauk, did you simulate acquisition extinction in the models?

MAUK: We haven't done that, because since some animals were quite a bit calmer when we were training them, we haven't had to include anxiety circuits in our simulation yet.

ROBERT MCCRAE (*University of Chicago, Chicago, IL*): I'd like to repeat the question that Ebner asked the panel on the first day. What can we learn from eye-movement modeling and eye-movement research that applies to arm movements and other cerebellar functions?

KETTNER: It's a model system that people have selected because the motor plant output is simple; I used to work in arm-movement control, where you have myriad muscles and so on. I think that confuses the control issue because of the issue of inverse models, so I have moved to oculomotor to try to get at predictive/fancy control issues. Also, if you do recording, it is a nice preparation because the animal is not moving very much. I think that is all it is. But we like to imagine that it would have application beyond just oculomotor control or eye-blink control.

MAUK: I think another value of using these well-defined systems is that the first step is to understand the rules the cerebellum uses, so that we can predict inputs and outputs. We need that understanding before we apply those rules to situations where there are complicated inputs and complicated plants. From my perspective, eyelid conditioning is nice because it gives the control required to thoroughly understand inputs, outputs, and transformations.

LISBERGER: Of course, I am a believer, but one of the things that has troubled me about this very question since the first day of this conference is that the first two speakers gave us the clear impression that for arm movements the cerebellum was much more important for multijoint movements than for single-joint movements. About the only multijoint movement I've heard talked about at this meeting in terms of eye movements is the one tidbit that McCrea gave us with combined eye–head movements. So, I have been worrying myself about how we can think about and design experiments about eye movements that would play into this idea about compensating for interaction torques and the like.

STEPHEN M. HIGHSTEIN (*Washington University School of Medicine, St. Louis, MO*): Every eye movement involves all six extraocular muscles. You can train an animal to make an eye movement in the plane of say the horizontal recti, such that the firing rates of the vertical extraocular motor neurons do not change. Henn showed that that is a very unusual eye movement. So, just like a hand movement, I think you have interaction torques of the six muscles that move the eyes, so I would maintain that eye movements might indeed be a very adequate model to help us start thinking about multijoint movements.

THOMAS THACH (*Washington University School of Medicine, St. Louis, MO*): I would agree, but I would say that they somewhat trivialize the question in that the physical properties of moving the globe are much less than moving a limb. By the same token, the fact that all of the eye muscles can work together, and do so nicely, allows you to do an experiment where you don't worry about the eye muscles. I think this is one of the things that bothers me about the trivialization of eye movements: they work in such precise combinations that it seems to me that an important line of research would be the kinds of things that Scudder was doing, which have been done in the past, and that is to try to perturb the individual eye muscles to see what parts of the nervous system are required to compensate for that, or to organize that behavior.

JOHN STAHL: In terms of unifying concepts, one thing we learned here at this meeting in multiple talks and all the different sessions is that the cerebellum seems to be much more important for fast things than for slow things. I think that might be a unifying concept that most of us can agree with. Now, one thing about trace conditioning versus delay conditioning, independent of the interval between CS offset and US onset, is that trace-conditioning paradigms tend to be slower than delay-conditioning paradigms. It is also important to point out that not all forms of delay conditioning require the cerebellum. Classic delay appetitive conditioning, which doesn't have very brief CS–US intervals, usually occurs at long, slower intervals and doesn't require the cerebellum. I wonder if the difference is not trace and delay, but fast and slow, as we have seen with so many other things.

MAUK: I'm not sure about that, because, for one thing, trace conditioning requires the cerebellum, too, and it requires the hippocampus. Another thing is that the temporal properties of the trace responses come on with a longer delay from the onset of the stimulus than in delay conditioning, but their rise times and velocities can be comparable to those seen in delay conditioning.

STAHL: Then what about classic appetitive conditioning? Why is that not disrupted by cerebellar lesions? That's just as much as eye-blink conditioning as classic conditioning.

MAUK: I think that if the right combinations of mossy fibers and climbing fibers arrive at it, it does its business and sends an output, and for unknown reasons, that may not be the case for appetitive conditioning.

THACH: I would like to hear what Dr. Optican has to say about this.

LANCE OPTICAN (*National Eye Institute, NIH, Bethesda, MD*): First, I want to say that I agree that studying eye movements is easier than studying limb movements. But it is not that much easier. The limb-movement people have a lot of explaining to do, as to why they don't understand their system nearly as well as the eye-movement people. I don't think you can simply argue that the eye movements are simple rotations about one joint, and that somehow makes the way the brain controls them really trivial. I think the problem may be subtler, in that limb movements are controlled through the spinal cord, and that may introduce a layer of complexity that obfuscates what the brain is doing. But having worked on eye movements for quite some time, I can assure you that they are not that trivial. There is one point I would like to make about models in general. We all make models and whether we write them down mathematically or not, we use them to try and predict what will happen under different circumstances. We use them to lead our thinking and even to plan our experiments. So, a question I would ask is: What is missing from the experimentalists' work? Every experimentalist is, of course, very happy to tell the theoreticians what is missing from his or her models, but that's not to say that there is an absolutely complete literature out there that will allow you to learn anything you want to know about brain physiology. So can you identify for the experimentalist what kind of experiments or experimental data would really be helpful? What are the missing elements that they would need to make better models? I would like to start with Dr. Hirata.

HIRATA: To understand the system from a theoretical point of view, we have to know the response of the system to rich information. Rich in this context means in terms of frequency and amplitude, so from a theoretical point of view again, if we could know the response of the neuron to, let's say, Gaussian white noise input, then we can make a pretty precise model and predict any cell's behavior or monkey's eye-movement behavior in response to any kind of stimulus.

KETTNER: I think there is an extremely important issue of the learning rules associated with these plasticities. In my view of the cellular literature, there is a fixation on an existence proof of plasticity, and as soon as that existence proof is there, they then study it pharmacologically, try to understand second messenger systems, and so on. Theoretically, in the time flow you need to know the relationship between the parallel-fiber–climbing-fiber activation sequence. Because the teaching signal is delayed by 90 ms, you need to be able to somehow have a situation where parallel-fiber activation that occurs before climbing-fiber activation is optimal. There is some initial information about this. My second wish list is associated with other synapses that are plastic, and evidence is starting to come in. A related issue concerns LTP versus LTD at the synapse. We need to know information about that, so these are the kinds of issues that I am interested in.

MAUK: I think in the eyelid-conditioning world, first and foremost, we as a group are way behind the eye-movement people and the limb-movement people in terms of recording. We have done lots of lesions and reversible lesions, but we need to be better about recording. There are some recordings out there, but we need it to be as thorough as the others so that we can compare the innards of our models to data, and not just the inputs and outputs. I think the second thing is again a way that eyelid-conditioning data have lagged behind others; whereas in many forms of adaptation there has been an emphasis on increases and decreases of the responses, in eyelid conditioning there has been almost an exclusive emphasis on acquisition. To widen this approach, we have been trying to understand much better how the responses can be unlearned in the same way that they can be learned.

YAMAMOTO: I was studying eye-movement control systems, but now I am studying arm-movement control systems. I would like to construct a model for arm movements. In the arm-movement world, what kind of information is carried by the simple spike signals is still controversial. But in the eye-movement world what kind of information is contained in the outputs from the primary motor cortex is controversial. After the end of the controversy about primary motor cortex and arm movement and cerebellum, I can build a realistic model of the working cerebellum. But first I want to know about the physiological experiments.

LISBERGER: Dr. Optican, this is out of the blue, but do you have an answer to the question Bower asked after Mauk's talk? It was a question about realistic models, and he alleged that Mauk's model was not adequately realistic and at the same time lacked emergent properties because everything had been built into it. I think if he had bothered to say anything after the other talks, he would have said similar things, perhaps with somewhat different words.

OPTICAN: Our models are simplifications of the real system, and thus unrealistic. However, if they were as complicated as the subject matter, we wouldn't understand our own models! The point is whether or not the models, although simplistic, make usable predictions. I think the examples we saw today were very good in that regard. Many of them were so simple, although not simpleminded, that there is hope that we can understand them. And these models do make predictions: what we think is happening, where gain changes are occurring. It is hoped that these simple models will help us make progress in understanding the cerebellum, which will lead to more detailed, and thus more realistic, models.

The Nonvisual Complex Spike Signal in the Flocculus Responds to Challenges to the Vestibulo-ocular Reflex Gain

T. BELTON, M. SUH, AND J.I. SIMPSON

Department of Physiology and Neuroscience, New York University School of Medicine, New York, New York 10016, USA

KEYWORDS: **complex spike; cerebellar flocculus; vestibulo-ocular reflex**

Complex spike (CS) activity in the cerebellar flocculus in the rabbit modulates not only in relation to retinal image motion, but also in relation to rotatory vestibular stimulation in darkness. Because the vestibularly evoked nonvisual CS signal typically is approximately 180 degrees out of phase with the optokinetically evoked visual CS signal, they compete during vestibular stimulation in the presence of a stationary, illuminated surround when the gain of the compensatory eye movement is less than one. In contrast, when visual and vestibular stimuli are rotated together to suppress the vestibulo-ocular reflex, the two CS signals are in phase. As part of our investigation of the interaction of visual and nonvisual CS signals and the relation of the nonvisual CS signal to the vestibulo-ocular reflex, we recorded eye movements and the CS activity of horizontally responsive floccular Purkinje cells in three awake rabbits that were sinusoidally oscillated in the light (0.4 or 0.6 Hz, 12 or 25°/s) in phase with an optokinetic surround to suppress the vestibulo-ocular reflex.

During the 30 to 90 minutes of suppression stimulation, the mean gain of the vestibulo-ocular reflex in darkness decreased by 29%. The amplitude of the nonvisual CS modulation also changed, but not with the same time course as that of the vestibulo-ocular reflex gain. In 6 of 14 cells that initially showed the typical pattern of nonvisual CS modulation, suppression stimulation resulted in a phase shift in the nonvisual CS modulation of about 180 degrees. Among the other 8 cells that initially showed the typical pattern of nonvisual CS modulation, suppression stimulation resulted in either no change in modulation (3 of 8) or the disappearance of modulation (5 of 8). In addition, in 4 of 4 cells with initially no nonvisual CS modulation, suppression stimulation resulted in the advent of significant modulation, also about 180 degrees phase shifted from the typical non-visual CS modulation.

Address for correspondence: Dr. Tim Belton, Department of Physiology & Neuroscience, New York University School of Medicine, New York, NY 10016.Voice: 212-263-5428; fax: 212-689-9060.

beltot01@popmail.med.nyu.edu

Ann. N.Y. Acad. Sci. 978: 503–504 (2002). © 2002 New York Academy of Sciences.

In sum, both the amplitude and the phase of the nonvisual CS signal can be altered by stimuli that challenge the vestibulo-ocular reflex gain.

ACKNOWLEDGMENTS

This work was supported by NIH Grants NS-13742 and DC-03284.

Role of Caudal Fastigial Neurons during Head-Free Gaze Shifts in the Monkey

S.C. BRETTLER AND A.F. FUCHS

Department of Physiology and Biophysics, University of Washington, Seattle, Washington 98195-0005, USA

KEYWORDS: caudal fastigial neurons; gaze; monkey

The caudal fastigial nucleus (CFN) has been implicated in the control of saccadic eye movements in the head-fixed monkey. Unilateral muscimol inactivations of the CFN cause ipsilateral saccades to become hypermetric and contralateral saccades to become hypometric,[1] suggesting that the CFN may be helping to accelerate contralateral and decelerate ipsilateral saccades. This suggestion is supported by the timing of the burst in CFN neurons, which leads contralateral but lags ipsilateral saccades,[2] and the connectivity of CFN neurons to elements of the brain-stem saccadic burst generator.[3,4] However, correlations between number of spikes and saccade size and burst duration and saccade duration are not robust (r^2 ~0.4). A major input to the oculomotor vermis, the main source of CFN input, is the nucleus reticularis tegmenti pontis, which in turn is the major output of the superior colliculus.

Recent evidence shows that superior colliculus neurons encode overall gaze amplitude during a head-free gaze shift when the head is also free to move. This led us to consider whether the correlations would improve under the more natural condition when the monkey made head-free gaze shifts. We recorded extracellular activity of 35 neurons in the CFN of three rhesus monkeys trained to produce accurate gaze shifts to sequentially illuminated LED targets on a hemispheric array with maximal target steps of 160 degrees horizontally and 40 degrees vertically and obliquely. Many CFN neurons discharged during head-free gaze shifts.

Preliminary findings reveal stronger correlations than those found in previous head-fixed data between the number of spikes and saccadic amplitude and an elongation of the burst as revealed by strong correlations between burst duration and saccadic duration (e.g., r^2 to 0.9). The most compelling result was the precession of the burst in time towards the end of the saccade, which revealed a preburst pause and/or a postburst pause in many neurons. The postburst pause led to a clear detection of the end of the burst, which was strongly timed with the end of the gaze shift for both ipsilateral and contralateral gaze shifts, as shown herein (FIG. 1). These findings suggest that the CFN may be actively involved in the termination of gaze shifts.

Address for correspondence: Dr. S.C. Brettler, Department of Physiology and Biophysics, University of Washington, Seattle WA 98195-0005.

Ann. N.Y. Acad. Sci. 978: 505–506 (2002). © 2002 New York Academy of Sciences.

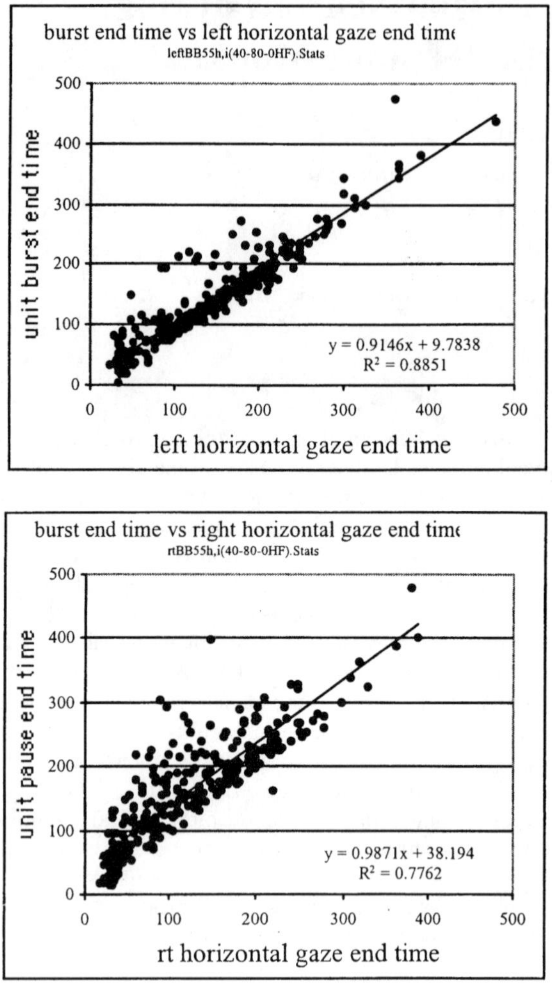

FIGURE 1.

ACKNOWLEDGMENTS

This work was supported by Grants EY00745, NS07395, and RR00166 from the National Institutes of Health.

REFERENCES

1. ROBINSON *et al.* 1993.
2. FUCHS *et al.* 1993.
3. NODA *et al.* 1990.
4. SCUDDER *et al.* 2000.

Behavioral Analysis of Gain Control for Smooth Pursuit Eye Movements

MEGAN R. CAREY AND STEPHEN G. LISBERGER

Keck Center for Integrative Neuroscience, University of California, San Francisco, California 94122-2747, USA

KEYWORDS: **gain control; smooth pursuit eye movements; eye velocity**

Previous studies from our laboratory have demonstrated that the visual-motor transformation for smooth pursuit eye movements is subject to gain control. Eye velocity responses to high-frequency, single-cycle perturbations of target velocity are enhanced during ongoing pursuit relative to fixation. The goal of our study was to determine what signals are involved in the modulation of pursuit gain control.

Monkeys were required to track a visual target for a liquid reward. Single cycles of 10 Hz, ± 5°/s sine wave perturbations of target velocity were used to probe the gain of the visual-motor transformation for pursuit while the monkey was engaged in one of four tasks: fixation, pursuit, vestibulo-ocular reflex, or vestibulo-ocular reflex cancellation. During vestibulo-ocular reflex cancellation relative to fixation, the response to perturbations of target velocity was increased to a similar extent as that during pursuit. This indicates that the pursuit system in engaged during vestibulo-ocular reflex cancellation and suggests that gaze velocity signals are capable of enhancing pursuit gain. By contrast, during the vestibulo-ocular reflex, the responses to perturbations of target velocity were similar to, or even smaller than, the responses during fixation. We conclude that gaze velocity signals, but not eye velocity signals, contribute to enhancement of the gain of visual-motor transmission for pursuit. Future experiments will investigate the possible role of gaze velocity signals in the cerebellar flocculus in pursuit gain control.

Address for correspondence: Dr. Megan R. Carey, Keck Center for Integrative Neuroscience, University of California, San Francisco, CA 94122-2747.

Ann. N.Y. Acad. Sci. 978: 507 (2002). © 2002 New York Academy of Sciences.

Coherence of Subthreshold Activity in Coupled Inferior Olivary Neurons

A. DEVOR AND Y. YAROM

Department of Neurobiology, Hebrew University, Jerusalem, Israel

KEYWORDS: **inferior olivary neurons; coherence; subthreshold activity**

In this study we investigated the spatiotemporal organization of the subthreshold oscillations in the inferior olivary nucleus using simultaneous double whole-cell patch recordings and optical imaging in brain slices of the inferior olivary nucleus. The frequency distribution in a sample of 94 neurons revealed two populations of frequencies with mean values of about 3 and 6 Hz. The frequency and amplitude in most neurons showed on-going modulations in time that were largely abolished after bath application of 40 μM CNQX.

The probability of finding a coupled pair of neurons decreased from 80% in immediate neighboring cells to about 30% at a distance of 40 μm. All electrotonically coupled cell pairs exhibited synchronous subthreshold oscillations. No coupling was detected in pairs, where the neurons oscillated with more than 20-ms phase shifts or oscillated at different frequencies. In all pairs with coherent oscillatory activity the frequency and amplitude modulations occurred simultaneously in both neurons. In agreement with pair recording, optical imaging performed on 40 inferior olivary nucleus slices using voltage-sensitive dye RH 414 revealed patches of oscillatory activity either occurring spontaneously or evoked by extracellular stimulation. The phase-shifted activity manifested as waves propagating within an oscillating patch.

These results suggest that subthreshold oscillations are a dynamic property, displaying continuous modulations in amplitude, frequency, and space. Synaptic inputs to the inferior olivary nucleus are likely to cause this instability.

Address for correspondence: Dr. A. Devor, Department of Neurobiology, Hebrew University, Jerusalem, Israel.

Ann. N.Y. Acad. Sci. 978: 508 (2002). © 2002 New York Academy of Sciences.

Comparing Continuous and Discrete Movements with fMRI

JÖRN DIEDRICHSEN,[a] ELIOT HAZELTINE,[b] RICHARD IVRY,[a] STEVE KENNERLEY,[a] AND BEKKI SPENCER[c]

[a]*Department of Psychology, University of California at Berkeley, Berkeley, California 94720, USA*

[b]*NASA Ames Research Center, Moffett Field, California 94035, USA*

[c]*Department of Health & Kinesiology and the Purdue Neuroscience Program, Purdue University, West Lafayette, Indiana 47907, USA*

KEYWORDS: **eye movement; fMRI**

Temporal control of repetitive movements may be achieved by distinct mechanisms depending on whether the movement is continuous (e.g., circle drawing) or involves discrete events, such as movement onsets or contact points (e.g., tapping). Patients with cerebellar damage exhibit increased temporal variability in tasks involving discrete timing, but not in tasks in which the movements are produced continuously.

We sought converging evidence for this neural dissociation in an fMRI study. Subjects performed continuous or discrete movements at two different speeds with either the left or the right hand. Within the cerebral cortex, activation in the contralateral motor cortex and SMA was similar for discrete and continuous movements and

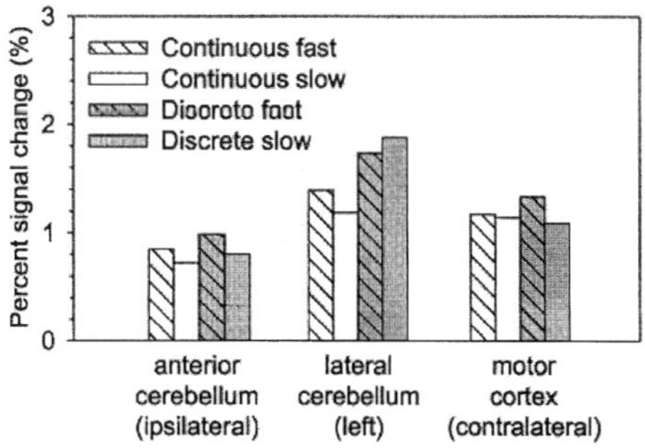

FIGURE 1. Percent change in BOLD signal in the anterior cerebellum ipsilateral to the moving hand, left lateral cerebellum, and motor cortex contralateral to the moving hand.

Address for correspondence: Dr. Richard Ivry, Department of Psychology, University of California at Berkeley, Berkeley, CA 94720. Voice: 510-642-0135; fax: 510-642-5293.
ivry@socrates.berkeley.edu

Ann. N.Y. Acad. Sci. 978: 509–510 (2002). © 2002 New York Academy of Sciences.

was strongly modulated by movement speed. A similar pattern was found within the anterior lobe of the cerebellum ipsilateral to the performing hand. By contrast, activation in the lateral cerebellum was bilateral and strongest on the left side independent of which hand was being used. Activity in the lateral cerebellum was higher in the timing of discrete movements, and no speed dependency was found.

These results provide supporting evidence that the lateral cerebellum may play a special role in the timing of movements involving discrete events. Whereas the anterior lobe and cortical motor areas seem to be involved with the implementation of movement, sites in the lateral cerebellum likely reflect the abstract timing demands of the task.

Is the Cerebellum Important for Podokinetic Adaptation?

G.M. EARHART,[a,b] F.B. HORAK,[b] G. MELVILL JONES,[a] E.W. BLOCK,[a] K.D. WEBER,[a] O. SUCHOWERSKY,[a] AND W.A. FLETCHER[a]

[a]*Department of Clincial Neurosciences, University of Calgary, Calgary, Alberta, Canada*

[b]*Neurological Sciences Institute, Oregon Health & Sciences University, Beaverton, Oregon 97006, USA*

KEYWORDS: **locomotion; sensorimotor adaptation; cerebellum; podokinetic**

BACKGROUND

After a period of stepping-in-place on a rotating circular treadmill, healthy subjects will inadvertently turn when asked to step-in-place without vision on a stationary surface.[1] This phenomenon, termed podokinetic after-rotation (PKAR), represents a remodeling of the relation between trunk rotation relative to the foot and the perception of trunk rotation in space. We investigated the role of the cerebellum in this podokinetic (PK) adaptation. Given the role of the cerebellum in many motor learning tasks, we expected subjects with cerebellar damage to show impaired PK adaptation. We compared subjects' ability to adapt to the rotating walking surface to their ability to adapt to throwing while looking through prisms.

METHODS

Eight subjects with diagnoses of cerebellar degeneration and/or spinocerebellar atrophy were compared to eight age-matched controls. Diagnoses were confirmed by genetic testing and/or MRI. Subjects completed a throwing task based on the design of Martin *et al.*[2] Each subject stood facing an 8×8 cm square target placed at shoulder level 2 m away. Each subject made 20 baseline throws before donning prism glasses (30 diopter 3M press-on Fresnel lens), 40 adaptation throws while wearing the prisms, and 20 postadaptation throws after doffing the prisms. The locations that each ball struck were marked and numbered sequentially during the task. After the throwing task, subjects were exposed to a PK stimulus that consisted of 15 minutes of stepping-in-place on the axis of a 76-cm disc rotating in a clockwise direction at 45°/s. Stepping frequency was maintained at 2 Hz by matching cadence to a metronome affixed to the trunk. Following the 15-minute period of adaptation, sub-

Address for correspondence: Gammon M. Earhart, Ph.D., Neurological Sciences Institute, OHSU West Campus, 505 NW 185th Avenue, Beaverton, OR 97006. Voice: 503-418-2605. earhartg@ohsu.edu

Ann. N.Y. Acad. Sci. 978: 511–512 (2002).

jects performed 30 minutes of stepping-in-place on the axis of the stationary disc while wearing a blindfold and earplugs. During this 30-minute period, PKAR responses were recorded by means of a rotational potentiometer.

DATA ANALYSIS

Data from the throwing task were analyzed using the methods of Martin *et al.*[2] The horizontal locations of the last eight baseline throws were compared to those of the first three postadaptation throws using a Mann-Whitney U ($p = 0.05$) to determine whether or not a significant aftereffect was present. PKAR data were fitted with three-parameter exponential decay curves to determine the initial velocity of each subject's response.

RESULTS

Five of the eight cerebellar subjects showed impaired PK adaptation, that is, their initial angular velocity during PKAR was more than two standard deviations below the mean initial velocity of control subjects. There was no correlation between adaptation on the PK task and adaptation on the throwing task.

CONCLUSIONS

The cerebellum appears to be important for PK adaptation, as over half the cerebellar subjects (but none of the control subjects) showed impaired locomotor adaptation. This suggests that PK adaptation is not strictly a peripheral phenomenon but is mediated, at least in part, by central structures. Subjects with impaired PK adaptation do not necessarily also demonstrate impaired adaptation to prisms. We hypothesize that different regions of the cerebellum may be responsible for adaptation in the two tasks, as the PK task involves processing of proprioceptive and vestibular inputs, whereas the prism task involves visual-proprioceptive adaptation. Further studies of subjects with localized cerebellar damage from infarcts and surgical procedures are planned to help elucidate which regions of the cerebellum may be particularly important for PK adaptation.

ACKNOWLEDGMENTS

This work was supported by National Institutes of Health Grant R01-DC04082 and Medical Research Council Grant MA-15639.

REFERENCES

1. WEBER, K.D. *et al.* 1998. Exp. Brain Res. **120:** 377–385.
2. MARTIN, T.A. *et al.* 1996. Brain **119:** 1183–1198.

What, Indeed, Do Climbing Fibers Signal?

Some Cerebellar Microcomplexes Appear to Monitor the Activity in Spinal Nociceptive Withdrawal Reflex Modules

MARTIN GARWICZ, ANDERS LEVINSSON, AND JENS SCHOUENBORG

Section for Neurophysiology, Department of Physiological Sciences, Lund University, BMC F10, Tornav. 10, S-221 84 Lund, Sweden

KEYWORDS: **climbing fibers; cerebellar microcomplexes; nociceptive receptive fields; spinal reflex modules**

Although the olivo-cerebellar climbing fiber system is pivotal to cerebellar function, to date there is little consensus with regard to what climbing fibers actually do. Here we test the hypothesis that climbing fibers defining individual cerebellar microcomplexes convey highly integrated information reflecting the activity of specific spinal reflex modules.

To this end, cutaneous nociceptive receptive fields of spinal withdrawal reflex modules acting on single forelimb muscles were mapped and compared to receptive fields of climbing fibers projecting to cerebellar pars intermedia (sagittal zones C1, C3, and Y) in barbiturate-anesthetized cats. In addition, the convergent muscle afferent input to individual climbing fibers was analyzed with reference to characteristics of the associated reflex module. Quantitative methods were used for both mapping and comparing receptive fields. Individual cutaneous receptive fields in the two systems could readily be matched, and matched pairs exhibited a high degree of similarity with regard to detailed distributions of sensitivity. Moreover, muscle afferents from a given spinal reflex module were found to preferentially target climbing fibers with cutaneous receptive fields similar to that of the module itself. These afferents also targeted climbing fibers in other microcomplexes, but to a lesser extent.

Our findings suggest that climbing fiber input to individual cerebellar microcomplexes in pars intermedia represents highly integrated information reflecting (1) cutaneous input to an individual spinal withdrawal reflex module, (2) muscle afferent input from the muscle that constitutes the output from that module, and (3) mus-

Address for correspondence: Dr. Martin Garwicz, Section for Neurophysiology, Department of Physiological Sciences, Lund University, BMC F10, Tornav. 10, S-221 84 Lund, Sweden. Voice: +4646-222-7755; fax: +4646-222-4546.
martin.garwicz@mphy.lu.se

Ann. N.Y. Acad. Sci. 978: 513–514 (2002). © 2002 New York Academy of Sciences.

cle afferent input from muscles that constitute the output from functionally related modules. This suggests that an individual climbing fiber signals cutaneous sensory events reflecting activity of a single muscle conditional upon the functional state of the muscle itself and that of functionally related muscles.

ACKNOWLEDGMENTS

This work was supported by the Swedish Medical Research Council Projects No. 10569 and 1013, the Medical Faculty of Lund University, Astra Hässle Inc., Knut and Alice Wallenberg's Foundation, Elsa and Thorsten Segerfalk's Foundation, Greta and Johan Kock's Foundation, the Crafoordska Foundation, and The Royal Physiographic Society in Lund.

REFERENCES

1. Garwicz, M., A. Levinsson & J. Schouenborg. 2002. Common principles of sensory encoding in spinal reflex modules and cerebellar climbing fibres. J. Physiol. **540:** 1061–1069.
2. Garwicz, M. 2002. Spinal reflexes provide motor error signals to cerebellar modules—relevance for motor coordination. Brain Res. Rev. **40:** 152–165.

Cell Morphology and Local Circuitry of the Central Lobes of the Mormyrid Cerebellum

VICTOR Z. HAN AND CURTIS C. BELL

Neurological Sciences Institute, Oregon Health and Sciences University, Beaverton, Oregon 97006, USA

KEYWORDS: cell morphology; circuitry; mormyrid cerebellum

Cells in the central lobes of the cerebellum of mormyrid fish were filled with biocytin after intracellular recording in an *in vitro* slice preparation. The labeling was visualized with both the ABC and the fluoresence methods.

The morphologies of Purkinje cells, efferent cells, Golgi cells, stellate cells, and climbing fibers were determined, as follows.

(1) *Purkinje cells.* The oval cell bodies are in the ganglionic layer. Both axons and dendrites arborize extensively in the parasagittal plane, but do so only minimally in the transverse plane. The spine-covered dendrites extend 408 ± 148 μm in the parasagittal plane, but only 28 ± 23 μm in the transverse plane. The thin axon arises from the soma and branches extensively in the ganglionic layer, forming en passant boutons and remaining in the cortex. The axonal arbor extends rostrally from the cell body for a distance of 484 ± 192 μm, and caudally only 105 ± 45 μm in the parasagittal plane, but extends only 78 ± 30 μm in the transverse plane.

(2) *Efferent cells.* The large fusiform cell bodies are situated at the base of the ganglionic layer. Two or more smooth dendrites arise from the cell body and extend symmetrically in the molecular layer for a total distance 680 ± 150 μm in the parasagittal plane, but only 55 ± 38 μm in the transverse plane. The axons arise from the soma and usually end within a short distance of the soma at the cut surface of the slice, without branching and without terminal boutons. Occasionally, a single myelinated axon can be followed for over 1 mm in the parasagittal plane.

(3) *Golgi cells.* These cells have thin dendrites in the lower molecular layer. Their axons extensively branch in the lower ganglionic layer and granule layer.

(4) *Stellate cells.* These cells have molecular layer dendrites and axon arbors that branch extensively and terminate in the ganglionic layer and molecular layers.

(5) *Climbing fibers.* Climbing fibers were labeled in the fixed brain by placing DII in the inferior olive. Anterograde labeling in the central lobes was found exclusively in the ganglionic layer, without extending into the molecular layer.

Address for correspondence: Dr. Victor Z. Han, Neurological Sciences Institute, Oregon Health & Sciences University, Beaverton, OR 97006.

Ann. N.Y. Acad. Sci. 978: 515–516 (2002). © 2002 New York Academy of Sciences.

The present results confirm and extend previous studies. Efferent cells, which are equivalent to cerebellar nucleus cells in the mammal, integrate parallel fiber input within a narrow parasagittal plane oriented perpendicular to the parallel fibers. The efferent cells combine this direct parallel fiber input with input from Purkinje cells that receive their parallel fiber input within the same narrow parasagittal plane.

ACKNOWLEDGMENT

This work was supported by Grant MH49792 from the National Institute of Mental Health.

Evidence of a Cerebrocerebellar Network

An fMRI Study of Functional Integration during Visuospatiomotor Sequence Learning

RON HEAL AND PAUL MATTHEWS

Centre for Functional Magnetic Resonance Imaging of the Brain, Department of Clinical Neurology, University of Oxford, John Radcliffe Hospital, Oxford, England, OX3 9DU

KEYWORDS: **cerebrocerebellar network; fMRI; sequence learning**

BACKGROUND

Middleton and Strick[1] provided evidence for an anatomic model of cerebellar connectivity involving neural regions arising in the cerebellar cortex projecting first to the dentate nucleus, then to the thalamus, and finally to the prefrontal cortex. Further findings suggest an anatomic connection between the prefrontal cortex and the pontine nuclei.[2] In a brief review of these findings, Ramnani and Miall[3] finalize the anatomic connection from the pontine nuclei back to the cerebellar cortex and suggest that these cerebrocerebellar loops may somehow be involved in cognitive functioning.

DESIGN

An opportunity sample was taken from the FMRIB Centre personnel. Eleven subjects (3 female) were all strongly right handed, had normal or corrected-to-normal vision, and had a mean age of 26 years. Results indicated block design active (27 seconds)/baseline (27 seconds), 20 repeats, with a total time of 18 minutes. All subjects had no abnormal cognitive or physical deficits and no history of neurologic trauma. Subjects completed a novel visuospatiomotor sequence learning paradigm developed by us as part of a larger doctoral study investigating functional integration during motor learning, which couples a learning task with a stimulus-matched baseline control task. A separate motor response control task was also carried out. Performance data were recorded.

Address for correspondence: Ron Heal, Centre for Functional Magnetic Resonance Imaging of the Brain, Department of Clinical Neurology, University of Oxford, John Radcliffe Hospital, Oxford, England, OX3 9DU. Voice: +44(0) 1865 222729; fax: +44(0) 1865 222717.
ron@fmrib.ox.ac.uk

Ann. N.Y. Acad. Sci. 978: 517–518 (2002).

fMRI

A 3T Varian INOVA MRI system (multislice gradient echo EPI sequence: TR = 3,000 ms, TE = 30 ms, 25 mm × 6 mm axial slices, FOV 256 mm × 256 mm, 64 × 64 matrix) was used. A high-resolution structural scan was acquired for each subject.

RESULTS AND SUMMARY

These initial findings taken from our larger study of motor learning revealed that the dorsolateral prefrontal cortex (ba9/46), basal ganglia, dentate nucleus, cerebellar posterior cortex, and cerebellar lateral cortex were all significantly more activated during learning than during postlearning trials. Linear trend analysis revealed that the dorsolateral prefrontal cortex (ba9/46), basal ganglia, dentate nucleus, and cerebellar posterior cortex showed a significant trend to decrease in activation over the period of the postlearning trials. However, this trend in decreasing activation was not found during learning trials. The learning-related activity within this experiment was distinguishable from motor response-related activity by using a statistical contrast to factor out the motor element using within-subject control of motor responses. Comparison of purely externally cued motor responses was further facilitated by the additional motor control experiment, which revealed a pattern of cerebrocerebellar activity distinct from that evident from learning-related trials. These new findings support the suggestion of a role for cerebrocerebellar regions in visuomotor sequence learning, and therefore the function of the cerebellum appears to be confined not merely to the motor element of the task itself, but also, more interestingly, to the cognitive element of learning a visuospatial sequence requiring motor responses.

REFERENCES

1. MIDDLETON, F.A. & P.L. STRICK. 1997. Cerebellar output channels. Int. Rev. Neurobiol. **41:** 61–82.
2. MIDDLETON, F.A. & P.L. STRICK. 2001. Cerebellar projections to the prefrontal cortex of the primate. J. Neurosci. **21:** 700–712.
3. RAMNANI, N. & C. MIALL. 2001. Expanding cerebellar horizons. Trends Cognit. Sci. **5:** 135–136.

Antibody Against a Putative Ligand-Binding Site Reveals Delta2 Glutamate Receptor Function

HIROKAZU HIRAI,[a,b] THOMAS LAUNEY,[a] SUMIKO MIKAWA,[a] DAI YANAGIHARA,[c,d] TSUYOSHI KASAURA,[c] AKIHIRO MIYAMOTO,[c] AND MICHISUKE YUZAKI[b]

[a]*Laboratory for Memory and Learning, RIKEN Brain Science Institute, Saitama 351-0198, Japan*

[b]*Developmental Neurobiology, St. Jude Children's Research Hospital, Memphis, Tennessee 38105, USA*

[c]*Toyohashi University of Technology, Toyohashi 441-8580, Japan*

[d]*CREST, Japan Science and Technology Corporation (JST)*

KEYWORDS: **delta2 glutamate receptor; ligand binding; antibody**

The delta2 glutamate receptor, predominantly expressed in postsynaptic sites of cerebellar Purkinje cells, is referred to as an "orphan" receptor because no ligands have been found. Whether it forms a functional receptor, especially in the adult, is unclear. Here we demonstrate that delta2 receptors are functional in adult cerebellum and critical for cerebellar function. To block the activity of delta2 receptors, we generated a polyclonal antibody against its putative ligand-binding region, which had significant similarity to that of both other ionotropic glutamate receptors and phylogenetically related, bacterial amino acid binding proteins. This antibody specifically reacted with delta2 receptors and modified its activity in transfected cells. It also attenuated the synaptic transmission and blocked the induction of LTD in cultured Purkinje cells. Moreover, injection of this antibody rendered adult mice ataxic and significantly impaired the establishment of motor memory. These results suggest that activities of delta2 receptors, regulated by its putative ligand-binding region, exert remarkable influence on motor function in the adult cerebellum through modulation of synaptic transmission and plasticity.

Address for correspondence: Dr. Hirokazu Hirai, Laboratory for Memory and Learning, RIKEN Brain Science Institute, Saitama 351-0198, Japan.

Ann. N.Y. Acad. Sci. 978: 519 (2002). © 2002 New York Academy of Sciences.

Electrophysiological Mapping of Eyeblink Microzones in the Rabbit

M. IVARSSON, L. MILLAR, AND C.H. YEO

Department of Anatomy, University College London, London, UK

KEYWORDS: mapping; eyeblink; microzones; rabbit

Classical conditioning of the eyeblink response of rabbits is a model system for analysis of motor learning. Lesion and reversible inactivation studies have shown that a region of cerebellar cortex within lobule HVI is critical for acquisition and storage of eyeblink conditioning. In other species, HVI is known to contain C1 and C3 eyeblink control microzones, but the distribution of microzones within HVI of the rabbit is not known. We have mapped eyeblink control microzones in the rabbit in HVI by recording climbing-fiber-mediated field and single-unit responses to periocular stimulation.

In nine Dutch belted rabbits, tungsten microelectrodes were used to record in cerebellar cortical HVI. Climbing-fiber responses were defined as C2 or C1/C3 zone inputs according to the following criteria from previous studies in other species. The C2 climbing-fiber field potential responses have onset latencies of 18–22 ms and large receptive fields activated by stimulation of the ipsilateral and contralateral periorbital area and the ipsilateral forelimb area. The C1 and C3 eyeblink microzone climbing-fiber field potentials have onset latencies of 12–15 ms and restricted receptive fields activated only by ipsilateral periorbital stimulation.

After fixation, the marked cerebellar recording sites were reconstructed from serial, 50-μm sections stained with cresyl violet. An eyeblink control region was found in the C3 zone ventrally in lobule HVI, below the C2 zone, and extending from medial to lateral in HVI and into Crus I.

The location of this C3 eyeblink control microzone zone corresponds very closely to a site critical for the acquisition, consolidation, and expression of conditioned eyeblink. Its localization will enable more detailed analysis of changes in the activity of the critical, control Purkinje cells during eyeblink conditioning and provides a more discrete map for analysis of our ongoing studies of learning-related changes in gene expression.

Address for correspondence: Dr. M. Ivarsson, Department of Anatomy, University College London, London, UK.

Ann. N.Y. Acad. Sci. 978: 520 (2002). © 2002 New York Academy of Sciences.

Extracellular Amino Acid Levels in the Posterior Interpositus Nucleus during Eyelid Conditioned Responses in Cats

A Push-Pull Study

L. JIMÉNEZ-DÍAZ,[a] A. GRUART,[a] F.J. MIÑANO,[b] AND J.M. DELGADO-GARCÍA[a]

[a]*Laboratorio Andaluz de Biología, Universidad Pablo Olavide, Seville, Spain*

[b]*Hospital Valme, Universidad de Sevilla, Seville, Spain*

KEYWORDS: amino acids; posterior interpositus nucleus; eyelid conditioned responses; cats

The eyelid motor system has been used for years as an experimental model to study control of reflex and learned responses. The posterior interpositus nucleus is supposed to be a place directly involved in the generation of learned responses. It was recently reported that posterior interpositus neurons fired in relation to conditioned blink responses in alert behaving cats.[1] Cats were prepared for the recording of the EMG activity of the orbicularis oculi muscle and of upper lid position and velocity, with the search coil technique. They were also implanted with a stimulating electrode in the contralateral red nucleus for the antidromic identification of recorded interpositus neurons. Animals were classically conditioned with a delayed conditioning paradigm. The unconditioned stimulus (US) was a 100-ms strong air puff initiated 270 ms after beginning the conditioned stimulus (CS). The CS was a 370-ms tone. A whole experiment consisted of two habituation sessions, eight conditioning sessions, and six extinction sessions. Animals were conditioned after the unilateral implantation of a guide cannula in the posterior interpositus nucleus. The area selected for implantation was previously found functionally related to eyelid blink responses. A push-pull probe was lowered through the guide into the nucleus and was perfused continually with artificial cerebrospinal fluid. Five-minute samples containing 100 ml of perfusate were collected during the whole session (2 hours). Perfusate was used for amino acid analysis with a HPLC-EC method. Extracellular levels of GABA, glutamate, taurine, and glycine were measured. Prelim-

Address for correspondence: Dr. J. M. Delgado-Garcia, Laboratorio Andaluz de Biologia, Universidad Pablo Olavide, Seville, Spain. Voice: +34-954-349374; fax: +34-954-349375. jmdelgar@dex.upo.es

Ann. N.Y. Acad. Sci. 978: 521–522 (2002). © 2002 New York Academy of Sciences.

inary results show a significant increase in glycine, glutamate, and taurine levels during conditioning in comparison with habitution and extinction periods. Apparently, levels of GABA were not modified during the acquisition of this new motor skill.

REFERENCE

1. GRUART, A. *et al.* 2000. J. Neurophysiol. **84:** 2660–2690.

In Vivo Whole-Cell Patch-Clamp Recordings from Purkinje Cells and Interneurons

Synaptic Integration of Peripheral Input

HENRIK JÖRNTELL AND CARL-FREDRIK EKEROT

Department of Physiological Sciences, Tornavägen 10, SE-221 84 Lund, Sweden

KEYWORDS: **patch-clamp recordings; Purkinje cells; interneurons; parallel fibers**

In extracellular recordings, peripheral activation of parallel fiber input in Purkinje cells and interneurons indicates that only a small percentage of parallel fiber synapses are electrically effective. These synapses are all driven from a small receptive field that is specific to the receptive field of the local climbing fiber. This suggests that the synaptic integration performed by these neurons may be much simpler than we generally assume. The scope of the current study was to analyze this synaptic integration using peripheral stimulation.

In cats decerebrated at the intercollicular level, whole-cell patch-clamp recordings in the current clamp mode were obtained from interneurons and somata and dendrites of Purkinje cells in the forelimb area of the C3 zone. Blood pressure, end-expiratory CO_2, and body temperature were continuously monitored and maintained within physiologic limits. Patch-clamp electrodes contained 140 mM K-gluconate and 7 mM Cl- as well as other electrolytes. Cells were generally held slightly below (–60 to –80 mV) resting potential to facilitate analysis of synaptic potentials without contaminating spike activity. Synaptic potentials evoked by manual peripheral stimulation and direct electrical stimulation in the inferior olive and in parallel fiber beams were investigated.

In the spontaneous activity of molecular layer interneurons, there are only a few large excitatory postsynaptic potentials (EPSPs). Peripheral stimulation of the receptive field activates many large EPSPs, including the climbing fiber EPSP, as well as inhibitory postsynaptic potentials (IPSPs). Outside the receptive field, few or no postsynaptic potentials are evoked. In Purkinje cell recordings from putative stem dendrites, all-or-nothing calcium spikes are evoked from both the climbing fiber and

Address for correspondence: Dr. Henrik Jörntell, Department of Physiological Sciences, Section for Neurophysiology, BMC F10, Tornavägen 10, SE-221 84 Lund, Sweden. Voice: +46 46 222 77 64; fax: +46 46 222 45 46.
henrik.jorntell@mphy.lu.se

Ann. N.Y. Acad. Sci. 978: 523–524 (2002). © 2002 New York Academy of Sciences.

the parallel fiber receptive fields, but not from any other cutaneous areas. Somatic Purkinje cell recordings are dominated by large excitatory events of various amplitudes. Several characteristics suggest that they are all attenuated calcium spikes generated in stem dendrites.

REFERENCE

1. EKEROT, C.-F. & H. JÖRNTELL. 2001. Eur. J. Neurosci. **13:** 1303.

Contribution of Background and Target Motion to the Induction of Motor Learning in the VOR

MICHAEL C. KE AND JENNIFER L. RAYMOND

Department of Neurobiology, Stanford University, Stanford, California 94305, USA

KEYWORDS: motor learning; vestibulo-ocular reflex; VOR; target motion

Different combinations of vestibular, visual, and oculomotor signals have been hypothesized to control the induction of motor learning in the vestibulo-ocular reflex (VOR). We conducted a series of behavioral experiments to examine the contributions of visual and oculomotor signals. We induced learning by pairing passive head movements with movements of two visual stimuli. A small visual "target" (0.75 degrees in diameter) was presented either alone or in combination with a large (40 × 25 degrees) visual "background." The target moved in either the same (×0) or the opposite (×2) direction as the head. The background moved in either the same or the opposite direction as the target, creating "consistent" or "conflicting" motion of the two visual stimuli

We evaluated the relative contributions of the target and background to the induction of motor learning in the VOR. The movement frequency was either 0.5 or 5 Hz with a peak velocity of 10°/s. At 0.5 Hz, the monkey tracked the target; therefore, the velocity of target motion on the retinae was low. By contrast, the background had little effect on eye movements, but it generated large-field retinal slip of substantial velocity. At 5 Hz, the center of gaze was maintained within 2 degrees of the target's small position excursion (± 0.3 degrees), but there was no effective tracking of the target. Therefore, the velocities of background and target motion on the retinae were similar. At both frequencies, conflicting background and target motion resulted in less learning than the target alone condition, indicating that both the target/tracking eye movements and the background contribute to the induction of learning. Nevertheless, the target alone condition at 5 Hz was similar in effectiveness to consistent motion of target and background, suggesting that neither large-field retinal slip nor tracking eye movement is necessary for the induction of learning.

Address for correspondence: Dr. Jennifer Raymond, Department of Neurobiology, Stanford University, Fairchild Building, D-200, 299 West Campus Drive, Stanford, CA 94305. Voice: 650-736-1066.

jenr@stanford.edu

Ann. N.Y. Acad. Sci. 978: 525 (2002). © 2002 New York Academy of Sciences.

Discharge Properties of Saccade-Related Neurons in the Oculomotor Region of the Primate Fastigial Nucleus

J.F. KLEINE, Y.F. GUAN, S. LANGER, H. STRAKA, T. TCHELIDZE, AND U. BÜTTNER

Department of Neurology, LMU München, 81377 München, Germany

There is controversy about the function of the cerebellar oculomotor vermis (OV) and its target zone, the fastigial oculomotor region (FOR), in saccade control. Lesion studies in OV or FOR on saccade metrics suggest an influence of initial eye position.[1,2] However, prominent eye-position dependencies were not observed in the saccade-related activity of vermal Purkinje cells (PCs) or of neurons in the FOR.[3,4]

On the other hand, for FOR neurons latency differences for ipsi- and contralateral saccade-related bursts have consistently been found,[4–6] which are not prominent in OV-PCs.[1] However, establishing a precise relation between saccade and burst properties was made difficult by the variability of the discharge patterns of different FOR neurons.

We recorded from single FOR units in monkeys executing horizontal saccades (16-deg amplitude) from different horizontally and vertically displaced starting positions. While only a minority of units exhibited significant influences of eye position on their saccade-related discharge patterns, most neurons showed prominent differences in burst latencies for ipsi- and contralateral saccades, as previously described. Burst latencies varied markedly across individual units (FIG. 1A). However, linear summation of the responses from different neurons revealed "population bursts" that were closely related to the average acceleration and deceleration profiles of contra- and ipsilateral saccades, respectively (FIG. 1B). Modifying the relative weighting of individual neurons with different burst latencies could alter the temporal profile of the FOR population output, and thereby influence saccade metrics.

Address for correspondence: U. Büttner, Department of Neurology, LMU München, 81377 München, Germany. Voice: +49 89 7095 2561; fax: +49 89 7095 5561.
ubuettner@brain.nefo.med.uni-muenchen.de

Ann. N.Y. Acad. Sci. 978: 526–528 (2002). © 2002 New York Academy of Sciences.

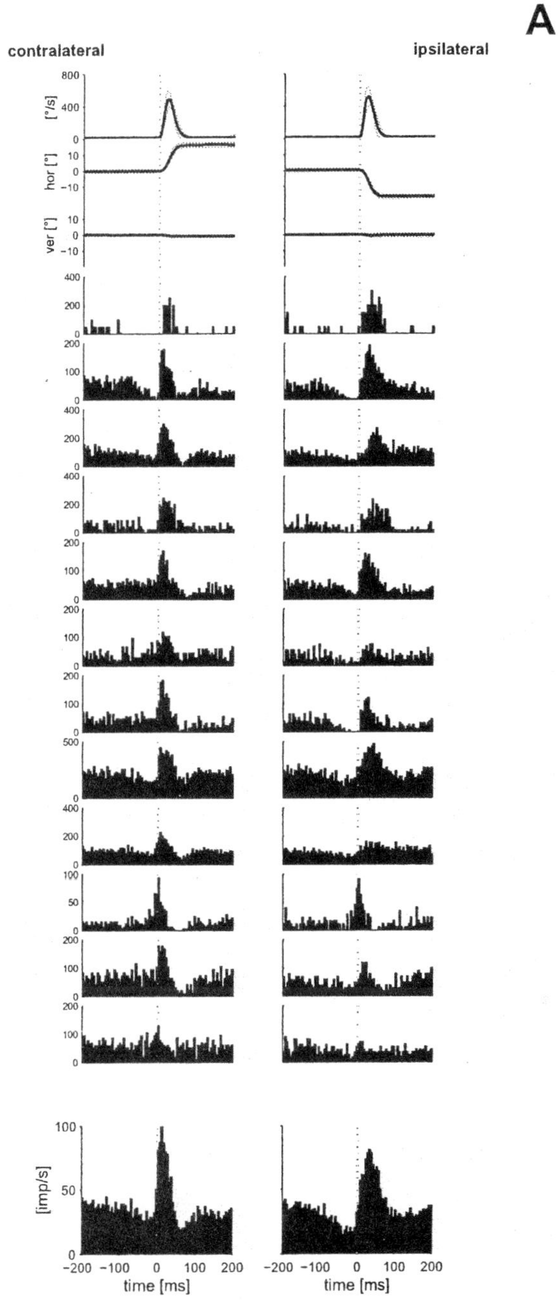

FIGURE 1 (A). *See following page for legend.*

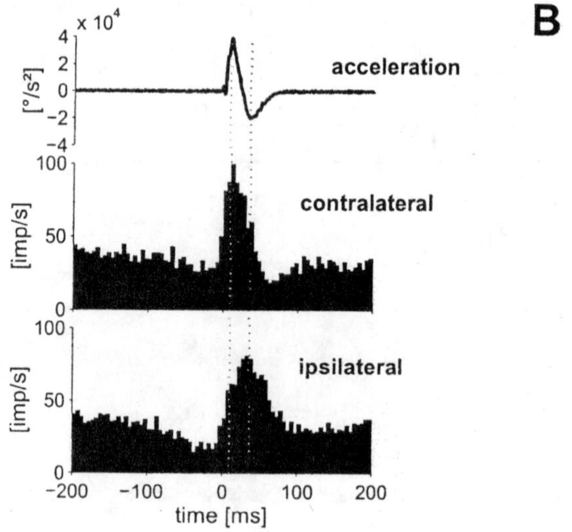

FIGURE 1. (**A**) Responses of 12 different FOR units during 16-deg horizontal saccades are shown with the average eye velocity and horizontal and vertical eye position signals (*upper traces*). *Bottom histograms*: population response (obtained by averaging across neurons). Burst latencies differ for ipsi- and contralateral saccades and vary markedly across units. (**B**) A close-up on the average responses [*bottom panels in part (A)*] indicates a close relation between average eye acceleration profile and population response.

REFERENCES

1. RITCHIE, L. 1976. Effects of cerebellar lesions on saccadic eye movements. J. Neurophysiol. **39:** 1246–1256.
2. ROBINSON, F.R., A. STRAUBE & A. FUCHS. 1993. Role of the caudal fastigial nucleus in saccade generation. II. Effects of muscimol in activation. J. Neurophysiol. **70:** 1741–1758.
3. OHTSUKA, K., H. SATO & H. NODA. 1994. Saccadic burst neurons in the fatigial nucleus are not involved in compensating for orbital nonlinearities. J. Neurophysiol. **71:** 1976–1980.
4. FUCHS, A.F., F.R. ROBINSON & A. STRAUBE. 1993. Role of the caudal fastigial nucleus in saccade generation. I. Neuronal discharge pattern. J. Neurophysiol. **70:** 1723–1740.
5. OHTSUKA, K. & H. NODA. 1991. Saccadic burst neurons in the oculomotor region of the fastigical nucleus of macaque monkeys. J. Neurophysiol. **65:** 1422–1434.
6. HELMCHEN, C., A. STRAUBE & U. BÜTTNER. 1994. Saccade-related activity in the fastiigial oculomotor region of the macaque monkey during spontaneous eye movements in light and darkness. Exp. Brain Res. **98:** 474–482.
7. THIER, P., P.W. DICKE, R. HAAS & S. BARASH. 2000. Encoding of movement time by populations of cerebellar Purkinje cells. Nature **405:** 72–76.

Spatiotemporal Profiles of Inferior Olivary Oscillatory Patterns

An *In Vitro* Voltage-Dependent Dye-Imaging Study

E. LEZNIK AND R. LLINÁS

Department of Physiology and Neuroscience, New York University School of Medicine, New York, New York 10016, USA

KEYWORDS: **dye imaging; olivary oscillatory patterns; spatiotemporal profiles**

Spatiotemporal profiles of ensemble subthreshold neuronal oscillation were studied in brainstem slices using high-speed voltage-sensitive dye imaging. Following local electrical stimuli, the overall voltage profile demonstrated coherent oscillatory waves that spread over the inferior olive. These oscillations were also observed in concurrently obtained intracellular recordings from inferior olivary neurons. Over the first few seconds after the stimuli, the optically recorded oscillations clustered into coherent groups comprising hundreds of neurons. Statistical analysis of the three-dimensional spatial profiles of these clusters revealed size fluctuation around stable core regions that were surrounded by a rim with phase-dependent dimensions.

The neuronal ensemble oscillations were calcium-deprived and had an average frequency range of 1–7 Hz. This rhythmic response demonstrated a different spatiotemporal distribution in the presence of picrotoxin, which induced the merging of neuronal clusters into larger areas of coherent activity.

ACKNOWLEDGMENT

This work was supported by the National Institutes of Health–National Institute of Neurological Disorders and Stroke (NIH-NINDS) Grant NS13742.

Address for correspondence: Dr. E. Leznik, Department of Physiology and Neuroscience, NYU School of Medicine, 550 First Avenue, New York, NY 10016. Voice: 212-263-5410; fax: 212-689-9060.

leznie01@popmail.med.nyu.edu

Ann. N.Y. Acad. Sci. 978: 529 (2002). © 2002 New York Academy of Sciences.

Oscillatory Properties of Inferior Olivary Neurons Modulate Transmission of Motor Cortical Activity to the Rat Cerebellum

S.P. MARSHALL, K.M. MAXWELL, AND E.J. LANG

Department of Physiology & Neuroscience, New York University School of Medicine, New York, New York 10016, USA

KEYWORDS: cerebellum; oscillatory properties; inferior olivary neurons; motor cortical activity; activity transmission

The intrinsic conductances of inferior olivary neurons allow the inferior olive to generate oscillatory activity. This rhythmicity may serve to gate transmission of afferent activity to the cerebellum. To investigate this issue, we used multiple electrode arrays to record crus 2a complex spike activity evoked by electrical stimulation of the motor cortex at various frequencies.

Autocorrelograms of complex spike activity during stimulation displayed peaks at intervals reflecting the stimulus frequency. However, the amplitude of the peaks fluctuated in a manner consistent with an approximately 10-Hz modulation of inferior olivary responsiveness to the stimulation. For example, successive peaks in autocorrelograms of activity during 20-Hz stimulation alternated in amplitude, whereas during 10-Hz stimulation, successive peaks gradually declined in amplitude (FIG. 1Ai and ii). If inferior olivary oscillations are responsible for these patterns, changes in these oscillations should alter the evoked complex spike response pattern. Charybdotoxin and apamin enhance and disrupt the rhythmicity of spontaneous complex spike activity, respectively.[1] Thus, the effect of injecting these substances on the motor cortex responses was investigated. Intra-inferior olivary injections of charybdotoxin enhanced the differences in the amplitudes of successive peaks in autocorrelograms of complex spike activity during 20-Hz stimulation relative to those in the controls. Conversely, injections of apamin resulted in a weakening of the alternation pattern during 20-Hz stimulation (FIG. 1Aii vs. 1Bii).

In 76% of cells (n = 11 rats, 210 cells), more complex spike activity was evoked in response to 8–12-Hz motor cortex stimulation than in response to stimulation at faster or slower frequencies, thus producing a bell-shaped response profile (FIG. 1Aiii). The number of cells displaying bell-shaped response profiles increased to 93% after intra-inferior olivary charybdotoxin injections (2 μM, n = 30 cells) and de-

Address for correspondence: Dr. E.J. Lang, Department of Physiology & Neuroscience, New York University School of Medicine, New York, New York 10016. Voice: 212-263-6638. lange01@popmail.med.nyu.edu

Ann. N.Y. Acad. Sci. 978: 530–532 (2002). © 2002 New York Academy of Sciences.

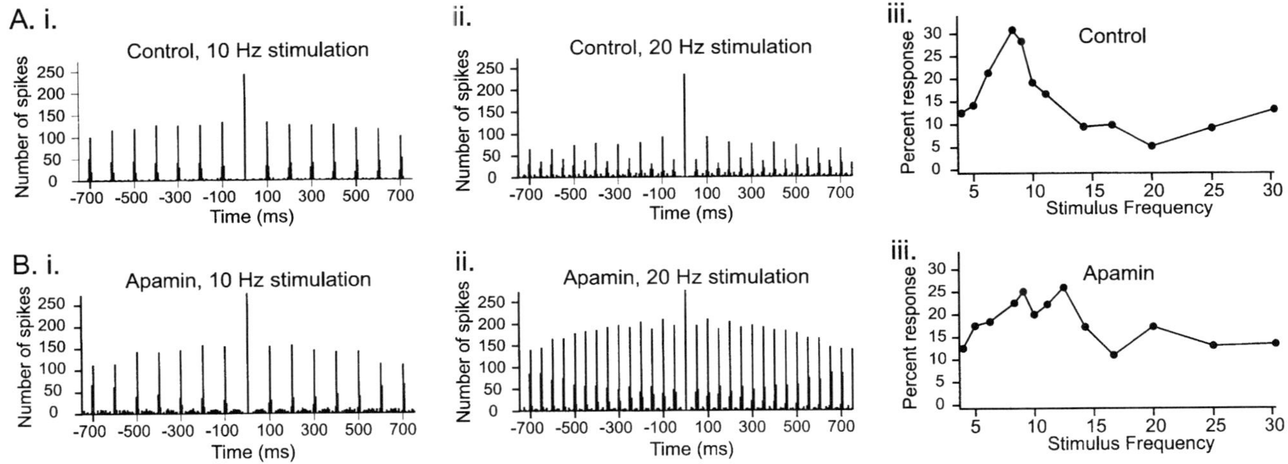

FIGURE 1. Ten-Hz modulation of complex spike (CS) responses to motor cortical stimulation. (**A**) Complex spike responses during motor cortex stimulation prior to drug injection. (i-ii) Autocorrelograms of CS activity during 10-Hz (i) and 20-Hz (ii) stimulation. (iii) Percentage of stimuli applied to the motor cortex that evoked CS responses for stimulus frequencies of 4–25 Hz. Stimulation at 8–12 Hz is most effective in eliciting CS responses. (**B**) Complex spike responses to motor cortical stimulation after intra-inferior olivary apamin injection. After injection, CS firing rates increase and responses show less modulation than in the control. (i) Responses continue to follow 10-Hz stimulation as in the control. (ii) At 20-Hz stimulation, successive stimuli evoke approximately equal numbers of CS responses, and the alternating peak amplitude pattern is virtually abolished. (iii) After apamin, a more uniform frequency response curve is found as stimulus frequencies faster and slower than 8–12 Hz evoke responses more effectively.

creased to 35% after apamin injections (10 μM, $n = 46$ cells), indicating that inferior olivary oscillations are responsible for the preferential responsiveness to approximately 10-Hz stimulation (FIG. 1Aiii vs. Biii). These results suggest that motor cortical input to the cerebellum via the olivocerebellar system is gated by the oscillatory properties of inferior olivary neurons.

ACKNOWLEDGMENT

This work was supported by NINDS and the National Science Foundation.

REFERENCE

1. LANG, E.J. *et al.* 1997. J. Neurosci. **17:** 2825–2838.

Inhibition of Climbing Fibers Is Necessary for the Extinction of Conditioned Eyelid Responses

JAVIER MEDINA

Department of Neurobiology and Anatomy, University of Texas Medical School at Houston, Houston, Texas 77030, USA

KEYWORDS: **climbing fibers; conditioned eyelid responses**

A central tenet of cerebellar learning theories asserts that climbing-fiber input to the cerebellum provides a teaching signal that drives the gradual adaptation of movements. Data from several forms of motor learning support this hypothesis. For example, in Pavlovian eyelid conditioning, where an initially neutral tone is repeatedly paired with a reinforcing unconditioned stimulus such as periorbital stimulation, the unconditioned stimulus promotes acquisition of conditioned eyelid responses by activating climbing fibers. Here we show that inhibition of climbing fibers can also serve as a teaching signal for the extinction of eyelid responses, where the signal for learning comes from presenting the tone in the absence, rather than in the presence, of the reinforcing unconditioned stimulus.

Cells of the inferior olive receive inhibitory input from the deep cerebellar nuclei. During eyelid conditioning, these deep cerebellar nuclei neurons show increases in activity that parallel the acquisition of conditioned responses. Therefore, as training proceeds, the increased excitation of climbing fibers provided by the unconditioned stimulus is presumably counterbalanced by increased conditioned response-related inhibition from the deep cerebellar nuclei. Consistent with this idea, the unconditioned stimulus activates climbing fibers reliably at the beginning of training, but the ability of the unconditioned stimulus to elicit climbing-fiber activity gradually decreases to normal levels after learning. Our hypothesis is that during extinction, increased conditioned response-related inhibition of climbing fibers, in the absence of unconditioned stimulus-related excitation, serves as a signal that drives the extinction of conditioned responses.

To test this hypothesis, we blocked inhibition of climbing fibers during extinction by infusing the GABA antagonist picrotoxin in the inferior olive of previously trained rabbits. Whereas conditioned responses were completely extinguished in approximately 50–60 trials in control sessions with saline infusions, rabbits that were

Address for correspondence: Dr. Javier Medina, Department of Neurobiology and Anatomy, University of Texas-Medical School at Houston, 6431 Fannin St., Houston, TX 77030. Voice: 713-500-5598.
javier.f.medina@uth.tmc.edu

Ann. N.Y. Acad. Sci. 978: 533–534 (2002). © 2002 New York Academy of Sciences.

infused with picrotoxin before the extinction session were still showing high levels of responding by the end of the session (108 trials).

These results demonstrate that extinction of eyelid responses requires inhibition of the inferior olive. This is consistent with the hypothesis that in addition to increasing their activity to signal the presence of an unexpected unconditioned stimulus, climbing fibers may decrease their activity to signal the absence of an expected unconditioned stimulus.

Contributions of Limb Incoordination and Postural Control Deficits to Cerebellar Gait Ataxia

S.M. MORTON[a] AND A.J. BASTIAN[b]

[a]*Program in Physical Therapy, Washington University School of Medicine, St. Louis, Missouri 63108, USA*

[b]*Kennedy Krieger Institute and Department of Neurology, Johns Hopkins University School of Medicine, Baltimore, Maryland 21205, USA*

KEYWORDS: postural control deficits; cerebellar gait ataxia

The cerebellum is involved in multijointed limb coordination and postural control, both of which are important components of walking. The purpose of this study was to determine whether particular features of cerebellar gait ataxia could be attributed to a specific deficit in either voluntary leg coordination or postural control. We compared the performance of subjects with cerebellar damage to age-matched healthy controls as they walked and performed specific tasks designed to measure isolated leg coordination and postural control. The leg coordination measures were a visually guided stepping task and the heel-knee-shin test. The postural control measures were a series of static standing tasks with eyes open or closed and a dynamic weight-shifting task. We measured three-dimensional position data from the leg and trunk during all tasks and center of pressure coordinates during the stepping and postural control tasks. We calculated foot placement errors and path ratios to measure leg coordination and sway amplitudes and weight-shifting distances to measure postural control.

Several subjects in our study had relatively isolated impairment of either leg coordination or postural control, whereas others had both types of deficits. Subjects with only a postural control deficit had more walking abnormalities than did those with only a leg coordination deficit. Subjects with both leg incoordination and impaired postural control generally walked the worst. Our data show that individuals with impaired postural control perform more poorly during walking than do individuals with impaired leg coordination. Thus, human locomotion on level surfaces may be more dependent on cerebellar outputs to vestibular and brain-stem motor regions than to cortical motor regions.

ACKNOWLEDGMENTS

This work was supported by National Institutes of Health Grants HD01199 and HD07434 and the Foundation for Physical Therapy.

Address for correspondence: Dr. S.M. Morton, Washington University School of Medicine, St. Louis, MO 63108.

Ann. N.Y. Acad. Sci. 978: 535 (2002). © 2002 New York Academy of Sciences.

Effects of Electrical Microstimulation and Muscimol Application in Selected Cerebellar Interpositus Sites on Reflex and Classically Conditioned Eyelid Responses in Alert-Behaving Cats

J.D. NAVARRO-LÓPEZ, L. JIMÉNEZ-DÍAZ, A. GRUART, AND J.M. DELGADO-GARCÍA

División de Neurociencias, Laboratorio Andaluz de Biología, Universidad Pablo de Olavide, 41013-Sevilla, Spain

KEYWORDS: cerebellum; nictitating membrane; eyelid response; interpositus; muscimol

The classic conditioning of nictitating membrane/eyelid responses is a very well-known experimental model for the study of neural mechanisms underlying the acquisition of new motor abilities. The cerebellum is one of the neural structures supposedly involved in those learning processes. In particular, the interpositus nucleus has been proposed as the site where memory for associative learning is acquired and stored. We prepared cats for the chronic recording of eyelid responses using the search coil in a magnetic-field technique, and of the electrical activity of the orbicularis oculi muscle. In a previous set of experiments, we identified interpositus sites with a firing activity related to reflex and conditioned eyelid responses.[1] For the present experiments, a microstimulation (cathodal square 50-ms^2 pulses) of those selected interpositus sites at different frequencies (5–60 Hz) for a maximum of 1 s was carried out. Muscimol (a GABAa receptor agonist) was injected in the same places through a guide tube (26-gauge stainless-steel needle) implanted up to 1.5 mm over the desired injection sites.

Preliminary results are as follow: (1) muscimol reduced the amplitude and mean velocity of reflex responses evoked by tones (600 Hz, 90 dB, 100 ms) or air puffs (3 kg/cm^2, 100 ms); (2) muscimol also reduced the amplitude of eyelid conditioned responses when applied in already-trained animals (following the fifth conditioning session); (3) electrical microstimulation of selected interpositus sites evoked a 20-Hz dominant, ramplike, small-amplitude eyelid response; (4) electrical microstimu-

Address for correspondence: J. D. Navarro-López: División de Neurociencias, Laboratorio Andaluz de Biología, Universidad Pablo de Olavide, 41013-Sevilla, Spain. Voice: +34-954-349054; fax: +34-954-349375.
jdnavlop@dex.upo.es

Ann. N.Y. Acad. Sci. 978: 536–537 (2002). © 2002 New York Academy of Sciences.

lation of interpositus sites increased the amplitude of eyelid conditioned responses evoked by a tone used as a CS; and (5) the effect of electrical microstimulation on conditioned responses was still evoked during muscimol application to the stimulation sites. These data suggest that the cerebellum reinforces the performance of eyelid reflex and conditioned responses.

REFERENCES

1. GRUART, A. *et al.* 2000. Cerebellar posterior interpositus nucleus as an enhancer of classically conditioned eyelid responses in alert cats. J. Neurophysiol. **84:** 2680–2690.

Neural Control of Discrete Movement Segments: The Role of the Cerebellum in the Control and Learning of Movements

KEVIN E. NOVAK, LEE E. MILLER, AND JAMES C. HOUK

Northwestern University Medical School, Chicago, Illinois 60611, USA

KEYWORDS; cerebellum; monkey; Purkinje cells; neuromuscularskeletal system; kinematics

In order to understand better how the brain adaptively controls voluntary movements, this work examined human and monkey subjects performing a knob-turning task, similar to tuning a radio dial. Subjects sometimes made these target-directed hand movements as a single, smooth motion, but often produced a series of discrete-movement segments. The central mechanisms responsible for the production of discrete-movement segments were investigated via three approaches: analysis of human kinematics, modeling the neuromusculoskeletal system, and single-unit recordings of Purkinje cells (PCs) in the cerebellum. A direct, objective algorithm was developed to detect when a movement was composed of multiple segments, and movements were separated into three segment types: primary movements, overlapping submovements, and delayed submovements. The kinematic properties of each of these segments were remarkably similar, suggesting they were generated by some common central process. Movements only became irregular and had shapes that differed from a bell-shaped velocity when they were made as a series of two or more overlapping segments. When torque perturbations were applied, the size and frequency of corrective submovements increased, but subsequently decreased as subjects adapted over time. As a result, the kinematics of the overall movement returned near normal.

A simplified model of the motor control system was used to show how changing pulselike commands affected kinematics, and how multiple pulselike commands could produce discrete-movement segments. Many PCs fired simple spike (SS) discharges related to both the primary movement and the corrective submovements, and most likely encoded for muscle activation levels. SS activity often changed when random perturbations were applied, showing that these cells were involved in the continuous monitoring and adjustment of movement commands. PCs fired complex spikes (CSs) after natural and electrical stimulation, and CS firing at movement on-

Address for correspondence: James C. Houk, Northwestern School of Medicine, Chicago, IL 60611.

Ann. N.Y. Acad. Sci. 978: 538–539 (2002). © 2002 New York Academy of Sciences.

set may have helped subjects make faster movements. Around 35–50% of PCs fired CSs after the monkeys made corrective submovements.

In summary, PCs monitor feedback and efference copy, and make discrete and continuous adjustments to movement commands. PCs can learn to generate single, accurate primary movements through changing synaptic efficacy of their inputs under the guidance of CS firing following discrete corrective submovements.

Quantitative Analysis of the Purkinje Cell Denritic Spines in the Voltage-Dependent Calcium Channel Mutant, Rolling Mouse Nagoya

IM JOO RHYU,[a] KEA JOO LEE,[a] S. ODA,[b] AND Y. S. SUH[c]

[a]*Department of Anatomy, Korea University College of Medicine, 126-1, Anam Dong, SungBuk-Ku, Seoul, Korea 136-705*

[b]*Laboratory of Animal Management, Nagoya University, Nagoya City, Japan*

[c]*Department of Anatomy, College of Medicine, Ewha Womans University, Seoul, Korea*

KEYWORDS: rolling mouse Nagoya; dendritic spines

The rolling mouse Nagoya (*Cacna1a*$^{tg\text{-}rol}$) is a spontaneous P/Q-type calcium channel mutant, which has been suggested as an animal model for human neurologic diseases, such as autosomal dominant cerebellar ataxia (SCA6), familial hemiplegic migraine, and episodic ataxia type-2. Altered synapses between parallel fiber varicosity and Purkinje cell dendritic spines were found in the molecular layer of the rolling cerebellum. Numerous spines from proximal dendrites of rolling mice were observed in our previous study.

After rapid Golgi preparations of the rolling and wild cerebellum, 4-μm sections were made and observed with high voltage electron microscopy. The height and density of the dendritic spines were analyzed quantitatively with an NIH image to in understanding the functional status of the Purkinje cell and the morphologic mechanism of the synaptic plasticity in rolling mice.

Purkinje cell dendritic spines were observed clearly enough for quantitative analysis. Many spines emanating from the proximal dendrites in rolling mice were observed. The average spine density was 2.12±0.18/10 μm in wild and 1.78±0.22/10 μm in rolling. The average length of the spines was 1.12±0.21 μm in wild and 0.82±0.21 μm in rolling. The density and height of the spines were significantly decreased in rolling Purkinje cells ($p<0.01$). Many hypertrophied parallel fiber varicosities were observed in rolling mice Nagoya.

Address for correspondence: Im Joo Rhyu, Department of Anatomy, Korea University College of Medicine, 126-1, Anam Dong, SungBuk-Ku, Seoul, Korea 136-705. Voice: +82-2-920-6149; fax: +82-2-929-5696.

irhyu@korea.ac.kr

Ann. N.Y. Acad. Sci. 978: 540–541 (2002). © 2002 New York Academy of Sciences.

Spine plasticity observed in this mutant may reflect the altered functional status of the Purkinje cell in this P/Q calcium-channel mutant.

ACKNOWLEDGMENTS

This work is supported by the Brain Korea 21 Project and the Korea-Japan Basic Scientific Promotion Program.

Does the Cerebellum Preferentially Control Discrete and Not Continuous Movements?

REBECCA M. C. SPENCER,[a] HOWARD N. ZELAZNIK,[a] RICHARD B. IVRY,[b] AND JÖRN DIEDRICHSEN[b]

[a]*Department of Health and Kinesiology and the Purdue Neuroscience Program, Purdue University, West Lafayette, Indiana 47907-1392*

[b] *Department of Psychology, University of California at Berkeley, Berkeley, California, USA*

KEYWORDS: **cerebellar damage; finger tapping; drawing tasks; timing**

Patients with unilateral cerebellar damage exhibit increased temporal variability in finger tapping and intermittent drawing tasks when performing with their impaired limb relative to their unimpaired limb. Surprisingly, we observed no difference in temporal variability between the limbs on a continuous drawing task. Together with previous work in our lab with normal individuals (Robertson *et al.*, 1999), it appears that the temporal control of movement can be achieved in two different ways. An explicit timing process may guide the movements by maintaining a representation of the temporal goal. Discrete movement tasks and temporal perception tasks appear to use such a timing process. Implicit timing, we posit, is an indirect or emergent process. The nature of the trajectory control processes can account for temporal consistency during continuous movements. In the present study the role of the cerebellum in discrete and continuous movements was examined further. Four groups, young adults (n=7), elderly adults (n=7), patients with unilateral cerebellar damage (n=6), and patients with bilateral cerebellar damage (i.e, spinocerebellar atrophy or cerebellar ataxia) (n=7), performed three tasks: discrete finger tapping, discrete circle drawing, and continuous circle drawing (FIG. 1). In the discrete tasks subjects made single movements (i.e, tapped once or drew one circle) following a variable interstimulus interval. The target cycle duration in all three tasks was 550 ms. We found increased temporal variability when patients with unilateral cerebellar damage performed the tapping and discrete circle drawing tasks with the impaired limb relative to performance with the unimpaired limb. The patients exhibited no impair-

Address for correspondence: Rebecca M.C. Spencer, Department of Psychology, University of California at Berkeley, 3210 Tolman Hall #1650, Berkeley, CA 94720. Voice: 510-642-9226; fax: 510-642-5293.

rspencer@socrates.berkeley.edu

Ann. N.Y. Acad. Sci. 978: 542–544 (2002). © 2002 New York Academy of Sciences.

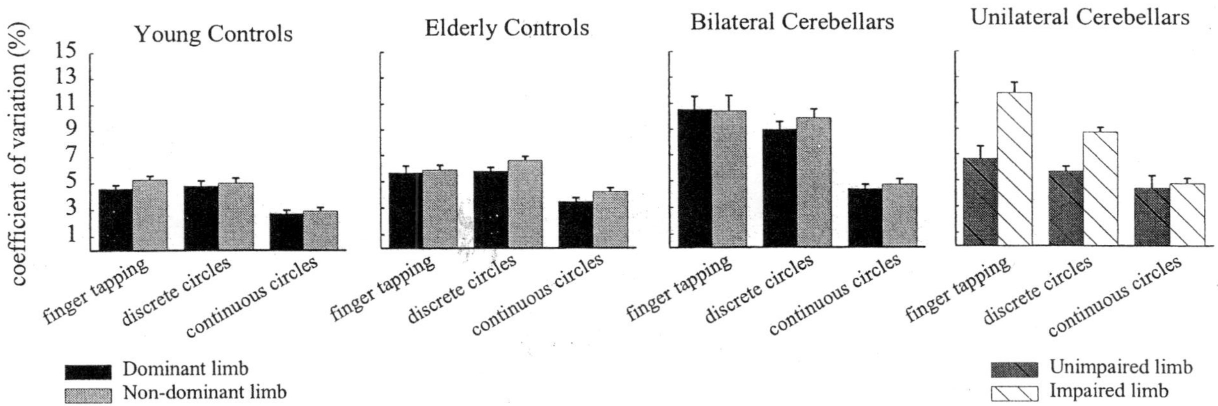

FIGURE 1. Bilateral cerebellar patients were more variable than control groups on the discrete tasks, yet there was no difference in performance between these groups on the continuous task ($F(4,36) = 6.12$, $p < .04$). Unilateral cerebellar patients exhibited increased temporal variability on the discrete tasks when performing with the ipsilesional limb relative to the contralesional limb ($t(5) = -4.23$, $p < .01$). There was no difference in performance across limbs on the continuous drawing task ($t(5) = -1.90$, $p < .12$).

ment on the continuous circle drawing task. Furthermore, patients with bilateral cerebellar damage had higher temporal variability in the discrete tasks than elderly control subjects, yet these patients were not more variable in the continuous drawing task. These data support the hypothesized distinction between continuous and discrete tasks. Explicitly timed tasks use the cerebellum. Implicit timing is not affected by cerebellar damage.

REFERENCES

1. Robertson, S.D., H.N. Zelaznik, D.A. Lantero, *et al.* 1999. Correlations for timing consistency among tapping and drawing tasks: evidence against a single timing process for motor control. J. Exp. Psychol.: Hum. Percept. Perform. **25:** 1316–1330.

Kinematic Modeling of the Simple Spike Modulation of Rabbit Floccular Purkinje Cells

M. SUH, T. BELTON, AND J. I. SIMPSON

Department of Physiology and Neuroscience, New York University School of Medicine, New York, New York 10016, USA

KEYWORDS: simple spike modulation; kinematic modeling; rabbit floccular Purkinje cells

Multilinear regression techniques have been used to study the combined influences of different eye kinematic variables in shaping floccular Purkinje cell simple spike (SS) firing rates. Suh and colleagues[1] used these techniques in relation to smooth pursuit in monkeys and found that eye velocity is the dominant influence, although eye position also plays an important role. Additional evidence of eye-movement-related responses of floccular Purkinje cells has been reported in monkeys[2,3] and rabbits.[4] Leonard found a preponderance of "eye movement dominated" neurons during the optokinetic reflex (OKR) and the vestibulo-ocular reflex (VOR) along with a smaller number of "image velocity dominated" neurons. To address how the SS modulation of rabbit floccular Purkinje cells is affected by visual input during VOR, we recorded from three Dutch belted rabbits during VOR with and without visual input. Since some of the 40 recorded vertical axis Purkinje cells provided data at more than one frequency, 57 light–dark paired data cases were obtained. Eye position and table position were recorded concomitantly. The awake rabbits were horizontally oscillated during VOR in light (VORL) and VOR in dark (VORD) at three different frequencies, (0.4, 0.6, and 0.8 Hz) all with a velocity amplitude of 12.5°/s To estimate the relative influence of eye kinematic variables on the SS modulation, we used a simple kinematic model expressing the SS firing rate as a multilinear function of eye position and eye velocity: $R(t) = \beta + \rho P(t) + \upsilon V(t)$, where $R(t)$ is the SS firing rate at time t, $P(t)$ and $V(t)$ are eye position and eye velocity at time t, β is the model estimate of baseline firing rate, and ρ and υ are the model estimates of the position and velocity sensitivity coefficients. For the majority of the population (42/57 cases, 29/40 neurons), the model produced good fits for both VORL ($R^2 = .87 \pm .06$) and VORD ($R^2 = .79 \pm .11$). Due to the lack of improvement of fits following addition of eye acceleration as a variable, eye acceleration was not included in the model. The preferred eye velocity direction was ipsilateral for both VORL

Address for correspondence: M. Suh, Department of Physiology and Neuroscience, New York University School of Medicine, 550 First Avenue, New York, NY 10016. Voice: 212-263-5428; fax: 212-689-9060.
suhm01@popmail.med.nyu.edu.

Ann. N.Y. Acad. Sci. 978: 545–546 (2002). © 2002 New York Academy of Sciences.

(42/42) and VORD (41/42). In contrast, the preferred eye position direction was contralateral in many cases for both VORL (13/42) and VORD (21/42). Interestingly, for some cases (15/57, 11/40 neurons), the model failed ($R^2 < .50$) to fit the SS firing rate for VORD ($R^2 = .33 \pm .16$), while it did fit the SS firing rate for VORL ($R^2 = 79 \pm .11$). Nonetheless, the failure of the model for VORD does not imply an inappropriate use of a model based on eye position and eye velocity variables. Instead, the failure of the model for VORD indicates that there was a low correlation between the sinusoidally modulated eye movements and the neural responses. In fact, the cases in the minority group (15/57 cases) had no clear modulation during VORD, although they showed good modulation during VORL. This finding supports the proposal[4] that the visual input *per se* plays a significant role in determining the SS response of some floccular Purkinje cells during VORL.

ACKNOWLEDGMENTS

This work was supported by NIH grants NS-13742 and DC-03284.

REFERENCES

1. Suh, M., H.-C. Leung & R.E. Kettner. 2000. Cerebellar flocculus and ventral paraflocculus Purkinje cell activity during predictive and visually driven pursuit in monkey. J. Neurophysiol. **84:** 1835–1850.
2. Noda, H. & D.A. Suzuki. 1979. The role of the flocculus of the monkey in saccadic eye movements. J. Physiol. (London) **294:** 335–348.
3. Belton, T. & R. McCrea. 2000. Role of the cerebellar flocculus region in cancellation of the VOR during passive whole body rotation. J. Neurophysiol. **84:** 1599-1613.
4. Leonard, C.S. 1986. Cerebellar Purkinje cell signals during eye movements. Ph.D. Thesis, New York University, New York.

Three-Axis Eye Movement Abnormalities with Cerebellar Lesions

M. F. WALKER AND D. S. ZEE

Departments of Neurology and Ophthalmology,
The Johns Hopkins University, Baltimore, Maryland 21287, USA

KEYWORDS: **three-axis eye movement; surgical cerebellar lesion**

The cerebellum plays an important role in the coordination and calibration of eye movement accuracy and dynamics. We have focused on the function of the cerebellum in the control of the three-dimensional (horizontal, vertical, and torsional) axis of eye velocity in visual following responses and the vestibulo-ocular reflex (VOR). Here we present examples of three-axis abnormalities from patients with cerebellar disease and a monkey with a surgical cerebellar lesion.

First, in response to rapid head rotations ("head thrusts"), cerebellar patients may have marked abnormalities in both VOR gain and axis. A characteristic finding is inappropriate vertical and torsional eye velocities during yaw-axis rotation. The vertical component is always upward, and the torsional component changes direction with the direction of rotation. The pattern is consistent with what would be predicted if the contralateral anterior canal (normally excited during yaw rotation with the head upright) were disinhibited, perhaps through loss of Purkinje cell projections to secondary anterior canal neurons. This cross-coupling has important functional implications, because the relatively high vertical velocities induce a substantial retinal slip, impairing vision during head movement.

Another finding is that patients often have excessive torsion during horizontal optokinetic stimulation (OKN) and that the normal relationship of torsional eye velocity to vertical eye position during horizontal pursuit and OKN is disrupted. This implicates the cerebellum in the implementation of Listing's Law.

Finally, in one monkey with a bilateral lesion of the cerebellar flocculus (with some preservation of the ventral paraflocculus), we found VOR cross-coupling. Responses to yaw-axis rotation included vertical and torsional components that were disconjugate.

These findings add further support to an important role of the cerebellum in the calibration of the axis of eye rotation, including the control of ocular torsion.

Address for correspondence: M.F. Walker, M.D., Department of Neurology, The Johns Hopkins University, Pathology 2-210, 600 N. Wolfe Street, Baltimore, MD 21287. Voice: 410-614-1575; fax: 410-614-1746.
mwalker@jhu.edu

Ann. N.Y. Acad. Sci. 978: 547 (2002). © 2002 New York Academy of Sciences.

Anterior Semicircular Canal and Otolith Afferents Converge at Single Vestibular Neurons in Cats

XIALING ZHANG, MITUYOSHI SASAKI, HITOSHI SATO, HUI MENG, RISHU BAI, MIDORI IMAGAWA, AND YOSHIO UCHINO

Department of Physiology, Tokyo Medical University, Tokyo, Japan

KEYWORDS: anterior semicircular canal; otolith afferents; single vestibular neurons

Central vestibular neurons receive convergent inputs from the semicircular canals and otolith afferents. In this study we used an intracellular recording in the single vestibular neuron by selective stimulation of the two afferent pairs of the anterior semicircular canal (AC) and saccular (SAC) nerves, and the AC and utricular (UT) nerves in decerebrate cats. Neurons were recorded from the central part of the vestibular nuclei, where the otolith afferents mainly terminate. All recorded neurons were classified as vestibulospinal (VS), vestibulo-oculospinal (VOS), or vestibulo-ocular (VO) by antidromic stimulation from the oculomotor/trochlear nuclei and the spinal cord: neurons that could not be activated from any test sites were classified as vestibular (V) neurons. Of a total of 118 neurons activated by stimulation of the AC/SAC nerves, 51 received convergent inputs (27 VS, 4 VOS, 5 VO, and 15 V neurons). Forty-two of the convergent neurons received monosynaptic excitatory inputs from both afferents. The majority of the VS neurons descended through the ipsilateral lateral vestibulospinal tract (i-LVST). The AC/SAC convergent neurons were located in the rostral part of the descending, the medial, and the caudal-ventral part of the lateral and superior vestibular nucleus. In 105 neurons studied by stimulation of the AC/UT nerves, both inputs converged in 42 neurons (28 VS, 6 VOS, and 8 V neurons; no VO neuron was recorded). Thirty-eight convergent neurons had monosynaptic excitatory inputs from both afferents. About half of the VS and most of the VOS neurons projected the descending axons via the medial vestibulospinal tract. Most of the convergent neurons were located in the medial, the rostral part of the descending, and lateral vestibular nucleus. The percentages of AC/SAC and AC/UT convergence were higher than the convergence between the posterior semicircular canal and otolith, and the convergence of the horizontal semicircular canal and otolith, respectively.

Address for correspondence: Xialing Zhang, Box 1135, Department of Neurology, Mt. Sinai School of Medicine, New York, NY 10029. Voice: 212-241-3927; fax: 212-831-1610. xialing.zhang@mssm.edu

Ann. N.Y. Acad. Sci. 978: 548 (2002). © 2002 New York Academy of Sciences.

Index of Contributors